Conducting Polymers, Fundamentals and Applications

Prasanna Chandrasekhar

Conducting Polymers, Fundamentals and Applications

Including Carbon Nanotubes and Graphene

Second Edition

 Springer

Prasanna Chandrasekhar
Ashwin-Ushas Corporation
Marlboro, New Jersey, USA

ISBN 978-3-030-09885-8   ISBN 978-3-319-69378-1   (eBook)
https://doi.org/10.1007/978-3-319-69378-1

Printed on acid-free paper

This Springer imprint is published by Springer Nature
The registered company is Springer International Publishing AG
The registered company address is: Gewerbestrasse 11, 6330 Cham, Switzerland

Preface

This book is a revised edition of a book by the same author, *Conducting Polymers, Fundamentals and Applications: A Practical Approach*, published in 1999 and well received at the time. The earlier book addressed a critical need for a primarily pedagogical and instructional text in conducting polymers (CPs) at a basic level. It ended up being used by many academic institutions in graduate- or senior under-graduate-level courses, as well as by a wide variety of researchers and students in the varied and multidisciplinary fields in which CPs have been applied. In the nearly 18 years that have passed since that first publication, CPs have fulfilled some aspirations for their applications but have also disappointed in many others. Nevertheless, they remain a firmly entrenched, yet continually developing, class of materials used by many researchers in various fields, as the chapter titles of this book clearly illustrate.

Since the first edition in 1999, two new classes of conductive materials have received much attention and research. These are carbon nanotubes (CNTs) and graphene. Indeed, one may say that these two classes of materials have received even more attention than CPs ever received in the last 18 years. In one case, that of CNTs, aspirations for their applications have somewhat disappointed. In the other case, that of graphene, there is still much promise in future applications. Nevertheless, these two classes of materials bear many similarities to CPs, primary among them being that their unique conductivity properties give rise to a whole host of other properties with potential applications in areas as diverse as drug delivery, communication, and metamaterials. Thus, an expansion of the subject material of the book to include these newer classes and the consequent, logical expansion of the title to *Conducting Polymers and Other New Conductive Materials* was in order.

As in the earlier edition, this book primarily addresses the need for an instructional text at a fundamental level, together with an introduction to potential applications for conducting polymers (CPs), carbon nanotubes (CNTs), and graphene. It can assist a variety of researchers from diverse fields, ranging from materials scientists, chemists, biologists, physicists, and environmental researchers to medical/pharmacological researchers and even medical doctors involved in research.

While there are several extant texts in each of these three fields, nearly all are "editor/contributor" texts of a high-level and specialized nature, addressing narrow subfields. They primarily present ongoing research in the contributors' laboratories. These texts oftentimes plunge directly into the thick of high-level research, with no pause for instruction, or even an explanation of terminology, for the novice or new researcher. Thus, they are sometimes quite difficult to follow for persons in areas that are different from the chapter author's research.

This book is separated into six parts, two each for fundamentals and applications for each of the three classes of materials, i.e., conducting polymers (CPs), carbon nanotubes (CNTs), and graphene. At the end of each chapter, there are problems/ exercises which should serve as a good test on whether the reader absorbed at least a majority of the material presented in the chapter.

As with the earlier edition, this book emphasizes a practical, "how-to" approach and is written in such a way that a new researcher can instructionally use only the parts relevant to his/her present research. The book also targets students, at the advanced undergraduate or graduate level, and could be included as part of a comprehensive introduction to these fields.

The author wishes to acknowledge the able assistance of Timothy Ambat in editing this text.

Marlboro, NJ, USA   Prasanna Chandrasekhar

Foreword to the First Edition

Like semiconductors, organic materials with extended π-electron conjugation (e.g., "conducting polymers") can give rise to novel electrical, optical, and magnetic phenomena. Like semiconductor materials, such phenomena can, at least hypothetically, be translated into a variety of useful devices. However, organic π-electron materials, unlike semiconductors, are not atomic solids but rather are typically amorphous polymeric materials. Phenomena such as charge transport in organic materials can be quite different (e.g., variable range hopping) from that encountered in semiconductors, and a range of mechanisms can be active depending on material processing. Possible "conducting polymer" structures and processing protocols are almost limitless. Certainly, the possibilities for "molecular engineering" of conducting polymers are very large indeed. The construction of devices such as light-emitting diodes and nonlinear optical switches involves quite different considerations when using π-electron materials compared to the construction of such devices from semiconductors. Not only are potential applications of π-electron polymeric materials very impressive in terms of anticipated economic impact of specific applications, but the anticipated range of applications is very large (e.g., light-emitting diodes, batteries, sensors, photorefractive devices, electro- and photochromic devices and materials, microwave absorbing materials, second- and third-order nonlinear optical materials and devices, etc.).

The field of "conducting polymers" has been very dynamic in its evolution. From simple-minded pictures of bond alternation defects and application to the development of lightweight batteries, the field has evolved to include an ever-widening area of topics and applications. In recent years, applications involving electroluminescence, photorefractivity, electrochromism, optical nonlinearity, and sensing have particularly attracted attention. The literature for each application area has become enormous (e.g., thousands of articles published on single topics such as organic light-emitting diodes). Because of the vastness and diversity of the journal and conference proceedings literature related to the topic of conducting polymers, a text written from the perspective of a single individual is particularly useful as an educational tool for acquainting scientists with various aspects of this topic.

The importance of the topics covered in the current text certainly makes this work useful to the scientific community

University of Washington,   Larry Dalton
Seattle, WA, USA

List of Common Abbreviations

The abbreviations listed below are classified into the following categories:

General
Common Conducting Polymers
Other Polymers
Monomers
Dopants
Chemicals, Solvents
Techniques, Methodology

Abbreviation	Explanation
General:	
AC	Alternating current
AFM	Atomic force microscopy
ASTM-	Testing or other standards issued by the American Society for Testing and Materials
CB	Conduction band
CON1/2/3	1st, 2nd, and 3rd configurations used in Ashwin electrochemical devices
CP	Conducting polymer
CV	Cyclic voltammetry
DC	Direct current
DPV	Differential pulse voltammetry
EA	Electron affinity
e-beam	Electron beam
E_F	Fermi level (primarily in the context of semiconductors)
E_g	E(gap), bandgap energy
EIS	Electrochemical impedance spectroscopy
EMI	Electromagnetic interference
EMI-SE	Electromagnetic interference shielding effectiveness

(continued)

Abbreviation	Explanation
EO, E/O, E-O	Electro-optic
ESD	Electrostatic discharge
ESR	Electron spin resonance
EXAFS	Extended X-ray absorption fine structure
FET	Field-effect transistor
FWHM	Full width at half maximum (peak half width)
HOMO	Highest occupied molecular orbital
H-T	Head-to-tail (coupling)
I_p, IP	Ionization potential
ITO	Indium tin oxide
ITO	Indium tin oxide thin film
I-V	Current-voltage (curves, etc.)
LB	Langmuir–Blodgett
LC	Liquid crystal(s)
LCD	Liquid crystal display
LEC	Light-emitting electrochemical cell(s)
LED	Light-emitting diode
LSV	Linear sweep voltammetry
LUMO	Lowest occupied molecular orbital
MIL-, MIL-C-, MIL-STD-	Military standards issued by the US Dept. of Defense
MWt, MW	Molecular weight
NPV	Normal pulse voltammetry
PV	Photovoltaic(s)
QCM	Quartz crystal microbalance
RCS	Radar cross section
S, S/cm	Siemen, Siemen/cm (Siemen, unit of impedance, $= \Omega^{-1}$)
SC	Semiconductor
SCALE	Symmetrically configured alternating current light emitting
SCE	Saturated calomel electrode
STM	Scanning tunneling microscopy
SWV	Square wave voltammetry
VB	Valence band
XPS	X-ray photoelectron spectrometry
Common Conducting Polymers:	
BBB	*see p. 423*
BBL	*see p. 423*
PBT	*see p. 423*
P(....)	Poly(....)
P(Ac)	Polyacetylene(s)
P(ANi), PANI, PAN	Poly(aniline)
P(DiAc)	Poly(diacetylene(s))
P(DPA)	Poly(diphenyl amine)
PPO	*see p. 423*

(continued)

Abbreviation	Explanation
P(PO), PPO	Poly(p-phenylene oxide)
P(PP), PPP	Poly(p-phenylene)
P(ProDOT)	Poly(propylene dioxythiophene)
P(PS), PPS	Poly(p-phenylene sulfide)
P(PV), PPV	Poly(p-phenylene vinylene)
P(Pyr), P(Py)	Poly(pyrrole)
PSS	Poly(styrene sulfonate)
P(T)	Poly(thiophene)
P(TV), PTV	Poly(thienylene vinylene)
P(3AT), P(AT)	Poly(3-alkyl-thiophene)
P(3DDT), P(DDT), PDDT	Poly(3-dodecyl thiophene)
P(3HT), P(HT)	Poly(3-hexylthiophene)
P(3MT), P(3MeT)	Poly(3-methyl thiophene)
P(3OT), P(OT)	Poly(3-octylthiophene)
P(4ABP)	Poly(4-amino biphenyl)
Other Polymers:	
HDPE	High-density polyethylene
LDPE	Low-density polyethylene
PAMPS	Poly(2-acrylamido-2-methyl-1-propane-sulfonic acid)
PE	Polyethylene
PEDOT	Poly(ethylenedioxythiophene)
PEG	Polyethylene glycol
PEMA	Poly(ethyl methacrylate)
PEO	Polyethylene oxide
PET	Polyethylene terephthalate
PMMA	Polymethyl methacrylate
PSS(A)	Poly(styrene sulfonate), poly(styrene sulfonic acid)
PVA	Polyvinyl alcohol
PVB	Polyvinyl butyral
PVC	Polyvinyl chloride
PVCz	Poly(vinyl carbazole)
PVS	Poly(vinyl sulfate)
Monomers:	
4ABP	4-Aminobiphenyl
ANi	Aniline
AT, 3-AT	3-Alkyl-thiophene
DPA	Diphenyl amine
DPBz	N, N'-Diphenyl benzidine
MT, MeT, 3-Me-T	3-Methyl thiophene
PV	p-Phenylene vinylene
Py, Pyr	Pyrrole
T	Thiophene
TV	Thienylene vinylene

(continued)

Abbreviation	Explanation
Dopants:	
PSS	Poly(styrene sulfonate)
PVS	Poly(vinyl sulfonate/sulfate)
Tos, TOS	*p*-Toluene-sulfonate (tosylate)
Trifl	Trifluoromethane sulfonate (tosylate)
Chemicals, Solvents:	
DMF	*N,N'*-Dimethylformamide (solvent)
Et	Ethyl
GBL, γ-BL	Gamma-butyrolactone (solvent)
Me	Methyl
PC	Propylene carbonate (solvent)
THF	Tetrahydrofuran (solvent)
Techniques, Methodology:	
AFM	Atomic force microscopy
CA	Chronoamperometry
CC	Chronocoulometry
CVA	Chronovoltabsorptometry
DFWM	Degenerate four-wave mixing
DPV	Differential pulse voltammetry
DSC	Differential scanning calorimetry
EELS	Electron energy loss spectroscopy
EH	Extended Hückel
EIS	Electrochemical impedance spectroscopy
EL	Electroluminescence
ENDOR	Electron nuclear double resonance (spectroscopy)
EPR	Electron paramagnetic resonance (spectroscopy)
EQCMB	Electrochemical quartz crystal microbalance
ESR	Electron spin resonance (spectroscopy)
FIA	Flow injection analysis
GC	Gas chromatography
GPC	Gel permeation chromatography
IR	Infrared
LB	Langmuir–Blodgett (film forming technique)
LWIR	Long-wave infrared
MWIR	Medium-wave infrared
MAS	Magical angle spinning (used in NMR spectroscopy)
MS	Mass spectrometry
Nd:YAG	Neodymium:yttrium-aluminum-garnet (laser, ca. 1.06 μm)
NLO	Nonlinear optic(s)
PV	Normal pulse voltammetry
NMR	Nuclear magnetic resonance (spectroscopy)
OCM	Open-circuit memory (optical memory retention)
PIA	Photo-induced absorption

(continued)

Abbreviation	Explanation
PIB	Photo-induced bleaching
PL	Photoluminescence
QCM	Quartz crystal microbalance
(%-)R	% reflectance
RBS	Rutherford backscattering
SEM	Scanning electron microscopy
SPEL	Spectroelectrochemical data, spectroelectrochemical characterization curves
SSH	Su-Schrieffer-Heeger (Hamiltonian)
STM	Scanning tunneling microscopy
(%-)T	% transmission
TGA	Thermogravimetric analysis
THG	Third harmonic generation
TPA	Two-photon absorption
UV–Vis	Ultraviolet–visible (spectral region)
VEH	Valence effective Hamiltonian
Vis	Visible (spectral region)
VRH	Variable range hopping (conduction model)
XPS	X-ray photoelectron spectroscopy
XRD	X-ray diffraction
Z-N	Ziegler–Natta (polymerization process)

Contents

About the Author

Prasanna Chandrasekhar received his B.Sc. (Honors) in chemistry from the University of Delhi, Delhi, India, in 1978; his M.Sc. in inorganic chemistry/X-ray crystallography from Concordia University, Montréal, Canada, in 1980; and his Ph. D. in electroanalytical chemistry from the State University of New York (SUNY) at Buffalo, Buffalo, New York, USA, in 1984. He was a postdoctoral associate at the Department of Chemistry and Materials Science Center, Cornell University, Ithaca, New York, USA, in 1984–1985, and a senior research scientist at Honeywell, Inc., in Minneapolis, Minnesota, USA, and Horsham, Pennsylvania, USA, from 1985 to 1987. From 1987 to 1992, he was Manager of Electrochemical Programs at Gumbs Associates, Inc., East Brunswick, New Jersey, USA.

Dr. Chandrasekhar founded Ashwin-Ushas Corporation (see https://ashwin-ushas.com/) in October 1992 and has been president, CEO, and sole owner there since. Ashwin-Ushas Corporation is a research company active in the defense and aerospace fields that has pioneered technologies based on conducting polymers (CPs), including IR-region electrochromics for terrestrial military camouflage, variable-emittance skins for spacecraft thermal control, visible-region electrochromics for safety and sunwear, and, most recently, voltammetric electro-chemical sensors for chemical warfare agents. Dr. Chandrasekhar is the author or coauthor of over 100 peer-reviewed publications and a large portfolio of worldwide patents.

Dr. Chandrasekhar's other interests include music, ancient history, archaeology, linguistics, and languages. He is the author of *NAVLIPI*, a universal alphabet published and patented in 2012 and applicable to all the world's languages. This claims to be the first truly phonemic script in the world (see http://navlipi.org/ and http://research.omicsgroup.org/index.php/Navlipi). He is also the author of several forthcoming books, including a primer of Sanskrit directed specifically to students of Western classical languages (ancient Greek, Latin) and a summary compendium of ancient Hindu philosophy.

Part I
Carbon Nanotubes (CNTs), Fundamentals

Chapter 1
Introducing Carbon Nanotubes (CNTs)

Contents

1.1   Historical

Carbon nanotubes (CNTs) were first widely popularized following the report of Iijima, working at Japan's NEC (Nippon Electric Company) Corporation's research laboratories, in 1991 [1, 2]. Iijima first prepared multi-walled CNTs in the arc discharge of graphite rods at high temperature, apparently while attempting to produce fullerenes (C_{60}, see below) [1, 2], although this same author had published electron micrographs of "carbonaceous materials" which were clearly multi-walled CNTs earlier, in 1980 [3]. Their first discovery and recognition as a different form of carbon appears, however, to have occurred considerably earlier. Radushkevich et al. [4] published an account in 1952 that clearly appears to describe CNTs. And Schützenberger and Schützenberger published an account in 1890 describing the production of "carbon microfilaments" obtained by passing cyanogen over porcelain at very high temperature [5].

Subsequent to this early work, publications of Endo et al. and Oberlin et al. around 1976 showed TEM (transmission electron microscopy) images of multi-walled CNTs grown via a chemical vapor deposition method [6]. After Endo's work, Baker and Harris [7] reported "filamentous carbon" with the structure of multi-walled CNTs, and Millward and Jefferson also published data clearly depicting multi-walled CNTs, both in 1978. Abrahamson et al. reported CNTs produced via an arc discharge method in 1979 [8]. The structures of CNTs produced via catalytic dissociation of CO were discussed [9]. Tennent obtained a US patent for what he described as carbon fibrils with a diameter of up to 70 nm, which appeared to be multi-walled CNTs [10].

© Springer International Publishing AG 2018
P. Chandrasekhar, *Conducting Polymers, Fundamentals and Applications*,
https://doi.org/10.1007/978-3-319-69378-1_1

It is also worthy of note that Ebbesen and Ajayan [11] were among the first to call what were, earlier, variously called "graphitic microtubules" and "nanoscopic carbon tubules" by their eventual name, i.e., carbon nanotubes. And several research groups had theoretically anticipated the existence of CNTs; the most prominent among these that come to mind are groups at the US Naval Research Lab [12] and the Massachusetts Institute of Technology [13], both of which described in 1992 just around the time of Iijima's seminal 1991 publication what they perceived as "fullerene tubes," and Jones described "rolled-up tubes of graphite" in 1986 [14].

1.2 The Very Basics of CNTs, Including Nomenclature

Graphite is an allotrope of carbon, other allotropes being C_{60} (buckminsterfullerene or simply fullerene, in the form of a geodesic dome) and diamond. Graphite is comprised of hexagonal sheets of sp^2 C-atoms ("hexane minus the hydrogens") stacked in approximately parallel fashion on top of each other. The C-atoms are bonded with σ bonds along their bond axis and delocalized π-electron density above and below their plane. *Graphene (presented in other parts of this book) then is simply a single sheet of this stack of C-atoms. CNTs in turn then are simply rolled-up graphene: graphene rolled up into cylinders with seamless bonds.*

Quite evidently, CNTs can come in the form of single tubes, known as single-walled carbon nanotubes or SWCNTs, and multi-walled tubes, MWCNTs. The multi-walled variety can be concentric cylinders, one inside another, or like a rolled-up document; the former is much more common. These varieties are depicted in Fig. 1.1. Much of the CNT literature also uses the self-explanatory terms "double-walled," "triple-walled," "multi-walled," and "few-walled" with, in the latter case, "few" generally meaning less than five.

Depiction of single- and multi-walled CNTs (SWCNT, MWCNT), together with actual images. Images reproduced with kind permission from Refs. [15] and [16].

If we take a sheet of graphene, and "cookie-cut" sections from it to eventually fold into a cylinder and form a CNT, it is evident that we can cut or stamp out our cookie in different ways, as shown in Fig. 1.2. In fact, scientists working in the field have come up with a mathematically based nomenclature to describe the manner in which the graphene "cookie" is cut to obtain a CNT. This is depicted in Fig. 1.3. In this nomenclature, two vectors, a_1 and a_2 (sometimes also designated simply a and b), are defined; both can be considered to originate in the top-left corner of the hexagonal unit of graphene, with a_1 terminating at the top-right corner and a_2 terminating at the bottommost corner of the hexagon in which one can then combine these two vectors to determine how the graphene sheet is cookie-cut to form the CNT. For example, if one uses five units of a_1 and five units of a_2, one obtains the "(5,5)" form shown in the figure. To form the long CNTs, one would use unit cells comprising such cookie-cut portions, which would then simply repeat

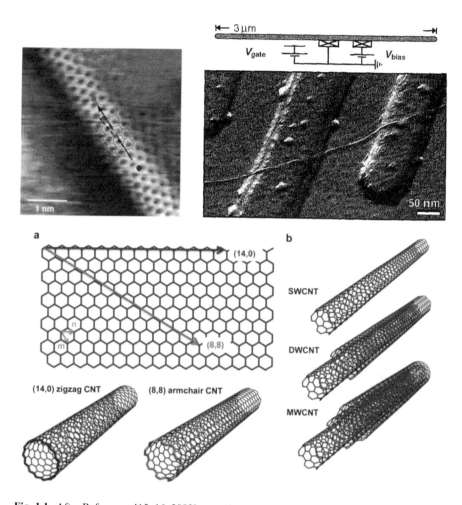

Fig. 1.1   After References [15, 16, 2083], reproduced with permission

along the length of the tube. The way the graphene is cookie-cut can then be defined by its *chirality vector* (also sometimes known as the *roll-up vector*):

$$\mathbf{C_h} = n\mathbf{a_1} + n\mathbf{a_2} \, \tilde{} \, (n, m) \tag{1.1}$$

where the *(n,m)* notation is used as shorthand, thus, e.g., defining a "(5,5) CNT." Certain values of *n* and *m* are in turn given further, shorthand names, *zigzag*, *armchair*, and the generic *chiral*, corresponding to the way in which the graphene edge appears to be cut. In the zigzag, the graphene is cut along the tops of the hexagons and $m = 0$, so this type of cut is designated $(n,0)$; in the armchair, it is cut along the centers of the hexagon and $n = m$, so this type of CNT is sometimes designated (n,n); and in the chiral, it can be cut in almost any way, defined by the

Fig. 1.2 Various ways in which a sheet of graphene can be "cookie-cut" or stamped out, to eventually fold into a cylinder and form a CNT (After Ref. [401] reproduced with permission)

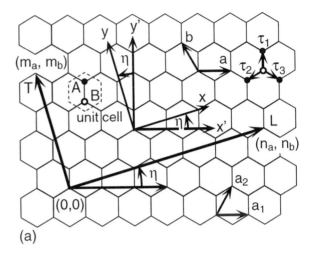

(a)

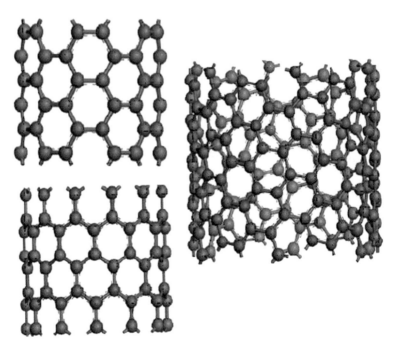

Fig. 1.3 Depiction of the *(n,m)* nomenclature used in "cookie-cutting" graphene to form CNTs (After Ref. [2084])

chiral parameters. This is also shown in Fig. 1.3. The vector C_h is always perpendicular to the axis of the cylinder of the CNT, denoted by the vector T, as also shown in the figure. A further, convenient nomenclature parameter is also defined as the angle, η, between the horizontal (zigzag) direction and the actual direction of the cookie-cut. As shown, for the zigzag direction, $\eta = 0$. For the armchair configuration, $\eta = \pi/6$.

The (n,m) values for an *ideal SWCNT* can be related to its diameter by the equation below; a caveat in this respect is of course that very rarely does one encounter an ideal CNT in real life:

$$d = \frac{a}{\pi}\sqrt{n^2 + nm + m^2} = 78.3\sqrt{\left((n+m)^2 - nm\right)}pm, \qquad (1.2)$$

where $a = 1.42(\sqrt{3}.)$ and Å, corresponding to the lattice constant in graphite sheet. The angle η can in turn be related to the n, m values by the equation:

$$\eta = \arctan\left[-\left\{(\sqrt{3} \div (2n+m))\right\}\right] \qquad (1.3)$$

It turns out that this (n,m) and η nomenclature is not just convenient for identifying and naming the type of CNT. It also determines, to a very significant extent, the conductive properties of the resulting CNTs (and, following from these, many other properties, though only applicable to ideal CNTs). As shown in a subsequent chapter in this part, the values of (n,m) can determine if an ideal CNT is metallic, semiconducting, or, in rare cases, nonconducting.

The CNTs encountered in experiment and in larger-scale production are not comprised of perfect *hexagons*, as depicted in our figures, but rather have many defects, including pentagons and heptagons, in their sidewalls; these of course cause properties of these more practical materials to be much inferior to those of pure materials or theoretically calculated properties 18]. Experimental evidence indicates that approximately 3% of the carbons in a CNT are at a defect site [19, 20].

The CNTs encountered in experiment are also most commonly *capped* at the ends. The cap may be an ideal hexagonal configuration of C-atoms, as shown in Fig. 1.4, or, more commonly, it may be an irregular configuration, usually a mix of C hexagons, pentagons, and heptagons. A commonly encountered type of defect is the "Stone–Wales defect," also referred to as a "7-5-5-7 defect"; this consists of two pairs of five-membered and seven-membered rings [19] and is also illustrated in Fig. 1.4.

1.3   PRIMER on Basic Properties of CNTs

When discussing the properties of CNTs as reported in the voluminous CNT literature, it is very important to realize that one is frequently presented with data that show extraordinary values of some physical or electronic property of CNTs, such as tensile strength or thermal conductivity, but many of these reports represent extrapolations to values for a single CNT (frequently a SWCNT), and some are even flash-in-the-pan observations difficult for others to repeat.

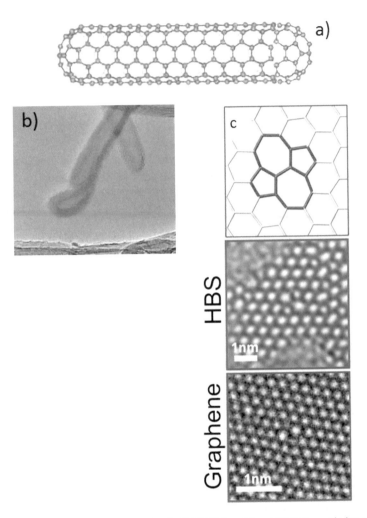

Fig. 1.4 Depiction of capping of the ends of a SWCNT (top (**a**) and TEM (transmission electron microscopy) image of an actual, capped MWCNT (**b**). Below (**d**), schematic representation of the Stone–Wales or 7-5-5-7 defect (After Refs. [2084–2086])

Nearly all CNTs display a C-C bond length of about 1.45 Å, representing an sp^2-C to sp^2-C bond. MWCNTs have intertubule spacings of about 3.4 Å, close to that of the inter-sheet spacing in graphite (3.35 Å).

Diameters of commonly produced SWNTs lie generally in the 0.8 to 2 nm range; those of MWNTs generally lie in the 5 to 20 nm range [18]. Diameters as small as 0.3 nm have been reported for an *armchair (2,2)* CNT [21]. Diameters of MWCNTs as high as 100 nm have also been reported [18].

A critical feature of most CNTs is that they have a very large length to diameter ratio, also known as an *aspect ratio*, sometimes as high as 100,000:1; in some cases, aspect ratios as high as 132,000,000:1 have been claimed, with lengths of up to 0.5 m (meter) [22].

CNTs are also said to possess exceptional physico-mechanical properties such as tensile strength (said to be greater than that of steel) and elastic modulus. However, it is also true that none of these exceptional mechanical properties have been converted to a "macro" product, i.e., a practical product that can be used in the real world, with anything close to the properties reported experimentally on nanotubes on a "nano-" or "micro-" scale, although incremental (20% to 60%) improvements to physico-mechanical properties have been obtained when CNTs are added to extant materials. Physico-mechanical properties and applications of CNTs are covered in separate chapters or parts in the sequel in this book.

1.4   CNTs, CNTs, and All That Hype

Very unfortunately, one of the issues that confronts all scientific research in the twenty-first century must do with what may be termed the "commercial–industrial funding complex" of modern scientific endeavor. This permits, increasingly, the unwarranted hyping up of discoveries that have attributes that make them stand out above the crowd. The ultimate and very prosaic and pecuniary objective is to simply obtain greater funding to further one's career path.

Putting this phenomenon in the perspective of CNTs, a recent cover story [23] in *Chemical & Engineering News*, the publication distributed weekly to all members of the American Chemical Society (the largest scientific society in the world with more than 100,000 members), gave recognition to the observation that much of the promise of CNTs from the time of their first popularization in the early 1990s may have been hyped. (Indeed, it noted that one of the proposed applications of CNTs had been an elevator to space, which was actually given serious consideration when it was proposed.) The number of scientific papers published in the field has grown nearly exponentially since the late 1990s. For example, a Google Scholar search of any field of CNTs confined to a single year (say 2012) will yield up to 100,000 papers. (The reader is welcome to try this.) Needless to say, a very large proportion of these papers represent "path-breaking" results that *cannot be reproduced*. Of course, the authors of such unreproducible work did not have to worry about this, because most of their colleagues were not interested in reproducing prior work, but rather in publishing new material in the same vein. Turning again to CNTs, a market research survey as recently as 2007 postulated that the commercial CNT market would amount to about US $500 million in 2015 [23]; in fact, as of this writing, the market is closer to $15 million and *declining*.

The sharp reader will of course recognize that this phenomenon is not confined to CNTs. It is everywhere in scientific research in the twenty-first century. Readers of a certain age will certainly recollect how often in the last few decades they may have heard such catchy phrases as "molecular electronics," "molecule-based sensors," "molecular motors," "molecular-scale heat pumps," "quantum nanowires," etc. Well, one might ask, decades hence, what became of these?

All the above having been said, it is nevertheless worth acknowledging that CNTs are today found in niche applications, such as additions to composite bicycle parts to make them slightly (about 25%) stronger; additions to polymer-matrix composites to make them slightly more (about 30%) thermally conductive and thus quicker to cure, increasing the specific strength of carbon fiber composites (about 50%); and the like [23].

1.5   Problems and Exercises

1. Briefly summarize the properties of CNTs, in terms of and including the following: (n,m) descriptors; C_h, chirality vector; zigzag, armchair and chiral CNTs, in terms of both general properties and (n,m,η) values; and Stone–Wales defect.
2. Briefly identify the typical ranges in diameters and lengths of SWCNTs vs. MWCNTs.
3. What in your opinion is the two most important reasons that CNTs have not borne out the original expectations of them since about the year 2000 in terms of applications?
4. Carry out an informal poll of several nonscientist friends, acquaintances, or colleagues for knowledge about what CNTs are and what kinds of products they are found in today.

Chapter 2
Conduction Models and Electronic Structure of CNTs

Contents

2.1 Conduction Models and Electronic Structure of CNTs

The electronic structure of CNTs is most easily discussed as applicable to ideal SWCNTs first and then conditionally extendable to imperfect (i.e., experimentally produced) SWCNTs and MWCNTs. The electronic structure of CNTs has been studied in detail theoretically, and some correlation with experiment has been observed. Theory predicts that CNTs can be semiconducting with a significant bandgap, semiconducting with a very small bandgap, and metallic, depending on their *(n,m,η)* parameters (discussed in an earlier chapter).

The underlying electronic structure of SWCNTs can be derived from that of a graphene sheet with neglect of hybridization effects due to the finite curvature of the tube structure, as discussed by Ouyang et al. and others [23]. Now graphene (the subject of later parts in this book) possesses valence and conduction bands which degenerate at only six corners of the hexagonal first Brillouin zone. (The Brillouin zone is a primitive cell in reciprocal space, whose boundaries are defined by planes and in turn defined by points on the reciprocal lattice. It is highly useful for treatment of periodic, repeating structures such as crystals or graphene sheets, since the solutions to the wave equation (the Bloch wave description) within the first Brillouin zone are easily extended to the entire, extended periodic structure.) The Fermi surface of the graphene sheet can thus be reduced to these six corners. Extending this to SWCNTs then, as a result of the periodic boundary condition represented in the equation below, *k* is:

$$k \cdot C_\mathrm{h} = 2\pi q$$
$$(k = \text{wavevector}, q = \text{integer})$$

(2.1)

© Springer International Publishing AG 2018

P. Chandrasekhar, *Conducting Polymers, Fundamentals and Applications*,

https://doi.org/10.1007/978-3-319-69378-1_2

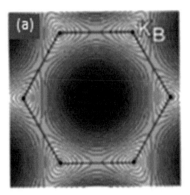
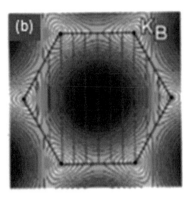
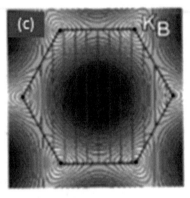

Fig. 2.1 Schematic representation of the electronic structure of the hexagonal graphene unit as applicable to the metallic or semiconducting character of ideal SWCNTs. Shown are (**a**) a 2D representation of the π and π^* energy bands of graphene, (**b**) allowed 1D wave vectors for a (9,0) SWCNT indicating metallic character, and (**c**) allowed wave vectors for a (10,0) SWCNT, indicating semiconducting character. The black dots at the hexagon apices are the $\mathbf{K}_B$ points, while the lines represent the allowed wave vectors in the first Brillouin zone (After Ref. [23], reproduced with kind permission)

Thus, only a particular set of states, which are parallel to the corresponding CNT tube axis with a spacing of $2/d_t$, are allowed (Fig. 2.1a–c). In consideration of this basic treatment, if one of the allowed wave vectors passes through a Fermi $\mathbf{K}_B$ of the graphene sheet, the SWCNT will be metallic; if it doesn't, the SWCNT will be semiconducting. The above equation thus reduces to the following, with the associated conditions for semiconducting or metallic behavior for an ideal SWCNT:

$$K_B \cdot C_h = 2\pi q$$

- Metallic when $(n - m) \div 3$ is an integer.
- Otherwise semiconducting.

The above simple treatment is depicted schematically in Fig. 2.1.

To further and more properly transfer the above 2D, graphene-based electronic structure treatment to 1D SWCNTs, the 2D graphene-based band structure can be *zone-folded* into the 1D Brillouin zone of an (n,m) SWCNT. A parameter called the *electronic density of states (DOSs)* can then be computed from the band structure by summing the number of states at each energy level. This π-only *tight binding model* yields several noteworthy properties for ideal SWCNTs [23]. As a first property, SWCNTs exhibit well-defined spikelike features in the DOS, denoted as *van Hove singularities* (VHSs). As a second property, the DOS at E_F is *zero* for semiconducting SWCNTs (where $(n - m) \neq 3q$, q being an integer), but it is *nonzero* for metallic SWCNTs (where $(n - m) = 3q$). As a third property, the VHS spacing has a characteristic "1-2-4" pattern relative to E_F (with spacing 1ξ-2ξ-4ξ, where $\xi = 2\pi/3|C_h|$) for semiconducting SWCNTs and "1-2-3" from E_F (with spacing 3ξ-6ξ-9ξ) for metallic SWCNTs. As a fourth property, the first VHS bandgaps for semiconducting and metallic SWCNTs are, respectively, $2\gamma_0 a/d_t$ and $6\gamma_0 a/d_t$; they are also independent of the chiral angle (η) to the first order.

The semiconducting and metallic character of SWCNTs as predicted and defined in the foregoing [24] was confirmed by experimental studies, e.g., those of Odom et al. [25] and Kim et al. [26]. This experimental corroboration served to show that the *π-only tight-binding model* of the electronic structure of CNTs is sufficient to predict experimental behavior at a basic level.

The finite curvature of SWCNTs is expected to influence electronic properties, due to the expected, significant hybridization of π, π^*, σ, and σ^* orbitals. One of its effects would be the opening up of small energy gaps at E_F. Additionally, due to the finite curvature, the Fermi points of "metallic" SWCNTs are shift away from original $\mathbf{K_B}$, from the $\mathbf{K_B}$ points along the tube circumference direction, while those of semiconducting SWCNTs shift along the tube axis direction. (Fig. 2.1c). The shifts in zigzag SWCNTs imply that wave vectors predicted to yield zigzag; "metallic" SWCNTs will no longer pass through the shifted Fermi points and will thus give rise to a small gap with $Eg = (3\gamma_0 a^2/d_t/16R')$. This is in agreement with experiment, which has indicated that the $(n,0)$ SWCNTs that were predicted to be metallic are in reality narrow-gap semiconductors, with bandgap magnitudes inversely proportional to the square of the CNT radius [23]. Figure 2.2 depicts the *calculated* DOS of CNTs as a function of energy for two configurations, the *armchair (5,5), considered metallic*, and the *zigzag (7,0)*, considered semiconducting.

Figure 2.3 depicts one example of agreement of theory and experiment, with calculated DOS based on the π-only tight-binding model appearing to agree with experimental observations

Figure 2.4 depicts an energy diagram comparing the DOSs in semiconducting vs. metallic SWCNTs.

A caveat concerning the above treatment of the electronic structure of *ideal* (meaning defect-free, among other properties) SWCNTs is that properties (such as electrical conductivity, semiconducting character, etc.) actually observed in real-life systems will be considerably "inferior" to those predicted. For example, the n-fold symmetry of an "armchair" SWCNT, which forms one of the elements of the

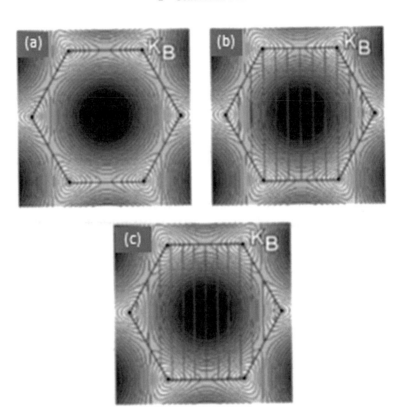

Fig. 2.2 *Calculated* DOS of CNTs as a function of energy for two configurations, the *armchair (5,5)*, considered *metallic*, and the *zigzag (7,0)*, considered *semiconducting* (After Ref. [24], reproduced with permission)

above treatment of electronic structure, can be broken by interactions between nanotubes in a tightly packed bundle of CNTs; in that case, a gap will open at E_F, resulting in a very significant reduction of the electrical conductance [23].

Thus, it has been observed that to realize the full potential of CNTs, one must produce "ideal," defect-free SWCNTs [23]. In practice, this is well-nigh impossible. Thus, electrical conductivities as high as 10^9 A/cm^2 are predicted theoretically for ideal SWCNTs [28, 29], more than 10^3 times that of Cu; in actual, experimental practice, nothing close to this value has been realized as of this writing, with the best conductivities observed in CNTs simply equaling those of Au [30]. Furthermore, conductivities of CNTs, and especially those of MWCNT bundles, strongly depend on the method of production and even the particular batch of CNTs produced in an identical process [16]. In typical experimental observation, batches of MWCNTs show both metallic and semiconducting behaviors, with observed resistivities, at 300 K, of about 1.2×10^{-4} to 5.1×10^{-6} ohm-cm and activation energies of <300 meV [16].

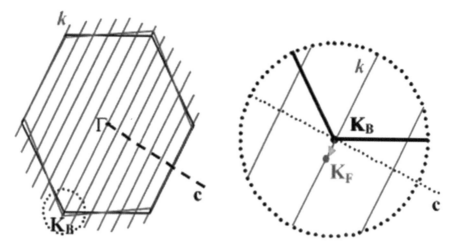

Fig. 2.3 One example of agreement of theory and experiment, with calculated DOS based on the π-only tight-binding model appearing to agree with experimental observations: tunneling conductance, dI/dV, for (15,0) zigzag SWNTs, with corresponding calculated DOS shown below. The new feature near E_F is highlighted with a dashed circle. Inset shows high-energy resolution normalized conductance, (dI/dV)/(I/V), for the (15,0) tube (After Ref. [23])

Fig. 2.4 Schematic energy diagram comparing the density of states (DOSs) in semiconducting (*left*) vs. metallic (*right*) SWCNTs. (Gray, valence bands; white, conduction bands) (After Ref. [27], reproduced with permission)

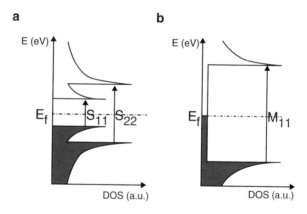

Other experimental results worthy of note with respect to the electronic (and related, magnetic) structure and properties of CNTs include: (1) The observation of "quantum nanowire" behavior in SWCNTs; in such "quantum wires," electrical *conduction occurs via well-separated, discrete electron states that are quantum-mechanically coherent over long distances.* Tans et al. [31]; necessarily, such work involved the painstaking isolation and electrical contacting of individual SWCNTs; how this was done in another, similar study is shown in Fig. 2.5. (2) The observation of the predicted *van Hove singularities* and the confirmation of the importance of the wrapping type and angle (i.e., the *(n,m,η)* parameters) in determining semiconducting or metallic behavior [24]. (3) Magnetoresistance measurements on

Fig. 2.5 An example of
methodology used to carry
out electrical measurements
on single CNTs (After Ref.
[2086], reproduced with
permission)

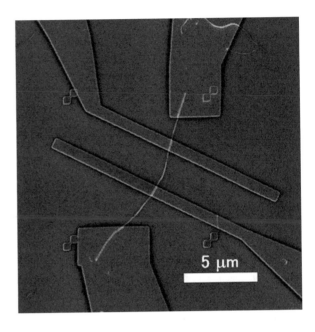

5 µm

individual MWCNTs in which the Aharonov–Bohm effect was observed [32]. The
Aharonov–Bohm effect is the dependence of the phase of the electron wave on the
magnetic field when electrons pass through a cylindrical electrical conductor
aligned in a magnetic field, where their wavelike nature shows up as a periodic
oscillation in the electrical resistance as a function of the enclosed magnetic flux.
This causes a phase difference, and hence interference, between partial waves
encircling the conductor in opposite directions. Such oscillations have been
observed in µm-scale thin-walled metallic cylinders and would thus be expected
to be observed in near-metallic, thin-walled CNTs.

2.2 Problems and Exercises

1. Briefly discuss the electronic structure of ideal SWCNTs as it relates to that of a
   single hexagonal graphene unit.
2. Discuss and compare the properties of ideal semiconducting vs. metallic
   SWCNTs in terms of their different structural properties and their (n-m) values
   and zigzag or other properties.
3. What is the effect of the finite curvature of SWCNTs on their electronic
   properties, especially in comparison with that of graphene?
4. Compare the DOS (density of states) for metallic vs. semiconducting CNTs.
   Sketch these on a piece of paper or electronic drawing board, if possible.

Chapter 3
Synthesis, Purification, and Chemical Modification of CNTs

Contents

3.1 Brief Enumeration of Methods of Synthesis

The methods of production of SWCNTs and MWCNTs comprise *electric arc discharge* (currently primarily of historical importance), *chemical vapor deposition (CVD)* (currently the most widely used), *laser ablation* and *plasma torch* methods, and several *miscellaneous methods.*

It is also important to recognize at the outset that CNTs (along with other carbon allotropes such as fullerenes (C_{60})) *occur naturally* in soot from hydrocarbon flames. These natural samples may be random mixtures of MWCNTs and a very few SWCNTs with widely varying diameter, length, and other parameters, and they may also be mixed intimately with graphite fibers and amorphous carbon.

Different methods produce differing proportions of MWCNTs vs. SWCNTs, with control of the relative proportions only within limited ranges. Diameter of CNTs produced is more controllable, although, again, even if CNTs of close to the same diameter are produced, they frequently have widely differing chirality. This and related factors lead to problems with the use of CNTs for potential electronic applications, since, for example, mixtures of semiconducting and metallic CNTs may be obtained in the same synthesis batch and be difficult to separate further. One of the well-established reasons that as-prepared CNTs occur in bundles which

© Springer International Publishing AG 2018
P. Chandrasekhar, *Conducting Polymers, Fundamentals and Applications,*
https://doi.org/10.1007/978-3-319-69378-1_3

require a lot of effort to further fractionate is that CNTs strongly interact with each other via van der Waals forces and thus have a tendency to "bundle."

Accordingly, methods to further differentiate and isolate SWCNTs from MWCNTs, or CNTs of a particular chirality, have been developed. These are discussed in the Purification section in this chapter.

Another issue with as-produced CNTs is that they are highly susceptible to effects of exposure to oxygen and water vapor in ambient air. Their electrical resistance, thermoelectric power, and local density of states (DOS) are extremely sensitive to exposure to air or oxygen, as confirmed experimentally by transport measurements, scanning tunneling spectroscopy, and other techniques [25]. Such parameters are said to be reversibly tunable by surprisingly small concentrations of adsorbed gases.

3.2   Electric Arc Discharge

Electric arc discharge is of historical importance for CNT production since it was the method reported in Iijima's seminal 1991 publication "rediscovering" CNTs [33], which first widely popularized CNTs. Needless to say, Iijima's group was attempting to produce fullerenes and rediscovered CNTs in the soot of the graphite electrodes used. However, it is also used (at this writing) for low-volume production of SWCNTs.

In typical application of this technique [14], a direct current (80–100 A) is passed through two high-purity graphite (typically diameter 6 to 10 mm) electrodes separated by about 1 to 2 mm, in a He atmosphere (500 Torr). Localized temperatures achieved can be as high as 1700 °C. During arcing, a deposit forms at a rate of about 1 mm per minute on the cathode, while the anode is consumed in the process. The deposit so obtained has a cigar-like structure, on the periphery of which a hard gray shell is deposited. The soft inner core contains CNTs along with other carbon structures. Closer examination of this cathode deposit using scanning electron microscopy (SEM) or other techniques reveals two quite different textures and morphologies: a gray outer shell, which consists primarily of graphene layers, and the soft inner core deposits. The latter, bundle-like structures reveal randomly arranged MWCNTs along with other carbon structures.

The above-described process produces MWCNTs almost exclusively. If it is desired to produce SWCNTs, a metal catalyst is added. Among the first such adaptations to produce SWCNTs were those carried out by the Iijima group [34] and Bethune et al., both in 1993 [34]. In the former work, Fe-graphite electrodes were arced in a methane-Ar atmosphere, yielding CNT diameters between 0.75 and 13 nm. In the latter work, Fe-Co-Ni-graphite mixtures were arced in a He atmosphere, yielding CNTs with an average diameter of 1.2 nm, found packed in bundles contained within spider weblike deposits formed inside the arc chamber. More refined catalysts developed since this work include those incorporating Gd, Co-Pt,

Co-Ru, Co, Ni-Y, Rh-Pt, and Co-Ni-Fe-Ce [14]. The Ni-Y catalyst in particular gives high yields of SWCNTs with diameters of about 1.4 nm [14].

Among issues associated with these production methods are the high purity of inert gases (He or Ar) required and thus the significant costs. Additionally, typical throughput of these electric arc discharge-based methods is about 50 g/day [16].

3.3 Chemical Vapor Deposition (CVD) and Variants and Refinements Thereof

CVD is as of this writing the most widely used method for high-volume CNT production. It also appears to offer the highest throughput, about 50 kg/day in continuous production [16]. Typically in this process [14, 15, 34], hydrocarbons such as methane, benzene, acetylene, naphthalene, and ethylene are decomposed over metal catalysts (e.g., Co, Ni, Fe, Pt, and Pd) and deposited on substrates such as silicon, graphite, Cu, or silica, usually with a catalyst support (alumina, magnesia, etc.). Substrates are typically heated to temperatures of about 700 °C. The fluidized bed reactors typically used allow for uniform gas diffusion and heat transfer to the catalyst particles. A process gas, e.g., NH_3, N_2, or H_2, is bled into the reaction chamber along with the hydrocarbon. The size of the catalyst particles can determine the diameter of the CNTs grown. Other conditions determining CNT properties and quality of course include the type of process gas used, the pressure of the deposition chamber, and the temperature (with lower temperatures generally producing a higher proportion of MWCNTs). As of this writing, the mechanism is still not well understood [36].

One of the drawbacks of this method is that the CNTs produced are frequently contaminated with the metal catalyst particles; this also leads to generation of defects in the CNTs and thus less-than-optimal electronic and other properties, as discussed in a previous chapter. This is somewhat circumvented by advanced, costly, and time-consuming chemical cleansing (including acid etching/cleansing), or thermal annealing, to remove the contaminants; in this respect, rudimentary acid treatments used to cleanse the catalysts in the early years of the technology were found to be deficient [36].

One of the advantages of the CVD technique in general is that it shows the most promise for potential very large-scale CNT production, since the CNTs grow directly on a selected substrate, whereas, in the other techniques, they must be isolated from odd growth products (such as cigar-shaped growths) which have other carbon structures as well.

One variant of the CVD technique is *plasma-enhanced* CVD, where a plasma is generated by the application of a strong electric field and where CNT growth then follows the direction of the electric field [37]. CNTs grown by this technique are generally oriented perpendicular to the substrate rather than randomly; such growth is generally termed a "carpet" or "forest" [38].

Another specialized version of CVD pioneered by the Smalley group is to use short, precut "seeds" of individual nanotubes to grow longer tubes of the same diameter [39]. Yet another specialized method that yields very dense, mm-high "forests" of CNTs perpendicular to the substrate is one in which water is added during the CVD process [40]; separation from the catalyst contaminants is also said to be easier in this method, which yields primarily SWCNTs.

Other more recently used variants of CVD synthesis appear to yield *aligned* CNTs without the need for further processing. These include production of self-aligned CNTs on catalyst-coated substrates, CVD systems with floating catalysts, and use of Si wafers with patterned, parallel line arrays in combination with Fe and $CoSi_x$ catalysts [41–44].

3.4   Laser and Plasma Methods

In the most basic version of the *laser ablation* method, a graphite target is vaporized using a pulsed laser (typically a YAG laser) in a high-temperature reactor [45–49], yielding temperatures of 1200 °C or higher up to 3000 °C in many cases [16]. An inert gas is then slowly introduced into the chamber. CNTs form on the cool surfaces of the reaction vessel. Variants of the method include the use of water or a refrigerant to cool the surface and the use of metal catalysts incorporated into the graphite target. The method is said to produce a major proportion of SWCNTs, with reactor temperature determining the diameter of the CNTs produced to a fair extent. For example, in one of the first such specialized syntheses, using graphite-Co-Ni-graphite targets yielded bundles of SWCNTs containing primarily (10,10) nanotubes with a ~ 1.4 nm diameter [50].

A method that is somewhat expensive but also produces a high proportion of SWCNTs is the *plasma torch* method, first reported by Smiljanic et al. [51]. This method also vaporizes carbon, like the laser ablation and arc discharge methods, except that the target is a carbon-containing gas rather than graphite. A gaseous mixture of an inert gas (usually Ar), a carbon-containing gas (usually ethylene), and a vapor-form catalyst (usually ferrocene) is subject to an intense flame at atmospheric pressure. SWCNTs are found in the "fumes" generated by this process and isolated when condensed on a suitable substrate. What results is a quicker and potentially continuous process, lower cost, and a larger proportion of SWCNTs. Variants of this method include the generation of *thermal plasma* via high-frequency oscillating currents in a coil in a flowing inert gas [52]; this method yields throughputs that are about two to three times those of laser ablation.

3.5 Other Methods

A unique method that received attention in the mid-1990s [53, 54] but has since been for the most part abandoned, and is worthy of mention for purposes of documentation only, used electrolysis of graphite electrodes immersed in molten LiCl immersed in a graphite crucible (serving as the counter electrode); it produced primarily MWCNTs of widely varying diameter.

Another method also receiving much attention in the early years of active CNT research (i.e., the 1990s) but less attention due to its less practical value is the substitution of solar energy for the heat source on graphite catalyst metal targets [55–57]. Concentrated solar energy can generate temperatures as high as 3400 K.

3.6 Functionalization of CNTs

Functionalization of CNTs serves various objectives, e.g., *dispersion*; due to their generally hydrophobic nature, CNTs tend to "stick" to each other in hydrophilic or amphiphilic solvents commonly used for processing. Functionalization with hydrophilic groups (e.g., carboxylate, see below) allows for greater dispersion and compatibility with such solvents. *Fractionation*, e.g., of SWCNTs from MWCNTs, of CNTs of particular diameters, and of CNTs of particular chiralities. *Analysis:* Functionalization is used for analytical purposes, e.g., with attached fluorescent tags for imaging [14, 15, 47], and *functionalization for particular uses*, e.g., sensors or batteries. Increase of mechanical strength or other physical properties. Functionalization of CNTs is sometimes broadly classified into two categories, *covalent* and *noncovalent*.

CNTs are widely offered for sale as suspensions in liquids [15]. To achieve stable suspensions typically requires chemical modification of the CNT surface and/or addition of surfactants. Prior to use, the surfactants are removed by copious washing in deionized water; in the case of chemical modification, thermal annealing removes the modification agent [15].

The curved nature of CNTs makes their sidewalls as well as end-caps more reactive than flat graphene sheet, and these features of CNTs are useful targets for functionalization. A currently very common method of functionalizing CNTs thus is to attach substituent groups to their walls or ends, with the most common substituents being carboxylate groups. A functionalization procedure finding wide use in earlier CNT work involves ultrasonic treatment in a mixture of concentrated nitric and sulfuric acid, which leads to the opening of the CNT caps and formation of holes in the sidewalls, followed by oxidative etching along the walls [16, 58]. The resulting CNT fragments of length 100 to 300 nm have ends and sidewalls decorated with a high density of oxygen-containing groups, primarily carboxyl groups.

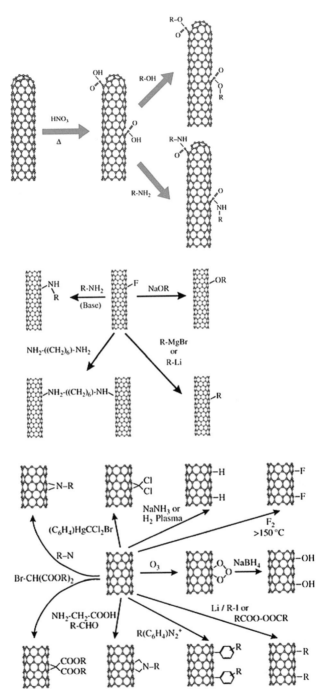

Fig. 3.1 Schematic illustration of various processes for functionalization of CNTs, including *carboxylate functionalization* followed by esterification or amidization, *addition*, and *fluorine functionalization* followed by *nucleophilic substitution* (After Ref. [16], reproduced with permission)

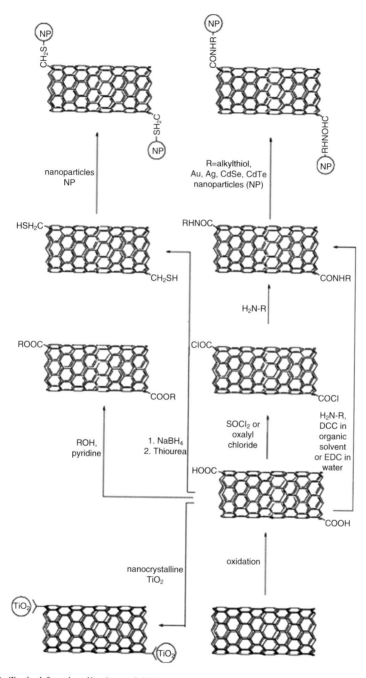

Fig. 3.2 Typical functionalizations of CNTs using carboxylate derivatives as a basis (After Ref. [25], reproduced with permission)

Fig. 3.3 Examples of
functionalizations of CNTs
with complex molecules for
potential drug delivery and
related biomedical
applications (After Ref.
[70], reproduced with
permission)

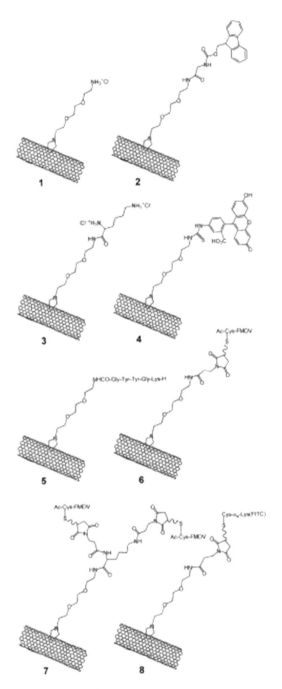

Carboxylate functionalization and *fluorine functionalization* of CNTs remain important stepping stones to many varieties of further functionalization of CNTs. Other functionalizations involve *addition, nucleophilic substitution*, and *electrophilic substitution* reactions. Some of these processes are illustrated schematically in Fig. 3.1. Variants of nucleophilic substitution reactions include those carried out with Li and Mg compounds which generated alkyl-substituted CNTs [59] and nucleophilic addition reactions with amines with Li reagents, yielding amino-substituted CNTs [60]. Acyl-substituted CNTs have also been obtained via Friedel–Crafts acylation [61]. Figure 3.2 shows typical functionalizations using carboxylate derivatives as a basis.

Noncovalent functionalizations of CNTs are said to minimally disrupt their structural features and electronic properties but at the cost of chemical stability, problems with phase separation, and other issues [62].

Other, miscellaneous functionalizations worthy of mention are those with sugars [63], those with phospholipids [64], those with pyrene derivatives and aryl thiols [65], and those involving addition of a pyrrolidine ring [66–69] (Fig. 3.3).

Nitrogen-doped CNTs have been studied for uses such as semiconductor devices, since they possess inherent doping [25], and Li battery electrodes [71], since they possess defects in their walls that are critical for diffusion and retention of Li ions. In the latter case, the resulting electrodes are said to possess storage capacities and cyclabilities superior to conventional intercalated graphite electrodes. N-doped CNTs are produced via modification of standard methods used for production of CNTs such as CVD. For example, in one method, NH3 is contacted with amorphous carbon at high temperature [72]. In another method, N-doped CNTs with low concentrations of nitrogen ($<1\%$) are synthesized via arc discharge in the copresence of melamine, nickel, and yttrium or via pyrolysis of pyridine and methylpyrimidine [73, 74]. Aligned arrays of N-doped CNTs (N-content $<2\%$) are produced via pyrolysis of aminodichlorotriazine and triaminotriazine over laser-etched Co thin films at 1050 °C [73, 74].

Electrochemically induced functionalization is another avenue adapted for functionalization of CNTs. Two examples are the covalent grafting of phenyl residues onto bulk SWCNTs via electrochemical coupling of aromatic diazonium salts [75] and the similar, claimed phenyl functionalization of single, "individually addressable" SWCNTs [76].

Figure 3.4 illustrates a more complex functionalization of CNTs, for sensing applications. A similar example is functionalization by phenylboronic acid moieties for glucose sensing by both bonding with poly(aminophenylboronic acid) and, alternatively, covalent attachment of phenylboronic acid using diazonium chemistry [28, 77]. Figure 3.5 illustrates other complex functionalization of CNTs, this time with porphyrin moieties. Here, photo- and redox-active porphyrins are linked to carboxylate-functionalized CNTs via esterification [25, 78, 79]. FTIR and thermal analysis indicate that amide or ester bonds are formed to bridge between the excited state electron donor and SWCNT. Other depictions in this figure include derivatizations with charge-transfer tetrathiafulvalene (TTF) derivatives, fluorescent dyes, polypyridyl moieties, and a polymer with a phthalocyanine derivative [25, 79–83].

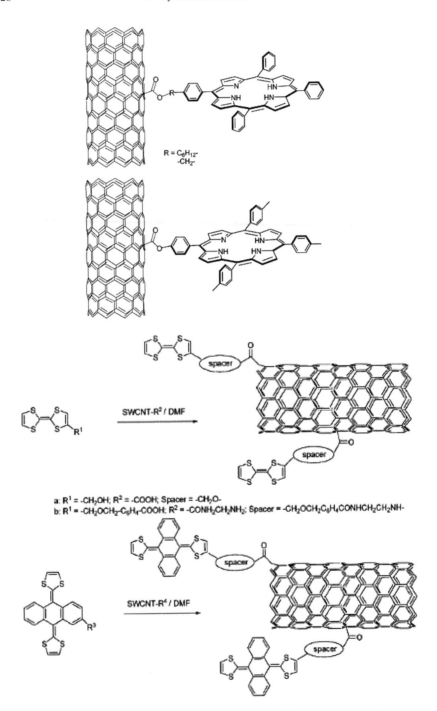

Fig. 3.4 Example of other complex functionalization of CNTs: porphyrin moieties, charge-transfer tetrathiafulvalene (TTF) derivatives, fluorescent dyes, polypyridyl moieties, and a polymer with a phthalocyanine derivative (After Ref. [25], reproduced with permission)

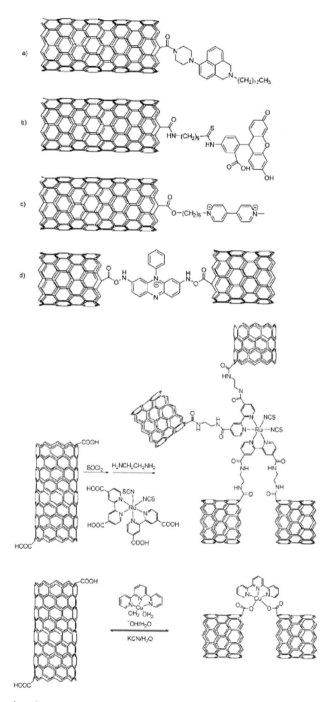

Fig. 3.4 (continued)

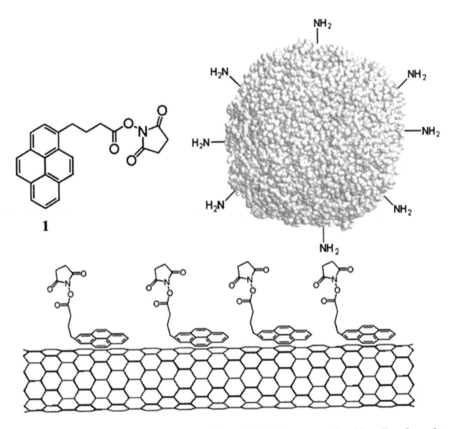

Fig. 3.5 Adsorption of 1-pyrenebutanoic acid succinimidyl ester onto the sidewall surface of a CNT, which provides a chemically reactive anchor for the immobilization of proteins. The ultimate application is sensing (After Ref. [28], reproduced with permission)

Photochemically induced functionalization of CNTs has been rarely used. One of the few examples of this is the sidewall osmylation of SWCNTs via exposure to OsO_4 under UV light, which appears to lead to photoinduced cycloaddition of the OsO_4 to C-C double bonds, giving rise to *reduced* conductivity due to the reduction in π-electron density in the CNTs [84, 85].

3.7 Brief Synopsis of Purification Methods

As noted above, "bulk" CNTs prepared using one of the above-described methods are frequently a combination of various diameters, a combination of SWCNTs with MWCNTs (even when methods known to give high yields of SWCNTs are used), and certainly a combination of various chiralities. Thus, following synthesis of "bulk" CNTs, various varieties of purification are used to further process and isolate particular fractions desired, such as bulk SWCNTs only or MWCNTs of a particular diameter or SWCNTs of a particular chirality (twist or wrapping mode). These methods are now very briefly described.

Among the common methods for purification of CNTs, especially those produced via CVD, which are contaminated with catalyst metal residues, are *thermal annealing* and *acid cleaning* and variants thereof. As an example of a combination method, microwave heating in air followed by treatment with HCl is claimed to reduce the level of catalyst metal residues to <0.2 w/w%; acid treatments followed by fluorination or bromination are claimed to cut, purify, and suspend the materials uniformly in desired organic solvents [86, 87]. Thermal annealing generally involves various methods of selectively oxidizing other carbon products (e.g., amorphous C particles), converting them to CO_2, while the CNTs remain unoxidized.

Methods used for *separation* of SWCNTs *by chirality* include density-gradient centrifugation combined with surfactant wrapping [88] and gel chromatography. Density-gradient centrifugation relies on the fact that a CNT's chirality determines how certain *surfactant* molecules arrange on its surface. The molecules functionalize CNTs with different chiralities differently. These small differences in surface chemistry create small differences in CNT buoyancy, which allows ultracentrifugation to sort CNTs into groups with closely matching chiralities [20].

A common method used to separate semiconducting from near-metallic CNTs is selective functionalization of SWNTs using 4-hydroxybenzene diazonium salts. Separation is carried out by deprotonation of the *p*-OH-phenyl group on the reacted, metallic CNTs in alkaline solution, followed by electrophoretic separation of these charged species from the neutral, semiconducting CNTs. This is followed by annealing, yielding separate groups of pristine semiconducting and metallic SWNT [89].

3.8 Common Methods Used to Characterize Synthesized CNTs

Among the most common methods for characterization of CNTs are Raman spectroscopy, high-resolution transmission electron microscopy (TEM), scanning tunneling microscopy (STM), and, less commonly, other, more specialized microscopies such as electrochemical microscopy.

The various microscopies yield dimensional as well as structural information. Spectroscopic techniques are useful for determining such parameters as the degree of semiconducting or metallic character and the extent of the sp^2-bonded framework of CNTs. As an example of the latter, in the Raman spectrum of SWCNTs, monitoring of the relative intensity of the D-line at 1290 cm^{-1} yields information on the disorder within the sp^2-bonded carbon framework [90]. Simple UV–Vis–NIR and IR spectroscopies can yield information on the bandgap via the valence to conduction band transitions of CNTs, which are found in the 400 to 650 nm, 830 to 1600 nm, and 600 to 800 nm spectral regions.

Semiconducting SWCNTs emit near-infrared light upon photoexcitation, which is absent in metallic SWCNTs; furthermore, the fluorescence is linearly polarized along the CNT axis. Emission wavelengths vary between 0.8 and 2.1 μm. These fluorescence (photoluminescence) properties of SWCNTs, in combination with their UV–Vis–NIR absorption and Raman scattering (as briefly described above), permit monitoring the orientation of SWCNTs without direct microscopic observation [91]. For this purpose, the CNTs are normally dispersed to reduce intertube quenching.

3.9 Prices and Production Volumes of CNTs as of This Writing

As of this writing, the above-described methods are able to produce MWCNTs for about US$100/kg, still about ten times the cost of specialized carbon fibers [15]. CNTs are currently offered for sale as powders or as suspensions in liquids.

Due to the predicted great promise (i.e., hype) for CNT applications in the 1990s, manufacturing capacities of about 3,000 tons of CNTs per year have been built up worldwide as of this writing; however, also as of this writing, the worldwide demand is said to be less than 600 tons per annum and declining [20]. Translating these figures to US$ values paints an even bleaker picture: as compared to about US $500 million revenues predicted for CNTs for 2014, actual values as of this writing are US$10 million [20].

Very recent reports have claimed breakthroughs in higher-volume production of CNTs. For example, the startup Molecular Rebar claims to have solved the problem of "unentangling" high-volume production CNTs, which are frequently "entangled" and thus largely useless for specialized applications, using a proprietary, multistep acid treatment and filtration process [92].

3.10 Problems and Exercises

1. Outline the five main classes of methods of synthesis of CNTs. Which are still used as of this writing (2016) and which are merely of historical importance?
2. For each of the methods you listed in Problem 1, briefly describe a typical synthesis. (You may describe these generally, e.g., omitting specific temperatures or time periods.)
3. For each of the methods you listed in Problems 1 and 2, identify the advantages and drawbacks.

4. Briefly outline the methods of chemical functionalization of CNTs. Specify the chemical groups used for such functionalization, the reasons for their choice, and the particular utility of each.
5. Briefly outline the methods used for purifying CNTs.
6. Briefly outline the most common methods used for characterizing CNTs and the advantages and limitations of each.
7. Briefly describe the current (2016) state of commercial production and cost of CNTs.

Chapter 4
Physical, Mechanical, and Thermal Properties of CNTs

Contents

4.1 Caveat on Experimental Versus Theoretical Versus "Practical" Properties

As with other properties of CNTs, their physical, mechanical, and thermal properties as reported in the literature must be taken with a strong caveat: properties of individual CNTs or isolated samples ("shells" or "micro-bundles") of CNTs do *not* usually translate into properties of as-produced, larger bundles of CNTs or anything close to a practical application. And of course, theoretically computed properties [96] nearly always differ from experimentally measured ones, substantially in many cases; thus, statements such as "nanotubes are indeed the strongest material known" [96] must be taken with a pinch of salt, as it were. One of the reasons for this, among many others, is that as-produced CNTs possess a significant population of *defects* which compromise mechanical properties such as strength. As-synthesized CNTs are not individual SWCNTs or MWCNTs; rather, they are randomly agglomerated CNT bundles [97].

It is additionally to be noted that there is also a significant *asymmetry* in the mechanical and thermal properties of individual CNTs in the axial (i.e., along the tube) vs. the radial direction, by factors of from 2 to 100 [98, 99]. And due to their high aspect (length to diameter) ratio, CNTs are subject to what is sometimes termed "structural instability" or buckling, rendering many experimental reports of very high strength and related properties practically useless [100].

© Springer International Publishing AG 2018
P. Chandrasekhar, *Conducting Polymers, Fundamentals and Applications*,
https://doi.org/10.1007/978-3-319-69378-1_4

4.2 Physical and Mechanical Properties

Taking into consideration the caveat outlined above, mechanical properties such as *tensile strength, specific strength, Young's modulus*, or *bulk modulus* reported for individual or small samples of CNTs almost never translate into properties of larger CNT samples or of composite materials in which CNTs are a component; this is amply illustrated by the discussion further below.

Thus, for instance, SWCNTs as well as MWCNTs have been reported with *tensile strength* in excess of 60 GPa, to be compared with that for stainless steel (a maximum of about 1.5 GPa) and for Kevlar (a very high strength poly(aramid) fiber used, among other things, in car tires, with a maximum tensile strength of about 3.6) [101–106]; indeed, estimates as high as 150 GPa have been made for the tensile strength of SWCNTs [101, 107], although values in the region of 12 to 52 GPa are more common [108]. Nevertheless, in spite of the above data on individual CNTs, the tensile strength of MWCNTs *bulk bundles* has been measured to only about 2 GPa [109].

Similarly, *Young's modulus*, a property defining how much strain or deformation occurs when a certain stress is applied to it (i.e., roughly, how rigid a material remains with applied stress), is comparatively about 0.2 TPa for stainless steel, 0.18 TPa for Kevlar, and 1.1 TPa for SWCNTs, with values up to 3.6 TPa reported for SWCNT and 2.4 TPa for MWCNT [100–107, 110, 111, 2073, 2074].

Values for another mechanical property, *elongation at break*, have been reported at 16 for SWCNTs vs. 20 to 50 for stainless steel [102–106].

Theoretically calculated values of these properties for CNTs (thus far not realized, needless to say) are even higher: Tensile strength of 126.2 for armchair-configuration SWCNTs and Young's modulus of 5 for SWCNTs [102–106].

Similarly, the *bulk modulus* (a measure of how resistant to compression, i.e., how hard, a material is) of a variety of CNTs known as "superhard" CNTs as high as 546 GPa has been claimed to be measured, a value exceeding that of diamond (420 GPa) [112], although bulk modulus values for more "conventional" SWCNTs are in the 38 to 40 GPa range [101].

As an example of the disconnect between "ideal" and "practical" mechanical properties, it is noted that yarns made from high-quality MWCNTs have been reported with a *stiffness* of 357 GPa and a tensile strength of 8.8 GPa. However, this is so only for a yarn length that is comparable to the mm-long individual CNTs within the yarn; as soon as one jumps to cm-scale yarn lengths, the tensile strength drops to about 2 GPa, close to that of Kevlar [18, 46]. Furthermore, larger yarn lengths yield even poorer tensile strengths.

4.3 Thermal Properties

It is worthwhile to remember, when discussing the thermal conductivity of CNTs (or, for that matter, graphene, later in this book), that the *thermal conductivity of other carbon allotropes*, e.g., *diamond (3D) or graphite (in-plane only), is also already extremely high*. The in-plane thermal conductivity of graphite at room temperature is about 3000 W/K. In contrast, some studies have yielded values for *bulk* MWCNTs of less than 50 W/K [113]. Values for room temperature-specific thermal conductivity for diamond, Si, and GaN are, respectively, >20 W/K-cm, 2 W/K-cm, and 1.5 W/K-cm [114].

Phonon transport is the primary means of thermal conductivity in materials such as graphite or CNTs. Many studies have shown that phonon transport in CNTs, including its temperature dependence, differs significantly from that in graphite [17, 115, 116]. For example, one study found that the phonon mean free path is about 500 nm in MWCNTs, and the temperature dependence of the thermal conductivity exhibits a peak at 320 K due to the onset of *umklapp* phonon scattering.

Experimental studies have shown that the *specific heat* of MWCNTs produced using pyrolytic methods varies linearly with temperature between 10 and 300 K [117–120].

In one study [118], it was found that the measured specific heat of single-walled CNTs differs from that of both 2D graphene and 3D graphite, *especially at low temperatures*, where 1D quantization of the phonon band structure is observed. The measured specific heat shows only weak effects of intertube coupling in nanotube bundling, suggesting that this coupling is weaker than expected. The thermal conductivity of nanotubes was found to be large, even in bulk samples: aligned bundles of SWNTs showed a thermal conductivity of >200 W/m K at room temperature. A linear K(T) up to approximately 40 K may be due to 1D quantization; measurement of K(T) of samples with different average nanotube diameters supports this interpretation.

In another study [120], it was found that the average thermal conductivity of MWCNT films (film thickness from 10 to 50 mm), prepared using a microwave plasma CVD method and studied using a pulsed photothermal reflectance technique, was about 15 W/mK at room temperature and independent of the tube length. However, the authors claimed that, taking a small volume filling fraction of CNTs into account, the effective nanotube thermal conductivity could be as high as 23,102 W/m K, which is still smaller than the thermal conductivity of diamond and in-plane graphite by a factor of 9 and 7.5, respectively.

A detailed study of CNT composites incorporating randomly oriented multi-wall carbon nanotubes (MWCNTs) and randomly oriented single-wall carbon nanotube (SWCNTs) from 300 to 400 K [119] showed that the specific heat of randomly oriented MWCNTs and SWCNTs showed similar behavior to the specific heat of bulk graphite powder. The specific heat of aligned MWCNTs was found to be smaller and had a weaker temperature dependence than that of the bulk above room

Fig. 4.1 Calculated specific heat (lines) of graphene, isolated SWNTs, graphite, and strongly coupled SWNT bundles. The solid points represent the measured specific heat of a bulk sample of SWNTs (After Ref. [119], courtesy of what-when-how)

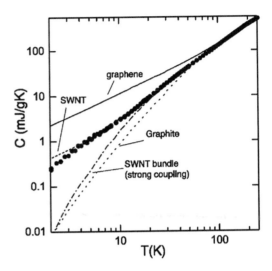

temperature. The effective thermal conductivity of randomly oriented MWCNTs and SWCNTs is similar to that of bulk graphite powder (about 0.73 J/gK at 300 K), exhibiting a maximum value near 364 K indicating the onset of phonon–phonon scattering. The effective thermal conductivity of the anisotropic MWCNTs increased smoothly with increasing temperature and was claimed by the authors to be indicative of the 1D nature of the heat flow. Figure 4.1 below shows representative data from this study.

4.4   Problems and Exercises

1. Briefly outline the factors present in "practical," as-produced CNTs that make their properties differ significantly from those predicted theoretically. Give examples of at least three such properties that differ.
2. Enumerate typical values of the following physical properties of SWCNTs and MWCNTs and compare them to those of typical, common materials (e.g., steel): *tensile strength, specific strength*, *Young's modulus*, or *bulk modulus*.
3. Enumerate and compare the thermal properties (e.g., *thermal conductivity, specific heat*) of all the carbon allotropes, at several different, exemplary temperatures.

Chapter 5
Toxicology of CNTs

Contents

The issue of *toxicity* of CNTs to humans and animals, primarily due to their nanostructural nature combined with their relatively high reactivity, started to be addressed only recently, more than a decade after their "rediscovery" in the early 1990s [116–119].

It is certainly relevant to note (and in some sense, quite self-evident) that the specific toxicity of a particular batch of CNTs will depend on its provenance – what synthetic method was used to produce the CNTs in that batch, their size, diameter, chirality, chemical functionalization relative concentration, and other properties and even the dispersant used when they are used in dispersed form [119]. Nevertheless, it is also now safe to generalize that *nearly all types of CNTs are toxic to humans* and animals, to some degree, although some researchers [18] have sought to draw a distinction between large bundles of long MWCNTs, which they believe are as or more toxic as asbestos, and shorter SWCNTs, which they believe are minimally toxic, but there is no published evidence thus far to support this contention (see below). Some studies have also indicated that the range of concentration of CNTs that is toxic to humans may be very large: from 5 ng/mL to 10 mg/mL [119], although it is also true that these studies were not "standardized" for type, dimension, chirality, etc. of the CNTs tested.

Firme and Bandaru [119] cogently summarize the large number of studies of toxicity of CNTs on a wide variety of cells carried out in vivo (primarily in mice) through 2010; the takeaway from their summary is that all these studies, taken together, are quite inconclusive, and in many cases conflict with each other, due apparently to the lack of standardized methods of testing toxicity CNTs. As an example of such conflict, one study [120] appears to indicate that the "majority of" intravenously injected CNTs in mice are simply passed through in urine, with "far less" found in the liver, spleen, and lungs. Another study however, indicates that the liver and spleen are the main sites of CNT accumulation [120]! The key then is in the interpretation of the words that we have put in quotes in the first study, i.e., "majority of" and "far less."

© Springer International Publishing AG 2018
P. Chandrasekhar, *Conducting Polymers, Fundamentals and Applications*,
https://doi.org/10.1007/978-3-319-69378-1_5

The primary method of ingestion into humans is understood generally to be *respiratory* [118, 121]. CNTs are known to cross membrane barriers, inducing inflammation, fibrosis, formation of granulomas (small, nodular growths), and, in some cases, cell death after accumulation in the cytoplasm [116, 117, 122–124]. CNTs are believed to align lengthwise in the human body (resembling asbestos fibers to some extent) and to react with metal ions in body fluids [116, 118, 121, 122]. Following ingestion into the human body, two mechanisms are further postulated whereby CNTs can breach the cell wall and enter cells: phagocytosis or endocytosis, which is the body's natural reaction to a foreign ("not-self") intruder, and nanopenetration [119]. Phagocytosis involves the encircling and containment of larger foreign intruders (e.g., bacteria, dimension ca. 1 µm); endocytosis is the counterpart of phagocytosis for smaller intruders such as viruses (dimension ca. 0.1 µm). Nanopenetration is the random, "passive" penetration, literally piercing, of the cell wall by electrically neutral, nanometer-size objects with high aspect ratios, such as very tiny metal needles or of course CNTs.

Although there are very limited, published epidemiological studies on the effects of CNTs in humans [119, 123–126], many other studies have hinted at potential carcinogenicity of CNTs [119, 127–129]. The greatest concern appears to be to workers in manufacturing and/or research who handle CNTs on a routine basis, with the postulated danger being through inhalation of aerosolization of CNT powder or small bundles. The US National Institute of Occupational Safety and Health (NIOSH) has issued specific guidelines for handling and working with CNTs [129].

An issue somewhat unique to CNT toxicology is what may be termed as "secondary" toxicity, i.e., toxicity of impurities or residues from fabrication methods used for synthesizing CNTs. Foremost among these are of course the metal catalysts used in most methods of synthesizing CNTs. Thus, it is noted that Fe and Ni catalysts used in CNT synthesis may constitute up to 40 w/w% of the product in some cases; such metal impurities may generate harmful free radical species or further products such as the superoxide ion in vivo [119, 130].

(Secondary toxicity arising from dispersants or carriers used to contain CNTs is self-evident. For example, quite toxic dichlorobenzene, pyrene, and tetrahydrofuran, to name just a few, are used commonly to disperse or carry CNTs [119]. Quite obviously, however, ample information on toxicity of such materials is available elsewhere, and it thus cannot be proposed to discuss such secondary toxicity in this chapter.)

Several mechanisms have been postulated for the toxicity of CNTs. One mechanism is said to involve activation of pathways that lead to DNA damage [119, 131, 132]. For example, it has been observed that mesothelial cells exposed to SWCNTs at concentrations ca. 25 µg/cm^2 activate DNA recovery along with changes in the cell cycle and generation of apoptotic signals and that cells incubated with CNTs halt at the G1 phase of the cell cycle [119, 131, 132]. It has also been experimentally observed that, somewhat perversely, normal mesothelial cells are more susceptible to DNA damage by CNTs than malignant mesothelial cells, as indicated by significantly higher levels of the DNA repair protein poly(ADP-ribose) polymerase

(PARP) [131]. Yet another potential mechanism, and one that may be very ominous as it indicates potentially very wide-ranging toxicity, is the ability of CNTs to adsorb chemicals and nutrients found in mammalian blood: for example, an in vitro study showed that common amino acids and vitamins (e.g., phenylalanine and folate) are passively adsorb onto CNTs, and concentrations of SWCNTs as low as 0.01–0.1 mg per mL of culture are able to deplete 2 nM of folic acid from solution [133].

Quite evidently, one of the most important consequences of the potentially high toxicity of CNTs to humans is that *applications* such as drug delivery, in vivo sensors, etc. *may be precluded* unless these toxicity issues are resolved.

5.1   Problems and Exercises

1. Briefly summarize (in one or two paragraphs) the toxicity of CNTs. Differentiate SWCNTs and MWCNTs as necessary.
2. What are the typical laboratory studies used to establish such toxicity (identify at least two distinct types of studies)?
3. After reading this chapter, what is your opinion on the likely in vivo medical applications of CNTs?

Part II
Carbon Nanotubes (CNTs), Applications

Chapter 6
Brief, General Overview of Applications

Contents

Since their "rediscovery" in the early 1990s, CNTs have been studied exhaustively from a theoretical and experimental as well as applications point of view. Applications researched have encompassed an extremely wide field. The applications below, while not exhaustive, are representative:

- *Specialized materials*, e.g., those possessing high strength, high conductivity (electrical as well as thermal), greater elasticity, high electromagnetic absorption (e.g., for radar cross-section (RCS) reduction), and other desirable properties. The high strength applications encompass not only postulated esoteric applications such as "elevators to space" but also more mundane ones such as yarns for more durable textiles and composite materials with higher strength and greater durability.
- *Batteries and energy devices*, including especially Li batteries, supercapacitors, solar cells, and materials for hydrogen storage. Many of the applications in such areas as Li batteries have originated in simple replacement of already established additives such as activated-C and graphite.
- *Sensors and actuators*, including electrochemically based sensors and biosensors (especially those coupled with DNA and RNA) but including other sensor types such as mechanical (e.g., strain) sensors, optical power detectors, acoustic detectors, and fluorescence-based imaging. It also includes actuators and "nanomotors."
- *Drug delivery*.
- *Microelectronics*, "*nanoelectronics*," and "*nano-bioelectronics*" *applications*, including "flexible electronics," "wearable electronics," and "biocompatible electronics" and computing applications.
- *Displays and transparent, conductive films/substrates*. This especially includes field emission displays and conductive replacements that seek to be replacements for transparent, conductive ITO (indium tin oxide) coatings.

© Springer International Publishing AG 2018
P. Chandrasekhar, *Conducting Polymers, Fundamentals and Applications*,
https://doi.org/10.1007/978-3-319-69378-1_6

- *Electrical conductors* (to replace metals), "*quantum nanowires*" or "*quantum wires*" (ideal one-dimensional conductors), potential *superconductors*, and other conductivity-based and one-dimension-based, specialized applications.
- *Environmental applications and materials used in separation science*. These include filters, membranes, and water treatment.

Miscellaneous Applications These include catalysis and esoteric or unconventional applications. As noted in Part I of this book, however, as of this writing (2016), CNTs are found [23] only in minor, niche applications, e.g., additions to composite bicycle parts to make them slightly (about 25%) stronger, additions to polymer-matrix composites to make them slightly more (about 30%) thermally conductive and thus quicker to cure, and increasing the specific strength of carbon fiber composites (about 50%). In spite of periodic reports of "breakthroughs" in mass production of CNTs [95, 139], commercially viable mass production for specialized applications remains distant as of this writing (2016).

Some companies, such as Zyvex Technologies [140], whose products are centered on specialized adhesives and composite materials for sport, marine, and other applications, have done well integrating CNTs into their products, frequently replacing more mundane materials such as carbon fibers. However, in such cases of presumed success as well, it is difficult to ascertain whether the reason for the success of the product is the claimed, incremental performance improvement or simply the "sex appeal" of carbon nanotubes in selling the product.

6.1 Problems and Exercises

1. Briefly list the nine (9) major categories of applications of CNTs, and cite one specific example in each.
2. From your reading of this chapter, as of 2016 what were the successfully commercialized applications of CNTs? And as of your reading today, if it is at least a year beyond 2016?

Chapter 7
CNT Applications in Specialized Materials

Contents

7.1 Overview

As noted in an earlier chapter, *specialized materials* include those possessing high strength, high conductivity (electrical as well as thermal), greater elasticity, high electromagnetic absorption (e.g., for radar cross-section (RCS) reduction), and other desirable properties. The high-strength applications encompass not only postulated esoteric applications such as "elevators to space" [141] but also more mundane ones (and ones that have actually found some commercial application) such as yarns for more durable textiles and composite materials with higher strength and greater durability. It is also to be noted that although SWCNTs show greater promise for CNT-based specialized materials in theoretical and limited, laboratory, experimental studies, most efforts at developing practical products have concentrated on MWCNTs, due to their considerably lower cost and greater ease of fabrication in bulk quantities.

7.2 Textiles

A first requirement for potential use in *textiles* is to successfully spin CNTs into *yarns* with improved bulk properties (such as tensile strength, elasticity, modulus, etc.) as compared to conventional yarns, even if these properties do not approach those for individual or a few CNTs. Such fabrication methods have included in situ chemical vapor deposition (CVD) spinning to produce continuous CNT yarns from CVD-grown CNT aerogels or drawing out CNT bundles from CNT forests and

© Springer International Publishing AG 2018
P. Chandrasekhar, *Conducting Polymers, Fundamentals and Applications*,
https://doi.org/10.1007/978-3-319-69378-1_7

subsequently twisting them into fibers using the "draw-twist" method [142, 143], or "co-spinning" with other fibers, such as cellulose-based fibers and other methods [144–149]; however, these processes usually yield very high strengths or other desirable properties only for short (about 1 mm) pieces of yarn, but strengths that are 1/10 or less of these for yarns are longer than 10 mm. In rare instances, SWCNTs with lengths up to several cm have been fabricated (in that case via catalytic pyrolysis of n-hexane) [150]. Spinning of MWCNTs produced in CNT "forests" has been used to produce two-ply yarns with the appearance of cotton or wool, a density of about 0.8 g/cm^3, and a tensile strength of 460 MPa [151, 152].

A process for pressure-induced interlinking of CNTs has been proposed for weaving them into high-strength *fibers* and, ultimately, high-strength cloth for applications such as bulletproof cloth [153]; once again, however, the practical (let alone commercial) implementation of this novel application is lacking to date. Applications sought have included textiles that have antibacterial, flame retardant, electrically conductive (e.g., for electromagnetic impulse (EMI) shielding), and other very wide-ranging properties [147–149].

Data have been published showing that yarns made from high-quality MWCNTs have reached a stiffness of 357 GPa and a strength of 8.8 GPa but only for a gauge length that is comparable to the mm-dimension CNTs within the yarn. On the other hand, cm-dimension gauge lengths displayed a strength of 2 GPa, corresponding to a gravimetric strength equaling that of commercially available Kevlar (DuPont). It has been noted that macroscale CNT yarns never achieve theoretically predicted strengths primarily because the probability of critical flaws increases with scale [18, 46, 154]. A process for the production of high-performance fibers containing *aligned* SWCNTs using coagulation spinning of CNT suspensions has been demonstrated in the laboratory [155], but scale-up of this process has not been demonstrated to date.

7.3   Composite Materials

Due to the high cost of producing SWCNTs, or MWCNTs with a specific diameter range and chirality, CNTs used in composite materials have, to date, primarily comprised bulk (bundles of) CNTs. Nevertheless, incremental improvements in mechanical, thermal, and electrical performance parameters have indeed been achieved in such materials. Examples of applications of this nature *currently implemented commercially* include:

- Use in bicycle components such as handlebars and seat bases for greater strength and in marine products, including small boats, implemented by Zyvex Technologies [140].
- The *Hybtonite* brand of carbon–epoxy composites manufactured by a Finnish company, Amroy Europe Oy [2058], wherein conventional carbon fiber is replaced by CNTs, yielding materials with up to 30% greater strength than

standard graphite–epoxy composites. Claimed applications for these composites are as varied as ice hockey sticks, surfboards, skis, and wind turbines.

Regarding mechanical properties of composite materials, it has been observed that the major difference of CNTs from conventional, fiber-reinforced composites is that the scale is narrowed to nm from μm and that CNTs can impart high stiffness as well as high strength [156, 157]. For example, it has been claimed in one study that, for load-bearing applications, CNT powders mixed with polymers or precursor resins can increase stiffness, strength, and toughness, and adding ca. 1 w/w% MWCNTs to epoxy resin enhances stiffness and fracture toughness by 6 and 23%, respectively, without compromising other mechanical properties [18, 158, 159]. However, it has not been clear from these studies whether these properties are improved enough from those of corresponding composites containing graphite or other carbon forms to warrant the extra cost involved in fabricating CNT-containing composites.

One of the critical issues in obtaining high-performance CNT-based composite materials is the achievement of good adhesion between the matrix and the individual CNTs, required for an effective transfer of the mechanical load onto the CNTs [19]. One approach to achieve this is polymer synthesis from covalently attached initiator sites, e.g., in radical polymerization of methacrylate esters on CNTs [160].

In other recent work, CNT-containing fiber composites have been created by growing aligned CNT forests onto glass, SiC, alumina, and carbon fibers. Fabric made from CNT-SiC impregnated with epoxy displayed in-plane shear interlaminar (mode II) toughness by 54%, and similar CNT-alumina fabric showed 69% improved mode II toughness (41). Applications proposed for such have included lightning strike protection and deicing for aircraft [18, 158, 161, 162], although, again, to date no commercial products have yet emerged.

Silica-coated MWCNTs have been prepared using sol-gel techniques combined with thermal annealing, which are claimed to be resistant to combustive oxidation up to about 1200 °C [163].

Aluminum-MWCNT composites with strengths comparable to stainless steel (0.7–1 GPa) at one-third the density have been demonstrated [18]. Again, however, to date no commercial product has emerged from this work.

MWCNTs have also been proposed as flame retardant additives to plastics (to replace halogenated flame retardants which have environmental issues), emanating from rheological changes induced by CNTs loading [164].

It has also been observed that, by replacing the carbon black in rubber tires with CNTs, improved skid resistance and reduced abrasion are obtained [165]; (needless to say, in spite of all these excellent experimental results, as of this writing, there are no automobile tires on the market that incorporate CNTs.) In other elastomeric applications, a CNT–[amine-terminated poly(dimethylsiloxane) (PDMS)] composite was developed with tensile modulus, and elongation at break vastly improved over corresponding PDMS control materials [166].

With respect to electrical conduction properties (e.g., when CNTs are added into resin precursors of composites to increase their electrical conductivity), CNTs' high

aspect ratio is said to allow them to form a percolation network at low concentrations (<0.1 w/w%); in some cases, polymer composites fabricated with MWCNTs
at a 10 w/w% loading have claimed conductivities of 10,000 S/m or higher
[18, 167]. Quite logically, highly electrically conductive CNT composites have
been proposed for use in electromagnetic interference (EMI) shielding and electrostatic discharge (ESD) prevention, e.g., as packaging for microelectronics and in
spacecraft (NASA's Juno spacecraft apparently used a partial CNT-based shield for
ESD prevention) [18].

Other, varied applications of CNTs have included scaffolding for bone cell
growth [168] and coatings for radar cross-section (RCS) reduction, where the
high absorption of MWCNTs in the microwave region combined with their "black-
ness" in the UV-Vis-NIR and IR spectral regions is an advantage. To date, however,
practical implementation has not been possible due to immunological and toxico-
logical issues in the former application and inability to compete with standard RCS
coatings such as the conducting polymer poy(aniline) (due to cost and related
issues) in the latter application [169].

7.4   Problems and Exercises

1. Briefly describe typical methods used to incorporate CNTs in textiles. Name at
   least two proposed applications.
2. Name at least one product type in the category of composite materials where
   CNTs have been commercialized as of 2016.
3. What are the advantages and drawbacks of the use of *coated* MWCNTs in
   composite materials vs. uncoated MWCNTs? Support your answer from the
   physical principles behind the strength of composite materials.

Chapter 8
CNT Applications in Batteries and Energy Devices

Contents

8.1 Overview

As noted in an earlier chapter, a large area of potential applications of CNTs are *batteries and energy devices*, including, especially, Li batteries, supercapacitors, solar cells, and materials for hydrogen storage. Many of the applications in such areas as Li batteries have originated in simple replacement of already established additives such as activated-C and graphite.

8.2 Batteries, Including Li Batteries

To date, the foremost area of effort for use of CNTs in batteries and energy devices has been in the field of Li batteries, mainly secondary (rechargeable) batteries. In this area, CNTs have found some limited commercial use as well, mainly as additives to cathode materials such as $LiCoO_x$ and, less frequently, additives to Li-intercalated graphite anodes.

For example, a 1 w/w% CNT loading in $LiCoO_2$ cathodes and graphite anodes has been shown to increase electrical connectivity and mechanical integrity of these electrodes, thus, ultimately, leading to higher battery rate capability and cycle life [165–167]. Many lab studies have shown very exceptional performance for Li secondary batteries incorporating CNTs in one or both electrodes. Examples of

© Springer International Publishing AG 2018
P. Chandrasekhar, *Conducting Polymers, Fundamentals and Applications*,
https://doi.org/10.1007/978-3-319-69378-1_8

such laboratory studies include MWCNT/CuO anodes with storage capacities up to 700 mAh/g [168], MWCNT-V_2O_5 composite electrodes with specific capacities of more than 400 mAh/g [169], and electrodes containing ball-milled SWCNTs with claimed capacities of up to 1000 mAh/g [170]. Unfortunately, such studies have not been translatable, to date, to enhance performance in practical, commercially viable Li batteries.

More recently, exotic batteries, such as "paper (Li) batteries," incorporating CNTs have been demonstrated wherein CNT films function as current collectors for anode as well as cathode [171–173].

Chen et al. [174] recently described a Li battery based on a 3D-interconnected, hybrid hydrogel comprised of a TiO_2 interspersed in a CNT-conducting polymer (CP) composite, with the CP being poly(ethylene dioxythiophene) doped with poly (styrene sulfonate). The flexible Li-CNT-CP electrode appeared to demonstrate excellent Li-ion intercalation therein, high cycling stability, a capacity of 76 mAh/g in 40 s of charge/discharge, and a high areal capacity (2.2 mAh/cm^2 at a 0.1C rate of discharge). Lee et al. [175] used a layer-by-layer approach to assemble a several-μ m-thick electrode that consisted of additive-free, densely packed, and functionalized MWCNTs. The electrode displayed a gravimetric capacity of ~200 mAh/g(electrode), power of 100 kW/kg(electrode), and a lifetime in excess of thousands of cycles. Guo et al. [176] described sulfur-impregnated, disordered CNT cathodes for Li-sulfur batteries with a unique stabilization mechanism for the sulfur in carbon and claimed superior cyclability and coulombic efficiency. However, to date, none of this work has been translated into a commercial, practical Li battery or supercapacitor.

CNTs have also been studied for use in *lead-acid batteries*. In one study, with incorporation of ca. 0.5–1 w/w% of CNTs into the cathode, the cycling characteristics of the battery were improved nearly twofold as compared to a control battery having no CNTs (1000 vs. 600 cycles to 40% discharge capacity) [177]; the proposed explanation for the improved performance was the ability of CNTs to act as a better physical binder than graphite.

8.3 Capacitors and Supercapacitors

Early work demonstrated that supercapacitors incorporating MWCNTs could achieve specific capacitances of up to 200 F/g and power densities of more than 8 kW/kg [178–181]. Some of the simplest methods of fabrication of CNT-based supercapacitors involve electrophoretic deposition from a CNT suspension or directly grown on a graphite or Ni substrate [182]. Attempts to improve performance via such methods as introduction of Ag nanoparticles, MnO_2, or the conducting polymer poly(aniline) have produced mixed or irreproducible results [183–186].

Excellent performance has been demonstrated for supercapacitors incorporating forest-grown, binder, and additive-free SWCNTs, specifically an energy density of

16 Wh/kg and a power density of 10 kW/kg for a 40-F supercapacitor with a maximum voltage of 3.5 V, with a forecast lifetime of about 15 years based on accelerated life testing [187]. Nevertheless, the high cost of the SWCNTs used was perceived as a major impediment to commercialization. Liu et al. [188] fabricated supercapacitors by casting SWCNTs from suspensions in several solvents on the surface of Pt or Au electrodes. They observed an increase in the effective capacitance of electrodes with SWCNT films at 0.5 V (vs. a Ag quasi reference electrode) of 283 F/g, approximately twice that of carbon electrodes in nonaqueous solvents.

Once again, however, as of this writing, none of the above supercapacitor work has been translated into a commercial, practical supercapacitor.

8.4   Fuel Cells

For standard fuel cells that combine H_2 and O_2 to produce water, CNTs have been studied for use as a catalyst support for potential reduction of Pt usage by up to 60% as compared with carbon black and, possibly in the future, complete elimination of Pt use [189–191]. They have also been studied for use in methanol fuel cells [197]; in this case, MWCNTs were grown on C-fibers using a CVD method, followed by deposition of Pt particles of about 1.2 nm particle size.

8.5   Solar Cells

CNTs have been studied for use in organic solar cells, as a means to reduce undesired carrier recombination and enhance resistance to photooxidation. Also for potential applications in solar cells, CNTs have been studied for incorporation into CNT-Si heterojunctions, due to the observation of efficient multiple-exciton generation at p-n junctions formed within individual CNTs [192, 193].

8.6   Storage of Hydrogen and Other Gases

One of the predicted properties of CNTs, said to arise from their cylindrical shape and geometry and nanoscale dimensions, is the potential to store gaseous H_2 at liquid-H_2 densities [96, 194, 195]; one of the postulated mechanisms for this is capillary action on a nanoscale in CNTs. Since the initial studies of H_2 storage in CNTs [196, 197], many strides have been made in this field. Nevertheless, experimentally measured H_2 storage parameters for "medium-purity" SWCNTs remain at a maximum level of w/w% of about 4.2%, at a temperature of 300 K and pressure of 10.1 MPa, as of this writing; these may be compared with corresponding values for graphite of 4.52% at 298 K and 11.35 MPa [96]. Clearly, significant improvement in the practical performance of CNTs is required before they can be considered seriously for use in H_2 storage.

It should also be noted that many irreproducible reports of high H_2 storage capacity in CNT-based materials have been questioned on the basis of the fact that they could arise due to metallic impurities (e.g., Ti) in the catalytically produced CNTs [12].

CNTs have also been studied for storage of other gases, e.g., Ar, He, and SF_6, with very mixed results [12].

8.7   Problems and Exercises

1. Briefly enumerate the method of functioning of secondary (rechargeable) Li batteries and identify their major components. Then identify which components and/or processes in such batteries are most amenable to improvement with the addition of CNTs, and identify how.
2. Briefly enumerate the method of functioning of supercapacitors. Then identify which components and/or processes in such supercapacitors are most amenable to improvement with the addition of CNTs and identify how. Enumerate the typical applications of supercapacitors.
3. In which component of fuel cells can the introduction of CNTs lead to improvement in their function and how?
4. Construct a schematic diagram of a solar cell and describe the specific function of CNT layers in its improvement (e.g., via prevention of carrier recombination).
5. What was the status of the use of CNTs in hydrogen storage, as of 2016? And as of your reading today?

Chapter 9
CNT Applications in Sensors and Actuators

Contents

9.1   Overview

As noted in an earlier chapter, extensive work has been carried out in *sensors and actuators* incorporating or based on CNTs. This includes electrochemically based sensors and biosensors (especially those coupled with DNA and RNA) and other sensor types such as mechanical (e.g., strain) sensors, optical power detectors and acoustic detectors, and fluorescence-based imaging. This also includes "nanomotors."

9.2   Actuators and Nanoscale Devices (Including "Nanomotors")

CNTs have been studied for such uses as nanoscale artificial muscles [198], nanoscale heat engines [199], and other nanoscale devices [200, 201]. In one study, starting with "aerogel sheets" of MWCNTs, actuators were fabricated that displayed ca. 180% actuation along their width with 5 ms delay time between applying the potential used to activate the actuator and observing the maximum stroke [202, 203], a rate slightly superior to that of the human muscle. However, claims of "artificial muscle stronger than steel and more flexible than rubber" appeared not to be substantiated by the results [202, 203].

© Springer International Publishing AG 2018

P. Chandrasekhar, *Conducting Polymers, Fundamentals and Applications*,
https://doi.org/10.1007/978-3-319-69378-1_9

Up to about 2010, there was much hype about *"nanomotors"* based on CNTs [204–207] with such phrases as "world's smallest synthetic rotation motor" [204] and "carbon nanotube electron windmill" [207].

An early (2008) publication from the Wang group at Arizona State University (USA) claimed to show that incorporation of CNTs into the Pt component of asymmetric "metal nanowire motors" leads to dramatically accelerated movement in H_2O_2 solutions, with average speeds (50–60 µm/s) approaching those of natural biomolecular motors [206]. Further acceleration – to 94 µm/s, with some motors moving above 200 µm/s – was claimed to be observed upon adding hydrazine to the peroxide fuel. According to these authors, factors influencing the accelerated movement included the CNT loading and fuel concentration.

However, to date, *there is no CNT-based nanomotor under further development and certainly none in actual commercial or other uses.*

9.3 Chemical Sensing, Including Gas Sensing and Electrochemically Based Sensing

One of the claimed advantages of CNT-based chemical (including gas) sensors is that since the sensor is nanoscale, the quantities of analyte required for sensing are also nanoscale [96, 47, 208–211]. Unfortunately, this has not been borne out by practical results to date, as the accounts below will demonstrate.

Some of the first sensing work carried out with CNTs was for sensing of gases such as NO_2, NH_3, CO, H_2, and CO_2 at the ppm level; this was primarily based on conductometric or capacitive sensing [212–223]. Sensors for toxic gases, including chemical warfare agent (CWA) simulants, and volatile organic compounds such as nitrotoluene (down to the hundreds of ppb) have been tested [224–227]. In some cases, low detection limits, e.g., 10 ppb for a conductometric NO_2 gas sensor operating at 165 °C [228], were reported, although the nongeneric nature of the conductometric response was not clearly established. Somewhat greater specificity has been obtained in some SWCNT-based sensors. For example, a conductometric sensor using a SWCNT film overlaid with Pd nanoparticles is claimed to be more specific toward H_2 than other gases [229–231]. Such functionalized CNT sensors have also been demonstrated for H_2S and CH_4 [232–234].

Microwave resonance has also been used for CNT-based sensing: Naishadham et al. described a unique, highly sensitive antenna-based "smart skin" ammonia gas sensor utilizing CNTs and inkjet printing [235], with the parameter measured being return loss corresponding to change in impedance. Others have also demonstrated microwave resonance-based sensing of NH3 using MWCNT and SWCNT sensors [236].

Since CNTs are essentially a carbon allotrope, they can potentially be used for *electrochemically based sensing* wherever carbon-based substrates, such as glassy carbon, activated-C, C-paste, graphite fiber, and other carbon substrates, are used as electrodes. Potentially enhanced performance is expected due to their nanoscale and higher conductivity. For example, enzymes may be immobilized on them for amperometric detection. A common example is glucose oxidase for glucose detection as used in commercially available, non-CNT glucometers for diabetes patients; in the case of CNTs, glucose oxidase can simply be mixed with MWCNT paste to fabricate the sensing electrode [237]. Examples of such enzyme-based sensing using CNTs abound, including those under test development for, besides glucose, estrogen and progesterone, organophosphates (simulants for chemical warfare agents, with the enzyme organophosphate hydrolase, OPH), proteins, and troponin [238–241]. Detection of dopamine with CNT-based amperometric as well as voltammetric sensors has been demonstrated [242]. Unique adaptations of enzyme-based electrochemical CNT sensors have included a sensor immobilized with the enzyme uricase (for uric acid detection) wherein the MWCNTs used were first functionalized with tin oxide nanoparticles, with the resulting sensor showing a greatly enhanced response over conventional MWCNT-based sensors [243]; a sensor with horseradish peroxidase and methylene blue immobilized on MWCNTs deposited on a glassy carbon surface, for detection of H_2O_2 [244]; and a sensor for glucose oxidase wherein, in place of a standard MWCNT paste, a CuO/MSCNT paste was used for reduced response time (down to 1 s) [245]. Catechol has been detected to a level of 0.01 mM using a sensing electrode fabricated from MWCNTs, tyrosinase, and epoxy resin [246].

Unusual or uncommon CNT-based electrodes tested for sensing have included those with MWCNTs incorporated into 1-octyl-3-methylimidazoline hexafluorophosphate gel, into layers of electropolymerized polyaniline, and into microelectrodes modified with (3-Mercaptopropyl)trimethoxysilane for amperometric detection of auxin in plant root samples [49, 247–249]. Other CNT-based electrochemical sensors studied have included those for the detection of homocysteine, brucine, hydrazine, uric acid, L-ascorbic acid, and carbohydrates [250–254].

A unique adaptation of electrochemical sensing to CNT-based electrodes has been *stripping voltammetry*. In conventional anodic stripping voltammetry, a metal ion such as Pb(II) in aqueous solution is first "plated" onto a thin-film Hg or Bi electrode. The Pb metal is then "stripped out" of the electrode in a reverse voltammetric scan. The "pre-concentration" in the first plating step allows for great sensitivity, with single-ppb detection routine. Such a stripping voltammetry technique has been adapted to CNT electrodes due to their observed, superior sorption properties for a variety of analytes. Frequently, the "plating" step does not even need to be at an active, applied potential – it can be at open circuit. Using such techniques, detection of xanthine, 6-benzylpurine, fluphenazine [57], and 4-nitrophenol has been apparently demonstrated [255–258].

Other CNT-based electrochemical sensors include those demonstrated for dopamine [259, 260] (in addition to the work cited above), ascorbic acid [260, 261], norepinephrine [262], NADH [263–265], and H_2O_2 [266, 267]. The advantage of CNT-based sensing electrode over conventional electrodes such as glassy carbon appears to primarily be in the lowering of the voltammetric peak potential. For example, in the Nafion-wrapped CNT on glassy carbon sensing electrode used by

Wang's group [268, 266], the voltammetric peak of hydrogen peroxide is at +0.2 V versus Ag/AgCl, whereas on a bare glassy carbon electrode, no H_2O_2 peak is observed up to +1.0 V. The high affinity of Nafion toward CNTs was said to play a part in this [268]. Similarly, at a CNT-paste electrode [32], the H_2O_2 peak is observed at +0.3 V versus Ag/AgCl [266].

It has, however, not been clear from all the above-cited prior sensing work whether the observed CNT (conductometric, capacitive, microwave resonance, or other) response is simply a generic response to any adsorbed chemical species and whether different analytes present together can be distinguished in any way from each other. The only exception to this is *voltammetric* sensors, since in that case the voltammetric peaks, if clearly distinct and resolved, inherently afford selectivity. In a rare case of good selectivity in an amperometric sensor, Qi et al. [217] showed that, with a polyethyleneimine coating, NO_2 could be detected without interference from NH_3, while with a Nafion coating, NH_3 could be detected without interference from NO_2 [217]; however, the sensor was extremely slow, with a response time in the tens or hundreds of seconds.

Additionally, the robustness and durability of CNT-based sensors, a strong requirement for practical field use, are also in question. And the advantages over established sensor technologies have also not been clearly established. A case in point in this regard is the glucose oxidase-based glucometer widely available commercially: Over a dozen different CNT-based glucose oxidase sensors for glucose have been reported, but none has thus far yielded a commercial product, apparently because, for all their advantages, their drawbacks, such as poor robustness, durability, and repeatability, are even greater. Wang's group has attempted to demonstrate that CNT-based electrochemical sensors can be practical by screen printing them in patterns using CNT-based inks, e.g., for the amperometric detection of organophosphate pesticides; however, it was not clear whether these sensors had any greater robustness or stability than conventional CNT-based sensors and whether they also responded to interferents [269, 270].

CNTs have been used for what are known as "*DNA hybridization*" electrochemical biosensors. These immobilize single-strand oligonucleotides onto the electrode surface labeled with an electrochemical indicator to recognize its complementary target [271–274]. Single-strand DNA modified SWCNT electrodes can be used to electrochemically detect the hybridization with target single-strand DNA strands [275]. Such "DNA-CNT" sensors have also been used for other analytes as well. For example, single-strand DNA drop cast onto a SWCNT-FET (field-effect transistor) sensor yielded improved response to methanol, trimethylamine, and organophosphates [276]; and surface-confined MWCNT electrodes subject to absorptive accumulation of guanine nucleobase have been shown to have greatly enhanced response to guanine over comparative non-CNT (e.g., glassy carbon) electrodes [277–279].

Another type of DNA hybridization reported is in a generic type of DNA biosensor [281–287]. In one example of this, amine-terminated probe DNA was immobilized onto the ends of aligned MWCNT arrays using carbodiimide chemistry. The probe DNA had inosine bases substituted for the normally occurring guanine bases (which are more easily oxidized). The target sequence was labeled with ten guanine bases at one end. Hybridization in the presence of $Ru(bpy)_3^{3+}$ allowed highly sensitive detection of DNA, the mechanism being the oxidation of the guanine bases on the target sequence by $Ru(bpy)_3^{3+}$ with concomitant reduction to $Ru(bpy)_3^{2+}$. The $Ru(bpy)_3^{2+}$ then diffused to the electrode where it was re-oxidized to $Ru(bpy)_3^{3+}$, thus forming a catalytic cycle. The DNA biosensor was estimated to have a detection limit as low as 6 aM (attomoles), much lower than conventional DNA electrochemical biosensors. A DNA probe-doped CP (poly (pyrrole)) film laid over a MWCNT layer was used for AC impedance-based measurements of DNA hybridization [2071]. DNA hybridization has also been detected using differential pulse voltammetry with a DNA-functionalized CNT array that uses daunomycin as a redox label, all intercalated into a double-stranded DNA-CNT assembly [2072].

Tran et al. [288] have described label-free and reagentless electrochemical detection of micro-RNAs using a conducting polymer (CP)–CNT composite "nanoelectrode" wherein the CNTs "interpenetrate" the CP, with specific application to prostate cancer biomarker miR-141. In this work, the electroactivity was believed to emanate from the quinone group in the CP backbone. Response to miR-141, and corresponding lack of response to miR-103 and miR-29b-1, was demonstrated. However, the precise origin of the response to miR-141, and whether other non-RNA interferents would also elicit a response, was not clear.

CNT-based transistors have been used for detection of protein binding, wherein a PEI/PEG (poly(etherimide)/poly(ethylene glycol)) polymer layer is used to avoid nonspecific binding, with attachment of biotin to the layer for specific molecular recognition; biotin–streptavidin binding has been detected using this sensor [289]. A field-effect transistor (FET) has been used for the selective detection of heavy-metal ions, wherein peptide-modified polymers are electrochemically deposited onto SWCNTs and selective detection is achieved by appropriate choice of peptide sequences [290, 291].

Gooding et al. [292] demonstrated protein electrochemistry using aligned SWCNT arrays wherein the enzyme microperoxidase MP-11 was attached to the ends of the SWCNTs aligned normally to the electrode substrate; the practical utility of this for sensing in a real environment was however not apparent from this work. Yu et al. [293] reported electrochemical sensors that used SWCNT forests with multi-label secondary antibody–CNT bioconjugates, for detection of a cancer biomarker in serum and tissue lysates. Greatly amplified sensitivity was attained by using bioconjugates featuring horseradish peroxidase (HRP) labels and secondary antibodies (Ab(2)) linked to CNTs at high HRP/Ab(2) ratio. This approach provided a claimed detection limit of 4 pg/mL) for prostate-specific antigen (PSA) in 10 µL of undiluted calf serum. Although this work was published in 2006 and the sensor showed promise, to date no practical sensor appears to have been developed either commercially or in the laboratory.

There has been some work on aptamer-modified CNT-sensing electrodes, along the lines of conventional (e.g., Au) aptamer-modified electrodes; however, this has for the most part yielded inconclusive or negative results [294–297].

CNTs have also been used for what are known as "immunosensors." For example, Wohlstadter et al. [298] reported an immunosensor for the detection of α-fetoprotein (AFP). Here, antibodies were immobilized on a MWCNT/EVA composite electrode. Capture of AFP resulted in binding of a $Ru(bpy)_3^{3+}$-labeled antibody. Since the $Ru(bpy)_3^{3+}$ is chemiluminescent, binding is transduced by the release of light, and the electrode is held at potentials more positive than +1 V. This immunosensor had a detection limit of 0.1 nM and a linear range up to 30 nM. Sanchez et al. [299] demonstrated an immunosensor incorporating CNTs with immunomolecules embedded in a porous polysulfone matrix, with five times the sensitivity toward anti-RIgG than a corresponding graphite-only sensor. The polysulfone embedding afforded good durability to the sensing electrode. Such polymer embedding was also used in other work, for enzymatic detection of human chorionic gonadotropin hormone [300].

An important observation in much work with CNT-based electrochemical sensors has been that electron transfer appears to occur at the ends of CNTs rather than along their walls [280, 301]. For this reason, it has been found that *aligned* CNT arrays yield somewhat sharper voltammetric responses than "random" CNT arrays.

Non-CNT-based amperometric biosensors in which an enzyme such as a dehydrogenase is co-immobilized along with its nicotinamide adenine dinucleotide (NAD+) cofactor on various solid electrodes are available commercially for sensing of substrates such as lactate, alcohol, or glucose. The oxidation of the NADH product serves to regenerate the NAD+ cofactor. Some problems encountered with such systems involve the large overvoltage for NADH oxidation at such electrodes and the fouling of the electrode substrate due to this. It has been shown that such problems can be overcome when the substrate is a CNT-based electrode [268, 302].

Some work has been done with electrochemical sensing wherein the sensor electrode is comprised of a *single* CNT [303, 304]; however, as of this writing, nothing practical has emerged from this work, which thus appears to be of academic interest only.

CNT-based sensors have also been tested for detection of such analytes as hydroxylamine (to 0.12 µM in 3 s) [305]. Detection limits of 50 ppb for chemical warfare agent (CWA) simulants have been claimed for certain SWCNT-based field-effect transistor (FET) sensors and other SWCNT-based sensors [306, 307] although these sensors appear to suffer from problems of irreversibility and/or slow response.

9.4   Fluorescence-Based Sensing and Imaging

CNTs have been used for biological imaging [308–310]. Cherukuri et al. [308] used NIR fluorescence microscopy for imaging in phagocytic cells. Welsher et al. [309] demonstrated use of CNT fluorophores for deep-tissue anatomical imaging in mice.

The well-known property of some biological systems to be highly transparent to near-infrared (NIR) light and the strong, intrinsic absorbance of SWCNTs in this NIR spectral window have been attempted to be put to use for optical stimulation of CNTs inside living cells, to arrive at "multifunctional nanotube biological transporters" [311, 312]. In this work, it was claimed that if SWCNTs could be selectively internalized into cancer cells with specific tumor markers, NIR radiation of the nanotubes in vitro could then selectively activate or trigger cell death without harming normal cells. This could be then used to develop SWCNT functionalization schemes with specific ligands for recognizing and targeting tumorous cells.

CNTs labeled with a fluorescent agent have been shown to be easily internalized and then tracked into the cytoplasm or the nucleus of fibroblasts using epifluorescence and confocal microscopy [313].

9.5   Other Sensors

Bhandavat et al. [314] recently described a *laser power sensor* based on a composite of CNTs and ceramic material. In this, the ceramic component provides the necessary resistance to high power, while the CNT component provides the very high thermal conductivity necessary to dissipate the heat generated. The sensor could measure powers up to 15 kW/cm^2 for short (<10 s) durations of time.

MWCNTs have been tested as tips for *atomic force microscopy (AFM)* and *scanning tunneling microscopy (STM) probes* [314, 315]; however, the claimed advantages (greater flexibility, greater strength) over more conventional tips were not clearly established. An adaptation of this probe or sensing application is the use of double CNT tips as "nanoscopic tweezers" [317].

Quartz crystal microbalance (QCM)-based piezoelectric CNT sensors have been studied for alcohol detection [318]. *Surface acoustic wave (SAW)*-based CNT sensors have been studied for detection of a variety of volatile organic compounds (VOCs) [319]. *Thermoelectric-based sensors* have been studied for detection of inert gases such as He and N_2 and other gases such as H_2 [320]. *Gas-ionization-based* CNT sensors have also been studied for sensing of mixtures of various gases [321]. These are based on the electrical breakdown of the gases at CNT tips, as the sharp tips of CNTs generate high electric fields at relatively low voltage, lowering breakdown voltages and allowing operation of such sensors at very low power.

A SWCNT film-based strain sensor was reported by Dharap et al. [322]. A linear change in voltage at a CNT electrode was observed upon tensile and compressive stresses. However, since many other materials (i.e., other than CNTs) would also show a linear change in voltage when subjected to tensile and compressive stresses, it was not clear why CNTs were better.

CNTs have also been studied for use as "nano-electro-mechanical sensors" [323, 238] and in *photoacoustic* imaging [324].

9.6 Problems and Exercises

1. Give at least two examples of the use of CNTs in actuators, including one on a nm-scale.
2. Describe at least two types of chemical sensors based on CNTs, including details on their principle of detection. Include method of functioning, typical analytes, and typical detection limits and times.
3. Describe at least two types of gas sensors based on CNTs, including details on their principle of detection. Include method of functioning, typical analytes, and typical detection limits and times. These should be different from your answers to Problem 2.
4. What are the advantages and disadvantages of a glucose oxidase-based glucose sensor that uses CNTs as opposed to others that do not use CNTs? Phrase your answer in terms of performance characteristics.
5. List at least five (5) analytes that have been the target of CNT-based sensors. Ensure that you select at least one from the following categories: gases, biologically important chemicals, toxic chemicals, or pollutants. Include method of functioning and typical detection limits and times.
6. Describe the principle of working of a CNT-based transistor and list at least one of its sensing applications.
7. Describe at least one CNT-based sensor using the following principles: fluorescence, laser-based, quartz crystal microbalance (QCM), and thermoelectric. Include method of functioning, typical analytes, and typical detection limits and times.

Chapter 10
CNT Applications in Drug and Biomolecule Delivery

Contents

10.1 Overview of Use of CNTs for Drug and Other Biomolecule Delivery

One of the intrinsic properties of CNTs relied on for drug delivery is the ability of these materials, with sub-cell dimensions, to quite easily penetrate various cellular and other barriers in the body [68, 313, 325]. In this respect, the behavior of CNTs has been compared to metallic "nanoneedles," some of which can perforate and diffuse through lipid bilayers of plasma membranes without inducing cell death [68]. Indeed, in vitro studies with magnetized CNTs and MCF-7 breast cancer cells have shown that CNTs do indeed cross the cell membrane like tiny needles [326]; this study was unfortunately extended to drug-bearing or functionalized CNTs.

CNTs can be functionalized with *bioactive peptides, proteins, nucleic acids, and drugs* [68]. These cargos can then be delivered to cells and organs. And functionalized CNTs have been found for the most part to be non-immunogenic.

10.2 Example Studies

Recent work [289, 311, 327] has demonstrated that various types of functionalized CNTs can be taken up by a wide range of cells and can penetrate cellular barriers: Intracellular trafficking of individual or small bundles of CNTs and CNT transport toward the perinuclear region were shown to occur a few hours after initial contact with the cells, even under endocytosis-inhibiting conditions, with presumed contributions from other mechanisms, e.g., phagocytosis. Overall, it could be concluded that functionalized CNTs possess a capacity to be taken up by mammalian

© Springer International Publishing AG 2018
P. Chandrasekhar, *Conducting Polymers, Fundamentals and Applications*,
https://doi.org/10.1007/978-3-319-69378-1_10

and prokaryotic cells and to intracellularly traffic through the different cellular barriers by energy-independent mechanisms [289, 311, 327]. It has also been shown that cellular uptake of CNTs, wherein short SWCNTs with various functionalizations are capable of the transportation of proteins and oligonucleotides into living cells, use a cellular uptake mechanism which involves energy-dependent endocytosis [311, 328, 329].

In other recent work [311, 330], advantage was taken of functionalized SWCNTs with antibodies in combination with the intrinsic optical properties of the CNTs to concomitantly target and destroy malignant breast cancer cells in vitro with the aid of "photodynamic" therapy.

Studies have been carried out on the use of CNTs as templates for targeting bioactive peptides to the immune system [68, 311, 331, 332]. Here, the B-cell epitope of the foot-and-mouth disease virus was covalently attached to amine groups present on CNTs, using a bifunctional linker. These peptide-modified CNT moieties apparently mimic the appropriate secondary structure for recognition by specific monoclonal and polyclonal antibodies. Immunogenic features of these peptide-based CNT conjugates were then assessed in vivo. It was found that immunization of mice with peptide–nanotube conjugates provided high antibody responses as compared with the free peptide, with the antibodies also displaying virus-neutralizing capabilities.

In work relating to CNT-mediated oligonucleotide transport inside living cells, the use of SWCNTs as nonviral molecular transporters for the delivery of short interfering RNA (siRNA) into human T cells and primary cells has been demonstrated [311, 333]. In this work, it was claimed that the delivery capabilities and RNA interference efficiency of CNTs were superior to existing nonviral transfection, such as various liposomes.

Plasmid DNA alone penetrates into cells and reaches their nucleus with great difficulty, according to some studies [334]. On the other hand, delivery of nucleic acids can enhance gene transfer and expression [335–337].

To explore the use of CNTs as gene transfer vectors, in one study [68, 2073, 2074], plasmid DNA pCMV-Bgal expressing b-galactosidase was adsorbed on CNTs carrying ammonium groups. Both single- and multi-walled cationic CNTs were shown to form stable complexes with this plasmid DNA. Subsequent gene transfer experiments showed a clear effect of the CNTs on the expression of β-galactosidase, with levels of gene expression up to ten times those with DNA alone observed. In another study [68, 340], it was shown that cationic CNTs were able to condense short oligodeoxynucleotide sequences and improve their immune-stimulating activity. It was shown that CpG interacts with ammonium-functionalized CNTs with no toxicity on mouse splenocytes. More importantly, high ratios of CNTs over a minimum stimulatory dose of a specific oligodeoxynucleotide CpG increased the immune-potentiating activity in vitro, while decreasing the secretion of inflammatory cytokine interleukin-6. In another study [68, 341], CNTs were used to deliver non-encoding RNA polymers into cells.

In another study, an antibiotic was linked to multiply functionalized CNTs, with the uptake of the CNT-antibiotic composite into mammalian cells in vitro then monitored fluorescently [68, 342]. It was found that the antibiotic carried by the CNT composites was easily internalized into mammalian cells without toxic effects in comparison with the antibiotic incubated alone; amphotericin B held in the CNT composites preserved its high antifungal activity against a broad range of pathogens, including *Candida albicans*, *Cryptococcus neoformans*, and *Candida parapsilosis*.

In another study [343], SWCNTs were functionalized with substituted carborane cages with the ultimate objective being a boron neutron capture therapy technique. These water-soluble CNTs were aimed at the treatment of cancer cells, with the studies showing that some specific tissues contained carborane following intravenous administration of the CNT–carborane composites and that carborane was concentrated mainly at the tumor sites. In a more promising drug delivery study [68, 344], SWCNTs were loaded with dexamethasone, and the binding and release of the drug then studied. It was found that dexamethasone could be adsorbed in large amounts onto oxidized nanohorns and, further, that it maintained its biological integrity after being liberated from the CNT composite; this was confirmed by activation of glucocorticoid response in mouse bone marrow cells and induction of alkaline phosphatase in mouse osteoblasts.

In a cancer drug delivery study, doxorubicin was loaded at ca. 60 w/w% into CNTs compared with only a ca. 10 w/w% loading achievable on liposomes; drug release was then being triggered using NIR radiation [345].

SWCNTs of very short length have also been studied for delivery of MRI contrast agents to patients [346], but, to date, nothing further has come of this.

In rather unique applications, CNT-based *electrodes* have been used to effect direct electron transfer to proteins, the ultimate objective of both being sensors for these proteins and, potentially, initiating or "actuating" the enzymatic processes embodied by many of them. Examples of these include peroxidases [4, 56, 86], cytochrome *c*-oxidase [280, 347–349], myoglobin [280, 350, 351], catalase [280, 352], and azurin [348].

In spite of all the above promising studies, however, *the use of CNTs for the delivery of anticancer, antibacterial, or antiviral agents has not yet been fully established, and we are, as of this writing (2016), certainly a long way from clinical trials* [68, 313, 325].

10.3 Problems and Exercises

1. Describe in detail at least two examples of the use of functionalized CNTs in drug delivery. Also, describe at least two *non*-CNT-based methods of drug delivery (you may use a quick Internet search for this). What are the relative advantages, if any, of CNT-based drug delivery over these two other methods that you found?

2. Describe in detail the use of peptide-modified CNT moieties in facilitating recognition by specific monoclonal and polyclonal antibodies, including any relevant immunogenic features.
3. Give one example of the use of CNT-based electrodes to effect direct electron transfer to proteins, for both sensing and actuating applications.

Chapter 11
CNT Applications in Microelectronics, "Nanoelectronics," and "Nanobioelectronics"

Contents

11.1 Overview

CNTs, and thereunder, mainly SWCNTs, have been promoted as the ultimate electronic material (whether in electronics, microelectronics, "nanoelectronics," or "nanobioelectronics"), possibly replacing even tried-and-true silicon in the near future [18]. Indeed, CNT transistors were expected to outperform Si transistors by about 2015 [18]. This was in view of the fact that CNTs hold an inherent size advantage over Si, where transistor elements are carved from bulk Si; this "top-down approach" appears to be limiting, especially in contrast to CNTs, which can be fabricated in dimensions that are a fraction of today's smallest Si feature [18]. Additionally, the combination of high intrinsic mobility (approx. $10^4 cm^2 V/s$), small capacitance (approx. 100 aF/μm), and nm-thick body channels makes SWCNTs supposedly promising for high-speed devices, with some potential for operation in the terahertz regime [353].

Nevertheless, to date, one does not see CNT-based electronics in even advanced laboratory trials, let alone commercial niche application. One of the features limiting application of CNTs in electronics is the fact that very high purity, very uniform chirality, very uniform diameter (also related to chirality), and appropriate length are required for CNTs to find application in practical electronics with performance anywhere near that predicted for them based on theoretical studies or laboratory studies on individual CNTs or small bundles of CNTs [18]. Selective synthesis of semiconducting-only and metallic-only SWCNTs must also be carefully carried out; then, for some applications, these two varieties must be incorporated together. As synthesized, SWNT thin films are typically approximately 1/3

© Springer International Publishing AG 2018
P. Chandrasekhar, *Conducting Polymers, Fundamentals and Applications*,
https://doi.org/10.1007/978-3-319-69378-1_11

metallic SWNTs and 2/3 semiconducting SWNTs [353]. Separation of semiconducting and metallic SWCNTs can be achieved via standard techniques described in an earlier chapter, such as ion-exchange chromatography (of functionalized CNTs) or ultracentrifugation.

Another issue limiting use of CNTs in transistors and other electronic circuit elements is high CNT-CNT contact resistances, which can be of the order of kOhms [12] and can have seriously deleterious effects on performance [354]. Thus, reducing or eliminating such CNT-CNT contacts can improve performance. Among techniques to achieve this, besides the attempted use of such materials as Pd for electrical contacts [355], are making the CNTs sufficiently long to directly bridge the source/drain electrodes or configuring the CNTs into dense, aligned arrays that avoid CNT-CNT overlaps; on the other hand, even with these mitigating techniques, "misalignment" of CNTs can lead to poor performance [354].

11.2   Microelectronics and "Nanoelectronics" Applications

CNT-based transistors include field-effect transistors (FETs) and variants thereof, such as metal–oxide–semiconductor field-effect transistors (MOSFETs). In early work in this field, room-temperature ballistic transistors with high-k gate dielectric were reported which displayed an ON current approximately 30 times greater than state-of-the-art Si MOSFETs [355, 356]. Thin-film transistors (TFTs) have been fabricated using SWCNTs with acceptable performance; Fig. 11.1 shows results from one such representative work.

A unique method demonstrated for fabricating CNT-based transistors has been the use of random networks of them [357]. This averages their electrical differences and allows production at the wafer level, a method patented and attempted to be commercialized (but without apparent success to date) by Nanomix, Inc. [358–362].

In more recent work, SWCNT FETs with sub-10 nm channel length have been reported with a normalized current density value of 2.41 mA/µm at 0.5 V, which is larger than that observed for Si-based devices [363]; and transistors that utilize horizontally aligned CNT arrays are claimed to have achieved mobilities of 80 cm^2/Vs, subthreshold slopes of 140 mV/decade, and ON/OFF ratios as high as 10^5 [364]. Most recently, advanced CNT film deposition methods are claimed to have enabled the fabrication of more than 10,000 CNT devices in a single chip using conventional semiconductor fabrication methods [365]. And in other promising recent work, a CNT FET yielded sufficient current output to drive OLEDs (organic light-emitting diodes) at relatively low voltages, enabling red-green-blue emission by the OLED through a transparent CNT network [366].

Figure 11.2 shows the fabrication and the performance characteristics of an inverter and a 4-bit decoder composed of CNTs, capable of operating in the kHz region, even at dimensions of 100 µm, suitable for macroscopic screen printing

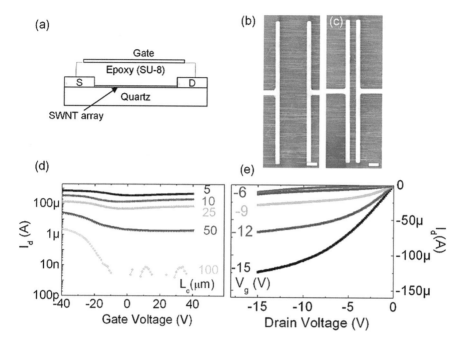

Fig. 11.1 Fabrication and performance of CNT-based thin-film transistors (TFTs) from early, representative work (After Ref. [354] reproduced with permission)

methods for fabricating circuits [367]. Figure 11.3 summarizes additional work carried out recently on SWCNT-based thin-film transistors (TFTs) [368, 369].

Common substrates used in general in the field of *"flexible electronics"* are poly (dimethyl siloxane) (PDMS, which is highly flexible) and polyimide (thin films of which are somewhat flexible). A typical scheme for grafting SWCNT-based electronic circuits onto flexible substrates is shown in Fig. 11.4 . These methods are based on CVD deposition above temperatures of about 450 °C [367, 369].

Due to their low scattering, high current-carrying capacity, and resistance to electromigration (at present, current-induced electromigration causes conventional metal interconnects to fail when their dimension becomes too small), CNTs have the potential of replacing copper in microelectronic *interconnects*. For such use, vias comprising tightly packed ($>10^{13}$/cm^2) metallic CNTs with low-defect density and low contact resistance are required. To this end, recently, complementary metal oxide semiconductor (CMOS) – compatible, 150 nm diameter interconnects with a single CNT contact hole resistance of 2.8 kOhm were demonstrated on full 200 mm diameter wafers [13, 187, 208].

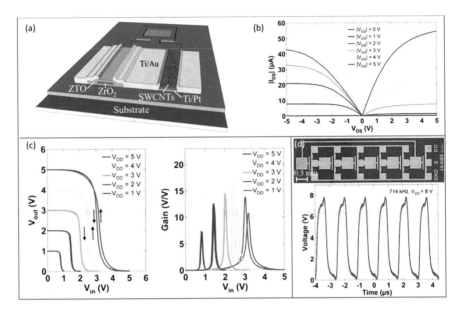

Fig. 11.2 Heterogeneous integration of CNT (p-type) and ZTO (n-type) semiconductors for printed complementary integrated circuits. (**a**) Schematic diagram of the CMOS inverter. (**b**) Output characteristics of the printed n-type and p-type TFT. (**c**) VTC (left) and voltage gain (right) of the printed inverter under various supply voltages. (**d**) Printed 5-stage ring oscillator using the complementary inverters. Top: optical micrograph. Bottom: output signal of the oscillator driven by a V dd of 8 V (After Ref. [2083])

11.3 Analog Systems

SWCNT-based TFTs are known to be able to produce larger than unity (1.0) power gain in the very high-frequency (VHF) range, allowing configuration as RF power amplifiers and integration together to form functional analog electronic systems. Using this capability, one group has constructed "CNT radios"; in these, SWCNT TFTs provide all active components [353, 368]. This is illustrated in Fig. 11.5. below.

11.4 Prognosis

As a parting thought, it may be noted that a review in 2004 [371] laid out the future for "integration of biomaterials with CNTs . . . . hybrid systems as active field-effect transistors (FETs) or biosensor devices" and "integration of CNTs with biomolecules . . . . generation of complex nanostructures and nanocircuitry of controlled properties and functions. The rapid progress in this interdisciplinary field of CNT-based nanobioelectronics and nanobiotechnology. . . ." However, *to*

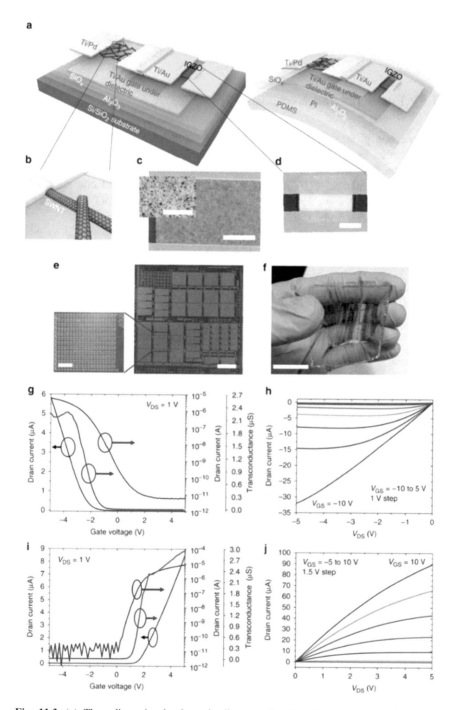

Fig. 11.3 (a) Three-dimensional schematic diagram of a CNT/IGZO complementary mode inverter on rigid substrate (left) and same circuit on flexible substrate (right). (b) Schematic diagram conceptually showing the interface between the Ti/Pd electrode and the CNT network. (c) Scanning electron microscopic (SEM) image of CNT network in the channel of a p-type TFT;

date (2016), nothing practical or commercial has emanated from the very large number of studies of CNTs for uses in microelectronics and "nanoelectronics."

11.5 Problems and Exercises

1. Name one promising and one actual application of CNT-based materials in electronics, as of 2016 and as of your reading today.
2. Describe the functioning of CNT-based field-effect transistors (FETs) and/or thin-film transistors (TFTs). Cite at least one proposed and one successful application. What are the purported/claimed advantages of CNT-based transistors over conventional transistors?
3. Describe the fabrication of a typical electronic circuit based on CNTs on a *flexible* substrate. (Name the substrate.)

Fig. 11.3 (continued) the scale bar in the low-magnification SEM image is 10 μm, and the scale bar in the high-magnification SEM image is 2 μm. (**d**) SEM image of IGZO in an n-type TFT; scale bar, 5 μm. (**e**) Optical micrograph of the hybrid CNT/IGZO ring oscillators, inverters, individual p-type and n-type transistors fabricated on a rigid Si/SiO$_2$ substrate. The inset shows a 501-stage ring oscillator on the rigid substrate. The scale bar in the rigid circuit chip is 500 μm. The scale bar in the 501-stage ring oscillator image is 600 μm. (**f**) Optical photographic image of the hybrid CNT/IGZO ring oscillators, inverters, and individual transistors on a flexible PI substrate laminated on a polydimethylsiloxane (PDMS) film. The scale bar in flexible circuit is 2 cm. (**g**) Transfer characteristic in linear and log scale and transconductance of a CNT TFT with gate bias varied from −5 to 5 V. Channel length (L_{ch}) = 20 μm, channel width (W_{ch}) = 100 μm, drain-source voltage (V_{DS}) = 1 V. (**h**) Output characteristic of the CNT TFT with V_{DS} varied from −5 to 0 V. (**i**) Transfer characteristic in linear and log scale and transconductance of a IGZO TFT with gate bias varied from −5 to 5 V. L_{ch} = 4 μm, W_{ch} = 12 μm, V_{DS} = 1 V. (**j**) Output characteristic of the IGZO TFT with V_{DS} varied from 0 to 5 V (After [2084], reproduced with permission)

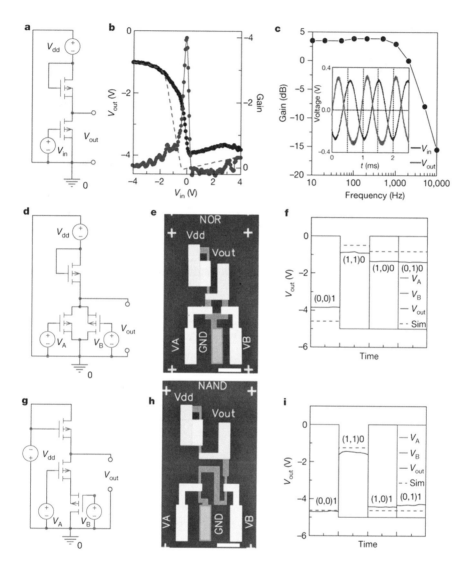

Fig. 11.4 Medium-scale carbon nanotube thin-film integrated circuits on flexible plastic substrates (After [2085], reproduced with permission)

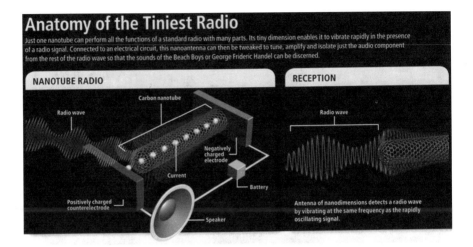

Fig. 11.5 An analog "CNT radio" wherein SWCNT TFTs provide all active components (After reference [2086]. Reproduced with permission)

Chapter 12
CNT Applications in Displays and Transparent, Conductive Films/Substrates

Contents

12.1   Overview

This chapter covers *displays and transparent, conductive films/substrates*. This especially includes field emission displays and conductive replacements for such industry workhorses as transparent, conductive ITO (indium tin oxide) coatings.

12.2   Displays and Light Sources

One of the first display applications of CNTs was in *field emission displays*. The principle behind CNT-based field emission displays is that when a potential is applied between a CNT surface and an anode, electrons are emitted from the CNT tips due to electron tunneling from the tips into the vacuum associated with the CNT curvature and defects therein such as dangling bonds. Applications of such CNT-based field emission sources potentially include flat-panel displays and intense light sources [12, 178, 372–375, 376]. The purported advantages of CNT-based field emission displays over other types include the possibility of much cheaper methods of fabrication than conventional field emission displays, which use conventional semiconductor fabrication methods, stable field emission over prolonged time periods, long lifetimes of the components, low emission threshold potentials, high current densities, no requirement for ultrahigh vacuum, and the possibility of achieving reasonably large current densities of about 4 A/cm^2 [12, 178, 377].

© Springer International Publishing AG 2018   73
P. Chandrasekhar, *Conducting Polymers, Fundamentals and Applications*,
https://doi.org/10.1007/978-3-319-69378-1_12

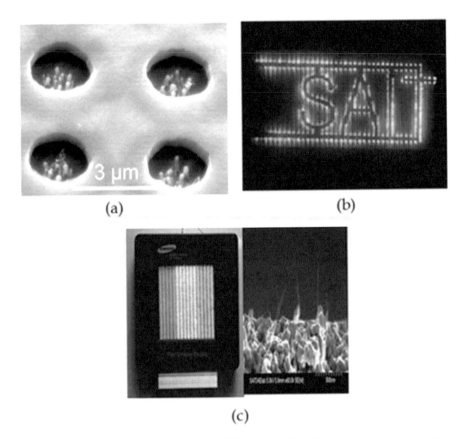

(a)   (b)

(c)

Fig. 12.1 Schematic and actual photos of a CNT-based field emission display, as fabricated at Samsung (After [378])

Samsung scientists reported CNT-based field emission displays with a threshold electric field of about 2 V/μm as early as 1999 [377, 378]. And as early as 2004, Samsung produced low-cost prototypes of 25 cm CNT color displays for the TV market [375, 379, 380]. Figure 12.1 shows a schematic and an actual such display.

CNT-based displays using the field emission principle have been fabricated into prototype, matrix-addressable, diode flat-panel displays; these used pulses of ±150 V switched between anode and cathode to produce images [194, 382].

However, in spite of all the above, promising work, to date (2016), there has been no successful CNT-based display product.

CNTs have also been studied for use as *bright light sources and X-ray sources*. It was demonstrated experimentally that CNT-based lamps were comparatively cheap to fabricate and possessed high efficiency and lifetimes of 8000 h. X-rays could be generated from these if the phosphorous screens used were replaced by metal targets and the accelerating voltage were increased. Imaging of human bones was demonstrated [376, 377, 384, 385]. Again, however, to date, nothing further has come of this pioneering work.

12.3 Conductive Coatings

In the early years of CNT research, there was much speculation that transparent, conductive CNT coatings would replace standard workhorses of the industry such as indium tin oxide (ITO) coatings [385]. One of the many purported advantages of CNTs was claimed to be that they would afford greater flexibility as compared to brittle ITO. In this regard, however, the best performance achieved to date for transparent CNT coatings has been a sheet resistivity of about 100 Ohms/square at 90% transmission on glass (integrated through the visible spectral region), still considerably poorer than ITO/glass, which can be fabricated at as low as 15 Ohms/square at 90% transmission [386]; it has been proposed that these best-performing CNT coatings may be adequate for such applications as thin-film heaters [13].

Nevertheless, there has been no dearth of studies of CNTs for transparent conductive coatings, especially to replace ITO *only*, *since some are theses*, etc. [387, 388]. For example, in an early study, Glatkowski [388, 389] showed that coatings could be deposited on various transparent substrates with a simple wet coating process that showed 90% T in the mid-visible with a 200 Ohms/square surface resistivity.

More recently (2014), Akhmadishina et al. [390] demonstrated a method for the large-scale growth of thin CNT films from solution on the surface of flexible, transparent substrates, via the preparation of a stable colloidal CNT solution using an aqueous surfactant. They observed that the optical transmittance of the films decreased linearly with increasing film thickness, whereas their resistance decreased quadratically; they attributed this to 3D CNT percolation in the films. They reported that, with increasing film thickness, the sheet resistance of the films dropped from 400 to 15 kΩ/square, while their % T decreased from 85 to 40%.

12.4 Problems and Exercises

1. Describe the principles of an LED based on CNTs. Sketch the components of the LED "sandwich" in schematic. Detail the active process that gives rise to the visible display. Consult figures in this chapter only minimally, if at all, for your answer.

2. Describe the similarities and differences between OLEDs based on CPs and those based on other organic molecules such as tetrathiafulvalene derivatives. Although such information is mostly proprietary, in your estimation, which of these two classes were actually found widely in OLEDs as of 2016? As of your reading today?

3. What in your estimation is the critical parameter that will allow CNT-based transparent conductive coatings to displace those based on ITO (indium tin oxide)? In your answer, include such parameters as surface resistivity values (in Ohms per square). Has this already happened as of your reading today?

Chapter 13
CNT Applications in Electrical Conductors, "Quantum Nanowires," and Potential Superconductors

Contents

13.1   Overview

Due to their electronic properties, as extensively discussed in a chapter in Part I of this book, CNTs have been considered for applications as *electrical conductors* (to replace metals), *quantum nanowires* or *quantum wires* (ideal one-dimensional conductors), potential *superconductors*, and other similar, conductivity-based applications.

13.2   Simple Electrical Conductors

Many laboratory-level or very small-scale studies have been carried out demonstrating the very high current-carrying capacity of CNTs [391–393]. CNTs doped with iodine or iodine monochloride have been shown to have specific conductivity exceeding that of common current carrier metals such as Cu and Al [391, 392]. And Cu–CNT composites have been shown to have among the highest observed specific conductivity of any electrical conductors [393].

Despite all the above-cited, very promising studies, however, to date there are no CNT-based electrical wires or other conductors in the commercial market or even under development. The reasons for this are the same as the reasons for the lack of success in a host of other potential applications for CNTs, as described elsewhere in this book: the inability to fabricate practical materials that are at least somewhat commercially competitive.

© Springer International Publishing AG 2018
P. Chandrasekhar, *Conducting Polymers, Fundamentals and Applications*,
https://doi.org/10.1007/978-3-319-69378-1_13

13.3 "Quantum Nanowires"

Individual SWCNTs (but not MWCNTs or bundles or groups of SWCNTs) have the potential to behave like "quantum nanowires" or "quantum wires" (ideal, one-dimensional conductors) because theoretical studies have shown that their molecular wave functions may extend over an entire nanotube due to their structural symmetry [11, 394–399].

Electrical transport measurements on individual SWCNTs have appeared to confirm this, with electrical conduction appearing to occur through well-separated, discrete electronic states that are quantum mechanically coherent over distances up to 140 nm [11]. Furthermore, studies carried out in a magnetic field indicate shifting of these electronic states due to the Zeeman effect [11]. Figure 13.1 shows an AFM image of one such SWCNT with observed "quantum nanowire" properties.

Current densities of up to 10^{10} A/cm^2 have been claimed for CNT nanowires, and the performance of CNT wires of about 9 nm diameter and about 3 μm length have been claimed to be equivalent to Au nanowires of the same dimensions [11, 25, 399, 400].

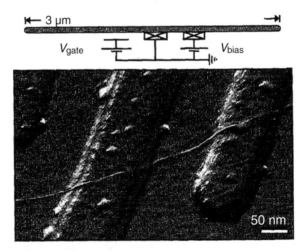

Fig. 13.1 AFM image of a SWCNT with observed "quantum nanowire" properties. The wire sits atop a Si/SiO2 substrate having two 15 nm thick Pt electrodes (seen in the image). The SWCNT shown is of ca. 1 nm diameter and ca. 3 μm long, with ca. 140 nm between the two Pt electrode contacts shown. The third Pt electrode at upper left in the photo is used to apply the gate voltage, as shown in the schematic (After Ref. [11] reproduced with permission)

13.4 Superconductivity

There have been just a few reports of superconductivity in SWCNT-based materials at low temperature, many of which were subsequently not reproducible. Among the most prominent and reproducible studies has been the early study of Tang et al. [401], where SWCNTs with very small diameter were embedded in a zeolite matrix. Among these, those with an approximately 4 Å diameter were found to display superconducting behavior at temperatures below 20 K (with a superconducting transition typically observed at 15 K), as seen via an anisotropic Meissner effect, accompanied by a superconducting gap and fluctuating supercurrent. The authors also showed that their experimental observations agreed with statistical mechanics calculations based on the Ginzburg–Landau free-energy function.

13.5 Problems and Exercises

1. What is, approximately, the highest current-carrying capacity recorded for CNT-based conductors? How does it compare to that of the best among other materials? What are the determinative factors for such high capacities in CNT-based materials? What are the factors preventing the practical implementation of CNT-based conductors, including nm-dimension materials, as of this writing (2016)? As of your reading today?
2. What is a "quantum wire" and how does it defer from a regular conductive wire? In your estimation, what are the factors preventing practical realization and commercial implementation of CNT-based "quantum wires"?
3. What would be the purported advantage of a CNT-based superconductor as compared to more established superconductors, including the higher-temperature "1–2–3" superconductors based on rare-earth materials discovered with much fanfare some decades ago?

Chapter 14
CNT Applications in the Environment and in Materials Used in Separation Science

Contents

14.1 Overview

CNTs have been considered for use in *environmental applications* and in *materials used in separation science*. Such applications include, e.g., filters, membranes, and water treatment.

14.2 Environmental Applications

In recent work [402] a porous "nanosponge" was constructed via a CVD process that used ferrocene as the catalyst precursor and added sulfur. This "nanosponge" was then used to absorb *o*-dichlorobenzene and oils from a water sample; the oils absorbed were more than 100x the weight of the sponge. In similar, earlier work, B-doped CNT "nanosponges" were fabricated from CVD-deposited CNT forests that were shown to behave somewhat like surfactants but with very high oil absorption capabilities [403].

A comparative study of the water treatment capability of granular activated-C, activated-C-fiber, and CNTs [404] found that the absorption capabilities of the CNTs (including SWCNTs as well as MWCNTs) for three aromatic organic compounds tested were somewhat better than those of the activated-C materials. Other work in water treatment has shown that molecular sieves constructed from CNTs have the potential of being activated and deactivated via an external electrostatic field [405]. CNTs have also been shown to be usable in a variant of *reverse osmosis* methods but operating at lower pressures, lower temperatures, and higher

© Springer International Publishing AG 2018   81
P. Chandrasekhar, *Conducting Polymers, Fundamentals and Applications*,
https://doi.org/10.1007/978-3-319-69378-1_14

flow rates than conventional (e.g., polycarbonate membrane based) reverse osmosis methods, although it also appears that only very specialized and very small-diameter SWCNTs function well in this application [406, 407]. In one study of desalination of water using self-supporting CNT membranes, a 99% salt rejection rate was claimed at a flux rate of water of about 12 kg/m^2 h [408].

In other work, CNT filters have been demonstrated that electrochemically oxidize organic contaminants, bacteria, and viruses or remove them through simple microfiltration mechanisms [409–413].

A company called Seldon Water (formerly Seldon Technologies and now owned by CB Tech) markets *Seldon Nanomesh*-brand portable water filters for water purification and claimed to "remove bacteria, viruses, and cysts, such as Cryptosporidium and Giardia, to US EPA Drinking Water Standards (99.9999% of bacteria, 99.99% of viruses, and 99.9% of cysts) also reduce sediment, total organic carbon (TOC), chlorine, color, bad taste, and odors also removes significant levels of harmful chemical contaminants, including: Lead and Cadmium, Organophosphates (pesticides and herbicides), Disinfection byproducts, radioactive contaminants such as cesium-137" [414].

14.3   Applications in Separations and Related Fields

CNTs have been studied for use in solid-phase extraction, as microextraction sorbents, in chromatography, and as laser desorption/ionization substrates [415].

One of the important parameters that must be considered in the use of CNTs in separation-related applications is the *point of zero charge (isoelectric point)*, at which CNTs' surface has *zero net charge* [415]. When the pH is higher than this point, the CNT surface is negatively charged; electrostatic interactions can be established to adsorb cationic species. Conversely, when the pH decreases below the point of zero charge, protons compete with cations for the same sites on CNTs, and the subsequent neutralization of CNTs provides a decrease in the adsorption. This fact can be used, for example, to retain at a certain pH value, in metals, and to later elute them using acidic solutions, allowing for facile separation of such materials. On the other hand, it is also estimated that the CNTs' covalent functionalization must not exceed about 10% of their surface [415, 416].

The use of CNTs in chromatographic stationary phases has been documented for the separation of the following analytes, as summarized succinctly by Herrera-Herrera et al. (type of CNT, if applicable, given in parentheses, and technique used given in brackets) [415, 417–442]: Ar and CO$_2$ (SWCNTs) {GC-TC}; alkanes and aromatic hydrocarbons (MWCNTs functionalized with –COOH and CONH$_2$) {GC-FID}; methanol, hexane, ethanol, carbon tetrachloride, benzene, six other alcohols, and two other esters (CNTs) {GC-FID}; benzene, toluene, dichloromethane, trichloromethane, acetonitrile, propanol, toluene, 1-butanol, m-Xylene, phenol, and naphthalene (SWCNTs functionalized with COCl, diameter < 2 nm) {GC-FID}; nine esters, nine aromatics, two alcohols, two ketones, and two alkanes

(MWCNTs with NH_2-R functionalization) {GC-FID}; chloro-substituted PCBs and terpenes (MWCNTs with NH_2 functionalization) {HPLC-DAD}; polyaromatic hydrocarbons (PAHs), aromatics, and amines (silica-MWCNT composites, CNT diameter < 8 nm) {HPLC-UV}; one nucleobase with seven benzene derivatives (variously functionalized MWCNTs) {HPLC-UV}; 12 peptides (MWCNTs with NH_2 functionalization) {HPLC-UV}; five nucleosides, a nucleobase, and four tetracycline antibiotics (silica-MWCNT composites, CNT diameter 20–40 nm) {CEC-UV}; five nucleobases, five nucleosides, eight flavonoids, and six phenolic acids (COOH-functionalized MWCNTs, diameter 20–40 nm) {CEC-UV}; five vitamins (COOH-functionalized SWCNTs, diameter about 1.2 nm) {CEC-DAD}; clenbuterol enantiomers (brominated MWCNTs, diameter 10–20 nm) {TLC}; thioamides (MWCNTs, diameter 8 nm) {micro-CEC-UV}; six nonsteroidal anti-inflammatory drugs (MWCNTs, diameter ca. 10–15 nm {CE}; and four flavonoids, four phenolic acids, and two saponins (MWCNTs, diameter 3–20 nm, and SWCNTs, diameter 0.7–1.1 nm) {CE} (Table 14.1).

CNTs have also begun to be studied for applications in laser desorption/ionization (LDI) time-of-flight mass spectrometer (TOFMS). These CNT-derived materials have been used as substrates in both matrix-assisted LDI (MALDI) [207–211] and surface-assisted LDI (SALDI) [415, 443–451]. However, this incipient work is still very much in the development phase, with no conclusive results as of this writing.

14.4 Problems and Exercises

1. Briefly describe the principles and practical construction behind the following applications of CNTs: water treatment and "nanosponges." Rate the applications using CNTs against the corresponding, more established applications using non-CNT materials (in terms of such parameters as performance, cost, size, and closeness to the market).
2. Enumerate and describe CNT applications in separations. Enumerate at least ten (10) chemicals whose mixtures have been separated using CNT-based materials. Define *isoelectric point*.
3. Describe the principles of the use of CNTs in mass spectrometry. What are their relative advantages and drawbacks in this application?

Table 14.1 After ref. [415]

Analyte	Methodology	Recovery (%)	LODs
Ag (I)	Flame atomic absorption spectrometry	96–108	0.35 µg L^{-1}
As (V)	Hydride generation-atomic fluorescence spectrometry	94–104	2 ng L^{-1}
As (III), As (V), Sb (III), Sb (V) (as APDC complex)	Hydride generation-double channel atomic fluorescence spectrometry	92–107	2.1–4.3 ng L^{-1}
As (III), As (V), Sb (III), Sb (V) (as APDC complex)	ETAAS	94–104	0.02–0.05 µg L^{-1}
Au (III), Mn (II)	Flame atomic absorption spectrometry	94–102	0.01–0.03 µg L^{-1}
Au (III)	Flame atomic absorption spectrometry	>95	0.3 µg L^{-1} for PANI/MWCNTs and 0.5 µg L^{-1} for PEDOT/MWCNTs
Au (III)	Flame atomic absorption spectrometry	>96	0.15 µg L^{-1}
Cd (II)	Flame atomic absorption spectrometry	98	0.22 µg L^{-1}
Cd (II)	ETA-AAS	97–100	0.010 µg L^{-1}
Cd (II)	ICP-OES	–	1.03 µg L^{-1}
Cd (II), Pb (II), Ni (II)	Flame atomic absorption spectrometry	97–104	0.04–0.23 µg L^{-1}
Cd (II), Co (II), Cu (II), Cr (VI), Pb (II), V (V), As (III)	ICP-MS	92–110	0.4–3.4 ng L^{-1}
Cd (II), Co (II), Ni (II), Pb (II), Fe (III), Cu (II), Zn (II) (as 8-hydroxyquinoline complexes)	Flame atomic absorption spectrometry	88–104	1.0–5.0 µg L^{-1}
Cd (II), Pb (II)		96–109	0.15 and 0.44 µg L^{-1}
Cd (II), Pb (II)	Flame atomic absorption spectrometry	97–101	0.3 and 1 µg L^{-1}
Co (II)	Flame atomic absorption spectrometry	101	50 ng L^{-1}
Co (II), Cu (II), Pb (II)	ETAAS	96–109	1.2–39 ng L^{-1}
Co (II), Cu (II), Ni (II), Pb (II), Fe (III), Mn (II)	Flame atomic absorption spectrometry		
Cu (II), Ni (II)	Flame atomic absorption spectrometry	81–100	0.31–0.63 µg L^{-1}
Cu (II), Ni (II), Zn (II)	Flame atomic absorption spectrometry	–	40–60 µg L^{-1}
Cu (III, Fe (III), Mn (II), Pb (II)	Flame atomic absorption spectrometry	23–106	3.5–8 µg L^{-1}

(continued)

Table 14.1 (continued)

Analyte	Methodology	Recovery (%)	LODs
Cu (II), Fe (III), Pb (II)	ICP-OES	97–105	0.15–0.26 μg L^{-1}
Cr (III), Fe (III), Pb (II)	ICP-OES	99–100	0.19–0.33 μg L^{-1}
Cr (VI)	UV spectrometry	98–12	8.5 ug kg^{-1}
F-, Br-, Cl-, No$_2^-$, BrO$_3^-$, NO$_3^-$, CO$_3^-$, PO$_4^{3-}$, SO$_4^{2-}$	HPLC conductivity detector	72–118	0.41–3.17 μg L^{-1}
BrO3-, NO3-, CO3-, PO43-, SO42-	HPLC conductivity detector	84–120	1.54–10.02 μg L^{-1}
Ga (III)	Flame atomic absorption spectrometry	87–94	3.03 μg L^{-1}
Hg (II)	Batch adsorption experiments	88–95	0.0123 μg L^{-1}
Ni (II), Pb (II)	ETAAS	–	10–30 ng L^{-1}
Pb (II)	ICP-OES	96–100	0.27 μg L^{-1}
Pb (II)	ICP-OES	99–109	0.32 μg L^{-1}
Pb (II)	Atomic absorption spectrometry	99–113	0.30 μg L^{-1}
Pd (II)	Flame atomic absorption spectrometry	99	0.3 μg L^{-1}
Pd (II)	Flame atomic absorption spectrometry	81–91	0.3 μg L^{-1}
Rh (III) (as PAN complex)	Flame atomic absorption spectrometry	96–105	0.010 μg L^{-1}
Total Tl, Tl (III)	STPF-ETAAS	>98	150 ng L^{-1}
Zn (II)	Flame atomic absorption spectrometry	>98	0.07 ng L^{-1}
Pesticides 8 multiclass pesticides	GC-MS	79–105	1.5–3.0 μg L^{-1}
Carfentrazone-ethyl	GC-ECD	74–80	10 ng L^{-1}
5 sulfonylurea herbicides	HPLC-DAD (236 nm)	79–102	1.1–7.2 ng L^{-1}
3 chloroacetanilide herbicides	GC-MS	89–1106	2–6 ng L^{-1}
7 OPPs and 1 thiadizine	GC-NPD	54–91	2.97–31.6 ng L^{-1}
Atrazine, methidathion on propoxur	UV–Vis spectrometry (200–300 nm)	84–104	0.31–0.41 mg L^{-1}
6 pyrethroids	HPLC-UV (210 nm)	71–118	1.3–5.0 ng L^{-1}
4 chloroacetanilide herbicides	GC-ECD	81–88	0.01 μg L^{-1}
11 triazines	UPLC-MS	73–98	<0.1 ng L^{-1}
4 triazines	μ-LC-UV (220 nm)	55–71	0.2–0.5 μg L^{-1}
3 triazines	GC-MS	–	–
Chlorpyrifos and phosalone	HPLC-UV (288 nm)	85–100	4.02 and 1.02 ng L^{-1}
Chlorsulfuron and metsulfuron	CE-UV (231 nm)	86–108	0.36 and 0.40 μg L^{-1}
Atrazine and simazine	GC-MS	87–110	2.5 and 5.0 ng L^{-1}

(continued)

Table 14.1 (continued)

Analyte	Methodology	Recovery (%)	LODs
Mefenacet and three photolysis degradation products	UPLC-UV-MS (220 nm)	70–120	0.02–0.04 µg L^{-1}
7 sulfonylurea herbicides	LC-MS/MS	82–110	0.01–0.20 ng L^{-1}
3 triazines and 2 dealkylated metabolites	HPLC-UV (220 nm)	86–102	4–30 ng L^{-1}
9 multiclass pesticides	Nano-LC-UV (200 nm)	36–101	16–67 ng L^{-1}
Pharmaceuticals			
Chloramphenicol	HPLC-MS/MS	96–102	Egg, 0.004 µg kg^{-1}; honey, 0.003 µg kg^{-1}; milk, 0.003 µg kg^{-1}
10 quinolones	UPLC-UV (279 and 319 nm)	70–100	5.8–14.5 µg L^{-1}
11 quinolones	Ce-DAD (250 and 280 nm)	62–114	0.028–0.094 µg L^{-1}
3 macrolides	HPLC-UV (210 nm)	85–96	–
Ursolic acid	HPLC-UV (210 nm)	80	–
Miscellaneous biological compounds			
Bovine hemoglobin and serum albumin	UV–Vis spectrophotometry (190–500 nm)	90	1.0 mg L^{-1}
Bovine serum albumin	Bradford assay	–	–
3 estrogens	MEKC-DAD (196 nm)	90–100	0.1–0.2 µg L^{-1}
3 estrogens	HPLC-FD ($\lambda_{ex}/\lambda_{em}$ = 280/310 nm)	88–112	1.21–2.35 µg L^{-1}
3 biogenic thiols	HPLC-FD ($\lambda_{ex}/\lambda_{em}$ = 360/510 nm)	92–113	0.004–0.080 nM
4 cobalamins	HPLC-DAD (265, 351 and 361 nm)	76–102	0.35–29 µg L^{-1}
3 albumins, bovine hemoglobin, and lysozyme	HPLC-UV (280 nm)	92–97	–
3 neurotransmitters	HPLC-FD ($\lambda_{ex}/\lambda_{em}$ = 326 nm/ 412 nm)	88–94 for SWCNTs; 82–94 for MWCNTs	–
PAHs			
3 PAHs	HPLC-FD ($\lambda_{ex}/\lambda_{em}$ = 274 nm/ 412 nm), 266 nm/ 436 nm, 270/416 nm	89–98	5–8 ng L^{-1}
16 PAHs	GC-MS	72–93	0.001–0.15 µg L^{-1}
16 PAHs	GC-MS	70–127	2.0–8.5 ng L^{-1}
8 PAHs	GC-MS	88–122	0.10–0.88 ng L^{-1}

(continued)

Table 14.1 (continued)

Analyte	Methodology	Recovery (%)	LODs
16 PAHs	GC-MS	72–99	4.2–46.5 ng L^{-1}
Chemical warfare agents			
9 chemical warfare agents	GC-MS	63–110	0.01–1 µgL^{-1}
9 chemical warfare agents	GC-FPD	55–96	0.05–1 µg L^{-1}
6 acidic degradation products of chemical warfare	GC-MS	48–112	0.05–0.08 µg L^{-1}
Other			
2 polychlorophenols and 2 tetrahalogenated biphenyls	GC-ECD	55–100	1.0–6.0 µg L^{-1}
2-Nitrophenol, 2,6-dichloroaniline and, naphthalene	HPLC-UV (-)	–	0.1–3 µg L^{-1}
27 amines, 2 anilines, 12 chloroanilines, 10 N-nitrosamines, and 3 aliphatic amines	GC-MS	–	–
8 low-molecular-mass aldehydes	LC-MS/MS	–	–
4 parabens	HPLC-C-CAD	85–104	0.5–2.1 mg L^{-1}
Sudan IV	HPLC-UV (510 nm)	89–95	2.3 ng L^{-1}
Pentachlorophenol	GC-ECD	92 for MWCNTs; 43–78 for o-MWCNTs	–
Trans- and cis-resveratrol	LC-MS/MS	95–108	0.02 µg L^{-1}
Melamine	HPLC-UV (240 nm)	90–93	0.3 µg L^{-1}
4 linear alkylbenzene sulfonates	HPLC-UV (224 nm)	82–110	0.02–0.03 µg L^{-1}
4 linear alkylbenzene sulfonates	HPLC-UV (224 nm)	87–106	0.013–0.021 µg L^{-1}
16 phthalate acid esters	GC-MS	63–119	0.0031–0.0038 µg L^{-1}

Chapter 15
Miscellaneous CNT Applications

Contents

15.1 Overview

Miscellaneous applications of CNTs include catalysis and esoteric or unconventional applications.

15.2 Catalysis

Catalysis relating to the use in *fuel cells* has been dealt with in an earlier chapter, on batteries and energy devices, quod vide.

One major issue still being grappled with relating to the use of CNTs for catalysis is the fact that as-produced CNTs contain residual chemicals (usually metal particles) from synthesis, in spite of attempts to remove these through post-synthesis washing and purification [452]. As a result of this, attempts to make CNT-based catalysts specific to targeted processes can become complicated.

Nevertheless, attempts to fabricate CNT-based catalysts for specific, targeted processes have shown some success. A Pt-based SWCNT system has been used to hydrogenate the aldehyde group in 3-methyl-2-butenal, with good selectivity [453]. CNTs have been substituted for alumina and silica supports for Pt and Pd catalysts used in the hydrogenation of o-chloronitrobenzene to o-chloroaniline [454].

MWCNT-supported Pt and Pd catalysts using MWCNT supports also showed high initial turnover rates in the hydrogenation of toluene and naphthalene as compared to silica, alumina, or ZrPSi supports [455]. More remarkably, Pd–Rh,

© Springer International Publishing AG 2018
P. Chandrasekhar, *Conducting Polymers, Fundamentals and Applications*,
https://doi.org/10.1007/978-3-319-69378-1_15

Pt, and Rh catalysts supported on MWCNTs have been demonstrated to hydrogenate benzene at room temperature, whereas analogous, carbon-supported Pd and Rh catalysts require significantly higher temperatures [456]. The catalytic metals are presumed to be adsorbed onto the external walls of the CNTs in such catalysts. Additionally, demonstrated uses of CNT-based catalysts have been in Suzuki coupling reactions and the oxidation of cellobiose [457, 458].

SWCNTs have been covalently coupled with an organic vanadyl complex to yield a heterogeneous catalyst for the cyanosilylation of aldehydes [459].

15.3 Miscellaneous Applications

CNTs have been studied for esoteric applications such as "nano-springs," fabricated from bundles or forests of MWCNTs, with claimed energy densities 10x that of steel and greater claimed durability [460]. However, again, to date, no practical or commercial or even niche applications have emerged from this work.

15.4 Problems and Exercises

1. Give at least two examples of the use of CNT-based materials as catalysts in organic synthesis using fuel cells. Describe the chemistry and electrochemistry involved in detail. What are their drawbacks preventing practical implementation as of this writing (2016)?

2. Propose and sketch the construction of a "nano-spring" based on CNTs, and describe its principles. What are its drawbacks preventing practical implementation as of this writing (2016)?

Part III
Graphene, Fundamentals

Chapter 16
Introducing Graphene

Contents

16.1   Historical

Graphene was first widely popularized following the publication of Geim and Novoselov, then working at the University of Manchester in the UK, reporting the isolation of individual sheets of graphene on a silica wafer using a very simple method involving peeling off layers of it from highly ordered pyrolytic graphite (HOPG) using cellophane adhesive tape (also called "micromechanical cleavage" or the "Scotch® tape method") [461, 462] in 2004. The name *graphene* itself was formalized by IUPAC (the International Union of Pure and Applied Chemistry) in 1995 [463] to avoid the use of other synonyms such as "graphite layers," "carbon layers," "carbon sheets," and "carbon exfoliates," although it was used by Mouras et al. [464] in 1987. Geim and Novoselov received the Nobel Prize in Physics in 2010 for their work on graphene [465].

However, as with all "discoveries" and "research," there is evidence that graphene was synthesized, observed, and even recognized much earlier than the above-cited work of Geim and Novoselov. For example, Boehm et al. reported on the adsorption properties of "very thin carbon sheets" in 1962, clearly recognizing what they were studying as individual sheets of graphite; they also first reported in detail the production of reduced graphene oxide (closely resembling graphene) from graphene oxide [466–468]. The term *graphene* was used by Mouras et al. in 1987, also to describe individual sheets of graphite [464]. As early as 1859, Brodie is said to have recognized graphene structurally in reduced graphite oxide [2067]. Wallace recognized graphene and discussed its structure theoretically in

© Springer International Publishing AG 2018
P. Chandrasekhar, *Conducting Polymers, Fundamentals and Applications*,
https://doi.org/10.1007/978-3-319-69378-1_16

1947 [470]. And Ruess and Voigt appear to have presented transmission electron micrograph (TEM) images of few-layer graphene ("highly lamellar carbon" according to their description) in *1948* [471]. Oshima and Nagashima produced "Ultra-thin epitaxial films of graphite and hexagonal boron nitride on solid surfaces" in 1997 [472]; however, in these films, the graphene appeared to be bonded to the substrates rather than free standing as in the later Geim/Novoselov product. Jang et al. [473] filed a US patent describing "nanoscaled graphene plates" in 2002, which was issued in 2006, in which medium-scale synthesis of graphene was described. (To their credit, however, Geim and Novoselov demonstrated an extremely facile and reproducible method of synthesizing it, thus opening windows to the detailed study of its theoretical and experimental properties and applications.)

Other work of historical importance for the record and preceding the above-described seminal work of Geim and Novoselov includes that of Enoki et al., who used high temperatures (>1600 °C) to convert "nano-diamonds" into nm-scale regions of graphene atop highly ordered pyrolytic graphite (HOPG) in 2001 [474].

16.2 Basics of Graphene

Here are some basic parameters and properties of graphene that may serve to usefully orient the reader:

- *C-C bond distance* of about 1.42 Å.
- *Thickness* about 3.35 Å (this is close to the interplanar spacing in 3-D graphite).
- *Weight per unit area* about 0.77 mg/m^2.
- *Specific surface area* of about 2600 m^2/g (similar to that of activated carbon and significantly larger than that of CNTs).
- *Mechanical properties – tensile strength, Young's modulus,* and *fracture toughness*:

  – Pure, defect-free graphene has a calculated *tensile strength* of about 130 GPa and a *Young's modulus* of about 1035 GPa. These values have also been experimentally measured in very small samples [475]; they may be compared with those for standard stainless steel, of about 1.5 GPa and about 200, respectively. On the other hand, the Young's modulus for graphite along the basal plane only is also about 1000 GPa [476].
  – However, very few as-synthesized samples are defect-free, and so their tensile strength and Young's modulus do not approach these calculated numbers; and very few samples of graphene are large enough for their exceptional strength to be utilized in any practical manner. Nevertheless, such values have been used to accord superlative predictions to graphene, e.g., that a 1 m^2 sample of graphene could "support up to 5 kg, but weigh only 0.77 mg, 1/10,000 of the weight of standard office paper of the same size" or that, in analogy to CNTs, it could be used to construct "an elevator to space."

- Conversely, pure, defect-free graphene is very brittle: it has a *fracture toughness* of about 4 MPa/m$^{1/2}$ vs. about 20–50 MPa/m$^{1/2}$ for most metals [477].

• Many forms of graphene are produced using *graphene oxide* (**GO**), which is relatively easy to synthesize from graphite, as the starting material. GO is then reduced. The reduction is however rarely 100% complete. Thus, the resulting *reduced graphene oxide* (**RGO**) frequently possesses not only impurities (such as remnants of catalysts, if used in the reduction process) but also some still-oxidized sites, frequently in the form of -C $=$ O or –COOH groups. Nevertheless, many researchers still use RGO in their work and believe their findings extend to pure graphene.

• In addition to single sheets of graphene, i.e., monolayers, graphene as studied in the laboratory also comes in forms such as *bilayer* or *few-layer*, *nanoribbons*, and even *quantum dots*.

• Pure, defect-free graphene has an *electrical conductivity* of about 3.6 $\times$ 10^8 S/m or about 6 X that of Cu or Ag.

• Pure, defect-free graphene has a *thermal conductivity* of 2000–4000 W/m^2 K. These values are typical for suspended monolayers, with the higher values typical of purer samples [477].

Figure 16.1 below shows the basic structure of graphene. Figure 16.2 illustrates the "Scotch® tape" method of fabricating sheets of single-layer graphene from highly ordered pyrolytic graphite (HOPG). Figure 16.3 shows a TEM image of one of the first free-standing graphene films ever produced, in 2007.

16.3 Toxicology of Graphene

While graphene is widely recognized to have much less toxicity than CNTs, it does still present some issues in this regard, especially with the potential of being incorporated into consumer products and in medical applications [486–489].

It is believed that the toxicity of graphene and graphene oxide (GO, including reduced graphene oxide, RGO) is mainly attributable to the *mechanical* damage of the cell membrane of bacteria due to the sharp edges of the nanosheets [490, 491]. -Hydrazine-reduced GO nanosheets were reported to be more toxic to the bacterial cell, possibly due to residual hydrazine. In in vitro mammalian cell line experiments, the cytotoxic effect of GO was found to be dose and time dependent, even at a concentration of less than 20 µg/mL. Above 50 µg/mL, GO appears to severely affect the cell survival rate and induce apoptosis in the human fibroblast cell line. Similar cytotoxic effects have also been observed in other cell lines such as MGC803, MCF-7, MDA-MB-435, and HepG2. Experimental data also demonstrate that GO enters into the cytoplasm of the cell and is found to be present in subcellular compartments such as in the lysosomes, mitochondria, endoplasm, and

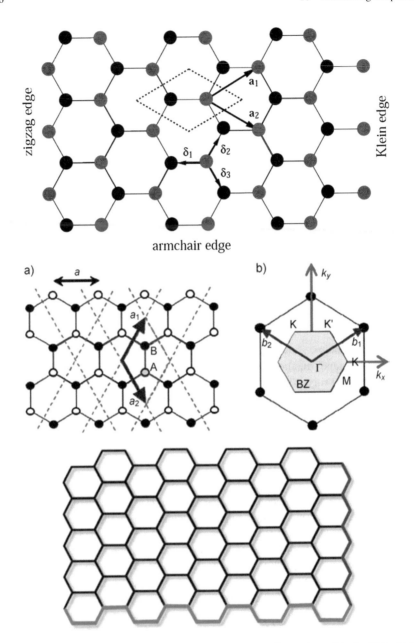

Fig. 16.1 Basic geometry of graphene lattice and the corresponding reciprocal lattice (After Refs. [478–480], reproduced with permission)

Fig. 16.2 Illustration of the "Scotch® tape" or mechanical exfoliation method of fabricating sheets of single-layer graphene from highly ordered pyrolytic graphite (HOPG) (After Ref. [481], reproduced with permission)

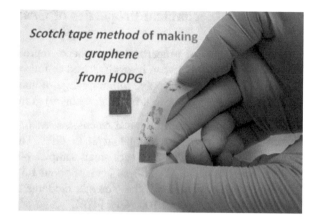

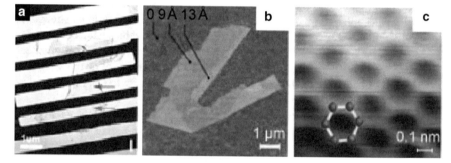

Fig. 16.3 Images of graphene: (**a**) TEM image of one of the first free-standing graphene films ever produced, in 2007. After Ref. [482], reproduced with permission. (**b**) AFM image of single-layer graphene. The folded edge exhibits a relative height of approximately 4 Å, indicating that it is single layer. (**c**) High-resolution STM image (After Refs. [483–485], reproduced with permission)

even nucleus. One hypothesis has been that GO attaches to the cell membrane and induces downregulation of the adhesion-associated gene [490, 491].

On the other hand, other studies have shown that forms of graphene, including graphene nanoribbons, have little or no toxicity up to a concentration of about 50 µg/mL, interpreted by the fact that they do not alter the differentiation of human bone marrow stem cells toward osteoblasts or adipocytes, although they do penetrate or pierce cell membranes and thus enter the interior or cells [486–489].

16.4   Basic Mechanical Properties of Graphene

The basic mechanical properties of graphene, reproduced here for convenience from earlier in this book, can be summarized as follows; many of the values cited for particular parameters apply, within a range of about $\pm 20\%$, equally to monolayer, bilayer, and trilayer (but not $>$ three layer) graphene [481]:

- Pure, defect-free graphene has a *calculated tensile strength* of about 130 GPa and a *Young's modulus* of about 1035 GPa. These values have also been *experimentally* measured in very small samples [492]; they may be compared with those for standard stainless steel, of about 1.5 GPa and about 200, respectively. On the other hand, the Young's modulus for graphite along the basal plane only is also about 1000 GPa [476].
- However, very few as-synthesized samples are defect-free, and so their tensile strength and Young's modulus do not approach these calculated numbers; and very few samples of graphene are large enough for their exceptional strength to be utilized in any practical manner. Nevertheless, such values have been used to accord superlative predictions to graphene, e.g., that a 1 m^2 sample of graphene could "support up to 5 kg, but weigh only 0.77 mg, 1/10,000 of the weight of standard office paper of the same size" or that, in analogy to CNTs, it could be used to construct "an elevator to space."
- Conversely, pure, defect-free graphene is very brittle: it has a *fracture toughness* of about 4 MPa/m$^{1/2}$ vs. about 20–50 MPa/m$^{1/2}$ for most metals [477].
- The *spring constant* of suspended graphene sheets, as determined according to Hooke's law, is in the region of 1–5 N/m [493].
- Other mechanical properties of graphene of note and their most recent reported values are as follows [494–496]: *second- and third-order elastic* stiffness of 340 and (-)690 N/m, respectively, *breaking stiffness* of 1 TPa (1000 GPa), and *intrinsic strength* of 130 GPa.

16.5   Hype on Graphene vs. CNTs Compared

Intense research and development on graphene are comparatively more recent on graphene as compared to CNTs, by approximately 10 years. As a result, as of this writing, there appears to be more promise and less "early disappointment" with regard to graphene's impact on science and commerce than with CNTs. A recent article in *Chemical and Engineering News*, the member publication of the American Chemical Society, gave a succinct account of the current status of graphene [497].

Nevertheless, and rather unfortunately, barring breakthroughs in economical, bulk production methods for highly pure, defect-free graphene in the near future, graphene appears to be on the same trajectory as CNTs in terms of expectation vs. realization, just delayed by a decade or so. As of this writing, about seven research papers are published *daily* on graphene, and more than 26,000 patents have been filed on subjects related to graphene, with a large proportion of these in East Asia [497]. The European Union recently launched an R&D initiative (similar to US initiatives on, e.g., the human genome) focusing on graphene with a budget of US $1.1 billion over 10 years [498]. There is even an exposition dedicated to graphene, denoted "Graphene & 2D Materials Europe," part of a larger, IDTechEx exhibit [499, 500].

With regard to commercial products or potential products, a very few have, as of this writing (2016), now made it to the market. A Spanish company started production of Li batteries incorporating graphene, with a claimed energy density about five times that of standard secondary Li batteries [501]. A New Jersey-based US company now offers RFID (radio frequency identification) tags with a bit of graphene in them to improve their high-temperature/high-pressure performance [502] as well as specialized antennae for cell phones (mobiles) that incorporate graphene and afford improved performance [503]. And there are other niche products already on the market as of this writing: some tennis rackets incorporating a bit of graphene (substituted for carbon fibers) to increase stiffness and reduce weight and a few conductive inks incorporating graphene [497].

In spite of all this development, however, as of this writing (2016), worldwide sales of graphene-related materials are about *US $30 million* [497]. Commercial products incorporating graphene can be counted on the fingers of one hand, as noted in the previous paragraph. The prediction that these sales will hit US $220 million in 2026 [497] seems, at least as of this writing, highly optimistic. And the expectation voiced by some in the initial, exciting years of graphene development, that graphene will replace Si in electronic circuits, seems, at this writing, or that it will appear in "low-cost touch interfaces for wearable electronic devices" [497] even more far removed, in this author's humble opinion.

16.6 Problems and Exercises

1. Briefly enumerate the major milestones in the historical discovery and development of graphene, starting with the year 1947.
2. Enumerate the defining, basic physical and structural properties of graphene, including such properties as C–C bond length, thickness, density, surface area, major mechanical properties, salient electrical properties, and salient thermal properties. Where relevant (e.g., for mechanical properties), compare graphene's values with those of other common materials (e.g., steel, copper).
3. Briefly outline the principle classes of methods of preparation of graphene. Under which class does the "Scotch® tape method" fall?
4. What is the principle mechanism of toxicity of graphene in vivo?
5. Briefly identify the known and established commercial applications of graphene as of 2016. Do the same as of your reading today.

Chapter 17
Electronic Structure and Conduction Models of Graphene

Contents

17.1 Electronic Structure and Conduction

As with CNTs, the electronic structure of *graphene* is most easily discussed as applicable to pure, defect-free graphene first, then conditionally extendable to "imperfect" (as-produced) graphene. Additionally, in the case of graphene, discussion of the electronic structure of *bilayer* and *few-layer* graphene as well is pertinent, since it gives some useful insights into the structure and properties of single-layer graphene.

The crystal structure of graphene is best considered as a combination of two, equivalent triangular sub-lattices (see Fig. 17.1) [476, 479, 481, 504]. In this figure, the high-symmetry points M, K, K', and Γ in the first Brillouin zone are marked, with K and K' being two inequivalent points in the Brillouin zone. The s, p_x, and p_y orbitals of the carbon atoms form σ bonds with neighboring carbon atoms. One p_z electron from each carbon contributes then to the π (bonding) and π* (antibonding) bands of graphene. The dispersion relation of these π electrons is approximated by a tight-binding model incorporating only the first nearest-neighbor interactions.

One of the peculiarities of graphene with respect to electronic transport and conduction is that they are better described by the *Dirac* equation rather than the *Schrödinger* equation, because its charge carriers mimic relativistic particles [462]. As noted by Geim and Novoselov particles [462], although there is nothing particularly relativistic about electrons moving around carbon atoms, their specific interaction with the periodic potential of graphene's honeycomb lattice gives rise to new quasiparticles that, at low energies, are accurately described by the (2 + 1)-dimensional Dirac equation with an effective speed of light, i.e., v_F approximately equals 10^6/m-s. These quasiparticles, called *massless Dirac fermions*, can be interpreted as electrons that have lost their rest mass or as neutrinos that acquired

© Springer International Publishing AG 2018
P. Chandrasekhar, *Conducting Polymers, Fundamentals and Applications*,
https://doi.org/10.1007/978-3-319-69378-1_17

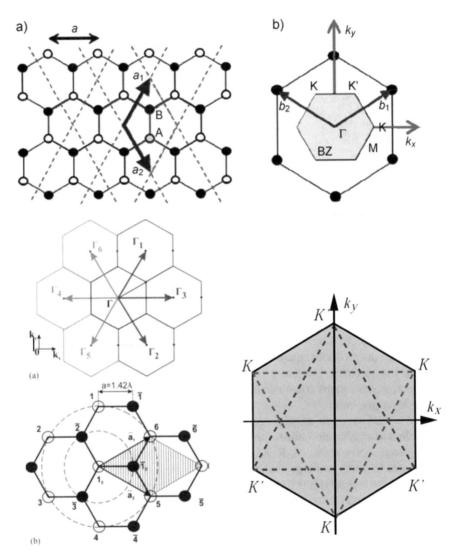

Fig. 17.1 Depiction of the graphene lattice in regular (left) and reciprocal (right) space (After Ref. [479], reproduced with permission). The Brillouin zone of graphene shown in more detail (After Ref. [478], reproduced with permission) (After Ref. [505], reproduced with permission)

the electron charge. Essentially, then, unlike in conventional semiconductors, the conducting electrons and holes in graphene are *massless* and behave like photons [506].

From symmetry considerations, the hopping of electrons between these two sub-lattices leads to the formation of *two* energy bands, which intersect at the K point. In the proximity of these intersection points, the electron energy is linearly

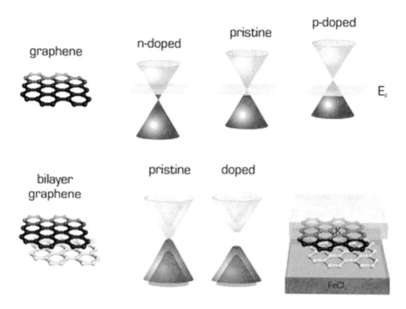

Fig. 17.2 Band structure of single-layer graphene showing p- and n-type doping with respect to the Fermi level and bandgap opening in bilayer graphene caused by doping (After Refs. [509, 510])

dependent on the wave vector. This linear dispersion results in massless excitons described by the Dirac equation. Now Dirac fermions (electrons or holes) have unusual properties when compared to ordinary electrons. For example, the anomalous *integer quantum Hall effect* can be observed in graphene even at room temperature (it is generally observed at very low temperatures for most other materials) and is generally ascribed to bilayer graphene [476, 481, 507–509].

Other unusual resultant properties of graphene include insensitivity to external electrostatic potentials (known as the *Klein paradox*), an instability in the wave function under confining potentials, and large mean free paths. Pristine graphene is a zero-bandgap semiconductor. Figure 17.2 comparatively shows the bandgap structure for single-layer and bilayer graphene. Single-layer graphene belongs to the D_{6h} point group which reduces to D_{3d} for the AB bilayer and ABC trilayer and to D_{3h} for the ABA trilayer.

Electron mobilities in excess of 200,000 cm^2/Vs at a carrier density of 2×10^{11}/cm^2 have been reported for mechanically exfoliated layers of graphene suspended above a Si/SiO$_2$ gate electrode [479, 481, 510], values that are several orders of magnitude higher than those for Cu. Furthermore, graphene's electron mobility appears to be nearly independent of temperature between 10 K and 100 K, implying that the dominant scattering mechanism in graphene is defect scattering [511, 512]. It has also been recognized that practically realizable mobility values for graphene are much less; for example, it has been observed that, with SiO$_2$ as a substrate, scattering of electrons by optical phonons of the substrate is a larger effect than scattering by graphene's own phonons, limiting mobility to about 40,000 cm$^{2/}$Vs [513]. It has also been shown that electrons in graphene can cover distances of several μm before being scattered, even at room temperature [514].

It is to be noted that many of the above-cited superlative electronic conduction and related properties of graphene are valid for pristine and defect-free graphene. Defects act as scattering sites; they inhibit charge transport by limiting the electron mean free path. The interaction of single-layer graphene with the substrate it sits on also affects conductivity; measurements must necessarily be made on insulators with no or limited π or van der Waals interactions, such as SiO_2. It has also been shown that charge transport, and hence electronic conductivity, of graphene is greatly affected by adsorption of ubiquitous "impurities" such as O_2 and H_2O on the highly surface-active graphene [515].

One of the hurdles in the use of graphene for electronics applications is that its *bandgap* is *zero* even at the charge neutrality point; this prevents its use in logic applications. The band structure of graphene can be modified by lateral quantum confinement (e.g., by constraining the graphene in nanoribbons), or in graphene quantum dots, or by biasing bilayer graphene (cf. Fig. 17.2) [481]. In this context, bandgap opening in both zigzag and armchair nanoribbons has been observed experimentally, including via processes such as doping and edge functionalization [516, 517]. In single-layer graphene, a bandgap can be induced either by spatial confinement or by lateral superlattice potential; with regard to the latter, sizeable gaps naturally occur in graphene epitaxially grown on top of crystals with matching lattices, such as BN or SiC [462]. Some researchers have directly observed a widely tunable bandgap in bilayer graphene, gate-controlled and continuously tunable bandgap up to 250 meV at room temperature [518]. In contrast, in trilayer graphene, no significant bandgap is observed in "ABA"-stacked trilayers, whereas "ABC"-stacked trilayers, whose structure is similar to that of "AB"-stacked bilayers, show a significant bandgap [519].

The extent of increase in bandgap in monolayer graphene can be gauged by the following figures [478]: 15-nm-wide graphene nanoribbons, 200 meV; graphene on BN substrates, 53 meV; graphene on Cu(111) substrates, 11 meV; graphene on SiC substrates, 250 meV; and graphene containing various adsorbed small molecules, 2000 meV.

It is to be noted that single-layer *graphene nanoribbons* have electronic properties significantly different from larger-area graphene sheets and from bilayer or few-layer graphene, and their electronic properties also vary depending upon whether their edges terminate in an armchair or zigzag pattern (see corresponding discussion for CNTs in Part I for definition of armchair and zigzag) [476]. Graphene nanoribbons are semiconductors with energy gaps which decrease as a function of increasing ribbon widths [476]. However, in this author's humble view, graphene nanoribbons have been shown to be extremely difficult to synthesize reproducibly and are thus largely a theoretical construct. Consequently, detailed discussion of their electronic properties is primarily of theoretical, academic interest and is thus omitted in this chapter.

Electrical conductivity in *reduced graphene oxide* (RGO, see above) has been shown to be consistent with a two-dimensional variable-range hopping model occurring in parallel with electric-field-driven tunneling, with the latter found to dominate at very low temperatures and high electric fields [524]. The model for conduction in RGO further indicates highly conducting graphene regions interspersed with disordered regions, across which charge carrier hopping and tunneling are promoted by strong local electric fields [524].

In work primarily of theoretical interest but also with potential implications for unique applications of graphene, Wang et al. [524] presented experimental data in 2008 that appeared to show room-temperature ferromagnetism in graphene materials and thus to corroborate theoretical predictions that suggested that one-atom-thick, two-dimensional graphene materials *may* show ferromagnetism, due to the existence of various defects or topological structures as the spin units and the possible long-range ordered coupling among them. These authors postulated that the observed ferromagnetism came from the defects on graphene.

17.2 Phonon Conduction and Thermal Conductivity

Single-layer graphene possesses a thermal conductivity of 3000–5500 W/mK at room temperature, based on a sampling of experimental measurements by multiple research groups [480, 525], although many researchers have expressed some skepticism of the values reported on the high end (near 5500 W/mK) [526, 527]. These values may be compared with that of high-purity graphite, which is about 2000 W/mK at room temperature; that for diamond, which is lower than that of graphite; and values for CNTs, which are generally in the 3000–4000 W/mK range. The wide range observed is ascribed to such factors as the relative purity of the graphene, presence of defects, and synthesis method. For example, the thermal conductivity of single-layer graphene exfoliated on a SiO_2 support is about 600 W/mK at room temperature, which is still higher than that of Cu; this value is said to be lower than that of suspended single-layer graphene due to the leakage of phonons across the graphene–support interface, as well as strong interface scattering of flexural modes, which make a large contribution to the thermal conductivity in suspended graphene [528].

Acoustic phonons appear to be the main heat carriers in graphene near room temperature. On the other hand, optical phonons find use primarily in Raman spectroscopic probing of graphene, yielding vital information such as the number of graphene planes in few-layer graphene [504]. It has been experimentally demonstrated that transport properties of phonons (such as energy dispersion and scattering rates) differ substantially in graphene as compared to the corresponding basal planes in 3D graphite or in other 3D bulk crystals [504]. It is this unique nature of two-dimensional phonon transport that gives rise to the observed, very high heat conduction in graphene. The 2D, phonon-assisted nature of heat conduction in graphene has been demonstrated over nearly the entire range of phonon frequencies [525]. It has also been demonstrated that the intrinsic, Umklapp-limited thermal conductivity of graphene grows with the increasing dimensions of graphene flakes and exceeds that of bulk graphite when the flake size is on the order of a few μm [525].

With respect to phonon-based thermal conduction, it is also noted that, specifically in nanostructures (which single-layer graphene qualifies as), the phonon energy spectra are quantized due to spatial confinement of the acoustic phonons. In most cases, the spatial confinement of acoustic phonons results in the *reduction* of the phonon thermal conductivity [529, 530].

17.3 Problems and Exercises

1. Sketch the graphene lattice in normal and reciprocal space, and relate the two.
2. Sketch the (theoretical) bandgap structures for single-layer and bilayer graphene, and compare the two.
3. Why is electron transport in graphene better described by the Dirac equation rather than the Schrödinger equation?
4. Describe the Klein paradox as it relates to graphene.
5. What are the typical theoretical and practically observed values of the electron mobility in graphene, and how do they relate to those of other conductors?
6. What are the methods used to work around the zero bandgap of graphene as it relates to applications in electronics? How successful have these been?
7. Have any practical electronic, microelectronic, or "nanoelectronic" devices resulted from graphene as of this writing? As of your reading today?
8. What is the main proposed mechanism for heat conduction in graphene near room temperature? What is the difference in this respect with graphite? How do spatial confinement and quantization affect thermal conductivity in graphene?

Chapter 18
Synthesis and Chemical Modification of Graphene

Contents

18.1   Brief Enumeration of Methods of Synthesis

Methods for the synthesis and large-scale production of graphene, including single layers, bilayers, and "few layers," can be broadly classified into two categories: (1) *top down*, generally starting from highly ordered pyrolytic graphite (HOPG) or some other form of highly pure, highly ordered graphite or its further derivative (such as graphene oxide, GO) and isolating graphene layers via some method such as mechanical exfoliation (peeling off), and (2) *bottom up*, generally involving building up carbon into graphene, e.g., by reduction of a starting material such as SiC or glucose.

The major methods currently used for the synthesis of graphene include:

1. *Micromechanical exfoliation or cleavage* (*Scotch® tape method* or "peeling off" method). This method affords the quickest samples still suitable for scientific studies without need for further purification or processing. It is also of historical importance since its first use by Geim and Novoselov in 1991 [531] set off intense study of graphene. This method is rather broad and includes, e.g., such methods as ultrasonic exfoliation, exfoliation using intercalation into graphite of various chemical moieties, and electrochemically based exfoliation.

2. *Chemical vapor deposition* (*CVD*), commonly on such surfaces as single-crystal Ni and Cu, although a very wide variety of substrates have been tested. CVD encompasses such variants as plasma-enhanced CVD.

© Springer International Publishing AG 2018
P. Chandrasekhar, *Conducting Polymers, Fundamentals and Applications*,
https://doi.org/10.1007/978-3-319-69378-1_18

3. *Reduction of graphene oxide (GO)*, with the GO itself being produced by a variety of methods. The GO is then exfoliated (it is easier to exfoliate than graphene) and then reduced to graphene or, more accurately, reduced GO. One of the drawbacks of this method is that the GO is almost never 100% reduced and thus always possesses some oxidized parts which are the source of defects, hence nomenclature reduced GO rather than graphene.
4. *Reduction of other precursors* such as glucose, SiC, or ethanol. Generally, this involves vapors of C being generated which then condense into graphene layers on a substrate.
5. *More specialized or exotic methods* such as electrochemical preparation, "unfolding" of CNTs, catalytic transformation of C_{60} (fullerene), microwave-assisted synthesis, and electron beam lithography.

18.2 Micromechanical Exfoliation or Cleavage ("Scotch® Tape Method")

As noted above and elsewhere in this book, Geim and Novoselov [531] first implemented this method in 1991, pulling off graphene layers from highly ordered pyrolytic graphite (HOPG) using Scotch® (cellophane) tape and transferring them onto thin SiO_2 on a silicon wafer for further study. In subsequent improvements to this method, graphene flakes of several mm^2 areas can be obtained [532]. Early work using this technique produced graphene flakes tens of μm in size; it is now possible to produce mm-sized flakes using variations of this technique. Figure 18.1, reproduced from an earlier chapter, illustrates the "Scotch® tape" or mechanical exfoliation technique.

From a technical and scientific point of view, mechanical exfoliation of graphite is a somewhat delicate process, for, to exfoliate a single sheet, van der Waals attraction between the first and second layers must be overcome without disturbing any subsequent layers, and, once the layers have been separated, reaggregation back into graphite must be prevented [479, 520]. Thus, assistive methods, e.g., *intercalation* of moieties between the individual sheets of graphite to help "pry them apart" and/or *ultrasonication*, are frequently used, and liquids or solutions are selected that will keep the layers apart and prevent reaggregation.

Figure 18.2 illustrates the intercalation/exfoliation method, for a particular case, potassium intercalation. *Ultrasonication* is generally used following intercalation with various chemicals to pry apart the graphene layers. This is generally followed by isolation of the resulting graphene flakes by centrifugation, by simple precipitation, or by other methods [533, 534]. In a typical solvent-based process, exfoliation of alkali-metal-intercalated graphite in *N*-methylpyrrolidinone (NMP) solvent gives rise to a stable solution of negatively charged graphene sheets which can then be deposited on various substrates [478, 535]. Liquids used for this purpose have included *N*-methylpyrrolidinone (NMP), cyclopentanone, *o*-dichlorobenzene, perfluorinated aromatics, several ionic liquids, and even molten salts such as LiCl

Fig. 18.1 Illustration of the "Scotch® tape" or mechanical exfoliation method of fabricating sheets of single-layer graphene from highly ordered pyrolytic graphite (HOPG) (After Ref. [480], reproduced with permission)

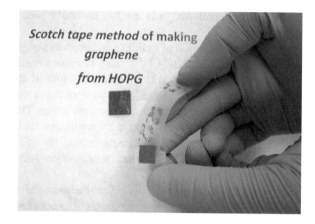

Fig. 18.2 Illustrative depiction of the intercalation/exfoliation method, here for K intercalation (After Ref. [544], reproduced with permission). The highly exothermic reaction of K with EtOH, as shown, causes the individual graphene sheets in graphite to be "blown apart"

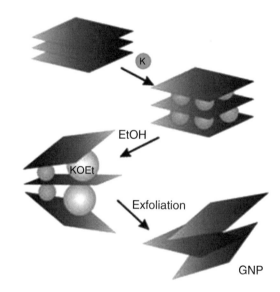

[533, 534, 536–539]; in this regard, NMP and cyclopentanone appear thus far to afford the highest proportion of single-layer graphene [479].

One of the problems encountered in early work in this area was reassembly ("restacking") of the graphene produced back into graphite; a number of solutions were attempted for this, including the use of surfactants in the solvent used and the use of immiscible liquids (e.g., heptane and water) [536, 537, 540]. Intercalation chemicals used have included $FeCl_3$, nitromethane, supercritical CO_2, and the abovementioned ionic liquids [541–543]. Ultrasonication solvents have included acetic acid, H_2O_2, and sulfuric acid.

Many adaptations of exfoliation have been implemented since the original Geim/Novoselov work, some quite unique or ingenious. For example, in one method, a diamond wedge is used to carefully peel away graphene layers from graphite [545].

In another adaptation of exfoliation, fragments of graphene that were soluble in tetrahydrofuran, CCl_4, and $CHCl_2$ were produced starting from graphite [546]. The graphite was first treated with an acidic mixture of sulfuric acid and nitric acids. Multiple oxidation and exfoliation steps were then used to arrive at small graphene plates with carboxyl groups attached to their edges. The carboxyl groups were then converted to acid chloride groups by treatment with thionyl chloride and then further to the corresponding graphene amide via treatment with octadecylamine. What resulted was small fragments of monolayer graphene soluble in the solvents delineated above.

In many nonmechanical exfoliation methods, solvents are used to prevent the graphene from reaggregating into graphite. Additionally, the exfoliation process typically generates small "platelets" of graphene between up to six layers thick, rather than monolayer graphene, with additional steps then required to the mono-layer content to an acceptable value such as 75% [496].

In *situ* produced CO_2 evolved in the space between sheets in graphene oxide (GO) during rapid heating has been used to exfoliate the GO, producing monolayer GO, which is then reduced to fairly large-area (1600 μm^2) monolayer graphene [547].

Electrochemically based exfoliation, starting with graphite immersed for a significant period of time in an appropriate solvent, is another scientifically inter-esting method that has been studied at some length, but has not been found to be competitive with other methods of mechanical exfoliation [548]. Variants of the method use surfactants and H_2SO_4-KOH solutions [549, 550], with single-layer or few-layer graphite then recovered from the solution by methods including centri-fugation and ultrasonication [551]. One unique method uses the property of Li ions utilized in Li-ion batteries, that of electrochemical intercalation into graphite [551, 552]; ultrasonication of the Li-intercalated graphite in a mixed dimethylformamide/propylene carbonate solvent, or in an aqueous solvent, sepa-rates the graphene layers, typically yielding about 80% few-layer graphene.

A unique exfoliation-based method for graphene synthesis with the potential for scale-up to industrial-level quantities was demonstrated in 2014; this method involved high-shear mixing of graphite in suitable stabilizing liquids and resulted in large-scale exfoliation to give dispersions of graphene "nanosheets" [553]. Char-acterization via X-ray photoelectron and Raman spectroscopies revealed the exfo-liated flakes to be unoxidized and free of basal-plane defects. It was observed that a local shear rate in excess of 10^4/s was needed for exfoliation to occur.

18.3 Chemical Vapor Deposition (CVD)

Chemical vapor deposition (CVD) has become an increasingly popular method for synthesis of graphene, although it remains more expensive than exfoliation. In its simplest rendition, methane (or another appropriate hydrocarbon) and hydrogen are heated in a chamber at temperatures up to 1000 °C, whereby graphene sheets are deposited on a substrate, typically a transition metal, and more particularly Ni or a special form of Cu (which has been painstakingly cleaned and from which traces of CuO have been removed) [496, 520]. In most cases, the metal substrates can be etched away to yield pristine graphene sheets that may be transferred to other substrates. Other varied substrates used have included Ru(111) for epitaxial deposition of graphene [554] and ZnS which is easier to remove than metal substrates [555]. Indeed, a very wide variety of other metals have been tested as substrates for CVD of graphene, including Fe, Ru, Co, Rh, Ir, Pd, Pt, and Au, as well as alloys such as Co–Ni, Au–Ni, Ni–Mo, and even stainless steel [479]. Cu has been the most commonly used substrate; Ni substrates have the problem that single-layer graphene regions are frequently intimately mixed with few-layer graphene regions and thus hard to isolate for further study [493].

Using ambient pressure CVD with a methane/hydrogen feed gas, very large-area films of monolayer and few-layer graphene have been grown on polycrystalline Ni films and then transferred onto substrates assisted by poly(methyl methacrylate) (PMMA) "wet etching" methods [556].

CVD has also recently been used to produce graphene on a potentially very large scale, using methods including very high heating of Cu foil [557–559] and using "resistive heating cold wall CVD" [560]. This work, carried out by the Yeh/Boyd groups at California Institute of Technology (USA), was claimed as a method to "grow large sheets of electronic-grade graphene in much less time and at much lower temperatures" [557]. Specifically, it claimed to demonstrate a plasma-enhanced CVD method that used a single step, at reduced temperatures (<420 C), and in a matter of minutes [559]. Growth on copper foils was claimed to nucleate from arrays of well-aligned domains, and the ensuing films possessed sub-nm smoothness, high crystallinity, low strain, few defects, and room temperature electrical mobility up to $(6.0 + -1.0) \times 10^4$ cm^2 V^{-1} s^{-1}, better than that of large, single-crystalline graphene derived from thermal CVD growth. However, subsequent to the last of these publications in 2015, the methods do not appear to have been scaled up further.

Very recently, a consortium of European universities and companies with US$8 million in funding has claimed to have developed a *continuous roll-to-roll* process for depositing graphene on a Cu substrate about 30 cm wide and several meters long [496].

Plasma-enhanced CVD carried out in a flow of a methane/hydrogen mixture has been used to produce µm-size flakes of few-layer graphene with four to six atomic layers on quartz and silicon substrates [561]. The plasma is commonly generated via microwaves or RF; typical conditions are 5–100% CH$_4$ in H$_2$ at a total pressure of 12 Pa, 900 W power, and 680 °C substrate temperature [562, 563]. Advantages of plasma-enhanced CVD over conventional CVD include shorter deposition times (typically <10 min) and lower temperatures (650–700 °C vs. >1100 °C for conventional CVD).

Variants of CVD methods are also used to synthesize *N-doped graphene*. Thus, e.g., introduction of NH_3 into the gas stream during CVD yields N-doped graphene [564].

18.4 Reduction Methods

A widely used method for synthesis of graphene is *reduction of graphene oxide (GO)*. Most typically, the *graphite* oxide is first exfoliated, as it is significantly easier to exfoliate than graphene, and then reduced [479].

GO can be thought of as a single graphitic monolayer with randomly distributed aromatic regions (sp^2 carbon atoms) and oxygenated aliphatic regions (sp^3 carbon atoms) containing hydroxyl, epoxy, carbonyl, and carboxyl functional groups [565]. The epoxy and hydroxyl groups lie above and below each graphene layer, and the carboxylic groups exist usually at the edges of the graphene layers.

GO itself is commonly synthesized by treating graphite with a mixture of concentrated nitric and sulfuric acids and potassium perchlorate at room temperature for about 5 days [566–568]. The GO obtained in this fashion, immersed in aqueous medium, is a layered pile of graphene oxide sheets, usually with "AB" stacking, ripe for exfoliation with thermal, mechanical, or other energy, usually delivered as a shock [520]; in effect, water and anions such as perchlorate intercalate between the graphene layers of the GO. Thus, a typical subsequent exfoliation method is to give a sudden thermal shock to the GO (at over 1000 °C, under an Ar atmosphere) in a suitable apparatus, such as a long quartz tube. From this, a stable suspension of this for later use (e.g., for reduction to graphene) can then be prepared by heating the exfoliated graphite oxide under strongly alkaline conditions at temperatures of about 50–90 °C. In many cases, the stability of the aqueous GO suspension may be enhanced by raising the pH [569].

Several methods are then available for reduction of GO to graphene, but it is important to recognize that *almost none appear to yield complete reduction* (i.e., *complete absence of oxidized sites*), with the degree of reduction achieved depending on the method used. Thus, reduced graphene oxide (RGO) has a significant proportion of –OH and –COOH groups, frequently along defect sites.

A common such method of reduction involves reduction with hydrazine followed by annealing in Ar and/or H_2 [570]. The hydrazine may be added directly to the aqueous suspension of GO as hydrazine hydrate (see above) [520], or it may have added in gaseous form. A common variant of hydrazine-based reduction of single-layer GO to single-layer graphene is refluxing with hydrazine; however, only about 20–30% of the oxidized groups (such as –COOH) are lost, leaving a significant number available for chemical attachment.

Among other reducing agents used to reduce GO to graphene are Na borohydride [571].

Another approach to reducing GO is to use simple *thermal reduction* at temperatures of >1000 °C, whereby a proportion of the carbon atoms, along with all oxygen moieties, are lost as CO2 [572, 573].

In the early days of work with GO, methods attempted for obtaining graphene (including single-layer graphene) from GO included spray coating the aqueous suspensions onto heated substrates [574] and use of Langmuir–Blodgett-based assembly methods [575]; to date, however, these methods do not appear to be scalable or highly reproducible.

A unique, room-temperature reduction method for production of stable *aqueous* dispersions of monolayer and few-layer graphene sheets has been reported which involves the use of a water-soluble pyrene derivative (1-pyrenebutyrate) as the stabilizer and hydrazine monohydrate as the reducing agent [576].

In another unique method, camphor has been pyrolyzed over Ni nanoparticles at about 770 °C in an Ar atmosphere, to yield large quantities of large-area, few-layer graphene [577].

Another unique method demonstrated involves production of poly(phenylene) using the Suzuki–Miyaura reaction, then dehydrogenating the poly(phenylene) using FeCl$_3$, and finally yielding graphene nanoribbons [475, 578].

Yet another unique method demonstrated recently [579–581] involves heating electrically insulating substrates, such as quartz and SiO$_2$/Si, spin coated with solutions of polyaromatic hydrocarbons (PAHs) in chloroform to temperatures of up to 1100 °C; this leads to fusion of the molecules and results in graphene films of varying thickness depending on the original concentration of PAH solution.

Nanoscale graphene oxide (GO) monolayers or few layers have been synthesized via a unique method that involves *reduction of sugars*, with the professed advantage being that the method is more environmentally friendly [582]. The GO produced can then be reduced to graphene.

Various methods involving *reductive pyrolysis* from varied precursors have the capability to produce graphene on a several-grams scale. These methods include, e.g., pyrolysis of SiC at temperatures up to 1100 °C at pressures of $<10^{-6}$ Torr [583, 584] and reduction of ethanol by Na with accompanying ultrasonication [583, 585]. It is believed that in these methods, the Si of the SiC "preferentially sublimes" leaving behind graphite layers and, ultimately, graphene [586].

Monolayer graphene sheets of relatively large area have been prepared by thermal decomposition of the (0001) face of a 6H–SiC wafer under ultrahigh vacuum [478, 587]. In a variant of this method, monolayer graphene has been grown on top of a 6H–SiC (0001) substrate in an Ar atmosphere by an ex situ method, yielding larger graphene areas which gives larger monolayer graphene [478, 588, 589].

Several rather exotic methods involving reduction have also been reported for the production of graphene or reduced graphene oxide (RGO), including supersonic spray [590, 591], microwave-assisted synthesis [592, 593], and ion implantation

[594]. The practical utility of these methods has however been questioned, and so, again, these are mentioned here for the record only.

Reductive "aerosol pyrolysis" using a mixture of ferrocene, thiophene, and ethanol has been used to synthesize highly crystalline graphene *nanoribbons*, of approximate dimension 30 mm length × 20–300 nm width in bilayer to few-layer form [595].

In relatively recent work as of this writing, the preparation of continuous, monolayer sheets of polycrystalline graphene from a liquid metal matrix has been described, with a claim that it could be extended to "industrial-scale production of low-cost poly crystalline graphene" [596].

18.5 More Specialized Methods of Synthesis

Arc discharge methods are more commonly used for the production of CNTs and fullerenes (see earlier PARTS of this book) but have also, less commonly, been used for the production of graphene [479, 597–601]. Arc discharge has been carried out in several different buffer gases. The inclusion of H_2 among the gases is believed to be important to terminate dangling carbon bonds and hence inhibit the rolling-up and closing of graphene layers into CNTs, etc.; inclusion of He is found to produce graphene of the highest crystallinity.

Variants of the above arc discharge methods have been used to synthesize *boron- and nitrogen-doped graphene*. Thus, e.g., *nitrogen*-doped graphene sheets have been synthesized by carrying out the arc discharge between carbon electrodes in the presence of hydrogen and pyridine or ammonia [602]. In contrast, to produce *boron*-doped graphene, boron-stuffed graphite electrodes or a mixture of hydrogen and diborane vapor are used [603]. The B-doped graphene shows a typical *p*-type semiconductor behavior. N-doped graphene has also been synthesized on a gram scale by a one-step solvothermal method, via the reaction of CCl_4 and LiN [604].

Methods involving *opening up or unzipping of CNTs* have included use of highly oxidative $KMnO_4$ and concentrated H_2SO_4 [605, 606], Li in liquid ammonia [475], laser irradiation [606], and plasma etching [607] to "open up" CNTs. These methods are generally used to produce only graphene *nanoribbons*, with the dimensions of the nanoribbons of course dependent upon the type and dimensions of the starting CNTs.

Several methods have been described which produce graphene along with other carbon allotropes, such as CNTs or fullerenes [608, 609]; however, such methods are not practical or useful for graphene applications, most of which require pure graphene only and so are mentioned here only for the record.

18.6   Common Methods Used to Characterize Graphene

In their early work, Geim and Novoselov took advantage of an interference effect at a specially chosen thickness (300 nm) of SiO_2 on Si to enhance optical contrast and absorbance under white-light illumination [520, 610].

Since this initial work, other techniques have been used more commonly to characterize graphene. These include conventional *optical microscopy, scanning probe microscopy (SPM), atomic force microscopy (AFM), scanning electron microscopy (SEM)*, high-resolution *transmission electron microscopy (TEM), fluorescence quenching microscopy (FQM)*, and, frequently, combinations thereof. SPM is a comparatively slow method which however has a typical resolution quite capable of measuring the ca. 0.34 nm step height for individual graphene layers and has issues such as poor recognition of substrate vs. graphene [520]. In FQM, the imaging mechanism involves quenching the emission from a dye-coated GO and RGO; the dye can be removed by rinsing without disrupting the sheets. The contrast arises due to the chemical interaction between the GO and the dye molecule on the molecular scale because of the charge transfer from dye molecule to GO that causes quenching of fluorescence [480, 611]. Figure 18.3 compares images of graphene flakes taken with various techniques.

Subsequently, however, *Raman spectroscopy* has been the *de rigueur* method of identifying graphene, most especially monolayer graphene. As discussed at some length by Allen et al. [520], the most prominent features of the Raman spectra of graphite and graphene are the G band at $\sim 1584\,cm^{-1}$ and the G′ band at $\sim 2700\,cm^{-1}$. The G band is assigned to the E_{2g} vibrational mode, and the G′ band is a second-order two-phonon mode. A third, D band at $\sim 1350\,cm^{-1}$, is not Raman active for pristine graphene; however, it can be observed where symmetry is broken by edges or in samples with a high density of defects. It is changes in the positions and relative peak heights of the G and G′ bands that serve to indicate the number of layers present in a particular graphene sample. The location of the G peak for *single-layer* graphene is at a point about 3–5 cm^{-1} greater than that for bulk graphite, while its intensity is roughly the same. The G′ peak shows a significant change in both shape and intensity as the number of layers is *decreased*. In bulk graphite, the G′ band consists of two components; the intensities of these are roughly one-fourth and one-half that of the G peak for the low and high shifts, respectively. *For single-layer graphene, the G′ band is a single sharp peak at the lower shift, with intensity roughly four times that of the G peak*. Thus, this method provides a very reliable method of determining the number of graphene layers in samples of graphene and thus verifying whether we have a single layer. Figure 18.4 shows representative Raman spectra for monolayer, bilayer, and few-layer graphene on a silica substrate.

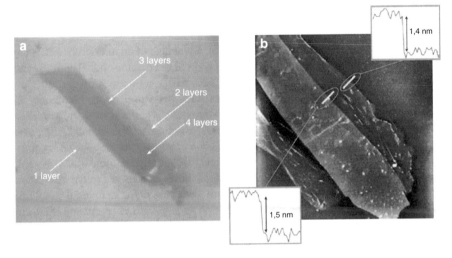

Fig. 18.3 Images of graphene flakes taken with various techniques: (**a**) optical microscopy, (**b**) AFM, and (**c**) FQM (After Refs. [612, 613], reproduced with permission) (After Ref. [614], courtesy of AzoOptics and NT-MDT)

Fig. 18.4 Comparison of Raman spectra at 532 nm for few-layer graphene. The position of G peak and the spectral features of 2D band confirm the number of atomic layer of the graphene devices (After Ref. [615])

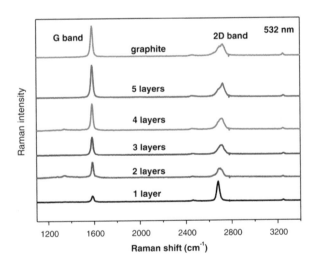

18.7 Brief Synopsis of Methods for Functionalization of Graphene

Functionalization of graphene and/or GO is used to make it more convenient for handling and derivatization for applications, such as sensors and potential electronic components. A very wide variety of methods have been demonstrated for functionalization (sometimes called "decoration") of graphene and GO with various chemical moieties. Georgakilas et al. have provided an excellent review of some of these functionalization methods for graphene and GO [565]; however, in this author's humble opinion, it is also clear from this review that much of voluminous work on functionalization of graphene and GO of the last 10 years or so does not appear to have a purpose or objective, beyond synthesizing ever more pretty and exotic materials. A *few*, *illustrative* (and by no means exhaustive) examples of these are now enumerated:

- Graphene has been heated with diazonium salts followed by further derivatization to yield graphene functionalized with nitrophenyl groups [615–617]; nitrophenyl covalently attached to graphene introduces a controllable, for potential for use in future semiconductors based on graphene.
- Hydroxylated aryl groups grafted covalently on graphene by the diazonium addition reaction serve as initiators for the polymerization of styrene, yielding polymeric chains that are covalently grafted on the graphene surface [618].
- Graphene sheets have been covalently functionalized with alkylazides, with the alkyl including hexyl, dodecyl, hydroxylundecanyl, and carboxy-undecanyl [619]. These functionalized graphene sheets showed enhanced dispersibility in solvents such as toluene and acetone.
- A method for the introduction of various functional groups, including oligomeric and polymeric chains, onto graphene sheets via nitrene cycloaddition has been demonstrated [565, 620]. The resulting functionalized graphene sheets show enhanced chemical and thermal stabilities compared with GO and can be further modified by amidation, surface-initiated polymerization, and reduction of metal ions, among other reactions. The resulting functionalized graphene sheets are claimed to have superior dispersibility in solvents.
- A method for aryne cycloaddition to a graphene surface has been demonstrated that uses 2-(trimethylsilyl)aryl triate as a precursor toward a reactive benzyne intermediate [565, 621]. A four-membered ring is produced that connects the aromatic arene rings to the graphene surface. These aryl-modified graphene sheets are readily dispersible in DMF, o-dichlorobenzene, ethanol, chloroform, and water. Additionally, the aryl rings can be substituted by groups such as -methyl and –F.
- Graphene–conducting polymer (CP) adducts have been synthesized with a view to arrive at advanced materials for solar cells and similar devices. For example, amine-terminated oligothiophenes have been grafted onto GO "nano-platelets" through covalent amide bonds [622]. In the product, strong interaction between GO and the oligothiophene is indicated by the almost complete fluorescence quenching observed in the graphene/oligothiophene conjunct in comparison with pure oligothiophene. And GO has been functionalized with $-CH_2OH$ terminated, regioregular poly(3-hexylthiophene) (P(3HT)) through the formation of ester bonds with the carboxyl groups of GO [623]. Due to the presence of an abundant number of hydroxyl groups in the added polymer, the conjunct is soluble in common organic solvents, allowing for its facile characterization and further processing in solution. The proposed end use of the P(3HT)-GO adduct is

in photovoltaics [623]. However, the demonstrated efficiency of a photovoltaic cell incorporating the P(3HT)-GO composite has been a rather poor 1.1% [624].

- With a view to potential, advanced dye-sensitization materials, chromophores such as porphyrins, phthalocyanines, and azobenzene have been covalently attached to graphene [625–628]. GO is functionalized with porphyrins through the formation of amide bonds between amine-functionalized porphyrins and carboxylic groups of GO.

- OH-functionalized graphene "nano-platelets" have been mixed with O_2 and poly (methyl methacrylate) (PMMA) to yield graphene reinforced PMMA "nanocomposites" [629]; these "nanocomposites" show an increase (of about 30 °C) in the glass transition temperature of PMMA at a proportion of just 0.5 w/w% of graphene content. In related work [630], amide-functionalized graphene "nano-platelets" have been prepared as reinforcements for PMMA, yielding an increase of 70% and 10% in the elastic modulus and the hardness, respectively, of the composites at a graphene content of just 0.6 w/w%.

- Amine-terminated poly(ethylene glycol) (PEG) has been grafted onto GO through amide bond formation [631]. The resulting functionalized GO is highly dispersible in water, as well as in serum, thus being a potential candidate for delivery of hydrophobic drugs in biological systems. As an example of this potential, a highly hydrophobic camptothecin analog, a potential anticancer agent which has very low solubility in aqueous media, has been immobilized on the surface of the PEG-GO adduct forming a nanostructure with excellent stability in biological solutions.

- Poly(vinyl alcohol) (PVA) has been grafted onto GO via ester bonds between the hydroxyl groups of PVA and the carboxylic groups of GO [632]. The resulting PVA/GO composite is dispersible in DMSO and hot water and can be reduced with hydrazine, yielding a unique affording the PVA/RGO composite.

- Graphene has been functionalized with 1,5-diaminonaphthalene (DAN) and 1-nitropyrene (NP) (electron-donating and electron-withdrawing molecules, respectively), although the objective of such functionalization was not clear [633].

- Functionalization of graphene "nano-platelets" with single-strand DNA (ssDNA) has been carried out by chemical oxidation of graphite to GO, followed by hydrazine reduction in the presence of ssDNA [634]. The resulting material is soluble in water with concentrations as high as 2.5 mg/L. Once again, however, the objective of this work was not clear.

- An RGO–TiO$_2$ composite has been shown to have a H$_2$-evolution efficiency about two times that of standard TiO$_2$ [635].

18.8 Problems and Exercises

1. Enumerate the five principle classes of methods of synthesis of graphene, and outline the basic steps of one example from each. Of these, which is the most commonly used for research? For larger scale production and/or commercial products?
2. Besides the Scotch® tape method, what are the other common methods of synthesis of graphene that fall under the exfoliation class?
3. Briefly describe the plasma-enhanced CVD method, and compare it with other CVD methods.
4. Briefly describe and compare the various methods of synthesis of graphene using graphene oxide (GO) as the starting material.
5. What are the practical methods proposed for large-scale manufacture of graphene (and which types of graphene), and what is their status as of this writing (2016)? As of your reading today?
6. Briefly describe one method each for synthesis of N- and B-doped graphene.
7. Describe the principal methods of characterization of graphene, and compare their relative advantages and drawbacks.
8. Enumerate the principal methods of functionalization of graphene, their applications, and relative advantages and drawbacks. What is the primary reason behind functionalization of graphene? What are the most commonly used forms of functionalized graphene (identify the functional groups used and the physical form of the resultant functionalized graphene).
9. (*Practical problem, contingent on availability of materials*): Procure a small piece of graphite, preferably highly ordered pyrolytic graphite (HOPG), a small sheet of copper, and some cellulose adhesive tape. Try out the Scotch® tape method by attempting to transfer graphene layers onto the copper sheet. Based on your results, describe how easy or difficult you find this.

Part IV
Graphene, Applications

Chapter 19
Brief, General Overview of Applications

Contents

As with CNTs, graphene has been studied exhaustively from a theoretical, experimental, and application's point of view since the renewed impetus for its study following the seminal work of Geim and Novoselov [462]. And as with CNTs, applications studied have spanned a very wide field. The *applications* below are *representative* and no means exhaustive:

- *Sensors*
- *Energy devices, including batteries*
- *Electronics and electrical conductors*
- *Displays and transparent films*
- *Drug delivery and biomedical*
- *Specialized applications*
- *Miscellaneous applications*

Again, however, as for CNTs, *actual, commercially implemented applications, as of this writing (2016), are still relatively few.* Some examples include:

- A graphene-containing printer powder, marketed as Graphenite WX by a US company called Noble 3D Printers and used for a technique called "lost wax casting" [636].
- Li batteries incorporating graphene, with a claimed energy density about five times that of standard secondary Li batteries [500]; these are currently produced and marketed by a Spanish company called Graphenano.
- A New Jersey-based US company now offers RFID (radio frequency identification) tags containing a small proportion of graphene to improve their high-temperature/high-pressure performance [501] and specialized antennae for cell phones (mobiles) that incorporate graphene and afford improved performance [502]. Both these products are offered by a New Jersey, USA based company called Vorbeck.

© Springer International Publishing AG 2018
P. Chandrasekhar, *Conducting Polymers, Fundamentals and Applications*,
https://doi.org/10.1007/978-3-319-69378-1_19

- Tennis rackets incorporating a bit of graphene (substituted for carbon fibers) to increase stiffness and reduce weight. These are at present marketed by companies including the tennis racket company Head, with the catchy ad-line "Once played with a Graphene racquet, you never want to miss the extra swing it gives you with each shot. . . . . . We inserted the world's strongest and lightest material in an even better way. . . . ." [637].
- A new version of Araldite® brand resins will be marketed by Huntsman Corp.'s Advanced Materials division in collaboration with Haydale Composite Solutions which incorporate graphene [638, 639]. Haydale is a UK company that has a production line for graphite nano-platelets.
- Miscellaneous, very low-volume applications, including "functional coatings" and composite materials, wherein graphene generally replaces activated carbon or carbon fibers in already established uses.

As noted in an earlier chapter, as of this writing (2016), worldwide sales of graphene-related materials are about *US $30 million* [496]. And as seen in the description above, commercial products incorporating graphene can be counted on the fingers of one hand.

Indeed, a company called Thomas Swan has, as of this writing, already put in place a high-volume manufacturing line for "non-oxidized graphene nanoplatelets (GNPs)" [640], although it is unclear who their current customers are and what kinds of end products these GNPs will be used in.

However, a prediction that these sales will hit US $220 million in 2026 [496] seems, at least as of this writing, highly optimistic. And the expectation voiced by some in the initial, exciting years of graphene development that graphene will replace Si in electronic circuits or that it will appear in "low-cost touch interfaces for wearable electronic devices" [496] seems, at this writing, even more far removed, in this author's humble opinion.

19.1 Problems and Exercises

1. Enumerate the five (5) classes of practical applications of graphene and cite one example from each.
2. Enumerate two (2) of the known commercialized products incorporating graphene as of 2016 and as of your reading today. Do a quick Internet search to determine the status, commercial availability, and pricing of these two products.

Chapter 20
Graphene Applications in Sensors

Contents

20.1 Gas and Vapor Sensors

In a manner similar to CNTs, pure, pristine, defect-free, monolayer, or few-layer graphene has very poor affinity for gas or any other molecules. It has no binding sites on its surface, and adsorption of molecules on its pristine surface is rather poor. However, functionalization of the graphene surface, using the various functionalization methods enumerated in PART III of this book, can lead to a dramatically enhanced capacity for sensing. Another method of rendering graphene more sensitive to gas-phase analytes is to use smaller portions of it, e.g., as graphene "nanoribbons"; in this case, the presence of a larger number of reactive edges improves sensing capabilities [475].

The most common sensing mode for graphene gas/vapor sensors is *conductometric*, i.e., via a change in resistance or conductivity. Other, primarily electrochemically based sensing methods include the *amperometric*, with *voltammetric* graphene sensors rarely encountered. Necessarily then, these modes of sensing have a tendency to be *generic*, i.e., lacking a specific response to specific analytes. Specificity can sometimes be obtained with carefully thought out functionalization of the graphene. (Other, optically and spectrally based sensing techniques for graphene can be more specific. These are discussed in more detail in a different section of this chapter.)

Unfortunately, much of the work on conductometric gas/vapor sensing using graphene sensors has not dealt with the issue of *specificity*, that is to say, specific response to the analyte of interest in the presence of interferents and not just a

© Springer International Publishing AG 2018
P. Chandrasekhar, *Conducting Polymers, Fundamentals and Applications*,
https://doi.org/10.1007/978-3-319-69378-1_20

generic response to any analyte, such as human breath or a common interferent. Indeed, this author finds it surprising that many of the publications in this field pass peer review. He finds it surprising that a reviewer does not ask the simple question, "yes, that's great that you're getting a response at 10 ppb, but does the sensor also respond to your breath, passing perfume, diesel fumes, or isopropanol?" The adsorption on a surface of just about anything will cause a change in the resistance of the surface! A case in point is a CO_2 sensor reported in 2011 and based on the conductometric response of a pristine graphene sheet synthesized using the mechanical cleavage technique [641]. In this work, linearity in response is claimed between 10 and 100 ppm CO_2. However, no data with other analytes, or with interferents present together with the CO_2, are presented.

One of the first and most promising studies of gas/vapor sensing with graphene was the claimed *single-molecule* sensing of NO_2 and NH_3 in 2007, wherein adsorption of what was postulated to be a single molecule of the analyte led to a decrease in resistance of monolayer graphene [520, 642–644] The claimed mechanism of action was that "The adsorbed molecules change the local carrier concentration in graphene by one electron, which leads to step-like changes in resistance. The achieved sensitivity is due to the fact that graphene is an exceptionally low-noise material electronically..." [642].

This and subsequent "single-molecule sensing" work however required either comparatively high voltages or high magnetic fields, a practical drawback.

A more recent (2016) claim of *single-molecule sensing* was a sensor reported for single CO_2 molecules, with the claimed, added feature that it was done at room temperature and did not require high magnetic fields and only modest voltages [645]. In this work, a bilayer graphene ribbon was angled between two electrodes at different heights, to lift the ribbon away from its SiO_2 substrate. The claimed mechanism of action was that, despite the negligible charge transfer from a single physisorbed molecule of CO_2, it strongly affects the electronic transport in the "suspended" bilayer graphene by inducing a charged impurity, which can shut down part of the conduction of the bilayer graphene with coulomb impurity scattering. The authors claimed they could thus detect *each individual-molecule* physisorption as a steplike resistance change with a quantized value in the bilayer graphene. They then used theoretical calculations to estimate the possible resistance response caused by coulomb scattering of one adsorbed CO_2 molecule and showed that it was in agreement with their experimental measurements.

With regard to NO_2 and NH_3 sensing, these analytes were said to induce p-type and n-type conduction, respectively. However, it was noted later [520] that lack of selectivity (i.e., a generic response to *any* adsorbate) significantly dampened interest in graphene gas/vapor sensors after this initial flurry of work. Subsequent work [646] using chemically derived graphene led to limited selectivity (but very slow response time, of the order of minutes) NO_2 and NH_3 sensors. Apart from the above-cited, claimed "single-molecule" detection, detection limits of about 100 ppm have been generally observed for these two analytes with graphene sensors [647]. Typical sensor response for various gases (including NO_2 and NH_3) at the ca. 1 ppm level is shown in Fig. 20.1.

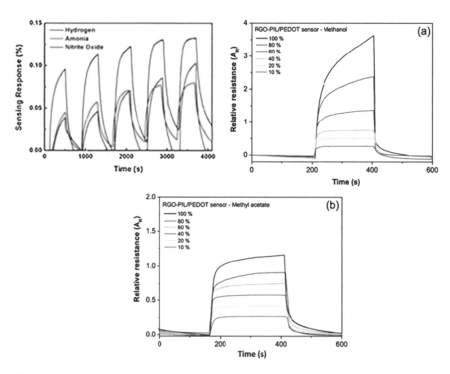

Fig. 20.1 Examples of varied generic response of graphene sensors (After Refs. [646–647], reproduced with permission)

Since the above-described work, graphene or graphene oxide (GO), including reduced graphene oxide (RGO)-based sensors, have been demonstrated for the analytes H_2, CO, dinitrotoluene, iodine, ethanol, and hydrazine hydrate [480, 648–652].

In work carried out in 2009, detection of gases/vapors of NO_2, NH_3, and 2,4-dinitrotoluene (DNT) has been demonstrated with sensors fabricated from chemically converted graphene dispersions using spin coating to create single-layer films on interdigitated electrode arrays, with dispersions of graphene in anhydrous hydrazine being formed from graphite oxide [653]. The response of these conductometric sensors appears to be generic, and it is not clear how the different analytes (and any possible interferents) would be differentiated.

One method of attempting to achieve more selectivity and specificity from the generic conductometric response of graphene vapor sensors has been through the use of intelligent algorithms looking for subtle patterns in the temporal and spatial profiles of the conductivity response. Thus, e.g., researchers at the US Army's Electronic Sensors Directorate in Virginia reported what they claimed was chemical discrimination with an unmodified (i.e., not functionalized in any way) graphene chemical sensor [654]. In this work, patterned Au-Ti electrodes were deposited on a 1 cm^2 graphene film on a SiO_2 substrate. When various gases or vapors deposited on the graphene surface, they generated a (quite generic) change in resistance in the composite electrode. However, subtle differences in the response were evaluated using principle component analysis (PCA) methods and were combined with prediction accuracy (PA) methods including linear discrimination analysis, k-nearest neighbor, random forest, and support vector classifiers, to yield the authors claimed fair discrimination between different adsorbed vapors and gases.

A purely theoretical study on graphene sensors based on "vibronics and plasmonics," carried out in 2010, purported to show that sensors could be fabricated using graphene nanoribbons which are "able to read molecular vibrations and molecular electrostatic potentials, acting as an amplifier and as a transducer converting molecular signals into current-voltage quantities of standard electronics" [655]. No prior or subsequent experimental work to corroborate the predictions has however emerged to date (2016).

20.2 Biosensors and Electrochemically Based Sensors

Graphene biosensors, including many electrochemically based ones frequently functionalized with various enzymes, have received much attention [656, 657]. One of the reasons for this is that graphene exhibits a wide electrochemical potential window, about 2.5 V in 0.1 M phosphate buffer (pH 7.0), which is comparable to that of graphite and glassy carbon (GC); on the other hand, the charge-transfer resistance on graphene, as determined from electrochemical impedance spectroscopy, is much lower than that of graphite or GC electrodes [659, 660].

Among the most-studied graphene-based, enzyme-functionalized sensing electrodes have been those immobilized with the enzyme glucose oxidase, the staple of commercial glucometers, with the form of graphene used including few-layer graphene and reduced graphene oxide (RGO) [656, 661, 662]. However, results from these studies indicate that the graphene-based glucose sensors are no better, and sometimes worse, in performance than commercially available glucometers, most of which are based on immobilized glucose oxidase sensing electrodes.

The presence of 8-hydroxydeoxyguanosine (8-OH-DeOxGuan) in body fluids such as blood, urine, and saliva is believed to indicate DNA damage and, potentially, developing cancers. A graphene-based biosensor for 8-OH-DeOxGuan has been demonstrated [663]. This uses graphene deposited epitaxially on SiC.

A graphene-based, carbon-layered electrode array device, implantable on the surface of rodents' brains for purposes of high-resolution neurophysiological recording; the claimed objective was in vivo imaging of the cortical vasculature via fluorescence microscopy and 3D optical coherence tomography [664].

Several studies have demonstrated DNA, protein, and bacterial sensors based on reduced graphene oxide (RGO) [480, 2070]. These have included a sensor claimed to be able to detect and differentiate a single bacterium. In the case of the claimed label-free DNA sensor, the presence of DNA on the RGO surface was observed to increase both the conductivity and the mobility due, it was claimed, to the interaction between the charged amine groups on the DNA and the RGO surface. In another study, single-strand (ss)DNA was tethered on the RGO surface and was able to bind its complementary DNA strand, reversibly increasing the hole density very significantly. In a similar biosensor, RGO-based FETs were shown to specifically detect biomolecules with high sensitivity using specific antibodies tethered to the RGO surface [480, 665]. And a sensor incorporating GO sheets into Au nanoparticle–antibody conjugates was demonstrated to detect a rotavirus [480, 666].

Researchers very recently (2016) reported on a graphene-based, wearable "diabetes patch," a dual-function device that monitors glucose levels (in body sweat) and also delivers drugs [667]. The device uses layers of the fluoropolymer Nafion to absorb sweat and carry it toward the device's sensors, which are built on functionalized graphene. The graphene is functionalized with Au atoms containing electrochemically active materials sensitive to glucose (e.g., the enzyme glucose oxidase, the staple of commercial glucometers). The wearable patch includes supplementary pH and temperature sensors. The drug delivery portion of the patch (not entirely relevant to the graphene-based sensor) uses polymeric microneedles that pierce the skin. With regard to glucose sensors, it was demonstrated in 2010 that glucose oxidase sensors using nitrogen-doped graphene (N-graphene) appeared to have much improved performance as compared to those employing pristine graphene [668].

A sensor specific to detection of prostate-specific antigen (PSA) was reported in 2011, based on "layer-by-layer self-assembled graphene" deposited on a poly (ethylene terephthalate) (PET) substrate [669]. The work used polyelectrolytes including poly(diallyldiamine chloride) and poly(styrene sulfonate) (PSS). The claimed sensitivity for label-free and labeled graphene sensors for PSA detection was, respectively, 4 fg/mL (0.11 fM) and 0.4 pf/mL (11 fM). A quick back-of-the-envelope calculation of the number of units or molecules of PSA that would be present in a sample weighing 4 fg reveals that this sensor would have absolutely extraordinary and not easily explained sensitivity specifically to this particular antigen! Since this publication, however, no further work on this very promising sensor, either developmentally or in terms of a commercial product, has been reported.

A "Graphene-based single-bacterium resolution biodevice and DNA transistor interfacing graphene derivatives with nanoscale and microscale biocomponents" was reported in 2008 [670]. In this work, fabrication and function of a chemically modified graphene-based "single-bacterium biodevice"/label-free DNA sensor and "bacterial DNA/protein and polyelectrolyte chemical transistor" were claimed. It was shown that the bacteria sensor generated about 1400 charge carriers in p-type graphene with attachment of a single bacterium. It was shown that single-strand DNA (ssDNA) tethered on the graphene hybridizes with its complementary DNA strand to reversibly increase the hole density by $5.61 \times 10^{12}/\text{cm}^2$. Since its publication, however, no further development or refinement and no practical or commercial sensor were reported emanating from this work.

20.3   Electrochemically Based Sensors

Graphene-based or graphene-modified electrochemical sensors have been demonstrated for analytes as varied as ascorbic acid (vitamin C), paracetamol, and rutin, with detection limits as low as 3.2×10^{-8} M [493, 671, 672]. However, most of this work involved graphene modification of established sensing electrodes (e.g., glassy carbon), and it was not clear whether the graphene improved sensor response or simply served as another mediation medium.

Graphene-based *electrochemical* sensors have also been reported for the analytes: H_2O_2 [673]; nicotinamide adenine dinucleotides (NAD+, NADH) [674]; dopamine [677]; and DNA and its components, including the free bases (G, A, T, and C) and single- and double-strand DNA, using RGO (reduced graphene oxide) sensors (see data in Fig. 20.2 [678]). All these studies have shown performance only slightly improved over corresponding electrodes with graphite or glassy carbon only.

Graphene-based electrochemical sensors have been studied for the detection of heavy metal ions (Pb(II), Cd(II)), with graphene-only or, occasionally, a Nafion®– graphene composite film comprising the sensing electrode and the method of detection being the tried-and-true anodic stripping voltammetry (ASV) [656, 679, 680]. Essentially, in this work, a graphene surface replaces the Hg/glassy carbon or Bi surface typically used in ASV. As with graphene-centric work, however, the. utility of these novel graphene-based sensing electrodes over standard ASV electrodes is not clear from this work; indeed, the detection limits and sensitivity appear to be less than those of conventional ASV.

20.4   Electronic, Optoelectronic, Photo-, and Magnetic Sensors

A wideband, frequency-tunable, *terahertz and infrared detector* based on graphene transistors was demonstrated in 2013 [681]. It was claimed that the graphene transistors, under a suitable magnetic field, were capable of detecting THz and IR waves in a very wide band of frequencies (0.76–33 THz) and that the detection frequency was tunable by changing the magnetic field. It was however observed that the electric potential distribution in the graphene transistor detector showed local step structure associated with impurities. The THz and IR photoconductivity properties of the graphene transistors were thus likely to be sensitive to such potential steps associated with impurities.

Other work in graphene-based sensors of this category, worthy of mention, includes:

- Graphene sensor based on the Hall effect, using either large graphene sheets or graphene encapsulated in hexagonal BN [682, 683]. In the BN work, a magnetic

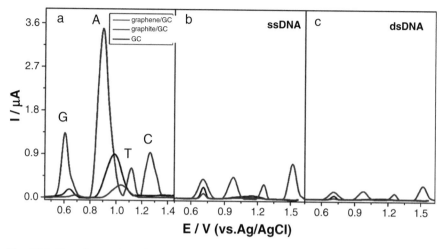

Fig. 20.2 Data for sensing of DNA and its components (After Ref. [678], reproduced with permission (*ss* single strand, *ds* dual strand))

resolution, limited by low-frequency electric noise, of less than 50 nT-(Hz)$^{-1/2}$ was demonstrated.

- A photodetector (photosensor) based on a graphene–Si heterojunction, showing strong rectifying behavior and high photo-responsivity (specific detectivity of up to 5.77×10^{13} cm Hz$^{1/2}$/W at the peak wavelength of 890 nm at room temperature, photocurrent-to-dark current ratio of about 107) [684].
- A position-dependent reduced graphene oxide (RGO) thin-film photodetector sensitive in the near-IR at about 800 nm [685].

The German company Bosch announced a BN–graphene-based magnetic sensor, based on the Hall effect, that was reported to have 100X the sensitivity of a conventional Si-based sensor (sensitivity of 7000 V/A-Tesla vs. 70 V/A-Tesla for the BN–graphene-based sensor) [686].

Graphene is a potential candidate for use as a magnetic sensor, based on the Hall effect. Now if we compare the room temperature Hall coefficient for a typical InAs sensor, $R_H = 4.3 \times 10^{-6}$ Ωm/T, with that for graphene, $R_H = 0.3 \times 10^{-6}$ Ωm/T, it appears that graphene would not be ideal for Hall effect sensing [493, 687–689]. However, the thickness of the conducting layer for the InAs device is taken to be 12 nm, but that for graphene is only 0.34 nm. This gives a Hall resistance of about 1000 Ω/T versus only 358 Ω/T for the InAs device. If we also include the facts that graphene can sustain current densities in excess of 10^8 A/cm^2 and is not buried beneath additional layers as is the case for conventional semiconductor 2 DEG systems, then it is clear that graphene has some significant advantages for Hall effect sensing. Based on this reasoning, a tunable magnetic sensor based on graphene and the extraordinary magnetoresistance (EMR) device concept has been demonstrated [688, 689]. The authors of this study have predicted that their device could have a 26 dB signal-to-noise ratio in a 1 GHz bandwidth [443, 687–689].

Monolayer graphene absorbs a *fixed* 2.3% of the illumination passing through it over a wide range of wavelengths (300–2500 nm), and this absorption is a linear function of the number of layers; thus, monolayer graphene has the capacity to be used as an optical standard, if problems in precise, reproducible fabrication of single- and few-layer graphene can be overcome [443, 690, 691].

20.5   Other Sensors

Graphene has been studied for *mechanical* sensors. A highly sensitive, flexible, *strain* sensor was reported based on large-area ultrathin graphene films prepared using single-step Marangoni self-assembly; this showed a gauge factor of 1037 at 2% strain, to be compared with values of about 4 for a common metal foil strain gauge, about 200 for a single-crystal Si-based strain gauge, and 100 for thick-film resistors (the gauge factor or strain factor of a strain gauge is the ratio of relative change in electrical resistance R to the mechanical strain ε) [684]. Graphene–rubber composites fashioned into bands have been used as body-wearable strain sensors, working at strains up to 800% and showing 10^4-fold increases in resistance, for athletic applications [692].

20.6   Problems and Exercises

1. Describe the three modes of *electrochemical* sensing using graphene-based sensors. Of these, which, if any, is capable of addressing the issue of *specificity* (i.e., specific response to the analyte of interest in the presence of interferents and not just a generic response to any analyte, including human breath)?
2. Give examples of two enzyme-based graphene sensors, and describe their mode of functioning and detection. What are the best detection limits and times achieved for these sensors, and how do they compare with other methods of detection of the analytes in question?
3. Describe the principle of operation of the prostate-specific antigen (PSA) sensor of Zhang et al. [517], and rationalize how its claimed sensitivity of 4 fg/mL could be achieved. How many units or molecules of PSA would be present in a sample of 4 fg?
4. What other varied analytes have been addressed by graphene sensors? (Consult the section entitled "Other Electrochemically Based Sensors.")
5. Enumerate one example each of an electronic, optoelectronic, photo-based, magnetic-based, and mechanically based sensor based on graphene. Describe the mode of detection and the known performance parameters (detection time/limit, ease of use, cost) of each.

Chapter 21
Graphene Applications in Batteries and Energy Devices

Contents

21.1 Overview

As noted in an earlier chapter, two large areas of potential applications of graphene are *batteries and energy devices,* including especially Li batteries, supercapacitors, solar cells, and materials for hydrogen storage. Many of the applications in such areas as Li batteries have originated in simple replacement of already established additives, such as activated-C and graphite.

Examples of applications that have *already been commercially realized* include Li batteries incorporating graphene with a claimed energy density about five times that of standard secondary Li batteries, currently produced and marketed by a Spanish company called Graphenano [500].

21.2 Batteries, Including Li Batteries

Graphene was studied early on as an additive for electrodes in Li batteries [693]. Flexible Li batteries incorporating graphene and where the anode acts as the active material as well as the current collector were demonstrated in 2013 [694].

Graphene has been incorporated into Li batteries containing the cathode materials Co_3O_4, Mn_3O_4, SnO_2, Fe_3O_4, and even Si, with demonstrated, enhanced energy densities in the 700 to 1100 mAh/g region [480, 695, 696, 699–704].

© Springer International Publishing AG 2018

P. Chandrasekhar, *Conducting Polymers, Fundamentals and Applications,*
https://doi.org/10.1007/978-3-319-69378-1_21

However, to date, performance of Li batteries incorporating graphene in their electrodes has been disappointing or mediocre: specific capacitances in the 540 to 1400 mAh/g range and stabilities to less than 1000 cycles [695, 705]. Thus, e.g., although graphene-based electrode materials can reach a high reversible capacity (1013–1054 mAh/g) at a low charge rate, it is still rate limited at a high charge/discharge rate ($\geq$500 mA/g) [706, 707]. In this respect, nitrogen-doped graphene (N-graphene) has been shown to increase battery capacities nearly twofold as compared to pristine graphene, due, presumably, to enhanced intercalation of Li ions from introduction of N-atoms [708]. N-graphene has also been suggested for use in fuel cells and other energy devices [709].

Very recent (2016) work reported on silicon oxycarbide glass-graphene composite paper electrode for long-cycle lithium-ion batteries [710]. In this work, a siloxane precursor was heat treated to form µm-sized particles of Si oxycarbide (SiOC) and then combined with reduced graphene oxide (RGO) to form freestanding sheets of an SiOC–RGO composite "paper." At a mass loading of 2 mg/cm^2, the paper electrode delivered a charge capacity of about 588 mAh per gram of electrode or about 393 mA per cm^3 of electrode at the 1020th cycle.

21.3   Fuel Cells

Fuel cells incorporating graphene, N-graphene and RGO, many as composites with Pt, have been demonstrated with power densities ranging from about 85 to 440 mW/cm^2 [480, 695, 711–715]. In these, the graphene support was claimed to maximize the availability of electrocatalyst surface area for electron transfer provide better mass transport of reactants to the electrocatalysts.

In a development of interest for the use of graphene in fuel cells based on proton exchange membranes (PEMs), though not quite leading to actual, functioning, practical fuel cells, the University of Manchester group reported in 2014 that pristine single layers of graphene and hexagonal boron nitride conduct protons quite well, with conductivities at about 100 °C that are adequate for use in PEM fuel cells [716]. Claimed advantages of these graphene-based PEMs over conventional PEMs include impermeability to ions and atoms other than protons, thus preventing mixing between fuel and oxidant. However, the study also noted that, at room temperature, the plain hexagonal BN is a better proton conductor than the graphene/hexagonal BN composite.

Metal-free, N-doped graphene (N-graphene) electrodes have been studied for use in fuel cells, with claimed enhanced catalytic activity toward the oxygen reduction reaction in the fuel cells and power densities of up to 440 mW/cm^2 [695, 705, 712, 717–719].

Interestingly, fuel cells using glucose (rather than the more standard H$_2$) as the fuel have also been studied (denoted as "enzymatic biofuel cells" in the literature), with current densities of up to 157 µA/cm^2 and power densities of up to 25 µW at 0.38 V, although their practical utility is not entirely clear [695, 711].

21.4   Capacitors and Supercapacitors

Graphene is a promising potential candidate for *supercapacitors*, due to its high surface area to mass ratio (2630 m^2/g for monolayer graphene) [720]. A number of graphene-based supercapacitors (sometimes also denoted "ultracapacitors") have been demonstrated in preliminary fashion. Just a few examples:

- Early work with graphene- and reduced graphene oxide (RGO)- based supercapacitors yielded poor performance (specific capacitances in the 75 to 187 F/g range and energy densities in the 6 to 32 Wh/kg range), even when ionic liquid electrolytes such as 1-ethyl-3-methylimidazolium bis(trifluoro-methylsulfonyl)amide, EMIM-NTf 2, were used [720–722]. The use of graphene with a curved structure to prevent reaggregation, prepared using a unique fluidized bed GO reduction method, yielded somewhat better performance, 100 to 250 F/g specific capacity at a current density of 1 A/g [723]; the further use of the ionic liquid, EMIM BF$_4$, in this system yielded an energy density of 85.6 W h/kg at 1 A/g discharge rate and a discharge voltage of 4.0 V at room temperature (or 136 W h/kg at 80 °C), values comparable to those of a Ni metal hydride battery [723]. In other early work, supercapacitors with capacitances of 135 and 99 F/g in aqueous and organic electrolytes, respectively, with good performance over a wide range of voltage scan rates were reported [724].
- Since this early work, supercapacitors incorporating graphene and graphene oxide (GO, mostly as reduced graphene oxide, RGO) have been demonstrated with the following variants: incorporating hydrous RuO$_2$ (38.3 w/w%); produced via microwave-assisted reduction of GO; produced via thermal reduction of propylene carbonate; produced via simple hydrazine reduction; as a composite with the conducting polymer poly(aniline), 50 w/w%; incorporating solid MnO$_2$; incorporating CNTs; incorporating MnO$_2$ nanowire; incorporating Co (II) hydroxide; incorporating Ni foam; incorporating SnO$_2$; and incorporating ZnO [480, 724–732]. These have shown specific capacitances in the 99 to 1046 F/g region (the latter, one of the highest values reported thus far, for graphene/poly(aniline) composites), power densities from 2.5 to 10 kW/kg, and energy densities up to 30 Wh/kg [480, 724–732, 733–737].
- 3D-stacked supercapacitors with increased energy density and cyclability in the thousands and mechanical robustness have been reported [738].
- Quick-charging, hybrid, very thin (one-fifth the thickness of paper) supercapacitors, using graphene coupled with MnO$_2$, have been reported with capacities claimed to be six times that of standard Li-ion batteries [739].
- A flexible boron-doped, laser-induced graphene "micro-supercapacitor," using boron-doped porous graphene, can be prepared in ambient air using a facile laser induction process from boric acid containing polyimide sheets and with reported areal capacitance values of 16.5 mF/cm^2 [740].
- Graphene/*poly(aniline)* "nanocomposites" (also mentioned briefly above) have shown high specific capacitance, up to 1046 F/g, and good cyclic stability, in potentially flexible supercapacitors [741–743]. In one of these studies, fibrillar

poly(aniline) doped with graphene oxide sheets was synthesized via in situ polymerization of the monomer in the presence of the graphene oxide [743].

- In other work with graphene/poly(aniline) composite electrodes [744], graphene/poly(aniline) composites were fabricated through in situ polymerization of aniline monomer in the presence of graphene oxide (GO) under acidic conditions, which was then followed by the reduction of the GO to graphene using hydrazine. Reoxidation and reprotonation of the reduced poly(aniline) followed to give the graphene/poly(aniline) composites. The composites that contained 80 w/w% of GO showed the highest specific capacitance, 480 F/g at a current density of 0.1 A/g. When the current density increased up to 0.5 A/g or 1 A/g, the specific capacitances still remained at above 200 F/g without a significant decrease upon charge/discharge cycling.
- In other work with other conducting polymer (CP)-graphene or CP-graphene oxide (GO) composites, specific capacitances of 510 F/g were obtained.
- GO–fibrous poly(pyrrole) composites [745]. Electrophoretically deposited graphene was used as a scaffold for poly(pyrrole) electropolymerization, yielding a composite electrode for potential supercapacitor applications, with a specific capacitance of 1510 F/g, an area capacitance of 151 mF/cm^2, and a volume capacitance of 151 F/cm^3 at 10 mV/s [746].
- Recently (2012), it has been shown that nitrogen-doped graphene (N-graphene) has enhanced performance in supercapacitors as compared to pristine graphene [705], but no further reports of N-graphene supercapacitors have since emerged.
- GO-MnO$_2$ composite electrodes have been studied for supercapacitors, with performance somewhat improved over MnO$_2$-only supercapacitors, with specific capacitance typically reaching about 211.2 F/g at a current density of 200 mA/g [747].

The takeaway from the above compilation of work on graphene-based supercapacitors however is that, as of this writing (2016), there are no graphene-based supercapacitors which possess markedly improved performance over conventional, non-graphene supercapacitors, to warrant their use practically and commercially.

21.5 Photovoltaics and Related Energy Devices

Among studies of note in areas relating to *photovoltaics and energy devices* are the following:

- Transparent, conductive, ultrathin graphene films as electrodes for dye-sensitized solar cells [748]. These graphene films, fabricated from exfoliated graphite oxide, followed by thermal reduction, showed high conductivity of 550 S/cm and >70% transmission in the 1000–3000 nm spectral region, together with high chemical and thermal stability.

- Large area, continuous, few-layered graphene as anodes in organic photovoltaic devices [749]. The graphene films in these devices were non-covalently functionalized with pyrene butanoic acid succidymidyl ester, which improved their power conversion efficiency to a still-low 1.71%. In contrast, a control device using indium tin oxide (ITO)/poly(3,4–ethylenedioxythiophene)‖-poly (styrenesulfonate)/poly(3-hexyl)thiophene‖- phenyl-C61-butyric acid methyl ester/LiF/Al electrodes showed a power conversion efficiency about 3.1%.
- Graphene–Si-based Schottky junction solar cells, wherein highly conductive, semitransparent graphene sheets were combined with an n-Si wafer to fabricate solar cells with power conversion efficiencies up to 1.5% at AM 1.5 illumination and an illumination intensity of 100 mW/cm^2 [750]. These solar cells were said to be extendable to other semiconducting materials in which graphene serves multiple functions as active junction layer, charge transport path, and transparent electrode.
- Higher-efficiency Schottky barrier solar cells using CVD-deposited graphene with efficiencies up to 15.6% with optimization of the oxide thickness on Si [2056].
- Low-temperature processed electron collection layers of graphene/TiO_2 composites in thin-film perovskite solar cells, with, again, efficiencies up to 15.6% [751].
- Continuous, highly flexible, and transparent graphene films by chemical vapor deposition (CVD) for organic photovoltaics [752].
- Organic light-emitting diodes using solution-processed graphene transparent electrodes [753]. In this work, the graphene electrodes were deposited on quartz substrates by spin coating of an aqueous dispersion of functionalized graphene, followed by a vacuum anneal step to reduce the sheet resistance. Small molecular weight organic materials and a metal cathode were directly deposited on the graphene anodes. The devices produced were claimed to have performance comparable to (but not better than) control devices on ITO-based transparent anodes.
- Recent (2015) work on *thermoelectric* power generation, from lanthanum strontium titanium oxide at room temperature through the addition of graphene [754]. This study noted that strontium titanium oxide-based thermoelectric materials are currently limited by their high operating temperatures of >700 °C and showed that the thermal operating window of lanthanum strontium titanium oxide could be reduced to room temperature by the addition of a small amount of graphene. The thermal conductivity of the nanocomposites decreased upon the addition of graphene, whereas the electrical conductivity and power factor both increased significantly. These factors, together with a moderate Seebeck coefficient, meant that a high power factor of ~2500 $\mu W/mK^2$ was reached at room temperature at a loading of 0.6 w/w% graphene. A thermoelectric figure of merit, ZT, of 0.42 and 0.36 was achieved at room temperature and at 750 °C at this graphene loading. A preliminary 7-couple device was produced using bismuth strontium cobalt oxide/graphene-LSTO pucks. This device had a

Seebeck coefficient of ~1500 μV/K and an open voltage of 600 mV at a mean temperature of 219 °C.

- Semiconducting nanoparticle quantum dots incorporating graphene combining semiconductors such as CdS, CdSe, ZnO, TiO_2, TSCuPc (Pc = phthalocyanine), and Co_3O_4 have been studied for use in energy devices as well as optoelectronic devices [480]. For example, a reduced graphene oxide/CdSe nanoparticle composite showed an enhanced photo-response under visible light [480, 755, 756], with this behavior being interpreted as the efficient and separate transfer of the photoinduced carriers from the CdSe nanoparticles to the RGO.

- TiO_2-modified graphene has been shown to enhance the photocatalytic reduction of CO_2 seven times as compared with pure TiO_2 [2074].

- A composite material comprised of $BiVO_4$ and RGO connected to a collecting electrode was shown to photochemically split water and evolve H_2 at a rate ten times that of pure $BiVO_4$, under visible illumination [2075].

- Photocatalytic H_2 production was demonstrated with a TiO_2/MoS_2/graphene composite electrode, with the possible indication that the positive synergetic effect between the MoS_2 and graphene could efficiently suppress charge recombination, improving interfacial charge transfer and providing a greater number of active adsorption sites and photocatalytic reaction centers for H_2 production [757].

- Functionalized graphene sheets with oxygen-containing sites, when applied on the counter electrode of a dye-sensitized solar cell, were claimed to perform comparably to Pt, with conversion efficiencies of 5.0 and 5.5%, respectively, at 100 mW/cm^2 with AM1.5G simulated illumination [758]. It was claimed that by increasing the amount of oxygen-containing functional groups in the graphene, the apparent catalytic activity of the material was improved.

Once again, it is worthy of note that all of the work cited in the preceding paragraphs has, as of this writing (2016), still not led to any practical, commercialized energy devices such as photovoltaics, in spite of highly hyped predictions that graphene-based photovoltaics could show an efficiency of 60% [759].

21.6 Hydrogen Storage

A graphene "nano-cage" or "nano-box," which could fold and unfold much like paper origami, has been demonstrated [760], with claimed storage capacities of 9.5 w/w% H_2; in this, the folding/unfolding is initiated by hydrogenation.

Graphene oxide (GO) has also been studied for hydrogen storage applications [565, 761]. Hydrogen can be stored between layers of GO, which are already somewhat separate. The GO layers can then be linked together to form a new layered structure, that is, a GO framework. Such GO structures are said to have tunable pore widths, volumes, and binding sites depending on the linkers chosen and have been claimed to adsorb hydrogen up to 6 w/w% at 77 °K and 1 bar pressure, a value higher than any other porous material known [565, 761].

A purely theoretical study of hydrogen storage in Al-doped graphene predicted a hydrogen storage capacity of 5.13 w/w% at 300 °K and 0.1 Gpa pressure, with an adsorption energy E_b of -0.260 eV/H_2 [762]. However, no prior or current experimental data were provided to substantiate the predictions.

21.7 Problems and Exercises

1. In which component of secondary (rechargeable) Li batteries is graphene generally incorporated? What has been its primary function or functions? What, if any, performance improvement has been seen with the use of graphene in Li batteries (cite parameters such as mAh/g, Wh/g, etc.)?
2. As of this writing (2016), are you aware of any commercially available Li batteries that incorporate graphene? And as of your reading today?
3. Enumerate the varied uses of graphene in fuel cells (hint: uses vary, e.g., in catalysts, in proton exchange membranes (PEMs), as electrodes). Cite one specific example of each.
4. How is graphene incorporated into supercapacitors (e.g., into which components of supercapacitors)? What are the advantages, if any, and drawbacks of graphene-based supercapacitors? Cite typical, comparative performance parameters (e.g., F/g) for graphene-based supercapacitors as compared with standard supercapacitors.
5. Give one example each of the use of graphene in conventional photovoltaics, dye-sensitized solar cells (DSSCs), OLEDs, and thermoelectric power generation. Cite comparative performance parameters (e.g., efficiency for photovoltaics and DSSCs) for the graphene- and non-graphene-based devices.
6. Outline the principles and method of use of graphene in hydrogen storage. Again, cite comparative performance parameters for graphene- and non-graphene-based hydrogen storage.

Chapter 22
Graphene Applications in Electronics, Electrical Conductors, and Related Uses

Check for updates

Contents

22.1 Conductivity-Based Applications

Graphene applications in electronics, electrical conductors, and specific conductivity-based applications abound. The conductive properties of graphene have been utilized in a niche but successful commercial product, RFID (radio frequency identification) tags. A New Jersey-based US company, Vorbeck, now offers RFID tags containing a small proportion of graphene to improve their high-temperature/high-pressure performance [500]. Vorbeck also uses the same principle to offer specialized antennae for cell phones (mobiles) that incorporate graphene and afford improved performance [501].

Graphene-based polymer composites have been studied for potential use in electromagnetic interference shielding (EMI-SE) [762]. More recently (2014), highly aligned graphene/polymer composites with excellent dielectric properties have been studied for high-performance EMI-SE [763]. In this study, self-aligned, reduced graphene oxide (RGO)/polymer composites were prepared using an all aqueous casting method, with a low percolation threshold of 0.12 v/v% in the RGO/epoxy system, using monolayer graphene sheets with extremely high aspect ratios (>30,000). Self-alignment into a layered structure occurred above a critical filler content yielded anisotropy in electrical and mechanical properties. High dielectric constants of >14,000 at 1 kHz at an RGO loading of 3 w/w%. EMI-SE and values of up to 38 dB were claimed in this work. In another study in 2009, EMI-SE of graphene/epoxy composites was reported [764]. These composites showed a low percolation threshold of 0.52 v/v.%. EMI-SE, tested over a frequency range of 8.2–12.4 GHz (i.e., X-band), showed a shielding efficiency of 21 dB for 15 w/w% (8.8 v/v%) loading, acceptable performance comparable to commercially available materials.

© Springer International Publishing AG 2018
P. Chandrasekhar, *Conducting Polymers, Fundamentals and Applications*,
https://doi.org/10.1007/978-3-319-69378-1_22

Since graphene is a very good conductor and can be put down as films or coatings, it was quite logically predicted that it may serve well for such applications as EMI (electromagnetic impulse) or radar shielding. In one study, graphene/quartz appeared to increase the absorption of radio waves in the 125 to 165 GHz region by over 90%, with a 28% fractional bandwidth from 125 to 165 GHz, with results indicating possible extension to the microwave and low-THz bands [765].

22.2 Applications in Electronics

In discussing graphene-based applications in electronics, microelectronics, and "nanoelectronics," the informed, intelligent reader must again, very reluctantly, attempt to evaluate reported results with a great deal of scrutiny and, yes, skepticism. It is important to note that the widely disseminated predictions of graphene-based "molecular-scale electronics," "nanoscale electronic devices," and "nanoelectronics" have, to date (2016), not come even close to being borne out. The reader will see this in the applications discussed below.

With respect to applications in electronics, it is important to recognize that a major, intrinsic drawback of graphene is that it possesses no energy bandgap, a fact that presents hurdles for its applications in electronic devices. As discussed in an earlier chapter, bandgaps can be induced or engineered in graphene via such methods as quantum confinement or surface functionalization; however, to date (2016), these efforts have not resulted in any practical success. However, an alternative route for use of graphene in microelectronics and information processing lies in the negative differential resistance observed in graphene field-effect transistors (FETs). This allows for construction of viable *non-Boolean* computational architectures using gapless graphene [766].

Chemical doping of graphene is another method used to create a bandgap in graphene; however, this is accompanied by significant Dirac point shifts, which give a low on/off current ratio and a poor switching property [564]. This has been somewhat overcome by dual doping of bilayer graphene [564, 767]. Thus, bilayer graphene (BLG) can be dual doped with $FeCl_3$ as acceptor and K as donor [564, 767]. Due to the high electronegativity of the $FeCl_3$ molecular layer, the $FeCl_3$-adsorbed bilayer graphene is hole doped, and the asymmetric composite of bilayer graphene with the $FeCl_3$ layer adsorption then exhibits an energy gap opening. By utilizing this asymmetric behavior, which increases the bandgap, the bilayer graphene with the dual $FeCl_3$ acceptor and K donor gives rise to a bandgap of about 0.3 eV, with, importantly, a proper Dirac point shift (-0.09 eV). It is to be noted that N-doping of graphene (i.e., with nitrogen) changes p-type semiconductor behavior observed in some monolayer graphene samples (due to effects of adsorbed molecules such as O_2 and H_2O) to n-type semiconductor behavior but also lowers the mobility, from about 1000 cm^2/Vs to about 200 cm^2/Vs at a 9 w/w% N-content, and even lower at a 2.4 w/w% content, according to published experimental data [704, 768].

In another chemical doping method, 2,3,5,6-tetrafluoro-7,7,8,8-tetracyanoquinodimethane (F4-TCNQ) molecules deposited onto bilayer graphene induce p-doping from the top, while NH_2-functionalized self-assembled monolayers (SAMs) constructed on a SiO_2/Si substrate induce n-doping from the bottom side [564, 769, 770]. These dual doping-driven perpendicular electric fields with opposite directions in FETs show that it is possible to open a bandgap with two molecular dopants (in this case, F4-TCNQ- and NH_2-functionalized SAMs).

The extraordinary expectations for electronics applications of graphene are predicated on the presumed high quality of its 2-d crystal lattice [519, 771, 772]. This necessarily assumes an unusually low density of defects, which typically serve as the scattering centers that inhibit charge transport. In such nearly defect-free, monolayer graphene, carrier mobility in excess of 200,000 cm^2/Vs has been experimentally measured [519, 773]. And another feature of defect-free, monolayer graphene is that carrier mobility remains high even at highest electric-field-induced concentrations, which implies ballistic transport on a sub-μm scale at 300 $°$K, leading to "dreams" of a room-temperature ballistic transistor [774]. Clearly, however, practical, commercial-scale use in electronics requires inexpensive, bulk production of defect-free graphene, still lacking as of this writing (2016).

To utilize defect-free graphene, large quantum dots, of diameter about 0.25 μm, have been used to demonstrate conductance that can be controlled by either a back gate or a side electrode also made from graphene [774]; however, again, such workarounds are not practical on a larger scale.

"Printed graphene circuits" and flexible and transparent electronic devices with extremely high mobilities (field-effect mobilities of up to 10,000 cm^2/Vs at room temperature) were reported (in 2007); these were fabricated from a graphene sheet by using a "transfer printing" method [775]. The claimed objective was "nanoscale electronic devices." However, to date, the work does not appear to have been further refined, and of course no such nanoscale electronic devices appear to have been produced thus far.

Another study with potential applications to "molecular-scale electronics based on graphene" used quantum dot devices "carved" entirely from graphene [776]. While not producing any practical device, this study did appear to demonstrate that at dimensions of >100 nm, the devices behaved as conventional "single-electron transistors," exhibiting periodic Coulomb blockade peaks. However, for quantum dots of dimension <100 nm, the peaks become strongly nonperiodic, appearing to indicate a major contribution from quantum confinement. Short constrictions of only a few nanometers in width remained conductive, revealing a confinement gap of up to 0.5 eV. Jia et al. [777, 778] claimed to have developed a "single-molecule switch" working at room temperature, as illustrated in Fig. 22.1, wherein the molecule shown is tethered to two graphene electrodes. UV light transforms the "switch" into a good conductor, whereas visible light retransforms it into an insulator.

Graphene-based field-effect transistors (FETs) with side and top gates were demonstrated in 2006 and 2007 [779, 780]. In the top-gated device, a conventional top-down CMOS-compatible process flow was applied, except of course for the

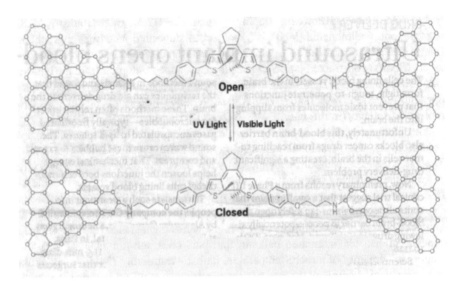

Fig. 22.1 Figure from [778]. After Ref. [778], reproduced with permission

graphene deposition. Carrier mobilities in graphene pseudo-MOS (metal oxide semiconductor) structures were observed to be comparable to those obtained from top-gated graphene field-effect devices. The extracted values exceed the universal mobility of Si and Si/insulator MOSFETs. As early as 2008–2009, researchers fabricated several hundred transistors from epitaxial graphene deposited on SiC substrates, using standard microelectronic fabrication methods [781, 782]. The graphene devices fabricated featured high-k dielectric, mobilities up to 5000 cm^2/Vs and, current on/current off ratios of up to seven. However, typical of graphene, the devices possessed negligible bandgaps and thus large leakage currents. In similar work, also in 2009, the operation of four basic two-input logic gates fabricated using a single graphene transistor was demonstrated [783]. Single-transistor operation was obtained in a circuit designed to exploit the charge-neutrality point of graphene, in order to perform Boolean logic. The type of logic function was selected by offset of the input digital signals.

Field-effect transistors (FETs) based on reduced graphene oxide (RGO) have been studied wherein few-layer RGO sheets serve as the semiconducting channel [784, 785]. The bandgap, which ranges from 2.2 eV to 0.5 eV, is made tunable by reduction treatments (i.e., different degrees of reduction of the RGO).

Self-assembled monolayers (SAMs) have been used as a buffer layer to induce instantaneous doping of graphene and tune the work function of graphene electrodes for high-performance *organic* field-effect transistors (FETs) [564, 769, 786]. In a pentacene-based organic transistor that utilizes graphene electrodes, the polymer residues have been shown to remain on the graphene surface and yield a stand-up orientation of pentacene formed on the polymer residue surface, giving rise to a molecular assembly that is optimal for charge transport [564, 786].

To further fabrication techniques for graphene-based electronics, an ink for 3D printing was developed by liquid-phase exfoliation of graphite in N-methyl-pyrrolidone (NMP) and used to print thin-film transistors, with mobilities up to about 95 cm^{2}/Vs, as well as transparent and conductive patterns, with about 80% transmittance and about 30 kOhm/sq. sheet resistance [787]. Graphene transistors have also been printed onto flexible substrates [788]. Another, solution-based fabrication method for few-layer graphene, based on first depositing layers of graphene oxide (GO) followed by their reduction, appeared to yield graphene deposits with mixed ambipolar and semimetal characteristics [789].

In 2008, scientists at IBM developed "high quality graphene transistors clocked running at 26 GHz"; these transistors were constructed of graphene monolayers, and the work was claimed to be "a significant step towards the realization of graphene-based electronics" [790]. Following this IBM report, in 2009, an "integrated complementary graphene inverter" consisting of one p- and one n-type graphene transistor integrated on the same sheet of monolayer graphene was reported [791]. Subsequently, in 2010, "100-GHz transistors from wafer-scale epitaxial graphene" were reported, also from an IBM team [792]. These FETs were fabricated on epitaxial graphene synthesized on the Si face of a SiC wafer, achieving a cutoff frequency of 100 GHz for a gate length of 240 nm. The high-frequency performance of these epitaxial graphene transistors was claimed to exceed that of state-of-the-art Si transistors of the same gate length. Further to this, the same IBM team reported a "wafer-scale graphene integrated circuit" [793]. In this work, a wafer-scale graphene circuit was demonstrated, in which all circuit components, including graphene FETs and inductors, were monolithically integrated on a single SiC wafer. The integrated circuit operated as a broadband radio frequency mixer at frequencies up to 10 GHz. The graphene-based circuits were claimed to exhibit outstanding thermal stability with little reduction in performance (less than 1 dB) between 300 and 400° K. Still further to this work, researchers described "integrated graphene oscillators" operating at 1.28 [794]. These graphene ring oscillators appeared to be less sensitive to fluctuations in the supply voltage compared with both conventional silicon CMOS devices and earlier, cruder oscillators made from the 2D materials.

Other demonstrations of potential electronics applications of graphene have included the following, which again is by no means an exhaustive list:

- Hybrid Conducting Polymer (CP)–graphene films with potential applications in solar cells, comprising (poly(3-hexylthiophene) on monolayer graphene [795]
- For potential applications in optoelectronics, a small (about 25 μm^{2}), graphene-based optical modulator operating at 1.2 GHz without a temperature controller, with a bandwidth from 1.3 to 1.6 μm [796]
- A frequency multiplier based on graphene films [797]

Again, however, as of this writing (2016), no electronic devices appear to have been developed and/or demonstrated from all of the work cited in the preceding paragraphs.

The electrostatic properties of graphene have been used to fabricate a graphene-based "electrostatic microphone and ultrasonic radio" for use in wireless communication [798]. Using as basis the fact that graphene-based acoustic transmitters and receivers have a wide bandwidth, from the audible region (20 ~ 20 kHz) to the ultrasonic region (20 kHz to at least 0.5 MHz), this study demonstrated efficient high-fidelity information transmission using an ultrasonic band centered at 0.3 MHz. The graphene-based microphone was also shown to be capable of directly receiving ultrasound signals generated by bats in the field, and the ultrasonic radio, coupled to electromagnetic radio, was shown to function as a high-accuracy range finder.

The nonlinear electromagnetic response of graphene was studied theoretically in 2013, with predictions of possible applications in "terahertz electronics" [799]. It was shown that the amplitude of the external electric field required for getting into the nonlinear regime for this material was of the order of several hundred V/cm for typical experimental parameters and that the operating frequency of a frequency multiplier fabricated from the graphene could vary in a broad range, from microwaves up to mid-infrared. However, no prior or current experimental data were cited in this work to substantiate the theoretical predictions.

22.3   Problems and Exercises

1. Describe in detail the uses of graphene in RFIDs (radio frequency identification tags) and electromagnetic interference shielding (EMI-SE). Quantify the performance improvements in each category as compared to non-graphene materials.
2. How and with what success is the lack of a bandgap in graphene addressed for its putative applications in electronic, microelectronic, and "nanoelectronics" devices? Cite specific materials and device examples.
3. Compare the construction, mode of operation, and performance parameters of graphene- and non-graphene-based FETs (field-effect transistors). Cite at least two examples of graphene FETs.
4. Describe the potential applications of the nonlinear electromagnetic response of graphene, citing at least one potential example.

Chapter 23
Graphene Applications in Displays and Transparent, Conductive Films/Substrates

Contents

23.1 Transparent, Conductive Films and Substrates

Like CNTs, graphene has been studied as a possible replacement for well-established *transparent conductors, transparent films, and conductive films* such as indium tin oxide (ITO), due primarily to its high electrical conductivity relative to its thickness. Among the professed advantages of graphene are the ability to achieve more flexibility. To be transparent, graphene films generally need to be <10 nm thick [479], a criterion of course easily satisfied by monolayer graphene.

The following studies are worthy of note in this field:

- One of the first demonstrations of graphene-based *transparent conductors* used films deposited by dip coating of graphene oxide (GO) and then reducing by thermal annealing [800]. Sheet resistances as low as 0.9 kΩ/sq. were obtained showing 70% transmission, to be compared however with typical ITO films (70 Ω/sq. at 90% transmission). The films' function was demonstrated in the anode in a dye-sensitized solar cell, which had a power conversion efficiency of (a rather low) 0.26%.
- Transparent, conductive, ultrathin graphene films as electrodes for dye-sensitized solar cells [801]. These graphene films, fabricated from exfoliated graphite oxide, followed by thermal reduction, showed high conductivity of 550 S/cm and >70% transmission in the 1000–3000 nm spectral region, together with high chemical and thermal stability. (This work was also cited in an earlier chapter.)
- Organic light-emitting diodes (OLEDs) using solution-processed graphene transparent electrodes [802]. In this work, the graphene electrodes were deposited on quartz substrates by spin coating of an aqueous dispersion of functionalized graphene, followed by a vacuum anneal step to reduce the sheet

© Springer International Publishing AG 2018
P. Chandrasekhar, *Conducting Polymers, Fundamentals and Applications,*
https://doi.org/10.1007/978-3-319-69378-1_23

resistance. Small-molecular-weight organic materials and a metal cathode were directly deposited on the graphene anodes. The devices produced were claimed to have performance comparable to (but not better than) control devices on ITO-based transparent anodes. (This work was also cited in an earlier chapter.)

- Continuous, highly flexible, and transparent graphene films by chemical vapor deposition (CVD) for organic photovoltaics [751].

23.2 Displays

Significant resources have been committed by organizations in South Korea toward development of transparent conductors and, eventually, displays, based on graphene. These organizations include the Korea Advanced Institute of Science and Technology (KAIST) and Samsung.

Among work of note in this area is the following:

- A prototype *flexible* display based on graphene was demonstrated by a group from the University of Cambridge in England [803].
- Researchers from Seoul National University reported flexible LED (light-emitting diode) displays based on graphene [804] in 2014. In this work, "carpets" of microscopic GaN wires were grown on an ultrathin mesh of graphene. These graphene-LED sheets were then peeled off a copper backing and placed on a flexible polymer substrate. The resulting LED displays were claimed to perform undiminished through 1000 bending cycles for the substrate. However, since the original 2014 report date, no further development of this technology has been reported to date (2016).

In an interesting application opposite to that of transparent, conductive films and coatings, 15-nm-thick "nano-textured, decoupled graphene multilayers" have also been claimed to absorb nearly 99% of the light incident on them from the mid-IR to the UV region [805].

23.3 Problems and Exercises

1. In a succinct summary, describe the advantages and drawbacks of transparent, conductive materials based on graphene. What are the specific factors that have thus far (as of 2016) prevented their practical implementation (e.g., to replace indium tin oxide (ITO) coatings)? What are the typical surface resistivities of the best graphene coatings produced as of 2016? As of your reading today?
2. How specifically is graphene used in displays? What are its advantages and drawbacks? What are the specific factors that have thus far (as of 2016) prevented its practical implementation in displays (e.g., as replacements for LED displays or even in niche applications)?

Chapter 24
Medical and Pharmaceutical Applications of Graphene

Contents

Among applications most prominent in medical applications and pharmaceutical applications of graphene are the following:

- Graphene has been tested for use as an electrode in vivo, to communicate with brain neurons [806]. These microelectrodes used non-functionalized, uncoated graphene and were an attempt to substitute W- and Si-based microelectrodes used for this purpose earlier.
- Polymeric "nanocomposites" based on poly(propylene fumarate) (PPF) and having some graphene content (including graphene and graphene oxide nano-platelets and nanoribbons) have been studied for use in bone tissue engineering, with claims of being biodegradable [807].
- A graphene-based, cellular protease (furin)-mediated system was demonstrated for co-delivery of a membrane-associated cytokine ("tumor necrosis factor-related apoptosis-inducing ligand, TRAIL") and an intracellular-acting small-molecule drug (doxorubicin, DOX). It was shown that these two drugs, "TRAIL" and "DOX," could be sequentially released toward the plasma membrane and nucleus, respectively.

Other, very diverse medical, pharmaceutical, biotechnology, and related applications of graphene and its derivatives have included the following, which again is by no means an exhaustive list:

- Use as MRI contrast agents, specifically, graphene nano-platelets intercalated with Mn(II) ions and then functionalized with dextran [808].
- A graphene-based multimodal resonance imaging – X-ray computed tomography contrast agent, in the form of graphene sheets with intercalated Mn(II) ions, with relatively low cytotoxicity [809].
- Contrast agents for photoacoustic and thermoacoustic tomography, based on "single- and multi-walled graphene oxide nanoribbons," with a claimed five- to

tenfold signal enhancement for photoacoustic tomography and a 10% to 28% signal enhancement for thermoacoustic tomography [809].

- Use of graphene nanoflakes to improve the efficiency of the polymerase chain reaction (PCR), attempting to use the high thermal conductivity and related properties of graphene for this purpose [810].

- A new method for ultrafast *DNA sequencing* using a graphene nanoribbon (GNR)-based "nano-channel" device has been *suggested* recently [564, 2075]. In this, while a single-strand DNA passes beneath a GNR, each nucleobase interacts with the GNR via $\pi - \pi$ stacking interactions. Because the conducting properties of a GNR differ when they interact with different nucleobases due to something called the Fano resonance phenomena, real-time DNA sequencing is possible by analyzing the real-time conductance of the GNR. Since the stable $\pi - \pi$ stacking interaction reduces stochastic motion of a nucleobase during the conductance measurement, the overlap between the signals from each nucleobase can be minimized. Thus, more reliable DNA sequencing is potentially realizable, employing statistical analyses including a data mining approach and a two-dimensional transient autocorrelation function [564, 811]. Needless to say, as of this writing (2016), this method remains hypothetical and has not been proven experimentally. Graphene quantum dots, synthesized by a one-step solvothermal method, have been demonstrated for potential use in cellular imaging, with comparatively lower toxicity and high photoluminescence [812]. In a similar vein, GO functionalized with CdSe/ZnS quantum dots have been demonstrated for potential use in visible light imaging and near-infrared phototherapy of cancer cells [813]. In yet another biological imaging application, graphene functionalized with *p*-sulfonated calyx[6]arene has been shown to be usable as a "turn-on" fluorescent probe for L-carnitine both in vitro and in living cells [814].

24.1 Problems and Exercises

1. Cite two specific examples of attempts to use graphene in biomedical applications in vivo. What are the advantages and drawbacks of graphene in such application as compared to established, competing technologies for the same objectives? Which specific factors prevent the use of graphene in such applications as of this writing (2016)? As of today?

2. Cite two specific examples of attempts to use graphene in biomedical applications which are NOT in vivo. What are the advantages and drawbacks of graphene in such application as compared to established, competing technologies for the same objectives? Which specific factors prevent the use of graphene in such applications as of this writing (2016)? As of today?

Chapter 25
Graphene Applications in Specialized Materials

Contents

25.1 Composite Materials

Graphene-based *polymer composite materials* (sometimes called "nanocomposites") have been studied at some length [822]. They have included those with the following components (this list is cited as illustrative only and by no means exhaustive):

- Conducting polymers (CPs) such as poly(aniline), poly(3,4-ethylene dioxythiophene) (PEDOT), poly(phenylene sulfide), and poly(diacetylene)
- Epoxies
- Polyurethanes
- Polystyrenes
- Poly(vinyl alcohol) (PVA)
- Poly(ethylene terephthalate) (PET)
- Polycarbonate
- Poly(vinylidene fluoride) (PVDF)
- Nafion®; poly(ε-caprolactone)
- Poly(lactic acid)
- Poly(methyl methacrylate) (PMMA)
- High-density polyethylene (HDPE)

A recent review [822] discusses methods of synthesis of these composites, which are straightforward and predictable, at some length. Suggested, potential uses for these graphene–polymer composites have been very varied, with some examples (by no means exhaustive) being organic LEDs, solar cells, transparent conductors, energy storage, sensors, biomedical devices, drug delivery, and EMI shielding

© Springer International Publishing AG 2018
P. Chandrasekhar, *Conducting Polymers, Fundamentals and Applications*,
https://doi.org/10.1007/978-3-319-69378-1_25

(EMI-SE). These applications are covered in separate chapters, elsewhere in this book.

In a manner similar to the reinforcement of concrete with steel bars for greater strength and toughness, graphene sheets have been claimed to have been reinforced with CNTs, in what has been designated "rebar graphene" [823]; applications proposed for this "rebar graphene" have included more environmentally durable photovoltaics and transparent, conductive films (to replace the likes of indium tin oxide, ITO).

Graphene has been used as an additive in composite materials and coatings, generally to replace carbon fibers or activated carbon, for enhancing strength, conductivity, and other properties and combinations of properties. For example, as noted in an earlier chapter, a graphene-containing printer powder, marketed as *Graphenite WX* by a US company called Noble 3D Printers, has been employed for a technique called "lost wax casting" [642]. As another example, graphene has been incorporated in small amounts (substituted for carbon fibers) to increase stiffness and reduce weight of tennis rackets. These have been marketed by companies including the tennis racket company Head [643]. As yet another example also cited earlier, a new version of Araldite® brand resins (which include epoxy adhesives) will be marketed by Huntsman Corp.'s Advanced Materials division in collaboration with Haydale Composite Solutions which incorporate graphene [644, 645]; Haydale is a UK company that has a production line for graphite nano-platelets. Recent work has shown that the *tensile strength* of epoxy resins can be increased more than two times when functionalized graphene is added to them.

A specialized, conductive adhesive based on a composite of graphene powder, CNTs, and poly(dimethyl siloxane) (PDMS) was shown recently (2016) to emulate the structure and van der Waals force-based adhesive power of gecko feet [824]. In this work, a mixture of graphene powder, CNTs, and PDMS was poured into a Si mold dotted with holes. After curing the elastomer at 120 °C for 2 h, the result was 10 by 10 cm square patch covered with 15-μm-tall golf-tee-shaped pillars. These shapes were claimed to mimic the microscopic, hairlike features on gecko feet that can cling to surfaces via van der Waals interactions. The patch was wearable as an electrocardiography recorder that could be submerged in water without coming off.

25.2 Environmental Applications

With regard to specialized environmental applications, freestanding, monolayer graphene with nm-scale pores was claimed to filter NaCl from water, i.e., desalinate the water, at high efficiency, in a study reported in 2012 [825]; the authors however did not discuss the issue of how precisely a single-layer graphene sheet would be suspended in water for any length of time in a manner suitable for desalination.

The graphene derivative graphene oxide (GO) has been shown to be promising for some *environmental* applications, such as remediation (cleanup). This stems from the properties of GO, including its non-toxicity, biodegradable nature, amphiphilicity, and capability of functionalization with reactive groups. Promising results showing the capability of GO to adsorb, from their solutions in water at concentrations of <0.1 g/L, radionuclides such as the radioactive isotopes of thorium, uranium, and neptunium and the radioactive elements plutonium, americium, and europium, were reported in 2013 [826, 827]; removal rates of these radionuclides from solution of 20% to 90% were reported.

Graphene has also been shown to be able to absorb and clean up pollutants such as bisphenol A, Pb(II)(aqueous), Pb(II)(aqueous), Hg(II)(aqueous), naphthalene, and 1-naphthol, with performance slightly better than existing, commercial methods [571, 828–832]. Some of these methods used graphene functionalized with moieties such as sulfonate and poly(pyrrole). Graphene sheets "decorated" with magnetite nanoparticles have been claimed to be effective in As(III) and As(V) removal, and RGO/Fe_3O_4 composites have been claimed to remove polluting dyes from aqueous solutions [571, 833, 834]. Graphene "decorated" with photocatalytically active SnO_2 and TiO_2 has been shown to be useful in the degradation of several polluting dyes in aqueous solution under visible light irradiation; however, in that case, the efficiency of the photocatalysts alone, i.e., without the graphene, was as high or higher for this process, and so it was not clear what the utility of adding graphene was other than novelty [571, 835].

The use of graphene for CO_2 capture has been demonstrated with N-doped, porous graphene [836], yielding a selective absorption capacity of 4.3 mM/g CO_2 claimed at 298 °K, and with poly(aniline)-RGO composites [837].

25.3 Other Material Applications

Since graphite is already used widely as an additive in formulation of *lubricants*, it was quite logical to substitute graphene for graphite to see if improvements in performance were observable. In one study, monolayer graphene used alone as a lubricant between a steel ball and a steel disk was claimed to last for 6500 cycles, as compared to 1000 cycles for conventional lubricants such as graphite or MoS_2 [838]. However, the relative cost factor (graphite or MoS_2 vs. graphene) would also need to be considered for such lubricant applications.

Utilizing its high *thermal conductivity*, graphene was studied as an additive to coolant fluids based on ethylene glycol, with a 5 v/v% addition to the fluid causing an 86% increase in the thermal conductivity of the fluid [839]; the thermal conductivity of graphene oxide (GO) and graphene in the fluid were estimated to be about 4.9 and 6.8 W/mK, respectively. In another application related to its high thermal conductivity, graphene was added to each side of a Cu film increasing its thermal conductivity by nearly 25% [840].

Few-layer graphene nanoribbons have been used to fabricate a "nano-electromechanical switch" with a demonstrated use as a mechanical OR gate [561].

25.4   Problems and Exercises

1. List at least five (5) different components *other than conducting polymers (CPs)* that graphene has been attempted to be combined with to form polymer composites.
2. Describe the mode of use and advantages of graphene in printing inks. What factors or properties prevented their commercialization as of 2016? As of today?
3. Enumerate at least two varied applications of graphene in the environmental field.
4. What are the advantages and disadvantages of the use of graphene in lubricants? Which specific factor or factors prevented their commercialization as of 2016? As of today?
5. What are the advantages and disadvantages of the use of graphene as an additive in coolant fluids? Which specific factor or factors prevented their commercialization as of 2016? As of today?

Chapter 26
Miscellaneous Applications of Graphene

Contents

A very limited number of applications for graphene and its derivatives have been studied that do not fall into the categories of the prior chapters and so are listed briefly in this chapter.

Anti-corrosion coatings for Cu based on Si-doped graphene have been studied recently [841]. The advantages over established, commercialized anti-corrosion coatings for metals and other materials have however not been clear from these studies.

A "molecular switch" was very recently (2016) reported, in which a diarylethene molecule was tethered to two graphene electrodes [785, 842]. In this switch's "open" state, when there is no direct bond between the diarylethene's thiophene units, the molecule acts as an insulator. However, shining ultraviolet light on it causes a bond to form between the two thiophenes, rendering the molecule an electrical conductor. Visible light breaks the bond and returns the molecule to its insulating state. It would however appear, in this author's humble opinion, that the claim that this "molecular switch" could be a "key component for ever-shrinking electronic devices" [785] is still quite far from realization.

26.1 Problems and Exercises

1. Describe applications of graphene in anti-corrosion coatings. Review the reference cited in this chapter for this. How does this compete with established commercial anti-corrosion coatings?
2. Describe applications of graphene as a "molecular switch." In your estimation, what are the hurdles or technical milestones that would have to be crossed before such applications would actually be incorporated into electronic devices?

© Springer International Publishing AG 2018
P. Chandrasekhar, *Conducting Polymers, Fundamentals and Applications*,
https://doi.org/10.1007/978-3-319-69378-1_26

Part V
Conducting Polymers, Fundamentals

Chapter 27
Introducing Conducting Polymers (CPs)

Contents

27.1 What Are Conducting Polymers (CPs)?

27.1.1 Definitions and Examples

When one thinks of polymers, one perhaps envisions common plastics, such as polythene, that one may encounter in everyday life. If one then conjures up a conducting polymer, one may perhaps envision these plastics filled up with conductors such as metal or carbon particles. The *conducting polymers* (*CPs*, also sometimes called *conductive polymers* or *conjugated conductive polymers* or *organic polymeric conductors*), which are the subject of this book, are quite a different beast, in the sense that they are intrinsically conducting and do not have any conductive fillers as such.

This unique intrinsic conductivity of these organic materials, which generally are comprised simply of C, H, and simple heteroatoms such as N and S, and the myriad of properties emanating from it arise uniquely from *π-conjugation*. That is to say, a sometimes fairly extended and delocalized conjugation originating in overlap of π-electrons. Such conjugation is illustrated somewhat simplistically in Fig. 27.1 for poly(acetylene), a prototypical CP.

This conductivity of CPs is achieved through simple chemical or electrochemical oxidation, or in some cases reduction, by a number of simple anionic or cationic species, called *dopants*. That is to say, the polymeric backbone of these materials needs to be oxidized or reduced to introduce charge centers before conductivity is observed, and the oxidation or reduction is performed by anions or cations somewhat misnamed *dopants*, a term borrowed from condensed matter physics.

© Springer International Publishing AG 2018   159
P. Chandrasekhar, *Conducting Polymers, Fundamentals and Applications*,
https://doi.org/10.1007/978-3-319-69378-1_27

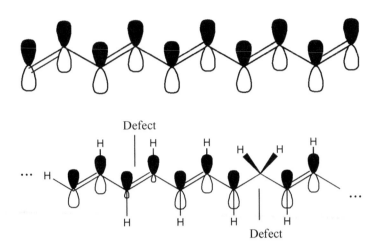

Fig. 27.1 Schematic representation of π-conjugation in the conducting polymer (CP) poly(acetylene). *Top:* Basic schematic. *Bottom:* Three-dimensional, including defects

Thus in an elementary fashion and if we consider the three broad material categories of *insulators, semiconductors, and conductors (such as metals)*, CPs differ from typical everyday organic polymers such as poly(ethylene), poly(vinylidene chloride) ("Saran" wrap), polyesters (used in textiles), and other everyday plastics, which are generally highly insulating, in that their unique π-electron properties impart electrical conductivity at room temperature, and many other derivative and interesting properties discussed at length in this book, on oxidation or reduction.

Typical representative CPs are shown in Figs. 27.2 and 27.3, where in the first figure the monomer building block is shown to the left of the polymer. In all cases, n is a large integer referring to the degree of polymerization. For the case of poly (aniline), the x's, y's, and n's are features of the structure which need not be worried about at this time. Among CPs, such as those illustrated below, poly(aniline), poly (pyrrole), poly(acetylene), and the poly(thiophenes) have been among the most studied, both scientifically and in terms of practical applications.

In the CP literature and among workers in the field, the monomers are generally represented by abbreviations, e.g., ANi for aniline, and the polymer also abbreviated in shorthand notation, following the custom in other polymer literature. For example, P(ANi) and PANi may be used to represent poly(aniline), P(Py) and PPy poly(pyrrole), and P(3MT), P3MT, or PMT poly(3-methylthiophene).

CPs continue to generate much interest and are continuing to find new applications, as covered in PART VI of this book [836–840].

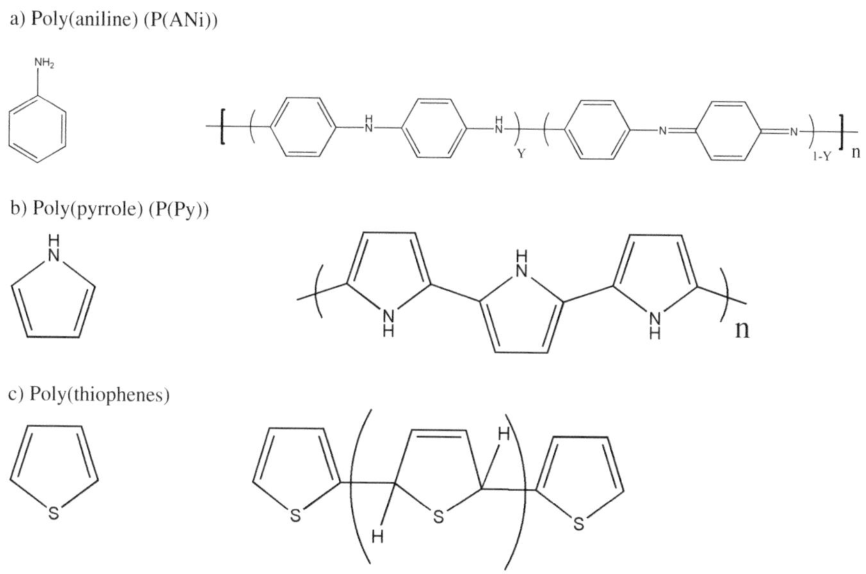

a) Poly(aniline) (P(ANi))

b) Poly(pyrrole) (P(Py))

c) Poly(thiophenes)

Fig. 27.2 Schematic illustration of monomer unit (*left*) and CP (*right*) for three. *(Over)* common CPs: (**a**) poly(aniline) (P(ANi)), (**b**) poly(pyrrole) (P(Py)), and (**c**) poly(thiophenes)

a)   b)   c)

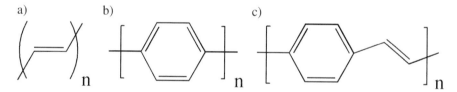

Fig. 27.3 Structures of three other common CPs: (**a**) poly(acetylene) (P(Ac)), (**b**) poly(para-phenylene), and (**c**) poly(para-phenylene vinylene)

27.1.2   Excluded Materials Classes (Those Not Treated as CPs in this Book)

An important point to note is that several classes of organic, inorganic, and organometallic materials which may fall in a gray area are *not treated as CPs and thus excluded from this book*. Specifically, these include:

1. *Charge-transfer materials based on metal–ligand charge transfer* and in which metals play an important part in conductivity, such as metallo-phthalocyanines (e.g., copper or cobalt phthalocyanines).
2. *Charge-transfer materials based on donor–acceptor complexes*, such as tetrathiafulvalene/−tetracyanoquinodimethanide (TTF/TCNQ), once the subject of intense study as the future organic conductors, much before CPs came on the scene.

3. *Poly(silanes)*, which share many of the attributes of CPs, such as conductivity, "dopability," and nonlinear optical properties, but which have the crucial distinction of having σ- rather than π-electron conjugation networks.
4. *Ion conductive polymers*, such as the prototypical LiClO$_4$-impregnated poly (ethylene oxide) (finding application as a solid electrolyte), which are many times called conducting polymers but which rely on conduction by ions as the basis of their conductivity and other properties.
5. *Organic polymers* such as (SN)$_x$, "poly(thiazyl)," which is highly conducting at room temperature and even superconducting below 0.26 °K, but whose different conduction mechanism yields high conductivity along the S-N chain but poor conductivity in other directions.
6. Lastly, we may mention that commercial materials comprising *insulating organic polymers filled with conductive fillers* such as graphite or metal powder (or, more recently, CNTs and graphene), which are sometimes also denoted "conducting polymers," are of course also excluded.

Representative structures for each of these classes of excluded materials are given, in the order mentioned above, in Fig. 27.4. It may be noted that CPs also differ from these five classes of materials in the feature that their polymer structure in general consists simply of a repetition of a well-defined monomer which is frequently the starting point for synthesis of the polymer.

27.2 Historical

Many conducting polymers (CPs) were well known in their nonconducting forms much before their conductivity and other features of interest were discovered. Some were also known in their conductive forms, but not well characterized and with not much interest paid to their conductivity. For example, poly(p-phenylene sulfide), PPS, has been commercially produced for thermoplastics applications under the brand name Ryton by Phillips Chemical Company since the early 1970s, and well-defined syntheses of poly(acetylene) have been reported since 1971 [841].

Chemical polymerization (chemical oxidative polymerization) of aniline was described by Letheby in 1862 [842] and studied in more detail by Mohilner et al. in 1962 [843]. Pyrrole was known to form a conductive "pyrrole black" [844] via spontaneous polymerization in air on the sides of pyrrole containers, and its chemical polymerization studied in some detail in 1916 [845].

Since 1957, studies of electrochemical oxidation of aromatic monomers, now widely used as one method of synthesis of CPs, have been reported under various descriptions such as "electro-organic preparations" and "electro-oxidations" [846–848]. More recently, in 1967, electrically conducting polymers from pyrrole, thiophene, and furan were characterized [849] and the electrical conductivity of poly(anilines) noted [850]. As early as 1968, dall'Ollio [851] described electropolymerization of poly (pyrrole). In many a sense, thus, CPs are "rediscovered" materials.

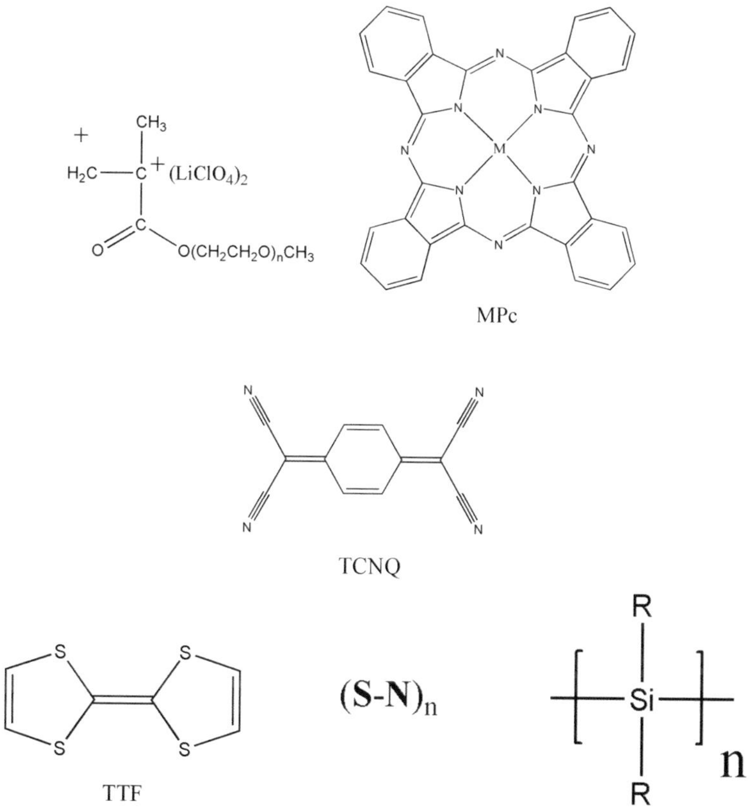

$$H_2C \overset{+}{\underset{}{\text{—}}} \overset{CH_3}{\underset{C}{\overset{|}{C^+}}} (LiClO_4)_2$$

MPc

TCNQ

TTF $(S-N)_n$

Fig. 27.4 Examples from the *excluded* materials classes (those not treated as CPs in this book)

The developments of importance that focused attention on CPs as potential novel materials with highly promising conductivity and other properties however started with the serendipitous discovery in a collaborative effort between the Shirakawa and Heeger/MacDiarmid groups [852] that poly(acetylene) exposed to iodine vapors develops very high and well-characterized conductivities (for an organic material). A chemical route to a new, conductive organic material was thus available. Further impetus to the field was given by the repetition and refinement, by the Diaz group at IBM [9], of the electrochemical polymerization of pyrrole originally ascribed to dall'Ollio [851], with a much better characterized polymer produced.

Poly(acetylene) was initially the most studied CP from both scientific and practical application points of view. However, due to its high chemical instability in air and related factors, interest in it has most recently been confined to its scientific aspects. Poly(aniline), poly(pyrrole), and the poly(thiophenes) remain the most extensively studied CPs to date, from both scientific and practical or commercial points of view.

For their work in CPs, the Nobel Prize in Chemistry for the year 2000 was awarded jointly to Alan J. Heeger, Alan G. MacDiarmid, and Hideki Shirakawa, officially "for the discovery and development of conductive polymers." Prof. MacDiarmid passed away in 2007. Prof. Shirakawa retired from the University of Tsukuba, Japan in 2000, becoming Professor Emeritus. Prof. Heeger continues at the University of California, Santa Barbara as of this writing (2016).

27.3 Basic Characteristics of CPS, Doping, and Structure

27.3.1 Conductivity Classification of Materials

Materials in the real world may be classified into three broad categories according to their room temperature conductivity properties: *insulators, semiconductors, and conductors.*

The overlapping of individual molecular electronic states in all these materials produces electronic bands. In crude and very simplistic schematic for purposes of our discussion, valence electrons overlap to produce a *valence band*, while the electronic levels immediately above these levels also coalesce to produce a *conduction band*. A gap, called the *bandgap*, generally denoted E_g, exists between these two, as shown in Fig. 27.5. If the gap is large, e.g., 10 eV, electrons are difficult to excite into the conduction band, and an insulator results at room temperature. If the gap is small, e.g., 1.0 eV, then electrons may be excited from the valence band into the conduction band by means such as thermal excitation, vibrational excitation, or excitation by photons; the electrons are then mobile in a sense, and the material is termed a semiconductor. For conduction to take place in

Fig. 27.5 The three conductivity classes (*metal, semiconductor, insulator*) (*left to right*) illustrated schematically in terms of their bandgaps

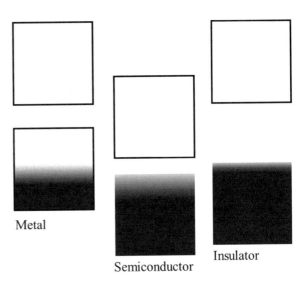

Metal

Semiconductor

Insulator

conventional, inorganic semiconductors, electrons must generally be excited from the valence to the conduction band. Normally, thermal excitation at room temperature gives rise to some conductivity in many such inorganic semiconductors. "Doped" CPs, when in an appropriate oxidized or reduced state as described in the next section, are semiconductors as a result of their unique, extended π-conjugation. Indeed, the extended-overlap π-bands become the valence band, and the π^* bands become the conduction band in CPs. The bandgap is generally greater than 1 eV in most CPs. Finally, if the gap vanishes, an overlap of the valence and conduction bands occurs, with the latter now partially filled, leading to metallic conduction.

These three conductivity classes are depicted schematically in terms of their bandgaps in Fig. 27.5. The electrons within the conduction band have unfilled energy levels (in the upper part of the conduction band) within easy reach, and in response to an electrical or other field can move into them and give rise to conduction. This is the situation in conductors such as metals. The highest occupied level in the conduction band is a level known as the *Fermi level*, with a corresponding *Fermi energy*, E_F.

A better idea of the place of CPs among the three broad conductivity classifications of materials, i.e., insulators, semiconductors, and metals, can be garnered from a comparison of conductivities, as shown in Fig. 27.6. In this figure, the conductivities are represented in Siemens/cm (1 Siemen $= 1 \, \Omega^{-1}$, S/cm $= \Omega^{-1} \, cm^{-1}$). The *conductivity* (generally denoted σ) is the reciprocal of the *resistivity* (generally denoted ρ, units of Ωcm), which can generally be defined as the potential drop across a given distance in a cross section of the material in question of given area, when a given current passes through this cross section, i.e., $\rho = (\Delta E/\text{distance})/(I/\text{area})$, keeping in mind Ohm's law, $E = IR$. A more detailed discussion of conductivity, including other units and measurement, is given in a later chapter.

27.3.2 Doping and Dopants

The semiconductor band structure of CPs permits electronic excitation or electron removal/addition, e.g., from the valence to the conduction band, leading to most of the properties that are of interest in CPs. Excitation of electrons from the valence band to the conduction band, e.g., by photons, yields typical excited state properties such as photoluminescence and nonlinear optical properties (e.g., third harmonic generation).

On the other hand, a chemical or other oxidation of the CP, i.e., essentially a removal of electrons from the valence band, leads to the presence of charges on the CP. These charges are in general strongly *delocalized*, over several monomer units in the polymer. These charges also cause a relaxation of the geometry of the (now charged) polymer to a more energetically favored conformation, a process which is discussed in more detail subsequently in this book. A charge may also be donated to the conduction band of the CP, causing reduction of the CP.

Ag, Cu	10^6	
Fe		
Bi	10^4	Graphite, doped poly(acetylene)
		TTF/TCNQ
	10^3	{Most
InSb	1.0	...
Ge	10^{-2}	... doped CPs}
	10^{-4}	
Si	10^{-6}	
AgBr	10^{-8}	
Glass	10^{-10}	
	(S/cm)	Units: S/cm

Fig. 27.6 Comparison of conductivities of various materials

Oxidation caused by a chemical species generates a positively charged CP and an associated anion. Reduction similarly generates a negatively charged CP and an associated cation. Examples of these processes, also designated "doping/de-doping," are, respectively, in very rough schematic, where M and A are any cation and anion, respectively:

$$P(Py) + MClO_4 \rightarrow P(Py)^+ ClO_4^- M^+ (\text{oxidation}) \qquad (27.1)$$
$$P(Ac) + NaA \rightarrow Na^+ P(Ac) + A^- (\text{reduction}) \qquad (27.2)$$

Both the above processes, oxidation and reduction, impart conductive properties to the CP. Because of the analogy with impurities which cause the removal or addition of charges from the valence or conduction band and thence impart higher conductivity in inorganic semiconductors such as silicon or CdSe, the chemical oxidation of the CP by anions, or its reduction by cations, was originally called *doping*. The associated anions/cations, i.e., the *counterions*, were called *dopants*. These terms, borrowed from condensed matter physics, have stuck. It should be noted from the outset however that the processes they refer to in the particular context of CPs are proper oxidation or reduction in the chemical and electronic sense, not doping/de-doping in the inorganic semiconductor sense.

Typical dopants are listed in Table 27.1. Dopants, i.e., chemical oxidants or reductants, are generally incorporated into the CP at the time of synthesis. They may also however be incorporated later by chemical, electrochemical, or a host of other means. Dopants may be small anions or cations, e.g., ClO_4^- or Na^+, or large polymeric species, such as the "polyelectrolytes" poly(styrene sulfonic acid) and poly(vinyl sulfonic acid). These are themselves polymers with pendant sulfonate groups.

A CP that is in its undoped, i.e., neutral, state is generally termed *pristine* (sometimes, *virgin*). The extent of oxidation/reduction, i.e., doping, is called the *doping level* and is generally measured as the proportion of dopant ions or molecules incorporated per monomer unit. This is then generally expressed either as a fraction or as a molar percentage. For instance, a CP with one dopant anion per four monomer units would have a doping level of 0.25 or 25%. Because of various constraints on the polymer, it is generally not possible to have a 1:1 doping, i.e., a 1.0 (or 100%) doping level. Increased doping level of course leads to increased conductivity, via creation of more mobile charges, and the maximum doping levels achievable vary for different CPs and different dopants. For example, the doping level for poly(acetylene) typically varies from 0.5% to 8%. Some representative doping maxima are also listed in Table 27.2.

Table 27.1 Typical dopants for CPs

Dopant	Structure/formula
Anionic	
Chloride	Cl^-
Perchlorate	ClO_4^-
Tetrafluoroborate	BF_4^-
Tos, *p*-toluene sulfonate	$CH_3\text{-}C_6H_5\text{-}SO_3^-$
Trifl, trifluoromethane sulfonate	$CF_3SO_3^-$
Hexafluorophosphate	PF_6^-
PSS, polystyrene sulfonate	$[\text{-}CH_2CH(C_6H_4SO_3)\text{-}]_n^{n-}$
Cationic	
Proton	H_3^+O
Sodium	Na^+
Potassium	K^+

Table 27.2 Typical maximum doping levels

Polymer	Maximum doping level (dopant)
Poly(pyrrole)	$33\%\left(ClO_4^-\right)$
Poly(thiophene)	$30\%\left(ClO_4^-\right); 6\%\left(PF_6^-\right)$
Poly(aniline)	$42\%\ (Cl^-)$
Poly(*p*-phenylene)	$44\%\ (Li^+)$

27.3.3   Doping Types

If the CPs are not already synthesized in the doped state, doping of pristine CPs can be accomplished chemically, e.g., by exposure to a solution or vapor of the dopant, or electrochemically, by subjecting the CP, generally in a solution, to an applied potential. When a positive potential is applied to a CP, for instance, one immobilized on an inert electrode, the dopant anion moves in from the solution into the CP toward delocalized charge sites on the CP, and anionic doping occurs. This anionic doping is termed *p-type doping*, again in erroneous analogy to solid state physics terminology. Similarly, if a negative potential is applied in solution to a CP immobilized on an electrode, a cation would move in from the solution into the polymer. This would be termed cationic, or *n-type doping*. Representative anionic and cationic dopings are depicted in Eqs. (27.3) and (27.4) in a simple way, for poly(pyrrole) (P(Py)) and poly(*p*-phenylene) (P(PP)), respectively.

Anionic doping:

$$P(Py) + nLiClO_4 \rightarrow P(Py)^{n+}(ClO_4)_n^{(n-)} \tag{27.3}$$

Cationic doping:

$$P(PP) + nLi \rightarrow (Li)_n^{(n+)}P(PP)^{(n-)} \tag{27.4}$$

In chemical doping, the dopant, i.e., the oxidizer or reducer, should have a redox potential suitable for oxidation/reduction, i.e., doping, of the CP, while the ionization potential (or electron affinity) of the CP should be small enough (or large enough) to facilitate this. For instance, in chemical doping of pristine (undoped or neutral) poly (acetylene) by I_2, the I_2/I^- redox potential is such that oxidation of poly(acetylene), i.e., p-doping, can take place (the actual dopant species in this case being I_3^-). Thus, simple exposure of pristine poly(acetylene) to I_2 vapor, or simple exposure of pristine poly(pyrrole) to HCl vapor, yields p-doped CPs with I_3^- and Cl^- dopants, respectively. During the doping/de-doping process, the dopant physically moves in and out of the CP lattice. This leads to physical, volumetric strains on the CP, i.e., swelling and shrinking. Besides the chemical and electrochemical doping described above, there are several other avenues available for doping of a CP, discussed in a separate chapter elsewhere in this book.

27.3.4   Real and Idealized Structures

An important caveat to remember regarding CPs, especially those synthesized via electrochemical or other less well-defined methods, is that structures that one draws in ideal representations, e.g., those in Figs. 27.2 and 27.3, may by no means be an exact or even an approximate representation in many cases. Many CPs are oligomers, with chain lengths ranging from as few as 8 monomer units to 50 units, much less than the typically portrayed 100,000 units of some poly(anilines), for instance. Polymer coupling is by no means head to tail as in ideal representations: there can be many branchings, e.g., 1,2-couplings in place of 1,4-couplings in phenyl rings, etc. Many oxidation centers, e.g., -C=O bonds, exist in the CP chain due to the effects of the environment or similar factors; some of these have been well characterized, for instance, for poly(pyrrole) [852, 853]. Doping may by no means be uniform; for example, if one chemically analyzes a 25% doping level, this by no means implies that for every four monomer units, there is one dopant molecule all along the polymer. Rather, there may be islands of very high doping level followed by polymer segments with no doping at all, as has been found for many CPs.

Even for the best CPs, morphology is poor, nowhere, for instance, approaching that routinely obtainable in common organic polymers such as poly(ethylene). If fibrillar, fibrils may be of widely varying length and width and may be highly disconnected. If globular, large void spaces may be routinely present.

Thus, the study of CPs becomes rather inexact in many instances, especially when highly defined theoretical models, e.g., of conductivity, are sought to be applied to them. In some instances, even theoretical models of "disordered" materials from condensed matter physics fail to be applied successfully to CPs. A good rule of thumb however is that if one pays consistent attention to the *bulk* properties of the material, and to the *reproducibility* of these properties, one is generally in good shape.

27.4 Basics of CP Synthesis

27.4.1 Categories and Classes of Syntheses

CP syntheses fall broadly into the two categories one recollects from elementary polymer chemistry or organic chemistry courses:

- *Condensation* polymerization (also sometimes called *step-growth* polymerization)
- *Addition* polymerization (also sometimes called *chain-growth* polymerization)

Syntheses also fall broadly into two other classes:

- *Chemical*
- *Electrochemical*

Nearly all electrochemical syntheses are *addition polymerizations*, and many chemical syntheses are *condensation polymerizations*.

Condensation polymerizations generally involve loss or elimination of a chemical species resulting from the reaction of end groups on monomer chains, the most famous example of course being the elimination of water from an alcohol and an acid group, yielding poly(esters):

$$..x\text{HO-R-R-OH} + x\text{HOOC-R}'\text{-R}'\text{-COOH}$$

$$\rightarrow \text{H-}\left(\text{O-R-R-OOC-R}'\text{-R}'\text{-CO}\right)_x\text{-OH} + (2x-1)\text{H}_2\text{O} \qquad (27.5)$$

Common plastics such as poly(carbonates) and poly(urethanes) are generally produced in this fashion. Addition polymerizations involve the well-known chain initiation, chain propagation, and chain termination steps, with chain initiation generally being through generation of a highly reactive, radical ion. Other common plastics such as poly(ethylene), poly(styrene), and the vinyl plastics are generally produced in this fashion.

27.4.2   Representative Syntheses: Chemical

In this chapter, several representative syntheses are presented as an introduction and to yield a feel for the "nuts and bolts" and "how to" of CP syntheses. A more thorough discussion of syntheses and polymerization mechanisms is presented elsewhere in this book.

The first CP syntheses, e.g., of poly(acetylene) or poly(p-phenylene sulfide), were chemical syntheses. From the descriptions, we will give in this book, it may be evident that chemical syntheses are simply adaptations of synthetic procedures well established in organic chemistry.

Poly(thiophenes)   The chemical synthesis of poly(thiophene) represents a typical condensation polymerization, employing a Ni catalyst:

$$X=\text{Br, Cl; dppp}= 1,3\text{-bis(diphenyl phopsphino)propane} \qquad (27.6)$$

Poly(acetylenes)   The most common chemical synthesis of poly(acetylene), as first effected by Shirakawa, involves use of a procedure common in the polymer world, viz., stereospecific Ziegler–Natta polymerization. In a typical procedure, a toluene slurry of AlEt_3 and Ti(OBu)_4 (4:1) is used to coat a reaction vessel (e.g., a Schlenk tube). Acetylene gas is then admitted at pressures ranging from 2 cm to 76 cm Hg. A well-formed poly(acetylene) film starts growing in a few seconds, with polymerization generally carried out to an hour. The polymer films are washed in solvent of polymerization, and dried, when they can be peeled off to yield free standing films, with thicknesses up to 0.5 cm.

A more refined poly(acetylene) polymerization which we may cite, one yielding a higher quality of polymer with higher conductivity and better morphology, is an example of one type of "precursor polymerization"; in this the monomer corresponding to the CP is not the starting substance. This is the polymerization known as the "Durham/Feast" route. The starting substance is a cyclobutene, 7,8-bis(trifluoromethyl)-tricyclo^{4,2,2,0}-deca-3,7,9-triene, which yields a *soluble precursor polymer* with the use of Ziegler–Natta catalysts such as $WCl_6:(C_6H_5)_4Sn$ and $TiCl_4:(Et))_3Al$, via ring-opening metathesis. This then undergoes an elimination, losing hexafluoroxylene, to yield poly(acetylene) (Eq. 27.7). A soluble precursor polymer is of importance because it allows preliminary processing of the normally intractable CP, e.g., casting of films or spinning of fibers, prior to preparation of the final polymer.

$$(27.7)$$

Poly(phenylene Sulfide) The common synthesis of poly(phenylene sulfide) is also a typical and simple example of a step-growth, condensation polymerization:

$$(27.8)$$

$$X = S, Se, Te$$

$$Y = Cl, Br, I$$

Poly(aniline) and Poly(pyrrole) Typical, simple, "in-a-beaker" syntheses of poly (pyrrole) and poly(aniline) may now be illustratively described. Both involve use of a *chemical oxidant to generate a radical cation* from the monomer, which initiates the polymerization. In the case of poly(pyrrole), 2.5 M of $FeCl_3$, the oxidant, is taken in MeOH. Pyrrole (liquid) is then added slowly with stirring and cooling, in a ca. 2:5 pyrrole/$FeCl_3$ molar ratio. The polymer so obtained is filtered, washed, and dried. In the case of poly(aniline), aniline (liquid) is taken either neat or on a substrate, and mixed with oxidant solution, which is typically 0.1 M ammonium peroxydisulfate in 2 M HCl. The solution is cooled in an ice bath to 0 °C, while stirring for several hours. The polymer obtained is washed, usually with MeOH, and dried. Variations of even these simple procedures involve use of other solvent media, e.g., acetonitrile/water mixtures, and other oxidants, e.g., $CuBF_4$.

27.4.3 Representative Syntheses: Electrochemical

Electrochemical polymerizations, in a similarity with chemical polymerizations, also use an initial electrochemical step, generally *oxidation* via *an applied potential, to generate the radical ion,* which then initiates the polymerization.

Typical, simple examples for poly(pyrrole) and poly(aniline) may again be illustrated. For the former, 0.05 M pyrrole monomer is taken, after molecular sieve/alumina purification, in 0.2 M $LiClO_4$ or Et_4NClO_4 in acetonitrile in an electrochemical cell. A three-electrode (working, counter, reference electrodes) or a two-electrode (working, counter) mode may be used. The counter electrode may be Pt, graphite, or a number of other materials, and the working electrode choice is dependent upon the end use for the polymer, for example, Au for reflectance studies, graphite for bulk CP powder, and a transparent electrode (in acetonitrile only) for spectral studies. For the former, an applied potential of +0.8 V vs. Ag/AgCl and for the latter +1.4 V are adequate for polymerization. The polymer, obtained on the working electrode, is washed in the solvent of polymerization and dried. In the case of poly(aniline), a 0.05 M aniline solution in 0.2 M aqueous H_2SO_4 also containing an additional dopant, e.g., Cl^-, if desired, is taken in the electrochemical cell. Polymerization is carried out at ca. +0.5 V vs. Ag/AgCl in three-electrode mode or +1.0 V in two-electrode mode.

The electrochemical polymerization has, in particular, become a widely used procedure for quick generation of CP when well-defined polymer structure or morphology is not imperative. Indeed, the all-pervasive paradigm applied to Ziegler–Natta polymerization, viz., "any metal alkyl, any transition metal halide and any monomer, when mixed together, will invariably yield stereospecific polymerization," can be adapted to electrochemical preparation of CPs; thus, "any monomer with a suitable ionization potential and when taken with any dopant in a suitable solvent will yield a polymer with some conductivity when a suitable electrochemical potential is applied."

27.4.4 Simple Representation of Mechanisms

To illustrate mechanisms of the polymerizations in a simple fashion, the polymerizations described above for poly(aniline) and poly(pyrrole) can be considered. *In both the chemical and the electrochemical cases, the initial step is the generation of the radical cation*, as illustrated for pyrrole polymerization in Eq. (27.9) below.

$$-\,e^-\qquad(27.9)$$

It is in the next step then that the chemical and electrochemical polymerizations differ, according to most studies [852, 853]. In the chemical case, the radical cation then attacks another monomer molecule, which are in abundance in the reaction sphere, generating a dimer radical cation. The polymer chain propagates in this fashion until termination (Eq. (27.10). In the electrochemical case, the concentration of radical cations is much larger than that of neutral monomers in the vicinity of the electrode where reactions are occurring, and radical–radical coupling leads to a radical dication. This then loses two protons, generating a neutral dimer, which is then oxidized to a radical cation, the polymerization thus progressing in this fashion to completion (Eq. 27.11). The scheme in Eq. (27.11), especially its features of radical–radical coupling and proton elimination, has been established from studies on varied CPs [852, 853] and is widely accepted as applicable to most electrochemical polymerizations generally.

$$(27.10)$$

Polymer $\longleftarrow$ radical dimer

radical dication

$$-2\mathrm{H}^+\qquad(27.11)$$

Polymer $\longleftarrow$ $\longleftarrow$ $\overset{-e^-}{\longleftarrow}$ dimer

During electrochemical polymerization, the current follows a typical catalytic pattern, and there is usually a small "gestation" period, for buildup of radical concentration, or in some cases of dimers and trimers, before the polymerization proceeds apace.

27.5 Problems and Exercises

1. Draw structures of poly(pyrrole) and poly(thiophene) showing 25% doping with ClO_4^- dopant and of poly(p-phenylene) with 33% Li^+ doping.
2. Define the following terms: conjugation, delocalization, semiconductor, insulator, and conductor. Draw in rough schematic a band diagram for each of the latter three.
3. Define the terms resistivity and conductivity, and list typical approximate conductivities, in proper units, of the following: Ag, Cu, Ge, a CP, and poly (propylene).
4. Draw in rough schematic the structures of poly(trans-acetylene), poly(aniline), and poly(p-phenylene vinylene).
5. Define the following terms in the context of CPs: dopant, doping, doping level, p- and n-doping, and pristine. List several common cationic and anionic dopants.
6. Enunciate the categories and classes of polymerization in the context of CPs.
7. Draw a schematic flow chart for chemical and electrochemical polymerization for 3-methyl thiophene, clearly noting the commonalities and differences in the mechanistics of the two.

Chapter 28
Conduction Models and Electronic Structure of CPs

Contents

28.1 Conventional Semiconductors and CPs

28.1.1 Conventional Semiconductors

Models from condensed matter (solid-state) physics for the treatment of semiconductors are the closest for arriving at a generalized understanding of CPs and thus have been adopted in broad measure by workers in the CP field. This of course includes conduction models. When one thinks of a semiconductor, one generally visualizes an inorganic semiconductor such as doped Si or Ge, widely used in electronics, or GaAs or CdSe, used for more specialized applications. These "conventional" semiconductors can for the most part be described well using the

© Springer International Publishing AG 2018
P. Chandrasekhar, *Conducting Polymers, Fundamentals and Applications*,
https://doi.org/10.1007/978-3-319-69378-1_28

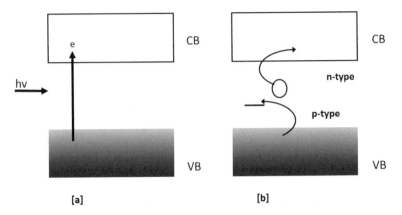

Fig. 28.1 Photoexcitation and doping in "conventional" semiconductor

semiconductor band representations discussed in an earlier chapter, possessing broad "valence" and "conduction" bands (cf. Fig. 1.5).

Removal of electrons in these conventional semiconductors can be accomplished by photoexcitation, as shown schematically in Fig. 28.1a. Removal or addition of electrons may also be accomplished through the introduction of impurities, known as "dopants." An impurity may contribute an electron to the conduction band, giving rise to electrical conduction through electrons in this band, and is then known as an "n-dopant." It may also abstract an electron from the valence band, giving rise to electrical conduction through "holes" in the valence band, which is then known as "p-doping." These processes are depicted in rough schematic in Fig. 28.1b. The electrons and holes are said to be the "charge carriers" for electrical conduction in these materials. Dopants may typically be elements from groups one position to the left or right of the central semiconductor atom in the periodic table, for instance, B for p-doping of Si and P for n-doping of Si.

An important point to note about these conventional semiconductors is that they generally have a rigid structure stemming from fourfold or higher coordination of the central metal atom. Removal or addition of electrons in such materials, whether by photoexcitation, electrochemistry, or some other means, leads to charged, ionized, or excited states which have nearly the same structure and geometry as the rest or ground states and in which the charge is usually delocalized over the entire material. These materials are treatable within the framework of what condensed matter scientists call "rigid band theory" or small deviations therefrom.

28.1.2   CPs as Semiconductors

The organic materials under discussion in this work, i.e., CPs, also develop broad valence and conduction bands stemming from extended overlap of the π-orbitals

Fig. 28.2 Ionization of an
organic molecule according
to standard Franck–Condon
treatment

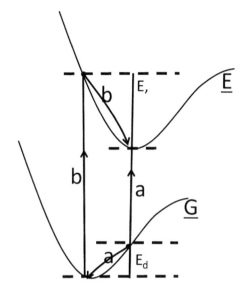

intrinsic to these systems, as described in an earlier chapter. The valence and
conduction bands can be broadly classified as π and π^* orbital bands, respectively.
As an important difference, however, the polymeric nature and twofold coordination of these organic systems allow for much greater flexibility and *susceptibility to
structural distortion*. The introduction of charges then leads to a *relaxation* or
structural distortion of the polymer structure about the charge, which is energetically favorable, and functions to stabilize the charge. These relaxations are intrinsic
to the development of states such as bipolarons discussed in the sequel. We note
also that in general these relaxations, for instance, when following photoexcitation,
are extremely rapid, of the order of 10^{-13} s (0.1 ps). They can give rise to other
interesting properties of the polymers, such as nonlinear optical effects, discussed
subsequently in this book.

The effects of these structural relaxations can best be visualized by first considering ionization in an organic molecule according to the standard Franck–Condon
treatment, as depicted in Fig. 28.2. In this, the rest or ground state, G, and the
ionized or excited state, E, have, according to convention, different geometries (i.e.,
a distortion of the geometry of E as compared to that of G), indicated by displaced
energy minima. Ionization can be thought of as occurring either via an initial
distortion of the ground state geometry to that of the excited state, followed by
vertical excitation of the electron (Path A), or it can be considered to occur via
Path B, i.e., a standard, vertical Franck–Condon excitation within the geometry of
the ground state, followed by relaxation to the equilibrium excited state geometry.
If the relaxation energy, Er, is greater than the distortion energy, Ed, then Path B is
preferred for ionization. The distortion in the ground state also causes shifting of the
highest occupied molecular orbital (HOMO) upward and of the lowest unoccupied
molecular orbital downward, as shown in Fig. 28.3.

Fig. 28.3 Photoionization
effects on HOMO, LUMO,
and E_F

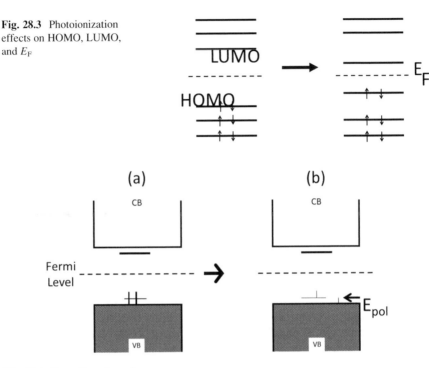

Fig. 28.4 Formation of a polaron

28.2   Structural Distortions: Polarons, Bipolarons, Solitons, and Excitons

A number of terms come up in relation to conduction models in CPs, including
polaron, bipolaron, soliton, and *exciton*. To understand these, let us briefly expand
on the discussion above, and consider a similar model as discussed above, but with
local, geometric distortion of the ground state in an organic semiconductor, i.e., a
CP, possessing a standard semiconductor band structure. The localized distortion,
requiring a distortion energy E_{dis}, gives rise to localized electronic states in the gap
region (the region between the valence and conduction bands), as depicted in
Fig. 28.4a, (where E_F, the *Fermi level*, is just an equilibrium energy level
corresponding to the chemical potential of the system): just as in the organic
molecule above, here too a small, local upward shift of the HOMO (i.e., the valence
band) and downward shift of the LUMO (i.e., the conduction band) occur, as
depicted in Fig. 28.3. If one were now to remove an electron from this localized
state, i.e., oxidize or "p-dope" the CP, the band structure would now look something
like that in Fig. 28.4b. It is clearly seen that the ionization energy, i.e., that required
to remove an electron from the structure in (a), is less than that required to remove
an electron from the valence band per se, by a quantity E_{pol}, as shown. If now E_{pol}
is greater than E_{dis}, the formation of the entity in (b) is favored.

The entity in Fig. 28.4b is called a *polaron*, a term again borrowed from condensed matter physics. The polaron as depicted in Fig. 28.4b is then just a *radical cation (one unpaired electron), which is locally associated with a structural distortion* in the CP. Put another way, a polaron is a charge in an extended lattice which is stabilized by a local distortion of that lattice.

We have, for argument's sake, shown the formation of a polaron occurring first through local distortion of the CP structure *followed* by removal of an electron. In practice, of course, the process is more likely to occur in the reverse direction, i.e., removal of an electron from the CP structure *causes* a local distortion in the structure which then serves to stabilize the charge associated with the electron removal. This distortion of the local structure by a charge in an extended lattice such as a CP chain is said in condensed matter terminology to arise from strong electron–phonon coupling, a phonon being a lattice vibration or "solid-state sound wave." The polaron is said to possess a binding energy, (which is just $E_{pol}-E_{dis}$).

It is important to note that because in polaron formation the conduction band remains empty and the valence band remains full, there is no resulting high conductivity – at least not as yet – as in the case of electron removal from the valence band in a "conventional" semiconductor. The newly generated, half-occupied, electronic level remains quite localized in the gap.

If we now look at removal of a second electron from the CP structure, we might consider that it would most likely occur through formation of another polaron at some other location in the CP chain, giving rise to two polarons. The second electron however could also be removed from the polaron itself, generating a *bipolaron*, and we need to decide whether this or the formation of two polarons is more favorable. If we revert to the lattice distortion argument, we note that the lattice distortion required to generate a bipolaron is likely to be much *larger* than that for a polaron, with correspondingly greater shifts in the localized HOMO (valence band) and LUMO (conduction band) states, as depicted in the hypothetical states of Fig. 28.5a.

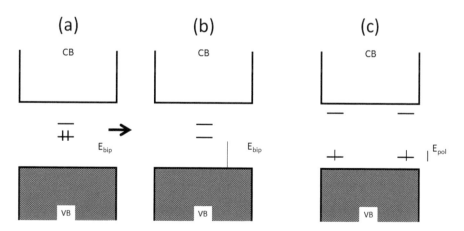

Fig. 28.5 Formation of a bipolaron

Let us now further proceed to the removal of two electrons from this state, generating the state in Fig. 28.5b, i.e., the actual bipolaron, and compare it with the representation for two polarons shown in Fig. 28.5c. The reduction in ionization energy for removal of two electrons in the formation of two polarons is $2E_{pol}$ (Fig. 28.5c), while that for removal of two electrons in the formation of a bipolaron is $2E_{bipol}$ (Fig. 28.5b), which is clearly larger. It has been found through extensive theoretical as well as experimental studies that the distortion energy, E_{dis} required to generate polaronic and bipolaronic distortions, is substantially the same. Thus, the bipolaron in such extended structures is significantly more stable than two polarons. This stability, despite obvious coulomb repulsion between two positive charges, can be considered to arise from coupling of the two charges to the lattice, via lattice vibrations (the electron–phonon coupling described earlier) in a fashion similar to Cooper pairs, i.e., two electrons coupled to a lattice vibration (phonon), in the older, BCS, theory of superconductivity. Again, we note that removal of electrons can be achieved through a number of methods, one of them being "doping" with a counterion, which in the case of CPs is really a redox process slightly different from doping in semiconductors in the conventional sense. At high doping levels, coulomb repulsion between the like charges of a bipolaron is also screened by the dopants, further contributing to their stability.

We have considered above the cases for generation of positively charged polarons and bipolarons, i.e., "p-doping" or oxidation, with an anion accepting the removed electrons from the CP chain. We may also of course have "n-doping," i.e., reduction or donation of electrons to the CP chain. We would then have negatively charged polarons and bipolarons, depicted in Fig. 28.6a, b, respectively. It is to be noted that apart from these positively and negatively charged polarons and bipolarons, the neutral polarons and bipolarons are just the corresponding structural distortions before removal of electrons and can be said to exist only hypothetically.

Another important feature to note regarding the bipolaron levels in particular is that they are either empty (p-type doping) or fully occupied (n-type doping) and thus *spinless*. An important reason why a model different from that of "conventional" semiconductors was sought for CPs is that in highly doped CP samples, which had high intrinsic conductivity, there was no evidence for unpaired electrons

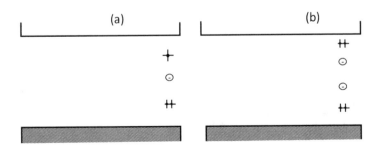

Fig. 28.6 Negatively charged polarons and bipolarons

from experiments such as electron spin resonance (esr) measurements, or correlation of conductivity with esr absorption, but rather, spinless charge carriers were indicated [853].

Figure 28.7 depicts some examples of actual structures that polarons and bipolarons represent in CP chains, illustrated for poly(*p*-phenylene) and poly (pyrrole). More such structures are presented further below. It can be seen that the region of the polaron/bipolaron in general represents a domain in which the normal, aromatic structure of an uncharged CP is interrupted in a significant way: for many cases, this aromatic-ring, *benzenoid* (or benzoid)-type structure of the CP is replaced in this region by a *quinonoid* (or quinoid) type of structure, as seen quite clearly in the figures.

Yet other features of such entities as bipolarons, observable from these chain structures, are that they can be *mobile* and can propagate along the CP chain. Thus, at sufficient concentration levels, they can function as charge carriers: They are essentially mobile charged structural distortions. It is also evident from the structures and the argument of mobility that the charges, e.g., the two charges in a bipolaron, are not isolated on specific monomer units but, rather, are delocalized over several (generally six to eight) monomer units, but *not* the entire CP lattice.

In general, for most CPs, two extreme geometric structures are possible, an entirely benzenoid one and an entirely quinonoid one, with the quinonoid structure generally possessing a slightly higher energy, and thus being less favorable, than the benzenoid structure. This is shown schematically in the energy diagram for poly (*p*-phenylene) (P(PPP)) and poly(acetylene) (P(Ac)) in Fig. 28.8a. For CPs, such as poly(isothianaphthene), due to the participation of an additional aromatic ring not part of the CP backbone, the energy difference between the two structures is less (Fig. 28.8b). The quinonoid structure in most CPs also possesses a lower ionization energy and a higher electron affinity and is thus favored on p-doping.

The terms benzenoid (or benzoid) and quinonoid (or quinoid) are encountered frequently in CP literature, in relation to such aspects as stability and effect (usually lowering) of bandgap. Another term frequently encountered, usually in theoretical discussions of stability and bandgap, is the degree of *bond length alternation*. This is frequently given a quantitative expression. It is defined as the difference, Δr, not just between double and single bonds in a CP chain but more precisely between bonds approximately parallel to the CP chain and the adjoining bonds at some angle, usually close to the perpendicular, to the chain. It is evident from Fig. 28.8a, b then that this quantity Δr is a measure, of sorts, of the relative degree of benzenoid (aromatic) or quinonoid character in a CP: $\Delta r < 0$ implies a benzenoid structure and $\Delta r > 0$ implies a quinonoid structure.

In trans-poly(acetylene), however, there exist two equivalents, i.e., degenerate, ground-state structures, differing only in the alternation of double and single bonds, with two equivalent-energy potential wells, as shown schematically in Fig. 28.9. Consequently, when cationic bipolarons, i.e., two positive charges, are generated in poly(acetylene), they can readily separate, as shown schematically in Fig. 28.10. In this structure, it can be seen that the structural element on one side of a charge

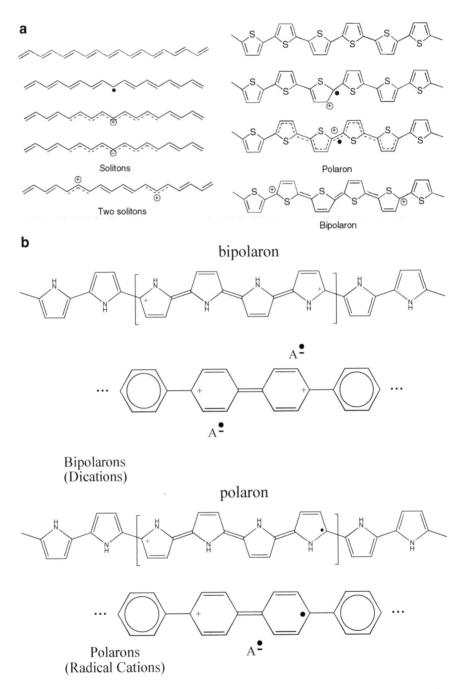

Fig. 28.7 (**a**) Illustration of solitons, polarons, and bipolarons in poly(acetylene) and poly(thiophene) (After Ref. [854], reproduced with permission) (**b**) Actual structures of polarons/bipolarons in poly(*para*-phenylene) and poly(pyrrole) (P(Py))

Fig. 28.8 *Top*: Benzenoid/ quinonoid structures for poly(*p*-phenylene), together with energy diagram. *Bottom*: for poly (isothianaphthene)

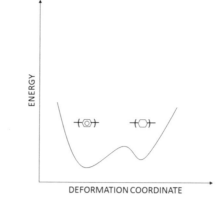

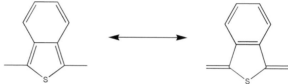

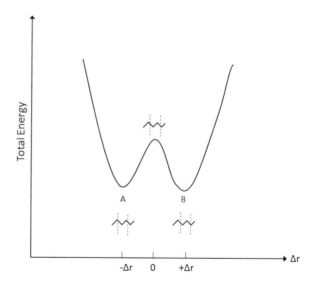

Fig. 28.9 Soliton potential well in poly(acetylene)

differs from that on the other simply by single/double bond alternation. In condensed matter physics, a charge of this nature which forms a domain wall, separating two different (here, degenerate) structural domains, is known as a *soliton*.

In a CP, such as trans-poly(acetylene), when there are odd numbers of carbon atoms in the CP chain, there remains an unpaired π-electron, and a radical, *neutral* soliton is said to result. There are also positively charged and negatively charged

Fig. 28.10 Formation of
two poly(acetylene) solitons
from bipolaron in poly-
(acetylene)

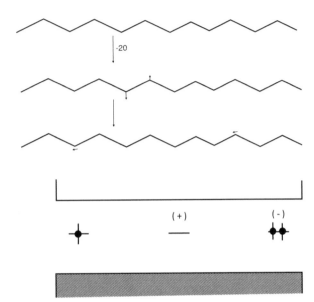

Fig. 28.11 Neutral,
negatively, and positively
charged solitons, e.g., for
poly(acetylene)

solitons. The schematic band structures for neutral, positive, and negative solitons are shown in Fig. 28.11. Just as discussed above for bipolarons and polarons, the soliton also is delocalized over several monomer units. The bond alternation change from one domain to another is not abrupt, as in the schematic figures above, but rather, gradual; as one progresses from one end of the soliton to the other, double bonds gradually lengthen until they reach single bond length and vice versa for single bonds, with near-equal bond lengths in the center of the soliton. The unique charge-spin relations of solitons are also again to be remembered: a neutral soliton has no charge but possesses spin; both positively and negatively charged solitons possess charge but no spin.

Finally, we are now in a position to look in more detail at how polarons, bipolarons, and solitons may actually appear in CP chains (Figs. 28.12, 28.13, and 28.14). In all cases, one may assume, based on many theoretical and experimental studies, that these entities are generally confined to, and delocalized over, about four to six monomer units. Furthermore, one must also visualize an appropriate counterion in the vicinity of the entity for charge neutrality.

For poly(p-phenylene) (P(PPP)) (Fig. 28.12), a polaron and a bipolaron are the simplest constructs. Among the former, one may have a positively charged polaron (a radical cation, (a), also called a p-type polaron or hole polaron) and a negatively charged polaron (a radical anion, (b), also called an n-type polaron or electron polaron); both these will also possess spin. One may also have a soliton (a cation, (c), without spin). Among bipolarons, one may have a positively charged bipolaron (p-type, hole bipolaron, dication, (d)) and a negatively charged polaron (n-type, electron bipolaron, dianion, (e)).

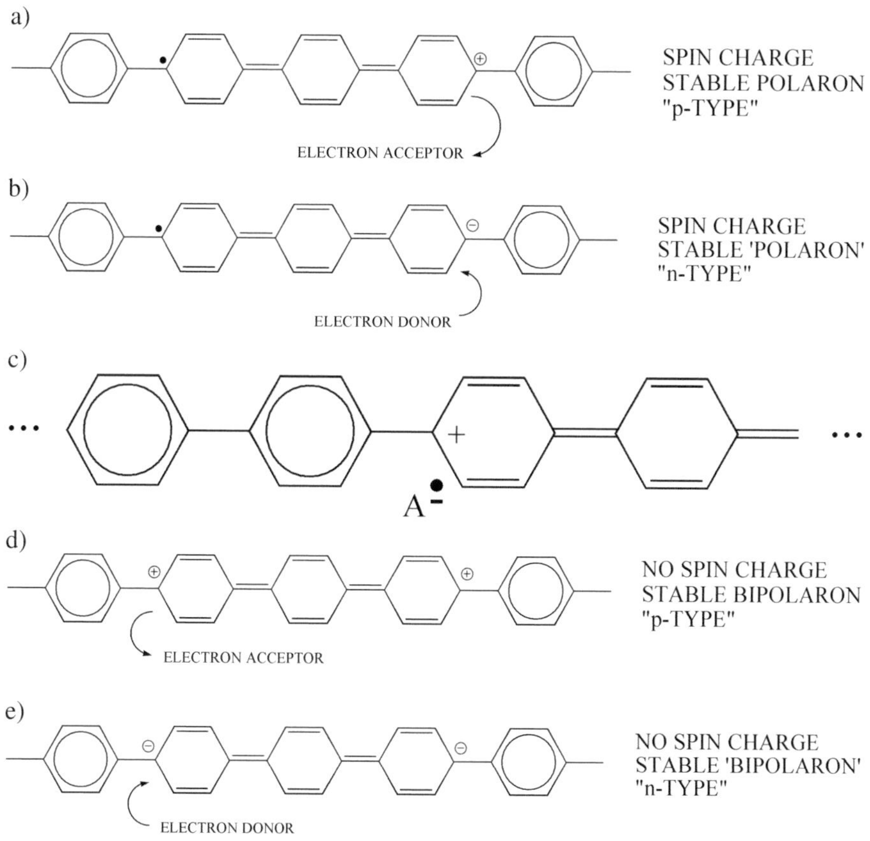

a)

SPIN CHARGE
STABLE POLARON
"p-TYPE"

ELECTRON ACCEPTOR

b)

SPIN CHARGE
STABLE 'POLARON'
"n-TYPE"

ELECTRON DONOR

c)

A•−

d)

NO SPIN CHARGE
STABLE BIPOLARON
"p-TYPE"

ELECTRON ACCEPTOR

e)

NO SPIN CHARGE
STABLE 'BIPOLARON'
"n-TYPE"

ELECTRON DONOR

Fig. 28.12 Actual structures of polarons/bipolarons in poly(p-phenylene)

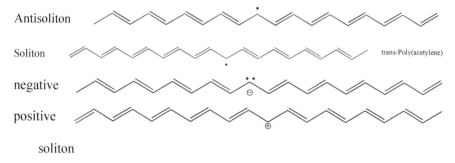

Antisoliton

Soliton   trans-Poly(acetylene)

negative

positive

soliton

Fig. 28.13 Actual soliton structures

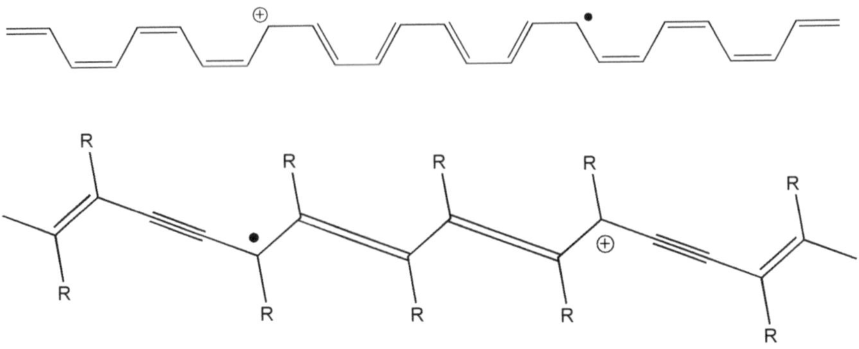

Fig. 28.14 Actual polaron structures in *trans*-poly(acetylene), *cis*-poly(acetylene), poly (diacetylenes)

For poly(pyrrole), positively charged polarons and bipolarons have already been illustrated above, (Fig. 28.7). Negatively charged polarons and bipolarons can be visualized which are analogous to those of poly(*p*-phenylene), but in practice they do not exist, as poly(pyrrole) cannot be n-doped stably. For trans-poly(acetylene) (P (Ac)) (Fig. 28.13), one may have a neutral soliton and "antisoliton" possessing spin (a) and spinless positive and negative solitons (b); two adjacent, positive solitons would be the equivalent of a bipolaron, which for reasons cited earlier degenerates into two solitons in this polymer. For cis-poly(acetylene) and poly(diacetylenes), one may have a positive (hole) polaron (Fig. 28.14).

Now electrons may exist in LUMO (lowest unoccupied molecular orbital) states, and holes may exist in HOMO (highest occupied molecular orbital) states; their *neutral bound* states are designated *excitons*. Although, technically, an exciton is any excitation above the ground state of a molecule, excitons are bound states confined well within an "effective Bohr radius" [855].

As an example of the practical application of excitons, for the phenomenon of electroluminescence seen in organic light-emitting diodes (OLEDs), free electrons and holes are injected from opposite contacts of the OLED. The electrons and holes then meet each other via diffusion, thus forming the excitons [855].

28.3 Band Structure Evolution

It is instructive at this stage to follow the evolution of the CP band structure as one progresses from an undoped, (i.e., unoxidized/unreduced and neutral), structure, to one of high doping. We take the specific example of poly(pyrrole), and follow the treatment (Fig. 28.15), based on semiempirical theoretical and experimental studies on this CP, first lucidly presented by Bredas, Street, and coworkers [856, 857]. At a hypothetical zero doping level, the polymer is neutral, and its band structure is that of a standard semiconductor, with a bandgap in the case of poly(pyrrole) of 3.2 eV,

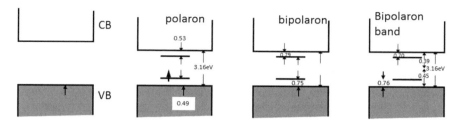

Fig. 28.15 Band structure evolution for poly(pyrrole) (P(Py)). The evolution from (**a**–**d**) is with progressively increasing doping

(Fig. 28.15a). Removal of one electron from the CP chain produces a polaron, as shown in (b) of the figure, with the two polaron levels about 0.5 eV from the valence and conduction band edges, according to the approximate calculations of Bredas et al. [856, 857]; these calculations also show that the distortion energy for polaron formation is ca. 0.37 eV, so that the net polaron binding energy is 0.13 eV. Since we have a species with a single charge, the lower bipolaron level is half-occupied, as shown, and the species would possess spin. The polaron has a structure like that depicted in Fig. 28.7 above, i.e., with the single positive charge delocalized over about four to six monomer units.

On removal of an additional electron, a bipolaron is produced, as shown in (c) of the figure, with now the bipolaron levels being further removed from the valence/conduction bands, as discussed earlier in this chapter. Bredas et al.'s approximate calculations show that the bipolaron levels are about 0.75 eV from the band edges. Calculations of the distortion energy associated with bipolaron formation show that the net bipolaron binding energy is ca. 0.7 eV, so that bipolaron formation is favored over formation of two polarons by ca. $(0.7\,\text{eV} - (2 \times 0.13\,\text{eV})) = 0.44\,\text{eV}$. Despite this, however, it is important to note that in an actual CP structure, the entire CP chain would first have to become nearly saturated with polarons before bipolaron formation would commence. This is also supported by spectroelectrochemical studies, as discussed below: at low doping levels, spectra corresponding to polarons are first seen, before bipolaron spectra occur.

As one progresses, further to very high doping levels (a maximum of about 35% is experimentally achievable in poly(pyrrole)), the individual bipolaron states of Fig. 28.15c coalesce into bipolaron *bands*, as shown schematically in (d) of the figure. Calculations by Bredas et al. and other groups have shown that for poly (pyrrole), the bipolaron bandwidths are 0.4–0.45 eV, as shown in the figure. It is also important to note that these bipolaron bands arise from electronic states "scavenged" from the valence and conduction band edges, and the gap between these bands consequently increases; according to calculations and some experimental data for poly(pyrrole), this bandgap increases from the 3.2 eV quoted above for a nearly neutral polymer to ca. 3.6 eV for a highly doped one.

At these high dopant concentrations, the bipolarons, which are spinless, can become mobile under the application of an electrical field, thus giving rise to the high conductivity observed in CPs concomitant with the absence of unpaired spins detectable by esr or other measurements.

It is also evident from the discussion above that if one were able to achieve even higher doping levels, the two bipolaron bands would gradually broaden enough to coalesce into the valence and conduction bands, producing metallike conduction stemming from the lower, half-filled valence/bipolaron band. This could occur at hypothetical, higher doping levels, near 100%, or if the intrinsic $\pi \rightarrow \pi^*$ (valence $\rightarrow$ conduction) bandgap of the CP was low. This is said not to happen in poly(pyrrole), with a fairly high bandgap (3.2 eV, 388 nm) and maximum experimentally observed doping level of ca. 35%, but evidence has been put forth for its occurrence in other CPs, notably poly(thiophene), which has a lower bandgap (ca. 2.0 eV, 620 nm). This merging of the bipolaron bands with the valence and conduction bands at hypothetical 100% doping levels is shown schematically in Fig. 28.16a–c for poly(pyrrole), poly(p-phenylene), and poly(thiophene). The numbers for the bandwidths and positions in these figures for these other CPs may also be compared to those of poly(pyrrole).

28.4   Densities of States and Wave Vector Representations

At this point, we are in a position to mention two representations of band structures, energy levels and electronic populations found commonly in the CP literature, if only for purposes of knowing their significance.

Schematics of band structures of the type illustrated in the figures above are only *rough* depictions of the population of energy levels, for example, those within the wideband known as the valence band. As may be observed, the illustrations in these schematic figures appear to indicate a homogeneous population distribution within width of the bands, appearing to show, for example, that the bottom of the valence band is as populated as the top. A more precise energy level distribution is however obtained in a representation called a *density of states* distribution, as, for example, illustrated in Fig. 28.17 for poly(pyrrole) (as a function of increasing doping level). Here, the energy levels are plotted as a function of their *relative* population. The bandgap is clearly visible as a region of no electron population. In approximately the center of this gap is the *Fermi level*, reflecting an equilibrium potential that represents the chemical potential of the system and which changes with oxidation state (doping) of the CP. Mid-gap states, such as bipolaron levels, are shown as spikes in the population distribution in this figure.

28.5   Correlation of Optical Spectra to Band Structure

With the above understanding of the band structure and semiconductor model for CPs, we may now turn to an elementary interpretation of the evolution of the spectra of a typical CP as a function of doping level. We first look at the spectra

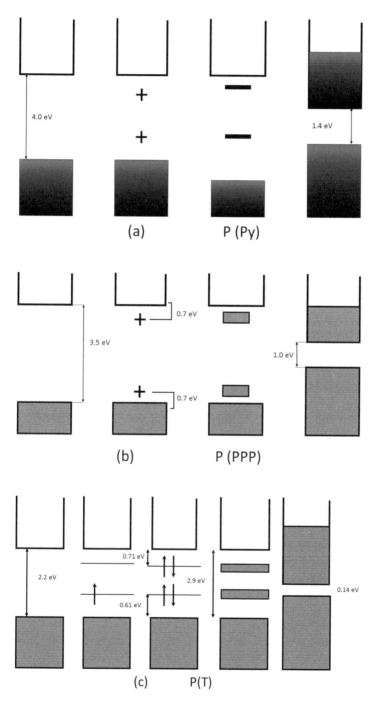

Fig. 28.16 Hypothetical evolution of band structure to metallic state for poly(pyrrole), poly (p-phenylene), and poly(thiophene)

Fig. 28.17 Density of states (DOS) band structure of poly(pyrrole) (P(Py)) (After Ref. [1326]. Reproduced with permission)

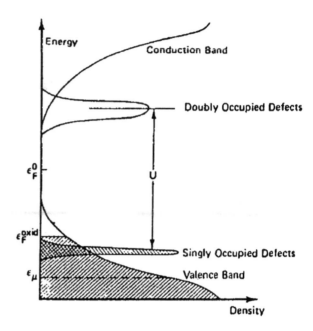

expected of CPs and then turn again to poly(pyrrole) as a prototypical example, following an earlier treatment first presented by Bredas, Street et al. [857].

If one looks at the schematic of the polaron energy level structure for p-type doping, Fig. 28.5c, one can see that there are four optical transitions possible, in decreasing order of energy: (1) from the valence band to the conduction band (the characteristic $\pi \rightarrow \pi^*$ transition), (2) from the valence band to the upper polaron level, (3) from the lower polaron level to the upper polaron level, (4) and from the valence band to the lower polaron level. For the case of bipolarons, however, it is evident from the schematic of Fig. 28.5b that there is one less transition possible: the transition between the bipolaron levels is eliminated. The lowest energy bipolaron transition, that from the valence band to the lower bipolaron level, has also been shown via theoretical and experimental studies to have the highest oscillator strength (i.e., intensity). Thus, as one proceeds with increasing p-doping of a typical CP, whether by electrochemical, chemical, or other means, one should progressively see one large valence $\rightarrow$ conduction band transition diminish and shift to higher energy, with three additional polaron transitions appearing, one of which gradually disappears as doping further progresses to yield two large "bipolaron band" absorptions.

Precisely such a spectral evolution is seen in poly(pyrrole), as first and most famously reported by Bredas, Street, and coworkers [857] and shown in Fig. 28.18, which shows the spectrum presented in a manner more familiar to physicists, in terms of energy (eV) rather than wavelength on the abscissa. The 3.2 eV (388 nm) valence $\rightarrow$ conduction band transition diminishes and shifts to higher energy, ca. 3.6 eV (344 nm) with increasing doping. The mid-energy, 1.4 eV (886 nm)

Fig. 28.18 Evolution of optical absorption spectra of poly(pyrrole) (P(Py)) as function of increased doping (After Ref. [856], Reproduced with permission)

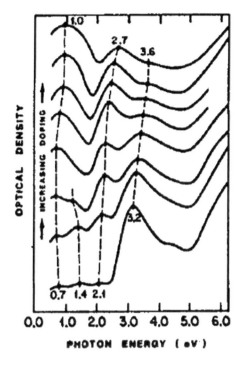

transition, between polaron levels, disappears at fairly low doping levels, as polaron populations gradually give way to bipolarons.

Important corroboration for the correlation of the 1.4 eV (886 nm) absorption with polarons is evidence from esr measurements of the existence of unpaired spins and also of the gradual disappearance of these spins with time (slow decay of polarons to more stable bipolarons) [856].

The lowest energy (0.7 eV, 1771 nm) transition (from the valence band to the lower polaron level) gradually transforms itself to the lowest energy bipolaron transition (from the valence band to the lower bipolaron level), acquires far greater intensity, and shifts to slightly higher energy, in agreement with the higher distortion energy for bipolarons and the schematic of Fig. 28.5. The higher energy polaron transition (from the valence band to the upper polaron level, 2.1 eV, 590 nm) gradually transforms itself to the valence band → upper bipolaron level transition, with intensity lower than the other bipolaron transition (valence → lower bipolaron level). The observed shift to higher energy of this transition for poly (pyrrole) is however not easily explained according to current theory.

With this discussion in mind, we can now examine optical transitions expected for other band structures found in CPs (Fig. 28.19). For neutral, positively, and negatively charged solitons such as those found in poly(acetylene), these are illustrated in Fig. 28.19a. For negatively charged polarons and bipolarons, they are illustrated in Fig. 28.19b, c, respectively.

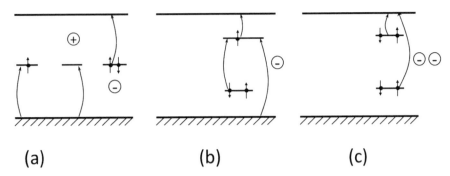

Fig. 28.19 Expected optical transitions for band structures found in CPs. (**a**, **b**), and (**c**) represent, respectively, solitons, polarons, and bipolarons

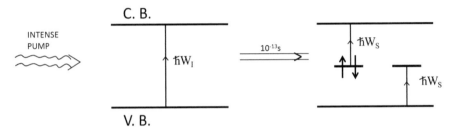

Fig. 28.20 Photogeneration of solitons in poly(acetylene) (P(Ac))

*Polarons, bipolarons, and other structural distortions in CPs, also denoted as mid-gap states, may be most commonly generated by electrochemical or chemical "doping", i.e., oxidation/reduction, with appropriate counterions then taking their place in the vicinity of these for charge neutrality. Another method of generating these mid-gap states is through intense irradiation, e.g., with a laser beam. This is illustrated schematically for the case of soliton generation in poly(acetylene) in Fig. 28.20. Following photoexcitation of an electron or electrons from the valence band to the conduction band, the electron-hole pairs generated relax within a very short time period (ca. 10^{-13} s) to soliton pairs, now with new absorptions, as shown. This process is in fact the source of *resonant* nonlinear optical (NLO) properties in CPs, discussed in a subsequent chapter.

Finally, another term frequently encountered in the CP literature in regard to excitation phenomena is an *exciton*. An exciton is nothing more than the short-lived species generated when excitations, generally an electron and a hole, combine. Depending on their spin characteristics, the excitons may then decay rapidly or with some time delay. They may decay thermally or radiatively. As an example, singlet excitons generated via electron-hole combination in CP devices consisting of sandwiches of CPs decay radiatively, giving out light of wavelength characteristic of the particular CP, the principle behind CP-based LEDs.

28.6   Theoretical and Practical Aspects

Semiconductor models for CPs are among the few theoretical or semiempirical models finding useful practical application to CPs, as attested to by the close correlation of optical spectra with CP semiconductor band structure discussed above. Even these models however encounter some limitations. As one example, the optical spectra of poly(dimethyl pyrrole) and poly(p-phenylene) as a function of increased doping do not show any polaron transitions at all but rather progress straight to the two bipolaron transitions. This is ascribed, with reason, to the immediate coalescing of polaron states to the more stable bipolaron states. However, the fact that this does not occur in poly(pyrrole) is attributed by workers in the field to poly(pyrrole) being more disordered, a somewhat tenuous argument at best.

   Another example of the sometimes poor application of semiconductor models [858, 859] can be found in the electrochemical properties of poly(aromatic amines): these typically show two redox couples with increasing oxidative potential in cyclic voltammetry, ascribed to formation of polaron and bipolaron states, respectively. However, analogous poly(aromatic amines) do not show identical behavior in this respect, and even with an identical poly(aromatic amine) prepared in an identical fashion, the two redox peaks may sometimes merge into one broad peak and sometimes may not, indicating that polarons sometimes degenerate into bipolarons, with no predictability as to when. This behavior may again be ascribed to increased or decreased disorder in the CP chain generating electrochemically inequivalent sites within the CP, but it also shows that CPs still remain generally rather poorly understood chemical and physical systems.

28.6.1   Experimental Measurements Substantiating Conduction Models

One of the first things to take serious note of when attempting a correlation of experimental aspects of conduction in CPs to conduction models is that to obtain a complete picture of conduction behavior; it does not suffice to merely measure DC conductivities. Rather, one must characterize the entire gamut of conduction behavior, i.e.:

- Conductivity as a function of temperature (e.g., temperatures from 500 K through ~0 K)
- Frequency (e.g., from DC through microwave frequencies)
- Magnetic field (whence the Hall effect, magnetoresistance, and Corbino resistance), thermal gradient (whence thermopower)
- CP processing and composition (whence conduction anisotropy, molecular weight, and dopant effects)

Additionally, many supplementary measurements assist in characterizing conductivity behavior and thus help in establishing models. These include, for instance, the complex microwave-region dielectric constants of the material and reflectance measurements from the visible through the far-IR.

It is noted at the outset in this chapter that the above experimental measurements or techniques for all aspects of conductivity are described in some detail elsewhere in this book, and this chapter is confined to a consideration of conduction *models* or, occasionally, a brief synopsis of the measurement technique when this is found necessary for the discussion.

We also note at the outset that while many models have been proposed for conduction in CPs, conduction has in fact been found to be a very complex phenomenon, and to date *no single model is comprehensively accurate*: in varying degrees, the various models are, for instance, able to account for conduction behavior within a specific temperature range or doping range or dopant type but then fail for other ranges or types or for other CPs. In one case, a 3D conduction model may be indicated, whereas in another case, it is not and so on.

Although a variety of measurement techniques have been applied, as noted above, one of the most important has been characterization of the conductivity (commonly DC, less commonly frequency dependent) of CPs as a function of *temperature*. This has been used in a very large number of studies to attempt to validate this or that conduction model, sometimes with a selective or incomplete representation of data.

28.6.2 Nature of Conduction and Relation with CP Morphology

Conduction in CPs, as opposed to that in "conventional" inorganic semiconductors such as doped GaAs or Si, for instance, is often said to be a "redox conduction" rather than an "electronic conduction" or "conventional ohmic conduction," because it is ascribed predominantly to charge carriers, such as bipolarons, rather than to electrons.

This is substantially but not entirely correct for several reasons. Firstly, these bipolarons and other charge carriers are not truly mobile, coasting along from one end of an infinite polymer chain to another, as in the idealized 1D conduction model. Rather, they are localized or confined by features such as defects or discontinuities in the extended conjugation (e.g., an sp^3 defect in the idealized sp^2 extended conjugation of P(Ac) or a cross-link or ortho branch in the idealized head-to-tail chain of P(ANi)) and by attraction to dopant counterions which pin them down.

In fact, considering that entirely defect-free areas in even the most carefully prepared CPs are not thought to be more than ca. 20 nm in length (with ca. 3 nm the more common figure [860], the time taken for a truly mobile charge carrier to traverse this length is in the 10^{-12} s region, and thus such charge carriers would truly contribute to conduction only in the terahertz frequency range. Thus, it is perhaps more accurate to characterize conduction in CPs as hopping of *electrons* between these charged carriers, a basis for many of the proposed "hopping" models of conduction.

For purposes of most discussions then, CPs are considered *disordered semiconductors*. Another interpretation of the defects or breaks in conjugation is to say that scattering of electron transport by the defect site or phonons is occurring. From a band structure point of view, amorphous and crystalline semiconductors may have similar band structures, but in amorphous materials the band edges are smeared out into the bandgap, and charge carrier mobilities are much lower. It may be noted that even the most recent theoretical studies of conduction models in CPs, e.g., that of Laks et al. [861], do not yield a firm, unique conduction model, although they do lend support to disorder being the physical mechanism behind the "insulator–metal" transition in CPs.

The importance of the morphology, including the nanoscopic-scale morphology described above, to conduction can be better visualized from Figs. 28.21 and 28.22 and, to a lesser extent, Fig. 28.23. It is evident from even the simple representation of Fig. 28.21 that, practically, one must take into consideration not only conduction along a CP chain (*1*) but also conduction between chains (*2*), between fibers (or globules for globular morphology) (*3*), and of course, the superposition of *1–3*. Figure 28.23 shows one of the modes of charge hopping between chains, that of bipolarons, as developed by Chance et al. [862]. Typical interchain distances may range from ca. 0.4 nm in the well-characterized trans-P(Ac) to several nm in some cases of loosely packed CPs with poor morphology. Figure 28.21 shows another practical aspect to be considered in conduction, the actual situation obtaining in most doped CPs, that of highly conducting doped islands ("metallic

Fig. 28.21 Schematic of relation of CP morphology to conduction; here showing intra-/interchain and intra-/interfiber relations (After Ref. [865], reproduced with permission)

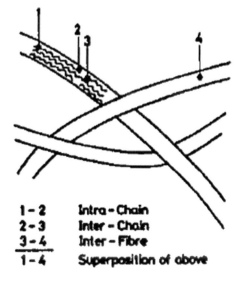

1 - 2	Intra - Chain
2 - 3	Inter - Chain
3 - 4	Inter - Fibre
1 - 4	Superposition of above

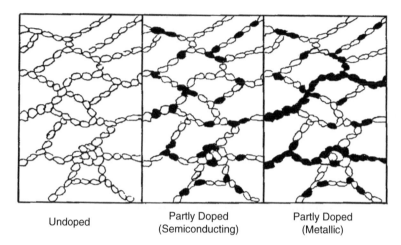

Undoped   Partly Doped   Partly Doped
                  (Semiconducting)   (Metallic)

Fig. 28.22 Schematic of relation of CP morphology to conduction, cont. Here, doped segments (dark) in an interpenetrating or cross-linked CP structure, showing how increased coping finally provides a complete conduction pathway (After Ref. [866], reproduced with permission)

islands") surrounded by insulating undoped regions: Such doping inhomogeneity has been substantiated in experimental studies on many CP systems [863]. Very recent theoretical work using the *Su–Schrieffer–Heeger* (SSH) model has appeared to show that interchain hopping of charged solitons in P(Ac) can occur via formation of pairs, i.e., bipolarons [864].

Most organic CPs have conductivities in the range of 10^{-6} to 10^2 S/cm. Well-prepared samples of CPs such as P(ANi) or P(Py) have conductivities in the 1 to 100 S/cm range; at very low doping levels, their conductivities can be as low as 10^{-6} S/cm. The highest known conductivities are those for "Naarmann-type" (using the Naarmann synthesis) P(Ac) [867], of the order of 10^5 S/cm, one order of magnitude below that of Cu. It is noted in this respect that short conjugation length is one of the key limiting factors for CP conductivity. Indeed, hypothetical computations have shown that conductivities of up to 10^7 S/cm, far exceeding that of Cu, could be obtained if one were truly able to synthesize a CP with extended conjugation along its entire chain [868, 869].

28.6.3   Temperature, Frequency, and Doping Dependencies

Finally, before we may consider conduction models, some essential experimental observations and givens need to be reviewed which deal with the temperature, frequency, and doping dependencies of conductivity.

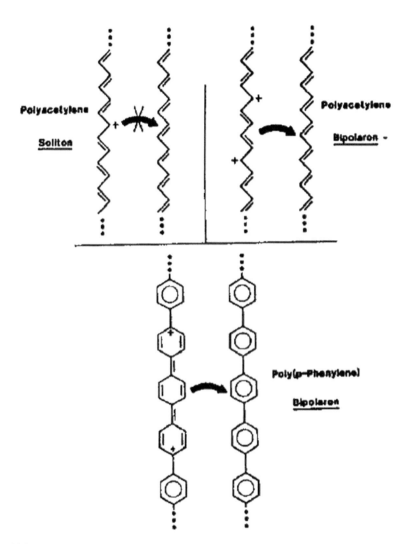

Fig. 28.23 Schematic of relation of CP morphology to conduction, cont. Illustration of interchain transport of solitons and bipolarons (After Ref. [862], reproduced with permission)

28.6.3.1   Temperature Dependencies

If one considers temperature ranges from slightly above ambient (400 K) to near absolute zero (several mK), metals, semiconductors, disordered semiconductors, and insulators have the following dependencies: (a) A metal's DC conductivity increases slightly with decreasing temperature and reaches a finite, limiting value near absolute zero (a caveat in this regard is that many disordered alloys and "dirty" metals sometimes show partial semiconducting behavior). (b) A semiconductor's DC conductivity decreases with decreasing temperature and remains finite at low temperatures; the conductivity of a crystalline semiconductor vanishes exponentially approaching absolute zero, while that of a disordered semiconductor vanishes more slowly (as a function of $T^{-1/n}$, with $n = 2$ to 4) or in some cases may not vanish at all. (c) An insulator's conductivity decreases with decreasing temperature and vanishes at temperatures near absolute zero. Figure 28.24 shows typical temperature dependence of conductivity for P(Ac) at a number of doping levels; it can be seen that the conductivity behavior appears quite different at different doping levels.

28.6.3.2   Frequency Dependencies

With respect to frequency, the following dependencies may be observed: (a) A metal's conductivity is independent of frequency at least up to frequencies of about 10 terahertz. (b) Conductivity of a crystalline semiconductor is independent of frequency across a large frequency range in which its carriers are capable of being excited across its bandgap. (c) Conductivity of a disordered, inhomogeneous semiconductor is substantially dependent on frequency and is generally higher at higher frequencies; indeed a marked frequency dependence is a characteristic of a disordered semiconductor. Again in this respect, CPs may be best considered as disordered semiconductors. Simply put then, DC (i.e., zero frequency) conductivity is the most "difficult" conductivity: for conductivity at high frequencies, charge carriers must hop or otherwise propagate between closely spaced sites only, possible when there are highly conducting metallic "islands," but for DC conductivity, charge carriers must "make it through" the entire sample, including insulating areas.

28.6.3.3   Doping Dependencies

For most CPs, conductivity increases dramatically with increasing doping across certain critical doping level ranges. These levels, sometimes also denoted doping percolation thresholds, can be very small, e.g., 1– 6% for appreciable conductivity to be observed. Figures 28.24 and 28.25 show conductivity vs. doping data for several CPs. Doping can be effected chemically, e.g., during CP synthesis, or, of course, electrochemically, by application of potential. The dramatic increases in conductivity within a narrow applied potential range have been documented for several CPs. Figure 28.26 illustrates the relation of conductivity vs. applied potential for P(3MT), showing the nearly 10 orders of magnitude variation over a narrow potential range. The conductivity is generally found to follow an exponential relation as a function of applied potential [870] of the form:

$$\sigma(E) = \sigma_0 \exp\{(E - E_0)/s\} \tag{28.1}$$

where σ_0 is the conductivity at a reference potential E_0, and s is the slope of a $\log(I_s)$

Fig. 28.24 Relation of doping level with conductivity for poly (acetylene) for the dopants AsF6(A), I2(B), and Br2 (C) (After Refs. [871, 872] reproduced with permission)

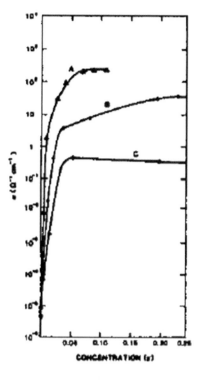

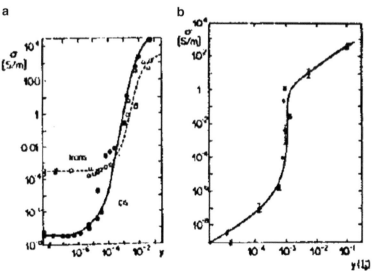

Fig. 28.25 (**a**, **b**) Relation of doping level with conductivity for (**a**) poly(acetylene), (**b**) poly (thiophene), both I2-doping (After Ref. [872], reproduced with permission)

Fig. 28.26 Relation of conductivity and applied potential for a poly (3-methyl thiophene) film, showing its nearly ten orders of magnitude variation (After Ref. [873], reproduced with permission)

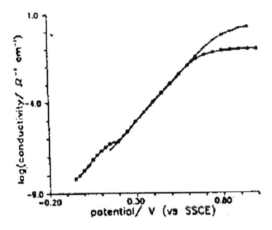

(I_s = steady state current at potential E) vs. E plot.

28.6.4   Practical Aspects

One may finally pose a question as to the importance of presenting conduction models in a purportedly practical book. The answer is that an appreciation of the various models and how they relate to actual experimental results provides an important handle on the possibilities as well as the limits for improving the conductivity of CPs, which is one of the primary properties of interest in so many applications for these materials. The author also notes in particular that in spite of the models and the large body of experimental data, actual improvements effected in conductivity have been minimal, and it has been largely found that the intrinsic conductivity of the CP/dopant system, together with parameters in the very first synthesis (i.e., not in subsequent processing), appears to be the overwhelming determinant of conductivity for a CP system.

28.7   Conduction Models

28.7.1   Mott Variable Range Hopping (VRH) Model

We are now in a position to consider conduction models, after which we may look at some of the experimental observations that these models attempt to predict. One of the first models to be considered for CPs, primarily because it appeared to apply to disordered "conventional" semiconductors and CPs were considered disordered, was that of Mott and coworkers [874]. This model correctly predicted the $T^{-(1/4)}$ dependence of conductivity observed in disordered semiconductors such as α-Ge. Lack of ordering in such materials was thought to produce localized electronic states similar to the localized states (structural distortions such as bipolarons) produced by short conjugation lengths in CPs. The model assumed that localization was not very strong and differed at different locations within a material and thus postulated that conduction occurred through *variable range hopping* (VRH) of electrons between these localized states. This electron hopping was *assisted by phonons* (lattice vibrations) and was somewhat dependent on the initial and final energies (overlap of wave functions) of the states between which hopping occurred, i.e., it was not isoenergetic. Another way to visualize phonon-assisted hopping is to picture charge hopping between two sites accompanied by a change of state of the lattice vibrations at both sites. This "*Mott VRH*," or subsequently, "Mott–Cohen–Fritzsche–Ovshinsky model" led to a general $T^{-(1/4)}$ (for three dimensions) dependence of the conductivity expressed by

$$\sigma = \sigma_0 \exp\left\{-(T/T_0)-(1/(n+1))\right\} \tag{28.2}$$

where,

$$\begin{aligned}
n &= \text{dimensionality of material,} \\
T_0 &= 24/\left\{\pi r_0{}^3 k N(E_F)\right\}, \text{and} \\
\sigma_0 &= 9/4q(3/2\pi)e^2\gamma_0 q\left\{\tau_0 N(E_F)/(kT)\right\}
\end{aligned} \tag{28.3}$$

and where σ_0 the "localization length" is the length over which the wave function decays, k is the Boltzmann constant, $N(EF)$ is the density of states at the Fermi level, and γ_0 is the phonon frequency (about 10^{12} Hertz). A slight modification of this Mott VRH model, equating localization length with conjugation length in CPs and considering hopping between these rather than between point-like localizations, was proposed by Schaefer-Siebert and Roth [875] and yields

$$\sigma = C(TT_0)^{-\frac{1}{2}}\exp\left\{-(T/T_0)^{-\frac{1}{2}}\right\} \tag{28.4}$$

with C as a constant.

The Mott VRH model agrees well with experiment for *moderate* temperatures and doping levels for CPs and is found to apply in many cases to materials such as amorphous carbon as well [876]. Indeed, many recent studies on a variety of CPs, such as those correlating ESR spins with DC conductivity [876], appear to confirm the Mott VRH model. Although there are two fitting parameters, σ_o and T_o, the latter is found to be directly related to room-temperature conductivity. In practice, one usually tries to fit temperature vs. DC conductivity data to a $T^{-(1/(n+1))}$ fit, with information on dimensionality of conduction then obtainable.

28.7.2 Sheng Model

A model on a different basis more in consideration of CPs was proposed by *Sheng* [876], found to apply to highly doped CPs. It assumes conduction between the "metallic islands" within CPs discussed above and yields the following temperature relation:

$$\sigma = \sigma_0 \exp\{-T1/(T + T_0)\} \tag{28.5}$$

with T_1 being an additional fitting parameter and the fitting parameters dependent upon the properties of the potential barriers (insulating islands) between the highly conducting islands. The potential barriers are also sometimes identified with interfibrillar spaces rather than insulating islands. The charge transport is said to result from tunnelling induced by thermal fluctuations. It is evident in this respect then that electronic overlap between different fibers, i.e., packing, would also affect conductivity. The above equation predicts a conductivity monotonically decreasing with temperature but still yielding a finite value at near absolute zero (due to quantum mechanical tunnelling) and a saturation value of conductivity at higher temperatures. The Sheng model applies well to highly doped or heterogeneous systems (such as PVC/CP composites), poorly at low temperatures and poorly for low doping levels. Several variations of Sheng's model have been presented with varying degrees of applicability and success; these include, for instance, the Paasch [877] and Naarmann [863] models. Indeed Naarmann et al. [863] have claimed that, using a phenomenological adaptation of the Sheng model to interpret extant experimental data, they are able to arrive at values for the potential barrier widths (effective conjugation lengths) close to those indicated by electron microscopic and other analysis of doping islands: ca. 1.5 nm, with barrier heights computed to be ca. 10 meV.

28.7.3 Kivelson and Other Models

A model found to correctly predict the steep increase in conductivity of P(Ac) on doping and dependence of P(Ac) on temperature is that proposed by *Kivelson* [878, 879]. In its essence, this "intersoliton hopping" model applied primarily to P(Ac) proposes *isoenergetic* hopping of electrons from a neutral soliton of one chain to a charged soliton at a neighboring chain via tunnelling. It is able to account well for conductivity behavior of P(Ac)s at lower temperatures. It predicts the following conductivity relation:

$$\sigma = 0.45 e^2/(kT) x_0 (1 - x_0) \tau(T)/(NR_0) r_0/R_0 \exp(-2.78 R_0/r_0) \tag{28.6}$$

where,
x_0 - Fraction of neutral solitons

(1-x_o) - Fraction of charged solitons
N - Average chain length
r_o - Average localization length
R_o -Average acceptor distance
$\tau(T)$ - Frequency factor (strongly dependent on temperature)   (28.7)

   Another graphical model of conduction, valid for very high doping levels where quasi-metallic behavior is observed for CPs such as trans-P(Ac), can be visualized if we return to the CP semiconductor band structure evolution discussed elsewhere in this book, but with the caveat noted above that CPs are disordered semiconductors which have band structures a bit distorted from these simplistic representations. At moderate doping levels, bipolarons (or uncorrelated charged solitons in the case of P(Ac)) are the dominant charge carriers, and conduction is from charge hopping among these carriers via any of the above hopping models of conduction. As the doping level is increased dramatically, however, the bipolarons merge into bipolaron bands, as is discussed further below.

   Microscopically, what this means is that the individual polaron charge carrier states in a polymer chain now overlap spatially, producing more continuous wave functions, along which true, conventional *electronic conduction*, rather than charge hopping, conduction can take place. In the first case, the charge carriers are spinless, whereas in the second case, they now have spin, exactly as observed experimentally (for instance, for trans-P(Ac) between 1% and 8% doping, with a transition at ca. 6%). In highly conductive CPs such as trans-P(Ac), the transition from spinless conduction to electronic conduction is analogous to a semiconductor-to-metal transition. In this case then, the conductivity increases dramatically starting at very low doping levels <1%, in which region conductivity is ascribed to spinless charged solitons. As doping increases further, the transition to the quasi-metallic state is observed near 6% doping, but this transition does not lead to a dramatic rise in conductivity per se – it is observed by other means which detect electron spin. That increased doping directly leads to this transition which does *not* parallel the conductivity that can be seen from the data such as that in Fig. 28.27a, b, which compare the onset of Pauli susceptibility (a magnetic susceptibility ascribable to mobile electrons) and conductivity as a function of increased doping.

28.7.4   Relationship of All Models

We note finally that *all* the above models are, in a manner of speaking, "one part of the whole truth," being valid under different ranges and conditions of doping, temperature, and other variables. For instance, in P(Py), P(ANi), and P(3AT) samples of varying homogeneity and doping level, the Mott VRH, Sheng, and other models are found to apply in the identical CP with differing dopant level and homogeneity [881].

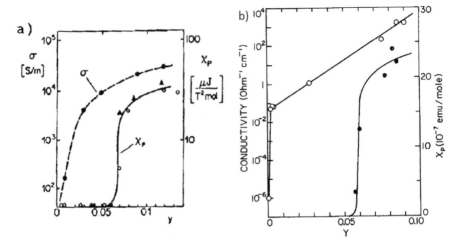

Fig. 28.27 (**a**) Onset of the Pauli susceptibility in poly(acetylene), doped with AsF$_6$ (triangles and open circles) and I3 (squares), compared with the variation of the conductivity with doping (broken lines and right-hand scale) (After Ref. [872], reproduced with permission) (**b**) Room-temperature electrical conductivity (left side, log scan, and temperature-independent Pauli susceptibility χp (right side, linear versus dopant concentration for [Na$_y$ + (CH)^y-]$_x$) (After Ref. [880], reproduced with permission)

28.8 Experimental Correlations

28.8.1 General

In practical application, the Mott VRH model is found to be widely applied to moderate doping levels and moderate temperatures for all CPs, while for high doping levels, the Sheng model (or variations thereof) is found to work reasonably well. The Kivelson model is applied in a very limited sense to P(Ac) within narrow doping and temperature limits as described above. We now discuss some experimental correlations.

28.8.2 Temperature Dependencies

The P(Ac) conductivity/temperature data of Fig. 28.28 can be fit to a $T^{-1/3}$ functionality [882], implying a two-dimensional characteristic for conductivity in this material. Since the two fitting parameters involved, σo and To (room-temperature conductivity and temperature-determined density of states), are found to be correlated, To can be plotted against σo to yield a "universal curve" [883] representing a wide variety of CPs (Fig. 28.29.) Figure 28.30 shows data for another sample of trans-P(Ac), now showing three-dimensional ($T^{-(1/4)}$) conduction with the Mott

Fig. 28.28 Temperature dependence of the DC conductivity of iodine-doped (CH)x fit to the VRH expression (Eq. 1)[848] (After Ref. [882], reproduced with permission)

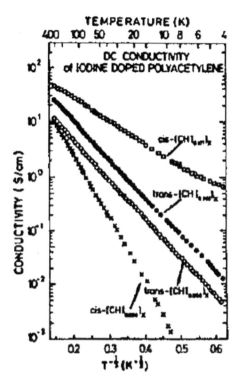

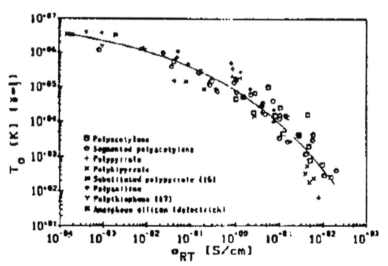

Fig. 28.29 "Universal" correlation between the room-temperature value of the conductivity and its temperature dependence for various conducting polymers (After Ref. [883], reproduced with permission)

Fig. 28.30 Plot of the log σ DC versus $T^{-0.25}$ for trans-(CH)x (After Ref. [886], reproduced with permission)

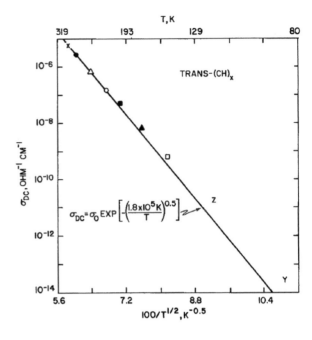

Fig. 28.31 Temperature dependence of the σ_{dc} of PA-PVB blends (After Ref. [887], reproduced with permission)

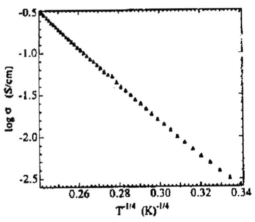

VRH temperature functionality. Such 3D Mott VRH conduction is also found in I2-doped and electrochemically prepared/doped P(Py) samples [884, 885].

CP composites also appear to show hopping mechanisms for conduction. Figure 28.31 shows the linear $\sigma - T^{-(1/4)}$ plot obtained for P(ANi)-poly(vinyl butyral composites), indicating a 3D Mott VRH mechanism. From the slope of these data, one can also obtain localization lengths (defined above) and density of states at the Fermi level, which in this case are found to be ca. 7 nm and $10^{20}/(\text{eV-cm}^3)$, respectively.

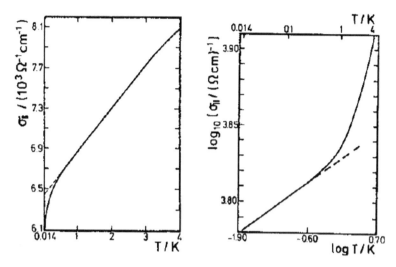

Fig. 28.32 Temperature dependence of σ between 14 mK and 4 K. The same experimental data are shown both in a linear and in a double logarithmical plot (After Ref. [892], reproduced with permission)

Substantiating the disordered semiconductor hopping models of conduction described earlier, Fig. 28.32 shows conductivities along the direction parallel to stretching in stretched films of P(Ac) at very low temperatures (to 14 mK), indicating finite and appreciable conductivities near absolute zero. For CP structures which are thought to be quasi-one-dimensional, a $T^{-1/2}$ dependence may also be observed, for example, as found for sulfonated P(ANi), P(o-toluidine), and even P(ANi) [888–890]. In DC and microwave studies of P(ANi) complemented by thermopower, ESR, and other studies, Wang et al. [890] found that overall 1D Mott VRH DC conductivity had a 3D quasi-metallic component for conduction within "crystalline" metallic islands, with surrounding amorphous islands wherein hopping conductivity predominated.

In isolated cases, very poor temperature dependencies have been found. For instance, in the case of the ladder CP BBL doped via ion implantation, the temperature dependence found (Fig. 28.33) [891], together with IR and other data, is thought to represent a disordered metal rather than a disordered semiconductor [891].

28.8.3   Frequency Dependencies and Microwave Measurements

The very elementary relation between total and DC/AC conductivities can be expressed as follows:

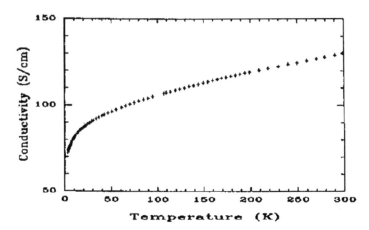

Fig. 28.33 Temperature-dependent DC conductivity of implanted PBO (After Ref. [891], reproduced with permission)

$$\sigma_{\text{TOT}} = \sigma_{\text{DC}}\{T\} + \sigma_{\text{AC}}\{T,f\} \tag{28.8}$$

That is to say, the total conductivity is a sum of the DC conductivity, which is a function of temperature, and the AC conductivity, which is function of frequency (f) and temperature. In this respect, the various conduction models predict varying relations for the AC conductivity. Figure 28.34 shows the AC conductivity at 1 KHz and 1 GHz for p(3-octylthiophene) (P(3OT)), as a function for P(3OT)-PE composites. For these materials, the AC conductivity is found to follow a simple relation, $\sigma = c \, X f^{2.6}$, where c is a constant and f is the frequency [891].

Although details of measurement methodology are provided in a subsequent chapter, we may note briefly here that for conductivities at frequencies less than 1 MHz (and fixed temperature), a capacitance–conductance bridge suffices. An impedance analyzer can be used to extend this range to 1 GHz. Microwave cavity perturbation techniques must be used in the range 10 GHz to 100 GHz. Kramers–Kronig transformation of infrared reflectance data can provide data for frequencies >100 GHz.

Figure 28.35 shows typical frequency vs. total conductivity data for trans-P(Ac) [893], while Fig. 28.36 replots these data on a log scale. Figure 28.37 shows similar data for poly(3-methyl thiophene).

The work by Javadi et al. [894] on a series of poly(anilines) of varying doping/ protonation levels has shown that microwave and DC conductivities follow different temperature relations, respectively, T^{-1} and $T^{-1/2}$, with the microwave conductivity larger than the DC conductivity by several orders of magnitude. These data appear to support a metallic island model for conduction, with a combination of hopping and electronic conduction occurring, depending on dopant level.

The recent work by the Heeger group [895] on P(Py)/PF6, using a Kramers–Kronig transformation of IR reflectance data to obtain conductivities and extinction

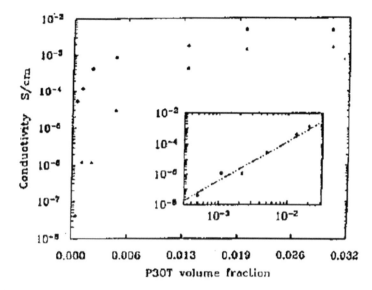

Fig. 28.34 Conductivity of poly(3-octylthiophene plotted as a function of the volume fraction of con polymer for two frequencies, 1 kHz sand 1GHz. Inset: log–log plot of the conductivity (at 1 kHz) vs. volume fraction of P3OT (After Ref. [892], reproduced with permission)

Fig. 28.35   Plot of log σ_{TOT} versus log f for trans-$(CH)_x$ at constant temperature. Note the break in the abscissa for DC. The solid lines are drawn as a guide for the eye. The data are for the sample used for Fig. 28.28 (After Ref. [886], reproduced with permission)

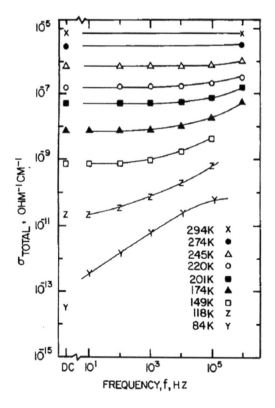

Fig. 28.36 Plot of log σ_{AC} versus log f for the data in Fig. 28.35. The solid lines are drawn as a guide for the eye. The temperatures represented by the different symbols are given in Fig. 28.35 (After Ref. [886], reproduced with permission)

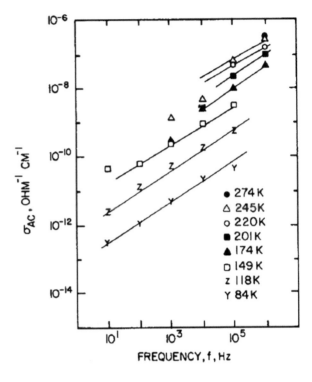

Fig. 28.37 Frequency and temperature dependence of the conductivity in poly (3-methyl-thiophene), synthesized electrochemically and being nearly completely reduced. The temperature varies from 290 K (a) to 200 K (h) in steps of 10 K (After Ref. [865] reproduced with permission)

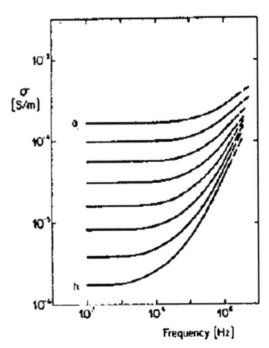

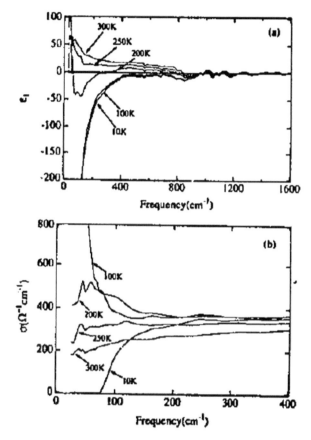

Fig. 28.38 Conductivity/frequency behavior for poly(pyrrole)/PF6: (**a**) $\varepsilon_1(\varpi)$ vs. ϖ and (**b**) $\sigma(\varpi)$ vs. ϖ at 300, 250, 200, 100, and 10 K (After Ref. [895], reproduced with permission)

coefficients at *IR frequencies*, has shown up an interesting property of this material: nearly all the ambient temperature DC conductivity of this polymer appears from a narrow conductivity peak (Fig. 28.38) near zero frequency, and there appears to be a gap in the spectrum of charged excitations at intermediate frequency that is not correlated to a similar gap in the spectrum of spin excitations.

28.8.4 Thermopower and Hall Effect Measurements

Thermoelectric power is the potential gradient generated between two points or faces of a material when they are subject to a temperature differential. Typically, it is large for semiconductors, while for metals it is small and decreases with decreasing temperature, vanishing near absolute zero. A small decrease of thermopower with temperature would thus likely indicate hopping conduction, while a very rapid

Fig. 28.39 Thermoelectric power (S) as a function of temperature for various I_2 doping levels. Dopant concentrations are $A = 0$, $B = 0.017$, $C = 0.03$, and $D = 0.22$ (After Ref. [871], reproduced with permission)

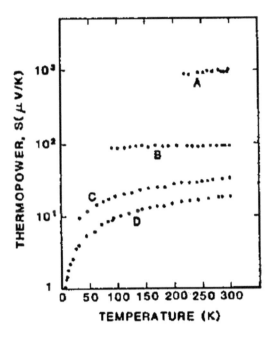

Fig. 28.40 Thermopower vs. concentration for poly (acetylene) (After Ref. [871], reproduced with permission)

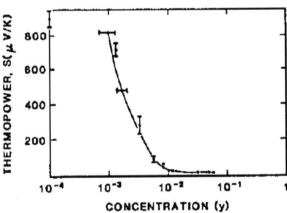

decrease and very low values (see below) for the thermopower would indicate a quasi-metallic conduction, as, for instance, observed for highly doped trans-P(Ac).

The techniques of thermopower measurements are described elsewhere in this book.

Figure 28.39 shows typical thermopower data for trans-P(Ac)/I_3 at various doping levels; it is clear that at the highest doping level (22%, D), metallike behavior is seen, while the transition to this metallike behavior is visible even at 3% doping (C). At ambient temperature (Fig. 28.40, thermopower for trans-P(Ac) ranges from nearly 1000 μV/K at very low doping to ca. 30 μV/K at high doping,

Fig. 28.41 Thermoelectric power of an iodine-doped polypyrrole film as a function of temperature (After Ref. [884], reproduced with permission)

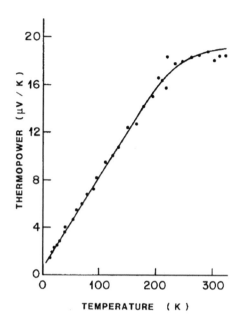

with the figure for other common CPs also being ca. 1000 μV/K and 100 μV/K for de-doped and doped states, respectively. The data for P(Py)/I_3 (Fig. 28.41 also show quasi-metallic behavior

The thermoelectric power behavior of most CPs parallels that of temperature and activation energy relationships.

Hall effect (magnetic field effect on conductivity) measurements carried out on trans-P(Ac) samples show that at very high doping levels, metallic behavior is observed, while for light doping, the Hall coefficient is larger than for highly doped samples and increases at low temperatures [896]. The work by the Heeger group [897] on P(Ac)/PF6, K/P(Ac), and P(Py)-camphor-sulfonate has shown that application of a magnetic field of 8 Tesla yields a negative temperature coefficient for the activation energy of conduction, with a semiconductor-to-insulator transition induced by application of this magnetic field.

28.8.5 Pressure Dependencies

The effect of pressure on conductivity can be visualized in a straightforward manner as closer packing of the material, enabling greater interchain contacts in both fibrillar and globular CPs. In theoretical terms, the electronic overlap between extended conjugation regions within chains is partially extended between chains and between fibers.

In nearly all pressure-dependence studies, actual conductivity gains realized with a small increase in pressure have been small, for example, ca. 25% between

Fig. 28.42 Pressure dependence of electrical conductivity for high-conductivity material showing return cycle to ambient pressure (crosses) (After Ref. [898], reproduced with permission)

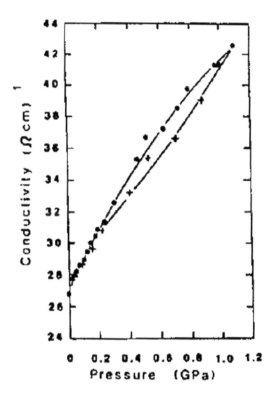

ambient pressure and 3 kbars found for trans-P(Ac) [898] or in some cases negligible [863]. In a study of P(Py)/tosylate, Maddison and Unsworth found an approximately 59% increase in conductivity over an added pressure range of 1.1 GPa and a hysteresis in the conductivity behavior (Fig. 28.42). Work by the Heeger [897] group on PF_6^- and K^+ doped trans-P(Ac) and P(Py)/camphor-sulfonate showed an apparent metal-to-insulator transition with increased pressure, indicated by a positive temperature coefficient for the activation energy of conduction.

At high pressures, however, large conductivity increases have been claimed: for example, in a trans-P(Ac)/I_3 sample, conductivity was found to increase nearly 5 X over that at ambient pressure at 30 kbar pressure but then to decrease by two to three orders of magnitude (i.e., becoming lower than at ambient pressure) at pressures up to 65 kbar. The latter effect was thought to be due to de-doping, cross-linking, chain scission, or other degeneration of the polymer [899].

In more recent work, Fukuhara et al. [900] have shown that application of about 13 kbar pressure to samples of P(3-Me-T)/PF_6 gives rise to a metallic temperature dependence of resistivity (conductivity) at low temperatures: below ca. 4 K, a $T^{1/2}$ dependence is observed. A study of thin films of P(ANi) at pressures up to 22 GPa at room temperature by Bao et al. [901] appeared to show enhanced interchain charge transport due to sample compression. Figure 28.43a–c shows some of the data obtained by these workers.

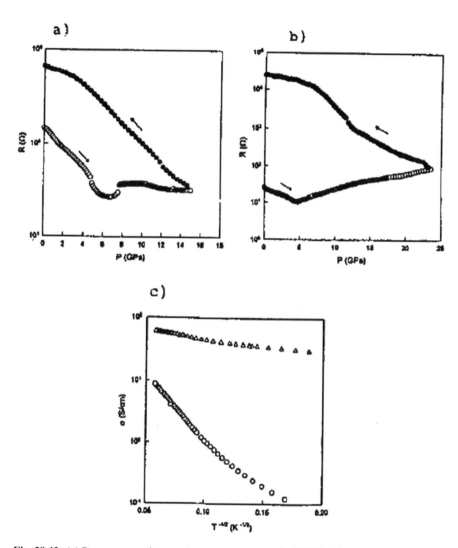

Fig. 28.43 (a) Pressure vs. resistance at room temperature of a P(ANi) thin-film sample prepared from N-methylpyrrolidinone (NMP) and protonated with HCl. Data shown are taken for a complete run and include loading (o) and unloading (·) (b) Sample made conducting via camphor sulfonic acid and cast from m-cresol (c) Temperature dependence of electrical conductivity of PANi thin films (o) prepared from N-methylpyrrolidinone (NMP) and protonated with HCl and (Δ) made conducting via camphor sulfonic acid and cast from m-cresol (After Ref. [901], reproduced with permission)

In the case of composites, such as that of P(ANi) with poly(styrene), increase of pressure on a 40% poly(styrene) sample from ambient to 10 kbar causes a conductivity increase from 10^{-3} S/cm to 1 S/cm.

28.8.6 Stretching, Anisotropy, Crystallinity, and Molecular Weight Effects

Since it is evident from the above hopping models that proximity of hopping sites, or more extensive delocalization, or both, would increase hopping conduction, measures such as stretching of CP films, or other morphological changes such as packing, have been undertaken by many workers. The simplest sort of anisotropy in conductivity is illustrated in Fig. 28.44, data along and perpendicular to the stretch direction in 1:4 stretched P(ANi)/Cl. For stretched films, conductivity parallel to the stretch axis is larger, sometimes by several orders of magnitude, to that perpendicular to the stretch axis. Figure 28.45 illustrates anisotropy of conductivity for stretched films of another CP.

The Heeger and MacDiarmid groups have carried out several studies on the dependence of conductivity on molecular weight (MWt) [902]. According to their studies on P(ANi), a dependence of conductivity is of the order of $MWt^{1.0 \text{ to } 1.2}$ or $MWt^{0.5-0.6}$ assuming, respectively, rigid rod and flexible chain conformations in solution. An illustration of the approximate relation of conductivity to MWt for P (ANi) observed by the MacDiarmid group is shown in Fig. 28.46a, while Fig. 28.46b shows the effect of increased stretching of P(ANi) ribbons on their crystallinity.

Fig. 28.44 Temperature dependence of DC conductivity of 50% doped 1:4 stretched PAN-ES (HCl) films in the directions parallel and perpendicular to the stretching directions (After Ref. [888], reproduced with permission)

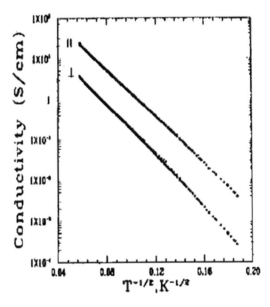

Fig. 28.45 Temperature dependence of DC conductivity of x = [Cl]/[N] 0.12 (o), 0.37 (+), and 0.50 (·) doped 1:4 stretched poly(aniline)-chloride films in the directions parallel and perpendicular to the stretching axis. The samples have been pumped overnight before the experiments (After Ref. [890], reproduced with permission)

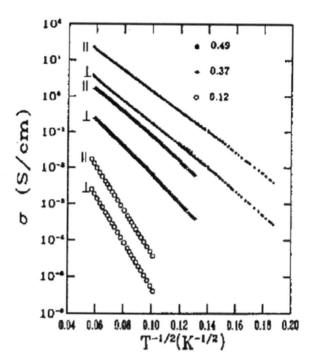

28.8.7   Activation Energies and Mobilities

Many workers [871] have assumed a crude Arrhenius relationship of the conductivity to temperature (Eq. 28.8), more correctly applicable near ambient temperature to crystalline semiconductors, to arrive at an estimation of the "activation energy of conduction." In such cases, typically, conductivity (σ) is proportional to n, the number density of charge carriers, which in turn has the following relation:

$$n \rightleftharpoons \exp\{-E_a/k_B T\} \qquad (28.8)$$

where E_a, k_B, and T are the activation energy, the Boltzmann constant, and the temperature, respectively. Figure 28.47 illustrates one such estimation as a function of doping level for P(Ac) with iodine and AsF_5 doping. While not accurate, such plots do show the correct trends, i.e., the dramatic increase in conduction with doping (here at 1 mole %). The wide variation in the absolute values of Arrhenius activation energies can however be seen by a comparison of the value reported by Burroughes and Friend [904] for a highly doped P(Ac), 0.3 to 0.4 eV (vs. the ca. 0.01 eV discernible from the figure).

There is a rather straightforward though simplistic relationship between macroscopic (bulk) conductivity (σ) and mobility of charge carriers (μ), viz.,

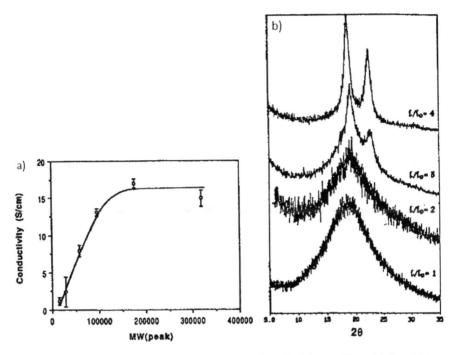

Fig. 28.46 (**a**) Dependence of conductivity of doped poly(aniline) (emeraldine oxidation state) on molecular weight (**b**) x-ray diffraction spectra of ribbons of emeraldine base of increasing draw ratio, $(l/l_o, l = $ final length, $l_o = $ original length before stretching) (After Ref. [903], reproduced with permission)

$$\sigma = ne\mu \qquad\qquad (28.9)$$

where e is the electronic charge and n the number of charge carriers per unit volume and has the Arrhenius temperature relation of Eq. (28.8). Thus, if the latter quantity can be estimated, as it frequently can from doping and other parameters, the charge carrier mobility may also be estimated. For example, several poly(3-alkyl thiophene)/I3s with estimated charge carrier concentrations of 10^{17} to 10^{18} per cm^3 have been estimated to have mobilities of between 10^{-6} and 10^{-4} cm2/Vs [906], these values being nearly in the regime of those in insulators. An apparent trend indicated in this series of P(3AT)s is that with increasing side chain length (at the 3-position of the thiophene unit), μ clearly decreases: the more cumbersome packing required for the longer-chain P(3AT)s would hinder hopping between chains for conduction. In contrast to these mobility values, values for the "metallic regime" of trans-P(Ac) have been claimed to be as high as 1000 cm^2/Vs [907].

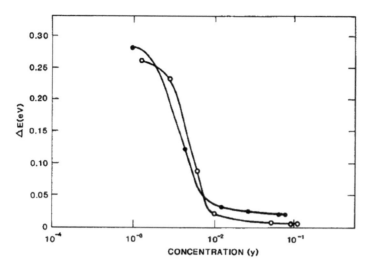

Fig. 28.47 Activation energy (E_a) as a function of the doping level of I_2 ($\cdot$) and AsF_3(o) for poly (acetylene) (After Ref. [905], reproduced with permission)

28.8.8 Minor Structural Effects

For the most part, a direct structure–conductivity correlation, predicting that P (ANi)s in general have higher conductivities than P(Py)s, for instance, is not established for CPs on structural grounds alone but, rather, depends strongly on empirical data. Some structural features or trends relatable to high conductivity can be found from semiempirical theoretical calculations, as discussed in a later chapter. Thus, it can be established from semiempirical calculations that the extended overlap and simple structure of trans-P(Ac) will lead to higher doping-induced conductivity than for P(ANi) and that P(3MT) will have a higher conductivity than its unsubstituted parent (P(T)).

However, even within individual CP classes, semiempirical or other theoretical treatments sometimes do not correctly predict even trends in conductivity; for instance, in the case of P(Py)s, the lower conductivities of N- as well as β-substituted P(Py)s are not correctly predicted by theory, and in the case of P (ANi)s, the lower conductivity of ortho-alkoxy substituted P(ANi)s is also not predictable from theory. In both these cases, the conductivity of the unsubstituted polymer is in the region 100 S/cm, while that of substituted homologues is in the 1 S/cm range. Another case in point with regard to theoretical/empirical correlations is that of the poly(pyrenes), which represent a structure of four fused benzene rings. It was thought that in these polymers, the expected extended orbital overlap would overcome steric effects in yielding a CP of very high conductivity, expectations which were belied [857].

As another example, in the series of P(3-alkyl thiophenes), for the electrochemically as-synthesized polymers with dopants such as BF_4^- or PF_6^-, the conductivities of P(T), P(3MeT), P(3EtT), P(3-hexyl-T), P(3-octyl-T), P(3-dodecyl-T), P (3-octadecylT), and P(3-eicosylT) are, respectively (in S/cm) 190, 500, 270, 95, 78, 67, 17, and 11 [907, 908]. Thus, for this class, conductivities peak for the methyl substituent, then steadily declining with longer alkyl chains (the longer chain polymers show, it may be noted, substantial solubility in many organic solvents). It thus appears to be the considered opinion of many workers that the only way of verifying potential high conductivity in a CP of hypothetical structure is to synthesize it and measure the conductivity.

28.9   Theoretical Studies of CPs

28.10   Preliminary Notes

With regard to theoretical studies of CPs, some familiarity with the very basics of quantum chemistry is assumed on the part of the reader: the Schrödinger equation, various Hamiltonians, the significance of ab initio vs. semiempirical methods and one-electron methods, and common parameterizations such as modified neglect of differential overlap (MNDO). These are found today in bachelor's level courses in nearly all the physical sciences and can be gleaned from any introductory quantum chemistry book. The terminology we will use is that of quantum chemistry rather than quantum physics. We attempt as far as possible to stay away from multitudinous equations, which can be cumbersome for the lay reader from another field, and difficult-to-understand representations, such as band structures illustrated in terms of wave vectors. We instead focus on a comparison of results of various methods in terms of which is most useful in interpreting experimental data and predicting CP properties. Equations and band structures are however cited in appendices at the end of the chapter for reference.

28.11   One-Dimensional Systems and Peierls Instability

Most early theoretical studies of CPs attempted to look at them as one-dimensional (1D) systems, since it made treatment easier and appeared directly applicable to P (Ac), one of the first CPs extensively studied.

A 1D polyene chain such as that of P(Ac), in which each CH fragment has an odd number of electrons, could from primitive electronic treatments be said to possess the band structure of a metal and thus might be expected to behave like a metal. This 1D metal would have equal C–C bond lengths along its chain. In fact, however, P(Ac) is a semiconductor. Peierls [909] was one of the first to recognize that the ground, equal-bond-length, metallic state of such a material was in fact unstable with respect to atomic shifts that would create alternating bond lengths (e.g., alternating double and single bonds). The *"Peierls instability theorem"* in essence states that a *1D metal cannot exist*. If one starts with such a 1D metal of equal bond lengths, with a filled (metallic) electronic band, and effects a perturbation leading to alternation of bond lengths, one finds that the total energy of the system declines (i.e., the new structure is more stable), and a splitting of the single metallic band into two bands with a gap in between occurs.

Figure 28.47 shows one of the ways of illustrating the formation of this Peierls gap. While the Peierls theorem has been found to hold well for experimental systems in general, a caveat is to be noted: It correctly predicts the formation of a bandgap only when no other mechanism exists to create one. (Among possible alternative mechanisms are electron–phonon interaction to create a superconducting gap or electron–electron interaction to give a metal–insulator transition, which are too cumbersome to deal with here). It will be realized here and in subsequent discussions that the property of bond length alternation, in various forms (such as, e.g., benzenoid vs. quinoid character), is one of the important properties determining CP behavior and characteristics such as bandgaps and conductivity.

28.12 Overview of Theoretical Methods Used

We must start with another important caveat, regarding all theoretical methods applied to CPs to date, that they form an inexact science at best and at worst can even yield misleading trends. This stems from the simplistic treatments that are necessary from the point of view of economy of time and resources, which force one to make some grave approximation or other, for instance, a one-electron method neglecting electron–electron correlations and using only valence electrons; a linear CP chain omitting conformational variations; an infinite chain omitting saturated, cross-linking, unusual-branching, or other defects; or any method explicitly omitting the presence of dopants. Some methods have however been more useful than others in establishing actual or predictive trends. The most accurate methodology would likely be an ab initio method treating CP chains of at least 20 monomer units with dopants and with conformational, defect, chain branching, and other variables, but this is of course still not possible due to computational restraints.

28.12.1 Historically Important Methods

Several theoretical methods have been applied to the study of CPs that are, at this writing (2016), predominantly of *historical* importance, since they have been superseded by more modern methods that are more appropriate for the greatly enhanced computing power available today. The application of the historically important methods is however worthy of note since they contributed to the first understanding of theoretical aspects of CPs. They are briefly cited here now in the following paragraphs.

More recently and as of this this writing (2016), *density functional theory* (DFT) appears to be the technique most commonly applied to CPs [854, 910, 911].

One of the first methods to be applied to CPs to elucidate structure/property or theory/experiment relationships was the semiempirical *extended Hückel* (EH) (Appendix 28.2) method [912, 913]. This method uses a one-electron approximation neglecting electron–electron correlations entirely and considers valence electrons only. Unfortunately, as will be shown below, it is able only to predict very broad trends – for instance, bandgaps – within chemical entities of similar structure or homologues and sometimes errs even with these. Extended Hückel methods yield overlaps between σ and π electrons that are too large and bandgaps that are generally too small vs. experimental values and appear to fail when highly polar substituents or dopants must be considered [914].

Ab initio (first principles) methods cannot of course be applied directly to CPs because of time and resource constraints but have been applied indirectly. Most commonly, they have been used to optimize geometries for oligomer chains of typically up to four monomer units, following which semiempirical methods are then applied. Occasionally, the reverse has been used, such as the work of the Marynick group in *which partial retention of diatomic differential overlap* (PRDDO) optimizations of geometries of oligomers up to the hexamer were followed by ab initio calculations [915, 916]. Ab initio methods have also been directly applied on a limited basis to P(Ac) to yield useful insights into its band structure as a function of such parameters as bond length alternation [913].

The most widely applied, and apparently the most practical in terms of theory/experiment correlation and predictive capability, has been the *valence effective Hamiltonian* (VEH) method propounded by Brédas and others [857, 917], which also commonly uses an initial ab initio optimization of monomer or oligomer geometry ("ab initio parameterization") or, less frequently, experimental geometries, prior to calculation. The VEH method has been used to arrive at ionization potentials, bandwidths, and, in the coplanar approximation, bandgaps, which reasonably follow experimental trends.

Other semiempirical methods used in early work have included the *AM1* method of Dewar et al. [918] and the *CNDO2/S3* method (a variant of the complete neglect of differential overlap method) applied with some success to P(Ac) and P (Py) [919, 920]. These have in nearly all cases been used in conjunction with initial or following ab initio calculations.

A method primarily applied to P(Ac)s and related CPs has been one using a special Hamiltonian, the *Su–Schrieffer–Heeger* (SSH) Hamiltonian [921, 922] (Appendix 28.2). This model omits all two-electron terms but includes many aspects of electron–phonon coupling, while neglecting some aspects such as the deformation potential coupling of phonons to the on-site electronic charge density.

It models the lattice (CP chain) potential in the form of elastic springs between neighboring chain sites. The method has yielded practical parameters such as density of state distributions and band structures that have been useful in interpretation of experimental data, primarily for P(Ac).

Methods expressly taking into account *electron–electron correlation ("coulomb interactions")* have found very limited application to CPs, although in the few such applications, they have clearly shown superiority to one-electron methods in terms of relation to experimental data. In this respect, the *Pariser–Parr–Pople Hamiltonian* [923] (Appendix 28.2), which again neglects many aspects of electron–phonon interaction while considering electron–electron correlation, has found application in predicting photochemical behavior of poly(phenylene-vinylene) (P(PV)) substantially more accurately than one-electron methods [924]. A *Peierls–Hubbard Hamiltonian* [923] (Appendix 28.2), which in addition to considering coulomb interactions also considers *electron–phonon interactions* in detail, has succeeded in accounting for such observations as the occurrence of the 2^1Ag state of polyenes below the optical gap, which SSH, VEH, or other methods are not able to account for with any parameterizations used [923]. It has been felt by many workers in the field of theoretical treatment of CPs that such an approach, encompassing electron–electron as well as electron–phonon interactions, appears to be the most complete, but for many reasons, the methodology has not been widely applied to CPs at least until recently.

28.13 Extended Hückel and Related Methods

In one of the early theoretical studies of CPs, Whangbo, Hoffmann, and Woodward [912] published an extended Hückel (cf. Appendix 28.2) study of P(Ac), P(DiAc), and P(PP) together with comparisons with poly(acenes), graphite, poly (cyanonitrile), poly(pyridinopyridine), and several other real and hypothetical molecules. As is customary, these used experimental bond lengths and angles. Band structures, bandgaps, and the presence/absence of partially filled bands for specific geometries constituted the primary comparative data obtained. As for this method generally, the authors cautioned that only approximate trends were indicated by the work, and even these would need to be corroborated by studies considering e–e correlation and experimental work. Figure 28.48 summarizes some of their salient results. Of particular note are the prediction of a zero bandgap for the Peierls unstable form of P(Ac) (4), the *difference in bandgap for the benzenoid and quinonoid forms* of P(PP) (5,6) (as will be seen subsequently, this is one of the important findings of theory well corroborated by experiment for nearly all CPs), and the nearly zero bandgaps of hypothetical structures 7, 9, and 10, the latter two due to the presence of partially filled bands found with nearly all the ladder-type structures studied. This work was one of the first studies providing a handle, however speculative, for the synthesis of potentially zero-bandgap CPs.

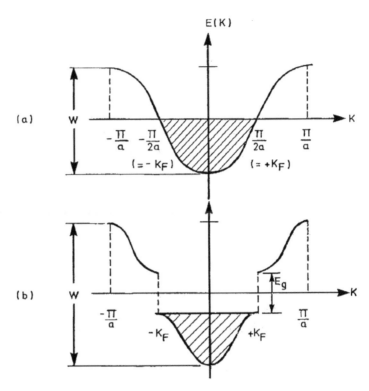

Fig. 28.48 Filled Peierls distorted band of semiconducting polyacetylene (After Ref. [866], reproduced with permission)

Grant and Batra [925] used a combination of first principles LCAO/extended tight binding and EH methodologies for a useful study of cis- and trans-P(Ac). Among the major findings were that even a small bond length alternation (0.07 Å) produces a substantial bandgap for the trans-isomer and that in the case of the cis-isomer even with equal bond lengths, there would be no metallic state (i.e., no vanishing gap).

Most other studies using extended Hückel methodology have been combined with ab initio or other semiempirical methods and are discussed in Sec. V.2.14 below. Among the important findings of EH and other Hückel-type methods has been the demonstration that bipolarons or other charge carrier structural distortions are confined to a width of ca. five to six monomer units [926]. Predictions based on Hückel-type calculations of zero bandgaps for this or that CP structure (e.g., the prediction that one isomer of poly(indenoindene), a CP with alternating fused benzene and cyclopentane rings, would have a zero bandgap [927]), have been difficult to substantiate either experimentally or using other, more involved theoretical techniques.

28.14   The VEH Method

To paraphrase Brédas, the advantages of the VEH method are that "it is completely theoretical, since the potentials do not contain experimental data, and it provides 1-electron energies of ab-initio double-zeta quality" [928]. Its simplicity appears to allow comparative studies of very varied *oligomers* and thus provides some guidance in CP design and a limited explanation of experimental data. Unfortunately, it also requires inclusion of several highly empirical "correction factors," for instance, subtraction of 1.9 eV from the original VEH-computed ionization potential, to Tables 28.1 and 28.2 that provide a summary comparison of selected VEH calculations with experimental data for several CPs (Table 28.3).

The VEH method appears to successfully estimate the experimental bandgap of P(Py) and the lowering of the oxidation potential of conductivity P(β,β'-di-Me-Py) relative to P(Py) [928]. However, it appears to fail for P(N-Me-Py), where the

Table 28.1 A sampling of VEH results for various CPs, obtained by the Brédas group (After Ref. [928], reproduced with permission)

Polymer	Ip(ev)	Bandwidth (ev)	Eg(ev)
Polyacetylene trans	4.7 (4.7)	6.5	1.4 (1.8)
Polydiacetylene acetylenic	5.1 (5.2)	3.9	2.1 (2.1)
Poly(*p*-phenylene) Twisted (22)	5.6 (5.5)	3.5	3.5 (3.4)
Polypyrrole bond length between rings = 1.45 Å	(~4.0) 3.9	3.8	(3.2)3.6

Table 28.2 Gas-phase ionization potentials (eV) for diphenylpolyenes (After Ref. [928], reproduced with permission)

Chain length (x)	VEH theory	Experiment
1	7.86 (7.84)	7.94
2	7.56 (7.57)	7.56
3	- (7.39)	7.33
4	- (7.27)	7.19
5	- (7.17)	7.05
6	- (7.10)	7.07
8	- (7.00)	

Table 28.3 Oxidation potentials (volts versus SCE) of conjugated polymers (After Ref. [928], reproduced with permission)

	Oxidation	Potential
Polymer	VEH theory	Experiment
Polyacetylene	0.4	0.2
Poly(p-phenylene)	1.2	0.9
Polythiophene	0.7	0.6
Polypyrrole	−0.6	−0.4

experimental conductivity is nearly five orders of magnitude less than that of P(Py), and the oxidation potential is substantially positive of that of P(Py), trends, and magnitudes in large disagreement with VEH calculations [929]. This appears to be due to the lack of consideration of twisting in the CP chain caused by steric effects of the N-Me substituent, and also lowered symmetry, an example of the caveat cited earlier. Nearly all VEH calculations, even when substantially in agreement with experiment, show variations substantial variations from experimental values; to cite an example, the experimental bandgap of poly(thienylene vinylene) is 2.2 eV, vs. the VEH calculated value of 1.6 eV [930].

The effect of benzenoid vs. quinonoid contributions is estimated quite well with the VEH method; for illustrative purposes, Fig. 28.48 shows the relation between the benzenoid and quinonoid forms of poly(p-phenylene). As with the EH and other semiempirical methods, the increasing coplanarity and decreasing bandgap of quinonoid as compared to benzenoid (aromatic) structures are well predicted. Perhaps the most explicit demonstration of the contribution of quinonoid structures has been the VEH calculations by Brédas et al. [931] on poly(isothianaphthene) (P (ITN)), a P(T) with a fused benzene ring on top of the thiophene moiety. They find in this that the nearly 1 eV experimentally observed lowering of the bandgap in P (ITN) vs. P(T) can be accounted for by *contributions to the electronic structure from quinonoid forms* which is enhanced by the presence of the fused benzene ring. They find that this effect is far greater than any effects due to substituents in the isothianaphthene rings.

In their calculations on P(ITN), Brédas et al. make use of a bond length alternation parameter, Δr, the difference between bonds approximately parallel and perpendicular to the CP chain direction, being negative for aromatic (benzenoid) geometry and positive for quinonoid geometry. Their calculations for the parent CP, P(T) (Fig. 28.49) show the effect of bond length alternation on the bandgap as well as HOMO/LUMO structures, with vanishing bandgap at a zero-bond length alternation, which is of course untenable from the point of view of the Peierls theorem and other factors. Figure 28.50 compares the VEH HOMO/LUMO structures for P(T) and P(ITN). An important finding of these VEH calculations is that in order to arrive at a CP with a very small bandgap, one must stabilize quinonoid contributions as far as possible in the undoped ground state itself (this stabilization occurs to some extent through doping of course).

Some of the most practically useful results of VEH calculations have been the band structures and the derivative band structure evolution as a function of doping, dealt with in some detail elsewhere in this book.

28.15   Ab Initio, Combination Ab Initio/Semiempirical Studies

Ab initio methods alone have been applied in very limited fashion to CPs other than P(Ac). A representative application to P(Ac) is the work of Tanaka and Tanaka (Appendix 28.2) [932], for Cl- and Na-doping. The computed band structure and

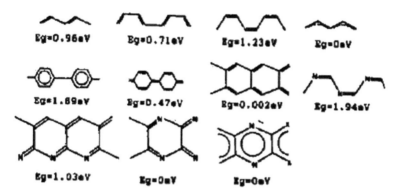

Fig. 28.49 Structures and bandgaps from work of Whangbo et al (After Ref. [912], reproduced with permission)

Fig. 28.50 Relation between benzenoid and quinonoid forms of poly(p-phenylene)

DOS representation, discussed at some length in Appendix 28.1 below, shows that for both (n-, p-type) extreme doping cases, HOMO and LUMO nearly merge, with quasi-metallic behavior indicated. The quasi-metallic states correspond to "polson–antipolson" lattices (polson = polaron-charged soliton pair). Shifts of the Fermi level on increased doping are found to agree well with experimental UPS spectra.

Brédas et al. [933] studied P(Py), P(PP), and P(T) using a combination of ab initio methods combined with VEH calculations (Appendix 28.2). The ab initio method, using an STO-3G basis set, was used to arrive at optimized geometries for neutral and Li or Na doped forms of the tetramers of the CPs. The optimized geometries were then used in the VEH calculations to arrive at band structures and band evolution upon doping.

Among the preliminary findings from Brédas et al.'s ab initio calculations was the result that the quinonoid form of P(PP) had an energy higher by 20 kcal/mol/ring than the benzenoid form. Using the STO-3G coordinates optimized on the undoped P(PP) tetramer and used as input to VEH calculations, a bandgap value for P(PP) of 3.5 eV was obtained, comparing well with the experimental 3.4 eV, and a bandgap value for P(T) of 2.2 eV, comparing well with the experimental 2.0 eV, was obtained. Charge-transfer calculations suggested that bipolarons were delocalized over ca. four monomer units, rather than the five to six units found with EH or other Hückel calculations. More importantly, the charge appeared to rest predominantly on the inner two rings.

The dramatic inducement of greater quinonoid structure and accompanying coplanarity by charge transfer via introduction of a dopant was clearly seen in these calculations (inter-monomer torsion angle for P(PP) changing from ca. $40°$ to ca. $2°$ for Li n-doping, $0°$ for Na-doping). Additionally, the pushing up of the HOMO and lowering of energy of the LUMO, i.e., lowering of the bandgap, were clearly seen upon doping, with the production of two states in the gap corresponding to bipolarons and states due to dopants appearing above these states.

In other work by the Brédas group [934], geometry optimizations up to the tetramer level using the semiempirical AM1 method were used, which were then input to restricted Hartree–Fock (RHF) 3-21G-basis-set calculations of the total energy of different rotational conformations for undoped and doped polymer forms. A "rigid rotor" approximation was used, i.e., not allowing for optimization of anything other than the conformational angles, which the authors claimed to affect the results only to an error of 10%. Among the important findings of this work were that rotational barriers are small, of the order of a few kcal/mole, in undoped CP, thus essentially permitting nearly free rotation at room temperature, *but dramatically increase (nearly 20 X) upon doping*. This is thought to be due to the inter-monomer bond acquiring double bond properties due to increased quinonoid character. The authors feel that their calculations up to the tetramer level adequately represent longer chains, which they feel would stack helically in their undoped form at room temperature. They cite small-angle neutron scattering results on undoped and doped CPs as substantiating their findings.

Pomerantz et al. [916] carried out calculations on a modified benzene-fused poly (di-thiophenevinylene) (Fig. 28.51), a new CP arrived at via a complex chemical synthesis.

Monomer structures were optimized via the PRDDO method (see above) for conformational analysis of trimers via a modified EH method. For band structure and related calculations, an ab initio program (GAUSSIAN 92) using the STO-3G basis set was employed for geometry optimization, used as input to EH calculations. Among their findings were that rotational barriers between monomer units were of the order of 1.5 kcal/mole or less (again for undoped polymer only as in the work of Lazzaroni et al. [934]). Bandgaps calculated using predominantly aromatic and quinonoid structures of the polymer were 3.6 eV and 1.0 ev, in comparison with the experimental value of 2.5 eV (absorption maxima) or 1.9 eV (band edges). The authors thus concluded that the actual CP structure most likely contained both aromatic and quinonoid segments, with the former in excess.

28.16 Methods Using SSH Hamiltonians

The SSH Hamiltonian has found widest application to P(Ac), and because of its active consideration of lattice vibrations, it has been used to attempt to account for phenomena associated with phonons, such as Raman scattering, IR excitations, and photo-induced IR [923, 935–938]. The SSH Hamiltonian includes many aspects of electron–phonon coupling, while neglecting some aspects such as the deformation potential coupling of phonons to the on-site electronic charge density. Figure 28.52

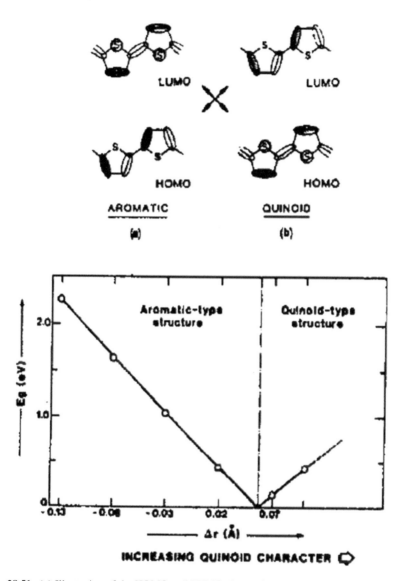

Fig. 28.51 (a) Illustration of the HOMO and LUMO electronic patterns in polythiophene for an aromatic-like geometry and a strongly quinonoid-like geometry (b) Evolution of the bandgap (in eV) as a function of increasing quinonoid character of the backbone for polythiophene (After Ref. [931], reproduced with permission)

illustrates local phonon DOSs computed using SSH Hamiltonians for undoped and soliton-containing P(Ac) chains, showing the most common phonon frequencies found in such soliton-dominated lattices.

Stafström [936] has employed a method that includes the SSH Hamiltonian with dopant and e–e correlation terms, which is discussed in the next section.

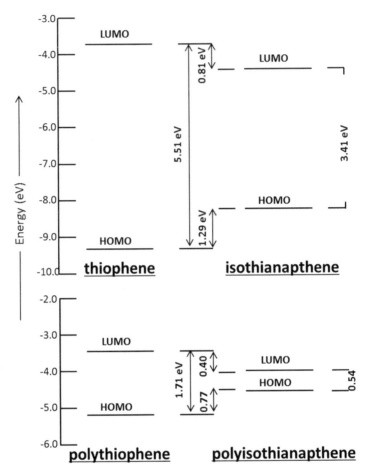

Fig. 28.52 Comparisons of the evolutions of the HOMO and LUMO one-electron energies when going from thiophene to isothianaphthene and poly(thiophene) to poly(isothianaphthene)

28.17 Methods Considering E–E and E–P Correlations

Among innovative methods combining one-electron Hamiltonians with consideration of dopant-induced and e–e correlation effects has been one put forth by Stafström [936]. The Hamiltonian used in his study includes the SSH Hamiltonian together with Hamiltonians accounting for dopants and e–e correlation (see Appendix 28.2). Figure 28.53 [936] shows the DOS at the Fermi energy as a function of doping level for ordered soliton, disordered soliton, and polaron lattices of trans-P (Ac), calculated using this method and considering a 1D polymer. Among the major findings of Stafström's study are that if one only takes a one-electron approach, the net charge of ca. 0.5 e- opposite to a counter-cation (for n-doping) is localized at the C-atom directly across the counterion, but that if one takes the dopant and e–e

Fig. 28.53 Poly
(di-thiophenevinylene)
structure studied by
Pomerantz et al. [916]

correlation effects into consideration, this localization is energetically unfavorable, and the soliton so generated is then found delocalized over a large segment of the lattice. Introducing weak disorder (e.g., by a slight arbitrary shifting of counterion positions) does not change the structural properties appreciably. The P(Ac) energy gap is found to decrease smoothly with increased doping. Disorder reduces the gap by only ca. 0.2 eV. The energy gap at a 6.67% doping level, close to the 6% level for the dramatic semiconductor-to-metal transition for P(Ac), is small enough to be caused to vanish by 3D interactions of the lattice. The calculations nevertheless do not fully explain the sudden onset of metallic properties, i.e., the semiconductor-to-metal transition, at 6% doping (they predict a gradual onset), and the observed saturation in the DOS above this doping level.

Mazumdar's group [924] has attempted a detailed correlation of experimental optical properties of P(PP) and P(PPV), showing that a theoretical method that considers e–e correlation among π-electrons (using a Pariser–Parr–Pople Hamiltonian, Appendix 28.2) is able to account for experimental optical spectra more accurately than one-electron methods. Figure 28.54a shows the optical spectrum of P(PV) predicted from one-electron calculations, which in particular predicts that peak (ii) will lie exactly at the center of peaks (i) and (iii). Figure 28.54b shows the experimental spectrum of a substituted P(PV) derivative; in this, except for the peak labeled Ib, the spectrum is nearly identical to unsubstituted P(PV). Figure 28.55 shows the spectrum of P(PV) calculated using the e–e correlation method, evidently corresponding much closer to the experimental spectrum. Based on correlations of the calculated spectra with experimental absorption spectra for P(PV), Mazumdar et al. are able to postulate that in the P(PV)'s studied charge carriers are delocalized over no more than eight to ten units. They postulate that the lowest optical state of the CP is an exciton with a binding energy of ca. 0.9 eV.

In more recent work from the Mazumdar group [937], a systematic characterization of excited states in CPs was performed, using full configuration interaction (CI) calculations in an exciton basis within which a long chain CP is considered as coupled molecular units. Among their findings, focusing primarily on P(Ac), were that $1B_u$ state was an exciton, and the fundamental two-photon states could be broadly classified into triplet-triplet (TT), charge-transfer (CT), and singlet-singlet (SS) excitations. The $2A_g$ state was classified as TT, and the mA_g state, an even parity state playing a strong role in NLO properties, was found to be a correlated CT state. The authors found that their work could be easily extended to poly(phenylenes).

Also in more recent work, Gallagher and Spano [938] carried out tight-binding, two-band Hamiltonian-based calculations in the formation and NLO properties of biexcitons in 1D CPs. Calculated two-photon absorption spectra were found to

Fig. 28.54 Local phonon densities of states projected onto carbon displacements for ideal polyene chains (dashed) and chains containing a soliton (solid) (After Ref. [935], reproduced with permission)

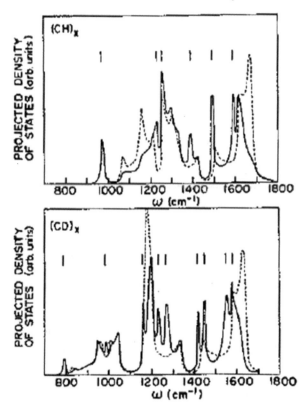

Phonons and the Peierls Instability

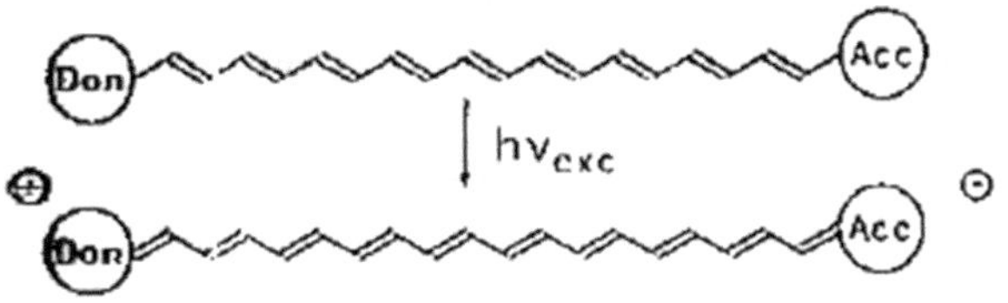

Fig. 28.55 Polyene chain with donor and acceptor end groups (After Ref. [936], reproduced with permission)

contain two types of peaks: a lower energy one due to a CP single exciton and a higher energy one due to a biexciton.

28.17.1   Density Function Theory (DFT) Methods

Since about 2005, theoretical (computational chemistry) methods, especially density functional theory (DFT), have been applied more frequently to CPs and monomers and oligomers thereof.

For example, Dai et al. [910] described a study of oligopyrroles, wherein first-principle calculations using DFT were used to study n-pyrrole oligomers ($n = 2$ to 18). They found that neutral oligomers were bent, whereas negatively charged oligomers became nearly planar, due to $(-)$ charge accumulation at the ends of the chains. They also found that doping of the $n = 12$ oligomer with Li atoms was possible, although they also found that Li-doped P(Py) was *not* metallic. Comparison between neutral and doped (P(Py)) showed that doped polymers displayed a substantial depletion of the bandgap energy and the appearance of dopant-based bands in the gap for a 50% monomer doping level.

Salzner [854] provided a useful summary of results of DFT calculations on CPs, discussing insights from time-dependent DFT studies. He noted that the polaron model that was developed from one-electron theories still remains the most widely used description of CPs.

Among his findings were that theoretical analysis of states has resolved several puzzles which cannot be understood with the polaron model alone, for example, the origin of the dual absorption band of green polymers and the origin of a "vestigial neutral band" upon doping of long oligomers. DFT calculations have also shown that defect localization is not crucial for spectral changes observed during doping and that there are hints that there may be no bound bipolarons in CPs. He also found that DFT can be applied to oligomers that are long enough to address crucial properties of CPs. To predict reasonable structures and electronic properties, inclusion of HF exchange is necessary.

Tretiak et al. [911] also used a time-dependent DFT method to predict exciton sizes of CPs, specifically the electronic structure and size scaling of spectroscopic observables in CPs. They showed that local density approximations and gradient-corrected functionals do not have an effective attractive coulomb interaction between photoexcited electron-hole pairs to form bound states and therefore do not reproduce finite exciton sizes. Long-range nonlocal and nonadiabatic density functional corrections such as hybrid mixing with an exact Hartree–Fock exchange are necessary to capture correct delocalization of photoexcitations in one-dimensional CP chains. Figure 28.56 shows representative results from this work.

Senevirathne et al. [939] reported a theoretical investigation of bandgaps of conducting polymers with heterocycles, wherein a new method was developed for calculating bandgaps of CPs. The method consisted of three major steps: In the first step, the geometry optimization of the oligomer structures was carried out. The optimized oligomer structures were then used as input in the second step. In the second step, bandgaps of oligomers were calculated by taking into account the energy difference of the highest occupied molecular orbital (HOMO) and lowest unoccupied molecular orbital (LUMO) levels. In the third step, the nearly free electron model and the extrapolation techniques on the oligomers were used to obtain the bandgap values of polymers.

Senevirathne et al. [939] presented an interesting study of bandgaps of heterocyclic CPs wherein the bandgaps were calculated using standard DFT methods, but then a single-parameter correction was applied that yielded calculated bandgaps that very closely approximated experimental values. Their results are summarized in Table 28.4 below.

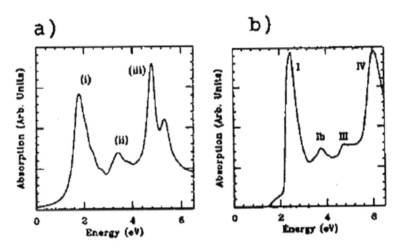

Fig. 28.56 (**a**) Optical absorption of PPV, cf. Eq. (1), in arbitrary units. (**b**) Experimental absorption spectrum of MEH-PPV (After Ref. [924], reproduced with permission)

Table 28.4 Bandgaps of selected heterocyclic CPs calculated using DFT techniques with a single-parameter correction, compared with experimental values (After Ref. [939])

Molecule	0.75	V_o (corrected) (eV)	Experimental (eV)
Polythiophene	0.75	1.8	1.8–2.21
Polypyrrole	1.4	3.3	2.9–3.2
Polyfuran	1.05	2.5	1.94–2.7
Poly 3-methyl thiophene	0.71	1.7	1.7–2.4
Poly 3-methyl pyrrole	1.23	2.9	2.93
Poly 3-methyl furan	0.9	2.1	

28.18   Appendix 28.1: Selected Calculated Band Structures

Figure 28.57 shows the band structure of trans-P(Ac) for varying bond length alternation computed via the one-electron ETB method. Figure 28.58 shows the corresponding densities of states (DOS). The lifting of degeneracy and the formation of a gap at Y for (b,c) (i.e., increasing bond alternation) are to be noted. These may be compared with Fig. 28.59 which shows trans-P(Ac) band structure and DOS computed using the VEH approach and with Fig. 28.60, which shows band structures computed on trans-, cis-, and idealized metallic P(Ac) using the EH technique.

The ab initio band structures/DOS calculated for P(Ac) for Cl- and Na-doping by Tanaka and Tanaka [932] are shown in Fig. 28.61. The center figure shows the one-electron band structure for neutral P(Ac) for comparison. The structures at the extreme edges, representing "polson–antipolson" lattices of the CP (polson = polaron – charged soliton pair), represent quasi-metallic behavior. In the Na-doping case (indicated in the top margins of the figure), the Fermi level shifts upward,

Fig. 28.57 Calculated
absorption spectrum of an
8-unit PPV oligomer with
$U = 8$ eV and $\kappa = 2$. The
arrow on the x-axis
indicates the location of the
continuum threshold within
SCI (After Ref. [924],
reproduced with
permission)

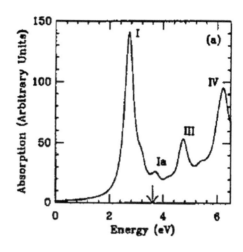

Fig. 28.58 Scaling and
saturation of calculated and
experimental bandgap s1B_u
stated energies (top panel)
and the respective oscillator
strengths with the poly(p-
phenylene vinylene) (P
(PV)) oligomer size, with
the structure of P
(PV) shown in the inset
(After Ref. [911],
reproduced with
permission)

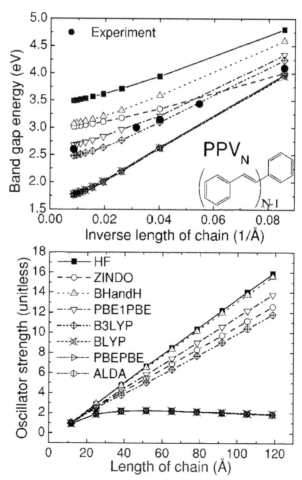

Fig. 28.59 Band structure
of *trans*-(CH)$_x$ for different
carbon–carbon bond lengths
(a) uniform (1.39 Å);
(b) weakly alternating
(C = C, 1.36 Å; C–C,
1.43 Å); and (c) strongly
alternating (C = C, 1.34 Å;
C–C, 1.54 Å). Note the
lifting of the degeneracy at
Y as bond alternation occurs
(After Ref. [925],
reproduced with
permission)

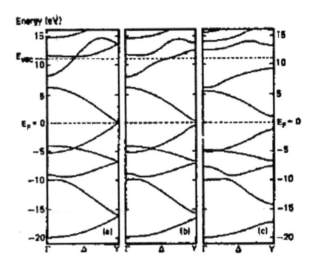

while in the Cl-doped case, it moves downward. The HOMO and LUMO make
effective contact in the very high doping cases (extreme left and right), leading to
the prediction of quasi-metallic behavior (Figs. 28.64 and 28.65).

Figure 28.62 illustrates the computed VEH band structure for P(Py), showing
12 occupied valence bands. Figure 28.63 in contrast shows a schematic, composite
band structure picture of P(Py), assembled from various theoretical and experimental studies. Figure 28.66 shows VEH band structures using ab initio STO-3G
optimized geometries for, respectively, P(PP) at 0% doping. Figure 28.67 shows
the VEH band structure for P(T) [940]. The π-band portion of this structure is
identical to that of poly(3-alkyl thiophene) at low temperatures, where greater
coplanarity between monomer units is observed. The sulfur 4p atomic orbitals do
not appear to contribute significantly to the lowest energy occupied band, labeled π_3
in the figure .

28.19   Appendix 28.2: Selected Methodology, Calculation
Details, and Relevant Equations

28.19.1   Extended Hückel (EH)

As applied to CPs, the EH method takes a set of basis orbitals for the atomic
constituents of a unit cell, $\{\chi\}$, and forms the set of Bloch basis functions:

$$b_\mu(k) = N - \tfrac{1}{2} \Sigma e^{ik \cdot R} t \chi_\mu (\mathbf{r} - \mathbf{R}_t), t \qquad (28.1)$$

where k is the wave vector and $\mathbf{R}t = \Sigma l_i a_i$ with ai being primitive vectors. With the
Bloch basis orbitals, l.c.a.o. crystal orbitals $\psi_n(k)$ are expressed as

Fig. 28.60 Densities of states of *trans*-(CH)$_x$ for the bond lengths of Fig. 28.59 (After Ref. [925], reproduced with permission)

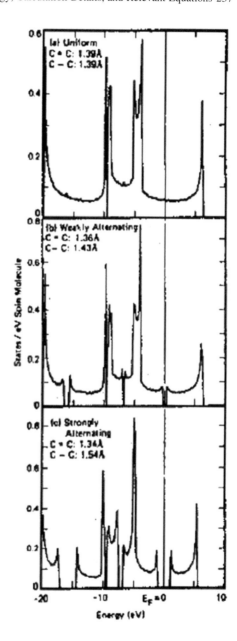

$$\psi_n(k) = \Sigma C_{n\mu}(k) b_\mu(k).\mu \tag{28.2}$$

The corresponding eigenvalues $\varepsilon_n(k)$ and coefficients $C_{n\mu}(k)$ are obtained from the eigenvalue equation:

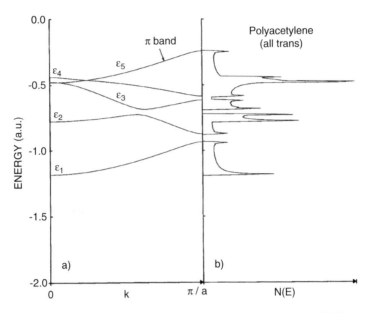

Fig. 28.61 (a) VEH band structure. (b) DOS, of poly(acetylene) (After Ref. [928], reproduced with permission)

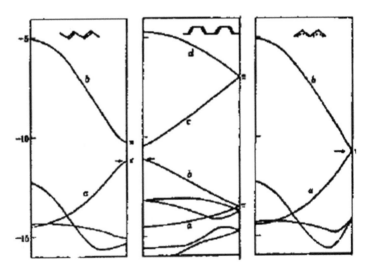

Fig. 28.62 Band structures of poly(acetylene) computed via the EH method. Units in eV (After Ref. [912], reproduced with permission)

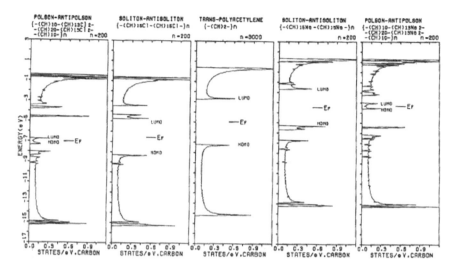

Fig. 28.63 Density of states of charged polyene chain showing the mid-gap (charged soliton lattice) and the metallic levels (polson lattice) (After Ref. [932], reproduced with permission)

$$H(k)C(k)=S(k)C(k)e(k), \tag{28.3}$$

where $H_{\mu\upsilon}(k) = <b_\mu(k)|H_{eff}|b_\upsilon(k)>$ and $S_{\mu\upsilon} = <b_\mu(k)|b_\upsilon(k)>$. Then band structures are determined by repeating the above calculation for various values of k (usually within the first Brillouin zone). The atomic parameters of the extended Hückel calculation are detailed in the appendix. Unless stated otherwise, lattice sums (i.e., the summation over l in Eq. 7.1) were carried out to first-nearest neighbors.

Equation 28.1 indicates that each Bloch basis $b_\mu(k)$ consists of the atomic orbitals $X_\mu(r-R_l)$ located at the various unit cells l, and each of them carries the phase factor $e^{ik\cdot R}l$. Thus the nodal properties of a crystal orbital $\psi n(k)$ at a specific value of k are constructed once the expansion coefficient $C_{n\mu}(k)$ is known for each $b_\mu(k)$.

28.19.2 LCAO/ETB (Extended Tight Binding)

This one-electron method uses a linear combination of Bloch-adapted Gaussian orbitals as its basis set, with double-zeta quality C s and p states. To construct the crystal potential, a superimposition is employed of atomic potentials generated from SCF-HF (self-consistent filed Hartree–Fock) charge densities using an approximation such as the Kohn–Sham–Gaspar local exchange approximation. Potential matrix elements are calculated to convergence in momentum space. Generally, overlap integrals > ca. 10^{-7} are retained.

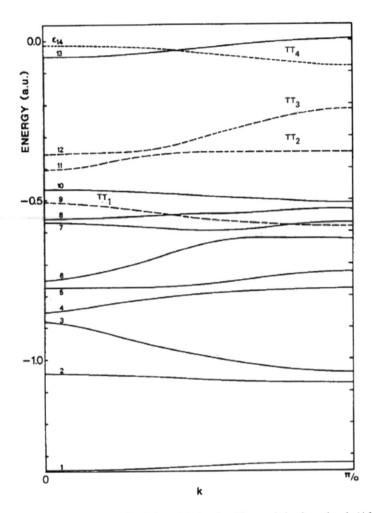

Fig. 28.64 VEH band structure of poly(pyrrole) showing 12 occupied valence bands (After Ref. [928], reproduced with permission)

28.19.3 VEH

The one-electron, nonempirical VEH method first developed by Nicolas and Durand and adapted to CPs by Brédas and others [941, 942] considers only the valence electrons explicitly, with Gaussian functions of appropriate orbital symmetry typically used for the C and H valence electrons. Coulomb interactions are simulated implicitly. In this, effective Fock operators contain the kinetic term and a sum of atomic potentials:

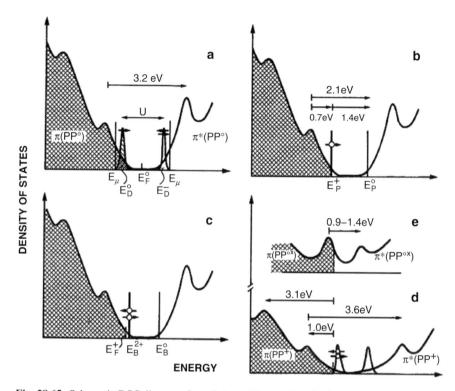

Fig. 28.65 Schematic DOS diagrams for poly(pyrrole) at various doping levels, (a) through (e), neutral to very doped (After Ref. [884], reproduced with permission)

$$F_{\text{eff}} = -\frac{\Delta}{2} + \Sigma_h \Sigma_A V_A^h \tag{28.4}$$

with the atomic potentials expressed as linear combinations of nonlocal Gaussian projectors:

$$V_A = \Sigma_l \Sigma_m \Sigma_i \Sigma_j C_{i,j,l,m}^A |\chi_{ilm}^A >< \chi_{jlm}^A| \tag{28.5}$$

The atomic potential parameters are the linear coefficients, $C_{i,j,l,m}$, and nonlinear exponents α_i, which are optimized for each atomic potential type on model molecules, thus explicitly including the effects of the chemical environment.

The numerical coefficients $C_{i,j,lm}$ are independent of m in the case of spherical symmetry, which we usually consider. In order to achieve double-zeta quality, the summations over i and j extend up to 2. The χ_{ilms} are normalized Gaussian functions:

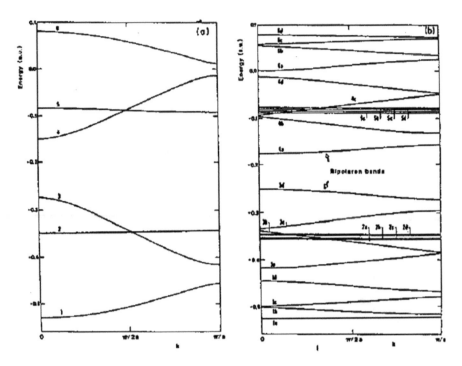

Fig. 28.66 VEH π-band structure for (a) a polyparaphenylene chain, with STO-3G geometry from undoped quaterphenyl corresponding to 0 mol % doping level, (b) a polyquaterphenyl chain, with STO-3G geometry from sodium-doped quaterphenyl corresponding to 50 mol % doping level, and (c) a polyquinoid chain, with STO-3G geometry from an inner ring of sodium-doped quaterphenyl corresponding to a 100 mol % doping level (After Ref. [933], reproduced with permission)

Fig. 28.67 The band structure (Ek vs k) for planar poly(thiophene) (After Ref. [940], reproduced with permission)

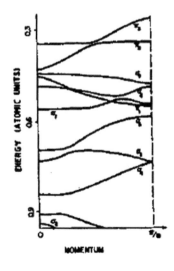

$$\chi_{ilm} = N_i r^l \exp\left(-\alpha_i r^2\right) Y_{lm}(\theta, \phi) \tag{28.6}$$

where N_i is the normalization factor, and $Y_{l,m}(\Theta, \phi)$ denotes usual spherical harmonics. Only ls and 2p Gaussian Cartesian functions are used.

The first step in the parameterization requires performing valence ab initio Hartree–Fock SCF calculations on the model molecules by a theoretical pseudopotential method [943] with first an STO-3G minimal basis set and then a double-zeta basis set. For both model molecules, we take the Fock operator to be

$$F = \Sigma_\nu \Sigma_\nu |\phi_\nu><\phi_\nu| \tag{28.7}$$

The sum is over all occupied levels. The valence orbitals ϕ_ν are those obtained from the STO-3G calculations, whereas the corresponding one-electron energies ε_ν are those determined from the double-zeta calculations.

In the second step, the atomic potential coefficients and exponents are determined by minimizing several variables.

28.19.4   Strictly Ab Initio Methods

Exclusively ab initio methods have been rarely employed on CPs other than P(Ac). An example is a calculation, by Tanaka and Tanaka [932]. For the very detailed methodology used, reference is made to the original source [932].

28.19.5   Representative Combination Ab Initio/ Semiempirical Methods

A typical ab initio/semiempirical combination method, used by Brédas et al. [933], may be cited. A restricted Hartree–Fock LCAO-MO technique is used as the starting point, with all core and valence electrons considered explicitly, with handling of up to 200 atomic orbitals. Two-electron integrals smaller than ca. 10^{-5} eV may be neglected, with convergence criteria for each of the density matrix elements of 5×10^{-5} atomic units. A Slater-type orbital three Gaussian (STO-3G) basis set is used, with each atomic orbital represented as a linear combination of three Gaussian functions, and standard molecularly optimized Slater exponents. Computation is limited to the *tetramer*, with only n-doping accounted for by introducing Na and Li atoms.

In other work of the Brédas group [936], geometry optimizations up to the tetramer level using the semiempirical AM1 method were used, which were *then* (i.e., in reverse order to the above) input to restricted Hartree–Fock (RHF) 3-21G basis set calculations of the total energy of different rotational conformations for undoped and doped polymer forms.

28.19.6   The SSH Hamiltonian

The Su–Schrieffer–Heeger [921, 922] Hamiltonian includes terms taking into account the π-electron hopping and σ-bond repulsion energy. As noted earlier, the SSH Hamiltonian neglects all two-electron terms and includes many aspects of electron–phonon coupling, but neglects some aspects such as the deformation potential coupling of phonons to the on-site electronic charge density:

$$H_{SSH} = -\sum_{n,\sigma} [t_o + \alpha(u_n - u_{n+1})] \left(C^\dagger_{n+1,\sigma} C_{n,\sigma} + C^\dagger_{n,\sigma} C_{n+1,\sigma} \right)$$
$$+ \frac{K}{2} \sum_n \left(u_n - u_{n+1} - C^2 \right) \tag{28.8}$$

The expression $t_0 + \alpha(u_n - u_{n+1})$ is the linearized bond length dependence of the hopping integral, u_n is the displacement of the n-th lattice site, t_0 is the hopping integral of adjacent sites when $u_n = 0$ for these sites, and α is the electron–phonon coupling constant. The compressibility of the σ-bonds is described by the classical, harmonic term in Eq. 28.8, where K is the effective spring constant. The approximate treatment of the distance dependence in the two terms in Eq. 28.8 is justified since the displacements, un, are small compared to the average C–C bond length.

The total energy will then be the sum of the nuclear kinetic energy plus an effective potential energy:

$$V_{eff}\{u_n\} = \frac{K}{2} \sum_n (u_n - u_{n+1})^2 + \sum_{\nu,s} \varepsilon_{\nu,s} n_{\nu,s} \tag{28.9}$$

where $n_{\varepsilon,s}$ is the occupation number of the eigenstate with energy ε and component s. The static solutions are those which minimize $V_{eff}\{u_n\}$.

28.19.7   Methods Considering E–E and E–P Correlations

Strafström's methodology [936] uses the Hamiltonian:

$$H = H_{SSH} + H_C + H_{e-e} \tag{28.10}$$

where H_{SSH} is the SSH Hamiltonian, and the Hamiltonians H_C and H_{e-e} represent dopant (counterion) and electron–electron correlation effects, respectively:

$$H_c = \sum_n \left(V_n^{imp} + V_n^{sol} \right) C_{n,\sigma}^\dagger C_{n,\sigma} \tag{28.11}$$

where V_n^{imp} is the coulomb potential at site n due to all counterions in the system, and V_n^{sol} is the potential at site n due to the charges on surrounding chains. These potentials are described by the following expressions:

$$V_n^{imp} = -e^2 \sum_j \frac{1 - v_{nnj}\left(1 - e^{-\gamma|n-n_j|a}\right)}{\varepsilon_1 \left[a^2(n - n_j)^2 + d^2 \right]^{\frac{1}{2}}} \tag{28.12}$$

$$V_n^{sol} = e^2 \sum_{n,\sigma} P_{nn'}^\sigma \frac{1 - v_{nn'}\left(1 - e^{-\gamma|n-n'|a}\right)}{\varepsilon_2 \left[a^2(n - n')^2 + 3d^2 \right]} \tag{28.13}$$

For P(Ac) here, $a = 1.222$ Å is the average distance between adjacent sites along the chain axis, $d = 2.4$ Å is the perpendicular distance of the impurity from the chain, and p3d is the perpendicular distance between adjacent chains. The lattice site opposite to an impurity is denoted n_j and a lattice site on an adjacent chain n'. Pn'n' denotes the charge density on the n'-th site of an adjacent chain. This quantity is taken to be identical to the self-consistently calculated charge density on the chain treated quantum mechanically, shifted however, in order to match the soliton sites on the adjacent chains.

The electron–electron repulsion term including diagonal elements only, often referred to as the extended Hubbard term, is

$$H_{el-el}^\sigma = \frac{1}{2} \sum_{n,m,\sigma,\sigma'} U_{nm} c_{n,\sigma}^\dagger c_{n,\sigma} c_{m,\sigma}^\dagger c_{m,\sigma} \tag{28.14}$$

Here, this interaction is treated within the self-consistent field Hartree–Fock approximation which gives the following mean-field spin-polarized electron–electron interaction Hamiltonian:

$$H_{el-el}^\sigma = \frac{1}{2} \sum_{n,m,\sigma} U_{nm} \left(P_{mm}^{\sigma'} c_{n\sigma}^\dagger c_{n\sigma} - P_{nn}^\sigma c_{n\sigma}^\dagger c_{n\sigma} \right), \quad \sigma, \sigma' = \uparrow, \downarrow \tag{28.15}$$

Here, P_{mm} is the charge density at the m-th site and U_{nm} the effective coulomb repulsion integrals between 2pz orbitals attached to sites n and m along the polymer chain. We adopt the screened Ohno expression [944] for these integrals:

$$U_{nm} = \frac{U_o e^{-\gamma |r_{nm}|}}{\left[1 + 0.6117 r_{nm}^2\right]^{\frac{1}{2}}} \tag{28.16}$$

The total energy of the system with the full Hamilton is

$$\varepsilon = \sum_{n,\sigma} -[t_0 + \alpha(u_n - u_{n+1})]2P^\sigma_{n,n+1} + \left[V^{imp}_n + V^{sol}_n\right]P^\sigma_{n,n}$$

$$+ \frac{1}{2} \sum_{n,m,\sigma,\sigma'} U_{nm}\left[P^\sigma_{n,n}P^{\sigma'}_{m,m} - P^\sigma_{n,n}P^{\sigma'}_{m,m}\delta_{n,m}\delta_{\sigma,\sigma'}\right] \tag{28.17}$$

where P^σ_{nm} are the elements of the density matrix, which is obtained from the solution of the Schrödinger equation as:

$$P^\sigma_{nm} = \sum_i^{OCC} B^\sigma_{n,i}B^\sigma_{m,i} \tag{28.18}$$

The recent work of Mazumdar and Chandross [945] on P(PP) and P(PPV) considers e–e correlation using a *Pariser–Parr–Pople*-type Hamiltonian:

$$H = H_{1e} + H_{e-e} \tag{28.19}$$

where the one-electron portion of the Hamiltonian is

$$H_{1e} = -\sum_{<ij>,\sigma} t_{ij}\left[c^+_{i,\sigma}c_{j,\sigma} + c^+_{j,\sigma}c_{i,\sigma}\right], \tag{28.20}$$

and

$$H = H_{1e} + H_{ee} \tag{28.21}$$

where H_{1e} is the same as in Eq. 28.20, and H_{ee} is the electron–electron interaction,

$$H_{ee} = U\sum_i n_{i,\uparrow}n_{i,\downarrow} + \frac{1}{2}\sum_{i,j} V_{ij}(n_i - 1)(n_j - 1) \tag{28.22}$$

Here, $n_{i,\sigma} = c^+_{i,\sigma}c_{i,\sigma}$ is the number of electrons with spin σ on-site i, $n_i = \Sigma_\sigma n_{i,\sigma}$, and U and V_{ij} are the on-site and intersite coulomb interactions (note that unlike t_{ij}, V_{ij} is not restricted to nearest neighbors and is long range). We use the same tij as in Eq. 28.20. For the V_{ij} we have chosen a parameterization similar to the Ohno parameterization of the Pariser–Parr–Pople Hamiltonian:

$$V_{ij} = \frac{U}{k\left(1 + 0.6117R_{ij}^2\right)^{1/2}}, \tag{28.23}$$

where R_{ij} is the distance between atoms i and j in Å, and κ is a parameter that determines the decay of the long-range part of the potential. Within the Ohno parameterization, $U = 11.13$ eV and $\kappa = 1$. Unlike the Ohno parameterization, however, they studied many different combinations12 of U and κ and used those most accurately fitting absorption spectra.

Mazumdar et al.'s calculations [945] are within the single configuration interaction (SCI) approach, where configuration interaction between all single excitations from the Hartree–Fock ground state is included, but higher states are not included.

   The Peierls–Hubbard Hamiltonian, which considers both e–e and e–p correlations explicitly, is given by

$$H_{PH} = H_{lat} + H_{1-e} + H_{e-e} \tag{28.24}$$

As discussed previously, H_{lat} describes the structural energy of the lattice arising from the σ-bonds, supplemented by the kinetic energy of the ions. Furthermore,

$$
\begin{aligned}
H_{1-e} &= -\sum_{l\sigma}\left\{\left[t_0 - (-1)^l(\alpha y_l + t_1)\right]\left(c_{l\sigma}^+ c_{l+1\sigma} + c_{l+1\sigma}^+ c_{l\sigma}\right) + \varepsilon_l c_{l\sigma}^+ c_{l\sigma}\right\} \\
&= H_{1-e}^{(1)} + H_{1-e}^{(2)} + H_{1-e}^{(3)} + H_{1-e}^{(4)}
\end{aligned}
\tag{28.25}
$$

is the one-electron Hamiltonian and

$$H_{e-e} = U\sum_l n_{l\uparrow}n_{l\downarrow} + \sum_{i,j>0} V_j n_l n_{l+j} \tag{28.26}$$

is the electron–electron interaction, where as usual $n_l = \Sigma n_{l\sigma}$. $y_l = (-1)^l(r_{l,l+1} - r_0)$ is the deviation of the bond length between the lth and (l + 1)th chemical moieties (CH in polyacetylene, for instance) from the average bond length, and $c^\dagger_{l\sigma}(c_{l\sigma})$ creates (annihilates) a π-electron at site l with spin σ ($= \pm \frac{1}{2}$) and

$$H_{lat} = -\sum_{l\sigma}[t_0 + \alpha_x(u_{l,x} - u_{l+1,x})]\left[c_{l\sigma}^+ c_{l+1\sigma} + c_{l+1\sigma}^+ c_{l\sigma}\right] \tag{28.27}$$

Evidently, the lattice portion is identical to that in the SSH Hamiltonian.

28.20   Problems and Exercises

1. Recount the significant features and the salient differences between the following: ab initio methods, semiempirical methods, one-electron methods, e–e correlation, e–p coupling, and Peierls instability.

2. Outline the major features, strong points, and drawbacks of the following theoretical methods: EH, VEH, methods using the SSH Hamiltonian, methods using Pariser–Parr–Pople-type Hamiltonians, and methods using Peierls–Hubbard-type Hamiltonians.

3. From a reading of this chapter, which methods do you find are best suited, in terms of accuracy and economy of time, for computation of the following properties of CPs: bandgaps, band structures, band structure evolutions, rotational barriers, UV-Vis-NIR absorption spectra, far-IR absorption and reflectance spectra, Pauli susceptibility, and optical transition probabilities (intensities).

4. From a reading of this chapter, list the major findings of studies using each method. What do you feel are the three most important properties for design of a CP with a vanishingly small bandgap upon doping at room temperature? Outline a possible chemical, electrochemical, or combined synthesis of one such candidate and its subsequent doping.

5. Outline all methods available for validating conduction models for CPs, and mention which aspects of conduction they shed light on.

6. Of the following – intrachain, interchain, intrafiber, and interfiber conduction – which would be relevant to hopping conduction models, and in what way? From the conductivity data cited in this chapter, what would you expect the packing and morphology to be like for trans-P(Ac)/I3 at 7% doping?

7. Delineate the salient points of each of the following conduction models: Mott VRH, Sheng, Paasch, Kivelson, and electronic. Identify which temperature and doping ranges each model is most correctly applicable to.

8. Differentiate DC and microwave (2 to 100 GHz) conductivities, and show their temperature functionalities for at least two CP systems.

9. Briefly outline effects of the following parameters on DC and AC conductivities of CPs: pressure, crystallinity, molecular weight, stretching along chain axis, substitution with alkyl groups of increasing length at the N-atom for P (Py), and the β-position for P(T).

10. What is thermopower and how is it related to conduction in CPs?

11. What are typical literature values reported for charge carrier mobilities in CPs for doped and undoped states? Compare these with values you can find from other literature sources for mobilities in solid (crystalline or amorphous) matrices.

12. Write out definitions of the terms polaron, bipolaron, soliton, and antisoliton. Identify commonalities and differences and spin-charge relationships. Discuss all possible optical transitions possible for each.

13. Draw expected positive bipolaron and positive polaron structures for poly (aniline).

14. Using the spectroelectrochemical data shown in the chapter in this book discussing electrochromics of CPs, for at least three different CPs, identify plausible $\pi \rightarrow \pi^*$ and bipolaron absorptions, and draw out the electronic transitions they represent in a band structure diagram for each CP.

15. Draw structures showing benzenoid and quinonoid segments alternating for four monomer units for each of the following CPs: poly(3-hexyl-thiophene), poly(3,3'-dimethyl pyrrole), and poly(isothianaphthene) (a thiophene with a fused benzene ring).

16. Schematically show and compare the results of n-doping of a "conventional" semiconductor (e.g., Si with P) and a CP (e.g., poly(thiophene) with tetrafluoroborate), using band structure diagrams, and molecular (for Si) and chain (for poly(thiophene)) structures.

17. Using band structure diagrams, identify, recapitulate, and compare all optical absorptions expected for neutral, $(-)$ and $(+)$ solitons, $(+)$ and $(-)$ polarons, and $(+)$ and $(-)$ bipolarons. Arrange these in terms of decreasing energy, and identify in which spectral regions you expect they will appear.

Chapter 29
Basic Electrochromics of CPs

Contents

29.1 Basics of Electrochromism and Spectroelectrochemistry of CPs

29.1.1 Basics

Electrochromism, i.e., change of color or spectral signature with applied voltage (or equilibrium potential or chemical potential), is one of the most prominent and fundamental properties of CPs. Electrochromic properties of CPs have aroused the most interest for practical applications initially. CPs possess the unique property that *their color changes with redox state*, which in turn is nearly always related to doping level of the CP. This doping level is altered in varied ways: electrochemically, by applying a suitable potential to the polymer; chemically, subjecting the CP to chemical oxidation or reduction and usually involving a dopant ion; and by secondary, indirect chemical means, such as alteration of the pH, which again affects the redox state of the polymer. Doping can, it must be remembered, be oxidative ("p-type", with the CP undergoing oxidation and the dopant being an anion) or reductive ("n-type", CP undergoing reduction, dopant cation), with the former far more common.

    Electrochromism is distinguished from *photochromism*, in which light, rather than electrical or chemical potential, causes the color change. Most current photo-chromics, e.g., those found in commercial sunglasses, need UV light to function and so do not function indoors or in automobiles.

© Springer International Publishing AG 2018
P. Chandrasekhar, *Conducting Polymers, Fundamentals and Applications*,
https://doi.org/10.1007/978-3-319-69378-1_29

Although it is most commonly considered in the visible spectral region (approximately 400 'to 700 nm), electrochromism of CPs can be found in many other spectral regions, including the *infrared* (IR, approximately 0.9 to 45 μm, of which the "near-infrared" (NIR) region is approximately 0.9 to 1.2 μm) and *microwave–mm-wave* (approximately 5 MHz to 50 GHz) spectral regions.

The visible-region color changes for most CPs are quite pronounced. For instance, the "p-doped" poly(pyrrole) undergoes a change from light greenish yellow in the reduced (pristine) state to gray black in the highly doped, oxidized state; p-dopable poly(aniline) undergoes a change from very light green in its highly reduced state, through green in a partially oxidized state, to blue black in its highly doped, completely oxidized state. Such color changes have been the basis for hopes for practical visible-region electrochromic devices based on CPs.

Other means to change chemical potential or redox equilibrium of a CP, and hence its color, include dilution in solvents (when the CPs are somewhat soluble) and change of temperature. These two effects, respectively, *solvatochromism* and *thermochromism*, are treated later in this chapter.

The change of color of a CP with doping is of course directly related to the creation of new, mid-gap states, and thus new optical transitions, as discussed at some length elsewhere in this book. New absorptions, generally at higher wavelength (lower energy) are added to the customary valence → conduction, ($\pi \rightarrow \pi^*$), transition of the pristine, undoped CP.

Besides redox state, other factors also influencing electrochromism include the following: the nature of the dopant (different dopants oftentimes yield different spectral signatures or colors), morphology (particle size, chain length/orientation and degree of folding, and degree of disorder can effect minute changes in color), and pH in aqueous systems (indirectly affecting redox state).

29.1.2 Spectral Regions

As noted above, electrochromism is observed not only in the *visible* spectral region but *also in the near-UV (300–400 nm), near-IR (0.7 to 2.5 μm), mid-IR (2.5 μm to 8 μm), far-IR (8 μm to 18 μm)*, and, to some extent, *microwave–mm-wave (5 MHz to 50 GHz) regions*. Properly speaking, conductivity, including microwave conductivity, is also an "electrochromic" property, since it can be changed with applied potential or CP redox state, and properties such as optical conductivity have a direct spectral connotation. These are discussed separately elsewhere in this book.

29.1.3   Elementary Electrochemistry of CPs as Basis for Electrochromism

Although electrochemistry of CPs is treated in detail in a later chapter in this book, we must digress to this subject here just sufficiently to give the unfamiliar reader a working knowledge adequate for interpretation of CP electrochromism discussed in the sequel. Electrochemistry of CPs can be carried out in *aqueous systems*, generally used for CPs such as poly(anilines) where protons participate in the doping processes. It can also be carried out in *nonaqueous systems*, applicable to the majority of CPs. Finally, it can also be carried out in exotic solvent systems such as *ionic liquid electrolytes*; these latter are essentially room-temperature molten salts which are organic (such as *n*-butyl imidazolium tetrafluoroborate) but behave somewhat like inorganic solvents.

Electrochemistry may be carried out in a *two-electrode mode* (working, counter electrodes) or a *three-electrode mode* (working, counter, reference electrodes); the latter is typically better for accurate potential control, while the former may more correctly emulate practical applications such as electrochromic devices.

The working electrode typically comprises a CP film on a conductive electrode substrate. These CP films are typically directly prepared on or cast onto visible-region-transparent, conductive substrates such as indium tin oxide (ITO)/glass and ITO/PET (poly(ethylene terephthalate), what soft drink bottles are made of); more recently used conductive substrates, used with less success, have included Ag nanowires, CNTs, and very thin (<60 nm typically) Au or Pt, all deposited on glass or plastics such as PET. The substrate can also simply be a solid substrate such as a Pt wire or flag.

CP film thicknesses can vary from ca. 50 nm to several microns; thicknesses, e.g., those measured coulometrically (by counting total charge used to deposit the CP), are usually only approximate, except when measured by absolute methods such as electron microscopy or stepper techniques. The counter electrode can be any electrode with the stipulation that it generally be larger or at least equal in area to the working electrode, so that electrochemical reactions on its surface are not limiting. A device known as a potentiostat/galvanostat is used for control of potential (potentiostatic mode) or current (galvanostatic mode) applied to the working electrode and monitoring of the resultant current or potential. For the three-electrode mode, the working electrode potential is controlled with respect to the reference electrode, which may be standard (e.g., saturated calomel electrode (SCE), wire-Ag/AgCl) or quasi-reference (e.g., Pt wire). An electrolyte of adequate conductivity, generally an ionic salt in a solvent, gel, or polymer matrix, is required for the charge transport necessary for observation of any electrochemical phenomena.

It is important to note that due to the nature of CP synthesis, usually involving the presence of dopants, most freshly synthesized CPs and especially those prepared electrochemically are not normally in their "pristine" (de-doped or virgin) state. Rather, these "as-prepared" CPs have a small doping level, of the order of <1% to up to 25%.

When a potential is applied to the CP which comprises the working electrode in an electrochemical system, it undergoes oxidation or reduction. Counterions, usually dopant ions, then flow from the electrolyte matrix to compensate this oxidation/reduction, and the CP undergoes additional doping or de-doping. The structural distortions that result in the CP are not only those accompanying charge generation which lead to generation of species such as polarons, as discussed at length elsewhere in this book, but also torsional and conformational distortions – twisting, turning, unfolding/folding, and swelling of the polymer – to accommodate entry of dopant ions and any associated solvent molecules into the CP matrix.

Thus, for example, when a negative potential is applied to a typical p-type CP with a 3% "as-prepared" doping level, the polymer will undergo reduction, and anionic dopants will be expelled from the CP into the solution or other medium. Its color will also of course change with this. Conversely, when a positive potential is applied to the polymer, it will undergo oxidation, additional anionic dopants or other anions will enter the CP matrix from the solution, and its doping level will increase, say to 10%, with concomitant change in color. Thus, for the more common "p-type" doping of a CP, i.e., one where oxidation occurs on doping and where dopant counterions are anions, the reduced state corresponds to the de-doped state and a negative applied potential, while the oxidized state corresponds to the doped state and a positive applied potential.

With this very elementary understanding of the basic electrochemistry of CPs, we can now proceed to an elementary treatment of their electrochromism.

29.1.4   Basic Methodology for Transmission- and Reflectance-Mode Electrochromism

Electrochromism can be monitored via standard transmission spectroscopy (*transmission-mode* electrochromism), applicable to thin films and to solutions in the very limited cases in which a CP can be prepared at different doping levels in solution. It can also be monitored by reflection spectroscopy (*reflectance-mode* electrochromism), useful, e.g., for IR measurements or if the final application envisioned is a display device.

For transmission-mode monitoring, CP films are typically directly prepared on or cast onto visible-region-transparent electrodes such as ITO/glass or Au or Pt/glass, as described above. For acceptable results, ITO/glass should be of resistivity 20 $\Omega/\square$ (Ohms per square, a dimensionless unit of surface resistivity) or lower, and Au/Pt on glass should be of at least 50% transmission (e.g., at 550 nm). Measurements can also occasionally be performed on free-standing films. The films' redox state is then changed to that desired, either chemically, via appropriate exposure to dopant or redox agent, or electrochemically, by application of the appropriate potential.

Spectral measurements can be done either ex situ, changing the doping level or redox state appropriately for every measurement, or in situ, the preferred method for greater accuracy. All such spectroelectrochemical data can be presented with absorbance (optical density) or % transmission on the ordinate and wavelength (nm, μm) or energy (eV, λ (in μm) = ca. 1.24/eV) or, sometimes, wavenumbers (cm^{-1}) on the abscissa. Occasionally, difference spectra, usually referenced to the pristine state of the CP, are also presented.

For reflectance-mode monitoring, CP films can be prepared or cast onto opaque substrates such as Au films evaporatively deposited on hard, flexible substrates or onto solid metal (Au, Pt, stainless steel) electrodes. Indeed, Au-based substrates provide some of the best substrates for one type of reflectance measurement (specular, see below), since Au possesses the highest reflectance known among metals, and the underlying substrate reflectance needs to be high for proper interpretation of reflectance data. In situ measurements are possible by ingenious design of samples, e.g., in the "attenuated total reflectance-infrared" (ATR-IR) method, or on sealed devices with special, proprietary designs, as has been carried out in the author's laboratories. In ATR-IR, for instance, Pt/Ge-crystal substrates are used. Reflectance measurements can be carried out from the near-UV through the far-IR.

Two types of reflectance measurements are of primary interest: *specular* and *diffuse reflectance*. All materials that reflect radiation can do so "specularly," like a mirror, in a fixed direction, or uniformly in all directions, i.e., "diffusely." The former case is that of an ideal polished, reflecting surface, while the latter case is that of an ideal matte, scattering surface. In practice, most materials scatter both specularly and diffusely. In the case of a CP which is deposited on a metallic surface and has high transparency in one redox state accompanied by high contrast between electrochromic states, the specular measurement is of greater interest; this case would typically find use in flat panel communication displays. Specular measurements can be carried out for various incidence angles, but a fixed angle measurement (typically at 16°) is preferred for practical and comparative reasons. In the case of a coarse CP coating which is to be used for camouflage applications, the diffuse measurement is of greater interest. A material can either *absorb, emit*, or *reflect* radiation, and a very crude relation between these three parameters, not taking into account directional and other effects, is absorptance = emissivity = 1 - reflectance. (Emissivity or spectral emissivity differs from emittance, which is an integrated measure). Thus, in very broad terms, % reflectance data can parallel % transmission data.

All specular and diffuse reflectance data need to be referenced. The reference material used for the former for the near-UV through far-IR regions is usually a mirror surface of some sort. The reference material used for the latter (diffuse reflectance) is usually a material based on BaSO$_4$, a near-perfectly diffuse reflector, for the near-UV–NIR and KBr powder or a special gold surface for the IR region. In nearly all the reflectance data presented in this chapter emanating from the laboratories of Ashwin®-Ushas Corp., the measurements have been performed on hermetically sealed, functional devices. The use of a proper reference, particularly for diffuse reflectance, is important, as unrepresentative references can give rise to reflectances that exceed 100%, a case of "comparing apples to oranges." It is also noted that all specular measurements presented here are for fixed incidence angles. Finally, we note that most manufacturers of commercial UV–Vis–NIR and FTIR spectrometers also supply adapters to their instruments for collection of reflectance data.

29.2   UV–VIS–NIR and IR Spectroelectrochemical Measurements

29.2.1   Transmission-Mode Spectroelectrochemistry

The simplest spectroelectrochemical measurement which yields information on electrochromic properties of CPs is the UV–Vis–NIR spectroelectrochemical curve, an in situ or sometimes ex situ measurement of the transmission-mode UV–Vis–NIR spectrum of the CP at various applied potentials. Such a spectroelectrochemical measurement is depicted in Fig. 29.1; this figure is a re-representation of the optical spectra of poly(pyrrole) (P(Py)) discussed in another chapter in this book, with an abscissa in terms of wavelength, and represents a

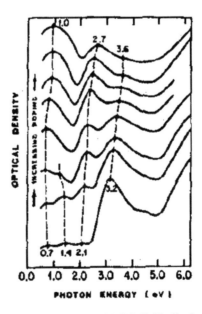

Fig. 29.1 Spectroelectrochemistry of poly(pyrrole) (P(Py)), idealized representation (Reproduced with permission from Ref. [856])

particularly well-behaved CP system. To recap again here, the single, prominent valence → conduction ($\pi \to \pi^*$) band transition in the pristine polymer (at ca. 388 nm) is accompanied by three additional polaron-based transitions at low doping level (ca. 590 nm, 885 nm, 1771 nm), which finally evolve into two bipolaron-based bands (ca. 459 nm, 1240 nm).

As one of the prototypical CPs, the color changes associated with the visible-region electrochromism of P(Py), subject to "p-type" (oxidative) doping, may be cited. For most "conventional" dopants, such as perchlorate, these are:

$$\begin{array}{c}
light\ green - yellow \\
(\text{reduced, dedoped, pristine state, negative applied potential}) \\
\longleftrightarrow \qquad\qquad\qquad\qquad\qquad\qquad\qquad\qquad (29.1) \\
dark\ grey - black \\
(\text{oxidized, doped state, positive applied potential})
\end{array}$$

The P(Py) system depicted in Fig. 29.1 represents a particularly well behaved and more "ordered" system, where the evolution from the polaron to the bipolaron bands is clearly visible. Figure 29.2 shows another such spectroelectrochemistry of an experiment P(Py) system, showing more clearly the disappearance of the polaron absorption (ca. 520 nm) with increased doping, to yield the broader bipolaron absorptions (beyond 700 nm). Figure 29.3a, b, and c shows more representative CP systems; Figure 29.3c also shows the alternative, % transmission representation.

In all of these, certain features, common to many but by no means all CPs, are of note. Firstly, all show a strong absorption at zero or low doping levels, usually in the near-UV, due to the valence → conduction band ($\pi \to \pi^*$) transition, separated from the polaron/bipolaron transitions by a characteristic isosbestic point. With increasing doping, this transition diminishes, with the intermediate polaron transition not always visible. At high doping levels, the bipolaron bands blend into one broadband stretching from the far-visible (ca. 0.6 µ) to the near-IR (up to 2.5 µ).

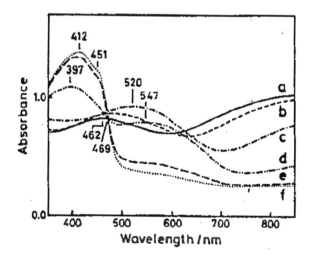

Fig. 29.2 Experimental spectroelectrochemistry of P(Py)ClO$_4$ (After Ref. [946], reproduced with permission). (f), neutral, through (a), increased doping

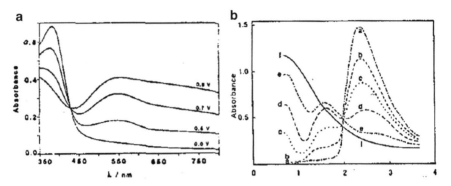

Fig. 29.3 (a–c) Spectroelectrochemistry of various CPs. (**a**) *Left*, poly(N-phenyl-2-naphthyl-amine) (Reproduced with permission from Ref. [958]). (**b**) *Right*, poly(3-methyl thiophene) (Reproduced with permission from Ref. [949])

The bipolaron or other mid-gap transitions are blended into one broadband primarily because in most experimentally produced CPs, there is sufficient broadening of energy levels caused by disorder or, put another way, sufficient electrochemical inequivalence of CP active sites. If one were able to arrive at a polymer with sharply defined chain length, chain packing, and conformational order, one would then probably see very sharp absorptions corresponding to these mid-gap transitions. This broadband absorption extending into the near-IR is indeed characteristic of CPs and is a sign of charge carrier mobility (sometimes called a "free-carrier" absorption signature). High absorptions in this region can frequently be associated with high conductivity of the CP as well. An additional feature of note is that although dopants do not generally interfere with characteristic CP absorptions in spectroelectrochemistry data, in rare cases, for example, with the dopant 1-amino-bromoanthraquinone-2-sulfonate used with poly(aniline) [947], dopant and CP can display absorptions independent of each other.

An obvious, defining electrochromic characteristic discernible from spectroelectrochemistry is the *dynamic range* or color contrast, that is, the difference between the absorptions between the pristine, completely de-doped polymer and the highly doped polymer, at a particular wavelength of interest. This can, for instance, be seen to be larger at 750 nm for P(Py) (Fig. 29.3a) than for P(N-Phe2-NaA) (Fig. 29.3b). A dynamic range of 0.5 in absorbance (optical density) units or 40% in transmission units can be considered acceptable for clearly perceptible visible-region color contrast. Another unit of measure of dynamic range used by some workers [948] is the absorbance/optical density difference divided by the charge capacity, discussed under charge capacities below.

Figure 29.4 shows time drive (i.e., %T vs. time as an electrochromic material is switched between different electrochromic states) data for an all-solid-state device fabricated by Kim et al. [950] which comprised N-butyl-sulfonate-doped P(ANi) as the anode, a solid polymer electrolyte comprising photo-cured PEG-methacrylate + tripropylene glycol diacrylate + LiClO$_4$, and a cathode comprising the inorganic

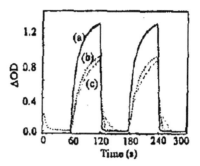

Fig. 29.4 Time drive data (i.e., %T vs. time as an electrochromic material is switched between different electrochromic states) for an N-Bu-sulfonate-doped P (ANi) system, applied voltages of ±1.5v. (a)–(c) denote various solid electrolytes (After Kim et al. [950])

electrochromic WO_3. It can be seen that significantly higher dynamic range can be obtained when one uses this complementary arrangement in which both electrodes are active electrochromics and they color in a complementary fashion.

Evidently, multicolor capability is a major feature of interest in electrochromic CPs. This can sometimes be achieved by combining several different CPs in bilayers or multilayers. For example, Yamasaki et al. [951] described an electrochromic system comprising poly(p-phenylene-terephthalamide) + poly(o-phenylene diamine) which showed multicolor capability encompassing a good representation of the visible spectrum: orange (−0.4 V), green (+0.4 V), and violet (+1.2 V) (all in two-electrode mode).

29.2.2 Reflection-Mode Data

The interpretation of reflection-mode data is a bit less straightforward than the transmission-mode spectroelectrochemistry above, since one has to contend with slight differences between specular and diffuse reflectance data and the reasons therefore. These data are nevertheless closely relatable to transmission-mode spectroelectrochemistry. If one remembers the crude relationship between the parameters, i.e., *absorptance* = *emissivity* = *(1 -reflectance)*, it is clear that reflectance data may closely parallel % transmission data. This is more true of specular reflectance than of diffuse reflectance, in which polymer morphology, particle size, and other factors appear to play a more important part. This is evident in the transmission-mode (%T) and specular/diffuse reflectance data for two poly (aromatic amines) studied in the author's laboratories, shown in Figs. 29.5 through 29.10. In the first polymer, which has a coarse-particulate, long chain-length morphology, the specular reflectance and %T data correspond, while the diffuse reflectance data are different (Figs. 29.5, 29.6, and 29.7). In the second, fine-particulate polymer (Figs. 29.8, 29.9, and 29.10), there is correspondence among all three measurements.

The correspondence between the fixed-angle specular and diffuse reflectance of yet another poly(aromatic amine) studied in the author's laboratories is shown in

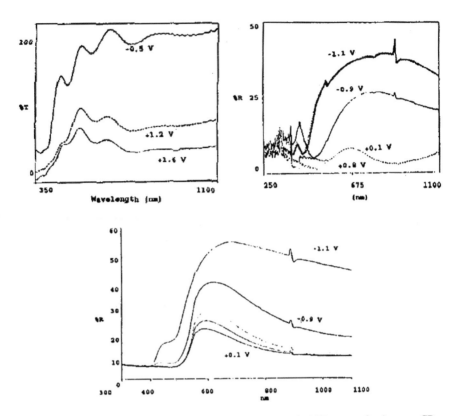

Figs. 29.5, 29.6, and 29.7 Transmission and reflectance UV–Vis–NIR spectra for the same CP, a coarse-particulate and a long chain-length poly(aromatic amine). *Left*, transmission-mode. *Right*, specular reflectance. *Bottom*, diffuse reflectance (Data courtesy of Ashwin®-Ushas Corp., Inc.)

Figs. 29.11 and 29.12. In both spectra, the major valence → conduction band ($\pi \rightarrow \pi^*$) transition of the polymer remains essentially unaltered. However, in the specular data, the highly oxidized, doped state of the polymer appears much more "blocking," essentially blocking out any reflection from the underlying metallic substrate in a particular (fixed-angle) direction, as is to be expected. Consequently, the bipolaron band absorption features are masked at such high doping level in the specular measurement.

29.2.3 IR-Region Data

The broad absorptions of CPs characteristic of charge carriers such as bipolarons tail off in the IR region, near ca. 6 μm, and subsequently, broadband variations in electrochromic signature are then absent in most cases for conventional dopants.

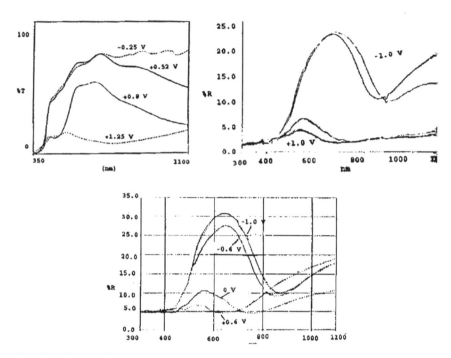

Fig. 29.8, 29.9, 29.10 UV–Vis–NIR spectra for a fine-particulate, long chain-length poly(aromatic amine). *Left*, transmission-mode. *Right*, specular reflectance. *Bottom*, diffuse reflectance (Data courtesy of Ashwin®-Ushas Corp., Inc.)

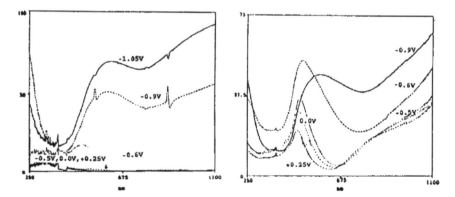

Figs. 29.11 and 29.12 UV–Vis–NIR electrochromism, cont.: specular and diffuse reflectance data for a poly(aromatic amine), showing near-complete blocking (~0% reflectance) at very high doping levels (Courtesy of Ashwin®-Ushas Corp., Inc.)

This can be seen from Fig. 29.13, showing specular reflectance for a poly(aromatic amine)/perchlorate; here the "tails" of the bipolaron bands from the far-visible and near-IR and the substantial dynamic range therein are clearly visible to ca. 6 μm.

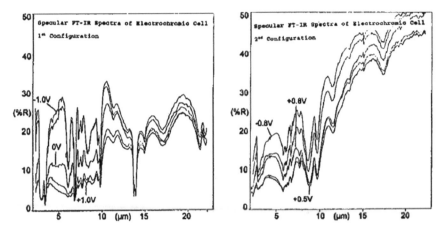

Figs. 29.13 and 29.14 IR specular reflectance electrochromic data for a poly(aromatic amine) using two differing device configurations (labeled 1st and 2nd) (Courtesy of Ashwin®-Ushas Corp. Inc.)

Beyond 6 μm, however, there is a nearly featureless absorption and very little broadband variation in dynamic range. This tailing of the bipolaron absorptions well in to the IR is a characteristic feature of most CPs (Fig. 29.14).

This kind of broadband variation in the signature with doping can however also be achieved in the mid- and far-IR with judicious tailoring of CP morphology and choice of dopant. This can be seen in the data in Fig. 29.14 (specular) and 29.15 (diffuse) for a poly(aromatic amine) with an altered configuration but the same dopant as in Fig. 29.13; here some far-IR broadband variation in dynamic range, through a shift of the entire baseline, is evident, with the bipolaron bandtailing also clearly visible. This bipolaron bandtailing with concomitant slight baseline variation in the far-IR is also visible in the spectra of poly(thiophene) as a function of doping level (Figs. 29.15 and 29.16).

More evidence that the broadband absorptions can be extended into the far-IR region is provided again by the specular (Fig. 29.17) data for a poly(aromatic amine) carrying a sulfonate dopant. Additionally, however, these spectra also show that the individual, "signature" absorptions characteristic of IR-active functional groups can still be preserved and be evident as superimpositions on these broadband absorptions. The effect of a radically different polymer morphology, effected during synthesis, and of large aromatic or polymeric dopants can be seen in the specular data in Fig. 29.18. Here, some characteristic functional group absorptions are evident for the higher doping levels, but more importantly, a very broad band absorption and large dynamic range is obtained even in the far-IR (e.g., nearly 94% at 10 μm). This wide "baseline variation," as it were, of CPs with a specific morphology and dopant type is also visible in the spectra of poly(3-Me-thiophene) (P(3MT)) in Fig. 29.19. Here, as in Figs. 29.17 to 29.18, the entire baseline shift toward higher absorption with higher doping, with characteristic "signature," IR-active functional group absorptions then superimposed thereon. It may be

Fig. 29.15 Diffuse
reflectance spectrum
corresponding to Fig. 29.14
(second configuration). The
ordinate units exceed 100%
due to the KBr powder
reference used (Courtesy of
Ashwin®-Ushas Corp.,
Inc.)

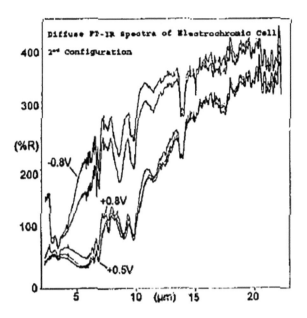

noted that these signature absorptions also shift to slightly lower energy with
increased doping; these characteristics are discussed in the next paragraph.

The above data and discussion bring forth several common features of
IR-reflectance electrochromism of CPs: (1) major, broadband, bipolaron band
absorptions, extending from the far-visible and near-IR, are preserved well into
the IR and are seen as tails up to ca. 6 µm; (2) IR signatures of many functional
groups are observable, but they are superimposed on broadband absorptions and not
changing very much with doping level, although shifting to lower energy; (3) broadband absorptions, with large dynamic ranges, do exist even in the far-IR, showing a
characteristic baseline variation on which the signature absorptions are then
superimposed but only for select CP morphologies and dopant types; (4) specular
and diffuse measurements follow trends more similar to each other than those found
in visible–NIR spectra, both with functional group signature absorptions visible;
(5) polymer morphology (determined primarily during synthesis), particle size, and
dopant characteristics can have a profound influence on broadband absorptions.

The broadband bipolaron-induced absorptions in the visible–NIR, which vary
with doping, are understandable and have been explained above as arising from the
characteristic lowest-energy bipolaron absorption. Because of disorder in the polymer, these absorptions tend to be broad, reflecting the merging of widely varying
energy levels, or again put another way, the inequivalence of active sites in the CP
lattice. These bands however taper off in the IR, near ca. 6 µm, reflecting the lowest
energy cutoff of the bipolaron transitions.

How then does one explain the broadband baseline variations with doping found
well into the far-IR, for some, but not all CP systems? The answer given to this by
many workers [952], including the present author, although not definitive, lies in

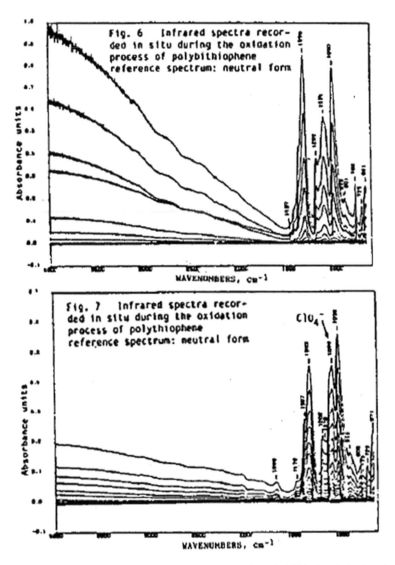

Fig. 29.16 ATR-IR (IR reflectance) electrochromic data for poly(bithiophene) (top) and poly (thiophene) (bottom) (Reproduced with permission from Ref. [953])

the unique properties of the organic CP systems: "vibronic coupling," i.e., mixing of vibrational and electronic states, an explain-all term sometimes used indiscriminately by many workers but fortunately very much valid for the CP systems: The unique vibrational coupling of the charge carriers – such as bipolarons et al. – to the CP lattice implies that the *characteristic functional group vibrations in the far-IR are coupled, through the CP chain/lattice, to these charge carriers*, and thus their net absorptions fluctuate with the charge carrier concentration, i.e., the doping

Fig. 29.17 Specular reflectance data for poly (aromatic amine) in a third device configuration (Courtesy of Ashwin®-Ushas Corp., Inc.)

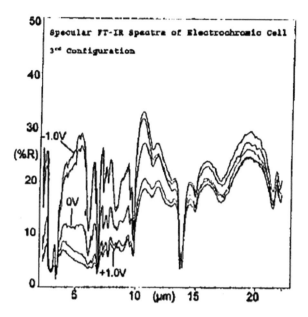

Fig. 29.18 Specular reflectance electrochromic data for thiophene-derivative CP (Courtesy of Ashwin®-Ushas Corp., Inc.)

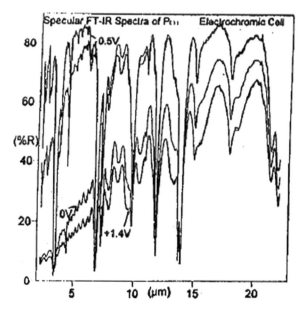

level. This coupling can also be used to explain the simultaneous shift of these characteristic absorptions to lower energy with increased doping. These broadband far-IR effects are moreover more observable in apparently more disordered CPs such as poly(thiophenes) than in poly(pyrroles). Additionally, large aromatic or

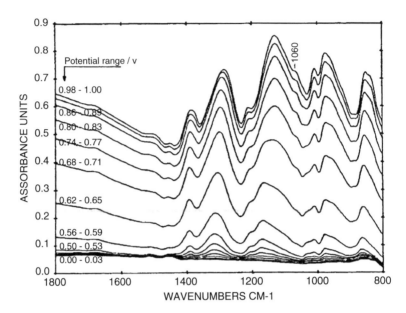

Fig. 29.19 ATR-IR (IR reflectance) electrochromic data for poly(3-Me-thiophene) (Reproduced with permission from Ref. [952])

macrocyclic dopants appear to enhance them, possibly by reducing conformational mobility in the CP chains.

The fact that enhanced absorptions in the late-mid- and far-IR are due to vibronic couplings in the CP lattice for certain CP systems is substantiated by results such as those in Fig. 29.20, showing the IR spectra for "neutral" (pristine) P(Py), P(Py)/ ClO_4, and P(Py)AsF$_6$. These spectra, besides showing the customary charge carrier bandtailing at the higher energies, illustrate that absorptions in the "signature" region (from 1600 to 700 cm^{-1}, 6.3 to 14.3 μm) are nearly identical for different dopants and for pristine polymer. This appears to indicate that many are due to "normal modes" associated with lattice defects such as bipolarons. This latter attribution has been supported by detailed interpretations of electrochromic data such as those in Figs. 29.21 and 29.22. In Fig. 29.21, the enhanced absorptions at ca. 1200 and 800 cm^{-1} are attributable to either vibronic coupling of charged solitons with the C=C stretch, C-C stretch, and C-H bend vibrations or due to the normal mode associated with a charged soliton, according to two different interpretations [944]. Also, in Fig. 29.21, vibronic enhancement of the ca. 1400 cm^{-1} band with increased doping is clearly visible. The additional absorptions in the 500–800 cm^{-1} region are ascribed to vibrational modes associated with charged solitons.

Some information in the IR region is also available from electrochromic Raman data, such as the spectra in Fig. 29.22 for P(Py). In Fig. 29.22 (left, middle), there are four sharp peaks in the pristine polymer. With increasing doping, one of these, at ca. 1564 cm^{-1}, shifts to lower energy (typical also for IR spectra above), while

Fig. 29.20 IR electrochromism, from ex situ spectra, for P (Py) perchlorate (top), neutral (middle), and hexafluoroarsenate (bottom) (Reproduced with permission from Ref. [953])

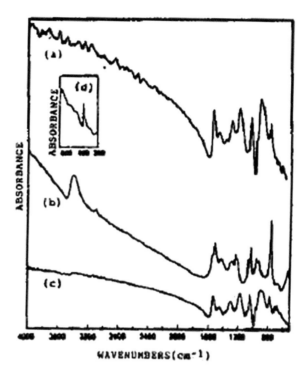

Fig. 29.21 Ex situ IR electrochromism of cis-P(Ac) with Br_2 doping (Reproduced with permission from Ref. [949])

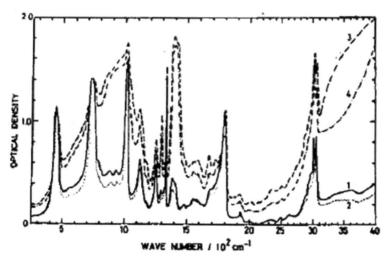

the other peaks are diminished. All peaks are broadened, and several additional broad peaks, in the 1500 to 1100 cm^{-1} region, are visible. These additional peaks, it

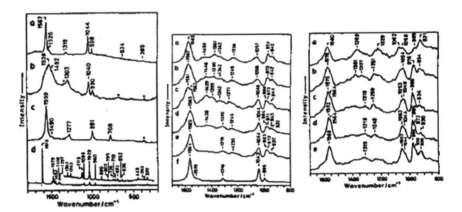

Fig. 29.22 Raman electrochromism for P(Py)ClO$_4$: *Left*, 457.9 nm excitation, (**a**), neutral P(Py). *Middle*, (**a**) through (**f**), with increased doping. *Right*, 632.8 nm excitation, (**a**) through (**e**), increased doping (Reproduced with permission from Ref. [953])

may be noted from other data with different dopants, are independent of dopant and ascribable to vibrational modes stemming from the localized, mid-gap lattice defects, i.e., bipolarons and polarons. Broadening and lower-energy shifting of peaks closely parallels that in the IR data cited above. In Fig. 29.22 (left), the transition between the polaron (radical cation) and bipolaron (dication) states is also quite visible, with, for instance, the peak at ca. 986 cm^{-1} ascribable to the polaron state while that at 1369 cm^{-1} is ascribable to the bipolaron state.

29.3   Other Electrochromic Parameters of Interest

Besides color contrast (dynamic range), there are several other characteristics of switching between various electrochromic states of interest, such as switching time and cyclability. A simple method of characterizing these is switching (usually with a square wave applied potential) between extreme electrochromic states while monitoring the optical absorption or % transmission at a particular wavelength as a function of time, which, for convenience, we shall denote as *time drive*. Figure 29.23 shows typical time drive data, for a poly(aromatic amine)-based electrochromic device (Fig. 29.24).

As seen from Fig. 29.23, among the information obtainable from such data is the switching time (hence speed) in each direction between highly doped and de-doped states, switching symmetry, dynamic range, and cyclability, i.e., degradation of dynamic range with cycling. In this respect, we note, importantly, that single-cycle representations of time drives such as that seen in Fig. 29.24 have by themselves very little meaning and are simply schematics of optical switching curves, reproducible by nearly all CP systems. For time drive (i.e., %T vs. time as an

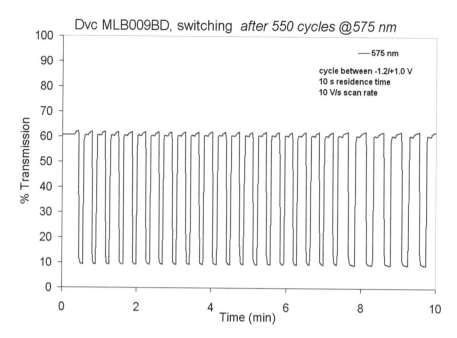

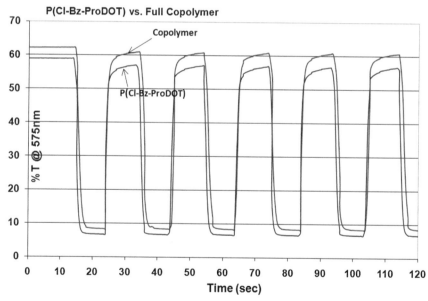

Figs. 29.23 and 29.24 Typical time drive data (i.e., %T vs. time as an electrochromic material is switched between different electrochromic states) for a poly(aromatic amine) (Courtesy of Ashwin®-Ushas Corp., Inc.; cycle numbers are indicated). Time drive data for FeBPs/P (Py) system (Reproduced with permission from Ref. [953])

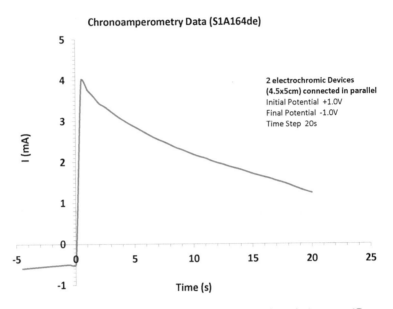

Fig. 29.25 Typical chronoamperometry data for a poly(aromatic amine) system (Courtesy of Ashwin®-Ushas Corp., Inc.)

electrochromic material is switched between different electrochromic states) data to have meaning, there must be (a) comparison of initial and extended cycles, e.g., the 1st and 1,000th cycles; (b) a reference run to steady-state (infinite time) optical transmission values as a baseline for comparison of all runs; and (c) runs between intermediate electrochromic states in addition to the extreme oxidized/reduced states.

The *switching time* between two electrochromic states has been defined in the author's laboratories as one analogous to rise times used in electrical measurements, i.e., the time for traversing between 10% and 90% of the total *steady-state* optical response. However, many workers in the field use different, entirely arbitrary definitions for this parameter. The steady-state optical response is determined by allowing a sufficient time for a CP system to attain a steady state, for example, 15 minutes, which suffices for most CPs. There are generally an oxidative (going from reduced to oxidized state) and a reductive switching time, τ_{ox}, τ_{red}, which are almost never the same. The switching speed is then just the inverse of the switching time. The switching symmetry is reflected in the difference between the oxidative and reductive switching times; as can be seen from Fig. 29.25, the two times can be clearly different. Typical switching times for standard thickness (0.1 to 1 µm) films of most CPs are on the order of seconds, although switching times of hundreds of microseconds have been claimed [954].

Electrochromic cyclability of CP films, i.e., number of cycles between extreme oxidized and reduced states, also requires precise definition. Again, we have in our laboratories defined this as the number of cycles possible between extreme oxidized

Fig. 29.26 time drive (i.e., %T vs. time as an electrochromic material is switched between different electrochromic states) and CA (chronoamperometry) data for poly(dithieno 3,2,-b:2',3'-d thiophene) (Reproduced with permission from Ref. [949])

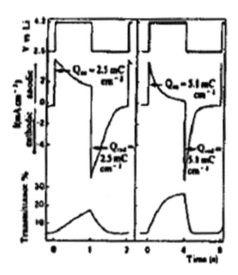

and reduced states of a CP with 10% or less *optical degradation*. The latter can be measured from time drive data (i.e., %T vs. time as an electrochromic material is switched between different electrochromic states), for a cycling time *at least* equal to the switching time, as defined above. A separate electrochemical or charge cyclability can also be defined, as in the next paragraph. Two important points are noteworthy in this respect. Firstly, electrochromic cyclability does not imply electrochemical/charge cyclability, although the two may be related or identical. Secondly, this definition of cyclability and the associated switching time does not in any way imply that an active potential must always be applied to the CP film. Switching, i.e., cycling, can be accomplished by whatever method is most convenient, including chemical methods. Thirdly, it is quite evident that switching times and cyclabilities in different cell types, in devices of different design, and with different dopants may be quite different for the same CP. Thus, a particular CP cannot be assigned a value for switching time or cyclability, although these usually lie in characteristic ranges for each CP. Cyclabilities of CPs can vary from tens of cycles to up to 10^6 cycles claimed for some systems [955].

Electrochemical methods such as *chronoamperometry (CA)* (monitoring of current decay with time upon application of a specified potential) and *chronocoulometry (CC)* (monitoring of charge decay with time upon application of a specified potential) can be useful methods for evaluation of the electrochromic properties of CPs. CA data of the type depicted in Fig. 29.26 may have little meaning, since they simply trace a generic, universally applicable current decay and indeed may not even correlate well with actual switching times. When applied through several thousand electrochromic cycles, however, or when integrated to yield *charge capacities* (see below), they can have meaning. An example of the former is shown in Fig. 29.25, where it is seen that the asymmetry between the reductive and oxidative segments of the cycle is enhanced on prolonged cycling.

Fig. 29.27 Corresponding
CA, CC, time drive curves
for poly(isothianaphthene)
(Courtesy of
Ashwin®-Ushas Corp.,
Inc.)

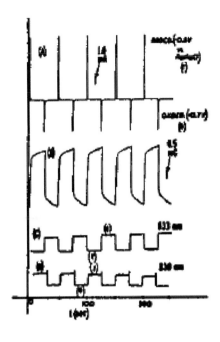

Fig. 29.28 Charge capacity
degradation with extended
cycling for poly(aniline)
(Reproduced with
permission from Ref. [949])

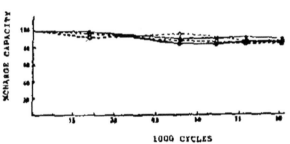

CA curves have some significance in data like that shown in Fig. 29.26, where time
drive data (i.e., %T vs. time as an electrochromic material is switched between
different electrochromic states) for two different switching times, 2 s and 8 s, are
compared; here, for the 2 s cycling period, both the optical response (time drive
data) and the current (CA data) do not return to steady-state values. CA curves also
have some significance when they are used to correlate optical response to current
decay, as in Fig. 29.27, for example, showing that current decay may be very rapid,
while optical response, dependent on factors including diffusion and dopant concentration, is delayed.

The *charge capacity* is simply the total charge needed to go from one
electrochromic state to another, usually between the extreme reduced and oxidized
states of a CP. Charge capacities are an extremely useful measure of such parameters as cyclability, sometimes more useful than optical measures for determining
causes of degradation. The data in Fig. 29.28, for example, show the degradation of

Fig. 29.29 Charge capacity degradation with extended cycling for a sealed poly (aromatic amine) electrochromic device (Courtesy of Ashwin®-Ushas Corp., Inc.)

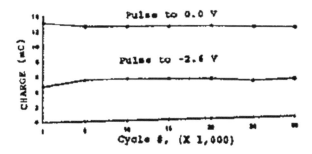

Fig. 29.30 Open circuit memory data for a poly (aromatic amine)/sulfonate system (sealed electrochromic device). At each arrow, the reducing/ oxidizing potential is applied for 60 s, and the cell then disconnected (Courtesy of Ashwin®-Ushas Corp., Inc.)

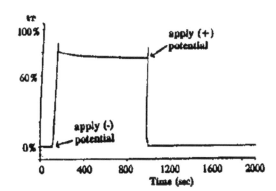

the charge capacity with extended cycling for a poly(aniline) electrochromic device. Charge capacity, in combination with time drive data (i.e., %T vs. time as an electrochromic material is switched between different electrochromic states), is also used by some workers to arrive at another measure of dynamic range: the optical dynamic range in optical density units divided by the charge capacity. A value of 0.1 OD/(mC/cm^2) is considered acceptable performance for electrochromic devices [948] (Figs. 29.29 and 29.30).

Charge capacity can be determined in a number of ways: by integration of areas under CA curves, directly via CC (not all commercially available electrochemical instruments have a coulometric capability), and by integration of area under CV peaks, which involves accurate estimation of baseline currents. If available, the CC method is the most direct. Charge capacities are a very important measure of performance for devices such as electrochromic windows, since the charge capacity is a direct measure of energy expended in switching. A charge capacity of 1 mC/cm^2 is a good, and achievable, target for most electrochromic systems.

Electrochemical or *charge cyclability* can then be defined exactly as for electrochromic cyclability above but with the stipulation that 10% *charge* degradation rather than optical degradation is measured. Frequently, electrochemical and electrochromic cyclability values are quite comparable [956]. Figure 29.31 shows typical charge capacity degradation with cycling for a poly(aromatic amine) system, determined from the data of Fig. 29.25. Charge capacities are more directly

Fig. 29.31 Alternative
interpretation of bandgap,
E_g, as intersection of
absorption ordinate with
abscissa axis on low energy
end, for neutral, undoped
CP

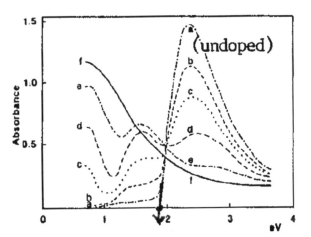

measurable using chronocoulometry, and can also be determined from cyclic
voltammograms (see the chapter dealing with electrochemistry of CPs elsewhere
in this book).

Another important electrochromic parameter of interest, especially in practical
applications, is the *open circuit memory*, i.e., the optical memory retention. For
practical applications, it is important to know how long a CP will retain its color,
i.e., electrochromic state, after an applied potential that brought it to that state is
disconnected (i.e., brought to open circuit). Quite evidently, the open circuit
memory of a CP in different electrochromic states may differ: For example, it
may have a long open circuit memory in its doped, oxidized state but a very short
one in its de-doped, reduced state, which is susceptible to atmospheric oxidation.

For our purposes, again, we use the definition adopted in the author's laboratories, defining open circuit memory for a particular electrochromic state (identified
by its applied potential), as the time during which there is less than 10% optical
degradation following disconnection of an applied potential. There is an added
stipulation that before disconnection, the system should have achieved a steady
state, as defined above. Thus, for example, one may apply a potential of +1.0 V
vs. SCE to a CP film for 10 min to bring it to its steady-state optical response
(monitored at a particular wavelength), then disconnect the CP film from the
applied potential, and monitor the optical response to the time that one sees a
10% degradation. This time is then the open circuit memory. The 10% stipulation is
necessary as there is always some optical fluctuation immediately on disconnection
of the circuit. Additionally, it is evident from our definition that for a CP even to
show an open circuit memory, it must be able to achieve a steady-state optical
response. Thus, for a polymer showing continued degradation on application of a
potential, the term has little meaning. It is evident again that open circuit memory
may be different for different (e.g., highly oxidized or highly reduced)
electrochromic states and dependent on factors other than the identity of the CP,
including electrochromic device or cell design and dopant. Typically, open circuit

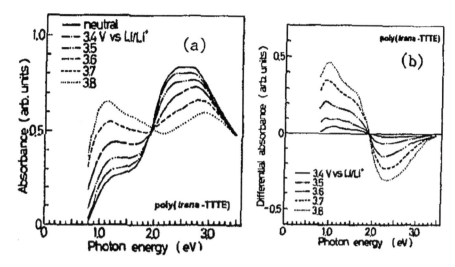

Fig. 29.32 Standard (**a**) and differential (**b**) spectroelectrochemistry plots for poly(trans-TTTE), a poly(thiophene) derivative (Reproduced with permission from Ref. [957])

memories can range from tens of seconds to infinite, i.e., total memory retention [950]. Open circuit memory can be measured with an experimental setup identical to that for time drive. Figure 29.32 shows a typical open circuit memory measurement carried out in the author's laboratory; the data indicate near infinite optical memory retention.

Another important parameter measurable from a spectroelectrochemistry is of course the *bandgap*, E_g, i.e., the energy between the valence and conduction bands of the CP corresponding to its main $\pi \rightarrow \pi^*$ transition. This is usually a characteristic parameter varying only slightly with polymer morphology and other factors. Any spectroelectrochemistry, but especially one which contains spectra of near-pristine and highly doped states of the polymer, clearly identifies this major absorption. Since the valence and conduction bands are broad, the absorption is usually broad. Consequently, the bandgap can be interpreted as the absorption maximum, or it may be interpreted as the commencement of this absorption on the low energy end, which is interpolated from the absorption spectrum as shown in Fig. 29.31. The latter method is preferred by many workers, but the former may be more accurate or practically relevant, since if one envisions a density of states in the valence and conduction bands, then the absorption maximum corresponds to the transition between the highest density of states in each band. Bandgap is usually expressed in eV, with the simple conversion $eV = \sim 1.24/\lambda (in microns)$.

29.4 Other Measurements

Several other derivative measurement parameters may also of interest in the electrochromism of CPs for specialized applications. The most straightforward of these are *differential measurements*, e.g., of the differential absorbance or trans-mittance of various doped states of a CP with reference to its pristine, undoped state. Figures 29.32 and 29.33 show two such differential plots. In the first, (a) shows the standard spectroelectrochemistry, while (b) shows its differential; the differential plot simply serves to enhance the trends observed in the spectroelectrochemistry but also serves to accurately pinpoint the isosbestic point. In the second, the differential absorbance monitoring a bipolaron absorption at 700 nm shows that up to an applied potential of ca. 0.6 V, the formation of bipolarons appears minimal; after 0.6 V, however, bipolarons form rapidly.

Other important derivative measurements are those of thermal *emittance*, ε and *solar absorptance*, α_s. Both can be derived directly from reflectance measurements, and both have importance in radiative loss applications, e.g., in space. For instance, satellites and spacecraft ideally require high reflection of solar heat during sunfacing, implying a low α_s, <0.2, while at the same time having high radiative capability to remove excess heat during darkfacing, implying a high thermal emittance, >0.8. At-will switchability – i.e., electrochromism – of these properties is highly desirable, with variation of α_s/ε ratios between 0.2 and 1.0 sought. Excellent treatments of these parameters are available elsewhere [959, 960]. Very briefly, however, emittance is given by:

$$\varepsilon = 1 - \left\{ \Sigma\{\text{from 1 to } n\}\left(\rho_{\lambda,j}\Delta F_j\right)\right\} \tag{29.2}$$

where the Δf_j are energy increments in the blackbody spectrum, obtained from a table giving the fraction of blackbody radiation energy between 0 and λT, and $\rho\lambda_j$ is derived from the reflectance (specular or diffuse) at wavelength λ, ρ_λ, through a summation procedure [959, 960]. Typically, 5 or 10 increments are adequate for most calculations. If these increments are equal, then the above equation becomes:

Fig. 29.33 Absorption difference (ΔOD) spectroelectrochemistry plot for poly(diphenyl amine) at 700 nm (Reproduced with permission from Ref. [958])

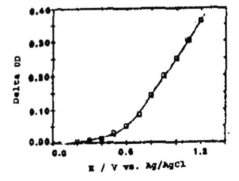

Table 29.1 Typical emittance (ε) and solar absorptance (α_s) data for variable-emittance "skins" used in spacecraft thermal control. Emittance and solar absorptance data are measured using specialized instruments called an emissometer and solar absorptometer, respectively (Data courtesy of Ashwin®-Ushas Corp., Inc.)

Device (sample) #	Light ε	Dark ε	$\Delta\varepsilon$
Skins with alpha(s)-reduction coating			
J3C056DD	0.325	0.771	0.446
J3A038FD	0.298	0.784	0.486
J3A038CD	0.234	0.676	0.442
J3A174BD	0.389	0.646	0.427
Skins without alpha(s)- reduction coating			
JB_011AD	0.257	0.771	0.514
JB_042AD	0.237	0.751	0.514
JA_168FD	0.335	0.835	0.500

Light ε	Dark ε	$\Delta\varepsilon$
Measurements with emissometer (at ambient pressure)		
0.376	0.777	0.401
Measurements with CalTVac (in space vacuum)		
0.376	0.730	0.354

$$\varepsilon = 1 - \left\{ (1/n)\Sigma\{\text{from 1 to } n\}\left(\rho_{\lambda,j}\right)\right\} \qquad (29.3)$$

The corresponding equations for the solar absorptance, α_s, are similar. Table 29.1 lists ε and α_s values obtained for a poly(aromatic amine) CP system, in a two-electrode mode, sealed device in the author's laboratory.

Electrochromism can be said to encompass any property of CPs that varies with applied potential or doping. This may include conductivity. An application for conductivity electrochromism may be envisioned, for instance, with an Au grid on a flexible substrate on which a CP is deposited. When the CP is cycled between conductive and nonconductive states, the dielectric permittivity of the material, affecting its microwave/radar signatures, changes dramatically. We thus have a CP "microwave/radar electrochromism" of sorts. Such uses are dealt with in the Applications section of this book (Fig. 29.34).

29.5 Thermochromism and Solvatochromism

Thermochromism is the property of change of color with temperature. When this occurs in solution, as in nearly all cases, it can be related to solvatochromism. Among CPs, the thiophene-based polymers have been exhibited the most pronounced thermochromism. Solvatochromism is the property of variation of color of a solution with the concentration of the solute in it. As for thermochromism, among CPs, the thiophene-based polymers have been exhibited the most

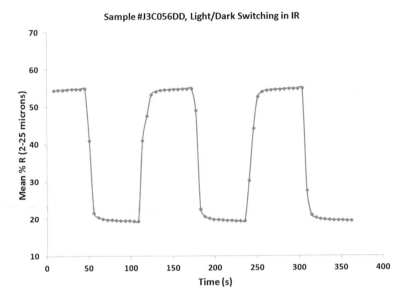

Fig. 29.34 Typical switching for variable-emittance "skins" used in spacecraft thermal control, measured as a variation in %R (specular reflectance, 16° incidence angle). (See Experimental Section for method of data collection)

pronounced solvatochromism. In CPs, both effects are due to *conformational* changes in the CP chain in solution.

In poly(3-hexylthiophene), a shift in absorption intensity maximum of ca. 0.5 eV occurs between room temperature and 180 °C [961]. In the case of poly (3-octylthiophene), a marked change in the signature of the proton at the 3-position is observed, as seen in the ^{1}H NMR spectra of Fig. 29.35 [962]. As can be seen in the comparative thermochromic and solvatochromic data for poly(3-hexylthiophene) in Fig. 29.36a, b, the actual magnitude of the effects is quite small.

In more recent work, Hirota et al. [964] discussed the thermochromism of a substituted-thiophene CP with a urethane bond and its application to reversible thermal recording (Figs. 29.37a–c). The color changes for both CPs were red-brown to yellow. A reversible alphabetical thermal recording on a CP-coated sheet was demonstrated.

Patil [965] observed a solvatochromic effect in a 3-butyl-sulfonate-substituted poly(thiophene). Both thermochromism and solvatochromism were also observed in a 3-urethane-substituted poly(thiophene) [966].

In more recent work, Lanzi et al. [967] studied solvatochromism of poly (3-(6-methoxyhexyl)-thiophene) in various solvents with the non-solvent component MeOH. Figure 29.38a–d summarizes their results.

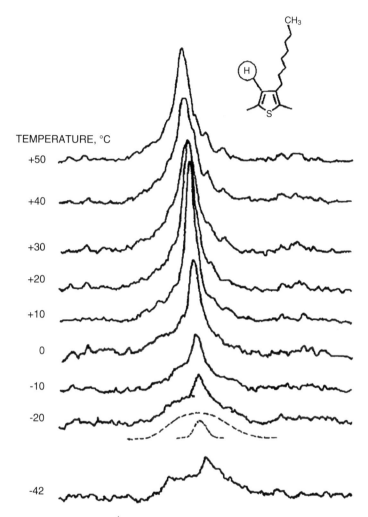

Fig. 29.35 Variable-temperature ^{1}H NMR spectra of the thiophene proton of P3OT (After Ref. [962], reproduced with permission)

29.6   Problems and Exercises

1. Identify all spectral regions and properties for which electrochromism of CPs is to be expected, also detailing the type of absorptions and spectra to be expected for each and the type of sample applicable to each for in situ and ex situ measurements.
2. In the UV–Vis–NIR region, how do transmission-mode and specular and diffuse reflection-mode spectra of CPs compare?

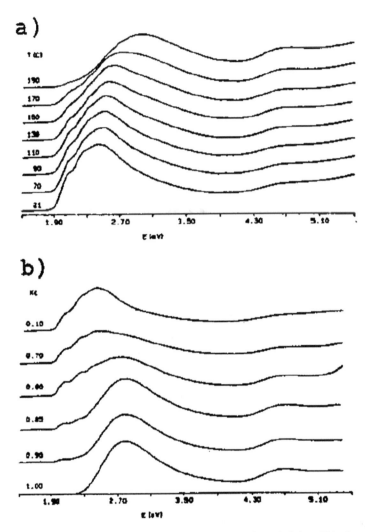

Fig. 29.36 Thermochromic (**a**) and solvatochromic (**b**) effects for poly(3-hexylthiophene). In (**a**), films were cast from chloroform solution. In (**b**), solutions were in chloroform/methanol, with: X_c = [CH Cl$_3$]/([CH Cl$_3$ + MeOH]) (After Ref. [963], reproduced with permission)

3. Clarify the differences and commonalities of broadband and signature transitions between the Vis–NIR and IR spectral regions. What one factor not found in Vis–NIR spectra is a determinant of the characteristics of IR-region spectra?
4. Write out detailed definitions, with illustrations if appropriate, for the following terms: time drive; oxidative and reductive switching times; electrochromic, electrochemical, and charge cyclabilities; charge capacity; dynamic range; open circuit memory; bandgap; solar absorptance; and thermal emittance.

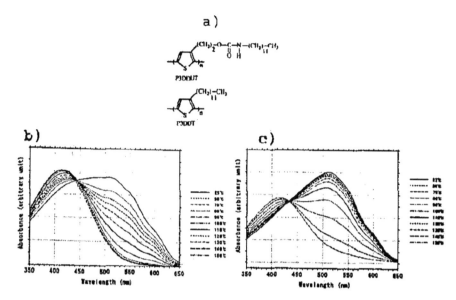

Fig. 29.37 (a–c) Thermochromism of urethane-substituted thiophenes; (a) CP structure; (b), (c) thermochromic data (After Ref. [964], reproduced with permission)

5. Design sample cells for three-electrode-mode in situ measurements of electrochromic properties in the UV–Vis–NIR and IR spectral regions applicable to transmission-mode (UV–Vis–NIR) and reflection-mode (UV–Vis–NIR, IR) data collection.

6. From an analysis of contributions in each spectral region, identify properties within a CP system most conducive to (a) a high-contrast, large-area visible-region electrochromic window, (b) a space satellite thermal coating, (c) a communication device in the far-IR, and (d) a camouflage panel in the visible–NIR, far-IR, and microwave regions.

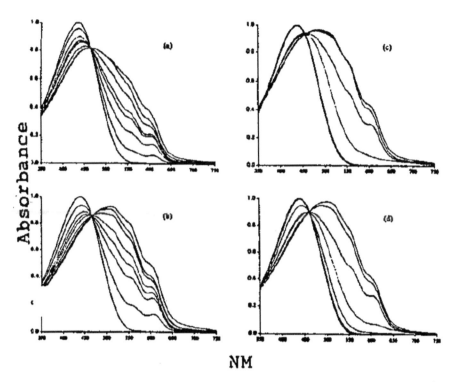

Fig. 29.38 (**a–d**) Solvatochromism for substituted poly(thiophenes), shown for increasing mole fraction of MeOH in mixtures with (**a**), (**b**) dioxane and (**c**), (**d**) toluene; (**a**), (**c**) and (**b**), (**d**) are different CPs (After Ref. [967], reproduced with permission)

Chapter 30
Basic Electrochemistry of CPs

Contents

30.1 Basics

30.1.1 Introduction

We note at the outset that this chapter deals specifically with the *electrochemistry of CPs* per se; electrochemistry and mechanisms associated with *polymerization* processes are discussed elsewhere in this book. It is also noted that it has been attempted to write this first section specifically for readers with little knowledge of electrochemistry and no familiarity with electrochemistry of CPs.

To briefly recapitulate from another chapter in this book, electrochemistry of CPs can be carried out either in aqueous systems, used, e.g., for CPs such as poly (anilines) where protons participate in the doping processes, or in nonaqueous systems and sometimes in combinations of the two. Besides standard liquid electrolytes, solid electrolytes may also be used, especially when solid-state device applications are sought to be emulated. Electrochemistry may be carried out in a two-electrode mode (working, i.e., actively studied, and counter electrodes) or a three-electrode mode (working, counter, reference electrodes). The latter is typically better for accurate potential control and is used for most characterization work, while the former may more correctly emulate practical applications such as electrochromic devices.

© Springer International Publishing AG 2018

P. Chandrasekhar, *Conducting Polymers, Fundamentals and Applications*,
https://doi.org/10.1007/978-3-319-69378-1_30

30.1.2 Basic Methodology

For most electrochemical studies, the working electrode is comprised of a film of the CP of some sort, deposited on a conductive, substrate electrode. This conductive substrate may be any of a number of choices, some classes of examples being:

1. Solid Pt, Au, or, occasionally, other metals (e.g., stainless steel or Al for corrosion studies)
2. Metal films on rigid or flexible substrates, for instance, Au/Pt vacuum deposited on flexible plastics (thickness anywhere between 50 nm and 1 μm), or rigid materials such as glass, quartz, or Ge crystal (the latter for IR studies)
3. Visible-region-transparent (or translucent) electrodes such as indium tin oxide (ITO) or very thin Au on glass or quartz or, more recently, Ag nanowires and CNTs, on the same substrates, used commonly for electrochromic studies
4. Ordinary graphite, pyrolytic graphite, or basal plane pyrolytic graphite, the latter for very high electrode stability, conductivity, and inertness

Generally, the resistivity of the substrate electrode *does* have an influence on the electrochemical results: for example, ITO electrodes with a resistivity in excess of 5 Ω/□ show some resistivity drop across distances in excess of a few cm, and thus the edge portions of a large-area CP film on such a substrate will show delayed response and larger "iR drops," leading to broader voltammetric peaks, slower switching times, and inhomogeneous switching of one section of a CP film with respect to another. For characterization purposes, solid noble metal electrodes or thick noble metal films (on glass, PET, etc.) yield the best results; adhesion of the CP to these electrodes is best. However, one accompanying problem is that remnants of the CP film may become difficult to remove from these electrodes without drastic means (e.g., dipping in conc. H_2SO_4). All substrate electrodes, including ITO electrodes, may be electrically contacted using standard means: alligator clips are usually sufficient, but if permanent contacts, e.g., with metallic epoxies, are desired, care should be taken that these do not contact the solutions, CPs, or any other active electrochemical surfaces. In some cases (e.g., for P(Py)), the CP may be obtainable as freestanding thin films which can be electrochemically contacted. These are usually prepared as very thick (several μm) films on noble metal electrodes which are then peeled off.

The CP film may be directly polymerized electrochemically (or sometimes, chemically) on the substrate electrode, the common method for most CPs. It may also be cast. Casting may be from de-doped, pristine (neutral) forms of the CP, which are generally soluble. For example, poly(aniline) (P(ANi)) can be cast from solutions of the pristine polymer in N-methyl-2-pyrrolidinone (NMP); afterward, it is re-doped either electrochemically or chemically and is then ready for further studies. One may also cast soluble precursor polymers which can then be converted (usually via chemical routes) to the CP. A number of other methods of casting CP films onto substrates are also available, including casting from colloidal suspensions (sols) and from suspensions containing surfactants or other solubilizing agents. In rare cases, CPs can be cast as already doped films directly from solutions (e.g., for some poly(3-alkyl thiophenes)) [1020]. All these methods are discussed in subsequent parts of this book. All the above casting methods can present some problems, for example, the solvent retention discussed further below.

Thicknesses of CP films for typical electrochemical studies vary from tens of nm to several μm, with the most common for acceptable results being ca. 100 nm; thinner films can sometimes give inadequate response or spurious results, and thicker films frequently start showing much slower switching, unwanted diffusion effects, and the like, as discussed in more detail below. Thicknesses are estimated coulometrically (counting total charge passed during electropolymerization and knowing polymerization stoichiometry), the most common method. They may also be estimated by more absolute means when available, e.g., scanning electron microscopy (SEM) of sections or surface texture or stepping methods.

30.1.3   *Electrolytes and Electrodes*

Electrolytes, whether solid or liquid, are comprised of the following components: an electrolyte salt, which can furnish ions for charge transport, and an electrolyte medium, such as a solvent (liquid electrolyte), polymer or other matrices (solid electrolyte), or an *ionic liquid* ("room-temperature molten salt") electrolyte.

For nonaqueous liquid electrolytes, solvents such as acetonitrile, propylene carbonate, γ-butyro lactone, methylene chloride, tetrahydrofuran, benzonitrile, ethylene glycol, dimethyl sulfoxide, nitromethane, and mixtures thereof have all been used. This list also gives the approximate order of preference, with acetonitrile being the most common for a number of reasons such as its general quality and reproducibility of voltammograms or other measurements in acetonitrile and its greater volatility and inertness, leading to less solvent retention within the CP matrix (a distortive feature in electrochemistry, see below). It is also the preferred electropolymerization solvent and is least likely to dissolve the more soluble, reduced forms of the CPs (a problem with other solvents).

For aqueous studies, usually carried out for CPs such as P(ANi) or other poly (aromatic amines), in which protons take part in the doping/de-doping processes, acids such as HCl or H_2SO_4 are preferred, at concentrations ranging from 0.01 M to several M. For solid electrolytes, varied matrices which usually have some combination of poly(ethylene oxide) and poly(ethylene glycol) are taken.

More recently, *ionic liquids* have been used for electrochemical studies, including those involving Visible- and IR-region electrochromism [975, 976]. Ionic liquids used as electrolytes for CP systems include the following, by no means exhaustive list: 1-butyl-3-methyl-imidazolium tetrafluoroborate, 1-ethyl-3-methylimidazolium tetra fluoroborate, 1-hexyl-3-methyl-imidazolium tetrafluoroborate, 1-hexyl-3-methylimi dazolium triflate, 1-butyl-3-methyl-imidazolium triflate, 1-butyl-3-methyl-imidazoli um hexafluorophosphate, 1-butyl-3-methyl-imidazolium trifluoroacetate, 1-butyl-3-methylpyridinium bis(trifluoromethylsulfonyl)imide, 1-butyl-4-methyl pyridinium hexafluorophosphate, 1-butyl-3-methyl pyrrolidinium bis(trifluoromethylsulfonyl)i mide, bis(pentafluoroethylsulfonyl)imide, and 1,3-dimethylimidazolium bis[tri fluoromethylsulfonyl]imide.

For three-electrode studies, the customary reference electrodes, such as the SCE (saturated calomel electrode) or SSCE (sodium-saturated calomel electrode), are used by most workers. For nonaqueous media, however, this is strictly not entirely correct, as water and ion leakage into the nonaqueous medium can and do occur. However, in the time frame of most studies – a few hours typically – this is usually not a problem. It can be a problem however in extended cycling studies, to cite an example. An entirely solid-state reference electrode, such as a Ag wire coated with AgCl (Ag/AgCl or, more correctly, Ag^+, electrode), may be better for nonaqueous media. Many workers also prefer to use a quasi-reference electrode such as a Pt wire and reference it accordingly; fluctuations of up to 75 mV in reference potentials are possible with this, but for most rough characterizations, it is adequate. A Pt or other metal wires may be the preferred reference electrode for sealed device applications. When performing two-electrode measurements with a standard potentiostat/galvanostat, it frequently suffices simply to shorten the reference and counter electrodes together as the single counter electrode.

In some form or another, the doping/de-doping, i.e., redox, process of the CPs ends up being the target phenomenon studied in most electrochemical work. This process in large part thus governs the choice of the electrolyte salt used. Thus, for example, electrochemical cycling or electrochromic switching of P(Py)/ClO$_4$ is preferably carried out with an electrolyte such as LiClO$_4$ or Et$_4$NClO$_4$; similarly, electrochemical cycling of P(ANi), which requires the participation of protons, is preferably carried out in acids such as HCl (where Cl$^-$ then becomes the dopant anion) or H$_2$SO$_4$ (SO$_4^{2-}$ dopant anion).

30.2 Basic Voltammetric Parameters and Information of Interest

30.2.1 Cyclic Voltammograms

We can now proceed to an examination of the most basic electrochemical datum obtainable from a CP film, the cyclic voltammogram, which is just a measurement of current resulting from application of a voltage (potential) function to the CP, i.e., just an I/E curve, with a fixed scan rate expressed in mV/s. The CV is correlated directly with the electrochromism of the CP, if it is electrochromically active, in the sense that the redox peaks observed in the CV usually coincide with the CP's color changes.

Figure 30.1 shows a CV of a prototypical CP, P(Py) doped with BF$_4^-$, i.e., P(Py)/ BF$_4$ from the early work of the Diaz group [977], while Fig. 30.2 shows scan rate dependence for the CP for a number of dopant ions. One of the first elements to note

Fig. 30.1 (left), CVs of poly(pyrrole)BF$_4$ as function of scan rate, with color change also shown (Reproduced with permission from Ref. [977])

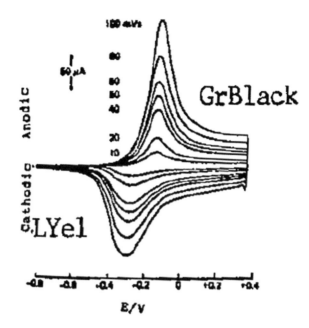

Fig. 30.2 Scan rate dependencies, as function of electrolyte used

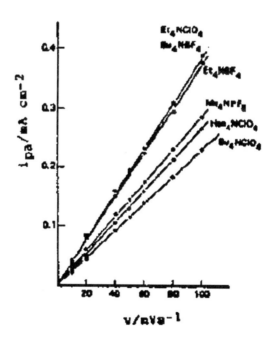

is that although one will find, in the literature, CVs presented with the (-) potential end on the left-hand side, as here, according to the predominant recent convention, CVs are now presented with the (-) potential end on the right-hand side. Scans preferably start on this end for CPs, also according to more common convention.

When scanning toward positive potential, the current is denoted "anodic" (with corresponding peaks, "anodic peaks") and when toward negative potential, "cathodic." Although many workers do not use "iR compensation," a method which allows the instrument to compensate for a potential drop caused by the resistance of the electrolyte, it is always preferable to do so to obtain more clearly defined electrochemical results.

30.2.2   Electrochemical Windows

Another element to note in the CV of a CP is that every CP has an electrochemical "window" within which the doping/de-doping (redox) process is reproducible (though not necessarily electrochemically *reversible*) to a large extent and beyond which oxidative, or reductive, decomposition of the CP occurs. Thus, one of the first measurements performed on a new CP system is to determine this window: Oxidation peaks unaccompanied by reduction peaks, or irreversible electrochromic changes, indicate that one is beyond the window. The window of P(Py)/BF_4, from Fig. 30.1, lies within -0.8 V $\leftrightarrow$ +0.4 V vs. SCE. Quite obviously, this window also represents the window for reproducible electrochromism of a CP; thus, for P(Py)/BF_4, -0.8 V and +0.4 V represent, respectively, light-yellow and dark gray-black electrochromic extremes of this CP system (as labeled on the CV). The oxidative decomposition peaks for most CPs lie only a little beyond the oxidative doping peak (or de-doping peak for n-type CPs), while the reductive decomposition peaks frequently lie well beyond the reductive de-doping (or doping for n-type CPs) peak, sometimes close to the electrolyte decomposition potential. The problem on the de-doped side for most p-doped CPs is more on that of dissolution of this more soluble, reduced form of the CP. Figure 30.3 shows a typical oxidative decomposition peak (indicated by arrow).

With regard to electrochemical windows of solvents and electrolytes usable for electrochromism of CPs, among the most versatile finding used recently are *ionic liquids*. An example of a commonly used such ionic liquid electrolytes is *n*-butyl imidazolium tetrafluoroborate. Figure 30.4 shows a typical cyclic voltammogram of a poly(aniline) CP in a variable-emittance device in this ionic liquid.

When electrochemical systems follow "Nernstian behavior," i.e., reversible thermodynamics and kinetics described by the Nernst equation of electrochemistry, CVs have well-established characteristics for such properties as anodic/cathodic peak separation, peak half width, and scan rate dependence.

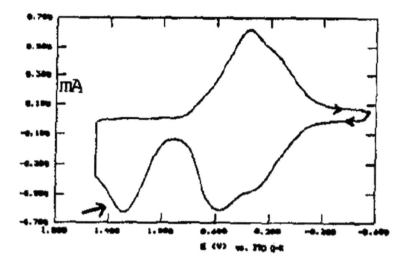

Fig. 30.3 CV of poly(aromatic amine)/-ClO$_4$, showing oxidative decomposition peak (arrow) (Courtesy of Ashwin®-Ushas Corp., Inc.)

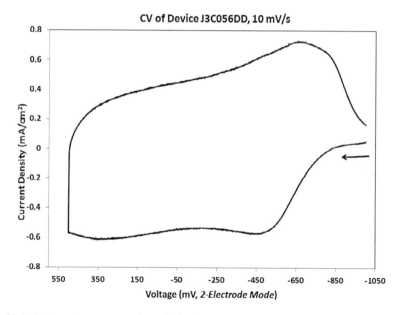

Fig. 30.4 Cyclic voltammogram of a typical variable-emittance skin incorporating a poly(aniline) CP in the ionic liquid electrolyte *n*-butyl imidazolium tetrafluoroborate. This is a two-electrode device, i.e., without a reference electrode. The IR light and dark states are at negative and positive applied potentials, respectively

30.2.3 Scan Rate Dependencies

One of the first characteristics of a CV of a CP film that one searches for is the dependence of the peak current (i_p) on the scan rate (ν). According to well-established electrochemical treatments, for a behavior dominated by diffusion effects, i_p is proportional to $\nu^{1/2}$, while for a material localized on an electrode surface, such as a CP film, i_p is proportional to ν. For $most$ CP films, the latter case obtains, thus indicating surface-localized electroactive species. For the P (Py) system of Fig. 30.1, i_p is proportional to ν. As more detailed analysis shows, however, this is so only for CP films that are not inordinately thick (which most are not), not inordinately compact (which most are not), and not doped with very large or sluggish dopant ions which have inordinately small diffusion coefficients (which most dopants do not). If any of the latter conditions prevail, however – i.e., wherever dopant diffusion effects can predominate – i_p can be proportional to $\nu^{1/2}$, as the case of poly(p-amino diphenyl amine) discussed below shows. Intermediate or transitional behaviors are also possible.

30.2.4 Other Parameters from CVs, Peak Broadening

Some other properties one may look for in a CP's CV are the symmetry of the peaks about the peak value, E_p; the background (residual) currents represented by the plateaux following the peaks; the cathodic and anodic peak separation $|E_{p,a}-E_{p,c}|$; the integrated current (area) under the peaks, which represents the total charge, Q; and the peak half width or full width at half maximum (FWHM). From an examination of Fig. 30.1, it is evident that the background currents on the anodic (+) side are much larger than those on the cathodic (−) side. This stems from an increase in capacitive current (a property of all real electrochemical systems) when the CP film is oxidized and conducting and thus able to augment the effective electrode substrate mass. Thus, such effects must be first discounted in interpretation of CVs. Figure 30.1 also shows that the peak in the oxidative (anodic) scan is more symmetrical than that in the reductive (cathodic) scan for this CP system, i.e., going from reduced to oxidized polymer appears more facile than the reverse for this CP system. This pattern is also evident in the slopes of the i_p vs. ν plots for this CP, which are much larger for the anodic peaks.

   Peak broadening also results from the electrochemical inequivalence of sites in the CP film. This phenomenon in turn can be associated with thick films as well as with other factors such as polymer morphology and chain length established during synthesis. For Nernstian behavior, a one-electron process, and a surface-immobilized electroactive film, the half width should be ca. 90 mV [2062, 2063]. For comparison, the half width of the narrowest anodic peak for P (Py)/BF$_4$ in Fig. 30.1 is 130 mV. Increased separation of the cathodic and anodic peaks frequently accompanies peak broadening. Finally, one must mention another important factor in peak broadening such as that seen for P(Py) (Fig. 30.1): The total electrochemical process involved in doping/de-doping of this CP can be said to encompass a two-step process – formation of polarons and formation of bipolarons. In P(Py), these two processes may be so close to each other in energy that electrochemically, a single, broad voltammetric peak is seen. In that case, the assumption of a one-electron process would also not be valid. Indeed, the slow evolution of the optical spectra with increased doping appears to substantiate this. Broad peaks in a CP's CV, which appear to be the result of coalescence of two peaks in nonaqueous media, can frequently be resolved in aqueous acidic media (see below).

If a CP film is switching very slowly (e.g., if it is extremely thick), the film will frequently not be able to "keep up" with the scan rate in a CV. In such instances, it is best to do a quick characterization of the approximate switching rate of the film via chronoamperometry (see Fig. 30.7 below), and then slow the scan rate accordingly to accommodate this switching speed. Thus, for example, a CP film that is showing unusually broad or otherwise distorted CV peaks at 100 mV/s may give well-defined ones at 10 mV/s.

30.2.5 Surface-Active Behavior

When in addition to having a surface-immobilized material, such as a CP film, one also has a very thin film, voltammograms similar to that obtained with polymeric-modified electrodes such as those based on poly(vinyl ferrocene) and poly(4-vinyl pyridine) are obtained, as shown in Figs. 30.5 and 30.6; here the cathodic and anodic peaks are nearly symmetrical above each other, with minimal separation.

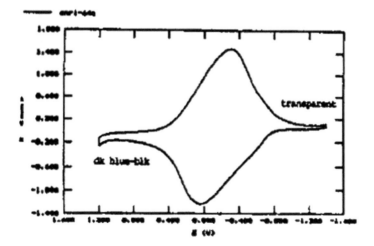

Fig. 30.5 *Left*, CV of poly(aromatic amine)/sulfonate (Courtesy of Ashwin®-Ushas Corp., Inc.)

Fig. 30.6 *Right*, poly
(N-phenyl-2-
naphthylamine)
(Reproduced with
permission from Ref. [965],
showing characteristic CV
of surface-localized
species)

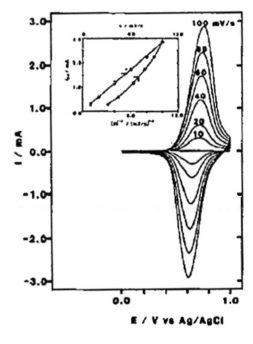

Fig. 30.7 Successive
cycles (as indicated) for
poly(oligothiophene)
derivative, showing
progressive "activation" of
film with increased cycling
(Reproduced with
permission from Ref. [978])

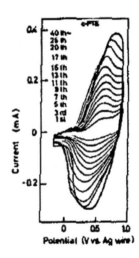

The cathodic/anodic peak separation then does not change, or changes minimally, with scan rate.

The thickness of CP films can also have other unexpected effects, one of which is the slow, continued "activation" of larger and larger portions of the film with increasing electrochemical cycling, as shown in Fig. 30.7. Here, peak areas appear to steadily increase without end with the number of cycles; for most CP films, CVs stabilize after the sixth to tenth cycle.

30.2.6   Charge Capacities

The area under the CV peaks can be integrated to yield total charge passed (*charge capacity*), if one has effectively provided a residual current baseline, which electrochemical software available today routinely does. This charge can then be used to estimate the proportion of the film that is electroactive. For instance, for all peaks in Fig. 30.1, the area, i.e., total charge, is ca. 0.7 mC/cm^2. This charge turns out to be ca. 9% of the total charge used to prepare the film; the latter includes the charge consumed in electropolymerization, which is a two-electron process, as well as additional charge to achieve the ca. 0.25, i.e., 25%, doping (oxidation) level of the P (Py) film, i.e., a total of 2.25 electrons per monomer unit. Thus, the charge under the CV peaks can be used as an indication of the electroactivity of the CP film, as follows: 9% ÷ {(100% ÷ 2.25) × 0.25} = 81%.

30.2.7   Dopant and Structural Relationships

Figure 30.2, the peak/scan rate dependence, also shows the effect of dopants on voltammetric behavior. The slope differs for different dopants. A lower slope usually implies slower dopant diffusion in and out of the CP and broader peaks. The effect of dopants is also seen in the voltammograms of Fig. 30.8. In general, the sharper, larger peaks are given by dopants which find diffusion in and out of the polymer facile. The effect of Bu_4ClO_4 vs. Et_4ClO_4 appears intriguing at first, since the dopant anion is the same in both cases; it is ascribed to ion-pairing effects where the cation also participates.

What happens if one introduces substituted analogs of the pyrrole monomer unit, e.g., N-methyl pyrrole, in place of pyrrole? This is seen in Fig. 30.9; the voltammetric behavior of the substituted analogs is very similar to that of P (Py) but with an anodic shift of the redox peak potentials. This is explained by the likely participation of protons in the facilitation of the redox of P(Py), which is not possible in N-substituted analogs. Another effect of N-substitution in poly (pyrroles) is of course increased environmental stability, as oxidants invariably first attack the exposed, electron-rich N-atom of unsubstituted pyrrole. That monomer structure can sometimes directly influence polymer behavior is also seen in the plot of oxidation potentials of monomer vs. polymer for poly(thiophene) analogs in Fig. 30.10; here the two oxidation potentials are exactly parallel to each other, i.e., a monomer that is more difficult to oxidize implies a polymer that is also more difficult to oxidize. On the other hand, alkoxy-substituted P(ANi) shows voltammetric behavior very close to that of the unsubstituted P(ANi), with peaks shifted less than 90 mV anodically (Fig. 30.11).

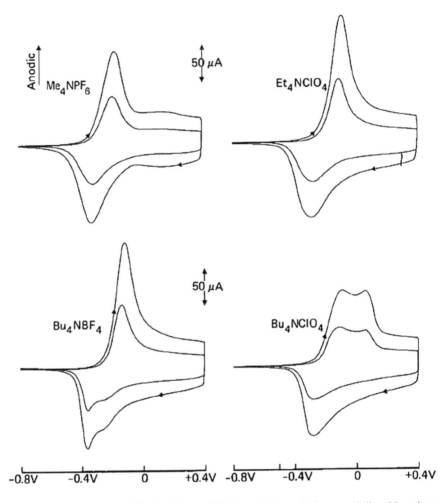

Fig 30.8 CVs at 50, 100 mV/s of poly(pyrrole) (20 nm thick on Pt) in acetonitrile with various electrolytes

30.2.8   *Reversibility*

The question of *reversibility*, *quasi-reversibility*, and *irreversibility*, according to the standard electrochemical definitions, is difficult to accurately estimate for CPs, since they frequently do not follow the rigid electrochemical criteria for these. Thus, for instance, to derive a kinetic electron transfer coefficient from electro-chemical data, as frequently attempted in the CP literature, is not entirely mean-ingful, in *practical* terms, for most CP systems. This is because of the very large number of variables involved. If during CP redox in a CV at a particular scan rate, a reduction peak is entirely absent or has 1/2 the peak height and half width of the

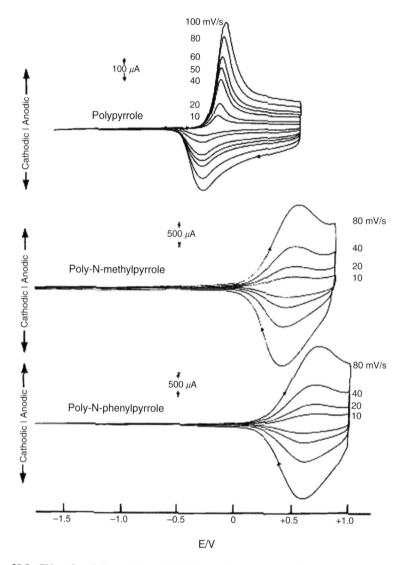

Fig. 30.9 CVs of poly(pyrrole), poly(N-Me-pyrrole), and poly(N-Phe-pyrrole) compared (Reproduced with permission from Ref. [979])

oxidation peak, then evidently the CP redox process is not electrochemically reversible for this scan rate. This does not preclude however that the peaks may become highly symmetrical and equal in parameters at very slow scan rates, in which case the CP redox would be considered reversible at those lower scan rates. Film thickness also quite evidently plays an important part in this behavior. Broadly speaking, the cathodic/anodic peak separations, half width ratios, peak ratios, and symmetries can cumulatively be used to arrive at a judgment on likely

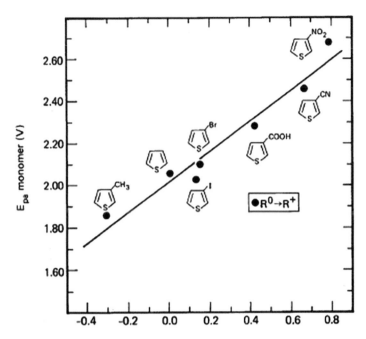

Fig. 30.10 Peak oxidation potentials of poly(thiophene) analog compared with thiophene monomers (Reproduced with permission from Ref. [979])

electrochemical reversibility, which would again be valid only for particular film
thicknesses, scan rates, and the like.

30.3   Solvent and pH Effects, Mixed Solvents, and Dopants

With regard to solvent effects in electrochemistry of CPs, solvent retention effects
can sometimes be pronounced in CPs. For liquid electrolytes, during the doping/de-
doping process, the polymer normally swells and contracts not only from minor
distortions accompanying charged carrier formation but also to accommodate
dopant ions and any accompanying solvent molecules. Solvent molecules can
subsequently be retained in the polymer even when dopant is expelled. Solvent
molecules can also be retained when films are cast from sols or solutions. Solvent
retention can then sometimes significantly alter the behavior of the CP film. This is
seen most clearly in the data in Fig. 30.12d, for P(ANi) films prepared via casting of
the pristine polymer from solutions in NMP (b.p. $> 200\ ^{\circ}$C). In both fresh
dip-coated and spin-coated films (a,b), one sees poorly defined CV peaks and,
sometimes, multiple peaks; anyone with some experience in CP electrochemistry
will immediately recognize that this is the case with most fresh CP films until the
CVs "stabilize" after 8–10 cycles. NMP is a solvent that bonds strongly at the more

a

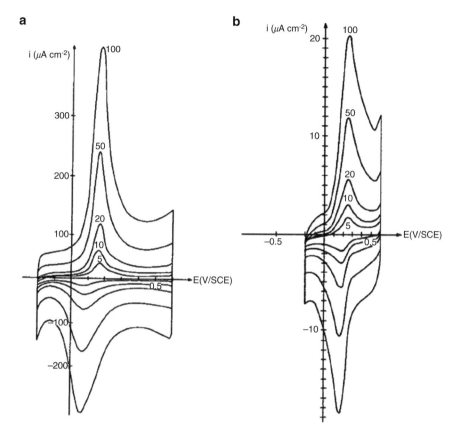

Fig. 30.11 Comparative CVs of poly(aniline) and poly(OEt-aniline) (Reproduced with permission from Ref. [980])

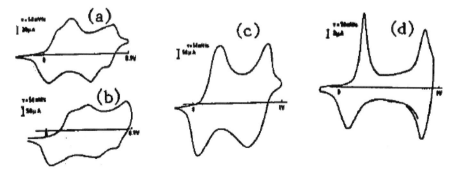

Fig. 30.12 CVs of P(ANi) recast in various ways, in 1 M HCl, scan rate 50 mV/s. (**a**) Dip-coated film; (**b**) spin coated; (**c**) spin coated after storage in water; (**d**) Langmuir-Blodgett processed film (Reproduced with permission from Ref. [981])

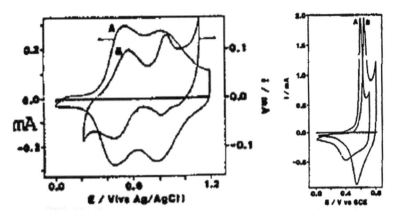

Figs. 30.13 and 30.14 Characterization CVs of poly(diphenyl amine) (**a**) and poly-(4-amino biphenyl) in LiClO$_4$/CH$_3$CN (30.13, *left*) and 2 M HCl (30.14) (Reproduced with permission from Ref. [982])

active sites, such as N-atoms, in a CP, and the poor definition of the CV peaks of Fig. 30.12a,b stems from the electrochemical inequality of CP sites with and without this solvent. As this retained solvent is removed by prolonged soaking in water, the CV sharpens considerably (Fig. 30.12c). With a very thin (Langmuir-Blodgett) CP film, this sharpening is even more pronounced (Fig. 30.12d). As noted earlier, acetonitrile appears to be the solvent showing the least solvent retention effects. An effective way of minimizing these effects is however simply to soak the CP film in low-boiling solvents that are inert to it, including acetonitrile and toluene, and then drying in a stream of warm air or under moderate vacuum.

For some CPs, both nonaqueous and aqueous media can be used for electrochemical studies. Poly(aromatic amines), such as P(ANi) or poly(diphenyl amine) (P(DPA)), are some of the CPs that are electrochemically active in both nonaqueous (e.g., acetonitrile) and aqueous (normally acidic) media (Figs. 30.13 and 30.14).

characterization CVs for P(DPA) and an analog, poly(4-amino biphenyl) (P (4ABP)) in nonaqueous and acidic (2 M HCl) media, respectively. It is seen that the peaks, especially the oxidation peaks, are much sharper in the acidic media. This is a characteristic of protonation processes and reflects active proton involvement (protonation of the active N-atoms during doping). The sharpness of the oxidative (doping) peaks can be again seen in Fig. 30.15, for poly(phenylene sulfide). Another important feature that may be discernible from Figs. 30.13, 30.14, and 30.15 is that when CV peaks are poorly resolved in nonaqueous media, they can frequently be sharply resolved in acidic media. Yet another curious property of many CPs is that if they are electropolymerized in one solvent medium, e.g., acetonitrile, they can be electrochemically cycled in another, e.g., propylene carbonate, but not vice versa.

The dependence of the peak height and position on pH is seen quite clearly in Fig. 30.16 for poly(1-amino-pyrene). This is a general behavior of such CP systems which are susceptible to protonation and in which protonation takes part in the

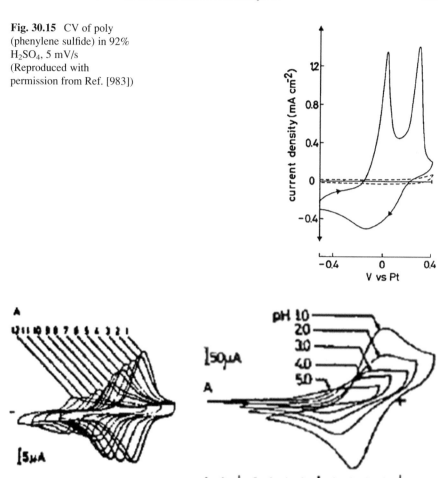

Fig. 30.15 CV of poly (phenylene sulfide) in 92% H$_2$SO$_4$, 5 mV/s (Reproduced with permission from Ref. [983])

Figs. 30.16 and 30.17 pH dependence of CVs for poly(1-amino pyrene) (30.16, *left*, with pH values indicated, reproduced with permission from Ref. [991]) and for poly(2,3-diamino-naphthalene) (Figure 30.17, reproduced with permission from Ref. [984])

doping/de-doping process: with increasing pH, the peaks diminish (and eventually disappear) and shift toward more negative potential, indicating increasing difficulty of protonation. For a system, such as poly(2,3-diamino-naphthalene) (Fig. 30.17), the shift of E_o with pH is nearly linear, indicating that electrons and protons take part in the doping/de-doping process in this CP in a 1:1 ratio [984].

Although mixed solvents, such as HCl-acidified acetonitrile, have been used for electrochemical cycling of CPs on some occasions [865], they are usually used to facilitate solubility and/or protonation of monomer or dopant during the electropolymerization itself and are thus discussed elsewhere in this book.

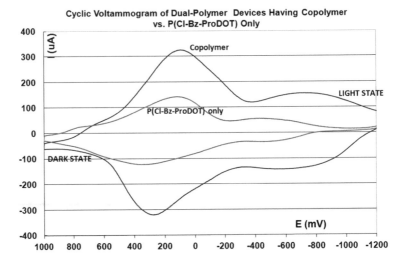

Fig. 30.18 Cyclic voltammograms for dual-polymer electrochromic devices comprising poly (Cl-Bz-ProDOT)/-poly(aromatic amine) and /[poly((Cl-Bz-ProDOT_-co-(Br-Bz-ProDOT)-co-(Bz-ProDOT))]/poly(aromatic amine), deposited with identical equivalent charge, in an ionic liquid mixture primarily comprised of *n*-butyl imidazolium tetrafluoroborate (After Ref. [985])

Dopants may frequently be *exchanged electrochemically* in CPs. For example, a CP may be electropolymerized in a medium containing perchlorate as the dopant, but repeated electrochemical cycling in tetrafluoroborate-dopant medium causes the eventual replacement of perchlorate in the CP by tetrafluoroborate.

Semisolid or gel electrolytes may also be used as effective electrolytes for studying electrochemistry, and, in a related fashion, electrochromism, of CPs. Figure 30.18 shows the cyclic voltammetry of an electrochromic CP in a gel electrolyte.

30.4   Electrochemistry in Ionic Liquid Electrolytes

Starting in about 2002, ionic liquids started being investigated as electrolytes for CPs, in applications involving electrochemistry, electrochromism, and synthesis [986]. Ionic liquids, a new class of materials first studied in detail starting in the 1970s, are essentially organic-based "room-temperature molten salts." They display high conductivities, up to 1/10 that of conventional organic electrolytes used in electrochemistry, and have other desirable properties, such as negligible vapor pressure and a wide electrochemical window. Low vapor pressure is especially useful in applications in the vacuum of space, where conventional electrolytes, including organic electrolytes with low vapor pressure, would rapidly evaporate and thus become nonfunctional. Figure 30.19 shows a typical CV of a CP in an ionic electrolyte, as functioning in a variable-emittance skin used for spacecraft thermal control.

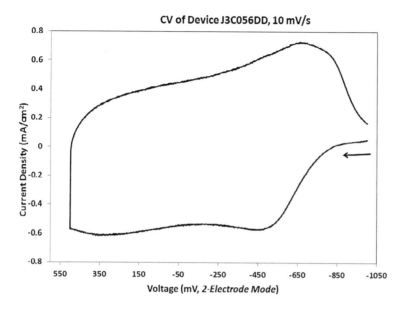

Fig. 30.19 Cyclic voltammogram of a poly(aniline) derivative contained in a variable-emittance skin (in the ionic liquid electrolyte *n*-butyl imidazolium tetrafluoroborate), used for spacecraft thermal control in the vacuum of space. This is a two-electrode device, with the working and counter electrodes comprised of the same CP. (*This figure was also presented earlier in this chapter in a different context.*) (After Ref. [975])

30.5 Relation with Semiconductor Properties

One of the first semiconductor-related properties evident from the CVs of the poly (aromatic amines) in Figs. 30.12, 30.13, 30.14 and 30.15 is the presence of *two* distinct redox peaks, with substantial separation. In the oxidative scan, the *first* corresponds broadly to removal of one electron from the polymer and thus to the generation of *polaron*-dominated mid-gap states. The second corresponds to removal of the second electron and the generation of *bipolaron*-dominated states.

Well-defined CVs can also be used to indirectly extract information such as CP bandgaps (E_g) and the relative strength and position of polaron and bipolaron absorptions. While the latter has been carried out effectively only for poly(aromatic amines), Fig. 30.20 below shows an example of the computation of the former from voltammetric data, using the simple formula $E_g = E_{p,ox} - E_{p,red}$. Figure 30.25 summarizes bandgaps for a number of CPs computed primarily from electrochemical data.

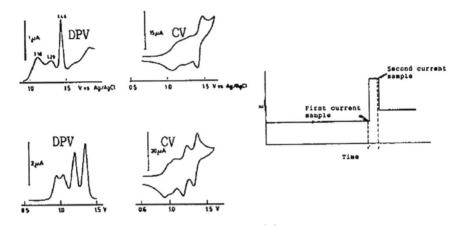

Fig. 30.20 Comparison of voltammograms from DPV (left) and CV (right) for novel CPs DMOP-OV. At right, waveform for DPV (Reproduced with permission from Ref. [984])

30.6   Other Voltammetric Methods

From the difficulties encountered with interpretation of CVs which the discussions above amply show, it would appear that other voltammetric methods, especially differential methods, would have found wider application to CPs. This has unfortunately not been the case. The results in Figs. 30.18a,b,c and 30.19 represent some of the few studies of this nature. In Fig. 30.18, the results of CV and of differential pulse voltammetry (DPV) are compared. The latter is a technique in which a small potential pulse is superimposed on a staircase potential function with the difference between the post-pulse and pre-pulse current measured (inset in Fig. 30.18). The differential method yields peak-shaped curves unencumbered by residual current tails, as in CVs, and thus a clearer identification of peaks and their widths. Figure 30.20 then shows DPV of poly(phenylene vinylene) used to compute the bandgap, as described earlier. Normal pulse voltammetry (NPV), in which a sort of digital pulse ramp is applied in place of the analog ramp of CV and the current sampled at the end of the pulse [50], has been applied to poly(1-amino pyrene) [991], yielding redox potentials as well as diffusion coefficients (Fig. 30.19). Other differential methods such as square wave voltammetry have been applied to poly (aromatic amines) in the author's laboratories. There is however little other extant work with pulse voltammetry of CPs, although the very brief results above clearly provide a strong indication for it.

30.7   CA, CC, and Diffusion Coefficients

Besides yielding useful electrochromic information, *chronoamperometry* (CA) can also help in otherwise characterizing CPs. They can crudely estimate switching rate and yield other information, such as diffusion coefficients. The quickest and simplest way to characterize an approximate switching rate for a CP – i.e., whether it is in the microsecond, millisecond, second, or tens of seconds range – is the potential step from its reduced to its oxidized states and back (i.e., using a double potential step) and monitor the transient current obtained to a time at which it is no more than ca. 5% of the peak value, which is then the approximate switching time.

The diffusion coefficients that are obtainable by CA and related methods are *apparent* diffusion coefficients, encompassing all diffusion-related processes relating to electrochemical "switching" (i.e., redox) of the CP. These processes of course primarily comprise diffusion of the dopant ion into and out of the CP structure, but may also include effects relating to the slow regeneration of conductivity in the CP, which impedes immediate electron flow to the surface of the CP film, an effect especially valid for the insulating → conducting transition. Thus, the diffusion coefficients are in reality *apparent diffusion coefficients for the entire charge transport process* associated with CP switching. Such diffusion coefficient calculations are valid for potential steps to diffusion-limited plateaux, where the *Cottrell equation*, describing the current-time response under such conditions, is valid. This equation, for the Ox → Red process, is:

$$i(t) = \left\{ n F A D_o^{1/2} C_o{}^* \right\} / \left\{ \pi^{1/2} t^{1/2} \right\} \tag{30.1}$$

where F is the Faraday, A the area of the electrode, $C_o{}^*$ the initial concentration of the active species being measured, and D_o its diffusion coefficient. Thus, if a plot of current vs. $t^{-1/2}$ is found to be linear, the slope of this plot will yield an apparent diffusion coefficient. Results for several initial concentrations should ordinarily be plotted.

Apparent diffusion coefficients for dopant ions in CPs have been in general found to be very small, of the order of 10^{-10} cm^2/s, vs. the "typical" diffusion coefficient of an ion in a liquid medium of ca. 10^{-5} cm^2/s. To quote some reported values, we have ClO_4^- in P(Py), 10^{-9} cm^2/s and H^+/P(ANi), 1.9×10^{-10} cm^2/s [859, 987–989]. A value as low as 10^{-6} cm^2/s has been reported for P(ANi)/SO_4^{2-}, but the calculation is in doubt [990].

NPV results may also be used to calculate apparent diffusion coefficients also employing Cottrell relationships. Oyama et al. [984] have used the NPV voltammograms of Fig. 30.20 (III.24. f10) to calculate apparent diffusion coefficients (with t now being replaced by τ, the NPV sampling time) for ClO_4^- in poly (1-amino-pyrene), of 1.3×10^{-10} cm^2/s.

A technique called *chronocoulometry* (CC) is much less frequently used in CP work. It can also be used to calculate, among other parameters, apparent diffusion coefficients. The relevant equation, again under the "Cottrell conditions" described above, is

$$Q = \{2nFAD_o^{1/2}C_o{}^*t^{1/2}\}/\{\pi^{1/2}\} + A + B \qquad (30.2)$$

where A and B are constants. Thus, if a plot of total charge, Q vs. $t^{1/2}$, yields a straight line, the slope yields the diffusion coefficient directly. Again, several initial dopant concentrations should be used. CC also yields charge capacities directly, allowing one to dispense with integration of CV peak areas, for instance.

Finally, many of the above equations may be used to estimate n, i.e., the number of electrons involved in the redox process of a CP.

30.8   Complex Film Thickness and Dopant Effects

The discussion above has thus far conveyed the impression that most CP films on conductive electrode substrates show properties typical of surface-immobilized species, such as peak currents (i_p) directly proportional to scan rates (ν) in CVs (rather than to the square root of the scan rate typical of diffusion-limited processes). As also noted above, however, these interpretations are somewhat simplistic, valid for the conditions of fairly thin and porous CP films and non-bulky dopants that usually prevail. It may be intuitively apparent that if diffusion of dopants during CP switching is unduly impeded by excessive film thickness, compactness, or dopant bulk, perhaps diffusion-limited or even other relationships may prevail.

This is precisely what is observed, for instance, in the case of poly(p-amino diphenyl amine) (P(ADPA)) [990]. Figure 30.21 shows the peak current vs. scan rate relationship under ordinary conditions at ca. 20 °C. The current is linear with

Fig. 30.21 *Left*, normal pulse voltammogram (NPV) of poly(1-amino pyrene) (Reproduced with permission from Ref. [991])

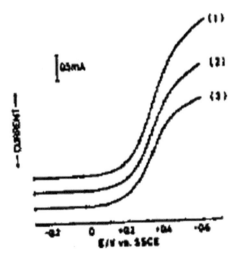

Fig. 30.22 DPV of poly (phenylene vinylene) on Pt, showing computation of bandgap, E_g (Reproduced with permission from Ref. [984b])

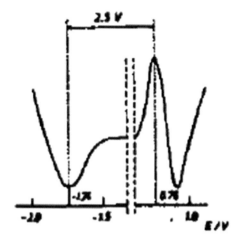

the *square root* of the scan rate, like diffusion-limited processes and very much unlike most CPs. However, when the temperature is raised to 70 °C, the current is proportional to ν; subsequently, if the film thickness is increased substantially, the current is proportional again to $\nu^{\frac{1}{2}}$ (Fig. 30.22). Clearly therefore, diffusion of dopant is enhanced at 70 °C so that it no longer limits the electrochemical processes but is again impeded if one has thicker films even at this temperature. This effect is visible in this CP because of its unusually compact structure as compared to other CPs, as other studies confirm [990].

The results of Fig. 30.21 also show some dopant diffusion characteristics, evident in the slopes of the lines (as for Fig. 30.2): Cl^-, ClO_4^-, and benzenesulfonate move more slowly in and out of the polymer (lower slopes) than NO_3^- and SO_4^{2-}. With the bulky benzenesulfonate dopant, the same, diffusion-limited behavior is observed (i proportional to $\nu^{\frac{1}{2}}$) in the initial electrochemical cycling. However, after prolonged electrochemical cycling of a film, the relationship reverts to i proportional to ν (Fig. 30.23). What appears to happen with this bulky anionic dopant is that after prolonged cycling, the dopant in effect becomes embedded in the polymer matrix, and subsequent doping/undoping processes involve transport of *cations* into the combined CP/dopant matrix. This is a characteristic phenomenon true for other bulky or polymeric dopants, such as poly(styrene sulfonate), as well as with the so-called "self-doped" CPs discussed later in this book. The magnified effect of bulky dopants in compact polymer matrices is also seen in Fig. 30.24, which show CVs for cycling of highly compact and nonporous P(Py)/poly(urethane) composites. Here, the CVs are quite reproducible in extended cycling for the agile Cl^- dopant ion but show diminution and cathodic shift of reduction peaks for the bulkier tosylate dopant (Fig. 30.26).

Fig. 30.23 *Left*, peak current vs. scan rate relationship for an aromatic amine CP, P(ADPA) at 20 °C and for various dopants

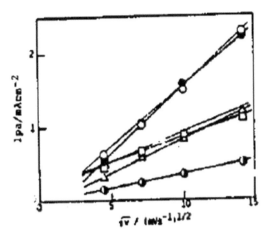

Fig. 30.24 *Right*, at 20 °C vs. 70 °C (Reproduced with permission from Ref. [866])

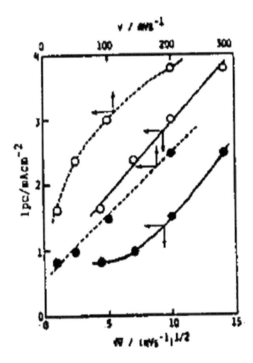

30.9   Modified Electrodes

We have until now discussed the electrochemistry of the CP films themselves, when immobilized on suitable substrate electrodes. These CP films can however also act as electrodes themselves when in their conductive states, if the redox activity of the

Fig. 30.25 *Left*, scan rate dependence of peak current as function of doping level of benzenesulfonic acid, for cycle numbers shown (Reproduced with permission from Ref. [866])

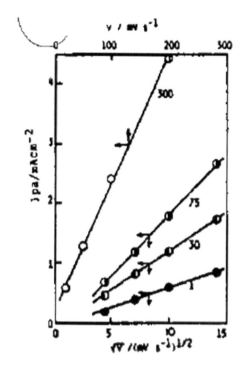

Fig. 30.26 *Right*, CVs of P (Py)/PU composite film in tosylate dopant (top) and chloride dopant (bottom)

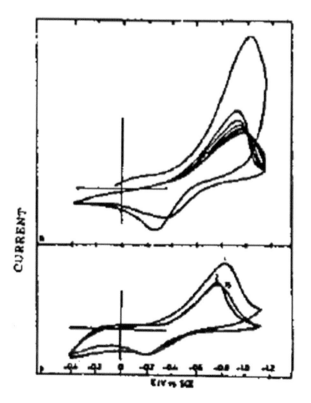

electroactive species to be studied lies within a potential range in which the CP remains conducting. We then have the customary modified electrode.

One of the first CPs to be studied as a modified electrode was P(Py), at which the electrochemistry of ferrocene ($FeCp_2$) was studied [992]. This $FeCp_2$ system (E^o + 0.42 V vs. SCE) showed near ideal reversible electrochemical behavior, with i_p proportional to $\nu^{\frac{1}{2}}$, indicating diffusion-limited processes, cathodic/anodic peak current ratios of near unity, and cathodic/anodic peak separations close to 70 mV. More recently, poly(bithiophene)-modified electrodes have been used to monitor the electrochemistry of $FeCp_2$ as well as p-benzoquinone and to probe the semi-conducting behavior of this CP [1481]. Ferrocene appears popular as a near-reversible probe material for CP-modified electrodes, having been studied with poly(thiophene) and other CP-modified electrodes as well [993].

From the brief discussion above, one may wonder as to the actual utility of such CP-modified electrodes, apart from scientific curiosity. The answer to this question may lie in preliminary studies of their catalytic activity, which is discussed at length elsewhere in this book. Here we may cite one brief, illustrative example of such activity. Bedioui et al. [985] have polymerized a Co(TPP) with a pyrrole pendant group, yielding a poly(pyrrole) of sorts in direct contact with the electrode substrate, which has catalytically active $Co^{I/II/III}$(TPP) pendant groups. These display all the normal Co(I/II/III) redox chemistry in both the conductive and insulating states of the poly(pyrrole) backbone. This modified electrode is able to catalyze the reduction of benzyl chloride to toluene.

30.10 Problems and Exercises

1. Identify typical parameters used for routine characterization and electrochemistry of CPs among the following: substrate electrodes, electrolytes, control instrumentation, CP thickness, and common techniques (voltammetric, amperometric, others) used.
2. Using one of the CVs from the catalog above, plot the scan rate dependence of peak current, and identify the major voltammetric parameters extractable. What does this tell you about diffusion processes, film thickness, dopant type, and other characteristics of this CP system?
3. Identify all possible causes of peak broadening in a CV of a CP.
4. Trace effects of substituents on electrochemical parameters for each of the following CP systems: poly(pyrroles), poly(thiophenes), and poly(anilines). Do a brief search of the recent CP literature to identify similar trends in other CP systems and between systems.

5. What are the commonalities and differences between electrochemical properties in acetonitrile and acidic aqueous electrolyte media for poly(aromatic amines)? For other CP systems?
6. What is the effect of pH on voltammetric parameters of protonable CP systems? The effect of dopant concentration in the electrolyte on other CP systems?
7. Carry out a brief literature search to identify studies on CPs employing pulse voltammetric methods within the last 5 years. Which information is available with greater facility and what additional information is available from these as compared to standard CVs?

Chapter 31
Syntheses and Processing of CPs

Contents

31.1 Electrochemical Polymerization

31.1.1 Mechanisms

31.1.1.1 Generic Electropolymerization Mechanism

We shall first very briefly discuss the broad division of CP syntheses into chemical polymerization and electrochemical polymerization and very briefly described the electrochemical synthesis and the mechanism thereof for P(Py). The clear difference in mechanisms between chemical and electrochemical syntheses of CPs described in an earlier part of this book appears to be preserved for most CPs.

Nearly all electrochemical polymerizations of CPs (which are nearly always oxidative polymerizations) appear to follow a *generic reaction pathway*. In Scheme 31.1, we illustrate this pathway, representing a generic monomer, for P(Py). This pathway shows the following common features:

1. The initiation step, (monomer) radical generation, via electrochemical oxidation

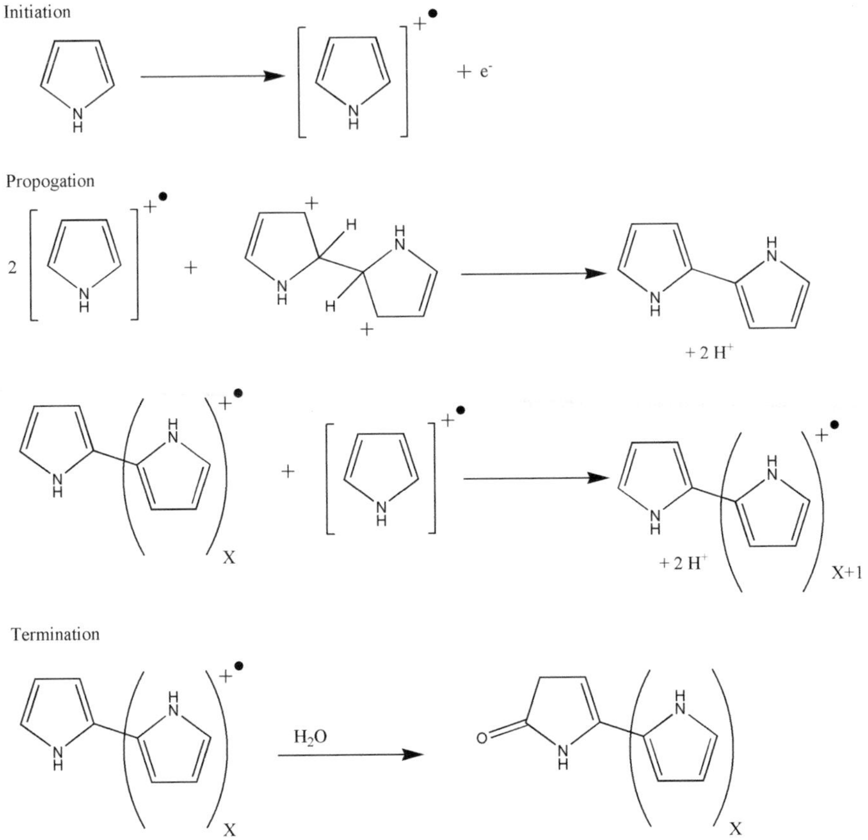

Initiation

Propogation

Termination

Scheme 31.1 Generic electropolymerization pathway valid for many CPs

2. Propagation via (a) *radical–radical recombination, not radical–monomer combination*; (b) loss of two protons from the radical–radical intermediate species, generating the dimer; (c) electrochemical oxidation of the dimer, generating another, "oligomeric" radical; and (d) combination of this or similar oligomeric radicals *with monomer radicals* and repeat of steps 2b and 2c, building up the polymer
3. Termination via exhaustion of reactive radical species in the vicinity of the electrode and accompanying oxidative (as shown for P(Py) in Scheme 31.1) or other chain termination processes

   Among other common features of the mechanism is the preferred polymer linkage at the α-position for most monomers, as illustrated in Scheme 31.1-(c). We now discuss the evidence for this generic mechanism among the many CPs studied heretofore.

One of the first features to note for an electrochemical polymerization is that once the initiating electrochemical potential is applied, the population of radical cations is likely to far exceed that of neutral monomer in the vicinity of the electrode surface. That is to say, a generated radical cation is far more likely to be surrounded by other radical cations than by neutral monomers or oligomers or other species. Among the several determinant causes for this is the usually rapid electron transfer kinetics for electrooxidation of the monomer, in comparison with the slower diffusion of monomer from bulk of reaction medium to electrode, thus causing a rapid depletion of monomer concentration at the electrode. This is in contrast to the situation in a typical chemical polymerization, where the concentration of monomer far exceeds that of radical, and thus radical–monomer combination (generating another radical) is more likely to be the next propagation step after radical generation. Apart from the usual spectroscopic structural evidence, other support for α-polymer linkages for most CPs is fairly strong: For example, 2,5-disubstituted pyrrole monomers fail to polymerize, and the polymerization of 2-monosubstituted monomers stops at the dimer stage; 1- and 1,3-substituted azulenes (Scheme 31.1) fail to polymerize [987]; theoretical electron density studies on azulene indicate that electrophilic attack, as with a radical cation on a neutral monomer, would occur at the 4-position, yielding a 1,4-linkage rather than the 1,3-linkage actually observed [988].

31.1.1.2 Factors Favoring Polymerization and No Polymerization

While it is true that electropolymerization is a nearly generic technique usable for quick and rapid production of CPs, not all organic monomers undergo electropolymerization of course. The keys to why certain monomers electropolymerize at all to yield CPs lie in a *combination* of factors of stability of the radical ions generated in the first step *and* the oxidation potentials for generation of these. In nearly all cases where successful electropolymerization is observed, the radical ions are found to be highly stabilized via mechanisms such as charge delocalization, and electrochemical oxidation was reasonably facile. For example, carbazole, furan, and indole, all molecules whose radical cations are less well stabilized than that of, say, pyrrole produce poorly formed polymers with low conductivity on electropolymerization [989, 990]; the oxidation potential of indole is however comparable to that of pyrrole (in acetonitrile +0.9 V, +0.8 V vs. SCE) and that of furan comparable to that of thiophene\ (+1.85 V vs. +1.6 V vs. SCE) [991, 992]. The electrooxidation of benzene, to produce poly(p-phenylene), requires a very high potential normally effected only in media such as HF or liquid SO_2 [993, 994], and thus chemical polymerization is preferred for this CP. Steric crowding, as illustrated, for instance, for the large macrocycles in Fig. 31.1, can also hinder electropolymerization or yield polymers of low conductivity. Table 31.1, which is based on a very large number or references, illustrates the wide variety of solvents and electrolytes (dopants) that can be used for

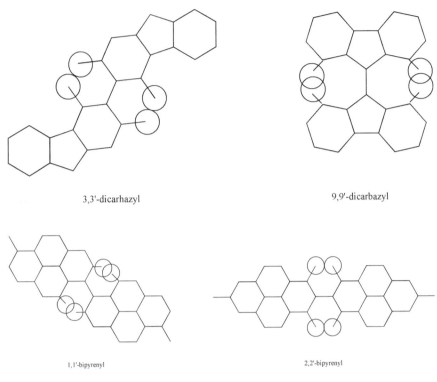

3,3'-dicarhazyl

9,9'-dicarbazyl

1,1'-bipyrenyl

2,2'-bipyrenyl

Fig. 31.1 Example of hindrance to electropolymerization caused by steric crowding

Table 31.1 Solvents and electrolytes used for Poly(pyrrole) polymerization. The list is representative and by no means exhaustive

Solvent/electrolyte	Polymerization
Acetonitrile/tetraethylammonium tetrafluoroborate	Good
Acetonitrile/toluenesulfonic acid	Good
Acetonitrile/tetraethylammonium tetrafluoroborate plus 1.0 M pyridine	None
Methylene chloride/tetrabutylammonium tetrafluoroborate	Good
Butanone/tetrabutylammonium tetrafluoroborate	Good
Propylene carbonate/tetrabutylammonium tetrafluoroborate	Good
Dimethylformamide/tetraethylammonium tetrafluoroborate	None
Dimethylformamide/toluenesulfonic acid	Good
Dimethylsulfoxide/tetraethylammonium tetrafluoroborate	None
Hexamethylphosphoramide/tetraethylammonium tetrafluoroborate	None
Ethanol/tetrabutylammonium tetrafluoroborate	Fair
Ethanol/toluenesulfonic acid	Good
Ethanol/sulfuric acid	Good
Ethanol/phosphoric acid	None
Ethanol/hydrochloric acid	Fair

Table 31.2 Well-known CPs (shown with structure) successfully electropolymerized as BF_4^- salts in CH_3CN

	Monomer		Oxidation potential
X	R_1	R_2	(V versus SCE)
NH	H	H	0.80
NH	CH_3	CH_3	0.35
NCH_3	H	H	0.80
S	H	H	1.60
S	H	CH_3	1.40
S	CH_3	CH_3	1.35
S	H	Phenyl	1.30
S	Phenyl	Phenyl	1.30

electropolymerization, here for P(Py). Table 31.2 briefly summarizes some of the more well-known CPs successfully produced electrochemically.

Proton removal renders the dihydro-dimer or dihydro-oligomer formation (Scheme 31.1) irreversible. Thus, it is expected that proton acceptors including water, bases such as pyridine or lutidine, and other nucleophiles enhance electrochemical polymerization. This is indeed found to be the case for a wide variety of CPs. It is also felt by many workers that this is one of the reasons that electropolymerizations are generally most successful in acetonitrile. A decrease in pH is sometimes observed during polymer film formation, as, e.g., for P(Py) [995].

Indeed, in many cases, a little water (about 1% v/v) is found necessary for electropolymerization: When highly dried materials are used, yields are invariably poor [999, 956]. Water or other nucleophiles also may serve another function, the provision of an effective counter electrode reaction during solution electropolymerizations: It is felt by some workers that in the absence of an effective counter electrode reaction in most such electropolymerizations, reduction of impurities such as trace metals or water serves as a poor substitute counter electrode reaction. This has indeed led some workers to explore deliberate use of counter electrode reactions (such as $Cu^{2+} \rightarrow Cu^o$ or $Ag^+ \rightarrow Ag^o$), but inexplicably, these procedures have not produced higher conductivity or improved morphology films [995, 996], perhaps indicating that impurities such as water may after all provide effective counter electrode reactions. Excess water may however lead to premature oxidative termination of the propagating polymer chain.

The dimers and subsequent oligomers produced invariably have oxidation potentials lower than (more negative of) that of the monomer. Indeed, this is a rule of thumb for CPs: As chain length increases, oxidation potential decreases. The variation of formal redox potentials for CP chains of different lengths is one of the factors contributing to peak broadening in CP cyclic voltammograms. Thus, in a typical potentiostatic or galvanostatic polymerization, the oligomer, once produced, is rapidly oxidized at the electrode to its radical cation. It then is more likely to react with monomeric radical cations, which are in abundance, rather than other oligomeric radical cations. It is however intriguing to note that attempts to electropolymerize dimers or trimers directly, e.g., bipyrrole or terpyrrole [997], are unsuccessful, as are attempts to polymerize these oligomeric pyrroles in the presence of monomer but at a potential at which the monomer is not oxidized [997]. A surfeit of *monomer* radical cation appears to be necessary for successful electropolymerization, a fact not easily explained.

31.1.1.3 Mechanistic Notation and Rate Expressions

The generic mechanism above is denoted in electrochemical terminology as an E $(C_2E)_nC$ mechanism, that is, an initial electrochemical step (the radical cation generation), followed by polymer propagation via successive chemical (combination with monomer radical cation, proton loss) and electrochemical (oxidation to generate oligomeric radical cations) steps, and likely termination via a chemical step. The chain termination steps have not been well characterized for most CPs; indeed, oxidative chain termination, such as shown for P(Py) in Scheme 31.1, may compete with combination with monomer radical and occur when the latter is depleted. Stronger evidence for the radical–radical coupling (vs. radical–monomer coupling) is available from many other studies, for instance, the detailed double potential step chronoamperometry studies of Oyama et al. [998], which are unfortunately too involved to treat in an elementary book such as ours.

Because oligomer and monomer radicals both figure in the rate expression for polymer propagation in most studies heretofore [956, 999–1004], an induction time – for the generation of sufficient concentrations of monomer radical cation as well as oligomers – is indicated for most electropolymerizations of CPs, before the polymerization can proceed at a useful rate. A fairly simple illustration of this is seen in the chronoamperograms and chronocoulograms of Figs. 31.2, 31.3, and 31.4: Electropolymerization current continues to increase in the initial stages of the polymerization, until it peaks and subsequently becomes limited possibly by monomer diffusion to the electrode; very similar behavior has been reported in chronoamperometric studies of a range of monomers, such as thiophene [952]. A "minimal CP nucleation" and nucleation times have also been cited by many workers for the deposition of an initial CP film on a fresh, highly polished electrode. Evidence for this emerges from data such as the electrodeposition CV in Fig. 31.5, which shows a crossover of the reverse cathodic scan over the anodic scan typical of metal deposition processes, where nucleation is required [1005]; once minimal nucleation has occurred in the initial anodic scan, CP deposition is much easier and takes place at lower potential on the reverse scan, hence the crossover.

31.1.2 Solvents and Electrodes

The solvent of electropolymerization is an important factor governing not only the quality of the CP obtained, but also its conductivity, morphology, and subsequent electrochemical and chemical behavior. Acetonitrile is one of the most versatile

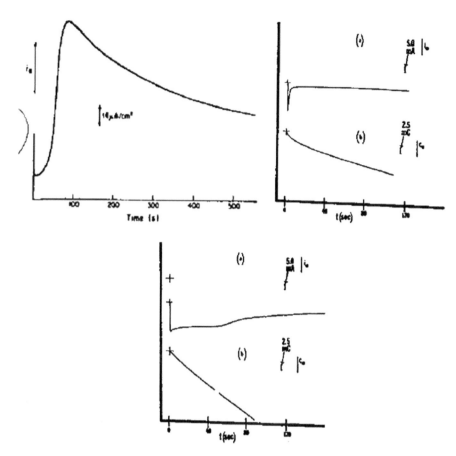

Figs. 31.2, 31.3 and 31.4 Representative electropolymerization chronoamperograms and chronocoulograms (Fig. 31.2: After Ref. [1002], reproduced with permission. Figs. 31.3 and 31.4: Courtesy of Ashwin®-Ushas Corp., Inc)

polymerization solvents for several reasons: good nucleophilicity, facilitating proton removal in the polymerization; poorer solubility or insolubility of nearly all CPs as well as of oligomers; and relative dryness, with sufficient water content for effective polymerization without early termination. Solvents such as propylene carbonate (PC), for instance, exhibit high solubility of oligomers and of reduced polymer and very high water content; thus poly(diphenyl amine) will electropolymerize in acetonitrile but not in PC [956, 1006–1008].

Solvent retention within the polymer matrix, as discussed elsewhere in this book for P(ANi), and, alternatively, retention of solvent "affinity sites" are important factors sometimes ignored by many workers. Thus, for instance, several poly (diphenyl amine) derivatives, when electropolymerized in acetonitrile or PC, will electrochemically cycle well only in the solvent of electropolymerization [956, 999, 1003, 1004]. Solvents may also be too nucleophilic: Besides its high solvation

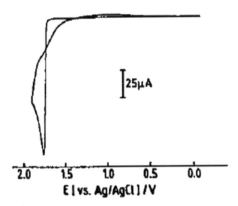

Fig. 31.5 Electropolymerization CV for thiophene in acetonitrile, showing crossover (see text) (After Ref. [1005], reproduced with permission)

capability for even doped CPs, dimethylformamide (DMF) is also a poor electropolymerization solvent due to high nucleophilicity; if this is reduced with addition of protic solvents, electropolymerizations are observed [972]. Table 31.2 cited above has [972] summarized electropolymerization data for P(Py) in various solvents.

In cases where the monomer has poor solubility in an otherwise ideal polymerization solvent such as acetonitrile, mixed solvents are used primarily to facilitate solubility. Thus, for instance, mixed acetonitrile + aqueous HCl solvents have been used for polymerization of the aromatic amine 4-amino-biphenyl [858], and hydrocarbons such as toluene may be mixed with acetonitrile to facilitate solution of monomers such as pyrene.

While almost any electrode which does not undergo oxidation at a potential equal to or less than the oxidation potential of the monomer can be used, several electrodes are preferred for specific applications. For electrochromic and initial characterization studies, a transparent electrode such as indium tin oxide (ITO) or Au/Pt on glass is preferred. For purely electrochemical characterizations, Pt wires, on which adhesion of nearly all CPs is excellent, are preferred. For preparation of freestanding films, graphite, special (e.g., basal-plane or pyrolytic graphite), or glassy carbon may be preferred. Stainless steel electrodes have been used for corrosion studies, and n-Si for semiconductor applications, both with success [1009]. Adhesion to such electrodes can be improved by using standard "semiconductor rinses," e.g., soap water, DI water, acetone, methanol, and acetonitrile.

As counter electrodes again, any of the above materials are suited, with, for instance, Pt coil or flag electrodes frequently used for electrochemical characterization studies, and Cu electrodes usable when the $Cu^+ \rightarrow Cu^o$ reaction is used at the counter electrode. As discussed at some length in an earlier part of this book, a three-electrode mode is used in most electrochemical work, while a two-electrode mode finds use in bulk syntheses or emulation of electrochromic or other end-use devices.

31.1.3   Potentiostatic, Cyclic, and Galvanostatic Polymerizations and Threshold Concentrations

A potentiostatic (constant potential) deposition mode generally yields polymer with the most consistent morphology. Repeated potential cycling, typically to a few hundred mV beyond the monomer oxidation potential, is also commonly used, yielding CP films almost comparable to potentiostatic deposition. Constant current, i.e., galvanostatic, deposition can be used when control of charge (e.g., for thickness) is desired, but it is almost universally accepted that it yields polymer of poorer morphology, conductivity, and general quality.

Typical monomer concentrations taken during electropolymerization range from as low as 1 mM to as much as 1 M. Dopant electrolyte concentrations must be from at least 2 X to up to 1000 X monomer concentration. Typical values are 50 mM monomer and 200 mM dopant electrolyte. The galvanostatic polymerization data in Fig. 31.6 show that threshold concentrations for polymerization can vary substantially for different monomers but are generally above ca. 1.0 mM.

31.1.4   Dopants

During a typical electropolymerization, it is nearly always noted that the net charge transfer is a little in excess of that indicated stoichiometrically, due to the additional oxidation (and doping) that occurs during electrochemical preparation of CPs. For example, while P(Py) and poly(azulene) both stoichiometrically require two electrons per monomer for electropolymerization, experimentally a charge of the order of 2.3 electrons per monomer is found to be consumed. The excess 0.3 is used to effect a 30% doping of the polymer.

In a similar fashion to differing solvents, when a CP is electrochemically prepared with one dopant, it may not always readily exchange with another, even if the latter is smaller. For instance, poly(4-amino-biphenyl) (P(4ABP)), when prepared with the large tosylate dopant, does not electrochemically cycle well in media containing the much smaller Cl^- or BF_4^- dopants [956, 999, 1003, 1004]. This appears to be due not to dopant size, but rather dopant affinity sites created in the particular morphology of the CP during polymerization. It is for this reason that it is best to simply incorporate the final dopant to be used during electropolymerization itself. Dopants can also cause a substantial shift in redox potential of a CP. For example, P(4-ABP) with perchlorate dopant has a redox potential (midpoint of $E_{p,c}, E_{p,a}$) of +0.6 V vs. +0.0 Ag/AgCl (in acetonitrile), which shifts to +1.1 V vs. Ag/AgCl with tosylate dopant [956, 999, 1003, 1004].

The selection of dopants usable in electropolymerizations is truly varied and large, as seen in a subsequent table in this chapter.

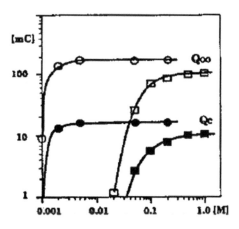

Fig. 31.6 Galvanostatic electropolymerization data for thiophene and bithiophene (After Ref. [1010], reproduced with permission)

31.1.5 Electrochemical Monitoring of Polymerization

Electrochemical monitoring of electropolymerizations is most effectively done using cyclic voltammetry (CV) and yields useful mechanistic information, which can be applied to interpretation of electrochemical and electrochromic behavior of new CPs. Figure 31.7 shows the most simple case of preparation of a CP film via potential cycling, using a slow sweep rate. On the first anodic (to + potential) scan, the predictable oxidation peak, corresponding to production of the radical monomer cation, is observed. On the reverse, cathodic (to (−) potential) scan, the reduction peak corresponding to this oxidation is not observed, indicating that in the time scale of the experiment (the sweep rate), the radical cation is rapidly consumed in a subsequent (chemical) reaction; furthermore, the oxidation peak progressively diminishes and shifts to higher potential on subsequent scans, indicating that monomer is rapidly depleted in the vicinity of the electrode and converted to radical cation, and thereafter its oxidation is limited by diffusion of monomer from bulk solution. Starting with the second potential scan, a redox couple is observed at ca. −0.4 V/ −0.1 V, increasing with every cycle, which corresponds to the redox of the steadily building CP film; as expected, this is at a much more negative potential.

The effect of going to a higher scan rate can be seen from Fig. 31.8a–d. In (a), at a fairly high scan rate of 10 V/s, the first few anodic scans show the expected monomer oxidation peak, ca. +0.65 V (the most positive peak), corresponding to generation of the radical cation and steadily diminishing with each scan. The reverse cathodic scans however also show the corresponding peak for the reduction of the radical cation, at ca. +0.55 V, indicating that at this scan rate this species is not depleted rapidly enough to not be detected. As one slows the scan rate (b through d), however, this corresponding reduction peak steadily vanishes. As in Fig. 31.7, the redox of the CP film is represented by a steadily increasing redox couple at a more negative potential (ca. +0.35 V), absent in the first scan.

Fig. 31.7 Representative electropolymerization via potential cycling, here for a poly(aromatic amine) (After Ref. [977], reproduced with permission)

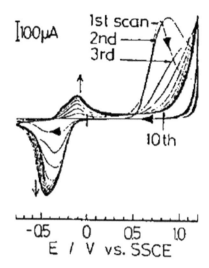

Fig. 31.8 Scan rate dependence of cyclic voltammograms obtained at a BPG disk electrode in an acetonitrile solution containing 0.2 M TBABF$_4$ and 5.0 mM PA. Scan rate: (**a**) 10, (**b**) 5.0, (**c**) 1.0, and (**d**) 0.2 Vs^{-1}. Electrode area: 0.20 cm^2 (After Ref. [998], reproduced with permission)

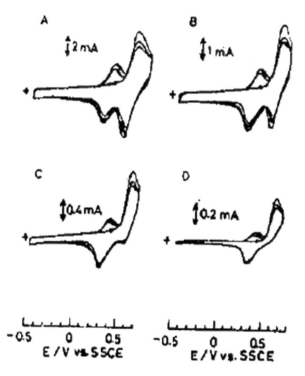

31.2   Chemical Polymerization

31.2.1 Mechanisms and General

The majority of chemical polymerizations of CPs, which are *addition* polymerizations, differ from electrochemical ones in a fundamental mechanistic step: After initial radical ion generation, rather than a radical–radical coupling (the principal propagation step) as in the generic electrochemical mechanism above, coupling occurs between radical and monomer. This directly follows from the fact that in the bulk of the reaction environment, where the radical initiators are generated, the concentration of monomer is in excess, unlike in electrochemical polymerizations. The second important difference with chemical polymerizations is that many of them are *precursor polymerizations*, i.e., involving synthesis of an initial, usually soluble and processible, polymer, which subsequently yields the final CP through relatively few and simple chemical steps. This precursor procedure is used to circumvent the usually poor processibility of the final CPs. The mechanistic difference is illustrated in the chemical polymerization mechanism postulated for P(T), Fig. 31.9.

As described briefly in an earlier part of this book, the more common typical chemical polymerization is relatively simple, involving monomer and an oxidant (usually also the dopant) in a suitable solvent medium, frequently aqueous, and temperatures in the region of -20 °C to $+80$ °C. Thus, for instance, a 0.25 M solution of an oxidant such as ammonium peroxydisulfate (($NH_4)_2S_2O_8$) is added with stirring to a solution of 0.5 M aniline monomer in 0.5 M aqueous *p*-toluene sulfonic acid solution (dopant electrolyte) at room temperature. As reaction temperature rises, the reaction bath is cooled with an ice bath. After 2 h of stirring, the polymer is filtered, washed with dopant electrolyte solution, and dried in vacuo at 60 °C. An optional procedure of Soxhlet extraction may be used for further purification. The polymer is then neutralized with 2 M NH_4OH to yield the emeraldine base form of P(ANi).

For P(Py) and P(3AT), combination oxidant/dopants, such as $FeCl_3$ and $CuBF_4$ (this in acetonitrile as well as aqueous medium), or oxidants such as ferricyanide may be used under similar reaction conditions. The polymers so obtained have conductivities equal or superior to those from electrochemical syntheses. Control of morphology, conductivity, doping, and related factors is however a little more difficult in chemical polymerizations, with slight changes in temperature, concentration, and other factors yielding substantial differences in polymer properties and even identical synthetic procedures never yielding exactly the same polymer each time.

Following polymerization, chemical doping can be used to incorporate new dopants not present in the original synthesis into the CP. For example, P(ANi) in its emeraldine oxidation state is convertible into Cl^-, SO_4^{2-} and acetate doped P(ANi) by simply soaking for 6 h. in the corresponding dilute (ca. 1 N) acid solution [1012]. Similarly, poly(3-Bu-thiophene), chemically polymerized in its near-

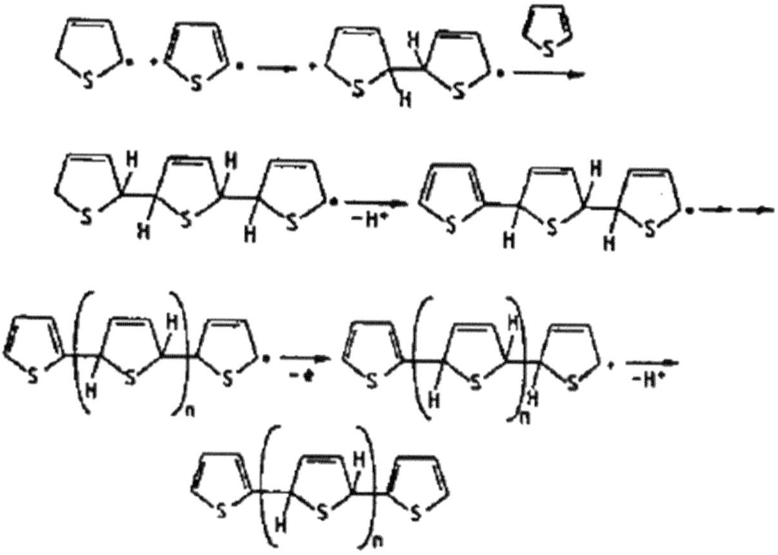

Fig. 31.9 One chemical polymerization mechanism postulated for poly(thiophene) (After Ref. [1011], reproduced with permission)

neutral state, can be chemically doped by I_3^- via simple immersion in methylene chloride solvent containing excess iodine and by SbF_6^- via simple immersion in a nitromethane solution of nitrosyl-SbF_6^- [1013].

We now deal briefly with typical effects of variation of factors such as concentration and oxidant redox potential on polymerization and follow this with a brief synopsis and representative examples of more involved chemical polymerizations. This will give the reader an idea of the wide variety of synthetic methods available, which as noted above parallel those available in synthetic organic chemistry. The reader is then referred to the sections on individual polymers elsewhere in this book for additional detailed treatment of syntheses.

31.2.2 Optimization in Solution Polymerizations

The effect of solvent and of the redox potential of the polymerization medium on CP quality and morphology can be very strong. For example, Miyata and coworkers [1014, 1015] investigated these for a typical chemical synthesis, that of P(Py) using anhydrous $FeCl_3$ as oxidant-cum-dopant in various solvents (monomer is simply added to the oxidant solution with stirring and the polymer recovered by filtration); they found that while methanol appeared to be the best solvent for optimal conductivity and morphology, the equilibrium redox potential of the solution, controllable via the relative concentrations of the monomer and oxidant and via addition of

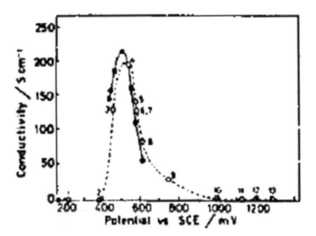

Fig. 31.10 Oxidation potential/conductivity relationship in poly(pyrrole) synthesis. Open circles show the conductivity of prepared poly(pyrrole) when the oxidation potential of the solutions was controlled by changing solvents. When the oxidation potential of methanol solutions was controlled by adding $FeCl_2$ at the initial stage, the conductivity of the prepared poly(pyrrole) is shown by the filled circles. (1) DMA, (2) DMP, (3) ethylene glycol, (4) Me-OH, (5) Et-OH, (6) water, (7) pentanol, (8) octyl alcohol, (9) 0.7 M $Fe(ClO_4)_3$, (10) benzene, (11) acetone, (12) acetonitrile, and (13) chloroform (After Ref. [1014], reproduced with permission)

$FeCl_2$, was a more important determinant of polymer quality. They found that at a redox potential of +0.5 V vs. SCE, optimal polymer of conductivity 220 S/cm was obtained Fig. 31.10. The polymer so produced also showed a more ordered, fibrillar morphology, and stretched films showed greater tensile strength and high flexibility.

31.2.3   Synopsis of Chemical Syntheses

We now cite representative chemical syntheses of CPs to impart an idea of the very wide variety of synthetic methods used, nearly all being adaptations of methods widely used in synthetic organic and conventional polymer chemistry.

Perhaps the greatest variety of methods has been applied to the synthesis of poly (acetylenes) (P(Ac)) and poly(diacetylenes) (P(DiAc)). The most well known is the Shirakawa method, which is an adaptation of Ziegler–Natta polymerization used for stereospecific polymers. In a typical synthesis, a solution of $Ti(OBut)_4$-$Al(Et)_3$ (4:1) is used to coat a reaction flask cooled in dry ice/methanol (-78 °C), with controlled thickness. Acetylene is introduced and removed at the end of the preselected reaction time. P(Ac) forms on the walls of the flask and can be peeled off [1015]. Other catalysts, used for substituted P(Ac)s, include $(C_6H_5)(CH_3O)$ C = $W(CO)_5$ (at 40 °C [1016], $Ta(Cl,Br)_5$ (80 °C, in toluene, [1017], and Mo $(CO)_6$ (in CCl_4 with UV light, [1018]). The synthesis of P(DiAc)s is typically a bit

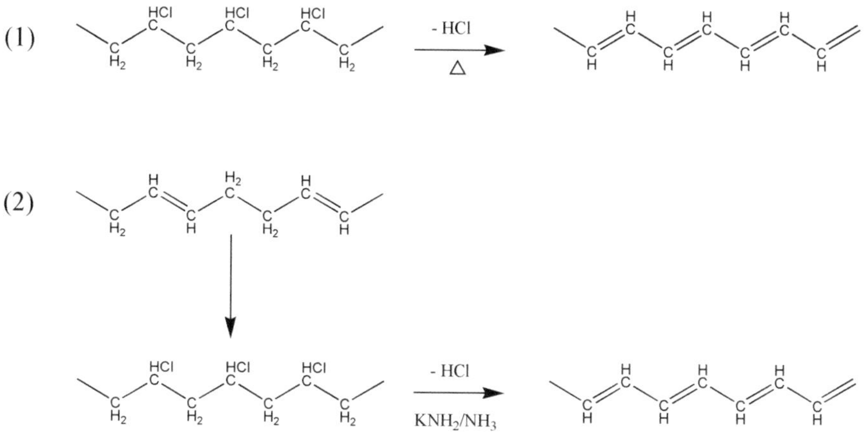

(1)

(2)

Fig. 31.11 Typical dehydrochlorination routes to poly(acetylene)

more involved, with various physical orientation methods used to align the diacetylene monomers prior to radiation or UV polymerization [1019]. One of the simplest precursor routes to P(Ac) is illustrated in Fig. 31.11, the precursor in this case being soluble, processible poly(chloroethylene). Attempts to polymerize P (Ac)s and P(DiAc)s via condensation polymerization routes have not met thus far with much success – nearly all methods are variants of addition polymerization.

The wide variety of coupling methods adapted from organic synthesis to condensation polymerization of just one CP can be appreciated from Fig. 31.12, for poly(phenylene). Typical condensations and eliminations adapted to syntheses of such CPs as poly(phenylene) and poly(phenylene vinylene) (P(PV)) are illustrated in Fig. 31.13. Figure 31.14 shows the variety of precursor routes available to P(PV). More recently, the Yu group [1023] has demonstrated application of Pd-catalyzed Stille and Heck reactions to the synthesis of poly(thiophene) (P(T)) derivatives (cf. Fig. 31.15). Besides Grignard couplings, P(T)s can also be prepared via a variety of other procedures, such as Friedel–Crafts alkylation [1020] and direct oxidation with $FeCl_3$ as for P(Py) above.

31.2.4 Unique Chemical Polymerization Methods

An example of the use of the use of novel polymerization techniques is provided by the direct photosensitized polymerization of pyrrole by $Ru(bpy)_3^{3+}$ (Fig. 31.16) carried out using a 490 nm dye laser in a matrix of Nafion, which also served as the dopant [1006].

A variation of the optimization of the equilibrium redox potential of the polymerization solution described above (31.2b) can be achieved by using solvent evaporation to change this potential (as solvent evaporates, concentration of

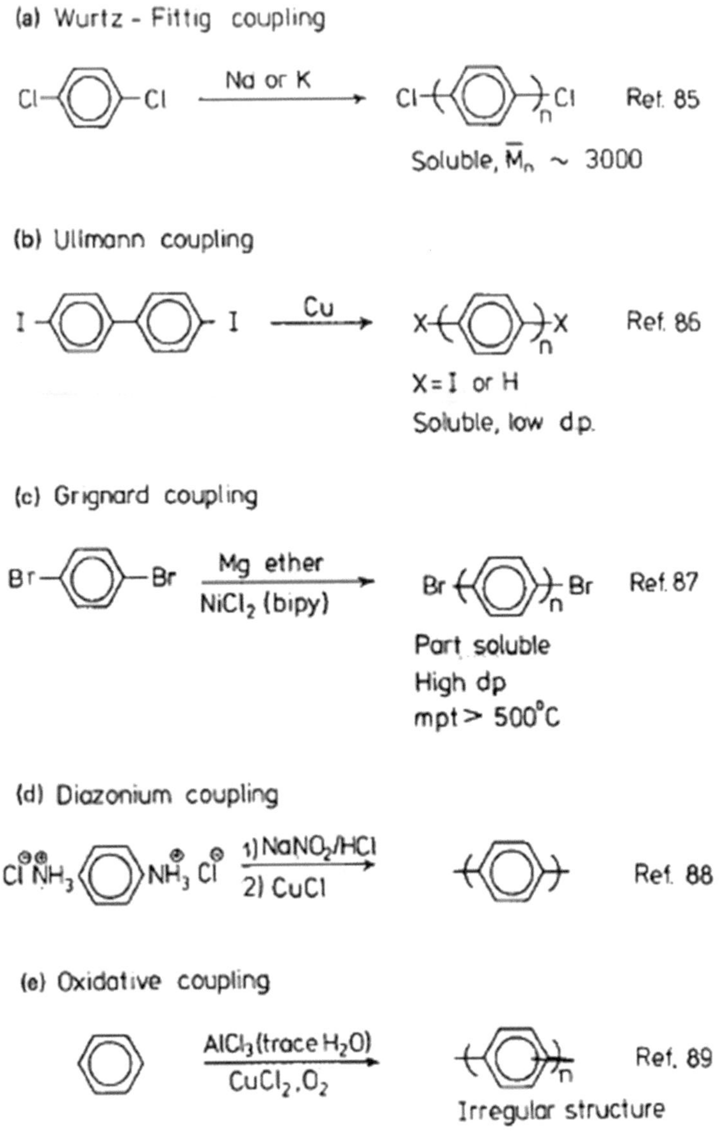
(a) Wurtz - Fittig coupling
Cl
Cl
Na or K
Cl
Cl
n
Ref. 85
Soluble, M̄n ~ 3000
(b) Ullmann coupling
I
I
Cu
X
X
n
Ref. 86
X = I or H
Soluble, low d.p.
(c) Grignard coupling
Br
Br
Mg ether
NiCl2 (bipy)
Br
Br
n
Ref. 87
Part soluble
High dp
mpt > 500°C
(d) Diazonium coupling
Cl NH3
NH3 Cl
1) NaNO2/HCl
2) CuCl
Ref. 88
(e) Oxidative coupling
AlCl3 (trace H2O)
CuCl2, O2
n
Ref. 89
Irregular structure

Fig. 31.12 Representative application of common coupling reactions to poly(arylene) synthesis (After Ref. [1021], reproduced with permission)

oxidant and electrolyte change, thereby changing redox potential). Thus, a polymerization solution (monomer, electrolyte, oxidant, solvent) can be cast onto an appropriate substrate, and as the solvent evaporates, polymerization occurs, usually indicated by a remarkable color change. Such a technique was demonstrated by the Miyata group [1008] for $FeCl_3$ polymerization of pyrrole in a poly(vinyl acetate)

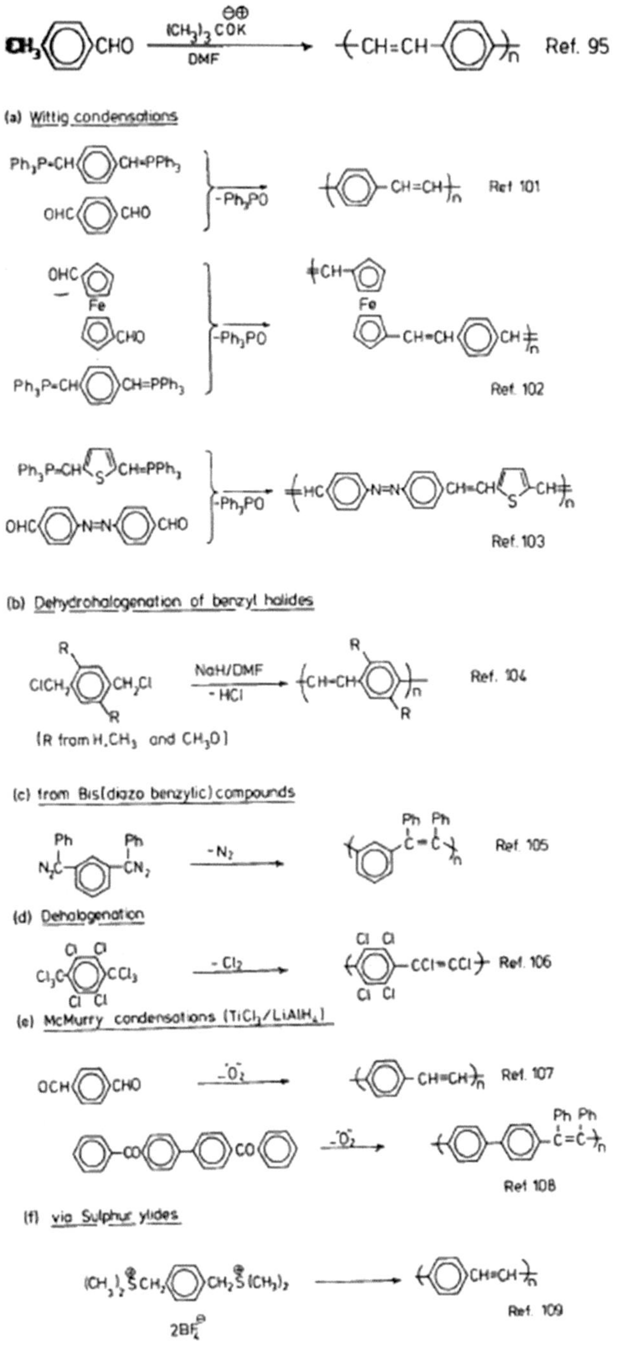

DMF
Ref. 95
(a) Wittig condensations
-Ph3PO
Ref 101
Ref. 102
Ref. 103
(b) Dehydrohalogenation of benzyl halides
NaH/DMF
- HCl
Ref. 104
(R from H, CH3 and CH3O)
(c) from Bis(diazo benzylic) compounds
-N2
Ref. 105
(d) Dehalogenation
- Cl2
Ref. 106
(e) McMurry condensations (TiCl3/LiAlH4)
Ref. 107
Ref 108
(f) via Sulphur ylides
Ref. 109

Fig. 31.13 Typical condensations and elimination reactions adapted to syntheses of various representative CPs (After Ref. [1021], reproduced with permission)

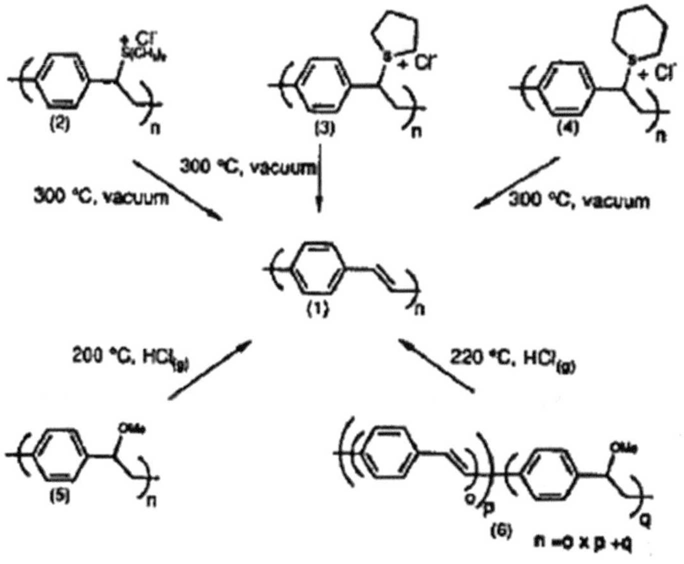

Fig. 31.14 Precursor routes to poly(phenylene vinylene) (After Ref. [1022], reproduced with permission)

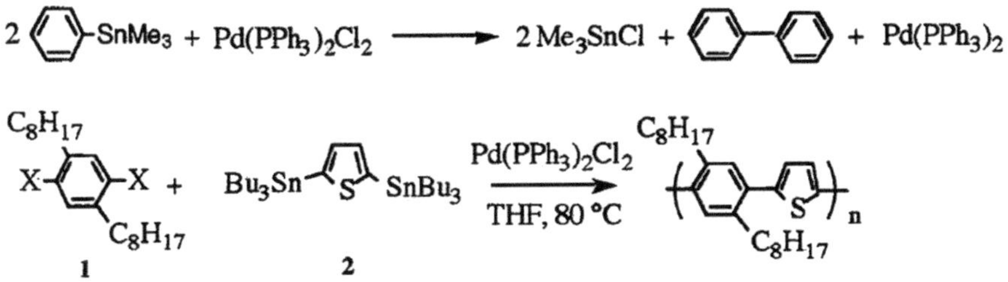

Fig. 31.15 Pd-catalyzed syntheses of poly(thiophene) derivatives (After Bao and Yu [1023])

(PVAc) matrix in methanol solvent, yielding a P(Py)/PVAc composite with a claimed conductivity of 10 S/cm at P(Py) proportions of only 5%. A similar technique was used later by the Lewis group [1024] to spin coat P(Py)/phosphomolybdate films on insulating substrates (phosphomolybdic acid as the dopant-cum-oxidant in THF), yielding ca. 100 nm thick P(Py) films with claimed conductivities in the 15–30 S/cm range.

A highly unique procedure for an "oriented polymerization" of acetylene has been described by Aldissi [1025]. In this technique for P(Ac), the usual Shirakawa-technique Ziegler–Natta catalyst (Ti(OBu)$_4$/AlEt$_3$) is aged in toluene, and the solvent then removed. The liquid crystal (LC) N-(p-OMe-benzylidine)p-butylaniline (MBBA) is then added to the dry catalyst in inert atmosphere, and reaction vessel degassed and subjected to a magnetic field of 4000 G at room temperature, orienting the LC molecules. The catalyst dissolved therein also

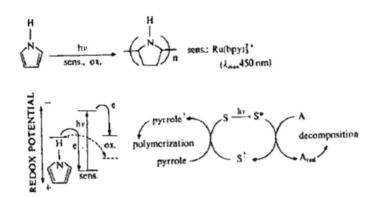

Fig. 31.16 Example of direct, photosensitized polymerization of pyrrole (After Ref. [1007], reproduced with permission)

orients, and when acetylene gas is introduced into the nonstirred reaction vessel, an "oriented polymerization" occurs. The P(Ac) so obtained shows no saturated defects (sp^3-Cs), a highly anisotropic conductivity ($\parallel/\perp = 4.25$), and other signs of chain orientation but a fibrillar morphology very similar to that of conventional Shirakawa P(Ac).

A truly unique *enzymatic synthesis* of a semi-soluble, sulfonated P(ANi), using the enzyme horseradish peroxidase, was carried out by Alva et al. of the Tripathy group [1026]. The enzyme, known to catalyze the oxidation of aromatic amines and phenols in the presence of H_2O_2 to generate their respective free radicals, was used to catalyze oxidative free radical coupling leading to the significantly water-soluble CP poly(2,5-diaminobenzenesulfonate). The reaction was found to be rapid, yielding average MWts of 18,000. A typical synthesis constituted dissolution of 0.1 g of the monomer in 50 mL of 0.1 M Na phosphate buffer of pH 6.0 containing 3 units of the enzyme. Reaction initiation occurred with addition of 100 μL of 30% H_2O_2 with stirring. After 3 h of stirring at room temperature, the reaction medium was dialyzed to remove unbound buffer, water evaporated off, and the resulting polymer extracted with MeOH. The authors noted that unlike the sulfonated P(ANi)s obtained through treatment of P(ANi) with fuming sulfuric acid, their P(ANi) appeared to be fully sulfonated. It was also noted that these CPs could be self-assembled into multilayer structures.

Cruz et al. recently described [1027] a plasma polymerization of P(ANi) films using RF glow discharges with resistive coupling between stainless steel electrodes, at a frequency of 13.5 MHz and pressures in the range 2–8 × 10^{-2} Torr. In this method, the electrons in the plasma generated by electric discharge give rise to ionization of monomer, and the reaction is initiated in the gas phase, with the growing polymer then attaching to the nearest wall. Conductivities obtained by Cruz et al. were however low, being ca. 10^{-4} to 10^{-11} S/cm with I_2 doping and varying strongly with relative humidity.

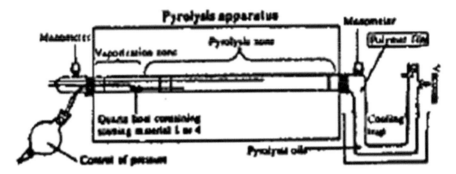

Fig. 31.17 Schematic of apparatus and process for CVD (chemical vapor deposition) polymerization of poly(phenylene vinylene) (After Ref. [1028], reproduced with permission)

Another unique polymerization method is embodied in the work recently described by Schäfer et al. [1028]. In this, P(PV) was prepared by chemical vapor deposition (CVD), specifically vapor-phase pyrolysis of α,α(')-dihalogenated-*p*-xylenes. The targeted application was LEDs, and unusually high rectification ratios were observed in these. Figure 31.17 shows a schematic of the polymerization apparatus and process.

31.3   Dopants and Alternative Doping Techniques

31.3.1   Common Dopants

Table 31.3 lists the most common and some uncommon dopants used heretofore with a variety of CPs, via both chemical and electrochemical doping. The actual form in which the dopant is introduced into the polymerization or in subsequent doping can vary. For example, in electrochemical doping, tetrafluoroborate or perchlorate can be introduced as Li^+ or Et_4N^+ salts used as electrolyte, while in a chemical polymerization, $CuBF_4$ may be used. AsF_6^- will be introduced as AsF_5 vapor for chemical polymerization (e.g., for poly(acetylene)), while for electrochemical polymerization, $LiAsF_6$ would be used as the electrolyte salt. $FeCl_4^-$ is invariably introduced chemically in solution as $FeCl_3$, functioning as dopant as well as oxidant. In the case of several poly(aromatic amines) such as P(ANi), dopants such as sulfate, chloride, and hydrosulfate are invariably introduced during electrochemical or chemical synthesis in the corresponding acids; in such cases, when an additional dopant electrolyte (such as perchlorate) is present, co-doping occurs, with the proportion of dopants closely related to their initial concentrations. Doping may also not be homogeneous; for instance, it is well recognized that doped poly (acetylene) may have a morphology of heavily doped "islands" surrounded by nearly undoped, insulating polymer [1029].

Table 31.3 Common dopants studied in prior work with CPs. When relevant, the typical CP or class of CPs to which the dopant is applicable is given in italics; this is by no means an exclusive list (References cited only where appropriate)

Dopant		Doping level (max or typical range)	References
P(Py), P(aromatic amines), P(bi/thiophenes):			
as dopant electrolyte, chemical/electrochemical prepn.:			
(Tetrafluoroborate)	BF_4^-	5–33%	[1447]
(Hexafluoroarsenate)	AsF_6^-	"	"
(Hexafluorophosphate)	PF_6^-	"	"
(Perchlorate)	ClO_4^-	4–30%	"
(Hydrogen sulfate)	HSO_4^-	2–30%	"
(Fluorosulfonate)	SO_3F^-	5–15%	"
(Trifluoromethane sulfonate)	$CF_3SO_3^-$	4–31%	"
(Benzene sulfonate)	$C_6H_5SO_3^-$	4–33%	"
(Br-benzene sulfonate)	$BrC_6H_4SO_3^-$	2–33%	"
(Trifluoroacetate)	CF_3COO^-	4–25%	"
(Sulfate)	SO_4^{2-}	5–22%	"
(Chloride)	Cl^-	3–42%	"
(Chloroferrate)	$FeCl_4^-$	"	"
(Poly(styrene sulfonate))	PSS^{n-}	3–15%	[1030]
Nafion (DuPont membrane fluoro-polymer)	Naf^{n-}	1–15%	[96]
Poly(acrylic acid)	$PAcrA^{n-}$	2–20%	[955]
Poly(vinyl sulfonate), poly(vinyl sulfate)		1–45%	[999]
Poly(di/acetylenes) (P(Ac)), anionic (p-type):			
from vapor phase:			
I_3^- (iodine), I_2Cl^-, Br^-, I_2Br^-		2–25%	[1032]
(Se,Te,o,W,U,Re,Os,Ir,Ru) $F_6^-BiF_5$		2–19%	[1033]
from toluene solution:			
$MoCl_5$, WCl_6, $(Ta,Nb,Zr)Cl_5$		2–7%	[1034]
$In(I,Cl)_3$, $(Sn,Ti,Zr)Cl_4$, TiI_4		0.3–7%	[1035]
HSO_4^- from $NOHSO_4$		1–12%	[1036]
(Ce,Dy,La,Pr,Sm,Yb) nitrates		0.9–2.2%	[1037]
P(Ac), P(phenylene), cationic (n-type):			
from THF or pentane solution:			
Na^+, from $NaC_{10}H_8$, Na_2Ph_2CO		4–28%	[1032, 1038, 1039]
Li^+, from $LiPh(CH_2)_6Ph$		6–44%	"

31.3.2 *Uncommon or Unusual Dopants*

There have been a large number of studies which have attempted to incorporate novel dopants into CP systems with a view to imparting special or unusual properties to them. All of these have unfortunately produced mixed results at best, with curious or interesting CP/dopant systems with however no clear-cut properties superior to existing CP systems. An example of this is the successful doping of P (ANi) with phosphotungstic acid (using the conventional persulfate oxidation in phosphotungstic acid), yielding scientifically interesting complexes with W (VI) active groups [1040]. While the CP is shown to have catalytic activity, it is poorer than that of the dopant alone, and the CP's conductivity, stability, and other properties are poorer than that of P(ANi) with a more conventional dopant. Doping of P(ANi), P(Py), and P(3MT) with other, similar, Keggins-type heteropolyanions, $XM_{12}O_{40})^{n-}$ (X = Si/P, M = W/Mo, n = 3,4) during electrochemical synthesis has been reported by Bidan et al. [1041]. Some catalytic activities for $H_3{}^+O$ and O_2 reduction have been reported for these CPs by these authors. Girard et al.'s [1042] synthesis of $P(Py)MoS_4$-MoS_3 with a 33% doping level via electrochemical polymerization from aqueous pyrrole and $(NH_4)_2MoS_4$. A sharp reduction peak observed in initial electrochemical cycling of the polymer was ascribed to release of $MoS_4{}^{2-}$ and neutral MoS_3. The utility of this dopant was purported to be in Li secondary batteries with composite Mo/P(Py) cathodes. Chemical synthesis of P (Py) with a phosphomolybdate dopant has been reported by the Lewis group [1024], with phosphomolybdic acid serving as dopant-cum-oxidant, and claimed conductivities up to 30 S/cm.

An interesting dopant studied by the Reynolds group [1043] is Cu-phthalocyanine-tetrasulfonate (CuPTS), for electrochemical synthesis of P(Py) CuPTS. As expected for very large macrocyclic dopants such as these, the redox process (doping/de-doping) occurring on electrochemical cycling of this polymer is cation (Cu(II)) transport, not dopant-anion transport. To date, the potentially interesting electronic conduction and photochemical properties of this CP system have however not been studied to our knowledge. Doping of P(T) with another potential organic electronic conductor, $TCNQ^-$, has been successfully achieved by Hotta et al. [1044], but the hopes of expected high conductivity have been belied in this CP (observed 10^{-4} S/cm).

This sort of cation doping with a large anionic dopant is also observed for "self-doped" CPs discussed in the sequel and for other large macrocyclic or macromolecular dopants such as the Keggins heteropolyanions discussed above, electrochemical hexacyanoferrate doping of P(Py) described by Sung et al. [1045], and P (Py)/PSSH complexes studied in some detail by Lim et al. [1046]. A large number of seemingly unique dopants, attachable to nearly all common CPs (P(Py), P(ANi), and P(3AT)) via electrochemical synthesis, have the sulfonate group as an effective dopant and also function via this sort of cation doping. These include anthraquinone-2-sulfonate, besides the commoner poly(styrene sulfonate) dopant cited in Table 31.3, Nafion (a DuPont fluoropolymer), alkane-naphthalenesulfonates, camphor sulfonates, dodecylbenzenesulfonate, and the FeBPS system discussed earlier [1031, 1047–1049].

Unique HF_2^- (from H_2F_2/water) and chlorometallate (MCl_4, M = Al, Fe, Tl, In, from $LiCl/MCl_3$/nitromethane) dopants have been reported for P(Ac) [1050, 1051]. Liu et al. [1052] reported a truly unusual dopant for poly(3-alkyl-thiophenes), C_{60}, although it was unclear if this functioned as a true dopant or just an admixture; the CP had to first be treated with a surfactant, arachidic acid, before the dopant could be mixed in.

31.3.3 *Alternative Doping Techniques*

Among the most common alternative doping techniques used have been *ion implantation, photochemical doping, heat treatment, solution doping, and "dry doping."* These are briefly discussed in turn.

Ion implantation, via bombardment of a substrate by high-energy ions which then implant in the substrate's lattice, forming covalent bonds therewith, is a technique widely used in the semiconductor and microelectronic fabrication fields. In the case of CPs, it has been successfully used to render a neutral, nonconducting CP conducting. In a typical ion-implantation study, for instance, the Epstein and Jenekhe groups [891] implanted a benzimidazobenzophenanthroline-type ladder polymer, BBL, and the rigid-rod polymers poly(p-phenylenebenzobisthiazole) (PBZT) and poly(p-phenylenebenzobisoxazole) (PBO) with Kr^+ ions of an energy of 200 keV and flux 4×10^{16}/cm^2, obtaining doped polymers with broad metallic-band behavior. Conductivities at room temperature were ca. 140 S/cm, and the temperature dependence of the conductivity was very small. The polymers however showed substantial carbonization (due to the ion bombardment) and environmental stability poorer than conventionally doped materials. It is noted that these ladder-type and rigid-rod polymers are rather unconventional CPs, as noted in the chapter on Classes of CPs in the sequel. Other examples of ion-implantation doping are the recent study by Tong et al. [1053], in which P(ANi) was bombarded with 100 keV Ar^+ and 24 keV I^+ ions at fluences of ca. 10^{16}/cm^2. While the Ar^+ ions effected a low, "damage-induced" conductivity, the I^+ ions effected a "permanent," environmental stable conductivity increase, from the undoped polymer, of 12 orders in magnitude.

Photochemical doping involves treating a CP with a dopant which is initially inert but rendered an active dopant by irradiation. These dopants are usable with common CPs. Examples are diphenyliodonium hexafluoroarsenate (in CH_2Cl_2) or triarylsulfonium salts (in aqueous medium), both of which are rendered active by UV radiation [1054,1055].

Solution doping has been briefly discussed in an earlier section in this chapter. Chemical solution doping can be used to incorporate new dopants not present in the original synthesis into the CP following polymerization. For example, P(ANi) in its emeraldine oxidation state is convertible into Cl^-, SO_4^{2-} and acetate doped P(ANi) by simply soaking for 6 h. in the corresponding dilute (ca. 1 N) acid solution [1012]. P(ANi) has also been doped by such dopants as camphorsulfonic acid (CSA) by simply mixing emeraldine base with CSA in *m*-cresol solvent [1048]. Similarly, poly(3-Bu-thiophene), chemically polymerized in its near-neutral state, can be chemically doped by I_3^- via simple immersion in methylene chloride solvent containing excess iodine and by SbF_6^- via simple immersion in a nitromethane solution of nitrosyl-SbF_6^- [1013]. Similarly, again, neutral poly (3-octylthiophene) can be doped by Cl^- via immersion in 0.1 M $FeCl_3$ in dry nitromethane [1056].

An interesting method of doping, that of heat treatment, has been applied mainly to ladder-type CPs such as BBL. Wang et al. [1057] found that previously wet-spun and mildly stretched BBL, when subjected to a heat treatment of 50 °C increments lasting 2 min in a calorimeter chamber, yielded a conductivity of ca. 10^{-8} S/cm with 100 °C treatment and a ca. 10^{-4} S/cm conductivity with 350 °C treatment, reverting again to ca. 10^{-8} S/cm at 600 °C. The enhancement of conductivity by heating has been ascribed by the authors to "improved structural order and thermally excited charge carriers" but may also be due simply to "impurity doping" of the classical sort, i.e., the creation of oxidation centers which then function as dopants (e.g., C=O or C^+-O^- bonds then yielding oxide dopants); the doping induced by inert ions such as Ar^+ in ion-implantation may also be due to a similar effect.

"Dry" doping is similar to the photochemical doping described above, with heat used to generate an active dopant from an inert species. Angelopoulos et al. [1058] have described such dry doping of P(ANi) (used in photoresists) with amine triflate salts, which are then thermally decomposed to yield the active dopant (triflate). This eliminates the need for solution, electrochemical, radiation, or other methods to effect the doping.

Solution doping is used here to denote direct doping of a solution of a CP by admixture of another solution, with the doped CP then remaining in solution for at least some time before precipitation occurs; it is thus distinct from solution doping in which a solution is used to dope an insoluble CP, e.g., HCl solution for P(ANi). The best example of solution doping is use of the dopant $NOPF_6$ in acetonitrile solution. This has been most successfully applied to doping of poly(3-long-chain-alkyl-thiophene)s, which are highly soluble in their de-doped or low-doped forms in organic solvents. The doped CP then usually partially or fully precipitates. As Hotta et al. note [1059], a simple exchange of ions and charge transfer occurs in solution.

31.4 "Composites"

Before going further in this chapter, we must pause briefly to define a term, *composite*, that is freely, and quite frequently erroneously, used in the CP literature. A *composite*, as defined by polymer and materials scientists, is a combination of two constituents, an adhesive matrix or resin base component and a fiber-reinforcing component. Natural wood, graphite-epoxy aircraft composites, and fiberglass (glass fibers in a thermoplastic or adhesive matrix) are examples of composites. However, the word "composite" is used freely by CP scientists to denote any combinations of CPs and conventional polymers, which are in fact either simple blends (mixtures), host–guest combinations (host polymer template with guest CP deposited therein), or genuine copolymers. None of these are really composites. In this chapter, we deal successively with blends, host–guest compositions, and copolymers. We shall however use the term "composite," without the quotes, loosely for all these compositions.

When dealing with blends having a conductive component (e.g., a CP) admixed into a nonconducting host matrix, as more of the conductive component is added, there is a point at which the conductivity of the blend experiences a sudden rise, eventually leveling off to a limiting or saturation value. This point is denoted as the *percolation threshold*. It is unfortunately used freely by workers in the field without a precise mathematical definition. For our definition, it may be taken loosely as the point at which the second derivative of a plot of the proportion of the conductive component (on the abscissa) vs. the conductivity (on the ordinate) is maximum. Percolation thresholds for CP blends generally lie in the 5–15% (w/w) region, although thresholds as low as 0.1% in simple blends have been claimed (see below). Thus, e.g., a blend of P(ANi) and nylon 12 [1060] shows a percolation threshold of 5% (w/w) P(ANi) with conductivity at 20% approaching that of pure P (ANi), values that are typical. These thresholds are generally lower than corresponding thresholds for "traditional" conductive polymer blends, i.e., those using carbon or metal as the conductive component, which are usually in the 15–30% w/w range. The reason for this is that CPs can form a completely interconnected polymer network within the host matrix at weight proportions significantly less than those required for physical connections between metal and carbon particles. Figure 31.18a, b [1061, 1062] are realistic examples of percolation threshold plots.

Transmission/scanning electron microscopy (TEM/SEM) and other measurements indicate that many composites are actually fine-particulate dispersions of the CP in the host polymer, with these dispersed particles achieving interconnection at the percolation threshold. A few composites turn out to be simply thin CP coatings on fine particulates of the host polymer. And a very few are simply aggregates of distinct CP and host polymer phases. As expected, these have poor conduction and other properties.

Now CP composites may be prepared via chemical or electrochemical polymerization.

Chemical polymerizations are usually in situ, with a host conventional polymer, such as poly(ethylene terephthalate) (PET), typically used to sorb monomer vapors, e.g., of pyrrole, then exposed to oxidant, e.g., $FeCl_3$, which results in polymerization of the sorbed monomer. The sorption may be reversed, with oxidant sorbed and exposed to monomer vapor or solution. Catalytic in situ chemical polymerizations, with, e.g., a transition metal catalyst sorbed into the host polymer and then exposed to monomer vapor, are used primarily for P(Ac) composites. One variant of

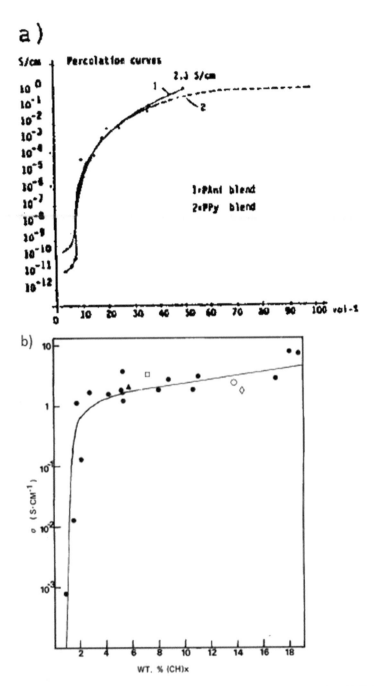

Fig. 31.18 (**a, b**) Examples of percolation threshold plots. (**a**) PVC-filled P(ANi) and P(Py) (After Ref. [883], reproduced with permission). (**b**) P(Ac) composites with various plastics (After Ref. [1063], reproduced with permission)

chemical polymerization uses a solution of the monomer and oxidant; as solvent evaporates, the oxidation potential changes to a value conducive to polymerization.

Interest in the above type of chemical polymerizations for preparation of CP composites continues unabated [1064] in view of the many potential applications of such conductive composites.

Electrochemical polymerizations usually involve polymerization from monomer solution into a host polymer that is adsorbed or coated onto a standard metal electrode such as Pt.

True copolymers are synthesized via conventional polymer and organic synthetic techniques and usually incorporate a CP monomer and the monomer of a conventional polymer (e.g., ethyl methacrylate or styrene). Interpenetrating polymer networks are similar to copolymers but use an extended cross-linking mechanism.

In the following sections, we cite detailed, illustrative examples of each of these types of syntheses, with a description of the properties of the resulting composites where appropriate. These should give the reader not only a general idea of each type of synthesis, but also hands-on details of how they are done.

31.5 Template-Based Polymerizations

There is a large variety of "specialty" CP polymerizations which appear distinct or novel in many ways but which in fact possess a unique, common feature: All of these polymerizations are based on *attaching or immobilizing a monomer onto a macromolecular or other template and then polymerizing this monomer*. They can thus be given the general appellation *template polymerization* or *template-based polymerization*. The attachment is usually, but not always, through a proper chemical bond; in a few cases, it is through van der Waals forces, hydrogen bonding, physisorption, or chemisorption. The template is chosen so as to afford some unique property to the resulting CP, such as water solubility (a polymeric sulfonate or surfactant used as template), radiation curability (onium macrocycles as templates), or catalytic activity (Co-porphyrin macrocycles as templates). We shall now briefly discuss these here. Polymerization using adsorption onto nanoscale (nanoporous) substrates is a special case of template polymerization discussed in the next section.

The simplest example of a template polymerization, one that has found the most common practical use (for water solubility), is that using the highly water-soluble poly(styrene sulfonate) (PSS^{n-}) as the template and dopant. PSSH is a "polyacid," exhibiting a large, winding, lanky polymer structure, with pendant acidic groups, to which a basic monomer such as aniline can readily attach, as illustrated in Fig. 31.19. Once the aniline monomers are attached, chemical polymerization is effected to yield a P(ANi)-PSS interwoven strand copolymer of the type shown in Fig. 31.20. Temperature and concentration are important parameters governing the type of product obtained in these syntheses.

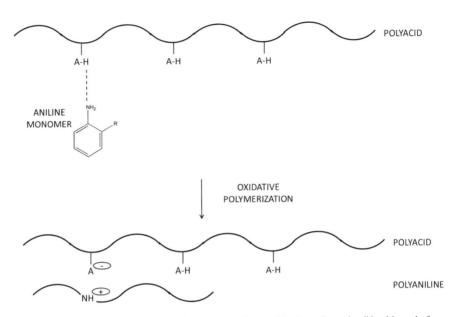

Fig. 31.19 Example of attachment of CP monomer (here aniline) to a "template" backbone before polymerization

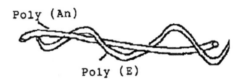

Fig. 31.20 Poly(aniline) and poly(electrolyte) (template backbone) structure resulting from polymerization of the template-pendant monomer shown in Fig. 31.19

To familiarize the reader, a typical, simple, synthesis procedure for P(ANi)-PSS, as described by Yang and coworkers [1030], may be cited: The stoichiometric amount of aniline liquid is added to 30% PSS acid solution. The solution is stirred for 30 min at room temperature, 0.1 M $FeCl_3$ and 3% H_2O_2 added, and stirring continued until a homogeneous, green solution is formed. Pouring the solution into isopropanol causes precipitation of the P(ANi)-PSS complex as a dark green gel. The complex can be redissolved in water. Unreacted aniline can be removed via dialysis in acidic medium and unreacted PSSH by repeated precipitation in isopropanol. The complex is soluble in water as well as solvents such as MeOH and DMSO, as a true solution (no retention through 0.2 μ filter paper). A similar procedure can be used to prepare a complex with poly(2-acrylamido-2-methyl-1-propane sulfonic acid) (PAMPSH) and with poly(vinyl sulfonate) [1065]. It is also to be noted that these complexes are also "self-doped" CPs in their own right, with PSS^{n-}, for instance, serving as dopant. Conductivities are however on the low side (ca. 10^{-2} S/cm compared to up to 10_2 S/cm for P(ANi)Cl). In the original synthesis of P(ANi)-PSS, Malhotra et al. [1066] used slightly differing procedures from the one above to arrive at a complex which they described as a "conductive rubber," with an elongation at break of 500% and stretchability up to 150%.

Template syntheses yielding complexes of poly(styrene)/poly(3-alkyl-thiophene) (P(Sty)/P(3AT)) claimed to be full interpenetrating polymer networks (IPN, all polymer components part of a cross-linked network) have been effected by the Rubner group [1067], with conductivities in some cases up to 0.5 S/cm. In this case, however, the P(3AT) actually forms the template: The pendant groups of a vinyl-derivatized P(3AT) are first cross-linked with styrene monomer. The resulting compound is then polymerized and cross-linked at the styrenes. As however the resulting complexes are insoluble, intractable, less conductive, and generally less stable than P(3AT)s, their utility is not made clear by the authors of the work.

In another template synthesis with an interesting potential application [1068], pyrrole monomers are bonded through an acetyl group to the dipeptide glycyl-D-phenylalanine. The resulting derivative is polymerized *electrochemically* on an electrode surface. As the dipeptide binds the enzymes carboxypeptidase A and trypsin, yielding stable complexes, so does the electropolymerized P (Py) derivative. This complex formation is detectable voltammetrically, the derivatized P(Py) thus serving as an enzyme-specific electrode. Electrochemical oxidation also causes dissociation of carboxyl groups in the polymer, releasing the enzyme and thus effecting a kind of "electrochemical delivery" of enzyme at will. In a similar application involving enzyme binding, Cosnier et al. [1069] incorporated the pyrrole-derivative monomer shown in Fig. 31.21 into a tyrosinase solution, dried the solution on an electrode, and electrochemically polymerized the resulting film. The catalytic activity of the tyrosinase was then utilized by employing the resultant P(Py) derivative film as a sensor for catechol, monophenols, and dopamine, with detection limits in the tens of nM, about 100 X lower than those for other tyrosinase-based sensors.

Stanke et al. [1070] have described chemical polymerization of poly(ethyl methacrylate) (PEMA, a common transparent plastic)-carrying pyrrole pendant groups, yielding I_2-doped conductivities of ca. 10^{-5}–10^{-2} S/cm (Fig. 31.22). However, since the objective of the work was to attempt to obtain processible and transparent CPs, and the products were unprocessible, insoluble, opaque, and with poor conductivity stability, the utility of these novel template CPs was in doubt.

The synthesis of pyrrole-styrene graft copolymers has been demonstrated by several groups [1072]. Pyrrole monomer is chemically grafted onto a template of styrene/Cl-Me-styrene copolymer (Fig. 31.23). The pendant pyrrole groups are then polymerized *electrochemically* in a solution containing additional pyrrole monomer. Conductivities of 0.4 S/cm are reported for the insoluble final product. However, the utility of this unique template polymer is again not indicated by the authors except for possibly providing "versatility for systematic modification of physical properties" [1072].

Bidan et al. [1073] synthesized novel Co/Cu 2,9-diphenyl-1,10-phenanthroline (DPP) derivatives having pendant pyrrole groups (Fig. 31.24). These were then

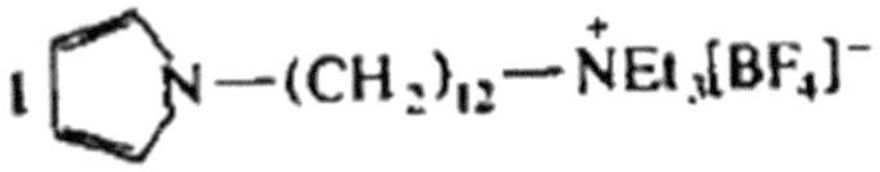

Fig. 31.21 Pyrrole-derivative monomer used by Cosnier and Innocent [1069]

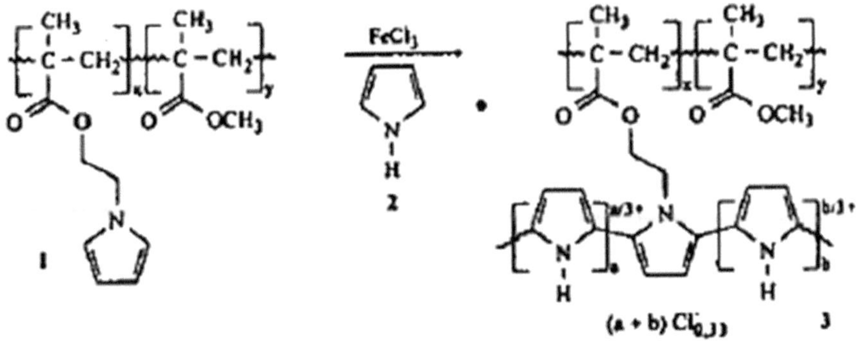

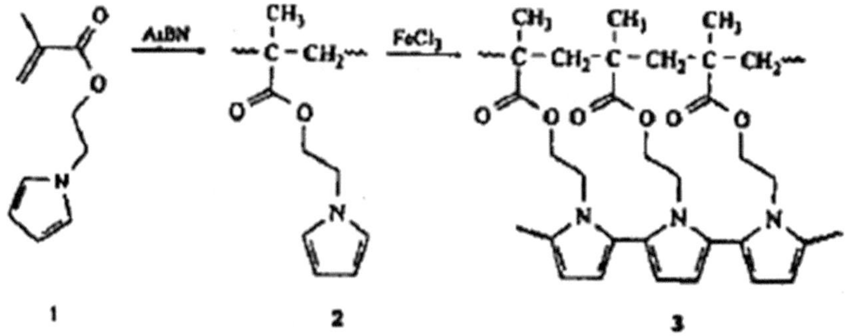

Fig. 31.22 Synthesis of methyl-methacrylate group containing poly(pyrroles) (After Stanke et al. Refs. [1070, 1071], reproduced with permission)

electropolymerized in acetonitrile yielding a novel template-based CP showing redox chemistry (Fig. 31.25) for both the metal redox couple and the P (Py) moiety. The potential utility of these complexes is claimed to be in electro-chemically mediated catalysis [1073].

A similar transition-metal-complex template, a Co-porphyrin with a single pendant pyrrole group also shown in Fig. 31.24, was used by Bedioui and coworkers [985] to prepare a Co-porphyrin-derivatized P(Py), immobilized on a carbon electrode, which exhibited all the Co(I/II/III) redox electrochemistry of the original Co-porphyrin. The new polymer film showed behavior typical of a surface-electroactive immobilized film, viz., peak current a direct function of scan rate and cathodic and anodic peak positions invariant with scan rate. The film was slightly

Fig. 31.23 Template/graft
synthesis of pyrrole-styrene
graft copolymers (After
Ref. [1072], reproduced
with permission)

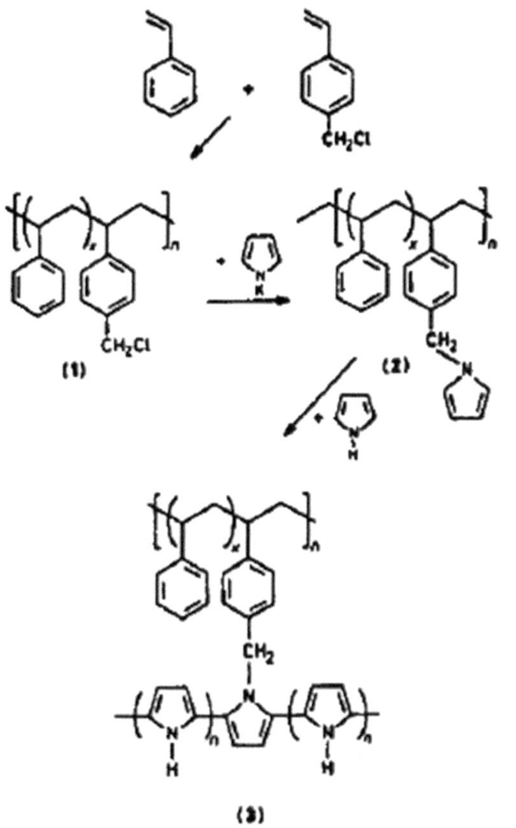

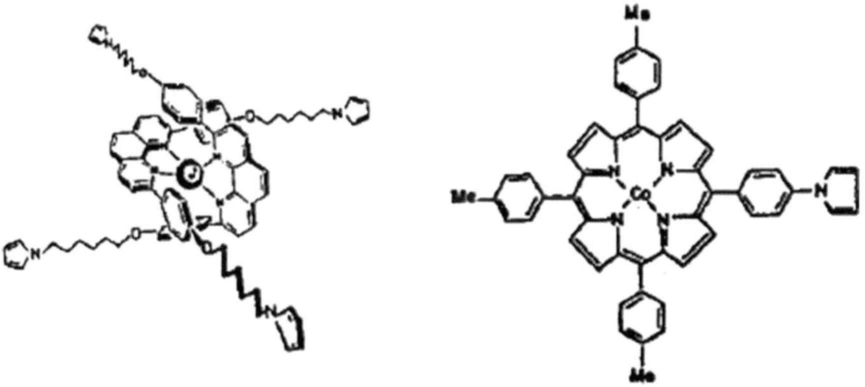

Fig. 31.24 *Left*: Novel metal-phenanthroline template used for CP synthesis, cited by Bidan et al. in Ref. [1073] (Reproduced from Ref. [1073] with permission). *Right*: Similar Co-porphyrin template cited by Bedioui et al., in Ref. [985] (Reproduced from Ref. [985] with permission)

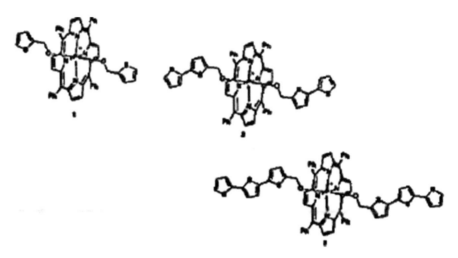

Fig. 31.25 Phosphorus porphyrins cited as templates by Shimidzu and coworkers in Ref. [1031] (Reproduced from Ref. [1031] with permission)

conducting in a potential region where P(Py) alone is nonconducting, and partial utility toward catalytic "electroassisted" reduction of benzyl chloride to benzene was demonstrated. In another template-based polymerization of a porphyrin derivative, Shimidzu et al. [1031] demonstrated successful electrochemical polymerization of *phosphorus* porphyrins with pendant thiophene and oligothiophene groups (Fig. 31.25. These unique poly(thiophene) derivatives then showed conductivity enhanced by photo-irradiation, although still near that of insulators (ca. 10^{-8} S/cm).

Other template-based synthesis methods for CPs have involved use of substrates comprised of *track-etched polymer* (e.g., polycarbonate) *membranes* and *alumina* [839, 840]. The principle behind the use of such substrates is shown in Fig. 31.26 below. Thus, e.g., poly(pyrrole) doped with perchlorate or poly(sodium 4-styrenesulfonate) anions has been electrochemically polymerized within a nanoporous polycarbonate track-etched membrane [1074, 1075] Poly(aniline) electrochemically synthesized within the pores of track-etched membrane has shown studied using scanning electron microscopy (SEM); here, it was shown that hollow tubules are formed, since polyaniline deposits during polymerization on the surface of the pore walls [839, 1076]. In a unique track-etched substrate polymerization of poly(2-methoxyaniline), the oxidized monomer was claimed to be first adsorbed onto a polycarbonate surface and then to subsequently polymerize to yield the polymer [839, 1077]. Track-etched substrate polymerization of other polymers, such as poly(3,4-ethylenedioxythiophene), has also been demonstrated [839, 1078]. A combined two-step synthesis reported for making poly(pyrrole) nanowires, comprised, firstly, electrochemical grafting of a thin polyacrylate film at a carbon electrode and, subsequently, electropolymerization of pyrrole; this yielded poly(pyrrole) nanowires, 600 nm in diameter and 300 μm in length, growing through the pores of the polyacrylate film [839, 1079].

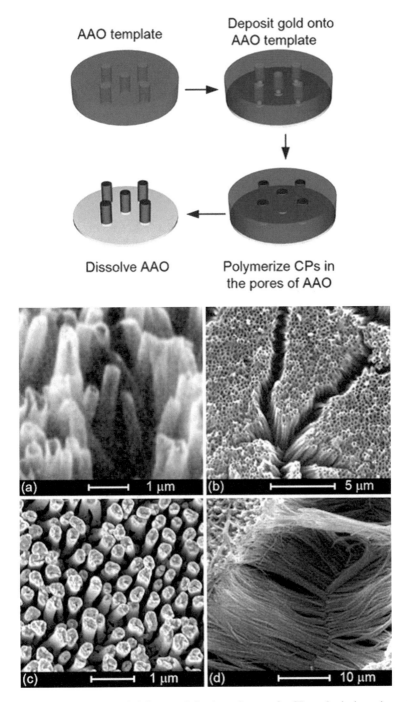

Fig. 31.26 *Top*: Principle behind the use of alumina substrates for CP synthesis, here shown for growth of CP nanowires in an alumina template. *Bottom*: Typical result – SEM images of CP nanowires synthesized using this method (After Ref. [840], reproduced with permission)

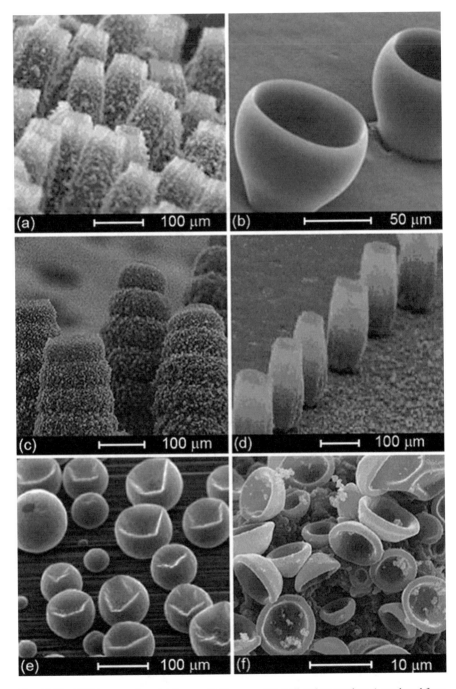

Fig. 31.27 SEM images of quaint "nanoart" ("nanocontainers" and nanospheres) produced from CPs (After Refs. [840, 1080, 1081], reproduced with permission)

Among other forms of nanoscale CPs whose synthesis has been demonstrated are "nanocontainers" and nanospheres. However, again, other than quaint interest as "nanoart," the precise application or utility of these structures is not immediately apparent, at least to this author. Some of these quaint structures are illustrated in Fig. 31.27 below.

With regard to alumina substrates, copolymers of pyrrole and thiophene have been prepared in the form of uniform and aligned nanoscale fibrils, via electrochemical copolymerization within the pores of alumina templates [839, 1083, 1084]. And aligned nanotubular heterojunctions of poly(p-phenylene) and poly(thiophene) have been prepared by successive electrosynthesis of both polymer layers in a nonaqueous solution within alumina pores 200 nm in diameter templates [839, 1085].

Galvanostatic polymerization has been used to grow poly(pyrrole) films in acetonitrile solution at a gold electrode modified with thiolated β-cyclodextrin self-assembled monolayers; this yielded nanoscale fiber structure for the resulting polymer. The thiolated β-cyclodextrin was shown to act as a molecular template restricting the polymer growth sites to within the β-cyclodextrin cavities, as was shown by scanning electron microscopy [839, 1086]. In a similar vein, poly(aniline) nanowires were synthesized at an Au electrode modified with self-assembled monolayers of thiolated cyclodextrin, embedded into an alkanethiol layer, wherein thiolated aniline derivative had been anchored to the surface within the cyclodextrin cavity, yielding an initiation point for aniline electropolymerization [839, 1087]. Nanosized dots and wires of polypyrrole and polyaniline have been obtained at a gold electrode modified with self-assembled monolayers of either β-cyclodextrin or p-[839, 1088].

31.6 Nanoscale Polymerizations

A series of CP syntheses on a nanometer scale fall into the category of template syntheses and are based primarily on adsorption of monomers within nanoporous templates. The concept is relatively simple: Include the monomer of a well-characterized CP, such as P(Py), within the pores of commercially available or easily synthesizable nanoporous matrices made of relatively inert material such as commercial polycarbonate or alumina membranes, or zeolites, and then polymerize it chemically via introduction of an oxidant or electrochemically. For the electrochemical polymerization, an electrical connection is made by metallizing the membrane pores and one side of the membrane. The result is conductive "nanofibrils," of quaint scientific interest if not, at least from presently available knowledge, of immediate practical utility.

The most well-known example of such polymerization has been the work of the Martin group [1089]. Commercial (Nuclepore, Poretics) polycarbonate, or sometimes alumina, filtration membranes of regular pore spacing, density (up to 10^9 pores/cm^2), and dimension (10 nm diameter typically) are used as templates. One side of these membranes is metallated for electrical contact. The membrane is then simply soaked in the polymerization solution and a potential applied to the electrode, initiating polymerization. The CP preferentially nucleates and grows on the pore walls, probably an effect of the initial adsorption of the monomer on these walls as well as subsequent electrostatic attraction between the oxidized CP and the anionic membrane walls, yielding hollow "nanotubules" for short polymerization times (solid fibrils for long polymerization times). Conductivities for large diameter fibrils (calculated from the transmembrane resistance using diameter and number of fibrils) are approximately those for bulk CP, but those for small-diameter fibrils are found to be 10 X higher. The template membrane can be dissolved and the CP nanofibrils or nanotubules collected. These nanofibrils and nanotubules can then be compacted for conductivity measurements just like bulk CP; electron microscopy shows that they apparently survive the high-pressure compaction process intact. However, the conductivities measured are not much higher than those of bulk CP. Most importantly, however, although the objective of the work was to obtain CPs with greater microstructural (or nanostructural) order, as Martin notes in a review article, the improvements in conductivity, CP chain and conjugation length, and other factors, are, disappointingly, but incremental.

Bein et al. chemically polymerized 2-ethyl-aniline contained within the 3D channels of dehydrated zeolites (commercial Linde A, Faujasite, Mordenite, others) [1090] and aniline within 3-nm-diameter pores of special mesoporous materials [1091], in both cases using $(NH_4)_2S_2O_8$ as the oxidant. The monomers were introduced via vapor or solution transfer. Properties such as conductivity were however not well characterized. In other, similar work, aniline has been polymerized within the voids of Cu-exchanged fluorohectorite [1092], and pyrrole polymerized within the 0.68 nm-diameter channels of an Fe coordination polymer [1093]. In all cases, the work is of much scientific interest, but does not appear to yield bulk properties, such as conductivity or conjugation length, of improved practical utility. Further extending this initial work more recently, Wu and Bein [1094] prepared P(ANi) filaments in 3-nm-wide hexagonal channel systems of the host aluminosilicate "MCM-41." Aniline vapor was first adsorbed into the dehydrated host and then oxidized with peroxydisulfate in the conventional manner. The product was obtained in the protonated, conducting, emeraldine salt form, and conductivity was confirmed by microwave measurements at 2.6 GHz. The application envisioned was "nanometer electronic devices" (cf. Fig. 31.28).

An interesting example of a template polymerization was provided by Wu et al. [1095]. Surfaces of SiO_2/Si substrates were first modified with a silane compound bearing pendant aniline groups via molecular self-assembly. This modified surface was then contacted with additional aniline monomer in the presence of the oxidant ammonium peroxydisulfate. The resulting CP films were claimed to be highly ordered with good substrate adhesion, and conductivities for 1-μm-thick films were found to be 0.5 S/cm.

Qi and Lennox [1096] recently described the fabrication os P(Py) nanostructures in a manner similar to the work of Martin et al. cited above but using lipid tubules as the templates. Nanofibers with diameters in the 10–50 nm range were thus obtained.

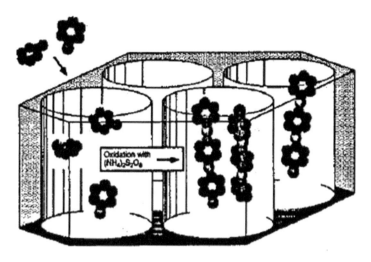

Fig. 31.28 Nanoporous synthesis of P(ANi) by Wu and Bein (After Ref. [1094], reproduced with permission)

Poly(aniline) *nanoparticles* have been synthesized at a highly ordered pyrolytic graphite (HOPG) electrode from aniline solution at low concentrations, using a potentiostatic method [839, 1097]. The disk-shaped poly(aniline) nanoparticles so obtained varied in dimension from 20 to 60 nm. In other works, large arrays of uniformly oriented poly(aniline) "nanowires" have been prepared via a controlled nucleation and growth process [839, 1098]. In this process, first at a high initial current density, a large number of poly(aniline) nuclei are deposited onto a substrate electrode; subsequently, growth of oriented nanowires occurs from the nucleation sites, achieved by stepwise reduction of the polymerization current density.

In similar synthetic processes applied to poly(pyrrole), nanowires were obtained by instantaneous 2D nucleation followed by 1D growth at a graphite–paraffin composite electrode [839, 1099]. In other works, nanostructures of poly(pyrrole) about 10 nm in size were prepared in this manner by electropolymerization at naturally occurring step defects and artificially formed pit defects of HOPG [839, 1100]. Under controlled potential electrolysis conditions, wire-shaped growth was observed for short electropolymerization times. In yet other work with poly (pyrrole), "nanotubules" with diameters ranging from 50 nm to 2 μm were claimed to be synthesized electrochemically in the presence of the dopants β-naphthalenesulfonic acid or p-toluenesulfonic acid [839, 1101]. The authors of this work proposed that the micelles of dopant or pyrrole–dopant clusters act like templates on the electrode surface, enabling poly(pyrrole) to grow in a tubular form.

Atomic force microscopy (AFM) and scanning tunneling microscopy (STM) have been used for synthesis of nanoscale poly(aniline) and poly(pyrrole) [839, 1102–1104]. Syntheses of nanoscale poly(aniline) and poly(pyrrole) have been claimed with the use of AFM, using tip–sample interactions; in this work, electropolymerization could be either blocked on the bare HOPG surface or enhanced at the as-polymerized film.

This approach is claimed by the authors of this work to allow one to use AFM as a tool for localized "nanolithography." Similarly, production of nanoscale poly (aniline) at a graphite surface, with the use of the Pt tip of a STM, has also been claimed. For this, a two-voltage pulse technique was employed, wherein the first pulse results in pit formation and, the second one, in oxidation and further nano-scale polymerization of aniline monomer, leading to poly(aniline) particles of dimension 10–60 nm × 1–20 nm.

A unique electrochemical technique was demonstrated for synthesis of nano-scale dot or line arrays of poly(o-phenylenediamine) at a bare Au(111) surface and at a p-aminobenzenethiol-modified Au surface [839, 1105].

Nanostructured composites of CPs with noble or semi-noble metal (Au, Pt, Pd Ag) clusters have been prepared by many groups [839, 1106–1113]. However, other than academic interest, the utility of these metal/CP composites is not immediately apparent from the publications in this area thus far.

Similarly, CPs have been coated onto CNTs and even monolayer graphene [839, 1114–1118], but again, other than academic interest, the utility of these composites is not immediately apparent from the publications in this area thus far; although one of the objectives claimed for them has been energy storage (e.g., in supercapacitors), they have performed rather poorly as compared to other, more established materials for such applications.

31.7 True Copolymerizations

As opposed to the template polymerizations described above, and composites such as those of CPs with plastics (e.g., with poly(vinyl acetate), poly(ethylene tere-phthalate), poly(vinyl alcohol)), *true copolymers* of CPs are those in which the CP and the other polymer components are synthesized *from their monomers* and also have all components part of a polymer structure, i.e., they are structural copolymers in the polymer chemistry sense of the term. Indeed, the common methods used to synthesize CPs, whether electrochemical or chemical, can also be used for copol-ymers through the simple expedient of starting with a mixture consisting of several monomers rather than a single monomer.

A good example of true copolymers is acetylene/ethylene copolymers prepared by the common Ziegler–Natta route described above but using mixtures of varying proportions of ethylene and acetylene as the starting monomers, as described by Raspopov et al. [1119]. At a 20% acetylene composition, these yield CPs with conductivities of 10^{-3} S/cm. Only the acetylene fragments appear to take part in doping and conductivity, however. In a similar synthesis using methyl methacrylate monomer but using Na-doped *poly*(acetylene) as the initiator for anionic polymer-ization, poly(methyl-methacrylate/acetylene) copolymers were synthesized

[1120]. Aldissi has described a copolymer of acetylene and isoprene which is castable from solution and dopable to ca. 5 S/cm [1121].

Block copolymers of styrene and acetylene which are soluble but poorly conducting have been prepared again using Ziegler–Natta catalysts [1061]. Another variant of poly(styrene/acetylene) block copolymers prepared by the Aldissi group [871] shows high conductivities for styrene proportions of less than 16%. For the 40%-styrene polymer, a conductivity (I_2-doping) of 10^{-3} S/cm increases to 1 S/cm simply by application of 10 kbar pressure, thus confirming that poly(acetylene) "islands" within the copolymer are responsible for the conductivity, and the styrene component simply plays the role of a kind of structurally incorporated "filler." Unfortunately, all of these copolymerizations show that attempts of the inert component impart properties such as solubility and durability to the CP (e.g., for the styrene copolymers) always show up huge trade-offs where conductivity or processibility is seriously compromised with no apparent advantage.

Electrochemical polymerization has been used to generate interesting copolymers which however again show no clearly evident utility. Pyrrole and naphthol have been electrochemically copolymerized [1122], as have Fe-bathophenanthroline-disulfonate (FEBPS) and pyrrole [1123] and Zn-tetra (4-sulfophenyl)porphyrin (ZnTPPS) and pyrrole [1124]. All of these pyrrole copolymers displayed some electrochromism but nothing superior to that of P(Py) or other CP. The ZnTPPS-P(Py) copolymer however did show photocurrents on irradiation with visible light. A more interesting true copolymer useful for tunable color LED applications has been described by Spangler et al. [1125] (Fig. 31.26). The Pickup group has used a true copolymer of Ru(2,2′-bipyridine)$_2$(3-pyrrol-1-ylmethyl)pyridine)$_2$] (Ru(BP)$_2$(pmp)$_2$) and 3-Me-thiophene (3MeT) to probe electronic conduction mechanisms in macrocycles. The copolymer was prepared by electrochemical cycling from solutions containing both monomers. As the thiophene/Ru molar ratio is increased above 1.0, conduction due to thiophene–pyrrole linkages is increasingly observed. There is however strong indication that the Ru sites contribute to the conductivity (Fig. 31.29).

Copolymers such as poly(2,5-dialkoxy-phenylene-vinylene/2,5 thienylene vinylene) have been synthesized by starting with the corresponding monomer mix for a standard Wittig synthesis [1126]. Similarly, a copolymer of pyrrole and naphthol was prepared by simple electropolymerization from a solution of both monomers [1122]. Hotta et al. [1059] describe a similar procedure for galvanostatic electrochemical copolymerization of 3-hexylthiophene and 3-benzyl thiophene from nitrobenzene solution with perchlorate electrolyte. It may be noted however that such procedures however typically produce CP copolymers of curious interest only, with no exceptional properties as compared to the respective homopolymers.

The Pickup group [873, 1127] has described interesting copolymers of P(3MeT) with transition metal polymer complexes, poly-[Ru(2,2′-bipy)$_2$(3-(pyrrol-1-yl-Me) Py)$_2$]ClO$_4$)$_2$ (poly-[Ru(bp)$_2$(pmp)$_2$](ClO$_4$)$_2$) [1127]. These copolymers were prepared simply via electrochemical polymerization from the monomer solutions in acetonitrile. The copolymer has much lower conductivity than P(3MeT) despite the presence of the metal atom. The Shimidzu group [1123] carried out a very similar

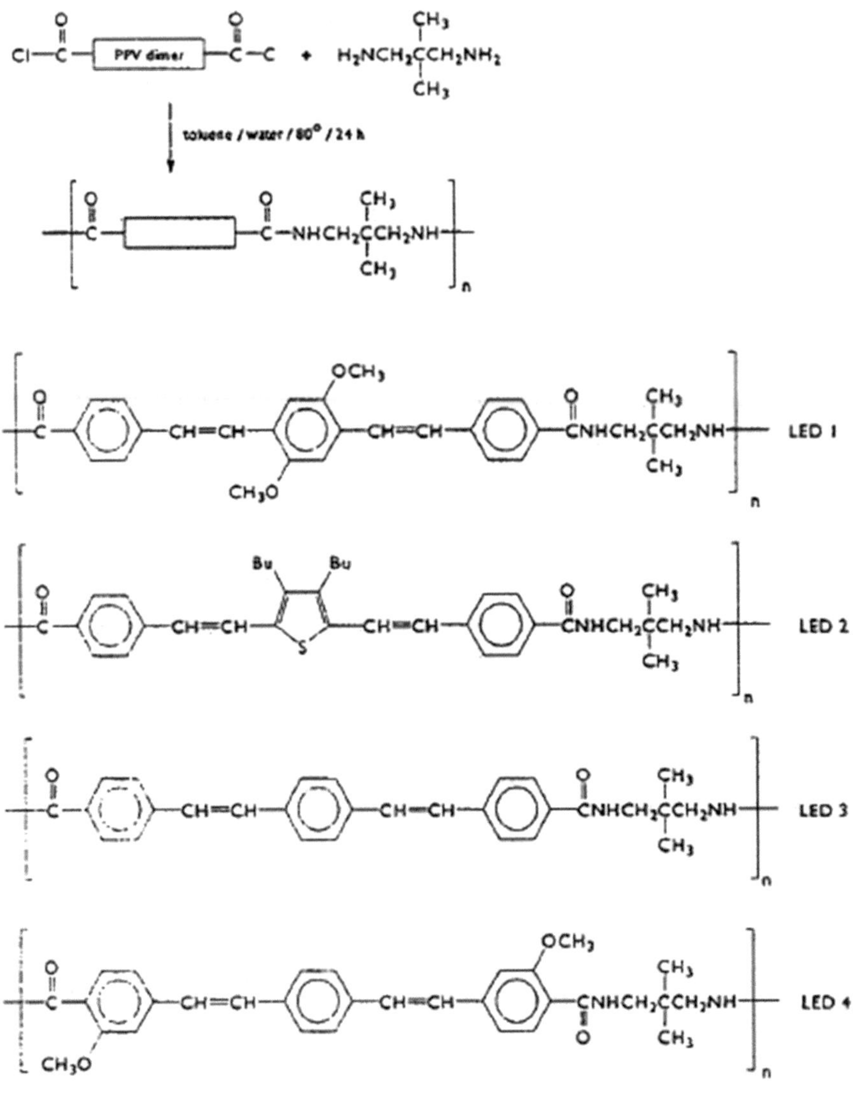

Fig. 31.29 Scheme for synthesis of a series of copolymeric CPs for use as LEDs (After Spangler et al., Ref. [1125], reproduced with permission)

electrochemical polymerization from a solution of pyrrole and Fe (bathophenanthroline disulfonate) (FeBPS), with the Poly(FeBPS)/P (Py) copolymer having a ca. 0.1 M P(Py) content and claimed conductivities as high as 7 S/cm. This copolymer showed electrochemical redox behavior intermediate between that of P(Py) and FeBPS, and blue-gray/red visible region electrochromism (vs. lt. yellow/gray-black for P(Py)).

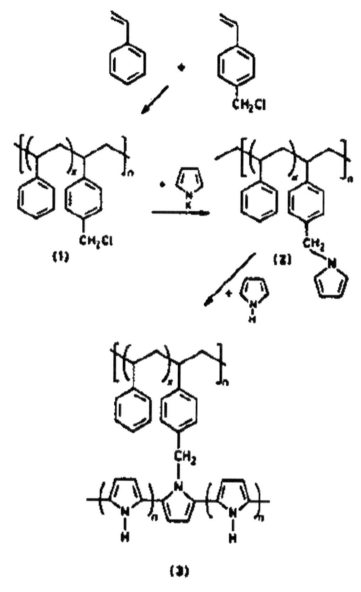

Fig. 31.30 Preparation scheme for poly(pyrrole-styrene) copolymers as adopted by Nazzal and Street [1072]

Perhaps the most well-known examples of a CP copolymer are the pyrrole–styrene graft copolymers synthesized by Nazzal and Street [1072]. As the preparation scheme in Fig. 31.30 shows, a polystyrene derivative is first derivatized with a pendant pyrrole group, and the derivative then copolymerized electrochemically with pyrrole monomer, yielding the final "graft copolymer." This is in fact an

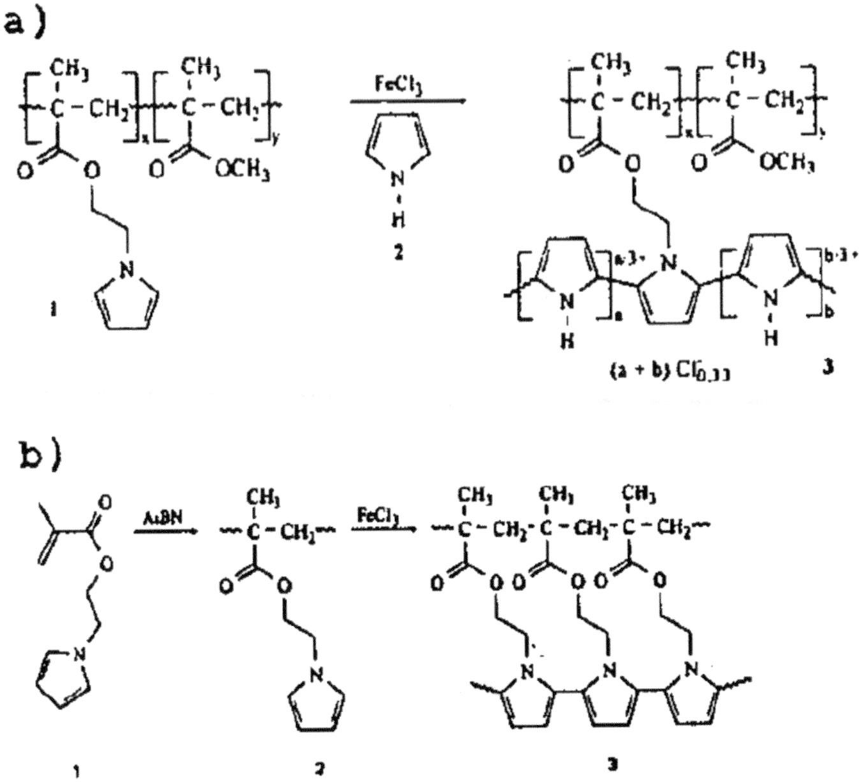

a)

b)

Fig. 31.31 Schemes for preparation of pyrrole-acrylate copolymers, after Stanke et al. (After Refs. [1070, 1071], reproduced with permission)

example of a "template polymerization" as discussed elsewhere in this book. Conductivities, as high as 0.4 S/cm, and acceptable mechanical properties are reported for these copolymers, but curiously, the work does not appear to have been pursued further for the synthesis of thermoplastic PStyr/P(Py) copolymers with desirable mechanical and conductivity properties. In similar work also leading to graft copolymers and using a template method, Stanke et al. [1070, 1071] prepared copolymers containing Py and MMA moieties via two different pathways (Fig. 31.31a,b). The first copolymer (a) showed limited conductivity, ca. 5×10^{-5} S/cm, when doped with I_2, and both copolymers showed poor mechanical properties and processibility. This belied hopes that copolymerization with a highly processible and versatile thermoplastic such as PMMA would produce a useful, processible conductive copolymer.

In other works of relevance, Raspopov et al. [1119] claimed to have synthesized a copolymer of acetylene and ethylene via a Ziegler–Natta catalytic method, claiming conductivities of 10^{-3} S/cm for a 20 mol% acetylene content. However, it was not clear whether this was a true copolymer or just a P(Ac) network with PE

Table 31.4 Structural and photoluminescence (PL) information for light emissive copolymers prepared by Cho et al. (After Ref. [1128], reproduced with permission)

Sample (AB/PL[a])(nm)	Structure[b]
PDHF (370/413)	
PDHFV (402/463)	
PDHFPP (370/406)	
PDHFVPHOD (386/434)	
PDHFMPV (382/422)	
PDHFPPV (412/477)	
PDHFDHOPPV (446/495)	

[a]$R = R_2 = -(CH_2)_5CH_3$
[b]PL spectra of solutions (conc. 10^{-5} mol l^{-1})

serving as a host polymer. Similarly, in work of Aldissi et al. [871] in the production of P(Ac)/PStyr "copolymers," it was not clear whether the end product was a true copolymer or a composite, since a percolation threshold, a term invalid for a copolymer, was claimed (ca. 16 v/v% P(Ac) for 10^{-3} S/cm conductivity).

Cho et al. recently described [1128] an interesting set of copolymers obtained by alternating units such as phenylene and vinylene with dihexylfluorenes; these were targeted for photoluminescence (PL) applications. Well-known organic reactions, such as the Heck, Suzuki, and Wittig reactions, were employed in the syntheses. Table 31.4 summarizes structural and PL information for some of these.

Beggiato et al. [1129] recently described a unique set of copolymers having the following monomer components: dithienopyrrole-dithenothiophene (DTP-DTT), DTP-thionaphtheneindole (DTP-TNI), and DTT-TNI. Simple electrochemical polymerization from acetonitrile solutions of the monomers in the desired proportions was used. However, only the first combination, DTP-DTT, yielded CPs showing conductivity and electrochromism.

Very recently [1125, 1130], a series of interesting copolymers (Fig. 31.32) of derivatives of P(T) and P(PV) have been synthesized, primarily with a view to

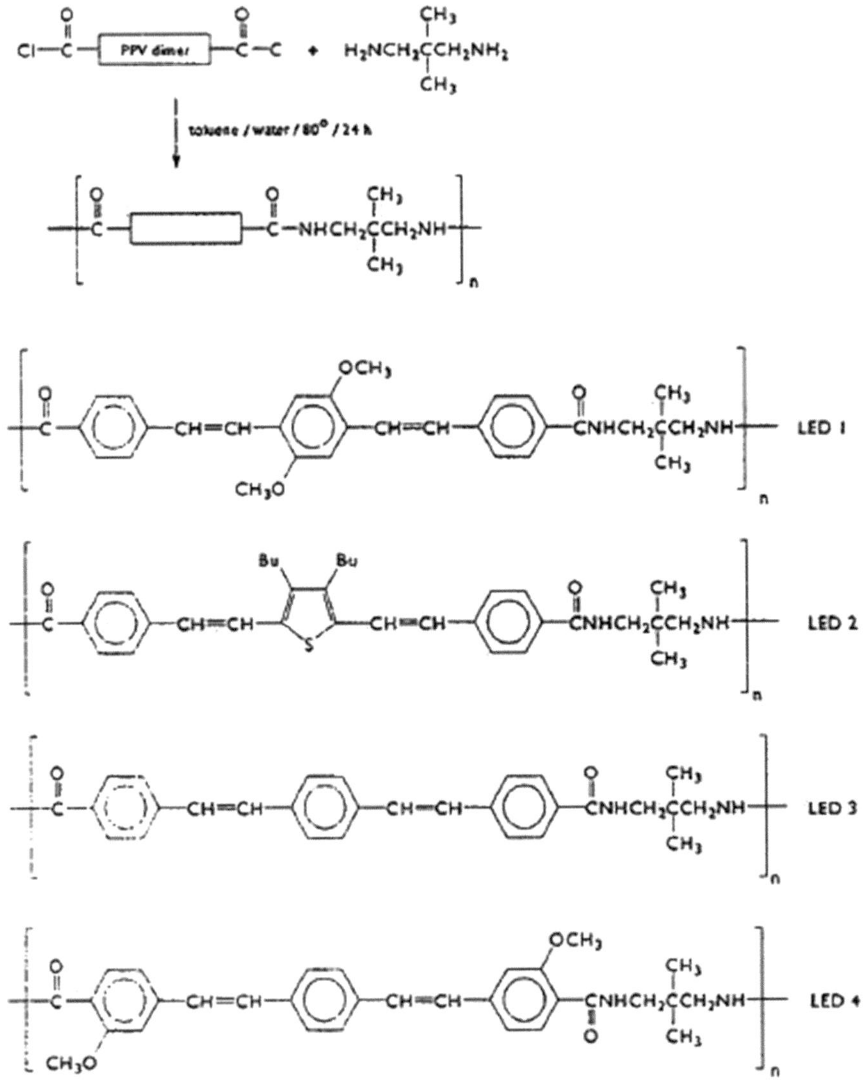

Fig. 31.32 New P(PV)-based copolymer CPs synthesized by Spangler et al. for LED applications (After Ref. [1125], reproduced with permission)

utilizing them in CP-based light-emitting diodes. These appear to impart tailorability of the LED spectral response, but their other properties do not appear to have been characterized in much detail to date.

Other applications of copolymers recently targeted have been conductor wires or films. For example, the patent application of Millan Rodriguez et al. [1131] describes copolymers of indole, thiophene, and pyrrole, prepared via conventional

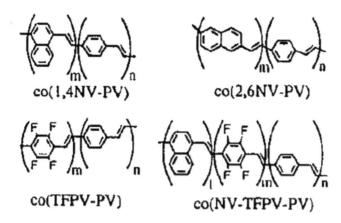

co(1,4NV-PV)   co(2,6NV-PV)

co(TFPV-PV)   co(NV-TFPV-PV)

Fig. 31.33 New P(PV)-based copolymer CPs synthesized by Benjamin et al. for LED applications (After Ref. [1134], reproduced with permission)

chemical copolymerization in solution using $FeCl_3$ oxidant; conductivities of ca. 1 S/cm are claimed.

Yet other recent examples of studies of copolymerizations have included that of aniline with *m*-phenylenediamine (MPDA) [1132], where copolymerization was found to occur only at low MPDA concentrations. Yang and Wen [1133] studied the electrochemical copolymerization of aniline with 2,5-diaminobenzenesulfonic acid (DABSA) in aqueous HCl solutions. The copolymer films obtained were electroactive, but did not exhibit significant conductivity (< 0.2 S/cm) nor electrochromism.

Benjamin et al. [1134] recently described a series of new P(PV) derivatives for potential LED applications, being in reality copolymers with such moieties as naphthalene; their structures are given in Fig. 31.33.

They used a modification of the "electrolyte precursor technique" described by Wessling and Lenz [1135].

31.8   In Situ Polymerization

An important method for processing of CPs as coatings – important because it is one of the few practical applications of CPs implemented successfully on a commercial scale – is *in situ polymerization of CPs* on varied substrates, ranging from individual textile fibers and cloth to thermoplastics. While not appearing to be a method representing true processibility or solubility of CPs, the final effect of the method, in terms of yielding highly durable and uniform CP coatings on varied substrates, is in fact superior to the many highly elaborate processing and solution methods described in this chapter which turn out ultimately to have little practical use.

On the other hand, several novels, generically applicable EP-coating methods, have been proposed and tested. For example, one novel method for in situ polymerization, but one quite logically obvious, has been apparently tested for several common CPs by the Genies group [1136]. In this, a spray-coating process is used in which separate monomer and oxidant sprays are mixed just before landing on the substrate. It is claimed that the process can be used to coat very large areas and odd shapes, but no subsequent data have been published on it, and to the best of available knowledge, the process has not thus far been commercialized.

A good example of this practical implementation is the series of "Contex"-trademarked P(Py)- and P(ANi)-coated textiles marketed by Milliken Research Corp., a textile company [890, 1137, 1139]. The basis of the deposition procedure is adsorption of monomer and/or dopant and/or oxidant on the substrate surface, at which chemical polymerization then occurs. In the case of P(Py), for instance, typically, fabric substrate is placed in a reaction chamber in water to which separate aqueous solutions of Py, dopant, and $FeCl_3$ (oxidant) are added. After reaction with stirring at an appropriate temperature, the product is washed, dried, and otherwise conventionally processed to yield coherent, uniform, well-adherent, smooth, and durable films of *highly conductive* CP on the substrate. CP thicknesses, in the region of 1 μm, are controllable. The key to the process seems to be use of high dilutions (e.g., 0.02–0.04 M monomer and oxidant and solution/substrate volume ratios of up to 50:1). The choice of doping agent also appears important; dopants such as methane sulfonate, benzene sulfonate, naphthalene sulfonate, and anthraquinone sulfonate yield progressively higher conductivities (1200–56 Ω/square). Polymerization rates are controllable via concentration and time of addition of oxidant, the former also being adjustable via addition of complexing agents such as sulfosalicylic acid. Evidently, the coating proceeds equally well on substrates ranging from rayon and cotton to polypropylene, polyester (e.g., poly(ethylene terephthalate)), nylon, Kevlar, and quartz fiber; coatings on all these substrates have been successfully produced. The Milliken researchers postulate a mechanism in which the free radical monomer or oligomers are the actual species adsorbed on the substrates but provide little strong evidence for this.

Similar methods have been used to coat P(Py) on nylon, C-fibers, and glass fibers by other workers [1140, 1141]. In one simple procedure for the production of P(ANi) coatings on glass fibers of diameters in the region of 7–9 μm, after acid wash and rinse of the glass fiber, it is kept for several hours with stirring in a solution of aniline, *p*-toluene-sulfonic acid (dopant), and ammonium persulfate (oxidant) in aqueous HCl [1141]. The result, after washing and post-processing, is a CP-coated glass fiber with CP thickness from 0.07 to 1.3 μm with surface resistivities from 1.4 to ca. 30 kΩ/square.

Another example of typical *in situ* polymerization is that which starts with sorption of the monomer on an appropriate substrate. A typical example, synthesis of a CP composite comprised of P(Py)/PET, may be described in this regard [1142]: Highly transparent (92%T @ 633 nm), dry (50 °C 24 h) PET films (30 μm thick, 1.4 g/cc) are dipped into (liquid) pyrrole monomer for 0.5–36 h. The pyrrole-sorbed film is washed with water and acetone and then dipped for similar times into aqueous $FeCl_3$ solution. The variables, determining conductivity and transparency of the composite, are dipping time in monomer and oxidant and concentration of oxidant. Conductivity and transmission achieve limiting values, ca. 0.1 S/cm and 20%T at 633 nm, respectively, after ca. 5 h dipping in pyrrole with 10 w/w% $FeCl_3$ solution. The percolation threshold is ca. 10 v/v% pyrrole. Optimal conductivity/transmission, ca. 0.05 S/cm and 60%T at 633 nm, is obtained with ca. 2 h dipping time in pyrrole, 10 w/w% aqueous $FeCl_3$, and 25 h polymerization time (dipping in $FeCl_3$). A very similar procedure [1143] can be used for P(ANi)/PET composites, with however a thinner PET film (10 μm), 0.1 M $(NH_4)_2S_2O_8$ (ammonium peroxydisulfate) + 2 M HCl as the oxidant, and slightly different dipping and polymerization times. The resulting P(ANi)/PET composites have claimed transmissions of 70% (450–700 nm) for a conductivity of 0.1 S/cm. The percolation threshold is ca. 5 w/w%. Both composites however are conducting only on one side of the PET host film.

Yet another example of typical *in situ* polymerization is that which starts with sorption of the *oxidant* on an appropriate substrate. A representative example is an oxidant-sorption procedure used by Ojio and Miyata [1144] for the fabrication of poly(pyrrole)-poly(vinyl alcohol) (P(Py)/PVA)) composites. In their work, PVA, of M_n 22,000, was dissolved with $FeCl_3$ (the oxidant) in water, a film cast from this solution onto a PET film substrate. This oxidant-saturated host polymer film was then exposed, in a desiccator at low temperature and in a deoxygenated atmosphere, to monomer (pyrrole) and water vapor for 0.5–24 h, with the resulting composite films, ca. 2 μm thick, dried in vacuum. Conductivities and transmission values saturated at about 1 h exposure time, to ca. 1 S/cm and 40% (at 550 nm) for a 70:30 w/w ratio PVA/$FeCl_3$. Transmission of the film was down to ca. 55% within 0.5 h (with conductivity ca. 0.1 S/cm) for the same PVA/$FeCl_3$ ratio. Higher PVA/$FeCl_3$ ratios (90:10, 95:5) gave not only higher transmissions with minimal conductivity reduction but also more homogeneous films as evidenced by SEM.

A similar procedure, using highly viscous liquid complexes formed by $FeCl_3$ (oxidant) with polymers such as poly(ethylene oxide) (PEO), poly(β-propiolactone) (PPL), and poly(1,5-di-oxepan-2-one) (PDXO), was employed by Rabek et al. [1145] to fabricate P(Py) composites. The complexes were first prepared in dry nitromethane (polymer/$FeCl_3$ ratio 7:3) and cast as films on glass. The films were then exposed to pyrrole vapor in a desiccator, yielding composites of 50–200 μm thickness over different reaction times. Excess oxidant and unreacted monomer were removed with a nitromethane wash. Conductivities were ca. 10^{-3} S/cm and transmissions below 50% in the mid-visible region.

A variant of the above procedures involving sorption of the oxidant is simple addition of monomer into a solution of the host polymer + oxidant, with the solution oxidation potential such that immediate polymerization occurs. The composite is immediately precipitated. As an example, Jousse et al. [1146] added liquid pyrrole monomer to a solution of $FeCl_3$ (oxidant) and PVC in solvents such as nitrobenzene, tetrahydrofuran (THF), or cyclopentanone. On shaking/stirring over 1 h, polymerization occurred, and the P(Py)/PVC composite was then precipitated by addition of ethanol. Interestingly, solvents such as THF and cyclopentanone, which solvate PVC well, yielded insulating composite due to the fact that the oxidant-cum-dopant, $FeCl_3$, is effectively complexed to the solvent, which are Lewis bases. On the other hand, nitrobenzene, a poor solvent for PVC and a poor complexing agent for $FeCl_3$, yields composite conductivities of ca. 10^{-2} S/cm. The composite had interesting microwave absorption properties.

Zinger and Kijel [1147] described a procedure for preparation of poly(ethylene) (PE) composites with pyrrole, which used initial sorption of poly(styrene sulfonic acid) (PSSA, dopant) in the PE followed by immersion into an Fe(II)/Fe(III) solution, with the Fe(II/III) ratio determining the oxidation potential of the solution. Into this solution, pyrrole was injected, yielding the composite films. The films showed appreciable conductivities and mechanical properties.

Wiersma et al. [1148] described a method for preparation of P(Py) or P (ANi) + polyurethane dispersions which may be applied to textile fibers. Aqueous solutions of pyrrole (or anilinium sulfate) and $Fe(NO_3)_3$ are added to a dispersion of polyurethane in water. After 20 h reaction time, a P(Py) (or P(ANi)) poly(urethane) dispersion is obtained, which can then be used to fabricate coatings with claimed conductivities up to 10 S/cm.

Lafosse [1149] described an oxidant-sorption method for synthesis of poly (pyrrole)-Teflon (PTFE) composites. A commercially available, surfactant-stabilized PTFE emulsion is mixed with aqueous Fe(III)-tosylate (oxidant-cum-dopant). Pyrrole is then added to this, yielding a finely divided P(Py)/PTFE composite after several hours, and precipitated with ethanol. SEM analysis showed the composite to be particles of 0.2 μm diameter which are presumed to be PTFE spherules coated with P(Py), rather than a homogeneous blend. Nevertheless, the composite showed conductivity of ca. 10^{-2} S/cm at a percolation threshold of ca. 16 v/v% P(Py) and acceptable microwave absorption properties. A similar emulsion method has been used by Sun and Ruckenstein [1150] to synthesize P (Py)/synthetic rubber composites, with $FeCl_3$ as oxidant, and several solvents (aqueous/organic combinations) and several nonionic surfactants being employed. Conductivities of the composites were ca. 3 S/cm, tensile strengths ca. 10 MPa, and elongation at breaks 38–166%.

Yet another variant of *in situ* polymerization is *solution evaporative polymerization*. To cite an example, Han et al. [1008] took a solution of $FeCl_3$ (the oxidant) and PVA (the host polymer) in methanol and then added pyrrole monomer thereto. The oxidation potential of the solution, and hence the polymerization, could be changed and controlled via the proportion of $FeCl_3$ to PVA as well as via evaporation of the solvent and was optimally ca. +500 mV vs. SCE. Thus, simple casting of a film from this solution yielded P(Py) composite. The percolation threshold, for a saturation conductivity of ca. 10 S/cm, was claimed to be very sharp, at 5 w/w%. A method which appears to be a combination of solution evaporative polymerization and monomer-sorption polymerization is that carried out by Morita et al. [1151]: A poly(pyrrole)-poly(methyl methacrylate) (P(Py))/PMMA) composite, with nonhomogeneous composition (a gradation of PMMA and P (Py) concentrations), was obtained in 0.1–10 μm thickness and 10^{-8}–10^{-1} S/cm conductivity from combination of 2-butanone solutions of pyrrole, PMMA, and benzoic acid with aqueous solutions of potassium persulfate.

A singular in situ chemical polymerization method for CP composites involves use of a *catalyst/host polymer complex into which monomer is introduced*. Thus, for instance, in a procedure used by Sariciftci et al. [887], a Co-based catalyst containing NaBH$_4$ is added to a solution of poly(vinyl butyral) (PVB) in ethanol or dimethylformamide (DMF). The solution is cooled to -30 °C and acetylene added thereto. After a polymerization time of 1–12 h, a P(Ac)/PVB composite results with conductivities up to 1 S/cm. The composite possesses significant third-order NLO properties. Very similar procedures, e.g., soaking of the host polymer in a toluene solution of the Ziegler–Natta catalyst Ti(OBu)$_4$/AlEt$_3$ and then exposing it to acetylene vapors, have been used to prepare composites of P(Ac) with poly (butadiene) and PE [1152]. The resulting composites can then be doped, but their environmental stability is not better than that of P(Ac).

Rubner et al. [1153] describe another such procedure for preparation of P (Ac) composites with Polybutadiene (PB). A solution of PB and Ziegler–Natta catalyst in toluene is used to coat the walls of a reaction vessel which is then used for polymerization of Ac at low temperature. The resulting PB/P(Ac) composite however had very poor structure, and a host of techniques show it to actually consist of phase-separated islands of PB and P(Ac), i.e., a true physical mixture and nothing more. Conductivity and other properties reflect this two-phase composition [1153].

Przyluski et al. [1154] recently described a stepwise chemical polymerization of aniline and pyrrole in the presence of aqueous vinylidene chloride-Me acrylate-Bu acrylate copolymer solutions to yield conductive "latexes." Contents were from 5 to 50% of P(ANi) and <15% of P(Py), and conductivities were in the 1 S/cm region. Another in situ preparation of latex composites was described by Xie et al. [1155] in which aniline was polymerized in the presence of poly(butadiene-co-styrene-co-2-vinylpyridine) (PBSP) to yield P(ANi)/PBSP latexes. Percolation thresholds were ca. 20 w/w% P(ANi), yielding conductivities of ca. 8 S/cm. Tensile strength and ultimate elongation at the percolation threshold were 14 MPa and 1300%, respectively.

Oh et al. [1156] recently described the preparation of P(ANi)/poly(styrene) composites via a combination of in situ chemical polymerization and blending. The polymerization was performed by adding oxidant and dodecylbenzenesulfonic acid dissolved in xylene to a xylene solution of aniline and poly(styrene). At a P (ANi) content of 12 w/w%, the conductivity reported was 0.1 S/cm.

Ramachandran and Lerner [1157] prepared a unique composite of P(Py) with the clay montmorillonite using a simple solution-blending procedure. The resulting "nanocomposite" was claimed to have the structure shown in Fig. 31.34.

Another, innovative variant of in situ polymerization is *in situ electrochemical polymerization*, especially for production of CP composites. One example of this is electropolymerization into a host polymer immobilized on a metal electrode, for example, electropolymerization of P(Py) composites with the liquid crystalline copolyamides poly(*p*-phenylene-terephthalate amide) (PPTA), poly(*p*-phenylene-terephthalate amide/diphenyl ether terephthalamide) (PPTA[O]), and poly(*p*-

Fig. 31.34 Schematic illustration of the nanocomposite structure for poly(pyrrole)– montmorillonite (After Ref. [1157], reproduced with permission)

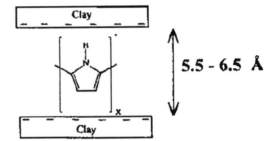

phenylene-terephthalate amide/diphenyl methane terephthalamide) (PPTA [C]) [1158]. After coating the PPTAs from solution onto a Ni electrode, P(Py) is deposited into the PPTA films via galvanostatic electropolymerization from 0.2 M pyrrole in CH_2CHOH/CH_3CN having 0.1 M tosylate dopant. Factors affecting the conductivities of the composites, typically 20 S/cm, and the tensile strength, typically 100 MPa, include pyrrole concentration and current density. An even more mundane example of an in situ electrochemical polymerization is the work of Niwa et al. [1159] in the polymerization of pyrrole into PVC films coated on a Pt electrode, where the composite is nothing but a physical web of P(Py) resting in the vacant spaces of the PVC matrix. A very similar procedure has been used by de Paoli et al. [1160] for fabrication of P(Py)/PVC composites.

Lindsey and Street [1161] describe an electropolymerization of pyrrole into a ca. 4-μm-thick PVA film previously spin coated on a metal electrode and cross-linked by heating to 150 °C for 0.5 h to reduce solubility, while permitting swelling in the solution used for polymerization, 0.1 M $CuSO_4$/0.1 M pyrrole in water. Control of polymerization conditions allows preparation of a composite which is conductive on one or both faces, with conductivities in the region 0.1–10 S/cm. In a similar procedure, Selampinar et al. [1162] described the in situ electropolymerization of P(Py) into bisphenol A-type PEK films yielding conductive composites.

In yet another very similar procedure, Bi and Pei [1163] have polymerized pyrrole from acetonitrile/ethylene glycol solution into films of polyurethane coated on Pt electrodes, obtaining composites of ca. 15 μm thickness and conductivity 1–35 S/cm (no percolation threshold up to a w/w% of 16). The composite films have much higher Young's modulus (i.e., are less rubbery) than pure polyurethane films.

Bozkurt et al. [1164] described the stepwise in situ electrochemical polymerization of P(Py) and poly(indene) from monomer solutions in acetonitrile, thus obtaining P(Py)/Poly(indene) composites.

31.9 CP Blends

Direct blending of CPs implies simply blending already synthesized CP and host polymer and then using that liquid phase blend for further processing. It can be done either in solution or in the melt phase.

31.9.1 Blending in Solution

An illustrative, simple solution-blending procedure is that used by Stockton and Rubner [1165] for P(ANi)/PVP composite (PVP = Poly(vinyl pyrrolidone), a hydrophilic plastic. Solutions of the emeraldine base (de-doped) form of P(ANi) and PVP were prepared in NMP (with some sonication, stirring, and filtering). Films of thickness ranging from 30 μm to 0.3 mm were cast directly from solution. Films unfortunately retained ca. 20 w/w% NMP (a consistent problem with use of this high boiling solvent) even after drying but had conductivities of up to 2 S/cm while showing no evidence of phase separation. Another simple example of solution blending is that of several poly(p-phenylene vinylene) (PPV-deriv) with host polymers such as poly(methyl methacrylate) (PMMA, a very common, household thermoplastic), polycarbonate, a common shatter-resistant thermoplastic, and poly (styrene) [1166]. Typically, a solution of I_2 (dopant) in THF was added to $CHCl_3$ solutions of the host polymer and the PPV-deriv with stirring. Films were cast on glass slides. The freestanding films, which were however opaque (black), had thicknesses of ca. 0.1 mm and conductivities of up to 2 S/cm.

In yet another example showing how simple solution blending of CP composites can be, Ogura et al. [1167] prepared composite films of poly(o-phenylenediamine) (P(oPD)) and PVA simply by casting from solutions of the two polymers in DMSO (dimethyl sulfoxide). They claimed percolation thresholds of 0.1 v/v% (sic) for saturation conductivities of ca. 0.1 S/cm, which however have not been duplicated anywhere else. These dark composites were used as humidity sensors.

Hotta et al. [1059] used chloroform solutions of a copolymer of 3-hexyl and 3-benzyl thiophenes and polystyrene (PStyr) to deposit films on glass slides. When solution doped with $NOPF_6$ in acetonitrile, they yielded conductive composites with a ca. 20 v/v% percolation threshold for a 1 S/cm saturation conductivity. Films cast on ITO/glass showed an acceptable electrochromic effect.

Pron et al. [1168] used novel phosphoric acid diesters to protonate the emeraldine base form of P(ANi) in solvents such as toluene, decaline, or m-cresol, yielding solutions of the conductive form of the CP. Solutions of the highly versatile thermoplastic acrylonitrile-butadiene-styrene (ABS) were then mixed with these solutions, and films cast from the mixture yielded composites with conductivity of ca. 10 S/cm (percolation threshold ca. 20 w/w% P(ANi)). Later work from the same group [1169] showed that P(ANi)/PVC composites which could be hot-pressed and had percolation thresholds of ca. 6 w/w% could be obtained with the same procedure.

Gonçalves et al. [1170] used a simple solution casting method to fabricate freestanding polyurethane-poly(*o*-methoxyaniline) composite films from a solution in DMF. The percolation threshold was 10 w/w%, and loadings of 65% CP gave conductivities of 10^{-3} S/cm. X-ray studies indicated significant crystalline sections in the composite.

A low percolation threshold, ca. 3 w/w% for a conductivity of ca. 10^{-1} S/cm, was claimed by Banerjee and Mandal [1171] for poly(aniline)/ PVC films prepared using a unique method. This method used sub-μ-particulate P(ANi)Cl, stabilized with ca. 5% poly(vinyl methyl ether), and a THF solution of PVC. The P(ANi) particles were dispersed in the PVC solution using ultrasound, and films were cast therefrom. TEM images at the percolation threshold indicated dispersed P(ANi) particles achieving interconnection.

In more recent work, Tan et al. [1172] described blends of P(ANi) with the rigid-rod polymer poly(*p*-phenylene-benzobisthiazole) (P(BZT)), wherein the P(ANi) content ranged from 2 to 90 w/w% and conductivities up to 124 S/cm were observable.

31.9.2 Melt Blending

The most well-known example of melt blending of CPs is that of P(ANi) in the molten state of host polymers such as PVC, nylon, poly(ethylene terephthalate glycol) (PETG), and PMMA [1173]. The molten form of the thermoplastic is blended with P(ANi) in its commercial, "Versicon" form (now from Monsanto Co., formerly from Allied Signal Corp.), which comes in fine-particulate, ready-to-disperse, doped, conductive form. Because of the high viscosities of the melt, a rheological control device known as a twin-screw extruder frequently needs to be employed for blending. Composites have thermoplastic and some elastomeric properties and conductivities of ca. 1 S/cm (the original CP is ca. 100 S/cm). Percolation thresholds are ca. 16 w/w%. Shacklette et al. [1174] describe similar work with blends of Versicon in PVC, PETG, and nylon 12, where the blends are compounded in a twin-screw extruder and then compression molded. TEMs of the composites at low loadings (0.1 w/w% P(ANi)) show ca. 50 μm particles evenly dispersed in the host polymer matrix. Percolation thresholds are ca. 10 v/v%.

A class of CPs that should be very amenable to melt blending in the inverse sense – i.e., using a melt of the CP rather than that of the host polymer – is the P(3-alkyl-thiophenes) with alkyl being an octyl or longer chain. Surprisingly, although composites of P(3-octyl-T) (m.p. ca. 160 °C) with ultrahigh MWt PE have been prepared and been claimed to show percolation thresholds as low as 0.6% v/v P (3OT) [1175] and stretching of P(3AT)/EP rubber composites to a ratio of 12 has been achieved [1176], there is apparently only one known study making use of the melt processability of P(3OT) itself: Laakso et al. [1056] describe melt processing of P(3OT) at near 200 °C with thermoplastics including poly(ethylene vinyl acetate) (PEVA), poly(ethylene butyl acrylate) (PEBA), PStyr, and PE. A special high-temperature blender cum twin-screw extruder is used. After blending, films and sheets of the composite are made using standard compression molding techniques at ca. 200 °C and 100 bar pressure. The composite films are then doped via immersion in $FeCl_3$/nitromethane. PEVA and PE yield the best composites with P (3OT), with conductivities at 20 w/w% P(3OT) loading of 1.0 and 0.04 S/cm, respectively. The composites are also apparently easily stretched, although no mechanical data are given.

Yang et al. [1177] prepared three-component composites, doped-P(ANi)/disperser/polyolefin via an extrusion process. They found percolation thresholds as low as 3 w/w% P(ANi) using the dopant bis(2-ethylhexyl hydrogen phosphate), the disperser lauryl gallate, and LDPE. A twin-screw extruder at 130–150 °C and a screw speed of 100 rpm for 10 min. were used. Films were prepared by hot-pressing at 135–155 °C and 3 tons pressure.

31.10 Interpenetrating Polymer Networks

Interpenetrating polymer networks (IPNs) are special cases of copolymers where selective cross-linking yields a composition that is an interpenetrating grid of several polymers. One of the illustrative examples of this with respect to CPs is the work of the Rubner group [1067] with IPNs of PStyr and P(T). In this work, a copolymer of 3-vinyl-hexylthiophene and 3-Br-octylthiophene was first prepared. Styrene monomer was then linked to the pendant vinyl groups of this polymer. The polymer was then swollen with solvent and styrene, and the styrene polymerized, cross-linking during polymerization. The full IPN so obtained, containing interpenetrating chains of the P(T) derivative and PStyr, could be doped with $FeCl_3$ in nitromethane after swelling in benzene. Maximum conductivities of 0.5 S/cm were obtained, and the IPN showed electrochromism closely resembling that of the P(T) derivative homopolymer.

In more recent work, Mandal and Mandal [1178] described unique IPNs of P (Py) and poly(methyl acrylate) or poly(styrene-co-butyl acrylate) which showed very low percolation thresholds for conductivity, ca. 0.023 volume fraction of P (Py).

Chiang et al. [1179] described an interesting series of IPNs with fullerenol-polyurethane elastomers interpenetrated by P(ANi). The aniline was in situ polymerized at the near-surface region of the fullerenol cross-linked poly(urethane-ether) networks. The resulting networks were claimed to exhibit conductivities of 2.6–5.4 S/cm depending on elongation, while retaining many of the mechanical properties of the parent elastomer. Liao et al. [1180] described a series of IPNs with P(ANi) interpenetrating maleimide-terminated polyurethanes. A sequential chemical polymerization procedure was used. Scanning tunneling microscopy (STM) images showed that the P(ANi), doped with dodecyl benzenesulfonic acid, was distributed as a continuous conductive network within the host. A moderate conductivity at a low threshold (percolation) concentration of P(ANi) was observed.

31.11   Fabrication OF CP-Based Fibers

Li and White [1181] have described an ingenious method of fabrication of thin fibers (10 cm long × 0.2 mm dia.) of CPs, also applicable to fabrication of CP composites. In this, the CP is electropolymerized on a very thin, long (127 μm dia., 1 mm long), Pt electrode suspended in a hydrodynamic flow stream of the polymerization solution. The flow causes a thin CP fiber to grow along the tips of the Pt electrode. CPs tested included P(3-Me-T) and P(Py). If a fiber of a host polymer such as Kevlar is placed on the point, a CP/Kevlar composite fiber results. Growth rates are up to 30 cm/h. The composites have conductivity of ca. 5 S/cm and are extremely durable and flexible, allowing repeated bending without cracking.

Andreatta et al. [1182] have given a detailed procedure for solution-phase blending of poly(p-phenylene terephthalamide) (DuPont Kevlar) with P(ANi). A 2 w/w% solution of Kevlar and a varied concentration solution of P(ANi) in conc. (98%) sulfuric acid are prepared. The two solutions are then mixed, stirred 5–12 h, and then let stand for 24 h. They are then used for wet spinning of composite monofilament fibers, coagulated in a 1 N H_2SO_4 bath. The fibers are drawn during coagulation itself, to a draw ratio of 7:10, and continuously washed with water to remove excess acid while being wound onto bobbins. The fibers show no sharp percolation threshold (maximum conductivities, ca. 10 S/cm, saturating only at ca. 50 v/v% P(ANi)) but show good mechanical properties, with Young's modulus ca. 10 GPa and tensile strength ca. 0.2 GPa at 40 w/w% P(ANi). As expected, the conductivity and Young's modulus are linearly related. In a variant of the same method, Hsu et al. [1183] have prepared P(ANi)/Kevlar fibers of 25 μm diameter with a lower P(ANi) content (ca. 1 w/w%) and conductivity but better mechanical properties (Young's modulus 62 GPa, tensile strength 2.8 GPa, vs. 76 GPa and 3 GPa, respectively, for pure Kevlar fibers).

Nemoto et al. [1184] describe a procedure for preparation of fibers of another Kevlar composite, P(Py)/Kevlar, but using the classic sorbed-monomer route. In their method, fibers are wet spun from Kevlar/pyrrole solution in DMSO into an aqueous $FeCl_3$ coagulating bath, which serves as the oxidant-cum-dopant as well. Conductivities as high as 10 S/cm are claimed for the fiber, but no mechanical data are given.

Chiang et al. [1185] describe an interesting variant of the catalytic in situ polymerization route described above for fabrication of P(Ac)/PE composite fibers. PE fibers are first spun from a PE/mineral oil solution. The fibers are then soaked in a standard P(Ac) Ziegler–Natta catalyst (Ti(OBut)$_4$/AlEt$_3$). Polymerization of P (Ac) from acetylene gas is then carried out in the standard manner. The excess catalyst and mineral oil in the fibers are extracted with toluene and HCl/methanol. The P(Ac)/PE composite fibers so obtained have conductivities of 1200 S/cm as prepared, or 6000 S/cm when drawn to a 2.2 draw ratio, and can have P(Ac) contents up to 82 w/w%. The Young's modulus, tensile strength, and elongation at break of as prepared composite fibers are 0.5 GPa, 0.1 GPa, and 170%, respectively. Unfortunately, the fibers also suffer from the usual environmental instability of P (Ac).

31.12   Bulk and Commercial Production

With regard to commercial production of CPs, apart from poly(phenylene sulfide) (P(PS)), which was produced in bulk long before the field of CPs was established and the dopability of P(PS) was understood, poly(aniline), and to a lesser extent, some of the poly(3-alkyl-thiophenes), have been the major CPs which have been bulk produced and marketed.

P(ANi) has been marketed for some time as a chemically polymerized, proprietary-sulfonate-doped CP under the tradename *Versicon*. Originally produced by the Allied Signal Corp. based in Morristown, NJ, USA, the product line was taken over in the mid-1990s by the High-Performance Materials division of Monsanto Co., St. Louis, MO, USA.

Around April 1998, Monsanto sold the product line to Ormecon Chemie, a subsidiary of Zipperling Kessler, based in Ammersbek (near Hamburg), Germany. As of this writing, this firm continues to produce bulk Versicon. The PVC dispersion, called "Incoblend," is claimed to be injection moldable. Neste Oy, Helsinki, Finland, and Hexcel Corp. also test marketed P(ANi) and P(3AT) powders; for example, one of the product lines of Neste Oy included poly(3-octylthiophene) chemically polymerized using $FeCl_3$ and with a weight-average MWt of ca. 89,000 [1186].

As of 1996, the production plant at Monsanto was still pilot scale. Typical properties of Versicon were compacted powder conductivity 4 S/cm, moisture 4% w/w, particle size 3–100 μ, density 1.36 g/cc, and surface area 5–20 m^2/g. Also, produced on a pilot scale by Monsanto was a "soluble P(ANi)," a green liquid which is a blend of 50% colloidal P(ANi), 40% xylenes, and 10% butyl cellulose. P (ANi) was at one time also marketed by AmeriChem, Inc., Cleveland, OH, USA, in collaboration with Monsanto, and by PolySciences Inc., Pittsburgh, PA, USA. Other dispersions of P(ANi) in PVC, nylon 12, and other thermoplastics were also, in the mid-1990s, in the test marketing stage in collaboration with Zipperling Kessler & Co.

Several other companies have bulk produced CPs in house, to be marketed as products: Milliken Research Corp, Spartanburg, SC, USA, P(Py)-coated textiles for microwave absorption, electrostatic dissipation, and related uses; Echema Co. has produced conductive electrodes incorporating P(ANi); Bridgestone-Seiko, Japan, and Siemens, Germany, have attempted unsuccessfully to market CP-based batteries; BASF, Ludwigshafen, Germany, also attempted unsuccessfully to market several CP (P(Py), P(ANi)) products such as films, particulate fillers, and batteries;

and Matsushita Co., Higashimita, Japan, apparently still markets P(Py)-based Al-electrolytic capacitors. This appears to be the only really successful marketing of a CP product thus far.

It is worth noting that nearly all commercial productions of CPs, whether they involve direct marketing of the CPs themselves, or incorporation of the CPs into a value-added product such as a battery, have suffered from one drawback: All of them use bulk production methods for CPs on little more than a pilot scale; these methods involve little more than direct, proportional scale-up of laboratory processes.

From the information, publicly available now, the only exceptions to this have been the following:

1. Attempts by the Kathirgamanathan group at South Bank University, London, UK, [1187] to adapt the bulk-scale PICHRE (Partially Immersed Cylindrical Horizontally Revolving Electrode) electrode used for the past 30 years in the bulk production of Cu and Ni foils to the bulk and continuous production of P (Py) film. Figure 31.35 shows a schematic of the PICHRE apparatus. In this, P (Py) is continuously peeled off (film) or scraped off (powder) as it is produced. The apparatus is also usable for production of continuous films of P(Py) on C-impregnated nonwoven polyesters [1188].
2. Work of the Naarmann group at BASF, Germany, who have also described and apparently patented a similar apparatus [1189].

31.13   Solubility and Processing of CPs

31.13.1   Truly Soluble and Processible CPs

It is pertinent to start our discussion by noting that *there are in fact very few CPs truly soluble and processible in their doped state*. Most elements of solubility and processibility are achieved through various ruses, such as dissolution in highly specialized and frequently impractical solvents such as conc. H_2SO_4 or AsF_3, processing of the soluble de-doped forms of the polymers followed by in situ re-doping, dispersion of the CPs as colloids, melt processing of powder-form CPs with thermoplastics, attachment of solubilizing groups such as urethane side chains which yield some processibility at the strong expense of conductivity and other properties, attachment of water-soluble groups such as sulfonates which yield poor solubility in the important protonated (conducting) forms of the CP, and in situ polymerization of adsorbed or otherwise preprocessed monomers. Among the very few CPs apparently soluble in their doped, conducting forms, for example, are "template-polymerized" P(ANi)/poly(styrene sulfonate) (PSSA) complexes which however have conductivity, thermal/environmental stability, and other properties seriously diminished from those of P(ANi) doped with more common dopants.

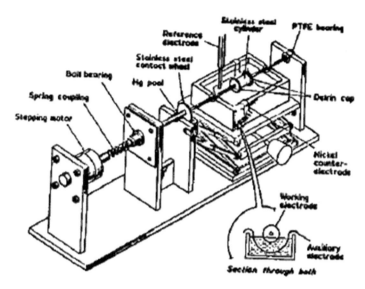

Fig. 31.35 Schematic diagram of PICHRE apparatus for bulk production of P(Py) (After Ref. [1187], reproduced with permission)

31.13.2   Most Common Processing Methods

Among the less elaborate and most common processing methods, one is still widely applied and deserves brief mention as a benchmark upon which to compare other methods described in this chapter. This method, most frequently applied to P(ANi), involves using a solution of the de-doped, neutral form of the CP (the emeraldine base in the case of P(ANi)), which is usually highly soluble in organic solvents, to prepare a coating on a desired substrate, followed by (usually chemical) doping. Thus, in the case of P(ANi), a typical procedure, after production of the conductive, emeraldine salt form of the polymer, involves reduction of this form to the reduced, emeraldine base form using a reducing agent such as hydrazine sulfate or ammonia. This reduced, highly soluble form is then dissolved in a solvent such as NMP (1-Me-2-pyrrolidinone) and the solution used to prepare a coating (some solvent is usually retained within the CP even after extensive drying). Re-doping of the coating to the conductive form of the CP of interest can then be effected in a number of ways. One relatively simple way, to cite an example, is to soak the emeraldine base for 6–12 h. in a solution of an aqueous acid. Thus 1 N HCl, 1 N H_2SO_4, and 80% HOAc solutions may be used to achieve Cl^-, SO_4^{2-}, and OAc^- doping, respectively.

Several problems have been recognized even with this simple procedure, however. Firstly, substantial retention of high boiling solvents such as NMP does occur, as documented in several studies [974]. Secondly, it also appears that the highest-molecular-weight (MWt) fractions of even the reduced CP do not dissolve and are left behind [1190]. The CP so obtained thus may not be truly representative of intrinsic behavior of the parent CP.

With regard to solvation of the de-doped forms of CPs, it may be perceived that this should be applicable to virtually all CPs as a means of processing. It is true that nearly all CPs show substantial solubility in organic solvents in their de-doped states. However, the highest MWt fractions frequently do not dissolve. Additionally, dissolution does not appear to be reproducible in all cases. For instance, if one dissolves a reduced P(Py) film in CHCl$_3$, recasts the film, and attempts to redissolve the recast film, one loses a bit of the polymer in each such cycle, and one observes substantial degradation in electroactivity and other relevant CP properties. Only a few CPs, such as long-chain alkyl-substituted poly(thiophenes) (see below), are *reproducibly* soluble in this manner.

Another very common processing method applicable to CPs with moderate thermal stability, such as P(ANi) in its de-doped (emeraldine base) and sulfonate-doped forms, is thermal processing through dispersion in thermoplastics above their melt temperature. This is discussed in more detail further below.

31.13.3   Solutions in Unusual or Difficult Solvents

Some success has been achieved in preparation of solutions of CPs *in the doped form* in unusual or very-difficult-to-handle solvents. AsF$_3$, conc. H$_2$SO$_4$, and *m*-cresol are some of these solvents. In an early study, Frommer [1191] showed that *doped* (with AsF$_5$) poly(*p*-phenylene sulfide) could be dissolved efficiently in AsF$_3$. Unfortunately, however, a film cast after solvent removal would not redissolve, i.e., the solution was not reversible.

The more common P(ANi) can be dissolved in *concentrated* (typically 97%) acids such as H$_2$SO$_4$, CH$_3$SO$_3$H, and CF$_3$SO$_3$H (methane and trifluoromethane sulfonic acids). Interestingly, P(ANi) concentrations in these solvents from a few % to 20% (w/w) are possible, and with precipitation (e.g., with MeOH or water), the P (ANi) comes out *in the conducting form* and shows some crystallinity [1190, 1192]; solutions of P(ANi) in these solvents also show bipolaron absorptions characteristic of fully oxidized P(ANi). Again, however, in spite of much work, the method has not been used practically to prepare P(ANi) films or otherwise process P(ANi), evidently because of the inconvenience of using highly corroding concentrated acid solutions.

In an adaptation of these concentrated acid methods, Pron et al. [1168] used dialkyl esters of phosphoric acid to protonate P(ANi), either in solution (solvents such as toluene, THF, *m*-cresol) or by mechanical mixing with neat diester. The resulting protonated P(ANi) was soluble *in the protonated form* in solvents such as decaline (up to 12 w/w%). Such solutions could then be used for solubilizing poly (methyl methacrylate), poly(acrylonitrile butadiene-styrene) (ABS), or other thermoplastics, and films or fibers could be cast or spun therefrom. Curiously again, however, practical applications of these promising conductive thermoplastic blends have so far not been seen.

Unusual solutions of camphorsulfonate (CSA)-doped P(ANi) in *m*-cresol which show liquid crystalline behavior have been described by the Smith group [1048]. These solutions contained P(ANi) in the protonated form, which has less inter-monomer rotational freedom due to increasing inter-monomer double-bond character and thus has a rigid conformation. The utility of this unique solution appeared to be that precipitation of the P(ANi) with acetone on a glass substrate (after shearing of the liquid crystalline solution thereon) yielded P(ANi) films which showed high orientation. In a similar procedure, Kim et al. [1193] doped P (Py) with dodecylbenzenesulfonate (using the Na salt during chemical preparation) and then found that the doped salt readily dissolved in *m*-cresol. Cast films however retained ca. 13% solvent, and conductivities were poor (ca. 10^{-3} S/cm at room temperature. There have however been no further reports of attempts to use these unique solutions for processing of P(ANi) or P(Py) for other applications to date.

31.13.4 Use of Solubilizing Agents

One of the few CP systems that may possibly qualify as one truly soluble in its more relevant, doped form is a system based on P(ANi) doped with poly(styrene sulfonic acid) (PSSA) or its salts. This and closely related systems were first described by Malhotra et al. [1066] and subsequently studied in more detail by the Yang group [1130, 1194, 1195], Angelopoulos [1012, 1196], and by others [1065].

The approach is simple: Take a highly water-soluble substance that is already a polymer, such as a "polyacid" (PSSA, poly(vinyl sulfonic acid) and poly(acrylic acids being some examples); use it as a template on which to attach monomer groups (such as aniline), usually via simple adsorption; and then polymerize the resulting monomer-enriched template. The CP then "piggybacks" on the solubility of the template, while the latter also serves as the dopant. The basicity of the monomer molecule (e.g., aniline) and the acidity of the template (e.g., polyacid) are used to assist in the initial monomer saturation. We now describe the best-known example of this, the P(ANi)/PSSA CP system.

Figure 31.36 shows the use of a template of PSSA to generate a P(ANi)/PSSA structure which incorporates strands of P(ANi) and PSSA intertwined or otherwise closely associated. The PSSA (or other polyacid template used) appears to then provide a hydrophilic outer surface to the aqueous solvent environment, facilitating solution. Such work was first described by Malhotra et al. [1066] who described the production of a dark green paste on addition of undoped P(ANi) to an aqueous solution of PSSA. On evaporation of the water from solution, a mass was obtained which had an elongation at break of 500%, a reversible-stretch ratio of 150%, and a conductivity of ca. 4×10^{-3} S/cm at 300% stretch, appropriately labeled a "rubber." The rubber could be redissolved reversibly in water, and the percolation threshold for a conductivity of ca. 10^{-3} S/cm was ca. 11.6% P(ANi).

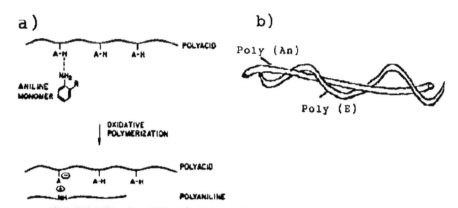

Fig. 31.36 Synthesis using solubilizing agents or templates: (**a**) use of poly(styrene sulfonic acid) (PSSA, "polyacid") for P(ANi) synthesis; (**b**) resulting intertwined strand of P(ANi)/PSSA (After Ref. [1196], reproduced with permission)

Yang et al. [1030, 1194, 1195] studied the system in more detail. In a typical synthesis effected by them, aniline was added to PSSA solution, and after stirring for 1/2 h. FeCl$_3$ and H$_2$O$_2$ (oxidants) were added, yielding a green solution. Pouring the solution into isopropanol caused precipitation of a dark green gel which redissolved in water. Dialysis to remove unreacted aniline and extraction with isopropanol to remove unreacted PSSA yielded pure P(ANi)/PSSA, which was dried to a powder. The powder could be redissolved and precipitated in water without limit, and its solution passed through 0.2 μm filter paper and withstood centrifugation at 10^{5}G. The CP system was claimed to be dopable and de-dopable in solution (with oxidant/reductant/acid), going through the normal pale/green/blue/ purple color transitions characteristic of P(ANi) and apparently showing the characteristic P(ANi) polaron bands in solution. Besides water, the P(ANi)/PSSA system was found to be soluble in MeOH, EtOH, DMF, and DMSO. At pH 7.0 in aqueous solution, the system continued in the "doped" emeraldine state, in contrast to other P(ANi)s which retain this state only at low pH. Perhaps more importantly from a processing point of view, Yang et al. claimed thin films with conductivities in the region of 10^{-2} S/cm could be cast from most solutions which were subsequently resoluble. No data on conductivity stability, electroactivity, or other properties of the films were however given. Yang et al. found that poly(2-acrylamido-2-methyl-1-propane-sulfonic acid) (PAMPS) yielded similar soluble P(ANi) complexes but poly(acrylic acid) yielded an insoluble colloidal suspension. Angelopoulos and coworkers [1012, 1196] have described similar, proprietary polyacid systems especially adapted to lithographic applications but with slightly lower conductivities. We note however that Yang et al., Angelopoulos et al., and other groups as well, nevertheless, give little data on whether it is the metal (e.g., Na) salt form of the CP that remains highly soluble or whether cast films need to be retreated with acid to yield the conducting, protonated form. Bates et al. [1197] have shown that P(ANi)/PSSA complexes are heat processible.

A popular method for effectuating water solubility in a CP, partially demonstrated for the PSSA systems above, is the affixation of the highly soluble sulfonate group to the CP backbone in some form. The two most common forms of doing this are the use of sulfonated polymeric dopants, as for PSSA above, or sulfonation of the main CP backbone itself, either at the monomer stage or subsequently. In the former category, besides PSSA, examples of sulfonated dopants include Nafion [1198] (a DuPont proprietary sulfonated fluoropolymer) and poly(vinyl sulfonic acid) [1065] and anthraquinone 2-sulfonate [*XIV.287*]. In the latter category, CP systems in which the sulfonate (or similar solubilizing groups) are part of the CP chain have been frequently called "self-doped" CPs, which, as the discussion below shows, is somewhat of a misnomer.

Apart from the template-based sulfonated systems above, very few examples abound of CP systems that may be truly soluble in doped and undoped forms. One of these is a poly(phenylene vinylene) (P(PV)) derivative reported by the Wudl group [1199]. The polymer, prepared via conventional precursor-polymer routes commonly used for P(PV)s, has protonated and metal salt forms (Fig. 31.37). Films can be cast readily from solution. The metal salt form yields films that readily redissolve in water, but films of the protonated form do not redissolve, apparently due to cross-linking. The conductivities of films of the protonated and metal salt forms are, respectively, ca. 10^{-6} S/cm and 10^{-3} S/cm.

Another sulfonated P(PV) derivative, which has an [O(CH$_2$)$_3$SO$_3$(Na/H)] group at both meta positions of every other phenylene ring in P(PV), has been described by the Reynolds group [1200]. This was synthesized via a complex route using 1,4-phenyl bisboronic acid and Pd catalysts. As is characteristic for many of the sulfonated CPs, however, it is again highly soluble and processible only in its metal salt form.

31.13.5 Soluble "Self-Doped" CP Systems

A typical example of "self-doped" CPs is shown in Fig. 31.38 [1201], produced via simple sulfonation of P(ANi) with fuming H$_2$SO$_4$. The metal (e.g., Na) salt form of this CP is highly soluble in water and can be used to cast films, which are however insulating. Conductivity and solubility data for cast films of the protonated form however apparently indicate that, while conductive (ca. 0.1 S/cm), these films do not redissolve, just as for the sulfonated P(PV) derivative (8.1) above. The "self-doped" nomenclature arises from the appearance that the sulfonate group, a typical dopant in many CP systems, lies apposite to the aniline NH group, appearing to effect doping of the latter. However, actual observation of this and other "self-doped" CPs shows that the actual dopant is in fact a metal ion (e.g., Na$^+$) or proton, which is actively transported in and out of the polymer lattice as are conventional dopants.

Other so-called self-doped CPs have been reported by several workers [1202, 1203] (Fig. 31.39). One such, poly(4-anilino-1-butanesulfonic acid) (P(ANBUS)),

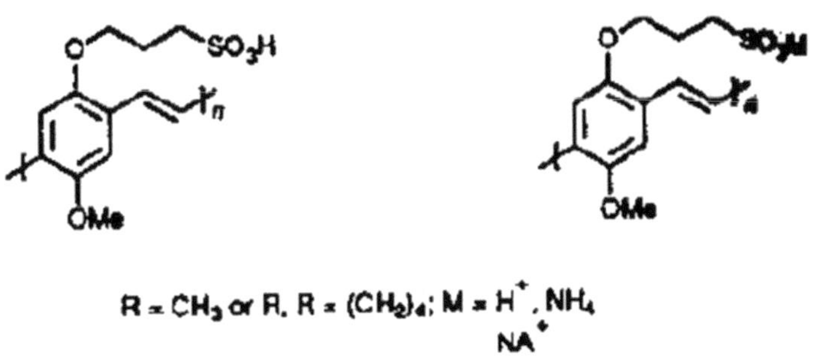
SO3H
OMe
SO3M
OMe
R = CH3 or R. R = (CH2)4; M = H+, NH4
NA+

Fig. 31.37 Protonated and metal salt forms of P(PV) derivative reported by Wudl et al. [1199]

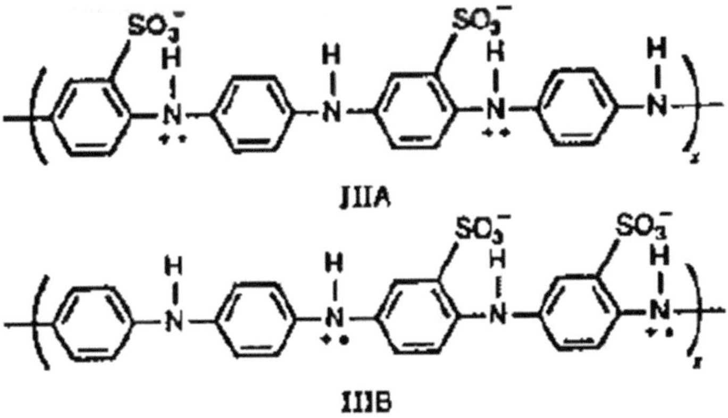
IIIA
IIIB

Fig. 31.38 Typical example of "self-doped" CP, (After Ref. [1201])

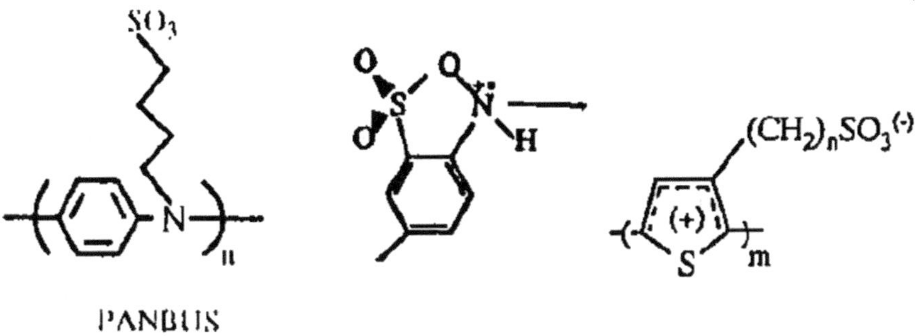
PANBUS

Fig. 31.39 Examples of other "self-doped CP" structures (After Refs. [1202, 1204])

with a sulfonate substituent at the N-atom of the aniline group (Fig. 31.39), is again highly water soluble only in its metal salt, nonconductive form [1202]. Conductivity data are not available for cast films. The Wudl and Heeger groups [1204] have demonstrated synthesis of sulfonated P(3AT)s (Fig. 31.39) which again show the characteristics of sulfonated CPs of reversible water solubility only in their metal salt (nonconductive) forms and poor conductivity of cast protonated form films.

31.13.6   Soluble Poly(thiophenes)

One of the first "soluble" CPs to be studied in detail were the poly(thiophenes) (P (T)s) substituted at the 3-position with long-chain alkyl groups. This rendered them amenable to *reproducible* dissolution (i.e., repeated solution/dissolution) in organic solvents (such as chloroform) and thus processible, *in their reduced, de-doped forms* [962, 1013, 1186, 1205, 1206]. The alkyl groups successfully implemented have gone all the way to up to octadeca, although the methyl group, as in poly (3-methyl) thiophene, remains the most common. The same procedures used for chemical synthesis of P(T)s can be used for these as well, if one starts with the appropriate monomer. For instance, 3-octylthiophene can be polymerized after iodination at the 2,5 positions, by conventional Ni-catalyzed Grignard coupling in THF [962, 1013, 1205, 1206]. The procedure can also be used to prepare mixed monomer (e.g., 3-Me- and 3-Oct-T) copolymers. Direct electrochemical synthesis from monomers was also demonstrated for alkyl chains as long as dodecyl [1205]. As expected, solubility increases as a function of increasing alkyl chain length. The de-doped CPs are soluble in solvents such as toluene, THF, and MeCl$_2$. MWts are moderate to high, from 8.3 K for Grignard-coupled P(3-Oct-T) to 150 K for the chemically oxidized P(3-Oct-T), for instance [1207]. They may be doped according to common methods, e.g., exposure to iodine vapor. Conductivities of these P(3AT)s are in the region 0.1 to 10 S/cm and apparently have only a very small dependence on alkyl chain length or dopant. Another important property of the long-chain P(3AT)s is their thermal stability and high m.p., rendering them melt processible, as discussed in a subsequent section below.

Interestingly, these poly(3-long-chain-alkyl-thiophenes) may also be doped in solution, by addition of solutions containing nitrosyl dopants such as NO-(BF$_4$/PF$_6$/ SbF$_6$) (e.g., in acetonitrile) to solutions of the de-doped CP (e.g., in toluene), yielding solution-like colloids that can be used to cast conductive films. These are however not true solutions, with colloidal aggregates visible even to the naked eye at polymer concentrations as low as 1.4×10^{-4} g/L [1186]. An important finding, however, was that doped colloidal gels obtained with high polymer concentrations in solution could be redissolved in certain solvents, such as THF.

31.13.7   Soluble Poly(Diacetylenes)

Poly(diacetylenes) (P(DiAc)) have the general formula

$$\left[= CR\text{-}C \equiv C\text{-}CR' =\right]$$

(31.1)

Substituent	Abbreviation
$(CH_2)_3OCONHCH_2CO_2C^4H_9$	P3BCMU
$(CH_2)_4OCONHCH_2CO_2C_4H_9$	P4BCMU
$(CH_2)_4OSO_2C_6H_4CH_3$	PTS12
$(CH_2)_9OCOCH(CH_3)_2$	
$(CH_2)_9OCOCH_2(CH_3)_2$	
$(CH_2)_9OCOCH_2(1\text{-naphthyl})$	

with R and R' being the substituents (side groups). While P(DiAc)s have been available since the late 1960s, their most common examples have been, like most common CPs, intractable, infusible, and insoluble.

More recent work [1208, 1209] however has shown that certain long-chain substituents render P(DiAc)s substantially soluble, in their de-doped form of course, in common organic solvents such as THF and toluene. Table 31.5 below lists some side groups (R,R' in the above formula) yielding such soluble P(DiAc)s. Once in solution, these can then be processed in the conventional manner, e.g., cast as films which are subsequently doped with I_2 vapor. Handling under inert atmosphere is however a strong recommendation for producing stable doped P(DiAc)s.

31.13.8 Use of Innovative Substituents for Achieving Greater Solubility of CPs

It is evident that the use of solubilizing groups, such as the long alkyl chains for poly(thiophenes) or the long-chain substituents for the P(DiAc)s described above, can render some measure of solubility or melt processability to CPs. Indeed, in one such example, Liu and Gregory [1211] describe the chemical polymerization of a 3-urethane-substituted thiophene, to yield a P(T) soluble, albeit *only in its de-doped form*, in common organic solvents such as chloroform as well as polar ones such as DMF and NMP. The CP had a glass transition temperature in the region of 60–70 °C. The CP could be blended in solution with common polymers such as polyurethane, polyacrylonitrile, polycarbonates, and polyesters. Cast films of the CP alone or of these blends could be doped with I_2, $FeCl_3$, etc.; conductivities of the CP films were ca. 1 S/cm.

It is important to note however that not all attempts at incorporation of solubilizing groups have been successful. For instance, Stanke et al. [1070] reported a very novel polymerization of a pyrrole with an ethyl methacrylate substituent, leading to a poly(pyrrole)-poly(ethyl methacrylate) (PEMA) (Scheme in Fig. 31.40). As PEMA is a highly soluble polymer processible from solution as well as from melt, it was expected that the resulting polymer would show some processibility or solubility. However, it was found that although it did form

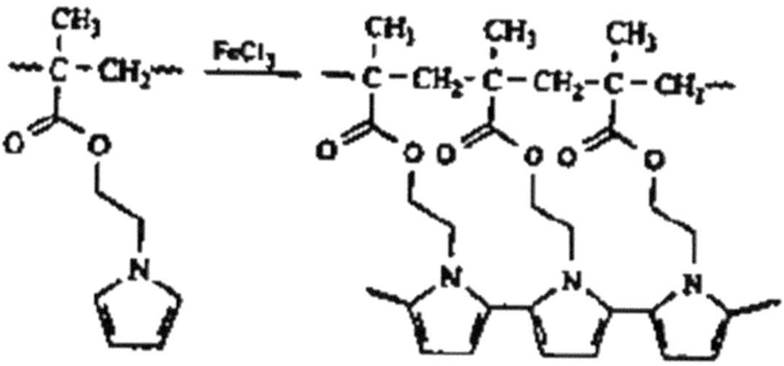

Fig. 31.40 Scheme for novel synthesis of ethyl methacrylate-substituent-P(Py) (After Stanke et al. [1070])

a true solution, upon precipitation or film casting (with conductivities on doping in the 10^{-5} S/cm range), the polymer cross-linked and became intractable, i.e., solubility was not reversible. The same was found true for melt processibility, although the glass transition temperature, at ca. 68 °C, was lower than that of PEMA itself.

31.13.9 Melt and Heat Processing of CPs

If a CP has reasonable thermal stability, e.g., to 140 °C without significant conductivity loss, a rather simple procedure for processing it into composites, laminates, or other conductive materials is to simply blend it with thermoplastics, either in the dry, powder form or as a solution. The thermoplastics can then be molded, extruded, etc., in the customary manner, yielding a formable, conductive material. As an example, such procedures have been applied with some success to P(ANi) and the thermoplastics poly(vinyl chloride) (PVC), nylon, and poly(ethylene terephthalate glycol) (PETG). Percolation thresholds of the CP in these thermoplastic matrices, yielding appreciable conductivities of ca. 20 S/cm, have been claimed to be as low as 30 v/v% [1173]. Apart from thermal stability of the CP, the key to successful CP processing in this manner appears to be proper, homogeneous dispersion of the CP in the thermoplastic.

Employing a similar molten-state dispersion procedure, Shacklette et al. [1174] have described the production of dispersions of P(ANi), in the commercial "Versicon" form supplied formerly by Allied Signal Corp. and now by the Monsanto Co. Dispersions of the P(ANi) are made in the melt of thermoplastics such as PVC, PETG, and nylon 12, with intense mechanical mixing used. The dispersion so obtained can then be used for compression molding in a conventional hot press. The percolation threshold appears to be ca. 12% (w/w) for all the thermoplastics tested, with conductivities of the dispersions at this threshold being in the 10^{-2}–1 S/cm range. The key to the apparently successful procedure is the thermal stability of the CP – in this case, the Versicon P(ANi), incorporating sulfonate dopants, is stable above the melt processing temperatures used.

Long-chain P(3AT)s, discussed above, have substantial thermal stability and moderately high m.p.s in their undoped states, rendering them melt processable. For instance, P(3AT)s with butyl, hexyl, and octyl substituents have m.p.s of 275, 190, and 150 °C, respectively [1212]. Common melt processing methods such as compression molding, extrusion, and fiber drawing/spinning can thus be applied to them. Melt extrusion and melt spinning of fibers [1212] and compression molding [1056, 1207] have indeed been demonstrated for several long-chain P(3AT)s. Evidently also, they can quite easily be blended with common thermoplastics such as PVC, PStyr, PE (polyethylene), and PEVA (poly(ethylene vinyl acetate)), to prepare various types of "composites" (blends really), dealt with elsewhere in this book. It is important to note however that these CPs are *not* melt processable in their doped form, where loss of dopant or other decomposition occurs at the temperatures required.

Among P(3AT)/thermoplastic blends, P(3-octyl-T)(P(3OT))/PEVA has been the most studied [1056, 1213]; the percolation threshold for appreciable (ca. 0.1 S/cm) conductivities for this has been as low as 20 w/w% P(3OT) [1056]. Laakso et al. [1056] described a detailed procedure for melt processing of P(3OT), having a m.p. of 160 °C, with thermoplastics such as PEVA, PE, and PStyr, using a commercial melt blender. Doping of the blends could then be carried out with iodine vapor or in nitromethane solution with $FeCl_3$. In spite of much promising work, however, actual commercial applications of such thermoplastic blends have still not appeared to date.

In more recent work, a composition which was claimed to be fusible and melt extrudable was developed in a joint effort of UNIAX Corp., Santa Barbara, CA, USA, and Neste Oy, Porvoo, Finland, given the tradename Panipol [1214]. In its typical preparation, a sulfonated P(ANi) salt, a plasticizer, and other "optional additives" are fed into a twin-screw extruder with a typical temperature profile along the screw of between 100 and 180 °C and residence time 5–10 min. The resulting conductive strands are cooled, pelletized, and packaged. The conductive pellets are then compounded for such end uses as injection molded articles or extruded films and fibers, or any other conductive thermoplastic products. The authors also claimed to have demonstrated conductive thermosets based on solution-processed Panipol.

31.13.10   Processing Using Colloidal Solutions

Another important way of getting around the comparative insolubility of CPs for processing applications is the use of colloidal solutions, i.e., sols. For this purpose, simple suspensions of fine-particulate CPs in inert solvents are of course usually not sufficient; they may simply precipitate out very rapidly or aggregate into unwieldy "macroparticulates" that lead to inhomogeneous coatings or lumpy blends with thermoplastics. Stabilization of the sols is desired, through various means, such as use of additives ranging from "film-forming materials" for P(ANi) [1215] to surfactants. We first describe here one of the best-known examples of a practical CP sol, applicable to P(ANi) and P(Py), and then dwell on several other examples of CP sols.

Bjorklund and Liedberg [1216] and Armes et al. [1049, 1217] have independently described apparently successful and practical methods for dispersion of P(Py) and, later, P(ANi) as sols in aqueous media. In Bjorklund et al.'s approach, pyrrole is oxidized (by $FeCl_3$) in an aqueous solution of methylcellulose (MeCel, MWt ca. 100 K). Washing with water at 70 °C, at which the MeCel gelled, enabled removal of unreacted materials. For very low Py concentrations, P(Py)-embedded-MeCel is seen by SEM, but at higher Py concentrations, the product is indistinguishable from P(Py). In place of MeCel, other water-soluble polymers such as poly(vinyl alcohol) (PVA), polyacrylamide, and poly(vinyl pyrrolidone) (PVP) are also usable, but are not as easy to wash in hot as the inverse-temperature-solubility MeCel. Most importantly, the sol could be used to prepare thin coatings of conductivity ca. 0.2 S/cm, but data on redissolution of the coatings were not provided. Similar syntheses of sols have been described by several other workers, for instance, the P(ANi)-poly(acrylic acid) (PAA) gel described by the Yang group [1030]; in this, a 90 K MWt PAA is used, and $FeCl_3 + H_2O_2$ is used as the oxidant, with fibrillar particles of ca. 50 nm dia. × 200 nm length obtained as product.

In the slightly more detailed work of Armes et al. [1049, 1217], particle morphology was characterized in detail for the similar polymerization of Py with $FeCl_3$ oxidant in aqueous solutions of PVP- and PVA-co-acetate (MWt greater than 40 K and 95 K, respectively). Poly(ethylene oxide), poly(acrylic acid), and several other polymers failed to yield usable sols. The P(Py)/"stabilizer" (PVA, PVP) particles had a spherical geometry with particle sizes of 50–300 nm not subject to aggregation. The stabilizer content was in the region 7–15% (w/w), and the authors believed that the particles consisted of a CP core surrounded by stabilizer which prevented aggregation. In a variation of this CP sol production technique, Armes et al. also described use of 20 nm silica particles in place of the polymeric "stabilizers," during chemical polymerization of P(ANi). With the correct concentration and other polymerization conditions, as the P(ANi) precipitated it was adsorbed onto the high-surface-area silica particles, yielding spherical colloidal particles of ca. 200–300 nm diameter which transmission electron microscopy (TEM) clearly showed as aggregations of smaller (ca. 30 nm dia.) P(ANi) particles around a (hidden) silica core.

In all the above cases, the sols obtained display a translucent nature due to the small particle sizes usually obtained.

Describing the adaptation of a room-temperature sol-gel technique, Mattes et al. [1218] demonstrated the synthesis of silica glasses containing P(ANi) and P(2-Et-ANi). In this work, silica gels of tetra(methoxy/ethoxy) silane (TMOS/TEOS) in aqueous HCl (catalyst) with a small amount of 1-Me-2-pyrrolidinone (NMP) were used as the starting silanol solution, added to P(ANi) or P(2-Et-ANi) solution in NMP. After filtration, the solution was set aside for several weeks for gelation. The utility of these unique CP-impregnated silica glasses was thought to be in nonlinear optical applications where transparency of the rigid host matrix to high intensity radiation is important.

Defieuw et al. [1219] describe the use of the naturally occurring sulfonated polymer lambda carrageenan as a polymeric dispersant (stabilizer) for the preparation of sols of poly(isothianaphthene) (PITN), in a procedure very similar to that used by Bjorklund et al., Armes et al., and others, described above. A PITN solution in sulfuric acid is added with stirring to a ca. 2 w/w% solution of lambda carrageenan, with subsequent dialysis used to remove unreacted materials. The sol so obtained could be used to coat a polyester (PET) film substrate, yielding a coating with a surface resistivity of ca. 9 Ω/square. The sulfonate groups in the carrageenan also served as dopants, maintaining the CP in the doped state, in which P(ITN), uniquely, is translucent. The coating could thus be used as a translucent antistatic coating on polyester.

Again, using a very similar chemical-polymerization-within-aqueous-polymeric-stabilizer procedure, Wiersma et al. [1220] describe a procedure for the preparation of P(ANi) and P(Py) colloidal suspension within polyurethane dispersions. The procedure follows the now-familiar technique of adding monomer and oxidant (Fe(NO$_3$)$_3$ in this case) solutions to the ready-made polymeric stabilizer (polyurethane) dispersion or solution in water. In this case, rather than using dialysis, centrifugation of the colloid followed by redispersion in water is used. The sol obtained can then be used to successfully coat various woven textiles.

31.13.11 Processing Using Soluble Precursors

Processing routes using a soluble, processible precursor which is then converted to the final CP are common. Some of the best-known examples involve P(Ac) and P(PV), P(TV), P(PO), P(PPS), and related polymers.

31.13.12 Preparation and Processing as Fibers

Processing of synthetic and natural polymers into fibers is an old and well-established science, and workers in the CP field have found that CPs can also be processed using these conventional techniques. We may commence our treatment of this field by very briefly outlining conventional fiber processing techniques and describing typical fiber processing methodology for prototypical CPs. We may then discuss the more involved aspects of CP fiber morphology.

Fibers are processed from polymer solutions or melts either via *wet spinning* or *dry spinning*. Typical solution viscosities for CPs are 0.5–1.5 dL/g. In wet spinning, the solution/melt is forced through one or more holes in an object called a jet or spinneret, into a suitable coagulating bath. The threadlike strands so obtained are called filaments. In dry spinning, the solution/melt is forced through the spinneret into a drying chamber where the solvent evaporates (or the melt cools). Variations to these techniques applied to CPs include polymerization of a precursor, CP, while being spun, in the coagulating bath or drying chamber. Fibers may then be *drawn*, to enhance crystallinity, order, or strength, either during or after spinning. Spinnerets may have tens of thousands of holes, producing multifilaments. Higher MWt polymers produce better fibers. The *draw ratio* (i.e., stretch ratio) is the multiple to which the fiber is subsequently elongated, usually less than 10 for CPs. Fibers are collected as they are produced on a bobbin or other collector. CP fibers are usually produced in the undoped form, which is soluble or melt processible. They may then be doped, annealed, or otherwise further treated. CPs are also produced in fiber form as composites or blends with thermoplastics or elastomers, a topic discussed in a later chapter in this book. It is important to note that although drawing at increasing ratios greatly improves conductivity and mechanical properties, it does not appear to greatly alter the overall microscopic morphology of the CP. For example, in threefold stretched P(Ac), it has been observed that the cross-linked/fibrillar morphology is not significantly changed [880].

We may now cite a few, detailed preparations and drawing procedures for representative CPs. P(3AT)s, in their undoped form, having MWts (M_w) typically 50,000–200,000, may be processed into fibers typically via wet spinning from 10 w/w% chloroform solutions into a coagulating bath of acetone. Typical solution viscosities are 0.6–1.1 dL/g; typical extrusion speed, 0.2 mL/min; and spinneret hole diameter, 0.4 mm [1221]. After winding on bobbins, the fibers may be dried in vacuum and then doped. P(3AT)s are typically not drawn (stretched) during spinning. Subsequent drawing is carried out in a tube furnace (e.g., at 105 °C) by varying the relative speed of the feed and take-up spools (up to ratios of ca. 10). P(ANi) fibers may be prepared using a similar procedure: Typically, a 20 w/w% solution of the emeraldine base form in 1-methyl-2-pyrrolidinone (NMP) is wet spun into a coagulating bath which can be a water/NMP mixture, yielding undoped fiber, or an HCl solution, yielding Cl$^-$-doped fiber directly. These fibers have diameters in the 30–70 μm range and contain some retained NMP, which acts as a plasticizer facilitating further drawing but is detrimental to properties such as conductivity or crystallinity and must be dried off. P(ANi) fibers have also been wet spun from conc. H_2SO_4 solution [1190]. This somewhat impractical procedure uses 20 w/w% solutions of emeraldine base in 86% H_2SO_4, extrusion at 60 °C, and wet spinning into a cold-water bath, yielding monofilaments of diameter ca. 300 μm.

Yoshino et al. [1222] describe a procedure for dry spinning of poly(3-dodecyl thiophene) fiber directly from the melt at ca. 250°–300° C, yielding enhanced conductivities (ca. 50 S/cm) over those of films. Tokito et al. [902, 1223] have described a singular method, a variant of the dry-spinning method, for preparation of fibers of poly(thienylene vinylene) (P(TV)) and poly(dimethyl-*p*-phenylene vinylene) (P(DMPV)) from methoxy precursors which on heating to ca. 240 °C eliminate methanol to yield P(TV) and P(DMPV). The precursor polymer fiber is first produced via wet spinning, typically from 7 w/w% solution in CHCl$_3$, and pumping through a single-hole spinneret of 0.5 mm dia. at 0.013 mL/min into a

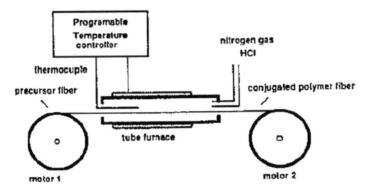

Fig. 31.41 Schematic of apparatus for simultaneous drawing and conversion of precursor polymer fiber (After Ref. [1223], reproduced with permission)

hexane coagulating bath [1223]). As seen in the schematic in Fig. 31.41, this precursor polymer is then simultaneously converted to final polymer as well as doped (with Cl^-) in one step. The draw ratio of the fiber is adjusted via the relative speeds of motors 1 and 2. Final fiber widths as large as 50 μm (for 16-fold stretching) are obtained.

Mattes et al. recently described a procedure [1224] wherein high MWt ($M_n > 30$ K, $M_w > 120$ K) P(ANi) in the emeraldine base form was dissolved in organic solvents for times sufficient to dry–wet spin solid fibers. A gel inhibitor was used in the solution, apparently disrupting H-bond formation between P(ANi) chains and thereby lowering solution viscosity and increasing gelation time. Doping with acids such as benzene phosphinic was found to preserve mechanical properties of drawn and stretched fibers, which were claimed to show better mechanical properties than conventionally drawn P(ANi) fibers. Figure 31.42 shows typical stress/strain curves for these fibers. Comparative mechanical properties of the 4X stretched P(ANi) fiber with Kevlar were as follows (Kevlar in parentheses): Young's modulus 2.21 (114) GPa; specific Young's modulus 27.1 (890) g/d; tenacity 0.77 (22.0) g/d; and elongation at break 6.04% (1.90%).

McBranch at Los Alamos National Laboratory, Los Alamos, NM, USA, described production of novel electrically conducting, *hollow* fibers of P(ANi), of diameters in the region of several μm, on an industrially meaningful scale; fibers of length greater than 1 m were shown during a presentation [1225]. Production of the fibers was enabled through improvements in solubility and processing of the high MWt form of the CP.

31.13.13 Preparation and Processing as Oriented Films

Oriented *films* are processed in a somewhat analogous manner to fibers, with subsequent stretching (orientation) of the films in particular directions through

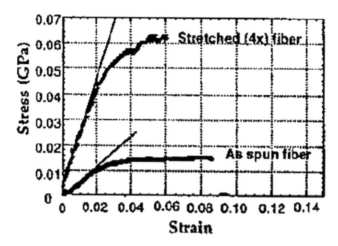

Fig. 31.42 Typical stress–strain curves for emeraldine base fiber samples (After Ref. [1224], reproduced with permission)

variable speed rollers. Films are prepared by conventional polymer casting methods (dip and spin coating). They may then be further oriented or stretched. To describe a typical film preparation [1065], that of P(ANi), a film of the emeraldine base form of the polymer, which is soluble, is cast from solution in NMP. The as-cast film retains about 15 w/w% NMP, which functions as a plasticizer. This film is then stretch oriented (stretch ratios typically 5:1) by passing it, at ca. 140 °C, between two (Teflon-coated) rollers rotating at different speeds. Quite evidently, this is a *uniaxial* orientation of a film, and films may be subsequently oriented along another axis, usually perpendicular to this one (biaxial orientation).

Concomitant with improvement of crystallinity parameters such as sharpness of diffraction pattern and coherence length with increased drawing or stretching of CP fibers and films is an enhancement of the conductivity. This is seen, for instance, in the representative data for poly(thienylene vinylene) (P(TV)) in Fig. 31.43a. Results for nearly every other common CP, e.g., P(Py) [1226], are similar, with increased conductivity leveling off at draw ratios of ca. 5.0.

It is quite evident that drawn fibers or films will display anisotropic conductivity, with conductivity parallel to the stretch direction ($\sigma(\parallel)$) higher than that perpendicular to it ($\sigma(\perp)$). Generally, conductivity increases with stretch ratio to values of the stretch ratio of about 5, after which it levels off. Figure 31.43a, b above shows typical dependencies of conductivity on draw ratio for two CPs. Anisotropy of conductivity and its relation to fiber draw ratio is most dramatic for certain forms of P(Ac). For instance, for stretch ratios of 3.0, 5.0, and 20.0, for Shirakawa, solvent-free, and Durham P(Ac), respectively, values of the anisotropy ($\sigma(\parallel))/(\sigma(\perp)$) obtained are 16.0, 98.0, and 100.0, respectively [882], again indicating a leveling off after a draw ratio of ca. 5.0. Evidently also, improved conductivity and mechanical properties go hand-in-hand, as shown below.

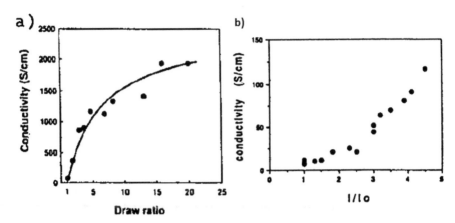

Fig. 31.43 (**a**) Relationship between draw ratio and conductivity of poly (thienylene) (P(TV)) (After Ref. [1223], reproduced with permission). (**b**) Conductivity of the HCl-doped P(ANi) drawn fibers vs. draw ratio ($1/1_o$) (After Ref. [903], reproduced with permission)

31.13.14   Processing Using Langmuir–Blodgett Films

A rather unique method of processing CPs is in the form of Langmuir–Blodgett (LB) films, and many studies have been carried out to date in this area. Before describing some of these, however, it is pertinent to note that they appear to be primarily of scientific interest, as the exact practical utility of such LB films is still unclear. Although many claims of greater molecular order, greater conductivity, and other such properties have been claimed, these have not been substantiated by the studies thus far.

The LB technique has conventionally been used to prepare monomolecular single-layer and multilayer films of surface-active chemical species, such as phospholipids or fatty acids. Conventionally, the films are first prepared on surfaces of highly purified water and then transferred therefrom to substrates of interest. Among common transfer techniques used are the vertical-transfer "Y-type" (vertical up- and downstrokes, commonly used for hydrophobic species), "X-type," and "Z-type" (vertical up- and downstrokes only), and the horizontal-transfer "lifting" technique (where the surface-active species simply adheres to the horizontally placed substrate).

The Shimidzu group [1227] has been active in the field of CP LB films and has demonstrated a number of novel approaches, one of which, for the production of P (Py) LB films on ITO substrates, is shown in Fig. 31.44. In this, electrochemical polymerization of P(Py) starts in the solution-immersed portion of the electrode, where it is disordered, but then progresses upward into the unimmersed portion of the LB film of Py monomer, where electropolymerization leading to a highly ordered LB film of P(Py) occurs (observable by a remarkable color change). In another unique approach of Shimidzu et al., a multilayer structure consisting of mixed monolayers of Py monomer and a surface-inactive hydrocarbon "spacer,"

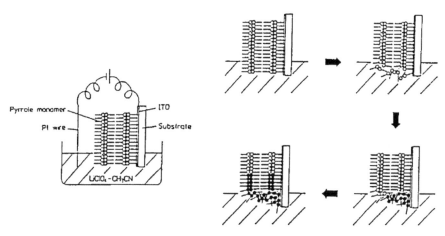

Fig. 31.44   Technique for synthesis of Langmuir Blodgett (LB) films of P(Py), (After Shimidzu et al., After Ref. [1227], reproduced with permission)

octadecane is first constructed. A Y-type transfer to an electrode followed by electropolymerization. The Py monomers used are amphiphilic pyrrole derivatives such as octadecyl 4-Me-pyrrole-3-carboxylate. A somewhat similar approach has been adopted by Rubner and coworkers [1228] to fabricate alternating multilayers of a surface-active but nonconducting pyrrole derivative (3-octadecanoyl pyrrole) and pyrrole on an $FeCl_3$ subphase, which then served to oxidize the monomers as well as dope the P(Py) to the conducting state, thus yielding alternating conducting/insulating polymer layers.

In yet another approach, Shimidzu et al. have shown that partially soluble surface-active CPs can be directly manipulated via horizontal lift LB techniques to construct 200-layer multilayer structures. Among such CPs are poly (2-octadecoxyaniline) (where the ortho octadecoxy substituent renders partial hydrophobic solubility) and poly(3,4-dibutyl pyrrole) doped with perfluorooctanoic acid (which imparts surface activity to the polymer). The multilayer LB films so obtained are then made conductive via standard I_2-vapor doping. Other attempts to manipulate CP films directly have yielded results with much painstaking effort, for instance, the difficult preparation of LB films of poly(3-dodecyl thiophene) (P (DDT)) films [1229]; using P(DDT)/chloroform solution poured onto an aqueous subphase yielded poor films using vertical lifting techniques and acceptable films using horizontal lifting techniques with some difficulty. Another example of direct lifting of CP films is the production of 6-nm-thick P(ANi) LB films from acidified solutions of emeraldine base in mixed NMP/CHCl3 solvent reported by Goldenberg et al. [974].

In a precursor-polymer synthetic approach [1230], amphiphiles which formed charged complexes with precursor polymers of poly(p-phenylene vinylene) (P (PV)) and poly(thienylene vinylene) (P(TV)) were spread onto solutions of these precursor polymers. This unique complex is then transferred to an appropriate substrate using standard LB techniques and then converted to P(PV) or P(TV) via heat treatment. In a similar but rather novel, in situ polymerization approach from the Rubner group [1231], LB films of ferric stearate are exposed sequentially to HCl vapor (generating $FeCl_3$ oxidant) and pyrrole monomer, yielding conductive LB P(Py) films.

It may be noted that many of the above methods use a *"piggyback" approach*, where the CP, if surface inactive, is used in its monomer or precursor form to piggyback on a conventional, LB, surface-active species. In yet another example of such an approach, an acetate-substituted P(T) (poly(thiophene-3-acetic acid)) is complexed to a monolayer of the common surface-active molecule dimethyl distearyl ammonium chloride, with the complex then transferred to an appropriate substrate by vertical lifting, and then doped to yield a LB CP film [1227]. Other examples of this generic procedure include the production of LB films from emeraldine-base-form P(ANi) piggybacked on stearic acid in aqueous media [1231]. Rubner et al. [1231] have carried this generic procedure one step further, by simply physically mixing a surface-active molecule (such as the common stearic acid) with a soluble CP (such as a P(3AT)) and then manipulating the resulting LB monolayers via customary techniques.

Among techniques used to characterize such LB CP films have been low-angle X-ray diffraction, comparative transmission-mode and reflection-mode IR spectroscopy, near-edge X-ray absorption fine structure (NEXAFS), and of course, conductivities, determined by transfer of LB films onto special four-electrode substrates. As expected especially for the LB CP films which include alternating nonconductive surface-active molecular layers, conductivities of such films are highly anisotropic. For example, the conductivity of Shimidzu et al.'s poly (3-octadecyl 4-Me-pyrrole-3-carboxylate)/octadecane alternating layer film, described above, is 10^{-1} S/cm in-plane and 10^{-11} S/cm across the film, yielding a conductivity anisotropy of 10^{10}.

LB CP films can be used to study many interesting CP properties such as charge transport, effect of nanoscale film thickness, and dielectric behavior. The presumed high molecular order, conductivity anisotropy, alternating conductive/insulating layer, and other properties of LB CP films have appeared to excite many workers in the field into finding unique applications for them, but these have still unfortunately been limited. Among the few active applications studied have been that of field-effect transistors (FETs). For example, FETs comprised of poly (3-hexylthiophene)/arachidic acid alternating layers have been constructed, with CP film thicknesses used being as low as 3.7 nm (a single alternating monolayer) [1227] and channel widths and lengths of 8 cm and 5 μm, respectively. The comparative advantage of such nanolayer FETs over macroscopic FETs also constructed from CPs has however not been clearly demonstrated. It has been proposed that the use of an insulating monolayer in chemFETs may be beneficial. Envisioned sensor applications of LB CP films have also not heretofore materialized. LB CP films have been studied for nonlinear optical applications, where it has been suggested that such thin films could show unique properties. While high third-order nonlinearities have been demonstrated, e.g., for LB P(DDT) films [1228], their fragility in the construction of rugged NLO devices, e.g., in optical computing, is an apparently serious drawback.

31.13.15 Processing Using Direct Vapor Deposition

Direct vapor deposition of CP films is illustrated by the work of Plank et al. [1233] who deposited P(ANi) films on substrates such as Cu(110) and Au in ultrahigh vacuum (10^{-8}–10^{-10} Torr). The source was a resistively heated quartz cell filled with emeraldine base powder. Thicknesses were controlled via control of pressure. Thin films, ca. 10 nm thick, doped with HCl were said to exhibit plasma frequencies in the far-IR, suggesting a high degree of ordering and associated high conductivity, while thicker films, ca. 100 nm thick, showed apparently lower conductivity.

31.13.16 Nanoscale Processing Using Other Innovative Methods

Zhou and Wipf [1234] described a singular method for "nanoscale" deposition of conducting P(ANi) patterns using the so-called "scanning electrochemical microscope" (SECM). Figure 31.45a summarizes this method, while Fig. 31.45b shows a typical result. Substrates used included Au, Pt, and C. The mechanism of deposition was a local pH change caused by proton reduction at the electrode tip. A 3 μm resolution was claimed. Conductivity of the resulting patterns was verified by SECM imaging.

Jang et al. [1235] described the production of "ultrafine" CP + graphite nanoparticles, of dimension as small as 2 nm. However, the precise use these would be put to was not immediately apparent from their work.

31.14 Problems and Exercises

1. Outline the plausible electrochemical polymerization mechanism for the monomer azulene (structure below). Outline a possible chemical polymerization method for this monomer, and identify mechanistic differences between the chemical and electrochemical methods.
2. Explain the term "$E(C_2E)_nC$ mechanism." What are the possible methods for termination in an electrochemical polymerization? What factors would determine the threshold concentration in an electrochemical polymerization?
3. Draw rough schematics of the first five cyclic voltammograms during electropolymerization of a typical CP by cycling.

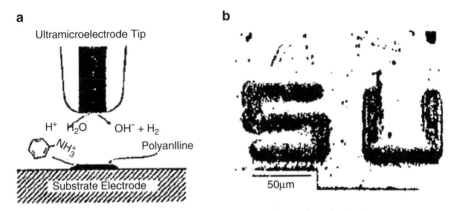

Fig. 31.45 (**a**) P(ANi) deposition using the "scanning electrochemical microscope"(SECM). (**b**) Typical result (After Ref. [1234], reproduced with permission)

4. Identify six common and three uncommon dopants, the CPs they would most likely be applied to, and the method of effecting doping of CPs with them.
5. List four factors that may be used to optimize chemical polymerizations of CPs.
6. Give as complete description as possible of one typical chemical synthesis for each of the following: P(Ac), P(DiAc), P(ANi), P(Py), P(3MT), P(PV), P(PS), and P(PP).
7. Give two examples (including procedures) for each of the following: chemical and electrochemical doping. Give one example (including procedures) for each of the following doping methods: ion implantation, photochemical doping, heat treatment, and solution doping.
8. Discuss two examples of template-based polymerizations, and outline how they differ from "conventional" chemical polymerizations.
9. Outline a procedure for a synthesis of nanotubules of P(ANi) in a polycarbonate matrix.
10. Give two examples of true copolymerizations of CPs, and clarify how they differ from CP composites and blends. Include as much detail of methodology as possible.
11. Design, in rough schematic, a flow-cell (flow-through) configuration for continuous electrochemical synthesis of a typical CP, such as P(Py), in a two-electrode mode. Estimate throughputs, production capacity, and likely doping level.
12. Why is the term "soluble and processible CPs" frequently a misnomer in describing many such CP systems? What do you think are the features of doped CPs that prevent their ready solubility and processibility? List all the methods gleaned from this chapter that circumvent these problems indirectly (such as in situ polymerization) and directly (such as solution in unusual solvents).
13. Identify two truly soluble CPs and describe their synthesis briefly. Describe processing methods used before the advent of the so-called processible CPs.

14. What are the properties of the sulfonate group that make it so popular for use in modification of CP systems for enhanced solution and melt processibility?

15. Cite one example each from P(3AT)s, P(DiAc)s, and other CPs with solubilizing groups, which you feel is the most processible CP in that category in both solution and melt. With reference to a rudimentary organic synthesis textbook, roughly sketch a synthesis which would attach a substituent group that might confer high solubility to P(ANi) in organic solvents.

16. Cite one example of a CP solution in three different "difficult solvents." Why do you think m-cresol may be classified as a "difficult" solvent?

17. What are the two keys to obtaining successful, well-adherent coatings on textile fibers during in situ polymerization of P(Py)? Using the textile chemical polymerization and the dual-spray (oxidant/monomer) polymerization as examples, devise another in situ polymerization methodology which might be successfully applied to large plastic sheets, such as those of a polyester like PET.

18. Mention three CPs that show good amenability to melt processing, and identify two thermoplastics that they form the most highly processible melts with.

19. What do you feel are the two most important keys to obtaining a stable colloidal solution (sol) of a common CP? Devise a process for making a stable P(3-MeT) sol in aqueous and organic media.

20. Outline three distinct methods of making LB films of CPs, with specific examples and as much detail as possible.

21. Define the terms composite, blend, host polymer, matrix, and percolation threshold. If the density of a host polymer matrix alone is 1.3 g/cc and 2.5 g/cc at atmospheric pressure and at 1 KBar, respectively, what is the maximum possible percolation threshold, in w/w%, in this host polymer for a CP with density 1.25 g/cc? Identify approximate percolation thresholds in all relevant figures cited in this chapter.

22. Outline the five major types of synthesis of CP composites, and give one detailed example of each. Which are more likely to give each of the following: monophasic blends, CP-coated nanospherules, and simple biphasic admixtures?

23. What kinds of characteristics and affinities (e.g., hydro(phil/phob)icity, void space) in a host polymer do you expect to be more important in determining sorption of $FeCl_3$ and Fe(III)-tosylate (as oxidant-cum-dopants) and sorption of Py, Ani, and 3-MeT monomers? What physical parameters would govern sorption efficiencies?

24. 200 mL of a $FeCl_3$ solution in a hypothetical protic solvent has a redox potential of +0.4 V vs. SCE. Using the Nernst equation, compute the redox potential as the solvent slowly evaporates from the solution (e.g., at 100 mL, 50 mL, 10 mL, and 2 mL).

25. In catalytic in situ polymerization of P(Ac) composites, which parameters do you feel are responsible for the production of homogeneous composites, as in the case of the P(Ac)/PVB composites of Sariciftci et al. [887], as against the

poorly structured, phase-separated PB/P(Ac) composites of Rubner et al. [1017, 1153, 1227], both described in this chapter?

26. Cite in detail one example each of what you feel are the most facile procedures for production of CP composites via blending in solution and in melt.

27. Outline in detail a procedure for the fabrication of a CP/Kevlar film stretched to a draw ratio of 4.0 using as bases the syntheses of Andreatta et al. [1182] described in this chapter.

28. Identify three points of difference between a true copolymer and a CP composite. From the structures and descriptions in this chapter, what is the difference between a "graft" and an ordinary copolymer, as it pertains specifically to CPs? List the distinguishing feature of an IPN as it pertains to CPs.

29. What do you feel are the reasons that the copolymer P(3MeT)/poly-[Ru $(bp)_2(pmp)_2](ClO_4)_2$ has poor conductivity?

30. The first CP copolymerization that one might think of, those of derivatized and non-derivatized monomers, has not been or has been unsuccessfully pursued for various reasons. Among these one might cite, e.g., N-Me-Py and Py, 3-alkyl-T and T, and o-alkoxy-ANi and ANi. What might be the reasons that make such copolymers either unviable or difficult to study when compared to the homopolymers? (Hint: Some factors could include electronic structure and redox potential.)

31. Without looking at Part VI of this book *(CPs–Applications)*, identify one important potential use each for CP composites in freestanding film, filler, coating, and fiber form.

Chapter 32
Structural Aspects and Morphology of CPs

Contents

32.1 General Considerations in Morphology

32.1.1 Idealized and Real Structures, Chain Defects, and Order

We must commence this chapter, which attempts to deal with the morphology and structural aspects of CPs, by reiterating the caveat cited in an earlier chapter regarding CP structure. In essence, the idealized, extended 1D structures of theoretical treatments are valid almost nowhere in the practical world of CPs. We may visualize this again by reciting a schematic structure (Fig. 32.1), where it is seen that short extended chains constituting the core of the CP are woven into fibers or globules, which are in turn constituents of larger fiber or globule bundles.

One of the first things that comes to mind then in comparing idealized with real polymer structures is that as opposed to the idealized head–tail-coupled, single-chain structures, a substantial degree of random coupling (head–head, side-chain, other) and cross-linking may occur in real CPs. This is in fact much more so than in more conventional, common polymers such as poly(ethylene) where polymerizations are more controllable. This is manifested, for example, in the "sp^3 defects" commonly observed in the normally sp^2 chain P(Ac), the 3-position branching observed in P(Py), and the ortho-linkages observed for poly(diphenyl amines) [956, 1218, 1235], all of which lead to cross-linking. It accounts for the ca. 70% or less H–T couplings observed in a typical CP such as poly(3-methyl thiophene) [1236]. Such cross-linking of course perturbs the extended electronic delocalization of CPs, leading to limitations in properties such as conductivity. Cross-linking can also lead, on a less microscopic scale, to highly intermingled and branched fibers, as is the case for the more common, Shirakawa-type P(Ac), for instance [863].

© Springer International Publishing AG 2018 389
P. Chandrasekhar, *Conducting Polymers, Fundamentals and Applications*,
https://doi.org/10.1007/978-3-319-69378-1_32

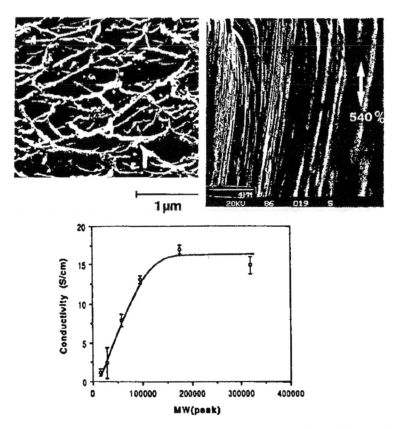

Fig. 32.1 (**a**) Morphology of *left*, "Shirakawa-type" poly(acetylene), which is highly cross-linked, and *right*, "Naarmann-type," which is said to be highly ordered (After Ref. [1240], reproduced with permission). (**b**) Dependency of conductivity of doped poly(aniline) (emeraldine oxidation state) on molecular weight (After Ref. [903], reproduced with permission)

It should also be apparent that in any polymerization batch, several regions or "islands" may be produced, of slightly differing structure (and molecular weight (MWt), see below), these different structures not being chemically or electrochemically equivalent. For example, Yang and coworkers [1237] have found that the broad 1.5 eV absorption band of P(ANi) is in reality an overlap of 1.52, 1.65, and 1.77 eV bands corresponding to chemically inequivalent segments within the polymer which are however produced in every polymerization of P(ANi). And the widely discussed "metallic islands surrounded by insulating regions" in CPs have been shown not only to be due to inhomogeneous doping but also due to structurally differing regions [894].

32.1.2 Polymerization Conditions

Polymerization conditions are of course an important determinant of these structural variations and inhomogeneities. For example, in a series of simple experiments, Whang et al. [1015] have shown that control of the oxidation potential during polymerization of pyrrole yields P(Py) with characteristics varying from fibrillar and highly elastic to globular and very brittle; at the optimal oxidation potential, a P(Py) is obtained which has a rubberlike elasticity and tenacity. This effect is also noticeable in the electrochemical polymerization of aniline, where, for instance, a change of dopant concentration from 0.5 M to 2.0 M leads to a transformation from globular to fibrillar morphology [1238]. Some studies [1239] have shown that variation of current densities in galvanostatic polymerization of P (ANi) allows tenfold variation in fibril radii.

Polymerization conditions have a most pronounced effect on the morphology of P(Ac), as Fig. 32.2a shows. At the left in the figure, the conventional, Shirakawa method P(Ac) clearly shows the microscopic effects of cross-linking, with a weblike fibrillar structure, while the Naarmann method P(Ac) at the right in the figure, prepared using a more refined method, shows highly oriented, thin, single fibrils; the conductivities of the Shirakawa and Naarmann P(Ac) reflect this microscopic order: ca. 10^3 vs. ca. 10^5 S/cm. Shirakawa and Naarmann P(Ac) are thought to have an sp^3 defect concentration of about 2% and nil, respectively [1240].

CPs that are more (or less) cross-linked can also be obtained via selective fractionation of the de-doped, soluble form; this method was, for instance, applied to poly(aniline) (P(ANi)) via fractionation, by gel permeation chromatography (GPC), of P(ANi) solutions in NMP [1241], with solubility of the various fractions in NMP being a measure of the degree of cross-linking.

32.1.3 Rotational Barriers

Undoped CPs, when not limited by cross-linking or other defects, have rotational freedom about the bonds linking the monomers. This freedom is however immediately lost on doping. The Brédas group has, in a simple set of calculations, shown that, while undoped P(Py) and P(T) have rotational barriers, about the monomer–monomer bonds, of ca. 3.0 kcal/mole, these rise to ca. 60 kcal/mole, i.e., 20-fold, in the doped polymers [1242]. Such predictions have been proven in experimental work, for instance, that of Cao and Smith [1048] showing that oriented films of P (ANi) are obtained from liquid crystalline solutions due to the chain stiffening that occurs on protonation (doping).

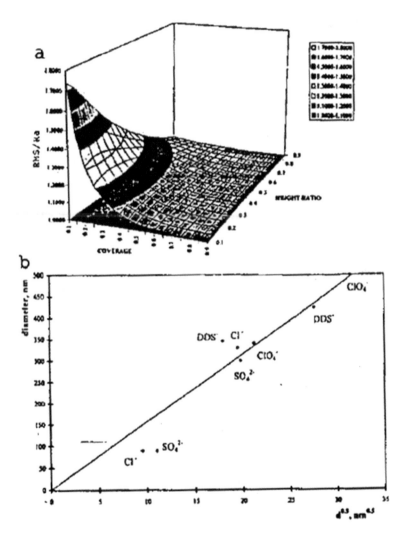

Fig. 32.2 Results from AFM studies of effect of film thickness and doping on surface roughness: (**a**) 3D graph of RMS/Ra ratio dependence on height ratio and degree of coverage. (**b**) Surface globule average diameter dependence on film thickness (*d*) (After Ref. [1252], reproduced with permission)

32.1.4 Molecular Weight (MWt)

Excepting for a few cases, most notably that of very high molecular weight (MWt) P(ANi), most CPs are more accurately oligomers, with chain lengths from 8 or 9 to no more than 2000 monomer units. Indeed, as noted earlier in this book, Garito et al. [1235] and others have shown that for most properties of CPs, such as optical effects, to be manifested, no more than eight or nine monomer units are in fact necessary. And the well-documented high disorder in most CPs entails that the effective delocalization length across which non-hopping conduction occurs is in fact no more than about eight or nine monomer units. A group of poly(diphenyl amines) synthesized in the author's laboratories, showing excellent nonlinear optical conductivity and other properties, were shown to average only about 35 monomer units in chain length [1235].

Before discussing MWt of CPs, we must very briefly digress to definitions of MWt. Three MWt parameters are commonly used by polymer chemists, the *number average*, *weight average*, and *Z-average* MWts, denoted, respectively, by M_n, M_w, and M_z.

Consider a sample of a polymer. It contains molecules of varying lengths and thus MWts:

- If the number of molecules with MWt M_i is denoted by n_i, the total weight of the sample is $\sum (n_i M_i)$.
- The weight fraction of the polymer having MWt M_i is $(n_i M_i)$, which can be denoted by w_i.

The three MWts are then defined as follows [1243]:

$$M_n = \sum n_i M_i / \sum n_i$$

$$M_w = \sum w_i M_i / \sum w_i$$

$$M_z = \sum w_i M_i^2 / \sum w_i M_i \qquad (32.1a, b, c)$$

Generally, techniques such as osmotic pressure, boiling point/freezing point change, and end-group titration yield M_n. Light scattering yields M_w. Dilute solution viscosity measurements yield an approximation for MWt as M^a, where the exponent a is ca. 0.5 for poor solvents and ca. 0.8 for good solvents. Light scattering is generally considered a more "absolute" method than solution-based techniques.

MWt of CPs is most commonly measured via GPC using a poly(styrene) standard. This yields good relative numbers but not strictly accurate MWt, one that is closer to M_n than to M_w. As an illustration, MWts measured by light scattering, a more absolute method, routinely show values ca. 2 X those obtained via GPC [1065].

For the prototypical P(ANi), the as-prepared (emeraldine base) form can show up to six distinct MWt fractions, having weight-average MWts (M_w) from typically ca. 22,000 to ca. 380,000. Each of these fractions shows distinct and differing properties, for instance, conductivities from 1.2 S/cm (for 22,000 M_w) to 17 S/cm (for 211,000 M_w) then leveling off at 15 S/cm (for 380,000 M_w, about 4050 monomer units); Figure 32.1b [903] shows this dependence for M_p of P(ANi). Heeger and Smith [902] have found that the conductivity of P(ANi) follows two functionalities with respect to M_w: $\sim M_w^{1.1}$ and $\sim M_w^{0.55}$ depending on whether the P (ANi) has, respectively, a "rigid-rod" or "flexible/uncoiled" configuration. Nazzal and Street [1244] have used a much more accurate MWt method employing tritium labeling to show that the absolute MWt of chemically synthesized P(Py) varied from ca. 100 to no more than 1000 monomer units, dependent on polymerization conditions.

32.1.5 Fibrillar and Globular Morphology

On a microscopic or nanoscopic scale, untreated, as-synthesized CPs generally fall into three categories in terms of outward topology: *fibrillar*, the most typical; *globular*; and more rarely, truly *filmlike* (which is actually extremely fine-structured globular). Typically, standard Shirakawa P(Ac) is produced in fibrillar form, electrochemically polymerized P(Py) in globular form, and electrochemically polymerized P(DPA) in filmlike form.

Especially in the fibrillar case, subsequent processing (e.g., drawing or stretching, see below) can improve the structural order substantially. And of course, quite expectedly, differing polymerization conditions can yield a variation of morphology from fibrillar through globular to filmlike, as noted earlier. Atomic force microscopy has been used [1245] to show that the "filmlike" conducting form of P(ANi) obtained via in situ chemical polymerization is quite smooth (RMS surface roughness ca. 3 nm) and is, as expected, actually composed of aggregations of globules of ca. 50 nm diameter. The Shacklette group at Allied Signal has demonstrated that dispersions of Allied Signal's commercial P(ANi), marketed under the trade name Versicon (now produced by Zipperling Kessler), in materials such as PVC, contain dispersed, single, globular particles of P(ANi) of ca. 250 nm diameter, each of which is supposedly an aggregation of smaller spheres [1174].

In the case of as-prepared fibrillar CPs, e.g., the Shirakawa P(Ac), fibril diameters can vary from as little as 3 nm to 1 μm [1246, 1247], very much dependent on polymerization conditions (time, solvents used, etc.).

Fibril formation in fibrillar CPs is generally through precipitation and coalescing of the growing polymer once a critical chain length is reached. It is important to note however that the bulk density of CPs appears to change little with morphology, or with other parameters such as nature of dopant, as shown below.

Some workers [1248, 1249] have claimed that CPs such as P(ANi) can show "folded/coiled" and "uncoiled" forms, interconverted through the presence of electrolytes in solution, somewhat like proteins. Evidence for this however is highly inferential and circumstantial.

32.1.6   Doping Effects

The effect of doping is first and foremost to produce a swelling of the CP, which generally leads to a large increase in the microscopically observable volume of the CP. The swelling effect however may be less pronounced if small dopants, such as BF_4^- or methane sulfonate, are used in a nonaqueous medium.

Swelling results from a combination of dopant penetration and solvent effects, including solvent molecules retained or brought in by the dopant. On extended electrochemical cycling (cycling between the oxidized and reduced electrochromic states), CP films may actually show an "annealing" effect, with morphology becoming more homogeneous and cracks or deformities disappearing. Doped and undoped CPs also differ in surface roughness; Buckley [1250] has used atomic force microscopy to show that doped P(ANi) has a significantly greater roughness parameter than undoped P(ANi).

While different dopants may appear to impart a very different topography to a CP, the physical properties of the CP may in fact change very little; this is evidenced by the case of P(Py), where different dopants show very different topographies [972], but the densities all lie in the range of 1.37–1.48 g/cc.

Another important feature of doping with respect to morphology, especially in the case of post-synthesis vapor phase doping, for example, re-doping with I_2 of a poly(acetylene) (P(Ac)) sample, is that it can be inhomogeneous. It is not only that diffusion of dopant through a thick fiber of already formed CP yields a gradient with high doping on the outside to nearly no doping at the fiber center, but that doping islands adjacent to undoped areas can form even at the edges of a CP fiber or globule [863, 1251].

Doping also imparts a rigidization to the CP chain, as discussed in Sect. 32.1.3.

Film thickness and doping affect surface roughness, which can be quantified to some extent by atomic force microscopy (AFM) studies. Figure 32.2a, b reproduces roughness data obtained by Silk et al. [1252] for P(Py) doped with various dopants.

32.1.7   Effects of Fundamental Structure and Substituents

A large number of studies have attempted to arrive at improvements in selected properties of CPs via alteration of their structure, either in a minor way through attachment of additional substituents or side groups or in a more fundamental way. Among the former, the best known examples are attachment of alkyl groups of increasing chain length at the 3-position in poly(3-alkyl thiophenes) (P(3AT)s) and of alkoxy groups at the ortho-position in poly(anilines) yielding P(AlkOANi)s. This has led to improved solubility of the P(3AT) and P(AlkOANi) in organic solvents with increasing alkyl chain length, and melt processibility for very long chain alkyl P(3AT)s. Figure 32.3 shows the reduction of the m.p. of P(3AT)'s as a function of increasing alkyl chain length; solubilities in organic solvents such as toluene,

Fig. 32.3 Melting point of
poly(3-alkylthiophene) as
function of the alkyl chain
length (After Ref. [1222],
reproduced with
permission)

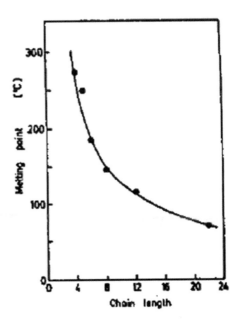

tetrahydrofuran (THF), and methylene chloride parallel the m.p. trend [997], while conductivities appear to be nearly the same for all P(3AT)s, ca. 1–10 S/cm. Dialkyl substitution on the thiophene nucleus however has been shown to lead to greatly reduced conductivities [997].

More fundamental studies of structural improvement have sought the use of oligomers rather than monomers as starting materials for polymerizations, with the expectation that this would lead to improved order and less chain branching/cross-linking. Studies have also addressed fundamental electronic delocalization features in CP chains. Most such studies have met with limited or no success. For example, Diaz et al. [997] showed that the use of bipyrrole or terpyrrole in place of pyrrole for electropolymerization yielded *no* polymer, a result discussed earlier in this book. Similarly, Onoda et al. [957] found that the use of thiophene–vinylene oligomers yielded CPs with poorer electrochromic and other properties as compared with the CPs obtained from the corresponding monomer. Jenekhe [1253] attempted a more fundamental structural modification, synthesizing CPs with alternating benzenoid and quinonoid structures, long predicted by theoreticians to lead to greater delocalization and lower bandgaps; much of Jenekhe's work however could not be reproduced [1254, 1255].

32.2 Mechanical Properties

Just as for "conventional" polymers, the standard mechanical properties of CPs can be characterized. Among the more important properties of interest are Young's modulus (also called tensile modulus, modulus of elasticity, and tangent modulus), tensile strength, and elongation at break. Other properties such as shear modulus, toughness, and flexural strength may also be of interest for some CPs. All such properties are generally measured using procedures established by ASTM (American Society of Testing and Materials). It cannot be proposed to give detailed definitions of these properties here, for which reference may be made to any introductory polymer chemistry book. One of the most important properties, the Young's modulus, is the tangent (slope) of the stress(x-axis)/strain(y-axis) curve for certain applied stresses and measures tensile/elastic properties. The elongation at break also measures elasticity and strength. As an illustration, typical Young's moduli for elastomers (rubbers) and thermoplastics (such as poly(ethylene), poly(styrene)) are, respectively, 1 MPa (150 psi) and 0.4 to 4 GPa (60,000 to 600,000 psi).

For P(TV) fibers prepared via the Yoshino precursor method described above, the Young's modulus and tensile strength are directly correlated to the draw (stretch) ratio and conductivity (Fig. 32.4a–c). The correlation of conductivity with tensile modulus is a general property observed for nearly all CPs, e.g., poly(3-octyl thiophene) (P(3OT)) [902]. The enhancement of the Young's modulus with doping for P(3OT) is shown in Fig. 32.5 and is thought to emanate from increased cross-linking with doping. 8:1 stretched, highly oriented, fibers of the ladder CP BBL were found to have Young's modulus of 120 GPa, tensile strength of 830 MPa, and compressive strength of 410 MPa after heat treatment [1057]. In fact, the Young's modulus and tensile strength of most CPs fall within this range, indicative of behavior closer to that of common thermoplastics than elastomers or thermosets.

Figure 32.6a, b shows Young's modulus and tensile strength for trans-P(Ac) films as a function of draw ratio. For P(Ac) as for other CPs, conductivity and mechanical properties closely parallel one another (with increased stretch ratios). P(Py), when obtained as free-standing, thick films polymerized electrochemically or chemically at optimal oxidation potential, has properties closer to those of elastomers than other CPs, e.g., Young's modulus of ca. 0.7 GPa and tensile strength of ca. 20 MPa [1015, 1256], values also obtained for P(ANi) fiber wet spun from conc. H_2SO_4 solution (Young's modulus 1 GPa, tensile strength 20 MPa, elongation at break 2.6%, conductivity ca. 60 S/cm [1257]).

32.2.1 Crystallinity

Crystallinity of CPs, usually in the form of fibers or oriented films, is most commonly measured via some variant of Debye–Scherrer method for X-ray powder

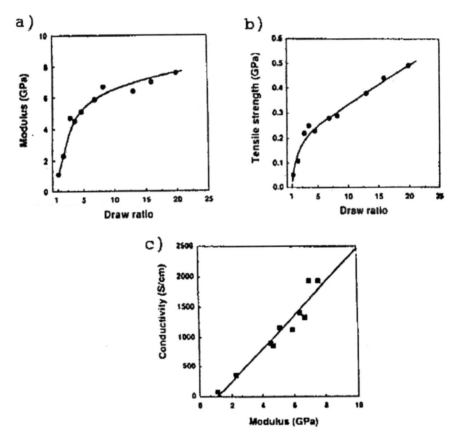

a)

b)

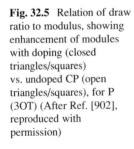

c)

Fig. 32.4 (a–c) Correlation of modulus, tensile strength, conductivity, and draw ratio for P (TV) fibers (After Ref. [1223], reproduced with permission)

Fig. 32.5 Relation of draw ratio to modulus, showing enhancement of modules with doping (closed triangles/squares) vs. undoped CP (open triangles/squares), for P (3OT) (After Ref. [902], reproduced with permission)

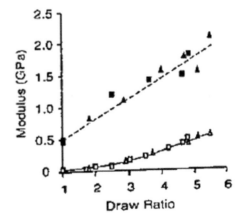

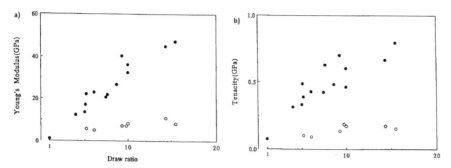

Fig. 32.6 (**a, b**) Young's modulus and tensile strength for trans-P(Ac) as function of draw ratio (After Ref. [902], reproduced with permission)

diffraction, although electron diffraction has also been employed. To recount very briefly the rudiments of such X-ray methods, a monochromatic X-ray beam impinges on the sample, whose reflections (diffractions) are recorded on a film cylindrically enveloping the sample (or a counter in modern-day diffractometers). Microcrystallites or microcrystalline regions in the sample yield strong diffractions when Bragg's law ($n\lambda = 2d\sin\theta$) is satisfied. An amorphous sample yields a diffuse halo. A partially crystalline sample may yield some high-intensity rings superimposed on such a halo. An *oriented* sample (e.g., with polymer chains oriented along a major crystallographic axis) would yield distinct diffraction spots betraying the orientation. One measure of the crystallinity taken by many workers is the *coherence length* (L_c) determined by Scherrer's formula, $\zeta = 2\pi\lambda/(\delta(2\theta))$ ($\delta(2\theta)$ = full width at half maximum (FWHM) for the reflection). Another measure is the *mosaic spread*, in degrees, also derived from the FWHM, with smaller values signifying generally better orientation.

Figure 32.7 shows the greater crystallinity observed for P(ANi) films with increasing stretch ratio (= final/original length). This is representative of a general finding that increased draw or stretch ratios for CP fibers or films increase all parameters relevant to crystallinity and orientation, such as coherence length, mosaic spread, and sharper diffractions. The work of the Epstein group with P(ANi) [1258] has shown that this CP, when prepared directly in its conducting form, shows ca. 50% crystallinity with a coherence length of ca. 5 nm. De-doping of this semicrystalline form of the CP yields amorphous emeraldine base, while re-doping is known to restore the original crystallinity [1259]. On the other hand, when P(ANi) is prepared in the emeraldine base (undoped) form, it is likely to show poor or no crystallinity, which does not improve even with doping. Results with the *o*-methyl derivative of P(ANi) are nearly identical.

Figure 32.8 shows an X-ray diffraction pattern for a form of P(Ac) prepared as microfibrils which have then been stretched eightfold and I_2-doped. The strong meridional reflections seen actually correspond to poly(iodine) columns of repeat distance 5 nm, which however betray a corresponding order in the P(Ac). A good example of the enhanced crystallinity observed with drawing is the case of the

Fig. 32.7 Greater
crystallinity observed for P
(ANi) with increasing
stretch ratio ($1/1_o$) (After
Ref. [1065], reproduced
with permission)

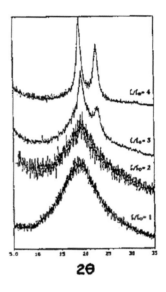

Fig. 32.8 X-ray diffraction
pattern of heavily doped and
eightfold stretched P
(Ac) (After Ref. [1260],
reproduced with
permission)

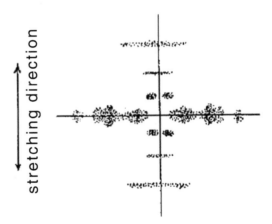

16-fold stretched poly(thienylene vinylene) (P(TV)) fiber shown in Fig. 32.4. An
X-ray pattern of a CP with some crystalline domains in an overall amorphous CP
matrix usually appears as diffraction spots situated in a diffuse halo [902], e.g., poly
(3-alkylthiophenes). A stark improvement in crystallinity has been observed in
going from undrawn P(TV)-precursor fiber to eightfold stretched P(TV) fiber.
Electron diffraction patterns closely correspond to X-ray patterns.

32.2.2 True Single-Crystal Character

Among all CPs, there is only one class that appears to display true single-crystal character. This is the poly(diacetylenes) (P(DiAc)s), discussed separately in a subsequent chapter in this book. Large (up to 1 cc) near-single crystals can be grown, with estimated coherence lengths of up to 1000 monomer units (ca. 1 μm in length) [1261].

32.3 Problems and Exercises

1. Draw structures of P(ANi) and P(3MeT) indicating possible cross-linking and deviations from idealized head–tail linkages. Try to include doping and charge defects and doping islands. Which types of defects would be fairly common in every batch of P(ANi) and thus lead to the three chemically inequivalent P(ANi) forms found by Yang et al. (Sect. 9.1a)?

2. Consult the references cited for this chapter to outline two different syntheses (polymerizations) yielding polymers with very different morphology for the following CPs: P(Py), P(ANi), and P(Ac).

3. Draw conjugated structures to show how rotational rigidity is imparted upon doping in the following CPs: P(ANi) and P(3-Bu-T). Use Cl⁻ as the representative dopant.

4. Look up cross-references cited for this chapter to arrive at estimated values for MWts or MWt fractions for two other CPs to supplement the data provided for P(ANi) in this chapter. From the information, available elsewhere in this book, do you expect that MWts of CPs will be greater, lesser, or the same when they are synthesized via chemical vs. electrochemical polymerization?

5. From your understanding of conventional polymers and CPs, what would fibrillar, globular, and filmlike morphologies tell about the degree of cross-linking and chain branching, and length of chains, for a CP?

6. What are the likely reasons for swelling of CPs with doping? What in your estimation would be the effect of dopants of different size and solvent affinity? Illustrate with examples of several dopants of radically different size and several solvents of different hydrophilicity.

7. What in your estimation would be the effect of a copolymerization, e.g., of 3-alkyl and unsubstituted thiophene, *o*-methoxy and unsubstituted aniline, and 3-alkyl and unsubstituted pyrrole, on the morphology, as compared to the corresponding unsubstituted homopolymers. Discuss in detail with structures.

8. Outline one procedure each for wet spinning, dry spinning, and melt spinning of CP fibers and drawing of each fiber to a draw ratio of 5.0.

9. Which among the following polymers do you expect to show the least and the greatest conductivity anisotropy when drawn as fibers to a ratio of 5.0: P(Ac), P(ANi), and P(3-octyl thiophene)? Why?

10. Consult an elementary polymer textbook and identify which (at least four) standard mechanical properties, besides Young's modulus, tensile strength, and elongation at break, would be useful measures for CP fibers and films. For each property, outline the definition, measurement method, synonyms, if any, and typical values for the three broad mechanical classes among polymers (elastomers, thermoplasts, and thermosets) for comparison with those measured for CPs. Consult the CP literature of the last 5 years to see if any of these have been measured for P(ANi) and P(Py).

11. Outline how you would interpret X-ray diffraction data, if any, obtained for the following CPs and using the following X-ray methods: P(ANi)Cl, P(Di-Ac)s, Debye–Scherrer powder camera, Weissenberg camera, precession camera, wide-angle X-ray diffraction, and automated diffractometer.

Chapter 33
Characterization Methods

Contents

33.1 Introduction: Outline of Skeletal Characterization

We mention at the outset that several important and routine characterization methods for CPs have been dealt with at length elsewhere in this book, quod vide, notably:

- *Routine electrochemical characterization*, including cyclic voltammetry, chronoamperometry, and chronocoulometry
- *Routine spectroelectrochemical characterization* and *electrochromic characterization*, including spectroelectrochemical data, *time drive* (%T or absorbance as function of time and applied voltage), and *open-circuit memory*
- *X-ray diffractometry* as a measure of crystallinity

© Springer International Publishing AG 2018
P. Chandrasekhar, *Conducting Polymers, Fundamentals and Applications*,
https://doi.org/10.1007/978-3-319-69378-1_33

403

- *Electron and other microscopies* for morphological characterization

A basic characterization of a new CP system (which may include, e.g., a common CP such as P(ANi) but with a new dopant) may encompass only a few rudimentary techniques. The following usually suffice for a comprehensive characterization:

- Elemental analysis
- Bulk DC-room-temperature conductivity
- Molecular weight
- Basic electrochemical properties (primarily via cyclic voltammetry, CV)
- Basic electrochromic properties (via spectroelectrochemical characterization)
- Composition (e.g., via FTIR and XPS)
- Crystallinity (e.g., via powder X-ray diffraction)
- Thermal properties (e.g., via thermogravimetric analysis (TGA) and differential scanning calorimetry (DSC))
- Morphology (e.g., via scanning electron microscopy (SEM) and transmission electron microscopy (TEM))

Whether the CP is producible in powder or film form can determine the course of characterization to a large extent. The specific examples cited in this chapter should give the reader an idea of the application of these common analytical and physical techniques to CPs in general and certain classes of CPs in particular.

33.2   Conductivity and Related Measurements

33.2.1   *Ex Situ DC Conductivity of Powders, Films, and Fibers*

The basic techniques for measurement of conductivity derive from the resistivity relationship in Ohm's law ($E = IR$, E the potential (voltage) difference, I the current, R the resistance). Across two points spaced at a distance d and with a cross-sectional area of A, the resistance is expressed as

$$R = \{E/I\} \times \{A/d\} \ \text{ or } R = \{E/d\}/\{I/A\} \tag{33.1}$$

Thus, knowing the cross-sectional area, the separation distance, the current flowing, and the potential difference between two points, the bulk resistance, R above, can be computed, expressed as *specific resistivity*, ρ, in units of Ω-cm. The bulk-specific conductivity σ, in units of Ω^{-1}-cm^{-1} or siemens (S) /cm, is then its reciprocal. The surface resistivity of a coating or film, in units of $\Omega/\square$ (square), is just ρ/t, with t the thickness of the coating/film.

When a conventional two-probe measurement instrument such as an ohmmeter is sought to be applied to a highly resistive sample, the contact resistance between

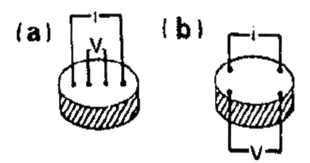

Fig. 33.1 Four-probe technique for measurement of ex situ DC conductivity. (**a**) For the case when thickness, t, of the sample $>> S$, the spacing between the probes, the conductivity, σ, is $(1/(2\pi S)) \times (i/V)$, where i = current passing through outer probes and V = potential drop across inner probes. When $t << S$, i.e., we have a thin film, then it is preferable to use line probes rather than point probes, placed in-line exactly as shown for the point probes. In this case, $\sigma = (S/(t \times w)) \times (i/V)$ where w = width of film. (**b**) For $t \leq S$, this probe placement is used, and $\sigma = (\ln (2)/\pi t) \times (i/V)$

the probe and the sample constitutes a major impediment. To circumvent this, four-probe techniques, such as those illustrated in Fig. 33.1 are used. In these, a current is applied between the outer probes, and the voltage differential between the inner probes is measured. Such techniques were proposed by van der Pauw, among others [1262].

A commercially available four-point probe, with sharp probe tips of very small diameter, and probe spacing typically ca. 0.2 cm, may be used in conjunction with a current source and a voltmeter (e.g., a digital multimeter). The probe may be lowered onto the sample through a mechanical device. Alternatively, the four-in-line configuration of Fig. 33.1a may be emulated in a permanent fixture into which the sample is lowered or otherwise affixed. If the sample is a powder, it is first compacted into the form of a pellet, using a laboratory hydraulic press or other device.

For the four-in-line configuration of Fig. 33.1a, the conductivity is given by

$$\sigma(\text{S/cm}) = \{\ln (2)/(\pi t)\} \times \{I/E\} \sim= \{0.22/t\} \times \{I/E\} \tag{33.2}$$

when t, the sample thickness (e.g., of a pressed pellet), is less than or about equal to the probe spacing. An alternative configuration in which all four-probe contacts are made at the periphery of the sample is used less frequently. For the same configuration (Fig. 33.1a), when the sample thickness is much greater than the probe spacing, s, the conductivity is given by

$$\sigma(\text{S/cm}) = \{1/(2\pi s)\} \times \{I/E\} \tag{33.3}$$

Measurements can also be carried out on a thin film on a nonconductive substrate. In such a case, the use of four-line probes, as shown in Fig. 33.1b, rather than point probes, is recommended, and the conductivity is then given by

$$\sigma(\mathrm{S/cm}) = \{I/E\} \times \{S/(tw)\} \tag{33.4}$$

where S, t, and w are the probe spacing, sample thickness, and sample width, respectively.

Yet another variation of this four-point method was proposed by Montgomery et al. [1263, 1264] to monitor anisotropic conductivity. In this, the four corner points of a face of the sample which is in the form of a rectangular block are used. Currents are passed and voltage differentials monitored along the short and long edges of the face, to yield the conductivity parallel and perpendicular to the face. Two components of a conductivity tensor are then calculable from this.

For very thin, Langmuir–Blodgett (LB) films, Logsdon et al. [1229] described an interesting technique for conductivity measurement. They transferred an LB film of poly(dodecylthiophene) onto an array of 40 pairs of Au electrodes of 3.2 mm length and 20 μm gap, which was previously coated with a hydrophobic phase comprising octadecyltrichlorosilane. Conductivity was then determined using a calculation similar to that for thin films described in the legend to Fig. 33.1 but with conductivities determined for each adjacent set of four electrodes averaged using a boxcar averager.

In the case of *fibers*, including very thin fibers of diameters as small as 1 μm, the same four-in-line probe methods described above may be used with a suitably proper mounting of the fibers. For example, Hermann et al. [1265] described a technique that uses standard Cu connectors on a printed circuit board, on which fibers are mounted using Ag paint contacts (Fig. 33.2). Wang et al. [1057] also used a standard integrated circuit socket having rows of Au-plated pins, where the fiber was physically pressed against the pins using a silicone rubber insert (Fig. 33.3) [1057].

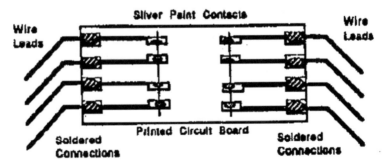

Fig. 33.2 Four-probe technique as adapted to very thin fibers, using printed circuit boards (After Hermann et al. [1265])

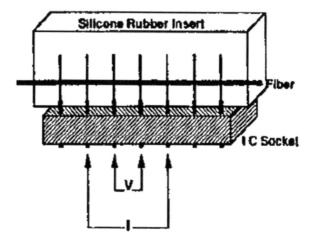

Fig. 33.3 Another variant of the four-probe technique for fiber conductivity measurement (After Wang et al. [1057], reproduced with permission)

33.2.2 In Situ DC Conductivity

On some occasions, it becomes imperative to determine the in situ conductivity of a CP while it is actively undergoing a chemical or electrochemical transformation, for instance, the conductivity as a function of applied potential. Several, varied methods have been used by workers in the field for this purpose [1266], all of which however incorporate the principles underlying Eq. (V..1). The Contractor group [1267] used an indirect, two-probe AC method to measure the DC conductivity of a P(ANi) film. They forced a five nA peak-to-peak AC current at 1.33 KHz across the two points while simultaneously monitoring the in-phase voltage across the same two points using a two-phase lock-in analyzer.

The Pickup group [870, 873] used a method employing a bipotentiostat (a potentiostat with two working electrodes) for monitoring conductivity of P (3MeT) as a function of applied potential. In this, the active CP film is sandwiched between two (working) electrodes. The applied potential at the two electrodes is simultaneously scanned, slowly (5 mV/s) to maintain steady-state conditions, and a small, constant potential difference, 10 mV, is applied between the two electrodes. The current flowing between them is monitored to yield the in situ resistivity, and hence conductivity, from Eq. (V.7.1). As noted in an earlier chapter, such measurements as a function of potential show an exponential rise in conductivity at the standard redox potential of the CP, indicating conversion from the insulating to the conducting form. Schiavon et al. [1268] and others [1269, 1270] used an essentially identical method, however using transistor–technology nomenclature (e.g., drain voltage and drain current for the voltage/current between the two working electrodes).

33.2.3 Electrochemical Impedance Spectroscopy (EIS) and AC Conductivity

Determination of real conductivity as a function of frequency, and the closely related technique of EIS which yields essentially identical information, can yield an important handle on processes relating to conduction mechanisms and other parameters not available from simple DC measurements. For example, in a metal, where charge carriers (here electrons) are in extended, delocalized states, conductivity is independent of frequency to ca. 10^{13} Hz or more. In crystalline semiconductors, with thermal excitation of charge carriers into the conduction band, again conductivity is independent of frequency across a large range but is dependent on temperature. However, in disordered, inhomogeneous or similarly structured semiconductors, to which class nearly all CPs belong, the conductivity is frequency dependent; at higher frequencies, it is generally larger than the DC conductivity. This behavior in CPs is usually due to "islands" of high conductivity (e.g., via localized doping) surrounded by more insulating (less doped) regions. Epstein [886] proposed a simple partitioning of total conductivity:

$$\sigma(\text{total}) = \sigma(\text{DC}) + \sigma(\text{AC}) \tag{33.5}$$

where $\sigma(\text{DC})$ and $\sigma(\text{AC})$ are functions, respectively, of temperature and temperature + frequency.

The techniques used for measurement of AC conductivity are varied and depend on the frequencies of interest. For $\nu <$ ca. 10^6 Hz, a capacitance–conductance bridge can be used [1271]. EIS, which yields both real and imaginary components of the complex resistance, i.e., impedance, can be used up to ca. 1 GHz. The cavity perturbation and related techniques in microwave measurements have been used for frequencies up to ca. 30 GHz [1272]. For higher frequencies, mathematical (Kramers–Kronig) transformation of IR reflectance data must be used [1273].

It cannot be proposed in a book such as the present to discuss common techniques such as EIS in detail; reference is made to excellent monographs on the subject [1274]. To summarize briefly, however, in a typical AC technique, a constant-amplitude (sine wave) current of a given frequency is applied across two electrodes connected via the CP alone or the CP in an electrochemical cell or device. The voltage that results is monitored via a lock-in amplifier which detects the in-phase and 90°-out-of-phase impedance components, which correspond, respectively, to the resistance and capacitance of the CP sample. If an EIS-type equivalent circuit analysis is not required, this information, as a function of several frequencies, is usually all that is sought for AC conductivity. Microwave and IR measurements are discussed separately within this chapter or elsewhere in this book.

Figure 33.4 shows a typical frequency/σ(AC) plot for P(Ac). Nogueira et al. [1275] recently carried out an AC conductivity study of P(o-MeOAni). Their results, as a function of frequency, temperature, and doping are plotted in Fig. 33.5a, b and show that for high doping levels, conductivity is essentially

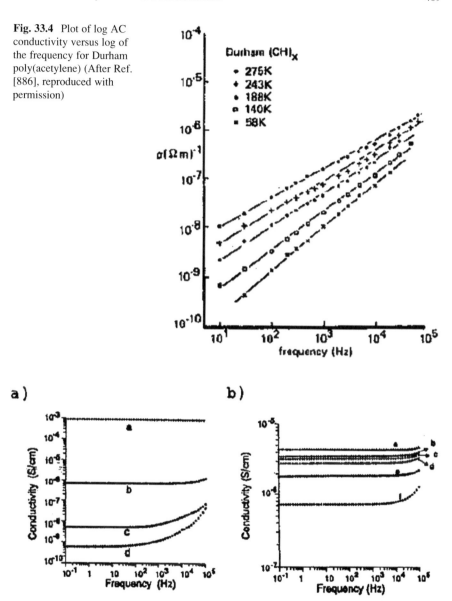

Fig. 33.4 Plot of log AC conductivity versus log of the frequency for Durham poly(acetylene) (After Ref. [886], reproduced with permission)

a)

b)

Fig. 33.5 Results of measurements of the AC conductivity of poly(o-MeO-aniline) as a function of frequency, temperature, and doping (After Ref. [1275], reproduced with permission)

frequency independent (quasi-metallic behavior, cf. discussion above), while at low doping levels, it increases at high frequencies. Their data fit the variable-range hopping (VRH, see elsewhere in this book) model of conduction very well.

We may cite representative EIS studies of CPs. Firstly, for reference, Fig. 33.6 shows a Cole–Cole (real vs. imaginary impedance, also called Nyquist) plot of an

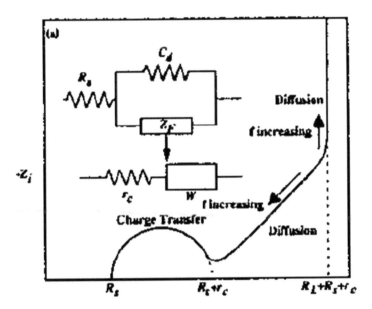

Fig. 33.6 Representative EIS Cole–Cole plot for idealized, planar, CP film electrode (After Ref. [1239], reproduced with permission)

idealized, planar CP film electrode, also showing the equivalent circuit; in a situation where the charge-transfer component of the impedance is negligibly small, the semicircle disappears. Kamamura et al. [1239], in a study of P(ANi) fibrils, found a rapid increase of the capacitive component of the impedance with decreasing frequency at lower frequencies. The charge-transfer resistance at the P (ANi) electrode was found to be considerably smaller than the impedance corresponding to the diffusion of counterions within the CP, indicating the latter to be the limiting process in conduction. Figure 33.7 shows typical Cole–Cole plots for P(ANi) films at two different applied potentials. Figure 33.8 shows the equivalent circuit of a P(Py) film in contact with an electrolyte solution arrived at by Amemiya et al. [1276] in their considerably detailed study of several P(Py) systems. Results from Cole–Cole plots for a ca. 1 μm film of P(Py)ClO$_4$ obtained by Smyrl and Lien [1007] showed that diffusion of counterions is dominant for a longer frequency span at low oxidation levels of the CP. In another typical study of interest, Passiniemi et al. [1277] found that the impedance of P(ANi) solution cast as a film had a very high capacitance component when compared to P(ANi) which had been melt processed in thermoplastic matrices.

Singh et al. [1278] measured the AC conductivity of lightly doped films of poly (N-Me-pyrrole) in the temperature range 77–350 K and frequency range 100 Hz to 1 MHz. In the low-temperature region, they found the relation $\sigma(\omega) = A\omega^s$, where s was found to be <1. At high temperatures, the conductivity was strongly frequency dependent but with the dependence weak at low frequencies. Fig. 33.9a, b show representative data obtained by these workers.

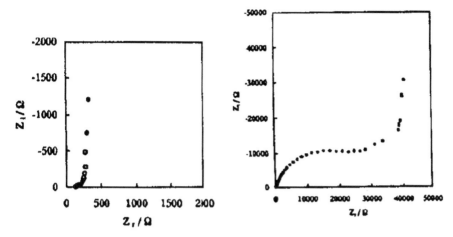

Fig. 33.7 Typical Cole–Cole plots for P(ANi) at 3.7 V (*left*) and 2.9 V (*right*) vs. Li/li$^+$ in propylene carbonate electrolyte (After Ref. [1239], reproduced with permission)

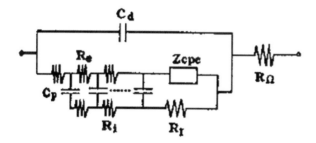

Fig. 33.8 Equivalent circuit for a P(Py) film in contact with a solution. C_d, double layer capacitance, R_e, electronic resistance, R_i, ionic resistance, C_p, distributed capacitance in the film, Z_{cpe}, impedance for electron, R_I, ionic resistance at the film/soln. Interface. R_Ω, soln. Plus ITO resistance (After Ref. [1276], reproduced with permission)

33.2.4 Thermopower Measurements

As noted in an earlier chapter in this book, thermoelectric power is the potential gradient generated between two points or faces of a material when they are subject to a temperature differential. Typically, it is large for semiconductors, while for metals it is small and decreases with decreasing temperature, vanishing near absolute zero. A small decrease of thermopower with temperature would thus likely indicate hopping conduction, while a very rapid decrease and very low values (see below) for the thermopower would indicate a quasi-metallic conduction, as, for instance, observed for highly doped trans-P(Ac).

To cite a typical thermopower measurement [889], the CP sample is mounted across two single-crystal quartz blocks maintained at two different temperatures, usually a difference of 1°K. A thermocouple is also used to monitor the temperature

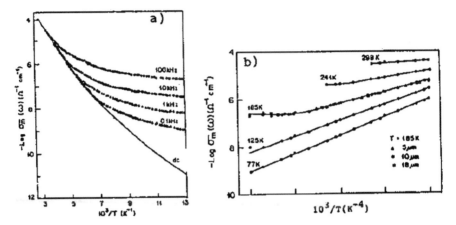

Fig. 33.9 Results from AC conductivity studies on lightly doped Poly(N-Me-pyrrole) films, after Singh et al. [1278], reproduced with permission

difference. During the measurement, the temperature difference is achieved slowly (e.g., over 10 min). The voltages across the sample and the thermocouple are monitored with a nanovoltmeter. The slope of the thermovoltage of the sample vs. that of the thermocouple yields the sample thermopower, after correction for contact contributions.

The thermoelectric power behavior of most CPs parallels that of temperature and activation energy relationships.

Selected thermopower data have been cited elsewhere in this book.

33.3   Infrared Measurements

Like other routine analytical diagnostic techniques, IR spectroscopy is applied no differently to CPs than to other materials. Techniques for infrared reflectance studies, including spectroelectrochemical studies, have been dealt with elsewhere in this book. Of the signature absorptions, the aromatic (six-membered and five-membered) ring and dopant absorptions are of particular interest in yielding immediate information on dopant incorporation and level and manner of polymer chain linkage and side branching. In this section, we cite representative studies yielding information on structural features particular to CPs.

As an example of a relatively simple structural determination, Catellani et al. [1279b] took a strong absorption at 920 cm^{-1} in the spectra of a family of substituted poly(2,5-thienylene vinylene)s to indicate a trans configuration about the vinyl bond. In studies of typical poly(aromatic amines) such as poly(diphenyl amine), poly(4-amino biphenyl), and poly(N-Phe-2-naphthyl amine), the Dao group [958, 975] observed the characteristic N–H stretching vibration at 3400 cm^{-1}, indicating that the polymer linkages were not at the N-group and thus likely on the aromatic rings. The heavily doped states of these CPs yield a band at 1555 cm^{-1}, corresponding to the N–H bending of amine salts and strong bands at 1040 and 1140 cm^{-1} corresponding to the ClO$_4^-$ dopant. In the aromatic region, several bands indicate *p*-disubstitution, validating the presumed aromatic ring head–tail polymer linkage. In Chandrasekhar et al.'s study of poly(diphenyl amine) and poly (4-aminobiphenyl) polymerized via a different route, it was found that in addition to the normally expected head-to-tail (para-para) linkages, there was considerable chain branching emanating from the ortho-positions and some from the N-atoms as well [956, 1235b,c].

For P(Ac), structural parameters are even simpler to deduce from IR spectra. For instance, strong peaks at 700 and 750 cm^{-1} are attributable to C–H out-of-plane bending for *cis*-P(Ac), yielding a clear indication of cis/trans ratio, while a peak at ca. 900 cm^{-1} due to an sp^3-C is a strong indicator of substantial cross-linking [1280]. In trans-P(Ac) of low doping, C–H stretching, wagging, and bending vibrations appear, respectively, at 3013, 1290, and 1015 cm^{-1} [1281]. The peak at 1290 cm^{-1} is substantially influenced by the charge on the CP backbone, via doping.

A fairly typical IR spectrum of undoped P(3AT)s is shown in Fig. 33.10 [1207]. Assignments are fairly straightforward: (1) thiophene C$_\beta$-H; (2) octyl; (3) thiophene C = C; (4) octyl -CH$_2$; (5)trisubstituted thiophene (a sharp, strong distinguishing band for all P(T)s [963, 1283]); (6) octyl -CH$_3$. For P(T)s generally, as for most other CPs, the aromatic region (600–900 cm^{-1}) gives useful information: for instance, 2,5 (α,α') coupling (strong band at ca. 790 cm^{-1}) in the electro-chemically made polymer can be distinguished from 2,5 combined with 2,4 (α,β) coupling (strong bands at 730, 820 cm^{-1}) in the chemically made polymer. P(T)s

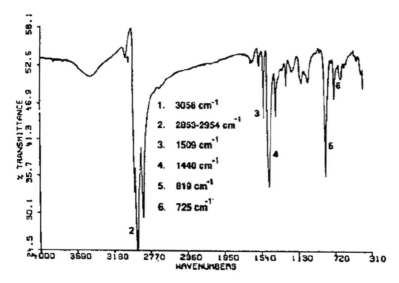

Fig. 33.10 IR spectrum of poly(3-octylthiophene) (After Refs. [962, 1207], reproduced with permission)

also sometimes show a broad peak at ca. 370 cm^{-1}, which has been ascribed to an "impurity-pinned translational mode" [1283]. IR spectra can also be used to identify approximate relative proportions in copolymers; for instance, Hotta et al. [1059] used relative intensities of hexyl and benzyl bands, as compared to monomer, to identify relative ratios in a copolymer of hexyl and benzyl thiophenes.

As noted in an earlier chapter, heavily doped CPs have a characteristic strong, broadband "free-carrier" absorption starting in the NIR region and extending through about 2000 cm^{-1}, sometimes obliterating key structural features in the 4000–2000 cm^{-1} region in the process. This is evident, for instance, in the IR spectra of P(Py)ClO$_4$ and P(Py)AsF$_6$ cited in an earlier chapter. While key N–H and C–H stretching, vibrations are masked in these spectra, within the small 1600 to 400 cm^{-1} region, the pyrrole ring peaks mask the ClO$_4^-$ but not the AsF$_6^-$ dopant peaks. In the case of P(ANi), the oxidation state, related to doping, can be approximately estimated by the relative intensities of the strong quinoid-ring-type (ca. 1600 cm^{-1}) and benzenoid-ring-type (ca. 1500 cm^{-1}) C–H stretching vibrations [1257], while the strength of the N–H stretching peak at ca. 3380 cm^{-1} can be used as an estimate of cross-linking, chain branching, or other defects at the N-atoms.

In most CPs, new bands indicative of doped polymer but fairly independent of dopant appear; for instance, for P(PP), the strong 1482 cm^{-1} absorption due to phenyl C–C stretching is shifted to ca. 1510–1535 cm^{-1} and is ascribed to the charged CP backbone. P(PP) also of course shows the characteristic strong vibration of p-disubstituted phenyl rings at ca. 820 cm^{-1} [1284]. In the related family of P(PV)s, IR spectra have also been used to determine E or Z configurations about the vinyl bond; for instance, in a study by Shi and Wudl [1199] of water soluble P(PV)s, the 960 cm^{-1} band, for out-of-plane bending of the trans-vinylene group, indicated an E configuration for the CP.

33.4 Molecular Weight

Molecular weight (MWt) of CPs can be determined by a number of direct and indirect methods, most direct methods unfortunately requiring some manner of solubility or processibility of the polymer. The three types of MWt applicable to polymers, denoted by M_n, M_w, and M_z, have been defined in an earlier chapter.

33.4.1 Indirect Methods

Of the indirect methods, perhaps the crudest and most approximate is one based on IR spectra and frequently applied to poly(aromatic amines) such as P(ANi). This involves comparison of the relative intensities of the aromatic-region bands. Thus, e.g., Nguyen and Dao [1285] observed strong peaks of 1,4-substituted phenyl (ca. 810 cm^{-1}) as compared to weak peaks for singly substituted (terminal) phenyl and used the relative intensities to compute MWts for poly(N-alkyl diphenyl amine)s. Similarly, for the case of P(PP), Jones et al. [1286] used the ratio of the log of the intensity of the phenyl CH vibrations corresponding to p-disubstitution (ca. $800\,cm^{-1}$) and monosubstitution (two bands ca. 770, 690 cm^{-1}, indicating terminal groups) to calculate a chain length of ca. 16 phenylene units. Among other reasons, this method is unreliable due to impurity or other effects on terminal groups. A more accurate but more difficult indirect method is one which has been applied to MWt determination for P(3,3′ dimethyl Py) [1244]. The 2,2′ positions in the monomer are tritiated, and when polymerization occurs, all except the terminal group tritium is eliminated. The relative radioactivity of the polymer and monomer then yield the chain length. A similar radiochemical method was applied by Chien et al. [1287] to P(Ac). Other indirect methods usually involve some derivatization. For example, one of the first MWt determinations of P (Ac), by Wegner [1288], involved chlorination to produce a soluble derivative, while Soga et al. [1289] used high-pressure hydrogenation of K-doped P(Ac) to produce processible poly(ethylene), whose MWt was then determined.

33.4.2 Gel Permeation Chromatography (GPC)

The most common MWt method applied to CPs, in their soluble, de-doped form, remains gel permeation chromatography (GPC) with poly(styrene) P(Styr) standards [1065, 1279, 1290], in solvents such as 1-methyl-2-pyrrolidinone (NMP), THF, or chloroform. GPC is a size-exclusion chromatography technique using ca. 10 μm silica or polymer particle columns, which have pores trapping large molecules for varying residence times depending on their size. Typically for P (ANi), a solution in NMP with 0.5 w/w% LiCl or 2 w/w% hydrazine hydrate added for improved solubility is used. Typical columns are "Ultrastyragel" (Waters) and "PLgel" (Polymer Labs). Eluent can typically be hydrazine hydrate or chloroform, flow rate typically 1 mL/min. For P(ANi), monomodal (single-component) or multimodal distributions are obtained, depending on P(ANi) morphology.

There are however a number of caveats regarding use of GPC with CPs. Davied et al. [1290] note that only the soluble portion of the sample is characterized, even though a good part of the CP in solvents such as NMP remains colloidal, and that the higher hydrodynamic volume of CPs such as P(ANi) as compared to P(Styr) make MWts too high. Indeed, when LiCl is added to NMP for better solubility of P (ANi), a MWt peak at ca. 970,000 disappears, indicating it is due to colloidal suspensions [903]. MacDiarmid et al. also note [1065] that GPC yields MWts that are roughly twice those obtained via the more absolute and accurate light scattering technique. For poly(3-octylthiophene), however, Hjertberg et al. [962] have noted, based on laser light scattering detection for GPC, that poly(styrene) may be an acceptable standard.

33.4.3 Viscosity

Viscosity is a method similar to GPC in terms of MWt determination of CPs, in that a reference polymer whose structure and void volume can be assumed to be similar to the CP being measured must be used, and a suitable solvent must be found. The most successful viscosity determinations for MWt purposes have been those carried out on P(ANi) in conc. (96%) H_2SO_4, using a standard viscometer [1279, 1291].

33.4.4 Light Scattering

Light scattering is possibly one of the most accurate and absolute methods of MWt determination. However, several problems remain with the use of this method for CP MWt measurement. Firstly, a solvent that yields a true solution of the CP, rather than a colloidal suspension, is preferred but not always found. Secondly, the high absorption of materials such as P(ANi) allow for only low concentrations to be studied well, whereas at these concentrations, the cumulative light scattering is poor. Thus, for instance, Plachetta and Schulz [1292] calculated MWts from light scattering measurements on soluble poly(diacetylenes) (P(diAc)) in chloroform solution, using standard methods, such as a Zimm plot, that we cannot go into detail here [1293] for an excellent reference). Davied et al. [1290] obtained unimodal particle size and MWt distributions for P(ANi) via dynamic light scattering studies in NMP solvent, which however does not yield a complete solution of P(ANi). The angular dependence of the scattering function measured in light scattering studies [1294] also gives information about the nature and extent of coiling, and the stiffness, of large CP chains. More recently, Seery et al. [1294] used dynamic light scattering to study the behavior of several P(ANi)'s in solution. Comparison with GPC data confirmed formation of supramolecular structures. Figure 33.11a, b summarizes some of the results of these authors.

33.5 Raman Spectroscopy

Like IR spectroscopy, Raman spectroscopy is a routine analytical technique that has been widely applied to CPs, and it is best to cite its use with specific examples. Most Raman studies of CPs to date have been under resonance conditions, since they have mostly used visible region frequencies which lie within the strong absorption bands of the highly colored doped state of the CPs or sometimes even beyond the bandgap. In certain cases, if the completely de-doped state of a CP is also studied along with doped states, then the former may correspond to non-resonance conditions, while the latter corresponds to resonance conditions. An important attractive

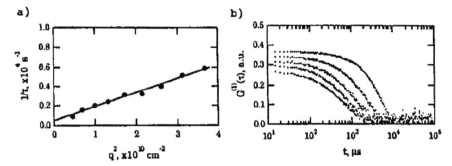

Fig. 33.11 Light scattering studies of P(ANi), after Seery et al. [1294]. (**a**) Inverse relaxation time vs. square of scattering vector, for 0.05 mg/mL P(ANi) in NMP at 30 °C. (**b**) Autocorrelation functions of polyaniline emeraldine base in m-cresol at 30 °C, concentration of 0.08 mg mL^{-1} (After Ref. [1294], reproduced with permission)

Fig. 33.12 Raman lines *trans*-(CH)$_n$ between 900 cm^{-1} and 1700 cm^{-1} at different exciting wavelengths. The lines marked with P$_1$, P$_2$, and P$_3$ show no shift, those marked with S$_1$ and S$_3$ are shifted with exciting wavelength (After Ref. [1281], reproduced with permission)

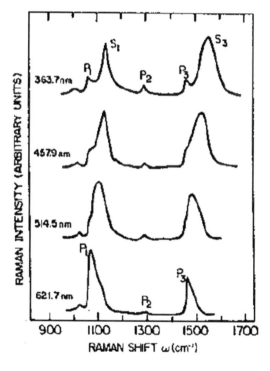

feature of Raman and resonance Raman spectroscopy is of course that it may be carried out in aqueous media and for powder samples as well.

P(Ac) shows Raman behavior strongly dependent on excitation wavelength (frequency dispersion), with resonant enhancement of many lines at energies above the bandgap due to electron–phonon coupling. Figure 33.12 shows Raman spectra of trans-P(Ac) at several excitation wavelengths, with the lines marked "P"

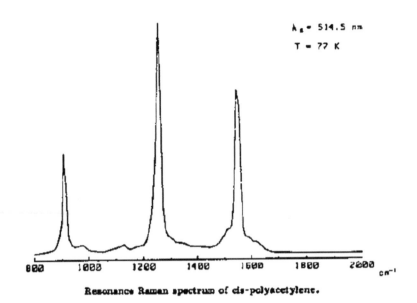

Fig. 33.13 Resonance Raman spectrum of cis-polyacetylene (After Ref. [866], reproduced with permission)

showing no frequency dispersion (the spectrum at 621.7 nm is not resonantly enhanced). Figure 33.13 shows a corresponding spectrum of cis-P(Ac) for one of the wavelengths. The lines at ca. 1460–1530 cm^{-1} are associated with the C=C stretching mode of P(Ac). The frequency dispersion of Raman lines in trans-P(Ac) (that in cis-P(Ac) is much less or nonexistent) has been the subject of much discussion, with one of the more popular explanations being that chains or regions with different conjugation lengths are excited at different frequencies [1295]. The cis-to-trans isomerization of P(Ac) with doping can be clearly observed with Raman spectroscopy. Lines characteristic of shorter chain lengths are typically evident with increased doping.

P(T) shows Raman active modes at 1500, 1458 (C=C antisymmetric stretch), 1045 (CH in-plane bending), and 700 cm^{-1} (ring deformations) [1296]. In P (3MeT), the C=C antisymmetric stretch is shifted to 1470 cm^{-1}, and overtone and combination bands associated with this mode, which are taken as an indication of a high degree of order in the CP [1297].

Tubino et al. [1298] have used Raman spectra at resonance and non-resonance wavelengths (454 and 1064 nm) of unique block copolymers containing thiophene and benzene rings connected with methyne units to arrive at an estimation of the proportion of the two monomer units, from the fact that only the thiophene ring absorption modes are resonantly enhanced. Hankin et al. [1299] have observed that resonance Raman spectra of several crystalline P(diAc) (Fig. 33.14 are predominantly a surface effect. Fong et al. [1300] have used Raman spectra to monitor the polymerization of pyrrole by aqueous ferricyanide and of aniline by persulfate in

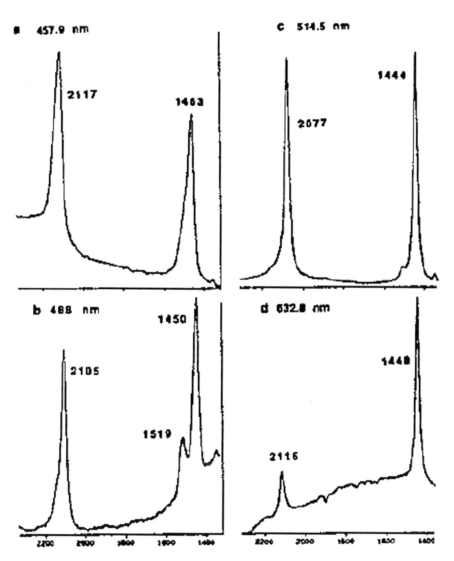

Fig. 33.14 Raman spectra for poly(diacetylene) for the 2200–1400 cm^{-1} region obtained with 457.9 (**a**), 488 (**b**), 514.5 (**c**), and 632.8 (**d**) nm excitation (After Ref. [1299], reproduced with permission)

aqueous perchloric acid, obtaining strong, distinct peaks for monomer and oxidizer, and determining the polymerization kinetics. Several Raman spectra of CPs as a function of doping level have been cited in the section on spectroelectrochemistry elsewhere in this book.

Fukuda et al. [1257] have carried out an interesting comparison of Raman spectra of P(ANi) in its three major redox states, viz., pernigraniline base (PB, completely oxidized), emeraldine base (EB, 50% oxidized), and leuco-emeraldine

a)   **b)**

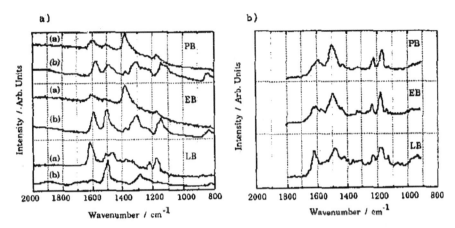

Fig. 33.15 (a) Nd:YAG laser-excited Raman (a) and infrared (b) spectra of each P(ANi)'s at different doping levels with resolution of 4 cm^{-1} and 2 cm^{-1}, respectively.(b) Ar^{+} laser-excited Raman spectra of each P(ANi)'s at different doping levels with resolution of 3 cm^{-1} (After Ref. [1257], reproduced with permission)

base (LB, completely reduced) under resonance (ca. 450 nm excitation) and non-resonance (1064 nm excitation) conditions (Fig. 33.15a, b). LB is the only species that does not absorb in the visible region and so shows no resonance Raman excitation. Under non-resonance conditions, the strength of the band at 1380 cm^{-1} for C=N stretching increased with degree of oxidation (Fig. 33.15a). Under resonance conditions (b of the figure) the spectrum for LB is nearly identical, but those for EB and PB are quite different: The C=N stretch mode at 1380 cm^{-1} nearly disappears. Additional peaks at 1175, 1230, and 1500 cm^{-1} appear, ascribed to polaron/bipolaron generation on photoexcitation, agreeing substantially with photoinduced absorption spectra (see below). Figure 33.16a, b gives higher-resolution Raman and resonance Raman spectra for the PB state of P(ANi), showing essentially the same effects.

Lefrant et al. [1301] carried out a detailed resonance Raman study of P(PV) and P(TV). They observed prominent bands at 966, 1174, 1330, 1550, 1586, and 1628 cm^{-1} for P(PV) and 654, 932, 1045, 1287, 1410, and 1586 cm^{-1} for P(TV) and claimed, uniquely, that there was no frequency dispersion effect for excitations from the blue to the violet region. Based on dynamical calculations, they assigned modes to four types of force constants associated, respectively, with the unperturbed benzene group, p-substituted benzene, benzene/vinylene coupling, and vinylene alone. Sakamoto et al. [1302] studied P(PV) doped with SO$_3$ vapors via resonance Raman and found evidence for both positively charged polarons and bipolarons, also observing that frequency dispersion in their data indicated that conjugation lengths for undoped "island" regions are shorter in doped polymer with conductivity ca. 10 S/cm as compared to that with 1 S/cm.

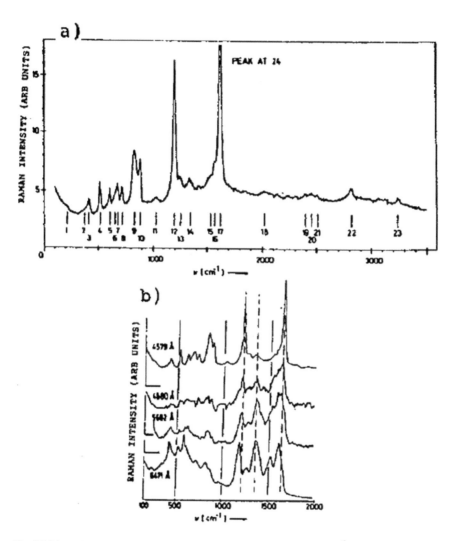

Fig. 33.16 (a) Raman spectrum of P(ANi), form D, excitation with 4579 Å as measured with an integration time of 4 sec and unpolarized scattered light. (b) Raman spectra of P(ANi), oxidized, pH 0 for excitation with various laser lines (not normalized and smoothed) (After Ref. [1301], reproduced with permission)

Kastner et al. [1303] recently carried out a resonance Raman study of poly (thienopyrazine), a low bandgap CP which has a pyrazine ring fused onto a thiophene ring. They observed two strong bands, 1520 and 1560 cm^{-1}, associated with the C=C stretching mode, which showed a frequency dispersion of 16 and 25 cm^{-1} from 742 nm through 458 nm excitation, associated again with varying conjugation lengths in the CP.

33.6   Thermal and Environmental Stability Measurements

33.6.1   Thermogravimetric Analysis (TGA)

Thermogravimetric analysis (TGA), a monitoring of the weight loss of a sample with increasing temperature under air, N_2, or other environments, is perhaps the simplest and most direct method yielding thermal stability data on CPs. TGA behavior in air and in inert gas is found to differ considerably, and the first decomposition mode in most CPs is thought to be dopant loss or decomposition.

Figure 33.17 shows TGA data in inert atmosphere for P(Ac) prepared by the conventional Ziegler–Natta route (A), via the AsF_5 route (B) and via the AsF_5/AsF_3 route (C). Curve (A) is fairly typical of many CPs under inert atmosphere, i.e., decomposition setting in only after ca. 400 °C. Figure 33.18 shows similar data, in inert atmosphere, for the commercially mass-produced CP P(PPS), showing exceptional thermal stability. Figure 33.19a, shows TGA data in air for as-prepared but neutralized P(ANi), THF-extracted P(ANi), and higher MWt, NMP-extracted P (ANi), showing stability to ca. 400 °C in air, and greater stability for the higher MWt CP (the apparent early decomposition in the NMP fraction is due to loss of retained solvent). Figure 33.19b shows TGA curves in air for doped P(ANi), illustrating that different dopants can have differing effects on stability. The two-step weight loss is thought by some workers [1304] to be due to dopant loss followed by dopant loss-induced polymer decomposition. Figure 33.20a, b shows that the stabilities in air of doped and undoped P(3-octyl-T) are considerably different. For undoped P(3-hexyl-T) in air, Inganäs et al. [963] observed substantial decomposition only above 350 °C. With undoped forms of a copolymer of benzyl and hexyl thiophene, decomposition under N_2 was observed at ca. 300 °C [1059]. P (Py) is, however, a much less thermally stable CP in air, even with more stable dopants such as tosylate. When air-unstable side chains are incorporated into

Fig. 33.17 TGA of (**A**) Ziegler–Natta poly (acetylene), (**B**) poly (acetylene) prepared using AsF_5, and (**C**) poly (acetylene) prepared using AsF_5/AsF_3 (After Ref. [1280], reproduced with permission)

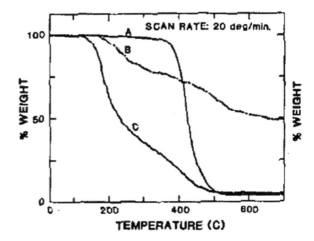

Fig. 33.18 TGA of (**A**) commercial poly(phenylene sulfide) and (**B**) the polymer prepared using AsF₅/AsF₃ (After Ref. [1280], reproduced with permission)

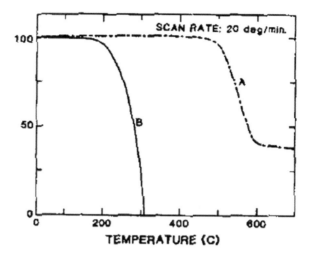

polymers such as P(T), as, for instance, in the three-urethane P(T) synthesized by Liu and Gregory, decomposition in air is extremely rapid (Fig. 33.21).

33.6.2   *Differential Scanning Calorimetry (DSC)*

Differential scanning calorimetry (DSC) measures heat flow of a sample with increasing temperature and is a determinant of exo-/endothermic transitions such as m.p. and glass transition temperature (T_g) and decomposition, which usually shows up as a broad endothermic transition. Figure 33.22 show DSC data for variously doped P(Py) under restricted oxygen ("closed pan"), open air, and N_2 conditions. The broad exothermic transition at ca. 150 °C for all samples appears to be due to a relaxation in the CP chain structure, while the dopant-dependent, sharp endothermic transitions at higher temperature appear to be due to dopant loss or decomposition. In the DSC (Fig. 33.23) of the P(3-urethane-T) discussed earlier, the small endotherm corresponding to T_g at ca. 62 °C and the broad endotherm corresponding to the polymer decomposition starting at ca. 220 °C are clearly visible, while the sharp endotherm in between is likely due to dopant or impurity loss. The highly thermally stable P(PS) shows a unique and characteristic DSC (Fig. 33.24). DSC analysis of undoped poly(3-hexylthiophene) [963] has shown that broad, endothermic decomposition beyond 220 °C is preceded by a sharp exothermic peak at ca. 193 °C.

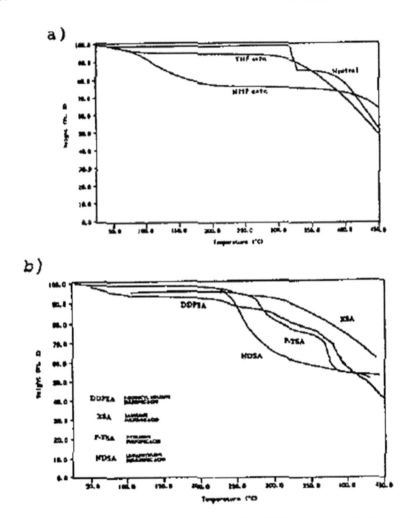

Fig. 33.19 (**a**) TGA curves of as-made neutral poly(aniline), THF-extracted and NMP-extracted polyaniline. (**b**) TGA curves of doped poly(anilines) (After Ref. [1304], reproduced with permission)

33.6.3   Other Thermal and Environmental Stability Test Methods

Dynamic mechanical tensile analysis (DMTA) has been used by several workers as a supplementary method to determine T_g and other critical data. The DMTA results for as-prepared, neutral P(ANi) and the THF and NMP soluble fractions (Fig. 33.25) show a minor transition at ca. -60 °C possibly due to a wagging mode of the benzene rings, and the major transition at ca. 150 °C, most likely associated with the T_g [1248, 1304]. The peak dispersion among the three fractions

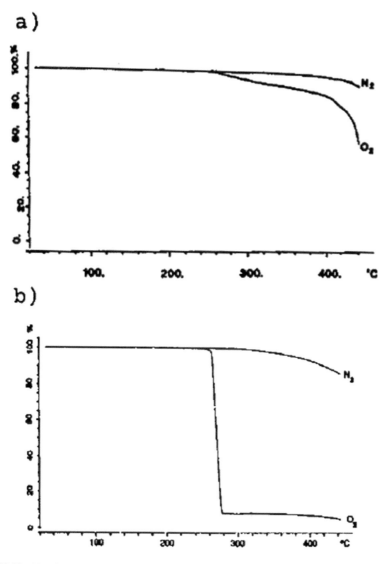

Fig. 33.20 (a) TGA curve of poly(3-octylthiophene) synthesized by the Grignard reaction. (b) The effect of Fe and Cl impurities (both around 1%) on the heat TGA stability of poly (3-octylthiophene) (After Ref. [1059], reproduced with permission)

indicates different chain lengths and morphologies. Doping induces a lowering of T_g, as the DMTA data of Fig. 33.26 show.

Perhaps the simplest thermal/environmental stability test with respect to one of the most important properties of CPs, the conductivity, is monitoring of this parameter as a function of prolonged exposure to high temperatures or extreme environments. A most intricate study of the effects of heat and humidity on P

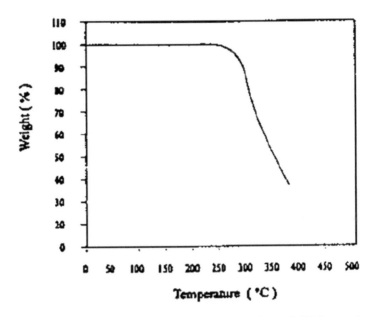

Fig. 33.21 TGA of a urethane-substituted poly(thiophene) (After Ref. [966], reproduced with permission)

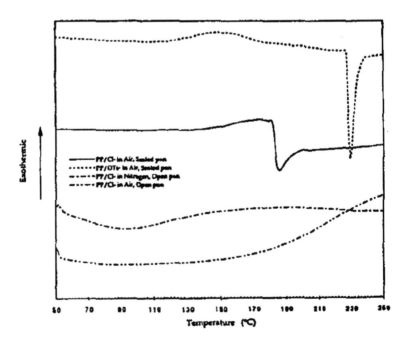

Fig. 33.22 DSC of chloride- and tosylate-doped poly(pyrrole) packed powders in nitrogen and air at 10 °C/min (After Ref. [1305], reproduced with permission)

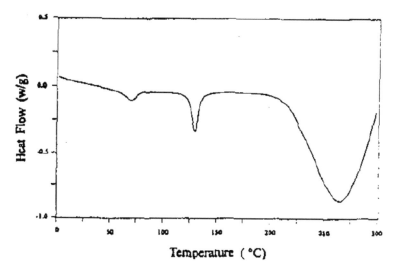

Fig. 33.23 DSC of the same urethane-substituted poly(thiophene) as in Fig. 33.21 (After Ref. [966], reproduced with permission)

Fig. 33.24 DSC of As-poly (phenylene sulfide). The scan rate is 20 °C/min (After Ref. [1280], reproduced with permission)

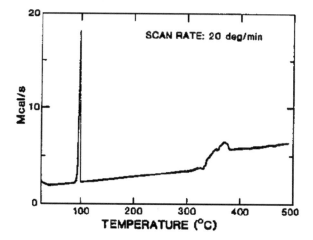

(3-octylthiophene) was carried out by Punkka et al. [906b]. Figure 33.27a shows that after the onset of a temperature of 70 °C for this CP, an initial, thermally induced conductivity increase is followed by steady decrease, even after heat is turned off, until the temperature reaches room temperature; the effect is more pronounced for a thinner film. Figure 33.27b shows that the relative conductivity decay is most pronounced for highly doped polymer. Wang and Rubner [1067] observed similar behavior for a poly(vinyl-hexyl-T). This behavior is however for doping with volatile $FeCl_4^-$, and is by no means universal. For instance, Chandrasekhar et al. [956, 1003, 1235b,c] have observed negligible conductivity decay for a poly(aromatic amine) doped with a unique sulfonate dopant. And

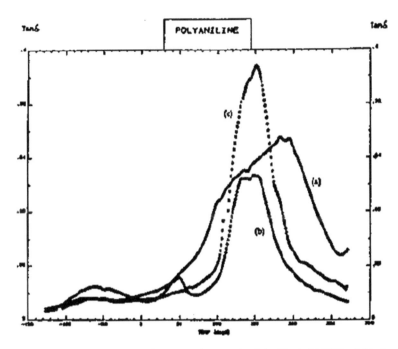

Fig. 33.25 DMTA from cast films of (**a**) as-made neutral poly(aniline). (**b**) THF soluble fraction. (**c**) NMP soluble fraction (After Ref. [1304], reproduced with permission)

Kulkarni and Mathew [1306] found that P(ANi) doped with tosylate shows a small conductivity decay on keeping in N_2 at 250 °C for 0.5 h, from 2.9 S/cm to 0.4 S/cm, but a larger decay, to 5.9×10^{-5} S/cm, on keeping in air under the same conditions; there is virtually no conductivity decay on keeping at 150 °C in air for 4 h.

Mathys and Truong [1307] described an FTIR spectroscopic and evolved-gas-analysis monitoring of the thermo-oxidative degradation of compressed P(Py) powder doped with $FeCl_4^-$ at 90 °C. In both air and N_2, they detected CO, CO_2, water, and NH_4Cl as evolved gases. They found that the chloroferrate dopant decomposes to ferric oxide and Cl_2/HCl. They surmised that atmospheric oxygen was probably present as an intrinsic dopant in the form of the highly reactive O_2^-. Thermal degradation was surmised to result in ring-opening reactions and a rapid loss of conductivity.

33.7 X-Ray Photoelectron Spectroscopy

X-ray photoelectron spectroscopy (XPS) or ESCA is another routine analytical technique whose application to CPs is best illustrated by examples. Wide-spectrum XPS is a simple tool for elemental analysis, useful, e.g., to see the presence of an atom belonging to a particular dopant to verify doping. Higher-resolution (core-

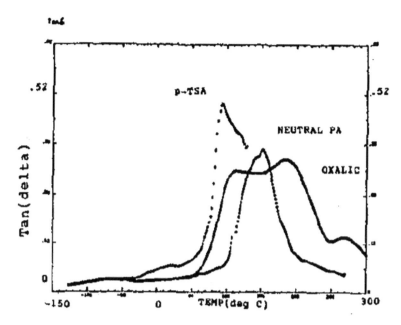

Fig. 33.26 DMTA from cast films of neutral poly(aniline), p-toluene sulfonic acid, and oxalic acid-doped polyaniline (After Ref. [1304], reproduced with permission)

level) XPS is typically useful for determining oxidation states of a central hetero-atom, e.g., the N-atom in poly(aromatic amines), hence elucidating the redox state of the polymer or at least of the central heteroatom. Finally, valence-level spectra, at very low energies, have been said to be usable for estimation of density of states, although their accuracy is questionable.

An example of wide-spectrum analysis is the spectrum of a copolymer of aniline with diaminobenzenesulfonic acid in Fig. 33.28 [1203]. As an example of the core-level spectra, the N 1 s spectra for neutral and oxidized (doped) P(DPA), and oxidized poly(N-phenyl-2-naphthyl amine) may be cited (Fig. 33.29). The spectral deconvolution shows that there is a single peak for the N-atom in the neutral polymer but two additional peaks appear in the oxidized polymers, ascribed to two distinct charge levels for the charged N-atom associated with bipolaron species, whose proportion of course remains smaller than that of the neutral N-atom [958]. The higher-energy peak (more charged N-atom) possibly represents the N-atoms directly in the vicinity of the dopant counterions [1308, 1309]. The deconvolution of the core-level N-1 s spectrum of doped P(ANi) in Fig. 33.30 is similar and shows three peaks, attributed in the same way. The N 1 s spectra for P (Py)ClO$_4$ similarly show three nonequivalent N-atoms (Fig. 33.31a). Intriguingly, the C 1 s spectra for this CP also show three peaks (Fig. 33.30b, c), and some workers have ascribed the highest-energy peaks in N-atom spectra to be due to "disordered" rather than dopant-proximate N-atoms [1009]. Interestingly, however, it has been observed in some samples of AsF$_6$$^-$-doped P(Ac) that no C 1 s peak

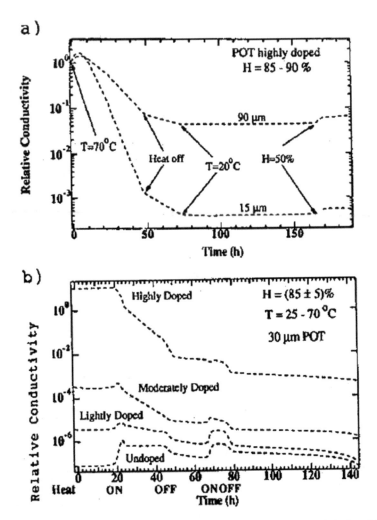

Fig. 33.27 (a) Relative conductivity decay in highly FeCl₃-doped poly(3-octylthiophene) for film thicknesses 15 and 90 μm with initial room-temperature conductivities of 1.1 and 2.4 S/cm, respectively. (b) Conductivity decay in 30 μm thick poly(3-octylthiophene) films at various doping levels. The temperature was cycled between 25 °C and 70 °C at a relative humidity level of (85 ± 5)% (After Ref. [906b], reproduced with permission)

splitting is observed up to doping levels of 10%, indicating that there may possibly be extremely efficient delocalization in this CP [1310].

Useful information on doping is available from core-level spectra of dopant atoms. For example, the Cl spectra for P(Py) prepared using FeCl₃ show two Cl equivalences, ascribed to $FeCl_4^-$ as also $FeCl_2^+$, and the Cl spectra of P(Py)ClO₄ show the presence of Cl^- as well as the expected ClO_4^- [1145]. Similarly, I-atom ($3d_{5/2}$) spectra have been used for estimation of extent, nature, and depth of

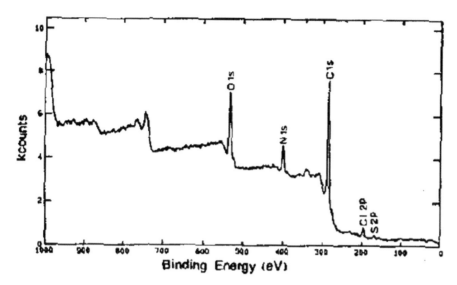

Fig. 33.28 An X-ray photoelectron (XPS) spectrum of ANi and diaminobenzene-sulfonic acid (DABSA). The film was grown (13 cycles) from 0.163 M ANi + 0.0185 M DABSA in 1 M HCl at a scan rate of 50 mVs^{-1} (After Ref. [1203], reproduced with permission)

Fig. 33.29 XPS N is spectra (**a**) neutral poly (diphenylamine), (**b**) oxidized poly (diphenylamine), and (**c**) oxidized poly(N-phenyl-1-naphthylamine). Dashed lines are the results of deconvolution (After Ref. [958], reproduced with permission)

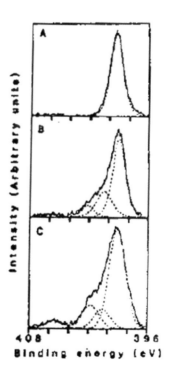

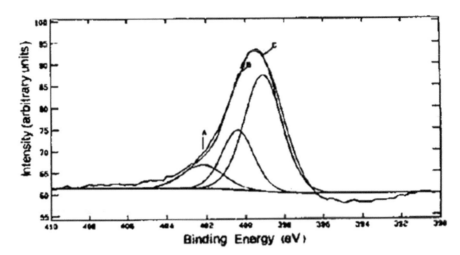

Fig. 33.30 N_{1s} XPS core-level spectra of P(ANi) (85 cycles). The finer lines are the fits to the data (After Ref. [1203], reproduced with permission)

I_2-doping of P(Ac) [1311, 1312]. XPS spectra can also be used to detect degree of decomposition or oxidation on surface films of CPs by monitoring core-level contributing peaks corresponding to C–O, C=O or N–O bonds, as, for instance, carried out by Nguyen et al. for P(PV) [1313].

33.8 Nuclear Magnetic Resonance Methods

Nuclear magnetic resonance (NMR) methods have somewhat limited application to CPs, due to requirements for solubility or constraints magic-angle spinning (MAS) or other adaptations to solid-state methods. Proton and ^{13}C solution NMR are of course confined to the soluble, de-doped forms of CPs, in solvents such as CDCl₃, and deuterated dimethylformamide. Solid-state methods have been applied with some success primarily to P(Ac). And nutation NMR studies have yielded some useful bond length information. Besides the illustrative studies cited below, many studies using NMR of CPs in conjunction with other analytic techniques for structural elucidation abound [1285, 1314, 1315].

Figure 33.32 illustrates the difficulty in studying CPs in the solid state. Here the low-temperature, 13C MAS spectrum of undoped poly(azulene) is shown together with that of a monomer derivative [1316]. The broad resonance for the polymer bespeaks delocalized charge or disordered polymer structure, but does not convey much additional information.

Figure 33.33a shows the proton NMR spectrum in CDCl₃ of de-doped poly (3-octylthiophene) [1317], while Fig. 33.33b shows the corresponding ^{13}C spectrum. As is evident, assignments here are much more straightforward. The 400 MHz

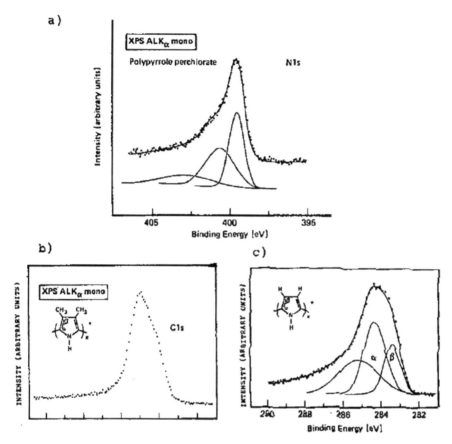

Fig. 33.31 (a) The N1 s XPS spectra of poly(pyrrole) perchlorate. The lower three curves show contributions from the three inequivalent nitrogens. (b) The C1s XPS core-level spectrum of poly (β, β'-dimethylpyrrole). (c) The C1s XPS core-level spectrum of polypyrrole perchlorate, where the lower curves show the contribution from the α-, β-, and disorder-type carbons (After Ref. [1009], reproduced with permission)

proton NMR spectra in $CDCl_3$ for de-doped P(3-butyl-T) and P(4,4'-dibutyl bithiophene) (Fig. 33.34) betray the greater structural order in the latter polymer [1318].

The effect of doping on a CP can be directly seen in Fig. 33.35. Here, the emeraldine salt form of P(ANi) is taken in conc. H_2SO_4, in which it is soluble (top spectrum), and then further oxidized (doped) in solution by adding a small drop of $(NH_4)_2S_2O_8$ (oxidant) solution in the same acid (bottom spectrum) [1192]. The single, broad resonance in the sample before oxidation, one observes resonances at ca. 130 and 145 ppm, attributed, respectively, to proton-bonded and N-bonded carbons in the benzene ring [1192], with no peaks corresponding to quinonoid-ring-type C-atoms. On oxidation, two additional lines, at 136 ppm and ca. 162 ppm

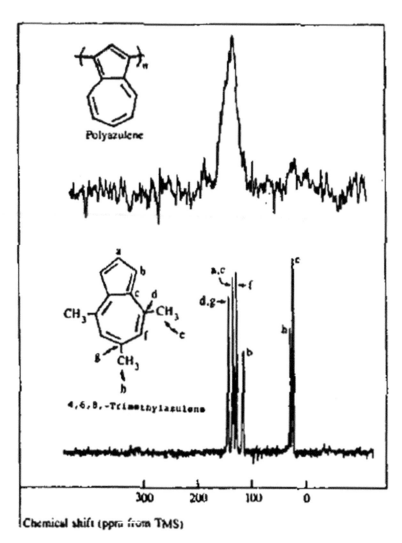

Fig. 33.32 Solid-state magic-angle ^{13}C NMR spectra of (**a**) poly(azulene)-ClO$_4$ and (**b**) trimethylazulene at 15 MHz and –150 °C (After Ref. [1318], reproduced with permission)

corresponding, respectively, to proton-bonded and N-bonded carbons in quinoid-type rings, suggesting heavy protonation for the oxidized CP.

The most useful application of solid-state cross-polarization MAS ^{13}C NMR studies to P(Ac) has been in the distinction between cis and trans isomers, and the isomerization. Figure 33.36 illustrates the distinction between undoped cis- and trans-P(Ac), while Fig. 33.37 illustrates the thermal isomerization process at 120 °C. There is a small downfield shift on doping, more pronounced in the cis-isomer, due to oxidation of the CP chain, and an accompanying broadening,

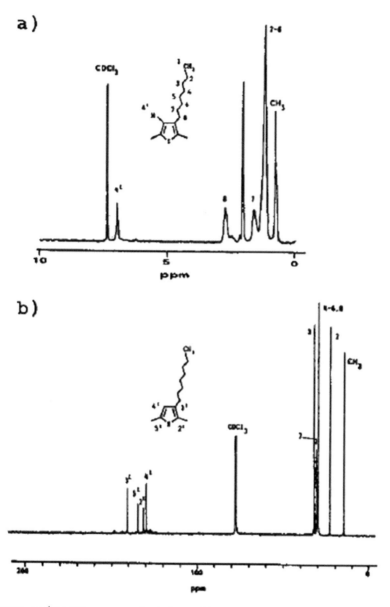

Fig. 33.33 (a) ^{1}H NMR spectrum of poly(3-octylthiophene), synthesized by Grignard coupling. (b) ^{13}C NMR spectrum of poly(3-octylthiophene), synthesized by Grignard coupling (After Ref. [962], reproduced with permission)

due to disorder effects (such as different degrees of oxidation along different chain segments).

The use of proton-decoupled ^{13}C nutation NMR studies of P(Ac) to model the bond length alternation in cis- and trans- P(Ac) is illustrated in Fig. 33.38

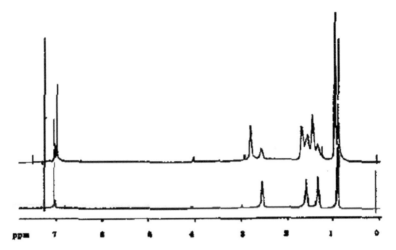

Fig. 33.34 400 MHz ^{1}H NMR spectra of solutions of poly(3-butylthiophene) and poly (4,4′-dibutylbithiophene) (lower) in CDCl$_3$ (After Ref. [1010], reproduced with permission)

[1321]. The spectral simulations indicate cis-isomer C–C bond lengths of ca. 1.37 Å and corresponding trans-isomer lengths of 1.36 and 1.44 Å in agreement with model X-ray crystallographic studies [1322]. The effect of doping on cis- and trans- P(Ac) on the solid-state cross-polarization MAS ^{13}C NMR spectra has been studied by Heinmaa et al. [1323]. They found that, peak broadening, downfield shift and other parameters are in agreement with the earlier cited studies.

33.9 ESR (EPR)

33.9.1 Basics of ESR (EPR) of CPs

Electron spin (or paramagnetic) resonance (ESR or EPR) and electron-nuclear double resonance (ENDOR) have been most useful when applied to CPs primarily for establishing the relationship of conductivity to the presence of unpaired electron density (and hence radicals) and, indirectly, the identity of charge carriers.

The ESR g factor for a free electron is 2.00232, and some CPs show g factors close to this value. The double-integrated intensity and related parameters can be used to fathom the concentration of unpaired electrons in the entire polymer system, yielding indirect information on charge carriers. The typical g factor for doped P (Ac) is ca. 2.0026, and the typical spin density ca. 10^{19} spins/g (with linewidth ΔH_{pp} ca. 0.5 G), pointing to 1 spin per ca. 1000 to 3000 -CH- (monomer) units [1324, 1325], while corresponding values for ClO$_4$$^-$-doped P(Py), which yields a narrow Lorentzian line, can typically be 2.0028 g factor, and 5×10^{20} spins/g with ΔH_{pp} ca. 0.3 G [1326], indicating a much lower proportion of monomer units per

Fig. 33.35 ^{13}C
n.m.r. spectra of the
emeraldine salt (**a**) before
and (**b**) after oxidation in
sulfuric acid (After Ref.
[1192], reproduced with
permission)

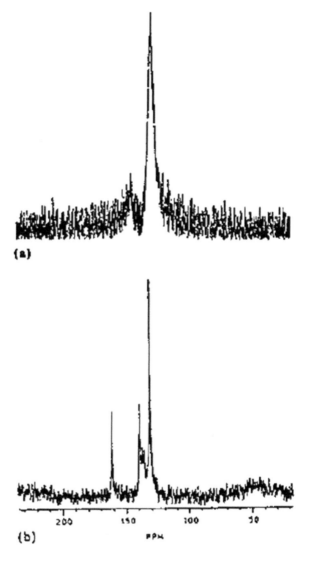

spin (one spin per ca. 200 monomer units). The radical (unpaired electron) concentration in most doped CPs appears to be of the order of 10^{19} spins/g, for example, it is ca. 5×10^{19} spins/g in the fused ring poly(1-aminopyrene) [984].

33.9.2   P(Py) and Poly(Thiophenes)

Perhaps the most significant ESR studies of CPs have been those on P(Py), which appeared to clearly establish the *absence of a relationship between conductivity and the presence of unpaired electrons* [1145, 1326]. Scott et al. [1326], in ESR studies

Fig. 33.36 Cross-polarization MAS ^{13}C-NMR spectra of pristine cis- and trans-poly(acetylene) (After Ref. [1320], reproduced with permission)

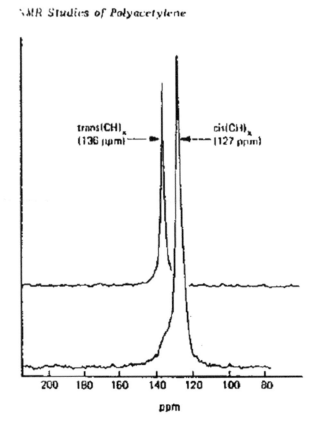

NMR Studies of Polyacetylene

trans(CH)$_x$ (136 ppm) — cis(CH)$_x$ (127 ppm)

of neutral and ClO_4^-- and O_2-doped P(Py), found a number of indications of a lack of correlation of unpaired electron density and conductivity: the spin concentrations of neutral, ClO_4^-- and O_2-doped P(Py) were of comparable magnitude, while the conductivity varied by a factor of 10^8; when going from 4 to 300 °K, there was no correlation between conductivity, which varied over many orders of magnitude and obeyed a 3D VRH relation (see discussion in an earlier chapter), and the linewidth, which increased by a factor of ca. 2; on doping with O_2, the conductivity changed by three orders of magnitude with little or no accompanying change in linewidth and intensity; and most importantly, little or no ESR signal could be observed in highly conducting, electrochemically cycled films. Figure 33.39 shows the variation of conductivity and ESR signal intensity in P(Py) as a function of O_2 doping found by these and other workers.

These results established early on that the charge carriers responsible for conduction in CPs may not be electrons and may be spinless bipolarons and gave an impetus to explanations of CP behavior based on condensed matter physics. Scott et al. [1326] postulated that the ESR signal in doped P(Py) arises from "accidental, neutral π-radical defects" which are more prominent in the oxidized state of the

Fig. 33.37 ^{13}C cross-polarization magic-angle spinning (CPMAS) spectra showing the *cis-trans* thermal isomerism in $(CH)_x$. The spectra *a* and *d* correspond to nearly pure *cis*- and pure *trans*-$(CH)_x$, whereas *b* and *c* correspond to thermal treatment at 393 K in vacuum for 10 and 40 min, respectively (After Ref. [1315], reproduced with permission)

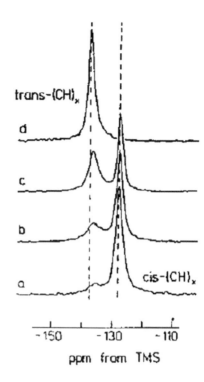

trans-$(CH)_x$

d

c

b

a

cis-$(CH)_x$

-150   -130   ~110

ppm from TMS

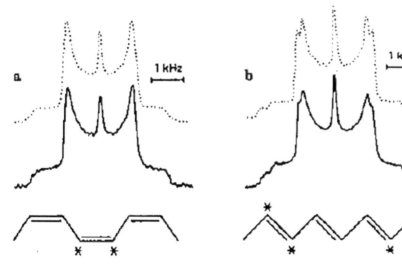

Fig. 33.38 Experimental and simulated (dotted line) mutation spectra at 77 K in (**a**) *cis*-$(CH)_x$ and (**b**) *trans*-$(CH)_x$ for the samples weakly enriched in ^{13}C pairs. The central peak corresponds to isolated ^{13}C spins. The deduced bond lengths are $r_{c=c} = 1.44$ Å and $r_{c=c} = 1.36$Åf or *trans*-$(CH)_x$ (After Ref. [1315], reproduced with permission)

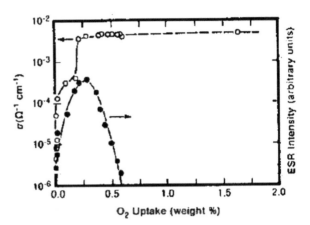

Fig. 33.39 ESR intensity and conductivity as a function of oxygen doping of neutral polypyrrole (After Ref. [1009], reproduced with permission)

CP. A similar explanation, i.e., that impurities (including metals and oxygen) in the oxidized state of the CP may be responsible for the ESR signal in most CPs where bipolarons (as opposed to polarons and solitons) are the presumed majority charge carriers, is now widely accepted.

Another example shedding light on the nature of charge carriers in CPs is that of P(3-MeT) [1327]. For neutral (undoped) polymer, there is no detectable ESR signal. At ca. 1% doping, e.g., with triflate ion, an ESR signal appears. At higher doping levels, however, there is a dramatic drop in spin concentration. Finally, at the highest doping levels possible, the ESR line shape changes from a symmetric Lorentzian to a Dysonian, resembling that of a metal (Fig. 33.40). These results taken together appear to indicate that charge carriers are paramagnetic polarons at low doping level, spinless bipolarons at intermediate doping levels, and possibly, electrons at the highest doping levels. Pure, neutral P(T) shows a symmetrical singlet ESR line with a g factor of ca. 2.0026, which has been taken to indicate that the unpaired electrons here belong to a defect in the C-backbone rather than to an S-atom containing species [1328].

33.9.3   Poly(Acetylene)

The ESR spectra of P(Ac) have also been studied in some detail. As-prepared, pure cis-P(Ac) shows no detectable ESR signal [1329]. Trans-P(Ac), which is generally obtained via thermal, chemical, or electrochemical conversion from the cis-form, shows an ESR signal which is the result of these conversions introducing fixed and mobile defects (such as solitons) or impurities [1330]. Indeed, cis-trans isomerization in P(Ac) can be conveniently followed by ESR (cf. Fig. 33.41).

At low levels of doping, ca. 1%, trans-P(Ac) shows a decrease in ESR signal due to some charge transfer from the paramagnetic solitons to dopants, yielding lower unpaired electron density [1331, 1332]. At intermediate doping levels for trans-P

Fig. 33.40 The ESR spectrum of P(MeT) doped with $SO_3CF^-_3$ ion at two levels, (**a**) P(MeT) $(SO_3CF^-_3)_{0.30}$ with a symmetrical Lorentzian line shape and (**b**) P(MeT) $(SO_3CF^-_3)_{0.50}$ with an asymmetrical Dysonian line shape (After Ref. [952], reproduced with permission)

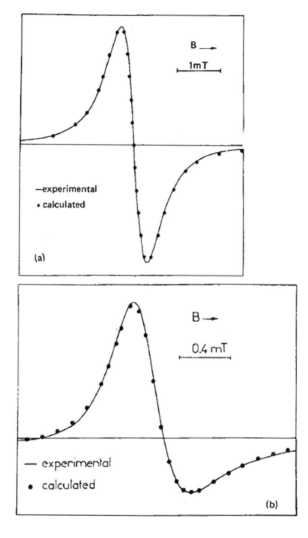

(Ac), ca. 1–6% corresponding to its characteristic dramatic increase in conductivity, the temperature-independent Pauli susceptibility, when compared to the conductivity, is a useful indicator of the plausible charge carriers (Fig. 33.42). While the conductivity increases dramatically at ca. 1% and plateaus at ca. 3% doping, the Pauli susceptibility does not increase dramatically until ca. 5% doping. These data have been taken to indicate that within the doping region to ca. 5%, charged solitons, which are nonmagnetic, are the predominant charge carriers. Only in the quasi-metallic regime, at doping >5%, does Pauli paramagnetism become observable. The magnetic behavior at these doping levels is not however completely that of a metal. For instance, although ESR line shapes come to be more Dysonian, like

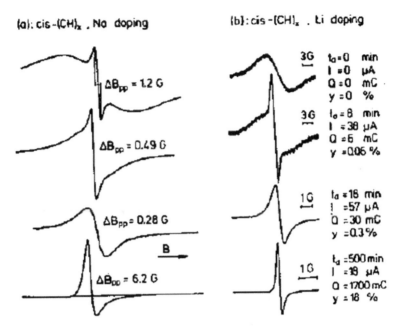

Fig. 33.41 Evolution of the ESR signal for various levels of doping in chemically (**a**) and electrochemically (**b**) doped $(CH \cdot yNa)_x$ [5.82] and $(CH \cdot yLi)_x$ [5.210], respectively. The arrow shows the increasing field direction. At high doping levels, the line shape is Dysonian in contrast to the Gaussian line shape for the pure *cis* material (After Ref. [1315], reproduced with permission)

that of a metal (cf. Fig. 33.41), other metallike behavior, such as temperature-independent Pauli susceptibility, is absent [1333]. Broadening of the ESR lines with decreasing temperature, ascribed to decreased mobility of solitons, is also observed.

33.9.4 P(ANi)s

Genoud et al. [1334] report a study of the ESR response of electrochemically cycled P(ANi)Cl as a function of pH and redox potential which appears to confirm the charge carrier behavior observed for poly(thiophene) and poly(3-methylthiophene) above. In the completely reduced state, the CP shows no ESR signal. As the CP is steadily oxidized (doped), the ESR signal reaches a maximum for an intermediate doping level at pH ~2, corresponding to a high population of polarons (Fig. 33.43). As the doping is further increased, the signal again falls off practically to nil (a residual signal ascribed to impurities and/or disorder is still observed). Hsu et al. [1183] report an interesting ESR study of a composite fibers of doped-P (ANi)/Kevlar. They found anisotropic g factor values, 2.00358 for magnetic field parallel to fiber direction and 2.00376 for perpendicular direction at room

Fig. 33.42 Variation with the dopant concentration y of the conductivity σ and the Pauli susceptibility X_P for iodine-doped polyacetylene (After Ref. [1279], reproduced with permission)

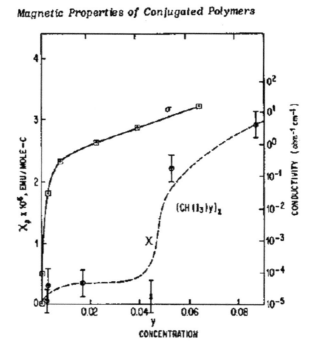

temperature, close to values for pure emeraldine-HCl (2.0038) [1335] and P(o-toluidine) (2.0034) [1336]. Decrease of line width with increasing temperature was taken to be due to motional narrowing, i.e., delocalization.

33.9.5   *Photoinduced ESR*

An interesting set of studies, of photoinduced ESR, has been carried out by Flood and Heeger [1337] and Moraes et al. [1338]. The former study provided indications that the ratio of photo-produced neutral to charged solitons in trans-P(Ac) was ca. 0.01. In the latter, very strong evidence of an increase in ESR intensity as a function of illumination was found for P(T) (Fig. 33.44). This and other evidence indicated that while (spinless) bipolarons may be the predominant charge carriers in the unexcited state of the doped CP, at least some of the charged excitations were (paramagnetic) polarons. This was in contrast to the situation with P(Ac), where at least some of the charged photoexcitations were seen to be solitons [1209, 1210].

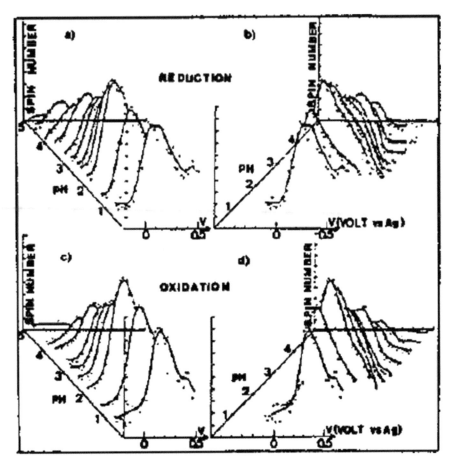

Fig. 33.43 ESR spin susceptibility as a function of the redox state and the pH of the electrolyte (increasing pH value), (**a, b**) during reduction of the polymer; (**c, d**) during oxidation. (**a, c**) representation in the low potential region; (**b, d**) in the high one (After Ref. [1334], reproduced with permission)

33.9.6   Other CPs

Although the ESR of many other CPs in the undoped as well as doped states has been studied [1290], many of these, for instance, poly(*p*-phenylene vinylene) (P (PV)) [1325] and poly(phenylene sulfide) (P(PPS)) [1339], display anomalous behavior not open to clear interpretation in terms of the expected charge carriers and doping/conductivity behavior in these CPs. There are also limitations to the interpretation of ESR data. For example, Wang et al. in their study of the emeraldine base for of P(*o*-toluidine), find a g factor of ca. 2.0054, close to that for an electron proximate to a N–H bond, which they take to indicate that the unpaired electron is delocalized over about one monomer unit only, a contention not entirely borne out by other experimental data [889].

Fig. 33.44 Photoinduced electron spin resonance in poly(thiophene). The ESR signal in the dark and under white light illumination are shown, as well as the photoinduced ESR absorption (After Ref. [1338], reproduced with permission)

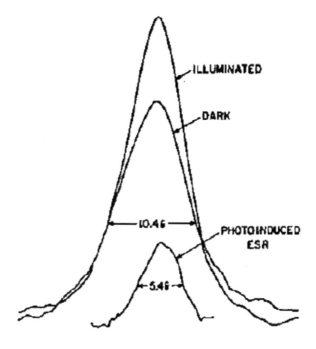

33.10 ENDOR

In ESR spectroscopy, e.g., of P(Ac), the hyperfine coupling is usually very small, while the linewidth is large (ca. 1 G). As a result, lines are not resolved, and much information is lost. This problem can be overcome by use of ENDOR (*electron-nuclear double resonance*) spectroscopy, in which nuclear spin transitions are induced in a paramagnetic species using a radio frequency and are detected by a change in the ESR signal intensity (an ESR detection of NMR transitions, so to speak). Figure 33.45 shows ENDOR data on pure and doped P(Ac) obtained by Raynor [1340]. Among other features, the broadness of the outer pairs of lines are adequately explained by the soliton model for P(Ac). The data also suggest that the ENDOR lines arise from undoped rather than doped trans-segments in the polymer [1340].

Triple resonance (double ENDOR) studies on cis- and trans- P(Ac) [1341] have shown that two separate sites are associated with the same paramagnetic center and that this defect is a highly delocalized π-radical interacting with sp^2 carbons and H-atoms bonded thereto, i.e., a trapped soliton. Comparison of the magnitude of the tensor elements with other data, e.g., that from MO calculations, indicates that this π-radical is delocalized over ca. 49 C-atoms [1320]. That the resonances in both cis- and trans- P(Ac) are due to the same type of defect is evidenced by studies that show that both isomers show similar ENDOR spectra at low temperature, ca. 2 °K (Fig. 33.46). Several ENDOR studies of P(Ac) and other CPs [1320, 1342, 1415] provide support to the solitonic (in P(Ac)) and polaronic (in other CPs) models of conduction in CPs.

Fig. 33.45 ENDOR data
on pure and doped poly
(acetylene), bottom to top,
increased doping (After
Ref. [1340], reproduced
with permission)

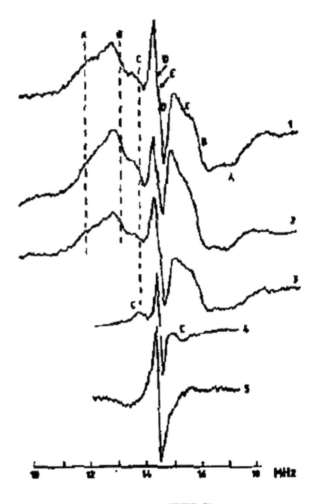

33.11   Electron Energy Loss Spectroscopy (EELS)

In transmission electron energy loss spectroscopy (EELS), the most common
technique, a beam of electrons (energy from 0 to kV region) is directed onto a
thin solid film, and the quantity, energy, and momentum of electrons transmitted,
i.e., inelastically scattered, through it are measured. The energy loss function
measured contains the most information obtainable about an electronic system
from a single experiment. In application to CPs, information such as the width of
the HOMO, yielding the degree of delocalization, is one of the important parameters measured. This has, for instance, been used to show that π-electrons are
increasingly localized in the CP series P(PP), P(PS), and P(PO) [1343]
(Fig. 33.47). Extensive EELS studies have been carried out for P(Ac)'s, P
(DiAc)'s, P(PP)-series, and P(Py) [867]. Salient results are cited.

Fig. 33.46 Proton ENDOR spectra of (**a**) cis-rich and (**b**) trans-rich (CH)$_x$ (After Ref. [1320], reproduced with permission)

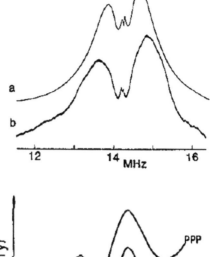

Fig. 33.47 EELS spectra of CPs as indicated (After Ref. [1343], reproduced with permission)

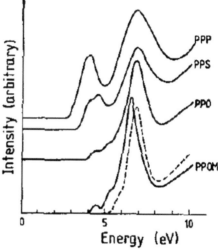

Early results on P(Ac) by Ritsko et al. and others [1344] showed two excitations (Fig. 33.48), assigned to interband transitions between π-π*, σ-σ* transitions. The EELS spectrum of AsF$_5$-doped (CH)$_x$ shows greatly increased absorption in the gap. Ritsko et al. [1345] have studied the EELS spectrum of poly(diacetylene)-10 L at various values of momentum transfer. Figure 33.49 shows EELS data obtained for P(Py) under various conditions by Ritsko et al. [1346].

33.12   Microwave Properties

33.12.1   *Interest in CPs, Properties Covered, Frequencies*

The strong interest in microwave properties of CPs has emanated from the fact that CPs are inherently strong absorbers of microwave and mm-wave electromagnetic

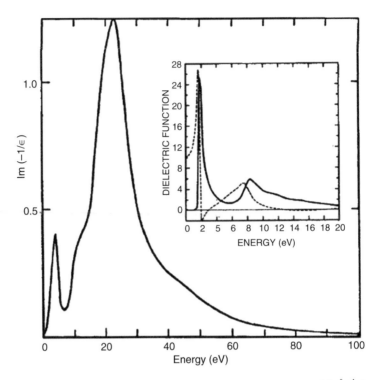

Fig. 33.48 Uncorrected energy loss spectrum for poly(acetylene) at $q = 0.1$ Å^{-1} from 0 to 100 eV. The insert shows the dielectric function of pristine PA ($\in_1$, dashed line; $\in_2$, solid line) (After Ref. [1343], reproduced with permission)

radiation, show high conductivity in these frequency regions, and when either alone or incorporated into composites display broadband microwave properties absent in metal- or C-fiber-filled materials. Applications of these microwave properties, which are covered elsewhere in this chapter, include aircraft structural materials capable of stealth against radar and other frequencies and withstanding lightning strikes and electromagnetic impulse (EMI) shielding materials.

Microwave properties encompass several, related individual properties of the CPs, including:

- *Complex permittivity (dielectric constant),* ε^*
- *Complex permeability,* μ^*
- *Conductivity* as a function of frequencies
- *Transmission, absorption,* and *reflection.* (Insertion loss or attenuation is derived from these)
- *EMI shielding effectiveness (EMI-SE)*
- *Radar cross section*

Frequencies of interest are generally in the range 5 MHz to 20 GHz (wavelength 60 m to 15 mm) but can have as wide a range as 1 KHz to 50 GHz (wavelength

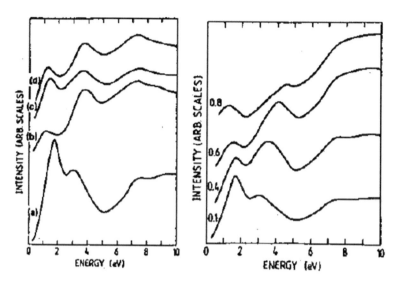

Fig. 33.49 (Left) Energy loss spectra at momentum transfer $q = 0.1$ Å^{-1} for (**a**)poly(pyrrole) doped with BF$_4$ (as-prepared), (**b**) heated at 10^{-8} Torr to 300 °C for 18 h., (**c**) exposed to 1 atm of AsF$_5$ for 18 h., and then (**d**) heated for 1 h. at 250 °C. (Right) Momentum dependence of valence excitations in poly(pyrrole) doped with BF$_4$ (as-prepared) for momenta indicated in Å^{-1} (After Ref. [1343], reproduced with permission)

300 km to 6 mm). For example, the higher frequencies are of greater interest in absorption studies, e.g., at the X (8 to 12 GHz) and Ku (12.4 to 18 GHz) bands of interest in communications, while frequencies in the GHz range are of greater interest for EMI shielding.

33.12.2 Parameters of Interest and Cavity Perturbation Measurements

One of the most widely applied early methods for characterization of the microwave properties of CPs has been the *cavity perturbation* method, wherein an electric cavity which resonates at a particular frequency is perturbed by insertion of a sample therein, normally as a flat sheet or film, and the resulting shift in frequency and intensity monitored. In Bauhofer's adaptation of the original method, a fitting procedure obviates the need to determine a sample depolarization factor which is difficult for samples of odd shape and required in the original method. The change in the resonant frequency, δ, and in the loss factor, Δ, caused by perturbation of the cavity field by insertion of the sample, are monitored:

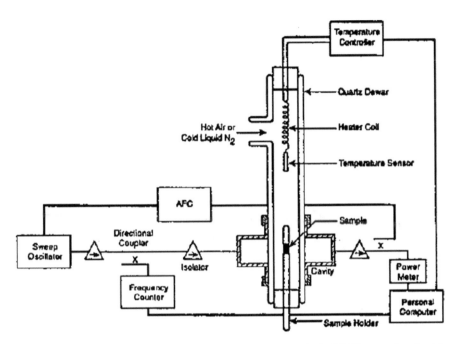

Fig. 33.50 Schematic of cavity perturbation method (After Ref. [1347], reproduced with permission)

$$\delta = \frac{f_o' - f_o''}{f_o'}$$

$$\Delta = \frac{1}{Q''} - \frac{1}{Q'} = \frac{1}{Qs} \tag{33.6}$$

Here f_0' and Q' are the resonant frequency and "quality factor" of the unperturbed cavity, and f_0'' and Q'' are those after perturbation of the cavity. In practice, it is usually necessary to measure δ and Δ over a wide temperature range to eliminate the depolarization factor.

A typical cavity perturbation setup is shown in Fig. 33.50. The sample is placed at the maximum of the electric field in the cavity. After temperature stabilization, the resonant frequency f''_0 and transmitted power P''_0 at resonance are measured. After sample removal, f'_0 and P'_0 are measured. δ and Δ are then obtained, respectively, from Eq. V.7.2 and

$$\Delta = \frac{1}{Q'} \left[\sqrt{\frac{P_0'}{P_0}} - 1 \right]$$

$$Q' = \frac{f_0'}{F} \qquad (33.7)$$

where F is the halfpower bandwidth of the unloaded cavity. δ and Δ are measured at various temperatures and $\delta(\text{max})$ and $\Delta(\text{max})$ determined. The complex permittivity

$$\varepsilon^* = (\varepsilon' - j\varepsilon'')\varepsilon_0 \qquad \left[\text{where } \varepsilon_0 = 8.854 \times 10^{-12} \text{F/m}\right] \qquad (33.8)$$

can then be determined as shown in Appendix 1, while the conductivity of the material at frequency f_0' is given by

$$\sigma = 2\pi f_0' \varepsilon'' \varepsilon_0 \qquad (33.9)$$

Another frequently quoted factor is the loss tangent, which is a direct measure of the absorption, and at a given frequency is

$$\tan(\delta) = \varepsilon''/\varepsilon' = \sigma/\left(f_0' \varepsilon^* \varepsilon_0\right) \qquad (33.10)$$

Loss tangents of the order of 10.0 indicate strong absorption of radiation. Both $\tan(\delta)$ and ε'' are good measures of microwave absorption, and the objective in most application-oriented studies is usually to obtain *non-resonant*, i.e., broadband, absorption across the desired frequency range.

33.12.3   Network Analyzer Based Methods

Since methods such as the cavity perturbation method do not easily permit wide-band (multifrequency analysis), several improvements, notably methods based on automated network analyzers (ANAs), are now widely used. In these methods, the primary measurement is that of the scattering parameters corresponding to a reflected wave (S_{11}) and a transmitted wave (S_{21}), as a function of frequency. In these, the sample is taken in the form of a waveguide. One illustrative formulation, used by Naishadham et al. [1347, 1348], is described in Appendix 1. In this, the reflection and transmission coefficients for the sample, Γ and T, are first calculated from the scattering parameters. Further analysis leads to the complex permittivity and permeability, ε^* and μ^*, respectively, for each frequency. The permittivity can then be used to calculate the microwave conductivity.

33.12.4   Transmission and Reflection Measurements

The complex transmission and reflection coefficients, Γ and T, and thence the absorption as well, can be determined indirectly according to the analysis described in the previous section. They may also be determined directly. We may describe a typical direct method, employed at Mission Research Corp., Dayton, OH [1349]. Samples are ca. 60 cm square. Transmission measurements are done in a transmission tunnel typically for normal, 30°, 45°, and 60° angles of incidence for two orthogonal linear polarizations, over a 1–20 GHz frequency range in increments of 25 MHz. The tunnel comprises a wideband horn radiating into an absorbing box, to eliminate stray radiation, with an aperture which radiates a planar field. This field impinges on a conducting septum covered with the sample. Opposite the septum is another absorbing box with a wideband horn receiving the transmitted energy. For reflection measurements, a sample of identical dimension is mounted in front of an absorbing wall and the reflected energy monitored via two broadband horns. The reflection and transmission coefficient data can then be used to calculate sample absorption. All measurements are carried out using a vector network analyzer. The complex permittivity of the sample can also be calculated from these transmission and reflection measurements (see Appendices 1, 2) and can be correlated to that measured using the waveguide format described in the previous section.

33.12.5 EMI Shielding Effectiveness (EMI-SE) Measurements

The shielding effectiveness of a material, S, is measurable from the intensity of an electric or magnetic field with and without a shield in place. It is also related directly to the complex transmission coefficient as defined above, as the equation below shows:

$$S = 20 \log^{10}\{A(\text{unshielded})/A(\text{shielded})\} = 20 \log_{10}\{1/\mathbf{T}\}$$
$$= \mathbf{A} + \mathbf{R} + \mathbf{B} \tag{33.11}$$

where A, electric field (E) or magnetic field (H), $\mathbf{T}$, the complex transmission coefficient, and $\mathbf{A, R, B}$ are, respectively, the absorptive loss, reflective loss, and multiple reflection factor components of the shielding. When the distance from the sample to the radiation source is $> \lambda/2\pi$, this is denoted as the far-field region, where E and H are not independent. When this distance is $< \lambda/2\pi$, this is denoted as the near-field region, where E and H are independent. Generally, it is useful to carry out both near- and far-field measurements. For commercial as well as defense applications, the frequency range of interest for EMI-SE measurements encompasses the 10 KHz to 1GHz range. As is evident, the microwave measurements described earlier readily yield the complex transmission coefficient, from which the EMI-SE can be calculated. There are however a number of standard methods for carrying out EMI-SE measurements. Among these are ASTM (American Society for Testing and Materials) ES-7-83 and MIL-STD (US military standard) -462.

33.12.6 Radar Cross Section (RCS)

Evaluation of RCS can also in principle be carried out using measurements similar to those above. However, additional factors such as the dimensions and shape of the sample and a strong directional component (incident and re-scattered radiation) come into play. It cannot be proposed to enter into a detailed discussion of these here; reference is made to excellent volumes available for the purpose, such as the *Radar Cross Section Handbook* [1350, 1351].

33.12.7 Salient Results

Most work to date has concentrated on a particular aspect of microwave properties, e.g., conductivity or dielectric constant, with few studies of the complete spectrum of properties over broad frequency ranges. For example, Fig. 33.51a, b show the DC vs. microwave (6.5 GHz) conductivity and the microwave (6.5 GHz) dielectric constant vs. temperature for a series of poly(anilines) measured by Javadi et al. [894]. The behavior observed – microwave conductivity greatly exceeding DC conductivity for higher doping levels and dielectric constant being independent of temperature for low doping levels – is typical of CPs. Buckley and Eashoo [1305] obtained relatively poor values for ε' and ε", ca. 90 and 60 (at the Ka band, ca. 33 GHz) for compacted P(Py)/Cl powder.

Figure 33.52 [894, 1336, 1352] shows the loss tangents measured by the MacDiarmid and Epstein groups for a series of poly(anilines) as a function of temperature at 6.5 GHz. These span an order of magnitude depending on polymer orientation, doping, and other factors. The same groups have observed a strong dependence of permittivity on polymer morphology, as the 6.5 GHz data in Fig. 33.53 show. These authors claimed to have observed an even more striking morphological dependence in data on the highly ordered P(ANi)/camphorsulfonic acid (CSA) prepared in *m*-cresol medium. Here, a negative dielectric constant similar to a metal and corresponding to a maximum in conductivity was observed in Fig. 33.54.

Many workers have measured microwave properties of blends of CPs with thermoplastics or with materials such as Teflon. Figure 33.55 shows ε" and ε' for P(Py)/Teflon blends measured at 2 GHz by Lafosse [1149]. It is seen that the rise in permittivity closely approximates the percolation threshold for DC conductivity and that for the more conductive blends (P(Py)> ca. 0.15), the loss tangent is of the order of ten, indicating good absorption. Hourquebie et al. [1354] studied P(Py) blends with a butyl elastomer, an epoxy, and Teflon emulsion and showed that a plot of log(ε") vs. log(frequency) was linear over the 130 MHz to 18 GHz range; their data however showed attenuation to be poor or moderate. In a study of P(Py) "latexes," Henry et al. [1355] obtained typical values of ε' and ε" at 5 GHz of ca. 945 and 1086. In a study of P(Py) blends with PVC, Jousse et al. [1146]

Fig. 33.51 (a)
l $\log_{10}\sigma$(6.5 GHz)
vs. $\log_{10}\sigma$(dc) at 295 K for
emeraldine polymers of
varying protonation level.
(**b**) $\log_{10}$(Conductivity) vs
$T^{-1/2}$ for emeraldine
polymer of composition
$x = 0.30$ and 0.50. The
symbols are the data at
6.5×10^9 Hz, the dashed
lines are data at dc (After
Ref. [894], reproduced with
permission)

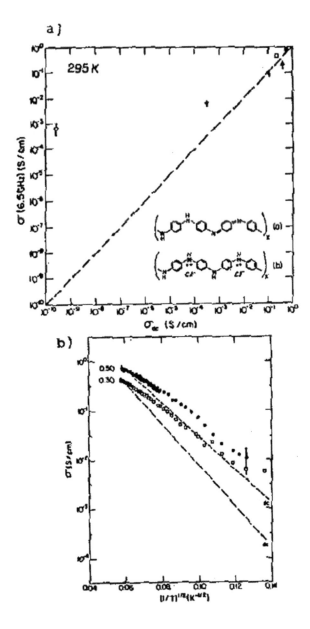

observed that for a pressed blend, "classical" microwave behavior, i.e., a monotonic fall of permittivities with frequency (Fig. 33.56a), is observed, while for injection-molded blends (Fig. 33.56b), a dielectric relaxation is observed (at ca. 5 GHz). This behavior is reflected in the corresponding reflection coefficient data, Fig. 33.57a, b, in which a resonance (at ca. 10 GHz, $\lambda = 3.3$ cm) is observed for the injection-molded material.

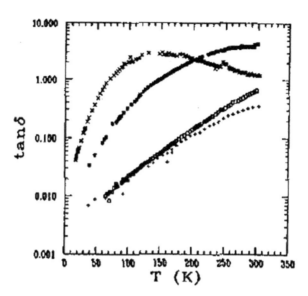

Fig. 33.52 Loss tangent at 6.5 GHz vs. temperature for four-folded stretched emeraldine hydrochloride parallel to the stretch direction (x), perpendicular to the stretch direction (*), unoriented poly (o-toluidine) hydrochloride (o), and unoriented self-doped sulfonated polyaniline (+) (After Ref. [888], reproduced with permission)

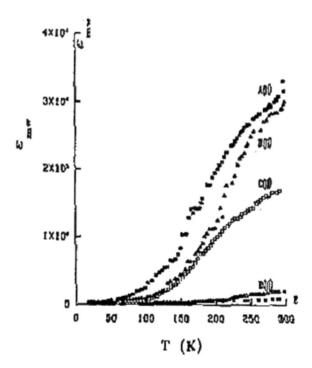

Fig. 33.53 Dielectric constant for P(ANi) HCl at 6.5×10^9 Hz for various morphologies (After Ref. [1353], reproduced with permission)

Work on CP-coated fabrics has been promising. Kuhn et al. at the Milliken textile company [1138, 1139] have demonstrated attenuations (Insertion Losses) as high as 28 dB at 8–10 GHz with P(Py)-coated glass fiber fabrics. Olmedo et al.

Fig. 33.54 Temperature-dependent microwave frequency dielectric constant of poly(aniline)-camphorsulfonic acid (PAN-CSA) prepared in CHCl$_3$ and *m*-cresol. Inset: microwave frequency conductivity of PAN-CSA (*m*-cresol). Data were recorded at 6.5×10^9 (After Ref. [1353], reproduced with permission)

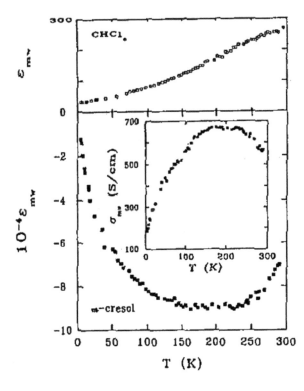

Fig. 33.55 Permittivity of P(Py)-Teflon composites (blends) at 2 GHz (After Ref. [1149], reproduced with permission)

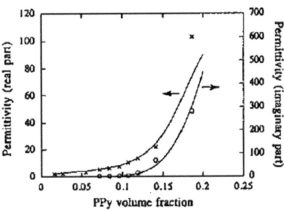

[1356] studied textiles into which P(ANi) and poly(alkyl-thiophenes) were incorporated and observed that the dielectric properties are not affected drastically by varying dopant and monomer chain length but do depend on doping level (Fig. 33.58a, b).

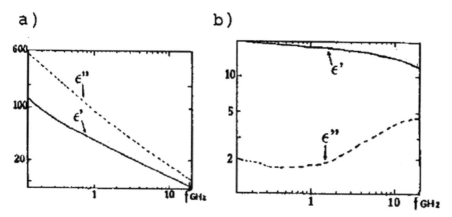

Fig. 33.56 Permittivity of P(Py)-PVC blends. (**a**) Pressed. (**b**) Injection-molded (After Ref. [1146] reproduced with permission)

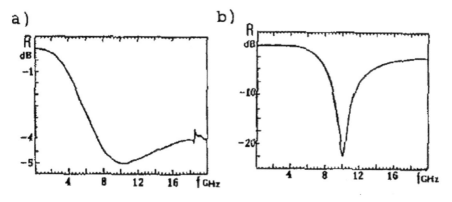

Fig. 33.57 Reflection coefficient data for P(Py) PVC blends derived from Fig. 33.56 (After Ref. [1146], reproduced with permission)

33.13   Photo-/Electroluminescence, Photoinduced Properties

33.13.1   Methods

A number of CP properties are directly related to function of devices such as light-emitting diodes (LEDs). These include *photoinduced absorption* (*PIA*) and the related *photoinduced bleaching* (*PIB*), *electroluminescence* (**EL**), *photoluminescence* (*PL*), and *fluorescence*.

In the most common implementation of PIA, a continuous wave (CW) laser is used to excite (pump) the sample. A white light source (probe), such as a Xe lamp, is then used to monitor the transmission (T). The laser is chopped at a frequency

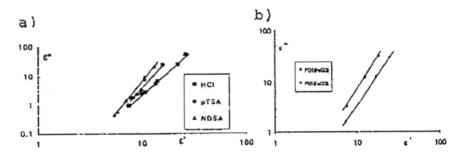

Fig. 33.58 (**a**) ε''_{10GHz} versus ε'_{10GHz} for polyaniline samples doped with three different acids, as indicated. (**b**) ε''_{10GHz} versus ε'_{10GHz} for poly(aniline) samples doped with three different acids: (HCL), p-toluene sulfonic acid (PTSA), and naphthalene 1,5-disulfonic acid (NDSA) (After Refs. [1356, 1357], reproduced with permission)

such as 100 Hz and the laser-induced change in transmission, ΔT, monitored using lock-in detection. The PIA is then expressed as

$$-\Delta T/T = l\Delta\alpha \qquad (33.12)$$

where $\Delta\alpha$ is the PIA absorption coefficient, and l is the laser penetration depth, $l = d$, the sample thickness, for weakly absorbing samples, and $l = 1/\alpha(l)$ for strongly absorbing samples, where $\alpha(l) = $ absorption coefficient for the laser in the sample. PIB occurs when the sign of $\Delta T/T$ is +, i.e., $\Delta\alpha$ is <0.

In EL, luminescence in a CP sample results from radiative decay of excited states (excitons), generally singlets, generated from combination of holes and electrons. These are generated (or "injected") in sandwich assemblies of two layers of (the same or different) CPs electrically by applying appropriate voltages at the opposing electrical contacts; a single layer of CP can also be used. In PL, the holes and electrons are generated photonically, usually by radiation at or beyond the bandgap of the CP.

Shim et al. recently studied [1358] the PL and EL properties of unique blends of poly(2-methoxy-5-(2-ethylhexyloxy)-1,4-phenylene vinylene] (MEH-PPV) with aklystyryl carbazole group containing poly(methacrylate) (CZ-PMA). These blends showed two isolated PL emissions at 440 and 560 nm corresponding to their component polymers, but only one EL peak, at ca. 580 nm, identical to that of MEH-PPV alone, but with very significantly enhanced intensity. Figure 33.59 shows some of their data.

33.13.2 Salient Results

Good examples of the close correlation of PL and EL spectra, i.e., photo- and electrically induced excitations in CPs, are provided by the data in Fig. 33.60a, b for

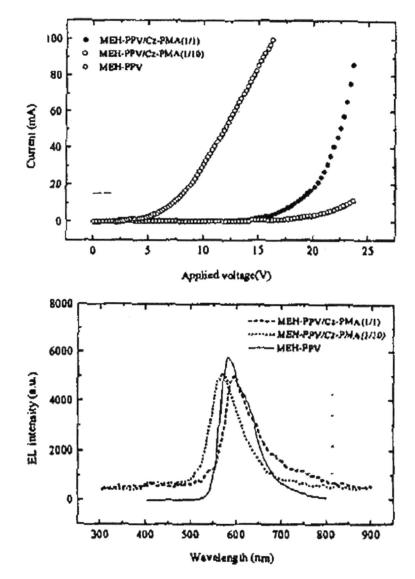

Fig. 33.59 EL spectra of MEH-PPV and the polymer blends, as indicated (After Ref. [1358], reproduced with permission)

P(PV), which together with its derivatives remains one of the most promising CPs for use in LEDs. EL spectra for P(PV) and several derivatives are given in Fig. 33.60c.

Figure 33.61a, b shows the PIA spectrum, and its dependence on temperature and laser chopping frequency, of a CN-derivative of P(PV); the frequency dependence gives an indication of the lifetime of the excitons.

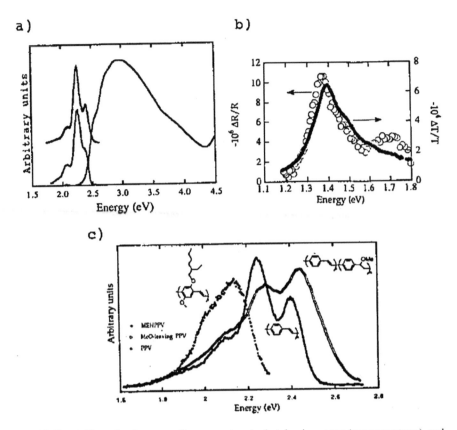

Fig. 33.60 (a) Electroluminescence (lower spectrum), photoluminescence (upper spectrum), and absorption spectra for poly(p-phenylene vinylene) (PPV) at room temperature. (**b**) PIA absorption in PPV (filled circles) and induced absorption measured in an LED formed with indium/tin oxide, PPV, calcium under forward bias (open circles). (**c**) Electroluminescence spectra from PPV and two derivatives (After Ref. [1022], reproduced with permission)

A comparison of PIA with PL for the same CP yields useful information. Tubino et al. [1359] studied a series of CPs with alternating thiophene (T) and benzene (B) units coupled with azomethine linkages. Their results for the PL and PIA of poly(T_3B_2) and poly(T_6B_2) polymers are given in Fig. 33.62a, b. Perhaps one of the most interesting comparisons is that between dopant- and photoinduced optical transitions. Just such a transition is seen, for P(trans-Ac), in Fig. 33.63a, b. Here, the broad, ca. 0.45 eV absorption is assigned to a charged soliton, while the sharp, ca. 160 meV and broad >200 meV absorptions correspond, and the ca. 110 meV transition induced by doping shifts to 60 meV in the PIA (Fig. 33.64).

The use of PIA to probe dopant-induced changes and the photogeneration of charge carriers is illustrated in the P(trans-Ac) spectra in Fig. 33.65. In (a), the strong IR absorption at ca. 1390 cm^{-1} results from positively charged soliton states bound to the dopant counterion. In (c), the analogous, photoinduced absorption,

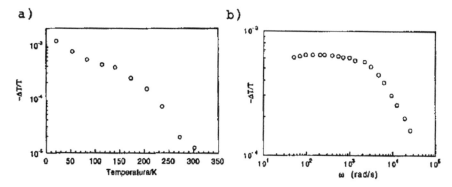

Fig. 33.61 (a) Temperature dependence of the photoinduced absorption for the cyano derivative of PPV, measured at 1.2 eV. (b) Frequency dependence of the photoinduced absorption for the cyano derivative of PPV measured at 30 K and at 1.2 eV (After Ref. [1360], reproduced with permission)

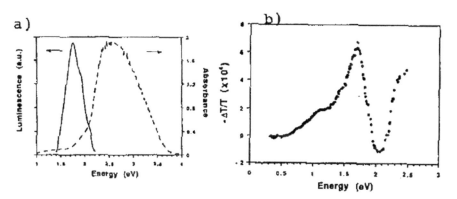

Fig. 33.62 (a) Optical absorption at 300 K (dotted line) and photoluminescence at 77 K (solid line) of T_3B_2. (b) Photoinduced absorption of T_3B_2 excited at 488 nm at 77 K (After Ref. [1359], reproduced with permission)

from above-gap excitation at 632 nm, is shown. The spectrum in (b) is obtained from excitation at the dopant-induced absorption band, i.e., mid-gap excitation. According to Orenstein [1361], the reduction in absorption at 1390 cm^{-1} indicates that photoexcitation at this mid-gap band induces releases a hole from a pinned +- soliton center. The resulting neutral soliton state is no longer IR active.

An example of PIB, for P(Ac), is shown in Fig. 33.65. Here, the ca. 1.4 eV PIA peak, ascribable to absorption from neutral soliton defects, progressively diminishes with increasing temperature, until it is negative, i.e., bleaching is occurring, at temperatures above ca. 150 K. This ca. 1.4 eV PIA is much longer lived in P(trans-Ac) than in P(cis-Ac), as ps-regime PIA decay data have shown. Sariciftci et al.

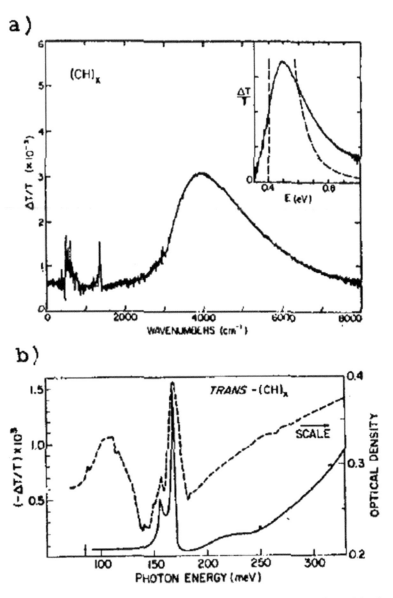

Fig. 33.63 (a) PIA for trans-(CH)$_x$ for photon energies <1 eV. (b) Comparison of the dopant-induced (dashed line) and photoinduced (solid line) IR-active modes (After Ref. [1361], reproduced with permission)

[887] studied PIA of P(trans-Ac)-poly(vinyl butyral) (PVB) blends and found a strong bleaching peak at ca. 1.9 eV (Fig. 33.66). The dependence of the PIA on the chopping frequency indicated a longer lifetime for the P(trans-Ac)/PVB blend as compared to the cis-blend (Fig. 33.67).

Fig. 33.64 Spectra which illustrate the process of generation of charge carriers by extrinsic photoabsorption. (**a**) shows the strong IR active mode near 1390 cm^{-1} due to dopant-induced carriers. (**c**) shows the optical absorption in this spectral region of photocarriers generated by above-gap excitation. (**b**) doped (CH)$_x$ photoexcited in the impurity-induced mid-gap absorption band (After Ref. [1361], reproduced with permission)

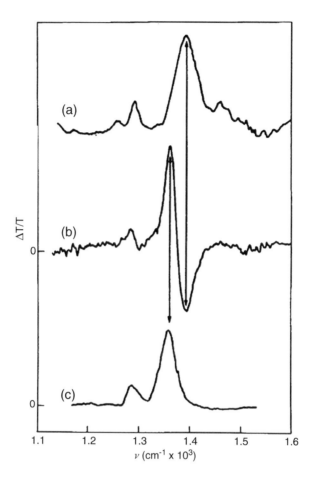

A comparison of EL, PL, and absorption for the same CP can be seen in the data in Fig. 33.61a, cited earlier. The EL in this case shows ca. 25% the efficiency of the PL. Figure 33.68 [1362] similarly compares the absorption and PL spectra of a poly (3-octylthiophene) film.

In the case of P(T), the PIA spectra are simple and their interpretation straight-forward [1363]. Two PIA bands, at ca. 0.45 eV and 1.25 eV, are observed together with several strong and weak IR bands. The 0.45 eV and 1.25 eV bands are assignable to photogenerated bipolarons, and their shift with respect to the dopant-induced bipolaron bands which occur at 0.8 eV and 1.8 eV is ascribed to the dopant binding energy. Later work by Moraes et al. [1364], where PIA data were obtained in combination with photoinduced ESR data, showed that at least some of the charged photoexcitations were polarons, with spin 1/2.

For an NMP solution of the leuco-emeraldine base form of P(ANi), the PL spectrum is compared with the absorption spectrum in Fig. 33.69. Here, the excitation is at the bandgap, and the luminescence decays over ca. 4 ns to

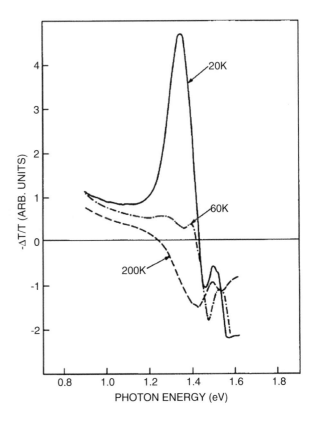

Fig. 33.65 PIB in poly
(acetylene) (After Ref.
[1361], reproduced with
permission)

practically zero [1365]. For the emeraldine base form of P(ANi), this luminescence
is absent.

Hilberer et al. [1366] have synthesized a series of new block copolymer derivatives of P(PV) (Fig. 33.70), which appear promising in LED devices (see Applications section).

Schulz et al. [1368] described a new series of poly(oxadiazole) derivatives of the
structures shown in Fig. 33.71 with promising properties for use in single-layer
LEDs. Some of their results are shown in Fig. 33.72.

Since P(PV) has now become a common building block or element for CPs
showing PL/EL properties and usable in LEDs, PL/EL studies of copolymers and
derivatives of P(PV) abound. For example, Dridi et al. [1369] recently described the
PL properties of a novel copolymer of P(PV) with an ether derivative of P(PV), and
Benjamin et al. [1134] described EL and PL studies of a new series of promising
copolymers of P(PV) and such moieties as naphthalene (structures shown in
Fig. 33.71). Figure 33.73 show EL and PL spectra of these polymers and P
(PV) in a comparative fashion. In a similar vein, Kim et al. [1370] described a
series of P(PV) derivatives containing organosilicon and carbazole units (see

Fig. 33.66 **(a)** Photoinduced absorption spectrum of *trans*-PA-PVB spin-cast film with 50 mW, 514.5 nm Ar ion laser excitation chopped with 273 Hz at 80 K (in-phase component). **(b)** PIA spectrum of PA-PVB solution in DMF at room temperature with 273 Hz chopped, 50 mW, 514.5 nm Ar ion laser excitation (in-phase component) (After Ref. [887], reproduced with permission)

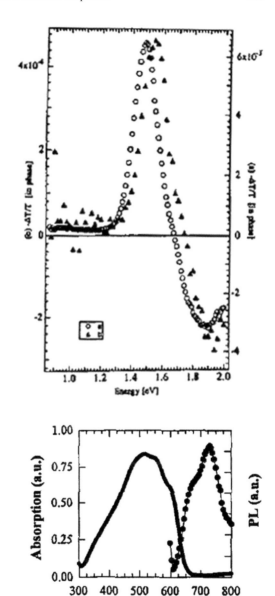

Fig. 33.67 Absorption and PL spectra of poly (3-octylthiophene) in film. The PL was recorded after excitation at 550 nm (After Ref. [1362], reproduced with permission)

elsewhere in this book for synthesis and structure). Figure 33.74 shows EL/PL spectra for one of these.

Cho et al. [1128] reported the syntheses of a series of copolymers containing dihexylfluorene units coupled with such moieties as phenylene, described in an earlier chapter in this book. Figure 33.75 shows the structure, absorption, and PL spectra for one of these.

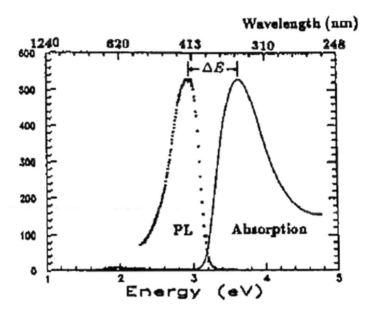

Fig. 33.68 Absorption and photoluminescence spectra for leuco-emeraldine base in NMP, the exciting wavelength for the PL spectrum is ~345 nm (After Ref. [1367], reproduced with permission)

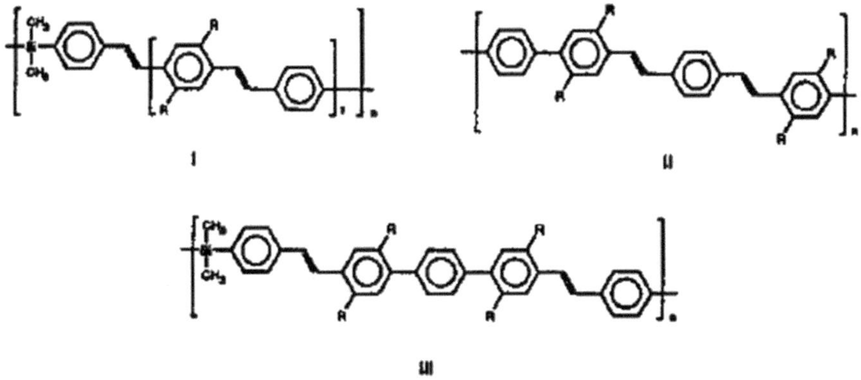

Fig. 33.69 Block copolymer derivatives of P(PV) synthesized by Hilberer et al. [1366] for LED applications

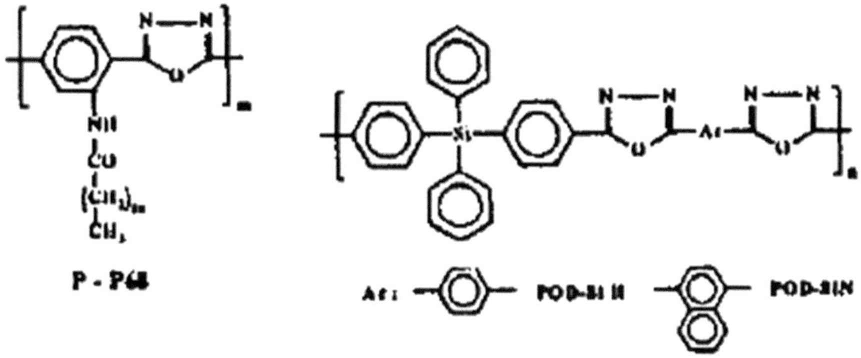

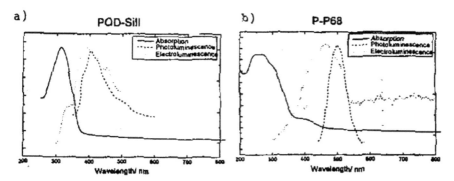

Fig. 33.70 Poly(oxadiazole) derivatives studied for LED use by Schulz et al. (After Ref. [1368], reproduced with permission)

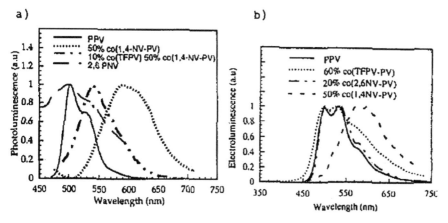

Fig. 33.71 EL and PL spectra of the (left) first and (right) second structure shown in Fig. 33.71 (After Ref. [1368], reproduced with permission)

Fig. 33.72   PL (left) and EL spectra of P(PV) derivative CPs of Fig. 33.71, compared with those of P(PV) only (After Ref. [1134], reproduced with permission)

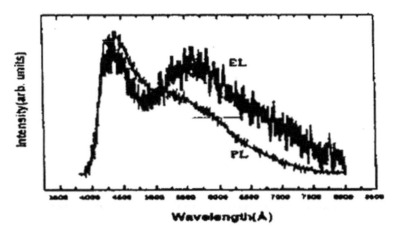

Fig. 33.73 PL and EL spectra of P(PV) derivatives containing organosilicon and carbazole units, after Kim et al. (After Ref. [1370], reproduced with permission).

Fig. 33.74 Absorption (dotted line) and PL (dashed line) spectra of PDHFV film and solution (10^{-5} mol l^{-1}) (solid line) (After Ref. [1128], reproduced with permission)

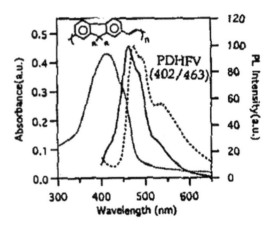

33.14   Third-Order Nonlinear Optical (NLO) Properties

33.14.1   The NLO Effect and Practical Requirements

Third-order nonlinear optical (NLO) properties of a material emanate from the susceptibility of the material to polarization by electromagnetic radiation and, more directly, from properties of excited electronic states of the material. They are of interest in many photonic applications in which change of an optical phenomenon as a function of intensity of incident radiation is sought. CPs have been shown to exhibit appreciable third-order NLO properties. For many practical applications, however, large third-order nonlinearities coupled (ca. 10^{-8} esu) with good stability of the material against thermal and photonic degradation (e.g., attenuation loss

Fig. 33.75 Measurements of three NLO parameters on urethane-derivative poly (diacetylene) films (After Kanetake et al. [1378], reproduced with permission)

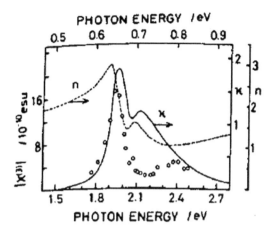

<1 dB/cm) are sought. This combination has so far eluded researchers in this area, but CPs remain one of the more promising materials in this regard. The only known, viable device implementation to date has been waveguides based on P(DiAc) single crystals. Indeed, in spite of the excitement generated for practical applications of NLO CPs [1371], industrial researchers have pointed out [1372] that perhaps a better indication of the practicality of the technology is the fact that all major industrial labs interested in using polymer NLO devices, such as IBM, Philips, DuPont, British Telecom, and AT&T Bell Labs, and many government-funded labs, such as the US Air Force Wright Laboratory, have quietly stopped research in the area, realizing it is futile.

In a dipole approximation, the bulk polarization, or *susceptibility*, of a condensed material is given by

$$P = P_o + \chi^{(1)} E + \chi^{(2)} E^2 + \chi^{(3)} E^3 \ldots \qquad (33.13)$$

Here, P is the bulk polarization, E is the electromagnetic field vector, and the *fill in* χ^{123} are, respectively, the first-, second-, and third-order *bulk* susceptibilities of the material. These bulk polarizabilities are correlated to the molecular-level *polarizability, hyperpolarizability,* and *second hyperpolarizability,* α, β, *and* γ, respectively. However, exact relationships between the bulk and corresponding molecular parameters are still not firmly established.

Because the direct correlation between the molecular and bulk parameters has been elusive, much work in this field has concentrated simply on empirical enhancement of the primary measures of practical performance of a material, i.e., its third-order bulk susceptibility and its resistance to photonic and thermal degradation resulting from exposure to radiation of high intensity that is required to effect the nonlinearity in the first place.

Table 33.1 Formalism for the three most common methods used for measurement of third-order NLO properties

Method	Formalism	Input beams	Output beams
Third harmonic generation (THG)	$\chi^{(3)}(-3\omega;\omega,\omega,\omega)$	ω	3ω
Degenerate four-wave mixing (DFWM)	$\chi^{3}(-\omega;\omega,\omega,\omega)$	ω	ω
Degenerate two-photon absorption (TPA)	$\chi^{(3)}(-\omega;\omega,\omega,-\omega)$	ω	ω

33.14.2   Methodology

A number of methods can be used to characterize third-order nonlinearities in materials, in solution, film, or bulk crystal form. Of these, we describe here only the most commonly used and most useful. These methods include *third harmonic generation (THG)* and *degenerate four-wave mixing (DFWM)* and two methods yielding useful supplementary information, degenerate *two-photon absorption (TPA)* and the *Z-scan* method (*Z-scan*). For the first three of these, the conventional formalism is outlined in Table 33.1. Techniques used to characterize response/decay times only include ps- or fs-regime pump-and-probe transient absorption/emission spectroscopy, where a laser pump is used to generate an excited state, and its decay is then monitored using a white light probe beam. Descriptions of these experimental methods are available elsewhere, and here we may confine ourselves to describing in simple terms what they do.

NLO measurements may be carried out at a wavelength (frequency) at which the active material does not absorb appreciably or where it has a strong absorption peak. These measurements are, respectively, called *off-resonant* (or *non-resonant*) and *resonant*. Necessarily, values of resonant NLO susceptibilities are expected to be larger than those of non-resonant susceptibilities for a given active NLO material.

33.14.3   Salient Results: Third-Order NLO Effects

Some of the first third-order NLO studies of CPs were carried out on the poly (diacetylenes) (P(DiAc)). One of the first such studies, on P(DiAc)-toluene sulfonate (PTS) single crystals, carried out by Sauteret et al. [1373], revealed third-order bulk NLO susceptibilities $\chi^{(3)}_{(zzzz)}$ (Z = polymer chain axis) of 1.6×10^{-10} esu at 2.62 μm and 8.5×10^{-10} esu at 1.89 μm. These values have as yet been exceeded only by one study, that of Drury on oriented Durham P(Ac) [1374], in which a value for $\chi^{(3)}_{(zzzz)}$ (Z = polymer chain axis) of 10^{-8} esu at 1.9 μm. However, this value of the susceptibility was inferred from THG data somewhat indirectly. Dennis et al. [1375] reported off-resonant $\chi^{(3)}_{xyyx}$ values of 7×10^{-12} esu for solutions of various P(DiAc), extrapolated to pure substance and determined via DFWM at 532 nm.

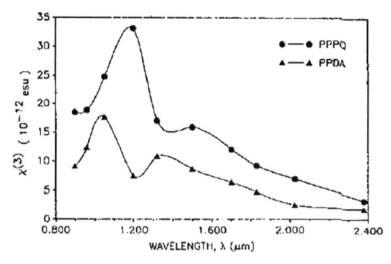

Fig. 33.76 Wavelength-dispersive $\chi^{(3)}$ measurements of rigid-rod poly(quinolines) (After Ref. [1384], reproduced with permission)

Sinclair et al. [1376] studied both P(Ac) and P(DiAc)-4BCMU. Their THG measurements on oriented, trans-P(Ac) parallel to the fiber direction at an energy of 1.17 eV yielded a value of $\chi^{(3)}_{xxxx}$ of ca. 4×10^{-10} esu, more than 15 times larger than the corresponding measurement on the cis-isomer. Their study on the P(DiAc) yielded a nonlinear refractive index (n_2) which was negative, with a value of 10^{-7} $(MW/cm^2)^{-1}$, with a decay time for this nonlinearity of ca. 2.5 ps. Chollet et al. [1377] studied the blue and red forms of several P(DiAc) films via THG and observed two and three-photon resonances, with $\chi^{(3)}_{xxxx}$ of ca. 5×10^{-11} esu at ca. 2 μm. Kanetake et al. [1378] measured THG on highly oriented, urethane-based P(DiAc) films, prepared by vacuum deposition, at a wavelength of 1.9 μm. They found that for a pump direction parallel to the polymer chain, a three-photon resonance to the ^1B_u exciton was observed, with a $\chi^{(3)}$ of 3.8×10^{-10} esu (vs. 0.2×10^{-10} esu for the perpendicular direction). Their frequency-dispersive measurements of three NLO parameters (Fig. 33.76) provided useful information. Neher et al. [1379] studied THG at 1.06 μm of ultrathin films of P(Phe-Ac). They found a value of $\chi^{(3)}_{xxxx}$ of ca. 7×10^{-10} esu, with a phase of ca. 152°, indicating that two- and three-photon resonances contributed to the observed nonlinearity.

The other common CP classes, such as the P(T)s and the P(Py)s, have also been the subject of much study. The Prasad Group reported bulk third-order susceptibilities, $\chi^{(3)}_{xxxx}$, in LB films of undoped P(3-dodecyl thiophene) at 1.06 μm of ca. 10^{-9} esu [1229]. In a subsequent study [1380], the same group found off-resonant (at 620 nm) values of $\chi^{(3)}_{xxxx}$, determined via DFWM for recast films of this partially soluble polymer, of 5.5×10^{-11} esu. Byrne et al. [1380], in a DFWM study of solutions of undoped P(3-Bu-T), found a three-photon resonance enhancement of the nonlinearity, with response times faster than 70 ps. Ghoshal [1381] carried out an elegant DFWM study of chemically synthesized P(Py) films of ca. 30 μm thickness at

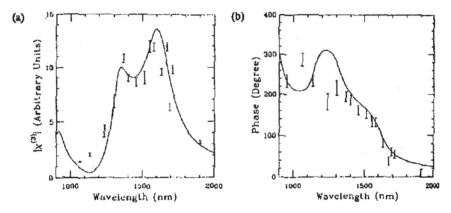

Fig. 33.77 Wavelength-dispersive magnitude (**a**) and phase (**b**) for a poly(diacetylene), poly (4 BCMU) (After Ref. [1385], reproduced with permission)

584 and 603 nm. He found nonlinear susceptibilities, $\chi^{(3)xxxx}$, of 3 to 3.5 × 10^{-12} esu for undoped films and films doped with SO_4^{2-} and Cl^-. The temporal response of the DFWM signal was of the order of single ps.

Ladder polymers of the benzimidazobenzophenanthroline type were studied using DFWM by Lindle et al. [1382], with susceptibilities of 1.5 × 10^{-11} reported at 1.06 μm. Cao et al. from the Dalton group [1383] studied poly(quinoxaline)-based conjugated ladder polymers in thin film form via DFWM at 532 nm. The off-resonant bulk third-order susceptibilities they obtained were ca. 10^{-11} esu, with a response time of less than 6 ps. Agrawal et al. [1384] studied third-order non-linearities, via THG at variable wavelengths, of a series of conjugated rigid-rod poly(quinolines). While their wavelength-dispersive $\chi^{(3)}_{xxxx}$ computations were of interest (cf. Fig. 33.77, no definitive structure–property relationships were found. Guo et al. [1385] also carried out a study of the spectral dispersion of the third-order nonlinearity and its phase for a series of CPs, results for one of which are shown in Fig. 33.78 .

Rigid ladder polymers of the P(PV) type were studied recently by Samoc et al. [1386] in chloroform solution. Figure 33.79 summarizes some of their data.

In a detailed study of structure–property relationships in off-resonant third-order NLO properties of several classes of CPs [1235b,c], as solutions of the very lightly doped CPs, Chandrasekhar et al. studied CPs synthesized from the monomers listed in Table 33.2 below. Their salient results are summarized in Table 33.3a, b, c, and d. Among the findings were that the nonlinearities of P(DPA) and P(4ABP) were the highest observed, and significantly higher than those of P(ANi), possibly due to the additional, unfused aromatic rings incorporated in the polymer backbone. When additional fused rings were incorporated in the polymer backbone, there appeared to be a dramatic increase in the nonlinearity, e.g., in P(1APyre) and P(ITN). Increased quinonoid character leading to a possible breaking of the π-delocalization along the polymer, as in P(NPhe2NaA) vs. P(DPA), appeared to lead to a complete lack of

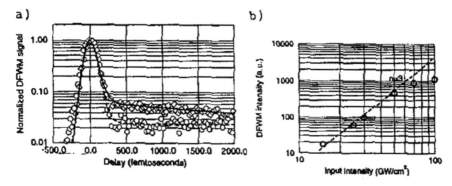

Fig. 33.78 :Summary DFWM data for 4 μm thick films of a rigid ladder derivative of P(PV) (After Ref. [1386], reproduced with permission)

nonlinear response in the former. The incorporation of an N-heteroatom into an aromatic polymer backbone, as in the poly(aminoquinolines) and P(3APAz), appeared to lower the bulk susceptibility. Chandrasekhar et al. showed that the values of the molecular-level third-order hyperpolarizabilities, γ_{ijk}, for the monomers, calculated using quantum chemical methods, confirmed the trends in the bulk susceptibilities observed, e.g., γ_{ijk} (X 10^{-36} e.s.u.): ANi, 8.67; DPA, 29.72; 1APyre, 41.15; 8AQ, 12.92.

Chandrasekhar et al. also found that for polymers with multiple heteroatoms in an aromatic ring, when N and S heteroatoms were coupled in the same aromatic ring in the polymer backbone, the nonlinearity practically disappeared, as was the case with P(2ATAz), P(2ABTAz), and P(4A213BTDAz). The study with variously substituted poly(aminoquinolines) showed the absence of a substitution position effect. Chandrasekhar et al. also studied the effect of dopant level and type. They found that low doping levels, e.g., 1% vs. 4.5%, yielded larger nonlinearities, possibly a solubility effect and that smaller dopants, e.g., BF_4^- and PF_6^-, appeared to yield larger nonlinearities. For all polymers, Z-scan and two-photon absorption measurements were used to demonstrate that the nonlinearity observed was truly non-resonant.

Several other third-order NLO studies of CPs of very varied structure have been carried out. Moon et al. [1387] recently reported DFWM studies, at 602 nm, of thin (ca. 15 μm), oriented films of a structurally complex P(PV) derivative containing alkoxynitrostilbene groups, finding $\chi^{(3)}_{xxxx}$ along the stretch direction of ca. 1.7×10^{-10} esu. Spangler et al. [1388] studied third-order nonlinearities, via DFWM at 532 and 1064 nm, of a class of bis-thienyl polyenes, having acetylene polymer chains terminated at either end by substituted thiophene rings, finding resonant $\chi^{(3)}_{xxxx}$ values not greater than 1.4×10^{-12} esu at 532 nm. Okawa et al. [1389] reported THG studies of a new series of P(PV)'s (whose synthesis using the Heck reaction was described in an earlier chapter.

These yielded non-resonant $\chi^{(3)}_{xxxx}(3\omega;\omega,\omega,\omega)$ at 1.907 μm of up to 1.2×10^{-12} esu. Meyer et al. [1390] carried out two-photon absorption measurements on poly

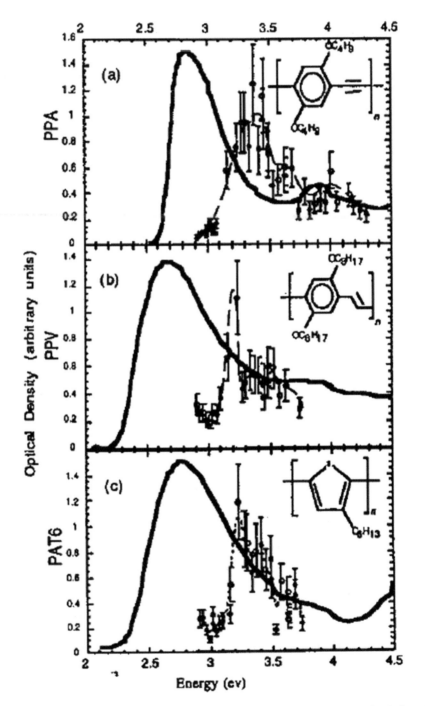

Fig. 33.79 Linear (solid line) and two-photon (open circles) absorption data for the CPs indicated
(structures inset) (After Ref. [1390], reproduced with permission)

Table 33.2 Monomer precursors of CPs used in NLO studies by Chandrasekhar et al. [1235b, c]

Monomer	Structure	Symbol
A. Poly(aniline) analogues		
N,N'-Diphenilbenzidine		NNPheBz
4-Aminobiphenyl		4ABP
Diphenlyamine		DPA
Aniline		Ani
N-Phenyl-2-naphthylamine		NPhe2NaA

(continued)

Table 33.2 (continued)

Monomer	Structure	Symbol
1-Aminopyrene	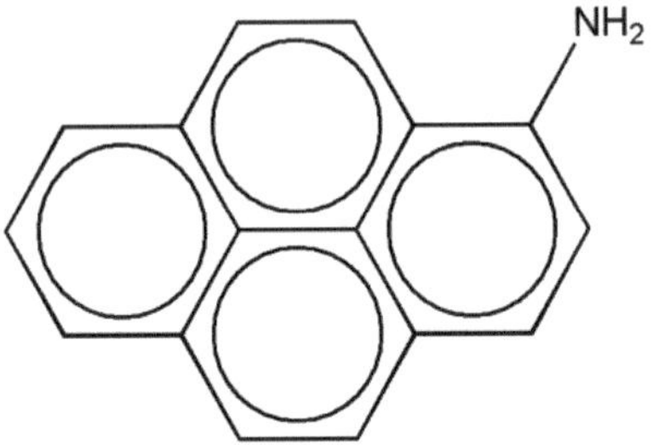	1APyre
B. Poly(aminoquinolines) and analogues		
3,5,6,8-Aminoquinoline (3-aminoquinoline shown)	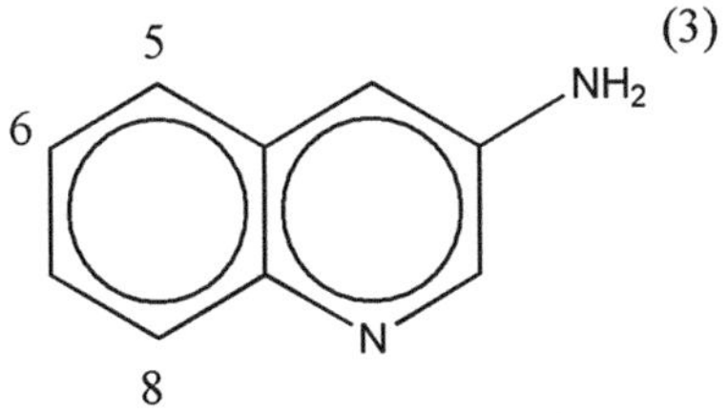	3AQ 5AQ 6AQ 8AQ
3-Aminopyridine	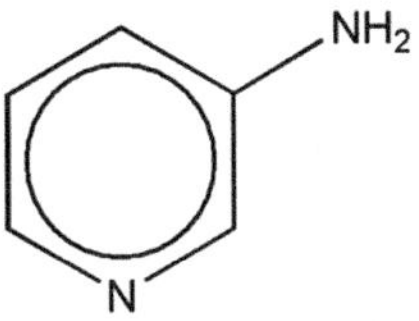	3APy
3-Aminopyrazole	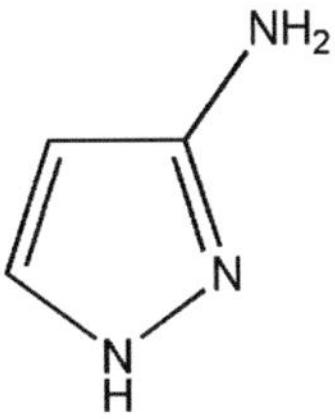	3APAz
C. S heteroatom-containing polymers		
Isothianaphthene		ITN

(continued)

Table 33.2 (continued)

Monomer	Structure	Symbol
2-Aminothiazole	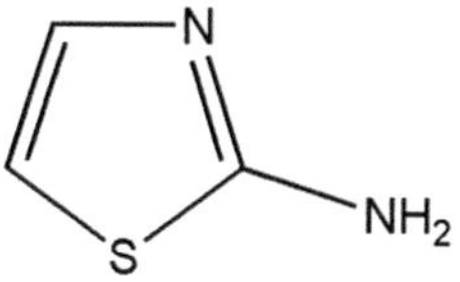	2ATAz
2-Aminobenzothiazole	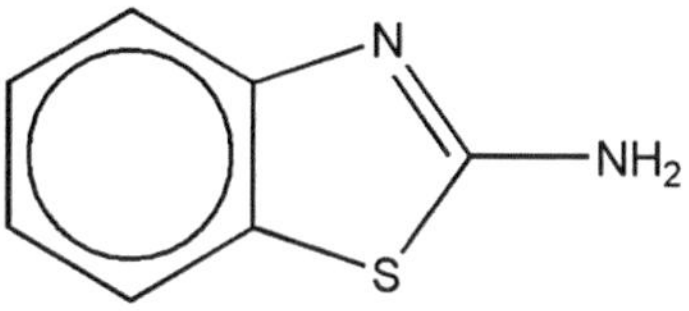	2ABTAz
4-Amino-2,1,3-benzothiadiazole	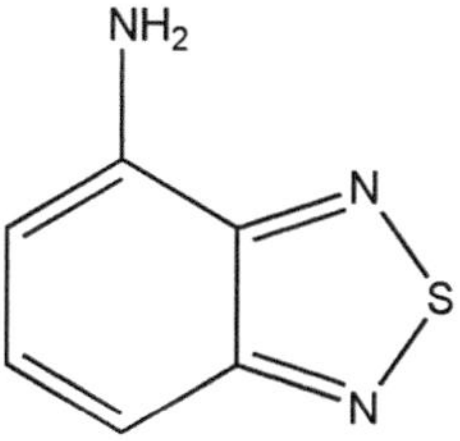	4A213BTDAz

Table 33.3a Comparative nonlinearities, $\chi^{(3)}_{xxxx}$(pure) for poly(aromatic amines) at identical doping (ca. 2.0% with BF^-_4) and concentration (20 mM) in DMF (After Ref. [1235b, c])

Polymer[a]	$\chi^{(3)}_{xxxx}$(pure) (e.s.u. $\times 10^{-10}$)		
P(DPA)	1.3		
P(4ABP)		1.4	
P(NNPheBz)			0.9
P(1APyre)		2.7	
P(3APAz)		0.40	
P(NPhe2NaA)			<0.04
P(Ani) (emeraldine base)			0.11

[a]For symbols see Table 12.2

Table 33.3b Comparative nonlinearities, $\chi^{(3)}_{xxxx}$(pure) for poly(aminoquinolines) and related polymers at identical doping (ca. 2.0% with BF^-_4) and concentration (20 mM) in DMF (After Ref. [1235b, c])

Polymer[a]	$\chi^{(3)}_{xxxx}$(pure) (e.s.u. $\times 10^{-10}$)	
P(3AQ)		0.37
P(5AQ)		0.19
P(6AQ)		0.26
P(8AQ)		0.23
P(3APy)	0.22	

[a]For symbols see Table 33.2

Table 33.3c Comparative nonlinearities, $\chi^{(3)}_{xxxx}$(pure) for S-heterocycle polymers studied in present report at identical doping (ca. 2.0% with BF^-_4) and concentration (20 mM) in DMF (After Ref. [1235b, c])

Polymer[a]	$\chi^{(3)}_{xxxx}$(pure) (e.s.u. $\times$ 10^{-10})	
P(ITN)	5.0	
P(2ATAz)	<0.01	
P(2ABTAz)		<0.01
P(4A213BTDAz)		<0.01

[a]For symbols see Table 12.2

Table 33.3d Effect of dopant type on nonlinearity, illustrated for P(DPA) at a constant doping level of 4.5% and concentration 40 mM in DMF (After Ref. [1235b, c])

Dopant	$\chi^{(3)}_{xxxx}$(pure) (e.s.u. $\times$ 10^{-10})	
BF^-_4		0.72
PF^-_6		0.74
ClO^-_4		0.35
Tosylate	0.37	

(2,5-dibutoxy-p-phenylene acetylene) and related polymers. Representative data of these authors are shown in Fig. 33.80.

Matsuda et al. [1391] reported the THG properties of a new series of poly (diacetylenes). In preliminary measurements, they obtained $\chi^{(3)}$ values in the region of 4×10^{-11} esu at 2.1 μm.

33.14.4   Salient Results: Second-Order NLO Effects

Because of symmetry considerations, CPs have generally shown less activity in second-order NLO properties. Among the approaches used to impart such properties to CPs have been the attachment of pendant, second-order NLO active groups, e.g., nitrobiphenyl and nitroazobenzene containing groups on P(DiAc), as done by the Tripathy group [1392]. Here, simultaneous second-order and third-order NLO properties were claimed to be imparted, with d coefficients that were 7 to 20 times smaller at 1.06 μm as compared to powdered urea. More recently, Cai et al. [1393] demonstrated large second-order NLO activity (45 pm/V) combined with thermal stability in rigid-rod poly(quinolines) by incorporating heteroatomic chromophores with large second-order nonlinearities therein. These polymers technically qualify as CPs, since they may be doped to yield conductive materials.

33.14.5   Salient Results: Decay Times of Excited States

While DFWM and other NLO measurement techniques yield information on response and decay times as well, as some of the data cited above indicate,

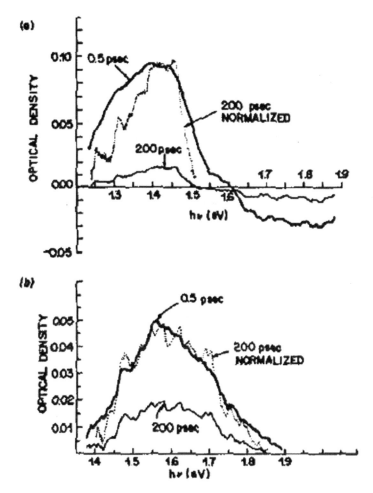

Fig. 33.80 Photoinduced difference spectra at 0.5 and 200 psec (**a**) for *trans*-polyacetylene and (**b**) for *cis*-polyacetylene, excited at 2.0 eV (After Ref. [1394], reproduced with permission)

additional data are obtainable from such techniques as pump-and-probe ps or fs transient absorption spectroscopy. Some of the first such data were collected by the Shank group at the erstwhile Bell Labs [1394]. Figure 33.81 shows their data for a 2.0 eV pump for cis- and trans-P(Ac), while Fig. 33.82 shows photoinduced absorption, at 1.4 eV, of P(cis-Ac) on a ps time scale. In the latter case, the risetime is less than 150 fs, the decay occurs over ca. 20 ps, and after 200 ps the spectrum is nearly the same as that at 1 ms. Femtosecond-regime risetimes have also been observed for P(DiAc)s [1395]. Such decay/response times appear to be typical of P(DiAc), P(Ac), and indeed a host of other CPs [1373, 1376, 1395, 1396]. Kobayashi [1397] gives a concise summary of risetime/decay data for P(DiAc) and some P(T)'s.

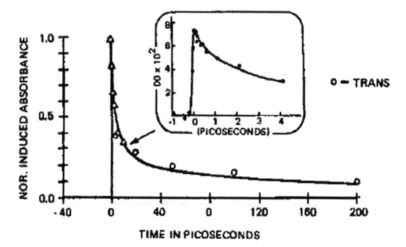

Fig. 33.81 The dependence of photoinduced absorption monitored at 1.4 eV in trans-$(CH)_x$ (circles) 1.6 eV in cis-$(CH)_x$ (triangles) (After Ref. [1361], reproduced with permission)

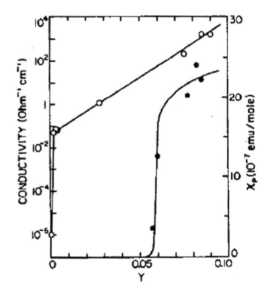

Fig. 33.82 Room-temperature electrical conductivity σ (left side, log scale) and temperature-independent Pauli susceptibility X_P (right side, linear scale) versus dopant concentration for $[Na^+_y(CH)^{y-}]_x$ (After Ref. [880], reproduced with permission)

In an interesting study on poly(3-dodecyl thiopene), Pang and Prasad [1398] showed the importance of the laser pump pulsewidth in the measured third-order NLO response. They found that with 60 fs pulses at 620 nm, this parameter is sixfold smaller than with 400 fs pulses, since the relaxation time of the photogenerated excitation responsible for the optical nonlinearity is comparable to the pulsewidth.

33.15   Magnetic Susceptibility

Apart from P(Ac), magnetic susceptibility measurements have been carried out for very few CPs. Values for the paramagnetic susceptibilities, $\chi(para)$ of cis- and trans-P(Ac), typically ca. 10^{-6} and 3×10^{-8} emu/mol, respectively, [1290], may be compared with that for a marginal semiconductor (PVC, irradiated, ca. 10^{-7} emu/mol), and the diamagnetic susceptibility $\chi(dia)$ for Si, ca. -3×10^{-6}. Undoped Shirakawa trans-P(Ac) shows adherence to the Curie law down to ca. 1.5 K, with ca. 10^{-4} spins/C-atom, while Durham trans-P(Ac) shows such behavior only to ca. 30 K, due to impurities.

Some of the first measurements of magnetic susceptibility of a CP, for AsF_5-doped P(Ac), were carried out by Weinberger et al., Epstein et al., and Shirakawa et al. [1204–1401]. The onset of Pauli susceptibility, indicating the critical concentration, was determined to be ca. 2 mole-%. The total susceptibility is dissected into three components, atomic core diamagnetism, localized Curie paramagnetism, and Pauli susceptibility. With increasing cis-trans isomerization, the Curie spin concentration, determining the number of unpaired spins in the CP, increased, indicating formation of neutral solitons with spin 1/2 during the isomerization. The spin susceptibility (Curie and Pauli susceptibilities combined) was observed to decrease with increasing temperature, up to 200 K, for undoped and I_2-doped P(Ac). Uniformity of doping appeared not be an important determinant of spin susceptibility (uniformly doped samples showing lower susceptibility than nonuniform samples). The varying onset of high conductivity and Pauli susceptibility as a function of doping (Fig. 33.83) has been rationalized using a soliton doping mechanism with charged solitons responsible for conductivity for P(Ac). The total magnetic susceptibility of cis-P(Ac) shows a peak at low dopant concentration (Fig. 33.84).

Mizoguchi et al. [1402] studied the magnetic susceptibility properties of P(Py) doped with a number of common dopants such as PF_6^-, AsF_6^-, ClO_4^-, BF_4^-, and Tos^-. They found that measurements of spin susceptibility using the Schumacher–Slichter (ESR/NMR) technique showed that the temperature-independent Pauli term ranged from 0.7 to 1.4×10^{-5} emu/mol-ring and the Curie term was 1 spin per 600 to 1500 rings, consistent with independent SQUID measurements in the same sample. Figure 33.85 shows representative data of these authors.

33.16   Miscellaneous Methods

Rutherford backscattering spectrometry (RBS) has been used to study doping levels of CPs having heavy-atom dopants or impurities, e.g., for P(3-alkyl-T)'s with I_2 [906]. Standard analytical potentiometry has been used, together with Raman scattering, to monitor the kinetics of aniline and pyrrole polymerization in situ [1300]. *Mössbauer* spectroscopy was used by Rabek et al. in a study of polymerization of pyrrole on polyetherester-Fe(III) coordination complexes, to show the

Fig. 33.83 Total magnetic susceptibility as a function of dopant concentration (mole percent) in 70% *cis*-$(CH)_x$ for the following dopants: MoF_6 (open circles), WF_6 (filled squares), ReF_6 (open squares), and UF_6 (filled circles) (After Ref. [1315], reproduced with permission)

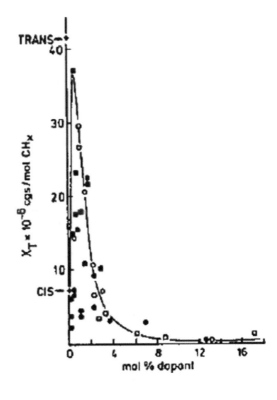

Fig. 33.84 Magnetic susceptibility χ_s for the as-grown P(Py) films synthesized at $-30 \sim -40\,°C$. PF_6^-, closed squares; AsF^-_6, open squares; BF_4^-, closed diamonds; ClO^-_4, open circles; ToS^-, closed diamonds. The closed circles and the open diamonds are PF_6^- and TsO $^-$ obtained by SQUID (After Ref. [1402], reproduced with permission)

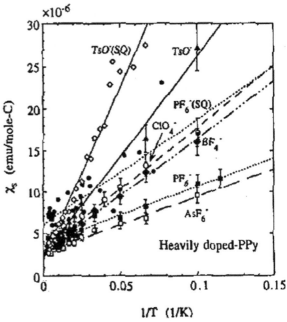

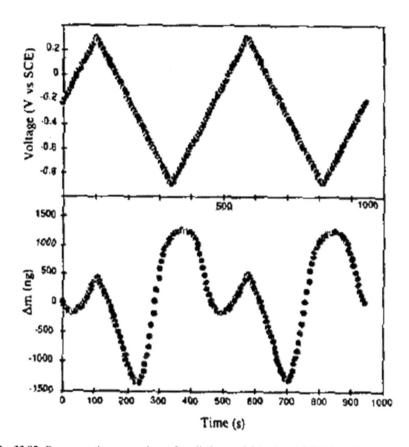

Fig. 33.85 Representative comparison of applied potential (top) and QCM data (bottom) for a P (Py)/PSS film in aqueous NaClO$_4$ (After Ref. [1007], reproduced with permission)

presence of FeCl$_4^-$ ions in the composite films so obtained [1145]. Kitao et al. [1403] used ^{129}I-Mössbauer to show that in I$_2$-doped P(T) and P(3MeT), the dopant was present as I$_3^-$ or I$_5^-$, confirming the results with P(Ac) of Suwalski et al. [1404], who also compared the ^{57}Fe Mössbauer signals of FeCl$_4^-$-doped CPs with those of quaternary ammonium chloroferrates.

The *electrochemical quartz crystal microbalance (EQCMB or QCM)*, a technique which measures microscopic mass changes on an oscillating quartz crystal via monitoring its oscillation frequency change, has been used in conjunction with voltammetry primarily in the study of doping processes in CPs. It has been used to demonstrate a very important concept in the doping/de-doping of CPs, namely, that ions physically move in and out of the CP matrix during such doping processes. This has been clearly demonstrated, e.g., in the case of P(Py) doping in aqueous solution by ClO$_4^-$ ions [1007]. Furthermore, in the case of P(Py) doped with a large polymeric dopant such as poly(styrene sulfonate) (PSS), QCM could be used to show that initial oxidation actually produces a decrease in mass, when cations are expelled from the P(Py)PSS matrix [1007]. This, and the high reversibility of cationic doping, can be seen in the data in Fig. 33.85. QCM has also been used to study metal electrodeposition on CP films [1405] and to study polymerization mechanisms in P(ANi) derivatives [1406].

Shilov et al. [1407] applied *neutron scattering* to the study of the magnetic properties of CPs, specifically addressing the problem of slow-neutron scattering by a singlet-triplet transition in the doubly degenerate Hubbard model. They found that the ground state was close too ferromagnetic.

33.17 Microscopic Evaluation

Standard microscopy techniques, such as optical, scanning electron (SEM), transmission electron (TEM), and the newer microscopies such as atomic force (AFM), scanning tunneling (STM), and scanning tunneling electron (STEM) have of course been routinely applied to CP samples wherever such samples are amenable to such analysis. A description of these techniques as applied to CPs would thus be somewhat redundant, much as a description of the application of standard elemental analysis to CPs would be. Thus, in this section, we will limit ourselves only to some studies that stand out.

Jang et al. [1235] presented TEM images of ultrafine (~ 2 nm), amorphous nanoparticles of P(Py) fabricated using microemulsion polymerization at low temperature, with supposed application to nano-electronic devices (though they did not indicate how exactly such application would be made). These are shown in Fig. 33.86 below.

SEM characterization of CPs, including fibers, "nanowires," "composites," and other derivatives, has been commonly used to yield synthetic and other information. A typical example is the SEM image of poly(ethylene dioxythiophene) (PEDOT) "nanowires" anchored on Au thin films, as seen in Fig. 33.86 . Figures 33.87 and 33.88 show a SEM image of polyaniline "nanotube junctions" [1408].

33.18 Problems and Exercises

1. Describe a series of simple analyses you would use for initial characterization of an unknown CP doped with an unknown dopant.
2. Outline three methods for determination of DC conductivity ex situ and one method in situ, and one method for AC conductivity, and indicate under which circumstances they are applicable.
3. Outline a simple technique, using IR, NMR, and UV–Vis spectroscopies, that you might be able to use to determine if chain branching exists, and what type it is, in P(ANi) and P(o-toluidine).

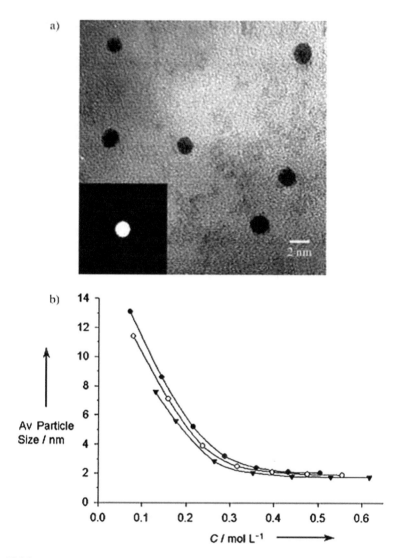

Fig. 33.86 TEM image of ultrafine (~ 2 nm), amorphous nanoparticles of P(Py) fabricated by Jang et al. [1235], using microemulsion polymerization at low temperature (After Ref. [1235], reproduced with permission)

4. Describe a simple electrochemical and spectroscopic procedure to determine level and type of doping with a P(Py) having a combination of Cl-, S-, and B-containing dopants.

5. If given a sample of an insoluble, intractable, doped P(Py), to which GPC may not be directly applicable, describe a method or methods for direct or indirect MWt determination.

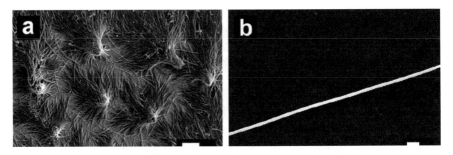

Fig. 33.87 SEM image of poly(ethylene dioxythiophene) (PEDOT) "nanowires" anchored on Au thin films (After Long et al. [1409], reproduced with permission)

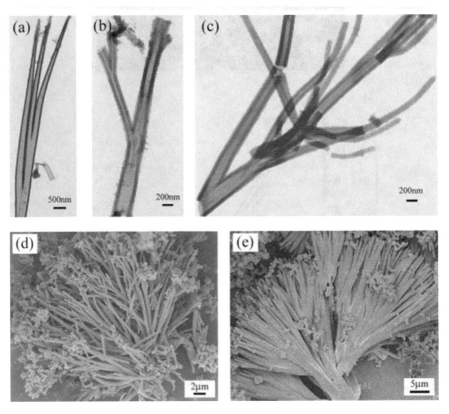

Fig. 33.88 SEM image of polyaniline "nanotube junctions" (After Ref. [1408], reproduced with permission)

6. Describe the principal features of the Raman spectra of both *cis-* and *trans-*P (Ac) and of P(3-alkyl-T).

7. What are the most likely modes of degradation of conductivity, macromolecular order and other related properties of CPs on heating? What do they tell you with regard to design and synthesis of thermally stable CPs?

8. Describe the use of XPS and NMR for study of doping effects in CPs, with one specific example for each.

9. Describe the relation of unpaired electron density, conductivity, Pauli susceptibility, and likely charge transport mechanism in CPs, and what techniques are used to elucidate these most clearly.

10. Identify the relationship and briefly outline the derivation of each of the following microwave-derived properties of CPs: *permittivity, permeability, frequency-dependent conductivity, EMI-SE, RCS.* In which application is each of these important, and what is the single largest constraint in each such application which is preventing their further use?

11. Discuss the differences and interrelation between the PIA, EL, PL, and fluorescence properties of a CP. Analyze the PIA and PL spectrum for one CP in detail, especially with regard to implications of charge carriers and lifetimes thereof.

12. What are the most important constraints preventing utilization of the third-order NLO properties of CPs in electro-optic (EO) devices? Which EO application do you think is the closest to fruition, and how would you attempt to design and synthesize a CP/dopant system to realize this application? Of all the CPs whose NLO properties are reported in this chapter, which do you think is the closest to practical implementation in terms of magnitude of NLO effects as well as beam stability, and why? Enunciate the strongest structure–property correlations you have gleaned from the various NLO structure–property NLO studies reported in this chapter.

13. Account in detail for the difference in the relation of Pauli susceptibility and DC conductivity to doping level in P(Ac) in terms of likely conduction models.

14. Besides the QCM method, which methods can also help demonstrate that dopant ions physically move in and out of a CP matrix during de/doping and that, in the case of large macromolecular dopants, the dopant just "sits there" while small ions move out of the combined CP/dopant matrix in an inverse manner?

Appendix 1: One Illustrative Network-Analyzer Based Calculation of Microwave Parameters

The scattering parameters S_{11} and S_{21}, which are automatically error corrected in the automated network analyzer (ANA), are measured as a function of frequency. The complex permittivity ε^* and the permeability μ^* are calculated from the measured S-parameters, using a modification of the Nicolson–Ross method, as summarized below.

Let Γ and T be the reflection and transmission coefficients due to the sample, of thickness d. The analysis is robust in that the thickness d need not be precisely known. Defining a parameter

$$K = \frac{S_{11}^{2}(\varpi) - S_{212}(\varpi) + 1}{2S_{11}(\varpi)} \tag{A.1}$$

where ω is the angular frequency, one can calculate

$$\Gamma = K \pm q\, K^2 - 1 \tag{A.2}$$

$$T = \frac{S_{11}(\varpi) + S_{21}(\varpi) - \Gamma}{1 - \{S_{11}(\varpi) + S_{21}(\varpi)\}\, \Gamma} \tag{A.3}$$

The sign ambiguity in (A.2) can be resolved by noting that $\Gamma < 1$ (passivity requirement). From rectangular waveguide analysis, one can relate the relative (complex) constitutive parameters $\mu_r{}^*$ and $\varepsilon_r{}^*$ to the reflection and transmission coefficients:

$$\frac{1 = \varepsilon_r{}^* \mu_r{}^* - 1}{\Lambda^2} = \frac{-\{(1/2\pi d) \ln(1/T)\}^2}{\lambda_0{}^2 \ \lambda_c{}^2} \tag{A.4}$$

where Λ is the guide wavelength, which is complex due to losses in the sample, λ_0 is the wavelength in free space, and λ_c denotes cutoff wavelength of the dominant waveguide mode. Since T is complex, the logarithm in (A.4) has infinite number of solutions, given by

$$\ln(1/T) = \ln|(1/T)| + j\{\mathrm{Arg}(1/T) + 2\pi n\} \quad (n \text{ an integer}) \tag{A.5}$$

The correct value of n can be determined by measuring the group delay using the ANA. From (A.4), the material parameters are calculated as

$$\mu_r{}^* = (1 + \Gamma)/\{\Lambda(1 - \Gamma)(\mathrm{Sqrt}[(1/\lambda_0{}^2) - (1/\lambda_c{}^2)])\} \tag{A.6}$$

$$\varepsilon_r{}^* = \{\lambda_0{}^2[(1/\Lambda^2) + (1/\lambda_c{}^2)]\}/\{\mu_r{}^*\} \tag{A.7}$$

In the original Nicolson–Ross procedure, an ambiguity in phase must be resolved at each frequency by comparing the measured and calculated group delay. We plan to modify this procedure by solving Eqs. (A.2) and (A.3) iteratively for ε^*, μ^*, from an initial guess supplied by Eqs. (A.6) and (A.7). The new formulation for ε^* and μ^* avoids phase errors resulting from group delay comparisons. The formulation is implemented on the PC controller to facilitate real-time display and on-line adjustments. The measurements are normally performed at room temperature, sweeping the whole frequency range of the waveguide.

Chapter 34
Classes of CPs: Part 1

Contents

34.1 Note on Chapter Focus

In this and the next chapter, we discuss each group or class of CPs in turn, starting with the poly(acetylenes). Much emphasis is placed on *synthetic aspects*, as properties and other aspects of the CPs are covered elsewhere in this book. Within this limited purview, more emphasis is placed further on giving more detail for the less well-known or less well-studied CPs, rather than the more studied ones such as P (ANi) and P(Ac).

34.2 Poly(acetylenes) (P(Ac)s)

34.2.1 Simple Syntheses and Basic Properties

The historical aspects of doped poly(acetylene) (P(Ac)) have been dealt with at some length in an earlier chapter. Much before Shirakawa et al.'s discovery of doped P(Ac) [841], Natta and coworkers employed Ziegler–Natta (Z–N) polymerization using Al(Alk)/organotitanium to obtain nonconducting P(Ac) [1410]. The commonest method for production of P(Ac), which is subsequently doped, remains Ziegler–Natta polymerization.

© Springer International Publishing AG 2018   489
P. Chandrasekhar, *Conducting Polymers, Fundamentals and Applications*,
https://doi.org/10.1007/978-3-319-69378-1_34

The cis-isomer, usually obtained in larger proportion in most direct syntheses, is converted to the thermodynamically more stable trans-isomer by heating, e.g., for 1 h at 150 °C. This isomerization can also be achieved during doping; indeed, electrochemical doping during electrochemical cycling via cyclic voltammetry also achieves this isomerization [1411, 1412]. The cis-isomer has a bandgap of ca. 2.0 eV (594 nm), while the trans-isomer has a bandgap of ca. 1.8 eV (700 nm). The trans-isomer has been determined by ESR and other measurements [1413, 1414] to have ca. 10^{18} spins/g or ca. 1 soliton defect per 15 CH units.

A typical, simple, Ziegler–Natta synthesis of P(Ac) may be described [1280]: the walls of a glass reaction vessel are simply wetted with a solution of $Al(Et)_3$ + Ti $(OBut)_4$ (4:1 M/M) in toluene. Acetylene monomer in gaseous form is introduced at pressures from 0.01 to 1 atm. A well-adhering film of P(Ac) grows on the walls in times ranging from a minute to an hour, dependent upon temperature, pressure, and other reaction conditions. The film so obtained, of thickness in the 0.1 μm to several cm, is washed and dried and then peeled off from the vessel walls.

When the reaction is carried out at room temperature, a mixture of 60%-cis and 40%-trans is typically obtained; at −78 °C, the product is nearly entirely cis-isomer; at 150 °C, using decane solvent base, the product is nearly entirely trans-isomer. The morphology of the product is entangled fibrils of diameter from ca. 5 to 100 nm (typically 20 nm), and no particular orientation (fibrils as large as 1 μm have been obtained with Al/Ti ratios of $>$10) [1247]. The undoped polymer has typical conductivities of 10^{-9} S/cm for the cis-form and 10^{-5} S/cm for the trans-form. Density of compacted fibrils is in the region of 1.2 g/cm^3. The extended sp^2 structure of the polymer is peppered with sp^3 defects, arising from cross-links, oxidation, and the like. As-prepared P(Ac) is highly air-unstable, turning brittle with dramatic loss in conductivity after 24-h exposure; this has been ascribed to an initial "oxygen doping" followed by formation of carbonyl groups breaking the polymer's extended conjugation [1247].

In the simple syntheses described above, trialkylphosphines can replace trialkyl aluminum [1415]; or $NaBH_4$ may be used as the reducing agent [1416]; or a composite, single-component Ziegler–Natta catalyst such as tetrabenzyl titanium may be used [1417]; or a polymeric component as poly(styryl-Li) may be used in place of the reducing agent [1061]. Suffice it to say that all these methods are variants of Ziegler–Natta catalysis polymerization. To date among widely used methods for P(Ac) synthesis, only the Durham–Feast route, discussed below, qualifies as being distinctly not a variant of Ziegler–Natta catalysis.

34.2.2   Doping of P(Ac)

As-produced P(Ac) may be doped, cationically or anionically, by chemical or electrochemical means. Common cations used have included the following (media used in parentheses): Li^+ ($LiClO_4$/THF), Na^+ ($NaPF_6$/THF), K^+ (KBu_3Py/THF), and Bu_4N^+ (Bu_4NClO_4)/PC) [1418–1420]. Common anions used have included the following:

Fig. 34.1 Dc conductivity
of iodine (I^-_3)-doped *trans-*
and *cis*-polyacetylene as a
function of relative dopant
concentration y (After Ref.
[872], reproduced with
permission)

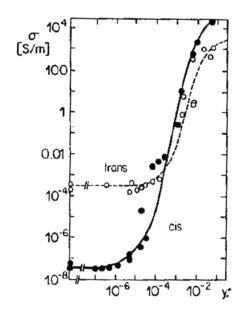

ClO_4^-, AsF_6^-, triflate, BF_4^-, I_3^-/I_5^-, HSO_4^-, HF_2^-, $AlCl_4^-$, $FeCl^{4-}$, $InCl_4^-$, $SnCl_5^-$,
and ReF_6^- [1033, 1035, 1050, 1051, 1421, 1426]. Maximum doping is generally
ca. 5% to 8%. The doping threshold for conductivity of trans-P(Ac) is very low,
ca. 1%, and even lower in some cases of I_2 doping; while beyond ca. 7%, conductivity falls with further doping, due probably to introduction of doping-induced
defects in the polymer chain (Fig. 34.1). From these and other data, the "semiconductor-to-metal transition" for P(Ac) is generally said to occur at a dopant concentration of ca. 6%. In the case of thin P(Ac) films, doping, e.g., by I_2 vapor, has been
shown to be nonuniform, with a gradient of high doping at the film edges and lower
doping in the center of the film, due to dopant diffusion effects [1427, 1428].

34.2.3   Orientation of P(Ac)

As-produced P(Ac) may be stretched for orientation, leading to anisotropic character (e.g., as measured by conductivity). Many other methods of preparing oriented P(Ac) exist, for example, carrying out the Ziegler–Natta polymerization in a
nematic liquid crystal with a 15 kG applied magnetic field [1429]. Stretching ratios
are generally in the region of 3, yielding anisotropy ratios for conductivity parallel/
perpendicular to chain direction of 3 to 40 [1430, 1431], although stretch ratios up
to 20 have been achieved (with anisotropy ratio 100) [1432]. Stretch orientation
even of cis-P(Ac) can yield polymer with highly crystalline segments yielding
sharp X-ray diffraction patterns and conductivities as high as 10^4 S/cm
[1260]. The entangled-microfibril morphology of the as-prepared P(Ac), with

void volume as high as 2/3, gives way in these highly-stretched polymers to highly aligned microfibrils, as seen in SEMs of conventional "Shirakawa" and Naarmann P(Ac). The oriented, stretched P(Ac) displays not only improved conductivity (along the stretching direction) but also improved mechanical properties such as Young's modulus and tensile strength [902]. Stretch orientation also leads to a change in the energies of the mid-gap and intergap transitions of P(Ac): for example, the $\pi \to \pi^*$ and $\pi \to S^+$ (S = mid-gap soliton state) for unstretched and stretched Durham–Feast P(Ac) are, respectively, 2.3 eV, 0.55 eV, and 1.9 eV, 0.45 eV.

34.2.4   Special Syntheses

In the Naarmann variant of the conventional Ziegler–Natta P(Ac) synthesis [867, 1433], first carried out at BASF, Germany, the $Ti(OBu)_4/Al(Et)_3$ catalyst is "aged" at 120 °C in silicone oil medium, with n-butyl-Li added as reducing agent. Acetylene vapor is introduced at room temperature. The reaction yields free-standing films of ca. 4 μm thickness, which are then stretched mechanically (ratios up to 6.5). The films are rinsed with toluene, methanol/HCl, and methanol. The films are then chemically doped in solution, yielding high crystallinity and a density of ca. 0.9 g.cm^2. The difference between the morphologies of conventional "Shirakawa" and Naarmann P(Ac) can be seen very clearly in SEMS, e.g., those in Fig. 34.2; fibril diameters in the Naarmann variant are typically 50 to 100 nm. Claimed conductivities for the Naarmann variant are as high as 10^5 S/cm along the stretch direction.

The Durham–Feast route (Fig. 34.3), the only common P(Ac) polymerization that is not a direct variant of Ziegler–Natta polymerization (although one common

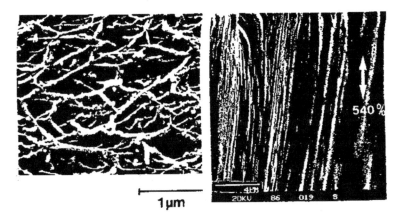

Fig. 34.2 SEMs of Shirakawa (*left*) and Naarmann (*right*) P(Ac) (After Ref. [1240], reproduced with permission)

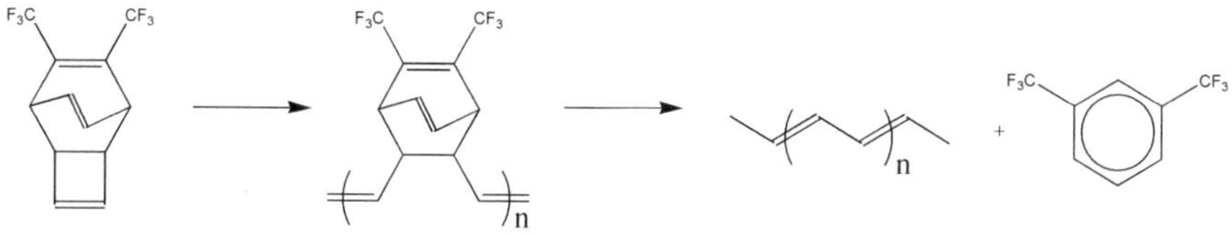

Fig. 34.3 Durham–Feast route for P(Ac)

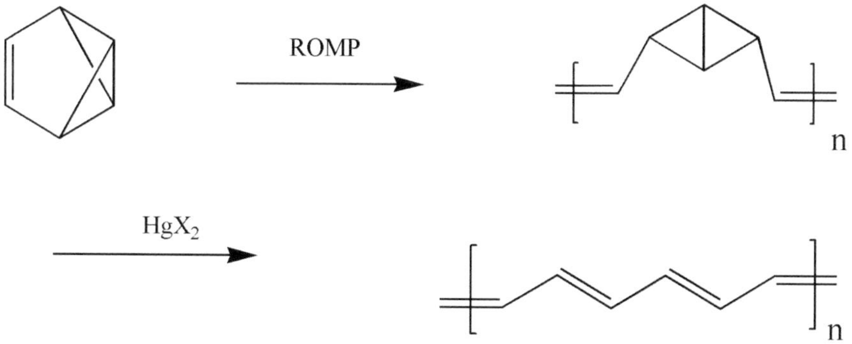

Fig. 34.4 Conversion of poly(benzvalene) to P(Ac) using ring-opening metathesis (ROMP)

of the catalyst used is a Ziegler–Natta catalyst), is a precursor polymerization route using ring-opening metathesis. The starting monomer is 7,8-bis(trifluoromethyl) tricyclo^{4,2,2,0} deca-3,7,9-triene, which undergoes a ring-opening polymerization with the catalyst system WCl_6:Phe_4Sn + $TiCl_4$:$AlEt_3$ (in 1:2 ratios). This precursor polymer is processible and thus can be cast onto any desirable substrate. It is then made to undergo an elimination, via vacuum pyrolysis during which it is simultaneously stretch-aligned, losing hexafluoroxylene to yield P(Ac) with properties distinctly improved in many respects from conventional, "Shirakawa" P(Ac). Interestingly, although this "Durham" P(Ac) is more highly ordered, it also has a higher proportion of sp^3 defects than Shirakawa P(Ac). A different ring-opening metathesis, demonstrated by Swager and Grubbs [1434], uses an exotic starting material, polybenzvalene, and is less practical (Fig. 34.4).

34.2.5 Substituted P(Ac)s

Substituted P(Ac)s may be synthesized using a wide variety of methods, primarily condensation polymerizations. Figure 34.5 shows a sampling of these. Poly (phenylacetylene) (P(PhAc)), a typical substituted P(Ac), is soluble in its undoped state in common organic solvents [1435].

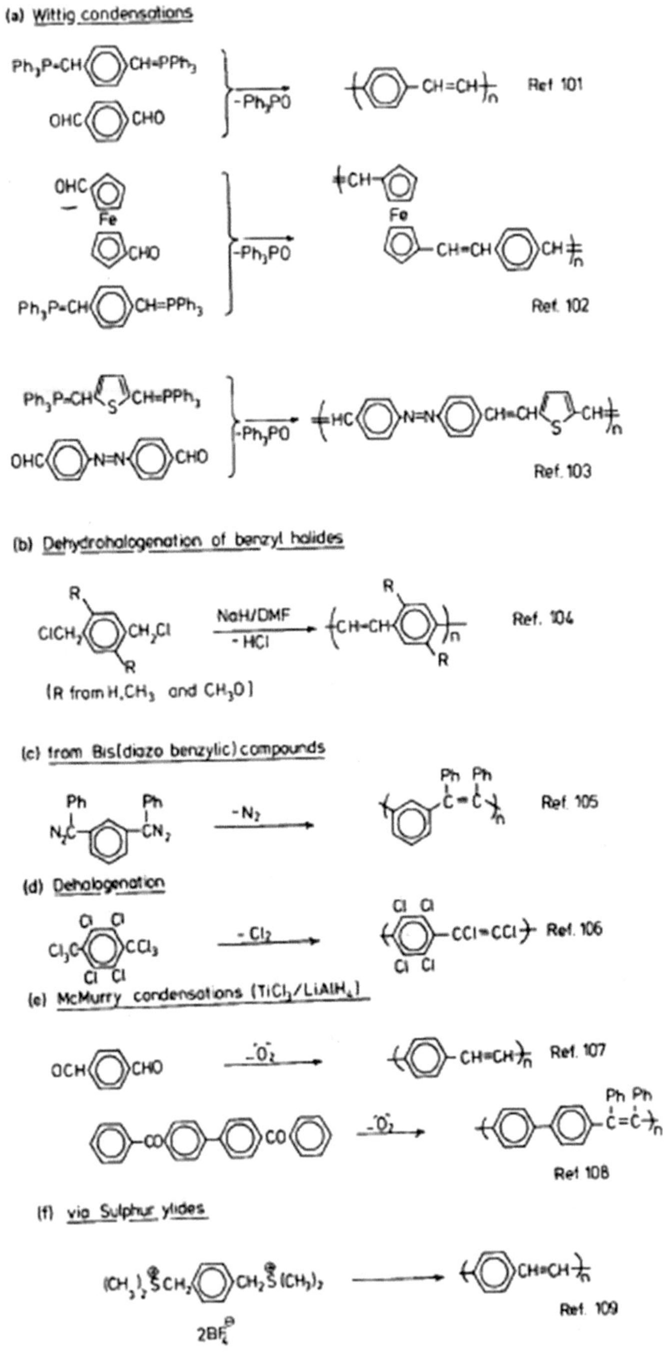
(a) Wittig condensations
-Ph3PO
Ref 101
Fe
Ref. 102
Ref. 103
(b) Dehydrohalogenation of benzyl halides
NaH/DMF
- HCl
Ref. 104
(R from H, CH3 and CH3O)
(c) from Bis(diazo benzylic) compounds
-N2
Ref. 105
(d) Dehalogenation
- Cl2
Ref. 106
(e) McMurry condensations (TiCl3/LiAlH4)
Ref. 107
Ref 108
(f) via Sulphur ylides
2BF4
Ref. 109

Fig. 34.5 Examples of variety of methods available for substituted P(Ac) synthesis (After Ref. [1021], reproduced with permission)

34.3 Poly(diacetylenes) (P(DiAc)s)

Typical schemes for polymerization of poly(diacetylenes) (P(DiAc)s), and the structural difference between P(Ac)s and P(DiAc)s, are shown in Fig. 34.6.

P(DiAc)s have alternating double and triple bonds, having the general formula [=(CR-C=C-CR')=]$_n$ (R, R' substituent groups). More importantly, they can be prepared as single crystals, the only CPs with this property. They are typically prepared via polymerization with light or gamma rays (^{60}Co) [1019, 1436]. Although like P(Ac)s they were known as ordinary polymers early on [1437], interest as CPs was aroused when it was discovered that with proper choice of the substituents R, R', they could be made soluble in their undoped form and that they could be doped [1209, 1438]. P(DiAc)s are typically named on the basis of the substituents R, R' used. Some of the most common P(DiAc)s, and several others, are listed in Fig. 34.7 below. Many P(DiAc)s display thermochromism and solvatochromism (color changes based on temperature and solvent composition changes), e.g., yellow to red or blue. [1439]. Indeed a thermochromic shift in Raman spectra has been observed. Photocurrent behavior in single crystals of PTS and MADF, two P(DiAc)s with different substituent groups (Fig. 34.7), is very different [1440], demonstrating the strong influence of the substituent groups in this class of CPs.

Controlled preparation of thin films of P(DiAc)s can be obtained by topochemical (solid state) polymerization of thin films of monomer previously deposited by evaporation under high vacuum, followed by UV or gamma irradiation [1441]; such thin films possess high order and alignment. Variation of the substrate crystallinity yields thin films or crystals of varying crystallinity [1441]. Innovative methods, such as application of pressure on a mobile monomer phase between two optical plates [1442], have allowed production of large-area thin-film single crystals of P(DiAc)s for potential use as waveguides in optical signal processing. P(DiAc)s display other interesting properties, such as triplet soliton-pair formation following UV excitation [1443]. Besides potential applications in waveguides, P(DiAc) single crystals have also been studied for second-order nonlinear optical effects [1392].

Matsuda et al. [1391] described a new series of P(DA)'s having directly bound sulfur atoms as electron-donating groups, as shown schematically in Fig. 34.8.

The monomers in this case were prepared in a four-step synthesis using 5,7-octadiyn-1-ol as the starting material, with the OH group of the substituted 8-thio-5,7-octadiyn-1-ol converted to the phenylurethane group in the final step. The monomers were then polymerized in the solid state using an appropriate dose of ^{60}Co γ-radiation (0.15 MRad/h).

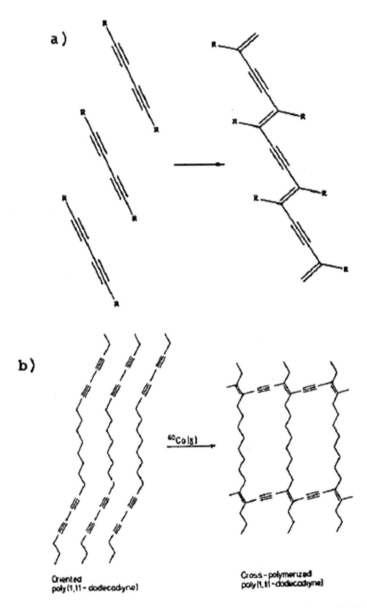

Fig. 34.6 Examples of P(DiAc) synthesis. (**a**) Topochemical-general scheme. (**b**) Using Co radiation (After Refs. [1021, 1444], reproduced with permission)

(CH₂)₃OCONHCH₂CO₂C₄H₉   P3BCMU
(CH₂)₄OCONHCH₂CO₂C₄H₉   P4BCMU
(CH₂)₄OSO₂C₆H₄CH₃   PTS12
(CH₂)₉OCOCH(CH₃)₂
(CH₂)₉OCOCH₂(CH₃)₂
(CH₂)₉OCOCH₂(1-naphthyl)

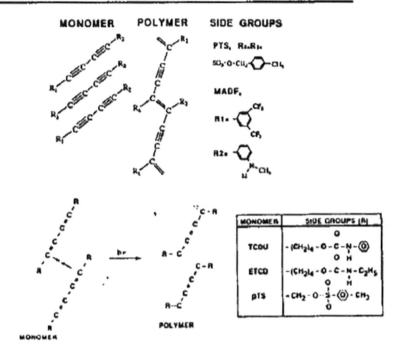

Fig. 34.7 Some common P(DiAc)s with their structures or formulations (After Refs. [1437, 1440, 1442], reproduced with permission)

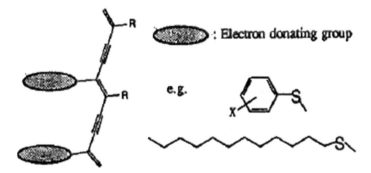

Fig. 34.8 New P(DiAc) having directly bound S-atoms (After Ref. [1391], reproduced with permission)

34.4   Poly(pyrroles) (P(Py)s)

Along with the P(Ac)s and P(ANi)s, poly(pyrrole)-based polymers (P(Py)s) are among the most well studied of the CPs. P(Py) as "pyrrole black" was chemically polymerized as early as 1916 [845]. The chemical and electrochemical routes for polymerization appear to be equally popular. In the chemical route, common oxidants such as $FeCl_3$ may be used. The electrochemical route generally uses a nonaqueous solvent medium such as acetonitrile and graphite, steel, or Pt electrodes.

34.4.1   Chemical Polymerizations

Machida et al. and Whang et al. [1014, 1015] have carried out a detailed study of chemical polymerization of P(Py), which also serves as a useful primer in this synthesis. In a typical synthesis, anhydrous $FeCl_3$ and pyrrole (pre-purified to remove chromophoric contaminants by passing through an alumina column) are taken in a 2.33:1 molar ratio (with 2.5 M $FeCl_3$) in MeOH solvent and stirred at 0 °C for ca. 20 min., yielding the polymer. The product is washed thoroughly with MeOH and dried in vacuo for ca. 12 h. The conductivity of the P(Py) so obtained has been claimed to be as high as 190 S/cm, although repetition of this synthesis in the author's and other labs has yielded product of conductivity not greater than ca. 20 S/cm. Machida et al. and Whang et al. compared various solvents and claimed product conductivities, under identical synthesis conditions, of greater than 90 S/cm for MeOH, EtOH, water, pentanol, octanol, and ethylene glycol, while conductivities were less than 20 S/cm for benzene, THF, chloroform, and acetonitrile and very small for acetone and DMF. This conductivity behavior would be consistent with the involvement of protons in the P(Py) polymerization mechanism.

Machida et al. and Whang et al. [1014, 1015] studied in detail the effect on the product conductivity of nearly all possible variables in the chemical synthesis. They found 2.5 M to be the optimal $FeCl_3$ concentration, yielding twice the conductivity of 1.5 and 3.5 M concentrations. A plot of product conductivity vs. polymerization temperature from −20 to 60 °C showed a maximum at 0 °C in protic solvents such as MeOH, with a decline to < half the value at 60 °C. After ca. 30 min., product conductivity rapidly fell with polymerization time, being ca. 1/3 of the peak value at 350 min. A most interesting finding was that when the reaction solution's oxidation potential was varied either by adding $FeCl_2$ or by admixture of other solvents, the peak product conductivity was obtained at an open circuit redox potential (vs. SCE) of +0.5 V, with steep conductivity declines on either side of this potential. SEM and X-ray (WAX) analysis of the products obtained under varying conditions showed an extremely smooth, aligned fibrillar structure and good mechanical properties, including flexibility of thick films, for product obtained under optimal polymerization conditions. $FeCl_3$ serves in this case as an oxidizer as well as dopant; a host of other oxidizer/dopants compatible with the oxidation potential of pyrrole can be used, for example, $CuCl_2$, $CuBr_2$, $Fe(NO_3)_3$, $Cu(BF_4)_2$, and ferricyanide [1300, 1355]; while other reaction conditions may vary, in nearly all cases, an oxidant/pyrrole molar ratio of ca. 2.5 must be used due to the stoichiometry of the polymerization. The polymerization of pyrrole in matrices of various types (e.g., PVC) to yield "composites" follows very similar procedures with the modification that the matrix is introduced in the solvent medium [1146, 1355].

34.4.2 Electrochemical Syntheses

Because of its simplicity and the frequently higher polymer conductivity obtained, electrochemical polymerization of P(Py) is many times preferred over chemical polymerization. In a typical electrosynthesis, 0.01 to 1 M monomer is taken with 0.1 to 2 M dopant/electrolyte (e.g., $LiClO_4$, Et_4NTos) in acetonitrile medium. Stainless steel, graphite, and Pt are most commonly used as electrodes. A potentiostatic polymerization, at ca. +0.8 V vs. SCE, yields the best results, although potential sweeps are also routinely employed [1022]. A small water content in acetonitrile, e.g., 2%, is found to be beneficial for the polymerization; some workers have even found that water contents as high as 50 v/v% in acetonitrile yield slightly improved P(Py) conductivities [1445]. With a Pt-working electrode, there is frequently a problem of exceptional adhesion of the polymer which is then difficult to remove entirely, spoiling the electrode. Working electrodes of stainless steel, e.g., AISI 304 [1446], and pyrolytic graphite can yield free-standing, thick films, of density ca. 1.5 g/cm^2, which are easy to remove. Ag/AgCl-reference or Pt-quasi-reference electrodes, or a simple two-electrode polymerization, may also be used for simplicity without any detriment to the quality of the product. Toluene sulfonate (tosylate, Tos) appears to be one of the best dopants for P(Py) in terms of oxidative stability, mechanical properties, and conductivity of the product; however, even P(Py)Tos is unstable on prolonged exposure to air, with the primary degradation mechanism thought to be formation of N $\rightarrow$ O bonds. Among other anionic dopants that have been used in electrochemical polymerization of P(Py) have been tetrafluoroborate, hexafluoroarsenate, perchlorate, hexafluorophosphate, triflate, trifluoroacetate, and hydrogen sulfate [1447].

P(Py) displays well-documented electrochromism (described elsewhere in this book) in the visible spectral region, going from light yellow in the reduced form (at negative applied potential) to nearly black in the oxidized form (at positive potential).

Ouyang et al. [1448] found in recent work that the quality of P(Py) films can be improved greatly when pyrrole is electropolymerized from aqueous solutions containing nonaphenol polyethyleneoxy(10) ether. Properties such as the tensile strength improved from 27.8 MPa to 127 MPa using this additive, possibly due to decrease in oxygen content (from increased oxidative stability) in the polymer.

34.4.3   Substituted P(Py)s

The most common substituted poly(pyrroles) P(Py)s include the N-methyl, N-phenyl, N-methoxyphenyl, 3-methyl, 3,4-dimethyl, N-ethyl, and N-propyl derivatives [1449–1451]. Conductivities of the substituted polymers are nearly always less than that of P(Py). Electrochromism of these P(Py) derivatives is not particularly noteworthy over that of the parent polymer.

34.5   Poly(anilines) (P(ANi)s)

34.5.1   Structure and Nomenclature

Poly(aniline) (P(ANi)) is a high MWt (typically 100,000) polymer. Its structure can be represented by the general formula shown in Fig. 34.9a. In this general formula, there are alternating reduced units, comprising the units contained in the leftmost parentheses, and oxidized units, comprising the units contained in the right parentheses. In the formula, n represents the degree of polymerization and $(1-y)$ the oxidation state, which can be varied from 0.0 to 1.0.

For $(1-y) = 0$, we have the completely reduced polymer, denoted as *leuco-emeraldine*, as shown in Fig 34.9b. With $(1-y) = 0.5$, we have the half-oxidized state of the polymer, denoted as *emeraldine*, depicted in Fig. 34.9c.

And finally, with $(1-y) = 1.0$, we have the completely oxidized state of the polymer, denoted as *pernigraniline*, a nonconductive form which in actual fact has very limited practical use, and its structure is shown in Fig. 34.9d.

Quite evidently, all states except leuco-emeraldine can be protonated. The states with no protonation are denoted as the "base" form, e.g., *emeraldine base* (which would be nonconductive). The reader may note the presence and alternation of "benzenoid" and "quinonoid" segments, as defined in an earlier chapter, in the structures of Fig. 34.9. Upon protonation, the polymer is denoted as the salt, for example, protonation of the emeraldine base form (Fig. 34.9c) with HCl would yield *emeraldine hydrochloride*:

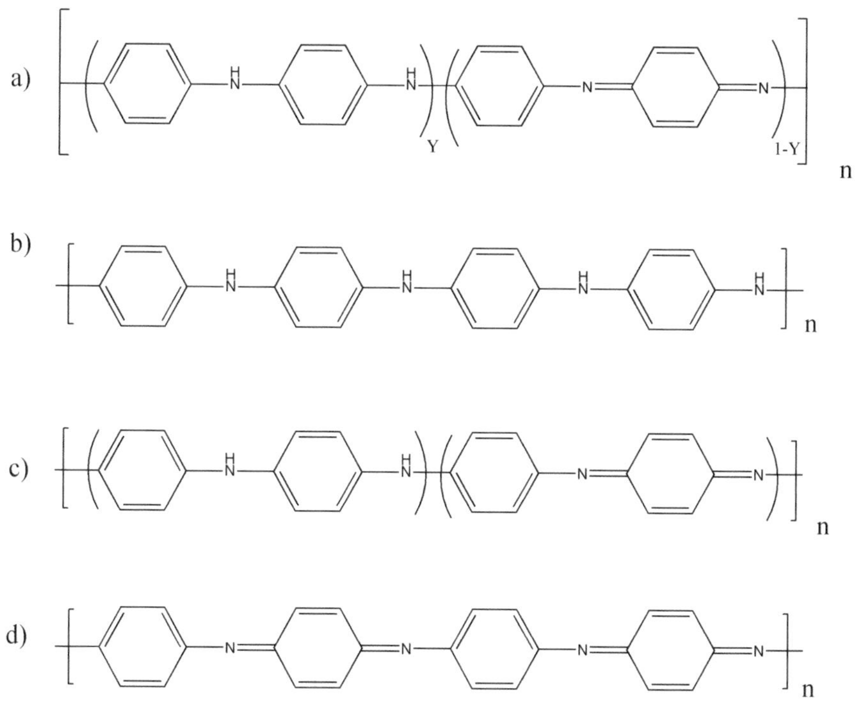

a)

b)

c)

d)

Fig. 34.9 Structures of various forms of P(ANi)

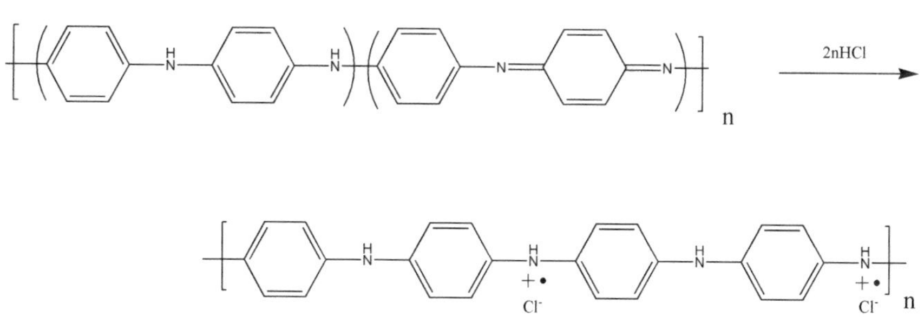

$$(34.1)$$

With full protonation as shown in the above equation, the polymer structure obtained is a radical cation, as confirmed by magnetic and other studies [995, 1060, 1452].

The emeraldine and in some cases pernigraniline base forms of P(ANi) are quite soluble in a variety of solvents; Shacklette et al. [1453], in a detailed solubility study, have measured solubility properties in 23 solvents, including such solvents as morpholine, butylamine, tetramethylurea, *m*-cresol, and diaminocyclohexane.

Because of its high MWt, P(ANi) in solution has been said to take on "close-coiled" or "expanded coil" configurations, studied in some detail by MacDiarmid and coworkers [1454].

34.5.2 Representative Chemical Syntheses

One of the attractions of poly(aniline) (P(ANi)) is its facile chemical synthesis. A typical such quick synthesis may be described. 10 mL of 2.5 M $(NH_4)_2S_2O_8$ (ammonium peroxydisulfate or persulfate, oxidant) is added with stirring to a 100 mL solution of 0.55 M ANi monomer and 0.55 trifluoromethane sulfonic acid (triflic acid, dopant) in an ice/salt bath (temperature <4 °C) over ca. 15 min. Stirring is continued for ca. 3 h. The precipitate is then filtered and washed with small amounts of 0.5 M triflic acid solution, then with methanol until colorless, and finally diethyl ether; the last step can also involve Soxhlet extraction with the ether. It is then dried at ca. 50 °C in air for ca. 3 h. The product so obtained is the conductive emeraldine salt form (typically 10 S/cm), i.e., emeraldine triflate (it is believed that pernigraniline is first formed but converted during reaction to this [903]. Aqueous HCl can be used in place of triflic acid, yielding emeraldine hydrochloride, but the Cl^- is a more fleeting dopant and is lost as HCl vapor with minor heating. The kinetics of this typical chemical synthesis have shown aniline consumption lags oxidant consumption at many stages [1300].

The as-prepared emeraldine salt form of P(ANi), which shows some crystallinity as seen by powder X-ray diffraction patterns, may then be reduced if required to the neutral, emeraldine base form, a blue–black powder with a metallic glint but with generally amorphous character as seen by X-ray diffraction. This can be accomplished using reducing agents such as aqueous NH_4OH (typically 2 M) or hydrazine. The substantially soluble, nonconductive, emeraldine base may then, after suitable processing, be re-doped (protonated) simply by treatment with the appropriate aqueous acid. Because of this capacity for re-doping, P(ANi) may be doped with a wide variety of dopant acids ranging from camphor sulfonic acid and oxalic acid to phosphotungstates [1040]; these dopants have been discussed elsewhere in this book. Re-doping is also used to enhance the conductivity of P(ANi) beyond a typical 10 S/cm produced in a first chemical synthesis. The conductivity of the greenish emeraldine salt form is directly and inversely related to its absorption bands at 420 nm and 580 nm, respectively [955].

The reprocessing of P(ANi) may include drawing into fibers or improving the crystallinity [1065] or fractionation of different MWts in solvents [1304]. P(ANi) substantially water-soluble in its doped form can be prepared by template-guided syntheses, which do not start from ANi monomer but rather polymeric dopants containing ANi pendant groups, as discussed at length in an earlier chapter; solutions of doped P(ANi) may also be obtained in conc. H_2SO_4 for spectroscopic and other studies [1192] or for spinning into fibers [1190]. In a detailed study of P(ANi) chemical polymerization, Cao et al. [1455] interestingly found no significant effect of parameters such as pH and relative concentration of reactants on the yield, although properties such as conductivity and viscosity in conc. H_2SO_4 were affected. They used a variety of oxidants, such as $K_2Cr_2O_7$, KIO_3, and $FeCl_3$, finding product conductivities similar for all of these.

P(ANi) has been marketed for some time as a chemically polymerized, proprietary-sulfonate-doped CP under the trade name *Versicon*. Originally produced by the Allied Signal Corp. based in Morristown, NJ, USA, the product line was taken over in the mid-1990s by the High-Performance Materials division of Monsanto Co., St. Louis, MO, USA. Around April 1998, Monsanto sold the product line to Ormecon Chemie, a subsidiary of Zipperling Kessler, based in Ammersbek (near Hamburg), Germany. As of this writing, this firm continues to produce bulk Versicon. P(ANi) has been produced in bulk in several other forms; see elsewhere in this book for a more detailed description.

34.5.3 Electrochemical Synthesis

Electrochemical polymerization is less frequently employed for bulk P(ANi) production and more frequently for production of thin films for spectroelectrochemical or similar characterization. In a typical procedure, ANi monomer is dissolved in an aqueous sulfuric acid solution (white ANi-sulfate first precipitated is shaken in solution until it redissolves). A potential of less than +0.9 V (vs. Pt quasi-reference) is then applied to a suitable electrode (e.g., Pt) on which the P(ANi) film is desired. Polymerization is rapid, and film thickness is controlled coulometrically. The polymer film is washed before further characterization. The cyclic voltammograms of such P(ANi) films display the dual-redox-peaks characteristic of poly(aromatic amines) P(ANi) that has also been electropolymerized in acetonitrile medium, but the overall properties of the polymer so obtained are considerably inferior to that obtained in aqueous medium.

An interesting electropolymerization of P(ANi), in aqueous (0.2 M) selenic acid rather than sulfuric acid medium, was reported by Tang et al. [1456]. The rate of polymerization of the aniline was however poor as compared to that in sulfuric acid and was complicated by the deposition of metallic Se.

34.6 P(ANi) Derivatives

Among the most common P(ANi) derivatives studied thus far are the alkyl and alkoxy derivatives, primarily *ortho*-substituted, with monomers, for example, being o-toluidine, *o*-methoxyaniline, and *o*-alkyl (up to dodecyl) aniline. Their syntheses from monomer are very similar to that of P(ANi), and in nearly all cases, they have been found to have conductivity, processibility, and other properties poorer than that of P(ANi). Gopal et al. [973] have even described a *o*-(2-methoxy)-ethoxy derivative, which has a maximum conductivity of ca. 4×10^{-3} S/cm. Their electrochemical, spectroelectrochemical, and spectral properties are predictable based on those of P(ANi) and do not shed additional light on the properties of P (ANi). From a practical aspect, therefore, these derivatives are largely of scientific interest only, to demonstrate that derivatives of P(ANi) can be synthesized.

In a comparative study of the polymerization of o- and m-toluidine, 2,6-diMeANi and ANi, Thyssen et al. [1406] found that m-toluidine comparatively showed an extremely long induction period and slow polymerization, but with no o-branching (100% para-coupling) characteristic of the other ANi derivatives.

Diaz et al. [1447] recently studied the synthesis and properties, including spectroscopy and morphology, of dihalogenated P(ANi)s, specifically poly((2,5)/(2,3)/(3,5)-dichloroaniline) and poly((2,5)/(2,6)-dibromoaniline). Chemical synthesis using Cu perchlorate, K dichromate, and K permanganate was employed.

Hwang et al. [1458] recently reported an interesting synthesis of N-alkyl P(ANi) derivatives in which $post$-alkylation is used, i.e., emeraldine base is first produced and then alkylated (rather than polymerization of N-alkyl anilines being carried out. They found that the resulting CPs have improved solubility in organic solvents, while doping up to 50%, as in P(ANi), could be carried out. Conductivities were higher than those of CP obtained through polymerization of N-alkyl anilines but were still poor, ca. 10^{-2} S/cm maximum, as compared to P(ANi).

A quite novel P(ANi) derivative, poly(2,2'-dithiodianiline), of postulated structure shown in Fig. 34.10 was synthesized recently by Naoi et al. [1459]. This CP was obtained via electropolymerization of the monomer, 2,2'-dithiodianiline, with the S-S bond preserved after polymerization. Tests of Li batteries incorporating this CP appeared to show enhanced charge capacity and energy density. The authors claimed that during reduction, the -S-S- linkages were broken, yielding -S$^-$...$^-$S- units, but that this process was reversible.

Fig. 34.10 Poly(2,2' dithioaniline) synthesized by Naoi et al. (After Ref. [1459], reproduced with permission)

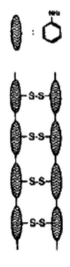

34.7 Other Poly(aromatic Amines)

Numerous CPs having a poly(aromatic amine) structure related to P(ANi) have been synthesized, most by electrochemical polymerization and a few by oxidants such as $Cu(BF_4)_2$. Indeed, the oxidation potential of aromatic amine monomers such as diphenyl amine is so low that their solutions are observed to oligomerize in the presence of air alone [999].

CPs including poly(diphenylamine) (P(DPA)) and poly(4-aminobiphenyl) (P (4ABP)) have been electrochemically synthesized by the Dao group [858, 975] and by Chandrasekhar et al. [956, 999, 1003, 1004, 1235]. The Dao group synthesized P(4ABP) in acidic aqueous/acetonitrile solution [865] (0.02 M monomer, 2.0 M HCl/CH₃CN, 2:1), via potentiostatic (+0.9 V) or cycling electropolymerization. The polymer so obtained, which these workers claimed showed only head-to-tail coupling, showed only a light-yellow to green to brown coloration when cycling in 2 M HCl solution. This is in contrast to the polymer obtained by Chandrasekhar et al. [956, 968, 999, 1003, 1004] via potentiostatic polymerization in acetonitrile solution with a variety of dopants. The latter polymer showed a reversible, glass-transparent to dark blue electrochromism and demonstrated evidence of substantial *ortho*-linkages. In a subsequent paper by the Dao group [975], it was claimed that DPA and 4ABP, when polymerized under such acidic aqueous/acetonitrile conditions, yielded the same polymer, "poly (4-phenylaniline)," although the cyclic voltammograms shown belied this. Chandrasekhar et al. [956, 999, 1003, 1004] showed that, at least in acetonitrile-medium-only polymerization, P(DPA) and P(4ABP) are distinct polymers with differing redox potentials although similar electrochromism (especially glass-clear transparency in the reduced state); both also show evidence of substantial *o*-linkages.

In a still later contribution by the Dao group [958], P(PNA) and P(DPA) polymerized in acetonitrile solution were studied. The electrochemical and spectroelectrochemical behavior (see earlier chapters) was more similar to that obtained by Chandrasekhar et al. except for lack of glass transparency in the reduced state and revealed only a single, reversible redox peak and poor electrochromism for P(PNA). Conductivities of 2 S/cm were claimed for P(DPA), comparing well with those obtained by the Chandrasekhar group (ca. 10 S/cm). Kitani et al. [859] carried out a detailed polymerization and electrochemistry study of poly(4-amino-DPA) yielding useful insight into the electrochemistry of CPs, discussed at some length in an earlier chapter. This CP was electrochemically polymerized in aqueous HCl solution using potential sweep or potentiostatic methods. The "polymer" so obtained was shown to be significantly soluble in solvents such as methanol and DMF, and this coupled with its low MWt, found to be ca. 2800 g, showed that it was an oligomer but with electrochromism and electrochemical behavior approximating that of the more extended polymer, if such could be obtained. This CP also had a fairly low conductivity, ca. 10^{-2} S/cm. Demonstrating that these low oxidation potential aromatic amines are very easy to polymerize, the Dao group subsequently reported [1285] chemical polymerization with $Cu(BF_4)_2 \cdot xH_2O$ oxidant/dopant for a series of poly(N-alkyl-diphenylamine)s. A 4–4'C-C'(phenyl-phenyl) coupling mechanism was claimed for this polymerization. These CPs showed poor conductivities (10^{-5} S/cm) and a yellow-to-violet electrochromism. In a variant of this synthesis, poly(N-alkyl-diarylamines), i.e., with a naphthalene group replacing one of the phenyls in DPA, were chemically synthesized by Dao et al. [1460]. These polymers however showed poor conductivity (10^{-5} S/cm) even in their highly doped form. The spectroelectrochemical characterization of these (see an earlier chapter) showed broadband responses characteristic of P(DPA) and its derivatives.

Ogura et al. [1167] demonstrated the use of electrochemically polymerized (0.1 M H_2SO_4 dopant solution, Pt electrode) poly(o-phenylenediamine) as a composite film with poly(vinyl alcohol) for conductometric humidity sensing. The electropolymerization of this CP was studied in some detail more recently by Wang et al. [1461]. The Oyama group [984, 998] carried out a detailed study of poly(1-aminopyrene), whose monomer unit possesses four fused rings with a pendant amino group (see above). The electropolymerization was standard, viz., from monomer + dopant solution in acetonitrile. A "polymer" of ca. eight monomer units was obtained which however showed spectroelectrochemical and electrochemical behavior (discussed in earlier chapters) typical of long-chain CPs. Work such as this and that of Ogura et al. cited in this paragraph appear to validate the contention of the Garito (Univ. of PA, Philadelphia, PA, USA) and other groups, based on theoretical calculations, that beyond ca. seven monomer units, much of the behavior of longer chain CPs which is based on extended π-delocalization is manifested [1235]. The poly(1-aminopyrene) showed low conductivity and electroactivity in both aqueous and nonaqueous media. An ESR response characteristic of a radical cation was observed. The polymerization was shown to be primarily along the N-atoms of the amino group. The study had useful implications for the electrochemistry of CPs, as discussed in an earlier chapter.

The Oyama group also electropolymerized and studied diaminonaphthalene polymers (2,3- and 1,8-derivatives) in acidic aqueous as well as acetonitrile media [977]. Conductivities were in the range 10^{-2} to 10^{-5} S/cm for highly doped polymer, and polymers served as scientifically informative poly(aromatic amine) analogues. More recently, Pham et al. [1462] studied the electrosynthesis and properties of poly (1,5-diamino-naphthalene), prepared in acidic aqueous, methanolic, and organic media. Spectroscopic evidence appeared to indicate that the polymer structure consisted of (-NH-C-) and (-N=C-) units bearing free -NH$_2$ groups. The CP obtained was electroactive and could incorporate Cu ions as inverse dopants. In a similar vein, Pham et al. also studied [1457] electropolymerization of 5-amino-1,4-naphthoquinone. The CP so obtained appeared to possess one quinone group per monomer unit. Such studies again confirmed the contention made in this chapter that virtually any monomer with an active amino group would lend itself to electropolymerization and yield a reasonably electroactive polymer. Mostefai et al. [1463] electropolymerized poly (5-amino-1-naphthol) from its monomer in aqueous perchloric acid solution. The CP films were characterized by Raman spectroscopy, among other techniques. The postulated structure of the final CP was as shown in Fig. 34.11.

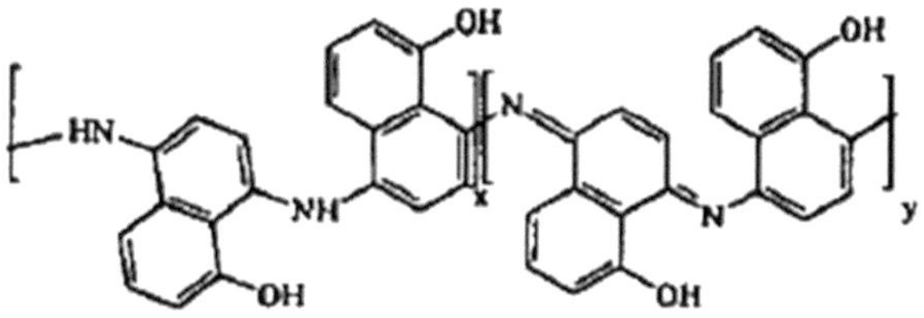

Fig. 34.11 Poly(5-amino-1-naphthol) synthesized by Mostefai, et al. (After Ref. [1463], reproduced with permission)

34.8 Problems and Exercises

1. Write down the structures of the doped and pristine forms of one member of the following classes (discussed in this chapter), and outline one chemical and one electrochemical (if available) synthesis for it: P(Ac), P(DiAc), P(Py), P(ANi), and P(ANi) derivatives and other poly(aromatic amines). In which cases are electrochemical (or chemical) polymerizations unavailable, and why?

2. Enumerate the most common dopants for one member of each of the polymer classes listed in problem 1 above and briefly outline the doping method.

3. Make a list of the classes listed in problem 1 in order of decreasing conductivity. Use conductivity ranges of the highest conductivity observed for a member of each class.

4. Outline one example of methods of processing – whether in pristine or doped form, in melt or in solution – for one member of each of the classes listed in problem 1.

5. For each class outlined in problem 1, briefly discuss the difference in all possible properties you can enunciate between the parent polymer and common derivatives (e.g., poly(N-alkyl pyrrole) vs. P(Py), poly(o-methoxyaniline) vs. P(ANi)).

6. Of the classes enumerated in problem 1, which are of practical utility? Cite all applications studied thus far for these.

7. Besides P(ANi), which of the CPs discussed in this chapter may qualify as long-chain (> 30 monomer units) CPs?

8. One method of introducing properties such as improved processibility in a derivative of a parent CP appears to be the introduction of substituents, such as long-chain alkyl/alkoxy, at appropriate positions in the parent monomer. What are the tradeoffs involved in this, in terms of other properties of the parent CP, such as conductivity?

Chapter 35
Classes of CPs: Part 2

Contents

35.1   Brief Note on Chapter Focus

In this chapter, a continuation of the subject matter of the previous chapter, we continue our discussion of each group or class of CPs in turn. Again, emphasis is placed on *synthetic aspects*, as properties and other aspects of the CPs are covered elsewhere in this book. Within this limited purview, more emphasis is placed further on giving detailed coverage of the less well-known or less well-studied CPs.

35.2   Poly(thiophenes) (P(T)s)

The term poly(thiophenes) (P(T)s) includes direct derivatives such as poly(3-alkyl-thiophenes) (P(3AT)s). Poly(thiophene) was one of the early CPs synthesized, but real interest in this class was augmented with the discovery that 3-substitution of the thiophene ring yielded a wide variety of practically very versatile CPs.

© Springer International Publishing AG 2018
P. Chandrasekhar, *Conducting Polymers, Fundamentals and Applications*,
https://doi.org/10.1007/978-3-319-69378-1_35

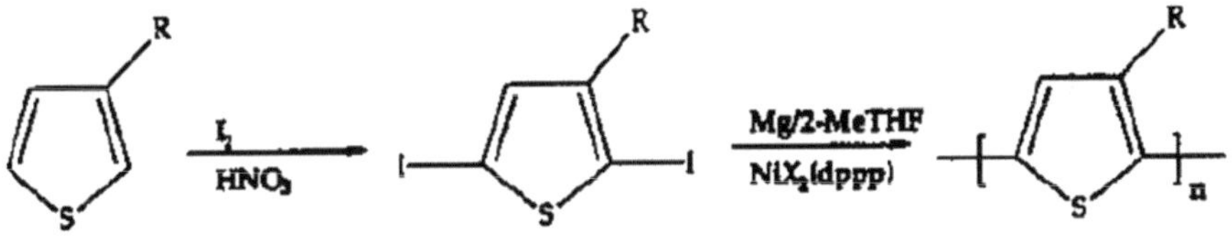

Fig. 35.1 Nickel-catalyzed polymerization of thiophene (After Ref. [1465], reproduced with permission)

35.2.1 Chemical Syntheses

Although electrochemical syntheses, discussed in the next section, are found to be the quickest way to obtain highly conductive, "clean" P(T)s, a very wide variety of chemical synthetic methods are available. The most common is a Ni-catalyzed coupling of the di-Grignard derivative of thiophene, shown in Fig. 35.1 [1328, 1464] and is advantageous because it is certain to give 2,5-coupling linkages.

Other methods employed include the relatively simple direct oxidation with $FeCl_3$ or $MoCl_5$ (dopant/oxidant) in chloroform and precipitation in MeOH [1466], Friedel-Crafts alkylation [1020], coupling of the dihalide using Ni catalysts [1467] and organocuprate coupling [1468]. In their preparation of a series of melt-spinnable long-chain-alkyl-3-substituted P(T)s, Yoshino et al. [1222] used a simple oxidation with $FeCl_3$ in chloroform. After 15 h of reaction, the polymer was poured into methanol and washed therewith.

Zhang and Tour [1469] recently described a chemical synthesis of amino/nitro-substituted P(T) derivatives said to possess very low bandgaps. Their synthesis is summarized in Fig. 35.2. The final product, **5** ($\lambda(max) = 662$ nm), was purified by dialysis due to its solubility in water and was said to possess significant zwitter-ionic/quinoidal character.

35.2.2 Electrochemical Syntheses

P(T)s (including derivatives) can be prepared using the standard electrochemical synthesis, viz., from monomer + dopant solution in organic electrolyte in 3-electrode mode [991, 1470]. Common dopant electrolytes used include Li or tetraalkyl ammonium tetrafluoroborate, hexafluorophosphate, and triflate (typically 0.5 M, monomer concentration typically 0.1 M). Solvent media include acetonitrile, THF, and CH_2Cl_2. Electrode substrates again depend on the end-use desired and can be Pt, Au, or ITO/glass. A potentiostatic polymerization at +1.6 V vs. SCE is adequate. As for P(Py)s and indeed any electropolymerization with mechanistic involvement of H_3^+O, a little water (0.01 M or less) in the electrolyte helps. P(T)s do not generally yield problems with severe adhesion onto Pt electrodes found in

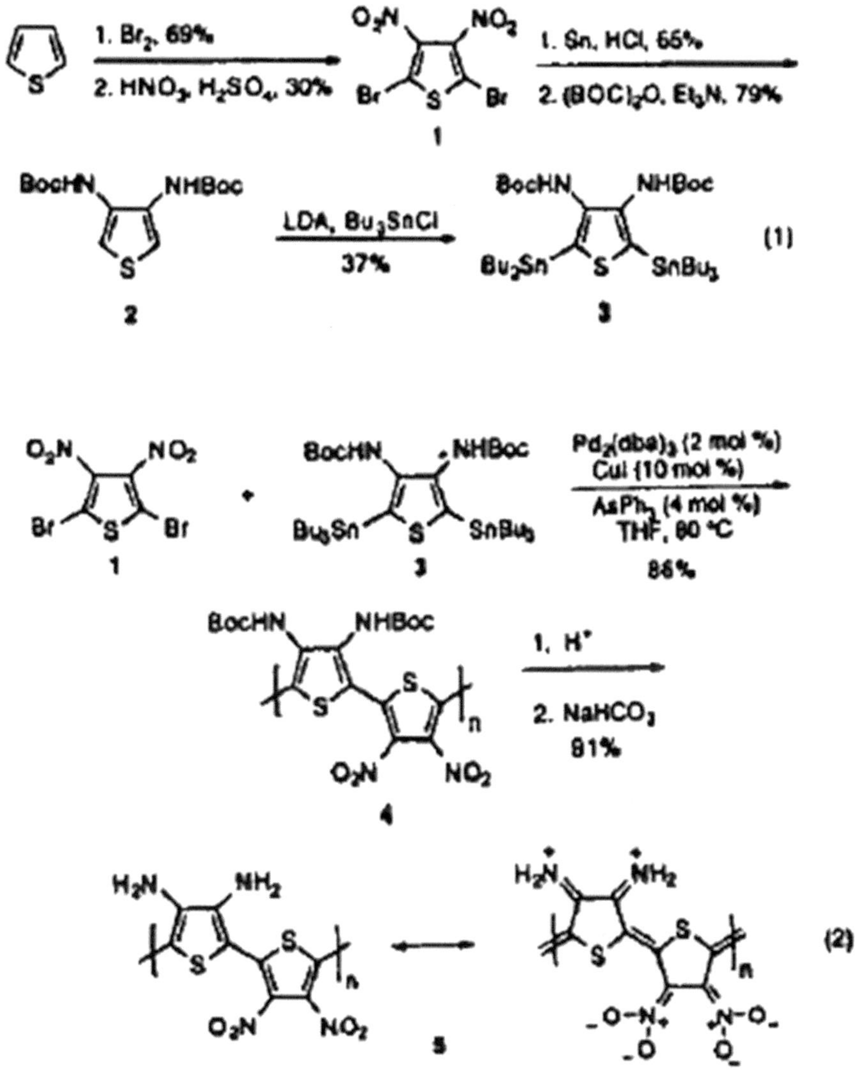

Fig. 35.2 Chemical syntheses of amino/nitro-substituted P(T) derivatives (After Ref. [1469], reproduced with permission)

P(Py)s, and free-standing films can be prepared on such electrodes. Quite evidently, and especially for unsubstituted P(T), 2,5-coupling cannot be ensured in electro-chemical polymerization although it is the dominant polymerization mode; the very close comparative IR spectra of electrochemically and chemically polymerized P(T) bear this out [1472, 1473].

A large variety of substituted thiophenes have been electrochemically polymerized [990, 1474], including many combinations of 3,4-(alkyl/halogen)-substituted thiophenes as well as 2,2′-bithiophene. For example, 3-Br-T, 3,4-di-Br-T, and 3-(CH$_2$CN)-T yield viable polymers on electropolymerization, whereas 3-CN-T and 3-NO$_2$-T do not. In one of the first preparations of soluble long-chain-alkyl-3-substituted P(T)s, from 3-hexyl through 3-octadecyl and 3-icosyl, Sato et al. used a galvanostatic (constant current) polymerization at 2 mA/cm^2 (ITO working electrode, Pt cathode, propylene carbonate or nitrobenzene solvent, Et$_4$NPF$_6$ electrolyte), monitoring the polymerization via charge. The doped polymer films could be peeled off from the working electrode and washed. An interesting preparation described by Hotta et al. [1059] is galvanostatic electropolymerization of a copolymer of 3-hexyl and 3-benzyl thiophenes from equimolar monomer solutions in nitrobenzene. The resulting copolymer could be "doped" in acetonitrile solution with NOPF$_4$.

Indeed, P(bi-T) on Pt has been used as an independent working electrode for the electrochemical characterization of the FeCP$_2$ system [1474].

Onoda et al. [957] carried out an interesting electropolymerization of oligothiophenes having a double bond (ethene or vinylene group) between thiophene chains, shown in Fig. 35.3a. In these, the conformational restriction introduced by the rigid vinylene groups is said to have some effect on the bipolaron absorptions. Figure. 35.3b shows a spectroelectrochemical data for one of these polymers: a reddish brown to dark blue electrochromism is observed, with calculated bandgaps ca. 0.2 eV smaller than that of P(T).

Welzel et al. [1475] carried out electropolymerization of various ether and ester derivatives of 3-(2-hydroxyethyl)-thiophene) in acetonitrile medium, in most cases copolymerizing them with 3-methyl thiophene since homopolymerization was unsuccessful. Some of the monomers polymerized included 3-(2-acetoxyethyl) thiophene) and 3-[2-((triphenylmethyl)oxy)ethyl)thiophene].

Zhang et al. [1476] found that when 3-fluoroalkoxy and 3-fluoroether thiophenes are electropolymerized, CPs with high MWt and good processibility resulted. These CPs also had comparatively higher thermal stability than non-fluorinated analogues.

35.2.3 Properties

Table 35.1 gives a measure of the property variation and versatility that can be obtained with a variation in the substituent of poly(3-substituted thiophene).

As discussed in an earlier chapter, dopant-induced swelling is especially apparent in P(T)s, with, e.g., undoped and doped (unsubstituted) P(T) having fibrillar diameters of ca. 25 nm and 80 nm, respectively, while both lack the X-ray-diffraction-manifested crystallinity characteristic of P(ANi).

Fig. 35.3 Structure (**a**) and spectroelectrochemical data (**b**) for unique poly(oligo-thiophene) (After Ref. [957], reproduced with permission)

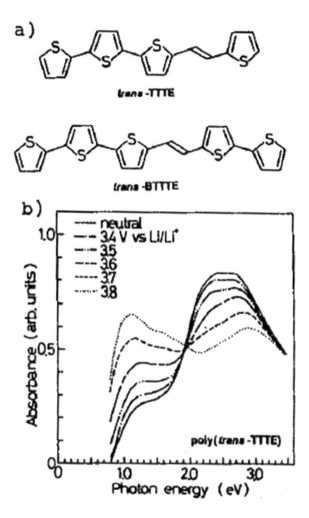

Table 35.1 Illustration of the variation of properties obtainable in 3-substituted poly(thiophenes) (P(T)s) with different substituents **R**

Substituent, **R**	Property imparted
Alkyl, C_nH_{n+1}, $n > 6$	Soluble in organic solvents
Alkyl, $n > 10$	Distinct m.p., melt processible
$(CH_2)_nY$, $4 < n < 10$, Y = halide	Useful for further syntheses/processing
$O(CH_2)_n$, $n > 5$	Some solubility in protic organic solvents, low bandgaps
$(CH_2)_nSO_3^-$, $2 < n < 5$	Water solubility
N-alkyl bipyridyl	Redox active group

Doped unsubstituted P(T) powder has conductivity in the region 1–50 S/cm (e.g., 10 S/cm with triflate doping), while P(3-Me-T) has a slightly higher conductivity (e.g., 50 S/cm for triflate doping); longer-chain-alkyl 3-substitution yields decreasing conductivity. Higher conductivities are obtained for drawn films of P(T), P(3-methyl-T), and P(3-ethyl-T) [907, 1477]: for example, heavy I_2-doping yields conductivities in drawn films of P(T), P(3-Me-T), and P(3-Et-T) of 190, 500, and 270 S/cm, respectively, with the trend from substitution also evident in these data. Stretch orientation in P(3-alkyl-T)s generally permits elongation of up to five times the initial length of films or fibers. It is also to be noted that with dopants such as I_3^- or $FeCl_4^-$, degradation in conductivity with air exposure is rapid, especially at elevated temperatures.

Among the most versatile substituted P(T)s have been the long-chain-alkyl 3-substituted P(T)s, which have yielded polymers which are thermally stable, and melt- and solution-processible (although only in the undoped and "pseudo-doped" states as described below) [1205, 1206]. The undoped 3-alkyl-substituted thiophenes have a clear melting point, which decreases with increasing alkyl chain length (in °C): butyl, 275; hexyl, 190; octyl, 150; decyl, 125; and dodecyl, 80. Thus these P(T)s in their undoped forms may be melt processed with thermoplastics, and melt spun into fibers [1222], as described elsewhere in this book. Because of conformational variations possible in such long-chain-substituted polymers, these P(T)s also display thermochromism and solvatochromism [962]. The undoped 3-alkyl P(T)s are also soluble in common organic solvents such as $CHCl_3$, CH_2Cl_2, THF, and toluene, with solubility obviously increasing with longer alkyl chain length. These 3-alkyl P(T)s are again in reality oligomers, with typical MWts in the 2500 region (i.e., ca. 20 monomer units), again confirming what has been observed many times in this book that such chain lengths are sufficient for CPs to display most of the characteristic properties of CPs with much longer chain lengths.

Yoshino et al. [1222] processed a series of 3-long-chain-alkyl-substituted P(T)s into fibers of ca. 30 μm diameter via simple melt-spinning (extrusion). After I_2-doping, the fibers had conductivities in the region of 50 S/cm. Ng et al. [1478] prepared a series of poly{3-(ω-OH-alkyl)-thiophenes} via standard chemical oxidation using $FeCl_3$. On I_2-doping, these gave conductivities of up to 38 S/cm, being maximum for the octyl pendant moiety.

Guillerez and Bidan [1479] recently described a new synthesis of highly regioregular poly(3-octyl-thiophene) based on the Suzuki coupling reaction. The stable polymerizable precursor which bore an iodo and boronic ester derivative in the 2 and 5 positions was obtained in two steps from the 3-octyl monomer. Polymerization of this in the presence of Pd(II) acetate gave a final CP said to contain up to 97% of head-to-tail couplings. This synthesis is summarized in Fig. 35.4.

A P(T) with a unique 3-substituent, (1-naphthyl), was electropolymerized by Dogbéavou et al. [1480]. The starting monomer was synthesized by a standard Ni-catalyzed Grignard coupling of 1-naphthyl with 3-Br-thiophene. The CP was then prepared via standard electropolymerization in acetonitrile with BF_4^- dopant, yielding a ca. 20% doping level.

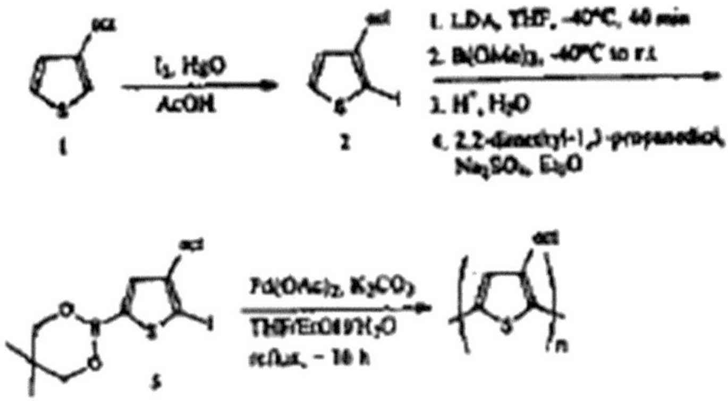

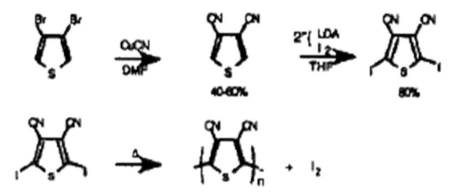

Fig. 35.4 Synthesis of highly regioregular poly(3-OT) (After Ref. [1479], reproduced with permission)

Fig. 35.5 Synthesis of new di-CN-substituted P(T) (After Ref. [1481], reproduced with permission)

Hide et al. [1481] described the synthesis of a new disubstituted poly(thiophene), poly(3,4-dicyanothiophene), schematically shown in Fig. 35.5. As a result of having two electron-withdrawing cyano groups, this CP was found to have depressed HOMO and LUMO levels (6.7 eV and 3.6 eV, respectively).

It has been claimed [1013, 1059, 1186, 1206] that these 3-alkyl P(T)s can be made soluble in their doped states in solvents such as nitromethane. However, closer examination reveals that such "doped" P(T)s, involving dopants such as nitrosyl-hexafluoroantimonate (NOSbF$_6$) in nitromethane or NOPF$_4$ in acetonitrile, are actually not true solutions. For example, they yield polymer, after solvent evaporation or casting, which has properties significantly different from that originally put in solution, e.g., a significantly lower conductivity. The solutions in reality appear to be sols or aggregated suspensions of dopant and pristine (undoped) polymer. During redissolution, it appears that the dopant is first solvated, yielding less doped polymer which is then more soluble and finally dissolves [1186]. In some cases recast films need to be treated further with oxidant/dopant such as $FeCl_3$ before showing acceptable conductivity [1206]. These data again show the difficulty in obtaining truly soluble, doped CPs.

35.3 Poly(alkylene Dioxythiophenes), Including Substituted Dioxythiophenes, the "PEDOTS" and "ProDOTS"

This class of CPs, of which the most well-known is poly(ethylene dioxythiophene) (PEDOT), has received much attention for varied applications, especially in electrochromics, since about 2002. We shall first briefly discuss PEDOT syntheses and then briefly address syntheses of another important member of this class, poly (propylene dioxythiophene) (P(ProDOT)), and derivatives thereof.

Several methods have been employed for *electrochemical polymerization* of PEDOT. These include potentiostatic, galvanostatic, and cyclic voltammetry techniques. Differences in film morphology were studied by Poverenov et al. [1482] by changing the EDOT monomer solvent, supporting electrolyte, and the electrochemical deposition technique (CV vs. potentiostatic). It was found that the solvent had the largest effect on film morphology, with acetonitrile giving rough "star"-shaped features, while propylene carbonate yielded mostly smooth films.[1]

Luo et al. [1483] investigated PEDOT polymerized from nonaqueous acetonitrile and aqueous microemulsion solutions. In the microemulsion, sodium dodecyl sulfate (SDS) was dissolved in water along with HCl and $LiClO_4$ as the supporting electrolyte. Films were deposited on ITO substrates by cyclic voltammetry, and comparisons were made between the nonaqueous and aqueous solutions. It was found that the microemulsion gave very smooth, crack-free films of PEDOT.[2]

Xiao et al. [1484] deposited PEDOT on –COOH functionalized multiwalled carbon nanotubes (MWCNTs) using a pulse potentiostatic method. After depositing the MWCNTs on ITO with electrophoresis, the MWCNTs were immersed in a solution of EDOT, sodium dodecyl sulfate, and $LiClO_4$ as supporting electrolyte. A pulse on potential of 1.2 V (vs. Ag/AgCl) was applied for 1 s, followed by a reverse pulse period of 0.2 V for 0.5 s, with a total duration of 500 s. SEM imaging shows deposition of PEDOT on MWCNTs which was used as a counter electrode in dye-sensitized solar cells.[3]

Chemical polymerization of PEDOT has been achieved through oxidation of the EDOT monomer with typically Fe^{3+}-based salts (i.e., $FeCl_3$ or $Fe_2(SO_4)_3$) or through persulfate salts (i.e., ammonium persulfate and sodium persulfate). McFarlane et al. [1485] have used phosphomolybdic acid hydrate ($H_3PMo_{12}O_{40}$, PMA) as an initiator for polymerization. The authors have reported that the use of PMA, which has an oxidation potential much closer to EDOT than $FeCl_3$, allows for crack- and pinhole-free films on glass substrates when deposited by spin-coating techniques [1485].

Cho et al. [1486] reported on the copolymerization of PEDOT with poly(styrene sulfonate-co-vinyltrimethoxysilane) (PEDOT:P(SS-co-VTMS)) and found that increased percentages of VTMS led to hydrophobic films with water contact angles as large as 96%. PEDOT:P(SS-co-VTMS) was polymerized with sodium persulfate (NaPS) and iron(III) sulfate $(Fe_2(SO_4)_3)$ and spin coated onto glass substrates [1486].

Kwon et al. used both PEDOT and hydroxymethyl EDOT (HEDOT) to detect organophosphate vapors to concentrations as low as 10 ppt. A ratio of PEDOT to HEDOT (3.6:1) was used and polymerized on PMMA substrates that were treated with $FeCl_3$. Depending on the temperature of deposition, different morphologies could be achieved, including nanonodules (90 °C, 760 Torr), nanorods (60 °C, 760 Torr), or a smooth layer (90 °C, 1 Torr). The lowest level of detection of DMMP (10 ppt) was achieved with the nanorod morphology.[6]

Zhang et al. [1487] synthesized 10–20 μm long PEDOT nanotubes using a reverse microemulsion solution. Sodium bis(2-ethylhexyl) sulfosuccinate (AOT) was dissolved in 70 mL n-hexane followed by the addition of 1 mL of aqueous 10 mmol $FeCl_3$. EDOT monomer was then added and allowed to stir for 3 h, after which the blue/black precipitate of PEDOT was filtered and washed with methanol and acetonitrile.[7]

Other unique synthetic approaches to PEDOT derivatives have included those described by Zhang, Wither-Jensen, and Paradee et al. [1488–1490].

Krishnamoorthy et al. [1491] of the Kumar group at IIT, Bombay (India), reported the synthesis and electrochromic performance of *dibenzyl*-substituted derivatives of P(ProDOT). They showed that the electrochromic performance of these materials was shown to be the best among all ProDOT derivatives reported before their work.

Chandrasekhar et al. [1492] reported in detail new syntheses of the new P (ProDOT) derivatives listed below, some of which are also depicted in Fig. 35.8. These authors also reported that these cathodically coloring electrochromic CPs yielded superior electrochromic properties when incorporated into electrochromic sunglasses with complementary, anodically coloring CPs.

- 2,2-(bis-4-chloro-benzyl)-3,4-propylenedioxythiophene ("Cl-Bz-ProDOT)
- 2,2-(bis-4-bromo-benzyl)-3,4-propylenedioxythiophene ("Br-Bz-ProDOT")
- 2,2-(bis-4-nitro-benzyl)-3,4-propylenedioxythiophene ("NO$_2$-Bz-ProDOT")
- 2,2-(bis-4-amino-benzyl)-3,4-propylenedioxythiophene   ("NH$_2$-Bz-ProDOT") (Figs. 35.6 and 35.7)

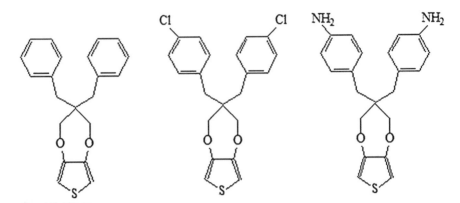

Fig. 35.6 Illustrative structures of two of the four new monomers synthesized by Chandrasekhar et al. [1492] and another base monomer, the monomer precursors of unique, cathodically coloring CPs. *From left to right*: 2,2-(dibenzyl)-3,4-propylenedioxythiophene ("Bz-ProDOT"); 2,2-(bis-4-chloro-benzyl)-3,4-propylenedioxythiophene ("Cl-Bz-ProDOT"); and 2,2-(bis-4-amino-benzyl)-3,4-propylenedioxythiophene ("NH$_2$-Bz-ProDOT")

35.4   Other Derivatives of Poly(thiophenes)

The most well-known derivatives of P(T), which are not simply substituted P(T)s, and the only ones discussed in this section, are poly(isothianaphthene) (P(ITN), also called poly(benzo[c] thiophene); its analogue poly(naphtho[c] thiophene); the poly (thienylene vinylene)s (P(TV)s), schematically shown in Fig. 35.8a–c; and certain poly(bithiophene) derivatives. Each of these is discussed in turn.

35.4.1   P(ITN) and Poly(naphtho[c]thiophene)

The original objective of the organic synthesis group of Wudl and Heeger et al. in synthesizing P(ITN) [1059, 1493, 1494] was the lowering of the bandgap through increasing "quinonoid" contributions. This lowering was realized in part (ca. 1.1 eV vs. ca. 2.1 eV for P(T)) and has been discussed elsewhere in this book. Here, we may briefly describe the common syntheses for this CP. All these center on an initial synthesis of the simple and dihydro-monomers, ITN, and DHITN (the latter with a saturated thiophene ring). DHITN is synthesized from *o*-dibromoxylene based on the detailed procedures of Cava et al., Oliver and Ongley, and others [1495, 1496], while other workers [1219] have used *o*-dichloroxylene as the starting material. DHITN can then be converted to the 2-oxide (S = O derivative). DHITN and this S-oxide can then be coupled to yield the somewhat unstable intermediate, ITN.

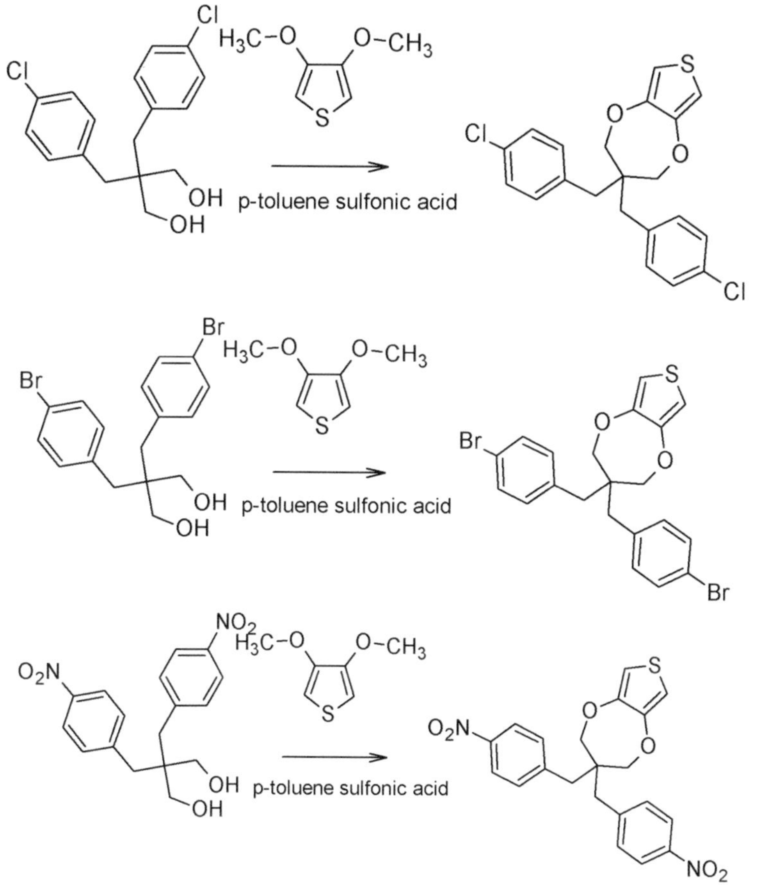

Fig. 35.7 Final steps in the syntheses of the monomers which are derivatives of propylene dioxythiophene ("ProDOT") used to electrochemically polymerize the corresponding electrochromic CPs (After Chandrasekhar et al. [968])

A chemical oxidation procedure has been used by many workers for polymerization of P(ITN) directly from DHITN: Jen and Elsenbaumer [1497] used O_2/FeCl$_3$ as the oxidant; Rose and Liberto [1219] used a multistep procedure; and Defieuw et al. [1219] used a modification of Wudl et al.'s procedure, employing conc. H_2SO_4/O_2 as the oxidant, the polymerization however taking as long as 2 weeks. Heeger and Wudl et al. [1498] polymerized P(ITN) electrochemically from the unstable ITN monomer. Perhaps the simplest and most practical synthesis of P(ITN) was that of Chandrasekhar et al. [1003], which used a direct electropolymerization from DHITN itself; this polymerization had important mechanistic implications, discussed in an earlier chapter).

Conductivities obtained by Chandrasekhar et al. for P(ITN) [1003] were in the 1–3 S/cm range, to be compared with the maximum ca. 4–8 S/cm obtained by

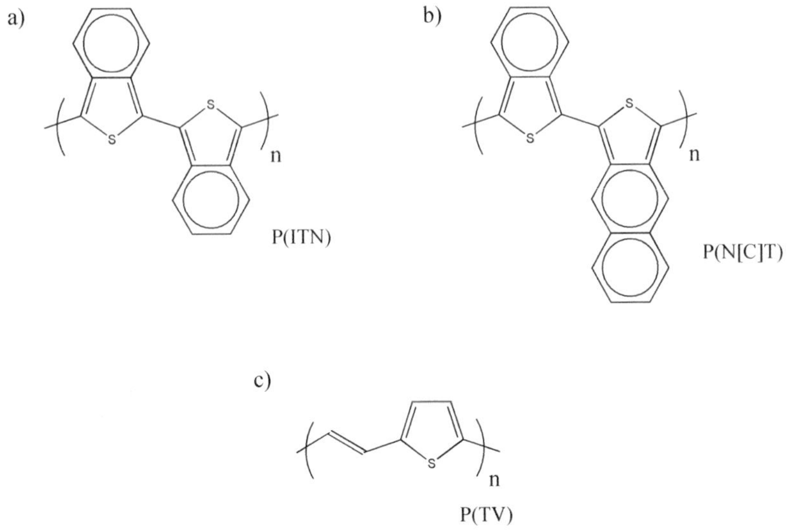

a)

P(ITN)

b)

P(N[C]T)

c)

P(TV)

Fig. 35.8 Structures of poly(isothianaphthene) and analogues

Defieuw et al. [1219] for the chemical polymerization. The spectroelectrochemical characteristics of P(ITN) have been presented in an earlier chapter. One of the unique characteristics of this CP is that it is darker in its reduced state (at negative applied potential), and very thin films can be nearly transparent in the oxidized (+ applied potential) state. This is said to stem from its low bandgap, ca. 1.1 eV (cf. 2.1 eV for P(T)). Defieuw et al. [1219] used this property to cast films of P(ITN) from suspensions in aqueous H_2SO_4/lambda carrageenan onto PET substrates for antistatic uses. Specialized P(ITN) syntheses and uses have been the subject of several patents [1498].

The CP poly(naphtho[c]thiophene) was synthesized by Wudl et al. [1003] using a method similar to that for P(ITN), with the objective of an even further lowering in bandgap. This objective was not favorably realized, and because of its lesser practical utility and more difficult synthesis, this CP has seen little work since.

Schlick et al. [1499] recently studied the electropolymerization of a number of monomers containing a variety of alternating thiophene, vinylene, and in one case isothianaphthene units, some of which are illustrated in Fig. 35.9. The only monomers successfully polymerized were those shown in part (b) of this figure.

35.4.2   Poly(thienylene vinylenes) P(TV)s

One of the major advantages of poly(thienylene vinylene) (P(TV))s is that they are generally prepared through a chemical precursor route, which allows for very practical processing prior to doping. Kossmehl [1500] first reported a Wittig

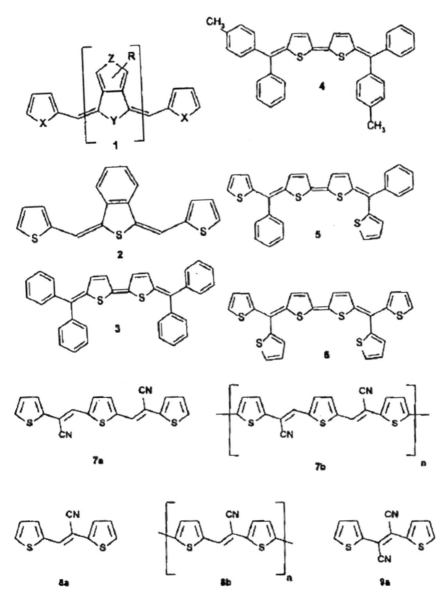

Fig. 35.9 Structures of monomers of unique poly(thiophene) (P(T)) derivatives of Schlich et al. (After Ref. [1499], reproduced with permission)

condensation synthesis of P(TV) yielding low MWt, low conductivity polymer. In one of the first useful syntheses, Jen et al. [1501] prepared P(TV) and substituted P (TV) through a polyelectrolyte intermediate route, two of whose variants are given in Fig. 35.10.

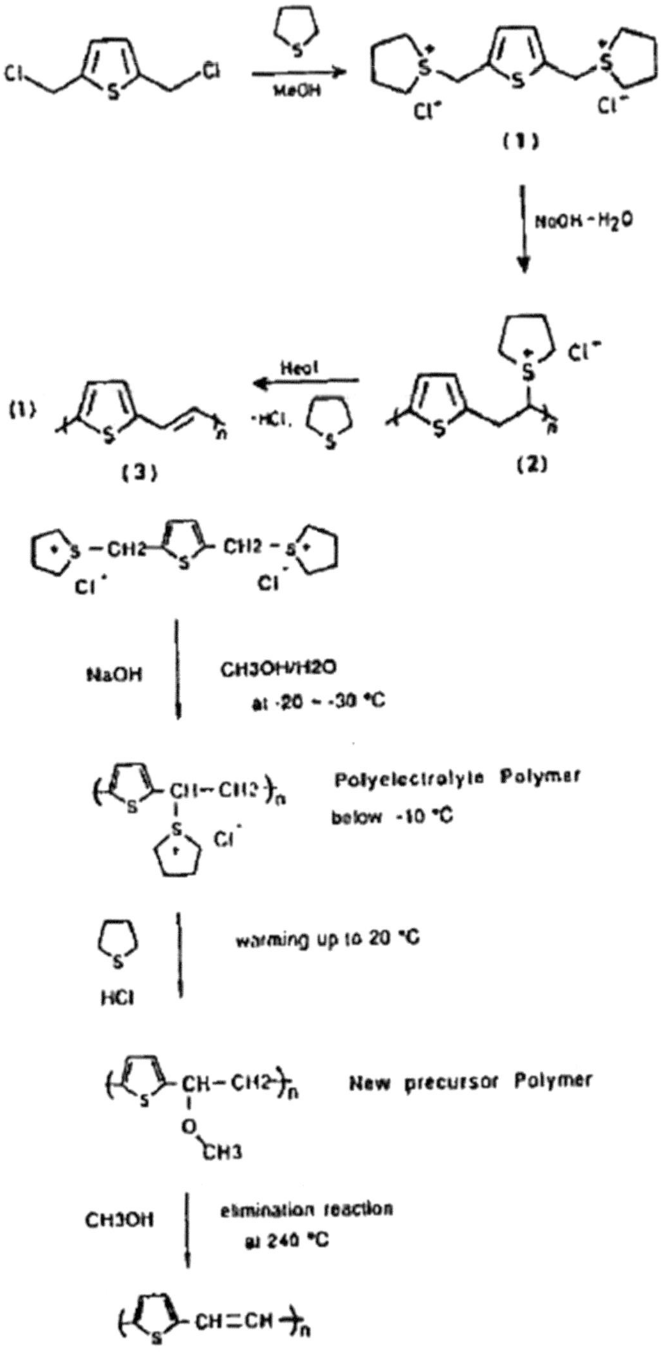

Fig. 35.10 Two variants of P(TV) synthesis (After Ref. [1501], reproduced with permission)

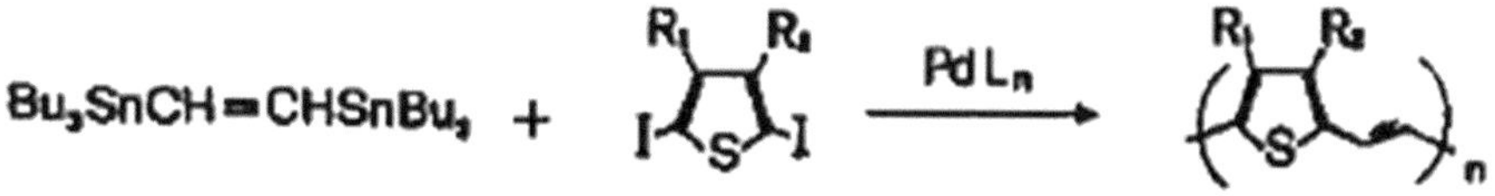

Fig. 35.11 Preparation of substituted P(TV)s, (After Catellani et al. [1502])

After casting films of the precursor polymer and thermal conversion to the final P (TV) as ca. 25 μm films, doping could be achieved through exposure to I_2, $NOSbF_6$, $FeCl_3$, $NaPF_6$, and other common dopants, yielding conductivities in the 37–62 S/cm range; expectedly with such dopants, however, conductivity deteriorated on air exposure or mild thermal stress. Catellani et al. [1502] used a different synthesis, shown in Fig. 35.11, to prepare substituted P(TV)s. These CPs had MWts (M_w) in the 11,000 region.

Yang and Geise [1126] synthesized alkoxy-substituted derivatives of P(TV) using the Wittig reaction (Eq. 35.1). The polymers they obtained were a mixture of cis- and trans-configurations, with high solubility in solvents such as $CHCl_3$ in their undoped state and conductivities of ca. 2 S/cm in their doped states. Using $CHCl_3$/THF solutions of the undoped polymer, I_2, and a host polymer such as polystyrene or poly(methyl methacrylate), composite films of conductivity ca. 0.1–2 S/cm were claimed to be castable.

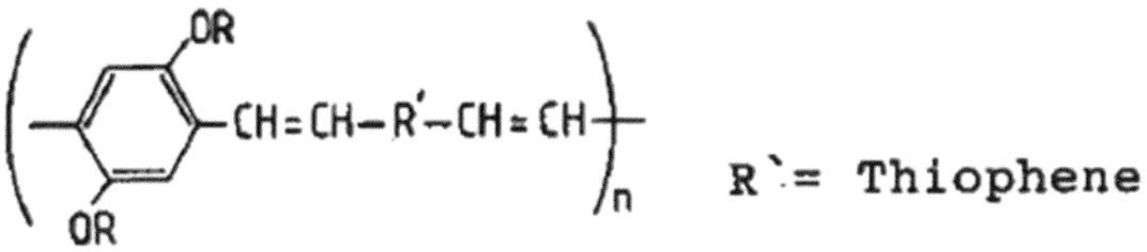

$$R' = \text{Thiophene} \tag{35.1}$$

P(TV)s have been investigated for a wide variety of versatile uses, including third-order nonlinear optical (NLO) applications [1503], corrosion protection [1504, 1505], and transmission-mode electrochromic devices [1506]. Heeger and Smith [902] have shown that fibers of P(TV)s drawn to ratios up to 20:1 possess enhanced conductivity (claimed up to 1200 S/cm) as well as enhanced mechanical properties, e.g., a Young's modulus of 8 GPa and tensile strength of 0.5 GPa.

35.4.3   Other Poly(thiophene) (P(T)) Derivatives

Certain poly(bithiophenes) qualify as P(T) derivatives. Thus, for instance, Ng et al. [1507] synthesized homopolymers of the symmetrical 3,3'-dibutylsulfanyl-2,2'-bithiophene (DBSBT) and its copolymers with thiophene and bithiophene via the standard Ni-catalyzed Grignard cross-coupling. CPs with bandgaps in the region of 1.7–1.8 eV were obtained. Conductivities with I_2-doping were however poor, a maximum of ca. 10^{-2} S/cm. Figure 35.12 summarizes this synthesis.

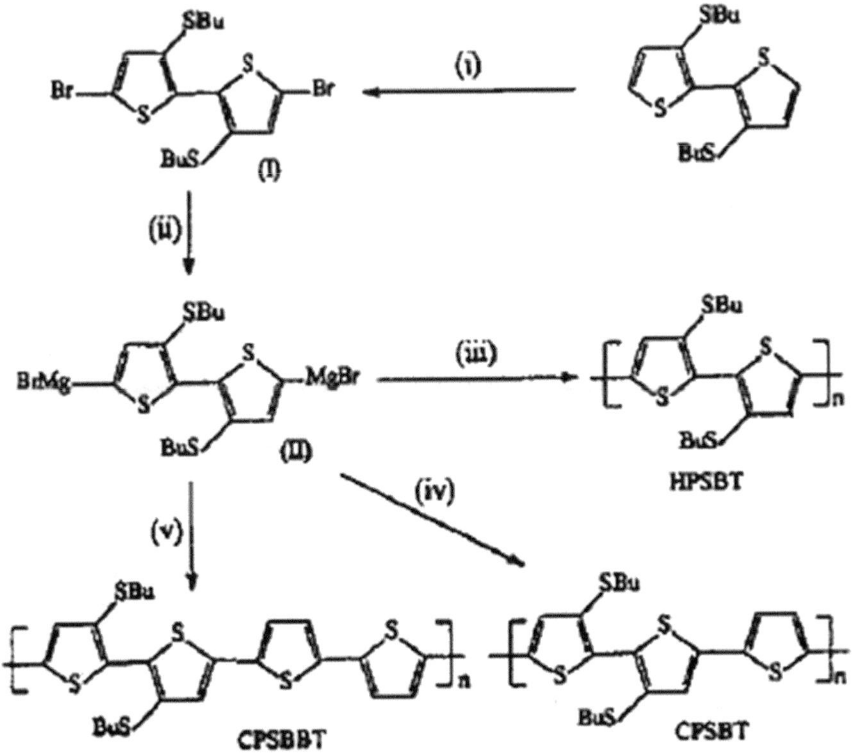

Fig. 35.12 Synthesis of unique poly(bithiophenes). Reagents and conditions: (i) Br_2, AcOH; (ii) magnesium (2-mole-equivalent), THF, refluxing; (iii) Ni (dppp)Cl_2, 5–5′-dibromo-3,3′-dibutylthio-2,2′-bithiophene; (iv) Ni(dppp)Cl_2, 2,5-dibromothiophene; (v) Ni(dppp)Cl_2, 5,5′-dibromo-2,2′-bithiophene (After Ref. [1507], reproduced with permission)

Inaoka and Collard [1508] successfully carried out electropolymerization of substituted dithieno[3,4-b:3′,4′-d]thiophenes, illustrated in Eq. 35.2, finding acceptable conductivities (ca. 5 S/cm with I_2-doping) and observing that the 2,5-dioctyl analogue was soluble in most organic solvents.

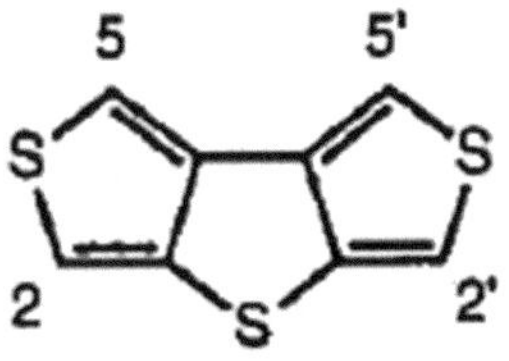

$$(35.2)$$

Pomerantz and Gu [1347] prepared poly(2-decylthieno[3,4-b]thiophene, through a multistep synthetic sequence. This CP had a $\lambda(max)$ of 738 nm, corresponding to a band-edge bandgap of ca. 1.2 eV, a MWt (M(n)) of ca. 30,000, and a conductivity on doping of ca. 7 X 10^{-2} S/cm. Their synthetic scheme and structure of the final CP (**5**) are shown in Fig. 35.13.

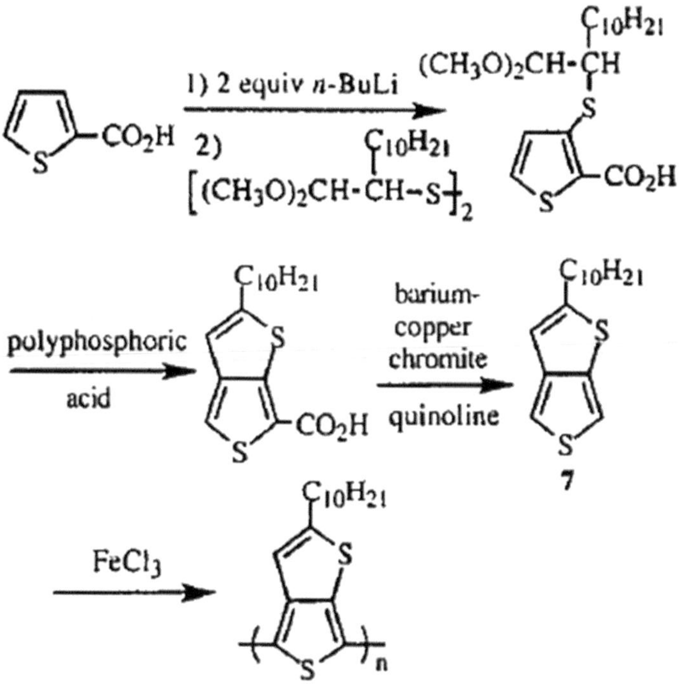

Fig. 35.13 Synthesis of novel poly(thiophene) (P(T)) derivatives, (After Pomerantz and Gu [1509], reproduced with permission)

Building on earlier work by Ferraris and Lambert [1510, 1511] and Gunatunga et al. [1512] found that an electropolymerized poly(thiophene) derivative possessed a bandgap below 1 eV, Beyer et al. [1513] electropolymerized a similar CP and confirmed that it also possessed a low bandgap, although not lower than 1 eV. In both cases, the parent monomers were synthesized using standard organic synthetic methods, such as Grignard couplings and Knoevenagel condensations.

Lima et al. [1514] recently studied the electropolymerization of 3,4-ethylenedioxy-thiophene and 3,4-ethylenedioxy-thiophene-methanol with dodecylbenzenesulfonate as dopant in aqueous acidic media and the morphology of the resulting CP films. Interestingly, Akoudad and Roncali [1515] studied the electropolymerization of the corresponding dimer, i.e., 2,2′-ethylene-dioxy-*bi*thiophene, and found that the resulting CP was nearly identical to that obtained from the monomer as in the work of Lima et al.; the advantage with the dimer was that, as expected, the electropolymerization occurred at a significantly less positive potential (−0.65 V as compared to the monomer).

An interesting series of CPs having alternating substituted thiophene and substituted phenylene rings were synthesized recently by Ng et al. [1516]. Their synthesis is summarized in Fig. 35.14. The CPs had a bandgap in the region 2.5 eV, and I_2-doping produced conductivities of ca. 0.2 S/cm. These CPs exhibited

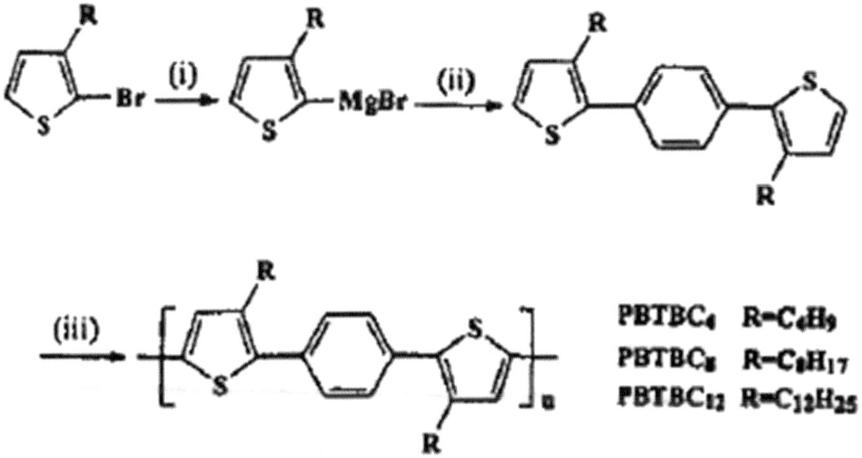

Fig. 35.14 Synthesis of fluorescent P(T) derivatives. Reagents and conditions: (i) Mg, Et$_2$O, (ii) 1,4-dibromobenzene, Ni(dppp)Cl$_2$, (iii) 4 equiv. FeCl$_3$, CHCl$_3$ (After Ref. [1516], reproduced with permission)

pronounced fluorescence, with quantum efficiencies of about 26% with respect to quinine sulfate.

Emge and Bäuerle [1517] synthesized novel uracil-substituted poly (bithiophenes), through electrochemical, potential-sweep polymerization of the monomer which was synthesized as shown in Fig. 35.15. These were used for voltammetric sensing of DNA/RNA bases, described elsewhere in this book.

35.5   Poly(*p*-phenylene)s (P(PP)s) and Derivatives

This class of CPs includes poly(*p*-phenylene) (P(PP)) itself; poly(*p*-phenylene vinylene) (P(PV)), which can also be considered a derivative of P(TV) discussed in the preceding section; poly(phenylene sulfide) (P(PS)), perhaps one of the earliest CPs to be manufactured in bulk; and poly(phenylene oxide) (P(PO)). This class of CPs comprises "older" CPs whose bulk syntheses were established some years before the recent surge in interest in CPs. They are singular and attractive in that many of them, e.g., P(PP), can undergo both p-type (oxidative) and n-type (reductive) doping, and nearly all have very high thermal stability in their undoped states; for example, P(PP) is stable up to 450 °C in air and up to 500 °C in inert gas. Many are also produced commercially: indeed, melt/solution-processible P(PS), by the trade name Ryton, has been available from Phillips Petroleum Co. since long before the surge in interest in CPs in the early 1980s. Each polymer is briefly discussed in turn.

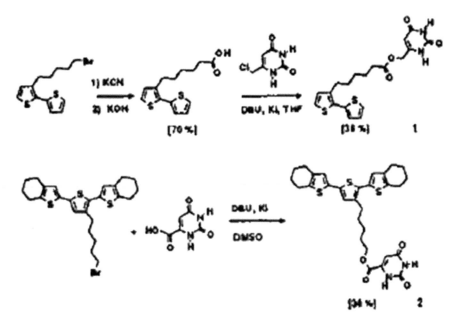

Fig. 35.15 Synthesis of novel uracil-substituted bithiophenes (After Ref. [1517], reproduced with permission)

35.5.1 Poly(p-phenylene) P(PP)

One of the first syntheses of P(PP), that of Kovacic et al. [1518], is shown schematically below:

$$\text{C}_6\text{H}_6 \xrightarrow[\substack{\text{CuCl}_2,\ \text{O}_2 \\ -35^\circ\text{C}}]{\text{AlCl}_3\ \text{(Trace H}_2\text{O)}} \left(\!\!\left(\bigcirc\right)\!\!\right)_n \qquad (35.3)$$

This method yields fairly high MWt P(PP) with little cross-linking. In theory, any of the standard organic aryl coupling reaction, such as the Ullmann, Grignard, Wurtz-Fittig, and diazonium couplings, can be used to synthesize P(PP)s, but these generally yield low MWt (< 35 monomer units) and much cross-linking. Two other syntheses are illustrated, respectively, by Yamamoto [1429] and Fauvarque [1519] syntheses (in the former, the Ni-Cat is NiCl$_2$ (bipy)):

$$\text{(35.4)}$$

$$\text{NiCl}_2\text{dppe} \xrightarrow{\;+2e\;} \text{Nidppe}$$

$$n\text{Br-C}_6\text{H}_4\text{-Br} \;+\; \text{Nidppe} \longrightarrow n\text{BrC}_6\text{H}_4\text{NiBrdppe} \qquad\qquad \text{(35.5)}$$

$$n\text{BrC}_6\text{H}_4\text{NiBrdppe} \xrightarrow{-2.6\text{vs Ag/Ag}^+} \left(\text{C}_6\text{H}_4\right)_n + \text{Nidppe}$$

A precursor route to P(PP) which involves a relatively low-temperature ($<300\ °$C) final step thermal conversion is that used by Ballard et al. and McKean et al. [1520,1521], starting with a cis-dihydrocatechol derivative, but this yields some *ortho*-linkages in addition to the 1,4-linkages.

$$\text{(35.6)}$$

Yet another route involves synthesis first of poly(1,3-cyclohexadiene), halogenation of the double bonds of this polymer with Br_2, and pyrolysis at ca. 350 °C to yield the final polymer [1522]. It is important to note that many of these differing synthetic routes yield polymer with significantly differing properties such as morphology, MWt, and conductivity.

Although it might be expected that the high oxidation potential phenyl group would prevent electropolymerization of P(PP), it has in fact been successfully done in exotic solvents having large potential windows, such as SbF_5-SO_2, Bu-Py-Cl/ $AlCl_3$ (room temperature melt), HF-SBF_5, HF-benzene, and triflic acid [1523–1527]. Most such products have poor MWt, conductivity, and other properties. The high oxidation potential however does have a correlation in the high bandgap of the pristine polymer, ca. 3.6 eV.

Li et al. [1528] recently showed that high-quality P(P) films could be prepared via electropolymerization of benzene at stainless steel electrodes, employing a BF_3-

diethyl ether medium. Films obtained were said to be flexible and cuttable into any shape. Their conductivity, modulus, and tensile strength were claimed to be, respectively, 2.7 S/cm, 1.2×10^{10} dyne/cm^2, and 600 kg/cm^2.

Curious derivatives of P(PP), e.g., poly(p-phenylene-co-2,5-pyrazine), have been synthesized [1529], but these appear to have little practical utility.

One of the most interesting properties of P(PP)s is that they may be either p- or n-doped. P-doping is generally carried out with AsF_6^-, through exposure to AsF_5 gas; a clear color change of the polymer to a dark, metallic green is visible in this. N-doping is generally carried out with alkali metal ions using their naphthalides in THF solution; expectedly, alkali-doped P(PP) is handled only under air- and moisture-exclusive conditions, limiting its practical utility. Doping can also be carried out electrochemically with these dopants in solution. Typical conductivities of AsF_5-doped and K-doped P(PP) are 500 S/cm and 50 S/cm, respectively [1284].

Maafi et al. [1530] described a truly novel P(PP) derivative, one containing a norbornadiene (NBD) substituent, prepared via electropolymerization. These authors obtained the parent monomer according to the scheme shown in Fig. 35.16, which on standard electrochemical polymerization in dichloromethane with tetrafluoroborate dopant gave the polymer, of predicted structure also shown in the figure. The polymer showed strong fluorescence emission with a maximum at ca. 395 nm.

Several polymerizations of benzene derivatives yielding CPs which may be considered derivatives of P(PP) have been carried out. For example, Martínez et al. [1531] recently electropolymerized poly(1,3-dimethoxy-benzene) from the monomer in acetonitrile with tetrafluoroborate electrolyte/dopant. Conductivities of the doped polymer at room temperature were however poor, ca. 2.5×10^{-4} S/cm.

An interesting series of CPs containing phenylene rings in series with pyrrole, thiophene, or furan rings was recently investigated by the Larmat et al. [1532], having the general structure shown below.

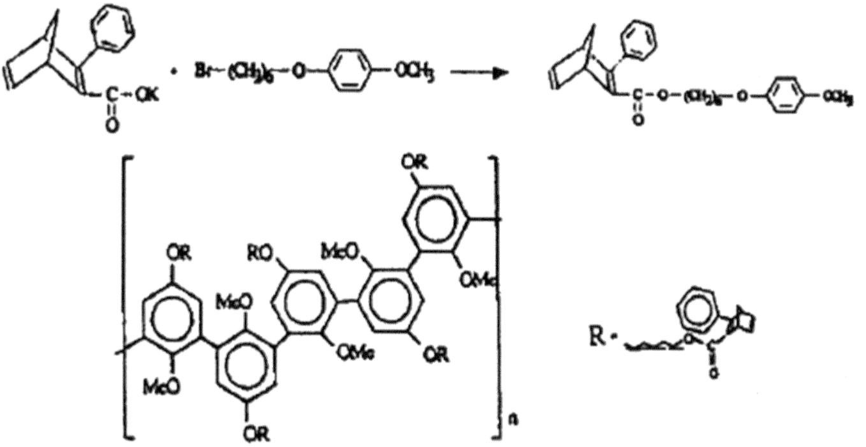

Fig. 35.16 Synthesis of novel P(PP) derivatives (After Ref. [1530], reproduced with permission)

$$X = O, S, NH \qquad (35.7)$$

Successful polymerization yielding a conductive, electroactive CP, was achieved only for the pyrrole derivative, in the easily oxidizable electrolyte Na tetraphenylborate.

35.5.2   Poly(phenylene vinylene) (P(PV))

This polymer, poly(phenylene vinylene) (P(PV)), bears some analogy to the P(TV) discussed above, both structurally and from the synthetic point of view; interest in it has been high recently due to its utility in LEDs. Indeed, one of the most popular syntheses for this CP is a precursor route closely analogous to that for P(TV) as shown in Fig. 35.17 (cf. Fig. 35.10).

The above is a variant of the route originally implemented by Wessling [1533], which yields high MWt polymer with claimed conductivities as high as 5000 S/cm and which also allows for production of 2,5-alkoxy-halo- and alkyl-substituted derivatives [1534]:

$$(35.8)$$

Wudl et al. [1536] have claimed that if in the precursor polymer in the scheme above, Cl$^-$ is replaced by, e.g., PF$_6^-$, a fully doped polymer results in the final thermal treatment step. Figure 35.18, after Brown et al. [1022], summarizes precursor routes to P(PV).

Because of the availability of precursor routes to P(PV), it is a relatively simple matter to prepare and orient or stretch fibers or films of the precursor polymer, and then convert them into P(PV) fibers or films, with fibers sometimes showing some crystallinity. It has been claimed [1199, 1536] that the 2,5-dialkoxy derivatives of P(PV) with long-chain alkyl (e.g., 2,5-dihexyloxy-) are soluble in their fully doped form.

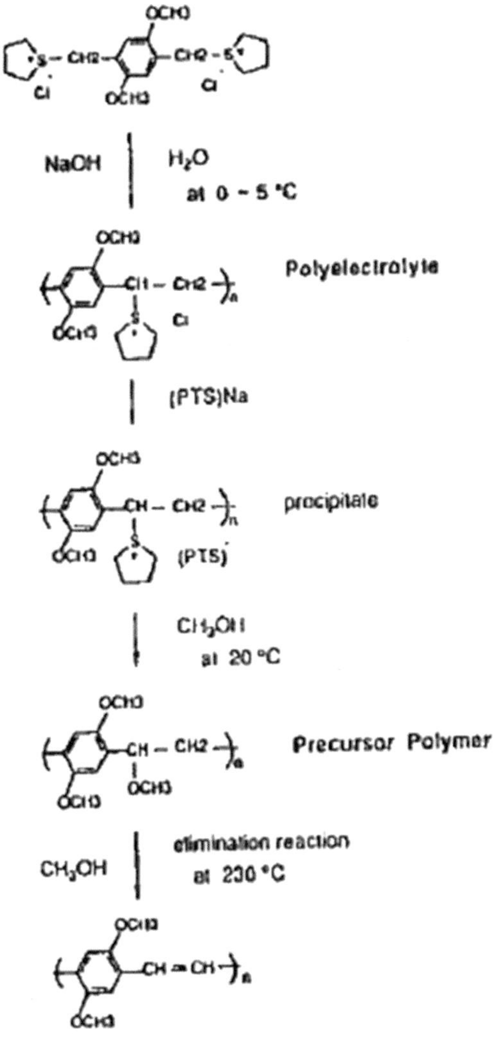

Fig. 35.17 One popular
synthesis of P(PV) (After
Ref. [902], reproduced with
permission)

Like their analogues, the P(TV)s, the P(PV)s have been studied for use in applications as varied as third-order NLO materials [1503] and electrochromic devices [1506]. Yang and Geise [1166] have prepared a series of P(PV) derivatives using a modified route that yields a mixture of cis-/trans-configurations which makes the polymers highly soluble in solvents such as CHCl$_3$. Table 35.2 summarizes their properties. Conductive (0.1–2 S/cm) films are said to be castable [1166] from solutions of these CPs with I$_2$ and a host polymer such as PMMA.

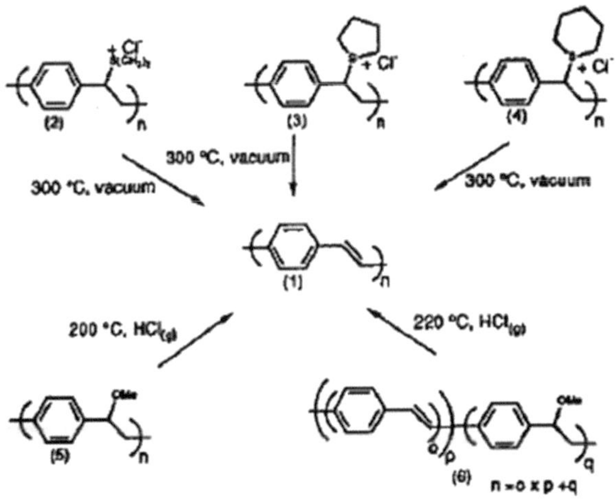

Fig. 35.18 Some precursor routes to P(PV) (After Ref. [1022], reproduced with permission)

Table 35.2 Caption

| Compound R' | R | Yield(g) | Melting range °(C) | | σ at 25 °C (Ω^{-1} cm^{-1}) |
			cis-trans	all-trans	
1 p-phenyl	CH$_3$	1.2	160–170	~300(dec.)	
2 p-phenyl	CH$_2$CH$_3$	1.4	117–140	~300(dec.)	0.1
3 p-phenyl	CH$_2$CH$_2$CH$_3$	1.9	115–135	~300(dec.)	0.2
4 p-phenyl	CH$_2$CH(CH$_3$)$_2$	1.9	135–145	~300(dec.)	0.2
5 p-phenyl	CH$_2$(CH$_2$)$_3$CH$_3$	1.8	95–110	~300(dec.)	0.1
6 p-phenyl	CH$_2$CH$_2$CH(CH$_3$)$_2$	2.1	78–92	~300(dec.)	0.1
7 p-phenyl	CH$_2$(CH$_2$)$_4$CH$_3$	1.9	76–90	~300(dec.)	0.1
8 p-phenyl	CH$_2$(CH$_2$)$_5$CH$_3$	2.0	75–85	~300(dec.)	0.1

Great interest has been directed recently (since about 1994) at P(PV)s with a view to their utilization in LEDs (a topic discussed elsewhere in this book). Hilberer et al. [1537] describe block copolymers for use in LEDs. The standard repeating conjugated structures of P(PV), i.e., the conjugated blocks, are linked with silanyl, aryl, or other groups, as shown in the schematic structures in Eq. 35.9. In these, the color is said to be tunable through control of the length of the conjugated block segments. These polymers are prepared via Pd-catalyzed Heck coupling [1538], an organic halide/vinylbenzene reaction. Structures of some of these are shown in Eq. 35.9.

$$(35.9)$$

Similarly, Spangler et al. [1125] have described the synthesis, via interfacial polymerization as poly(amides), of a group of four new P(PV)-based copolymers, LED properties of which are discussed in a later chapter; this is shown in Fig. 35.19.

Okawa et al. [1389] prepared a series of P(PV) derivatives with long alkyl side chains using the Heck reaction, for potential NLO applications. Their syntheses are summarized in Fig. 35.20. Work by Remmers et al. [1539] and others [1540] has shown that alkoxy-substituted P(PV)s (e.g., 2,5 di-ethoxy) yield CPs which are more solution processible in their undoped form than the parent P(PV)s.

An interesting derivative of P(PV), more strictly a copolymer with P(T), poly(*p*-phenylene[3-alkylthio)-2,5-thienylene]-*p*-phenylene), was synthesized by Ueda et al. [1541]. A NiCl$_2$-catalyzed polymerization of 3-(alkylthio)-2,5-bis(4-chloro-phenyl)thiophenes in NMP solvent in the presence of Zn, triphenylphosphine, and 2,2'-bipyridine was used. The CPs so obtained had MWts up to 47,000 and were soluble in and could be processed into films from such solvents as chloroform and THF. Figure 35.21 summarizes these syntheses.

Dridi et al. [1369] described synthesis of a novel P(PV) derivative which is actually a copolymer but is discussed here because of its similarity in properties with and relevance to P(PV). Equation 35.10 summarizes the synthesis and structure of this polymer. The synthesis was carried out in heterogeneous medium, either DMSO or DMSO/THF, in the presence of excess KOH powder, with yields from 28% to 62%, the polymer precipitating as neat yellow aggregates. These CPs had photoluminescence properties which were significant for LED applications.

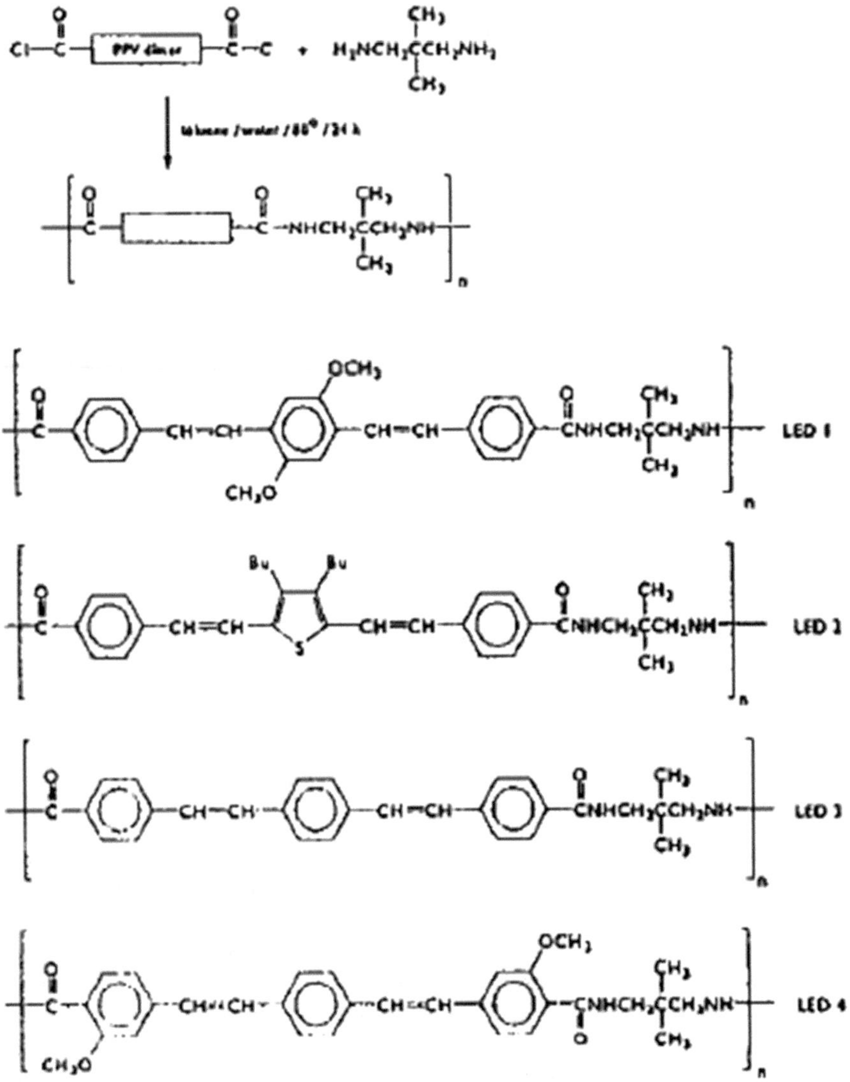

Fig. 35.19 Synthesis of group of P(PV) derivatives for LED applications by Spangler et al. [1125] (After Ref. [1125], reproduced with permission)

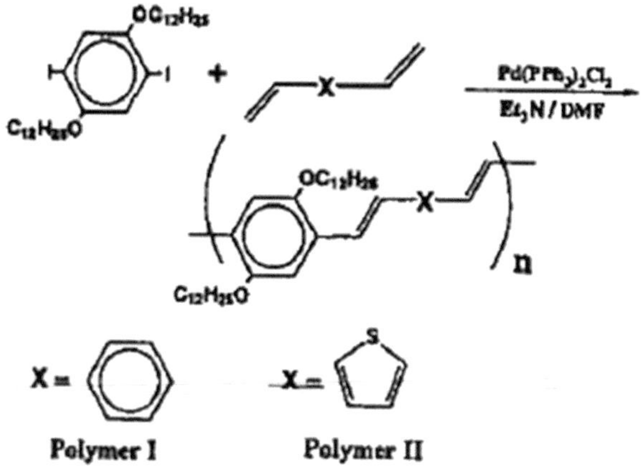

Fig. 35.20 Heck reaction synthesis of P(PV) derivatives (After Ref. [1389], reproduced with permission)

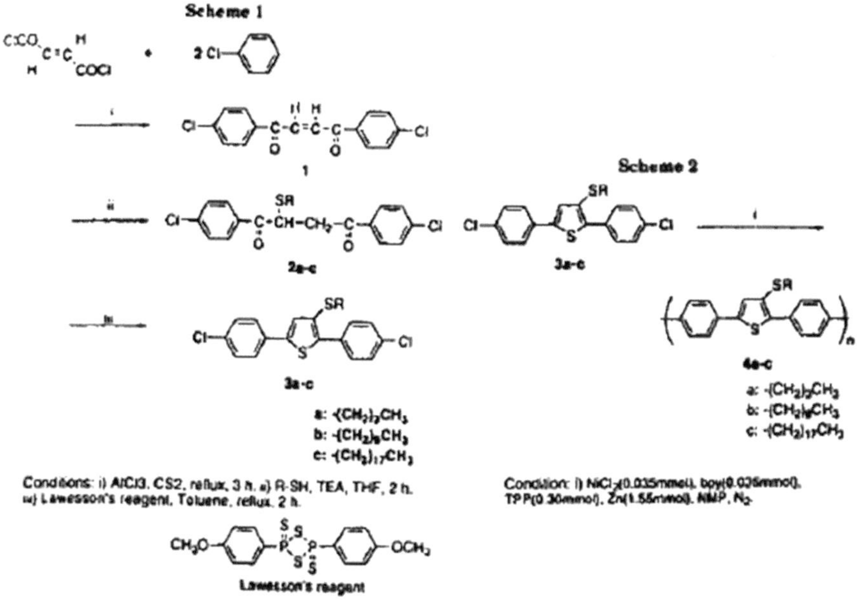

Fig. 35.21 Unique synthesis of P(PV)/P(T) copolymer (After Ref. [1541], reproduced with permission)

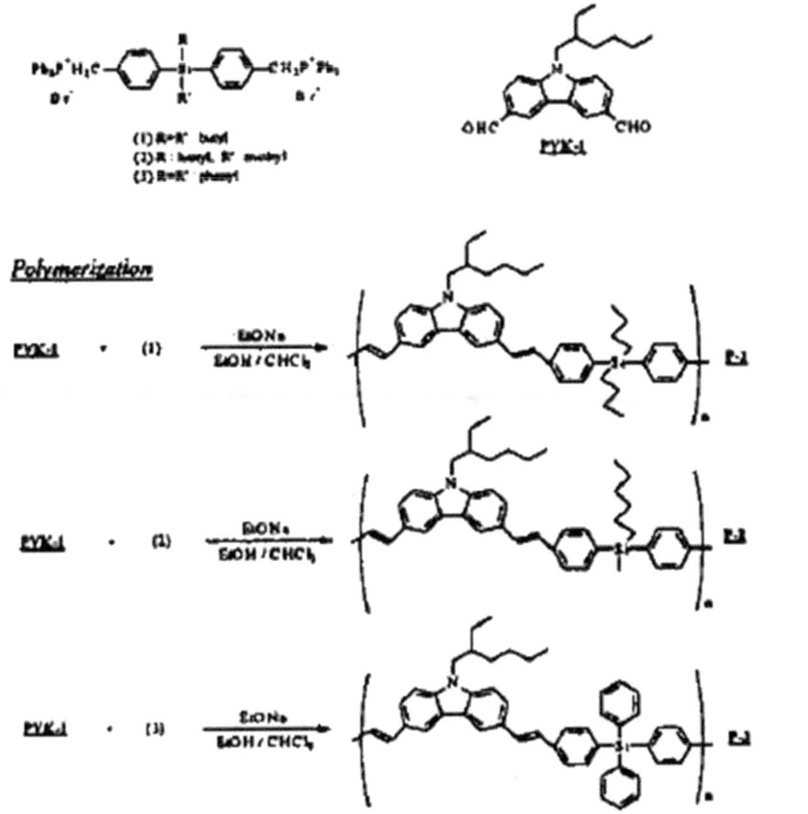

Fig. 35.22 Synthesis of poly(p-phenylenevinylene)-related polymers containing carbazole and organosilicon units in the main chain (After Ref. [1370], reproduced with permission)

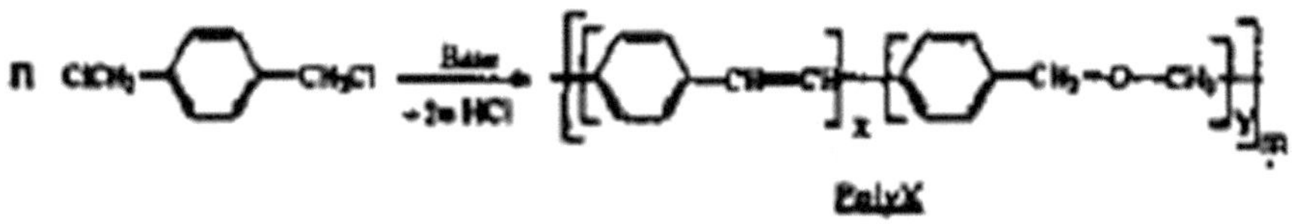

$$(35.10)$$

Kim et al. [1370] recently described the synthesis of a novel series of P(PV) derivatives containing organosilicon and carbazole units, as summarized in Fig. 35.22. The photo/electroluminescence properties of these have been cited elsewhere in this book.

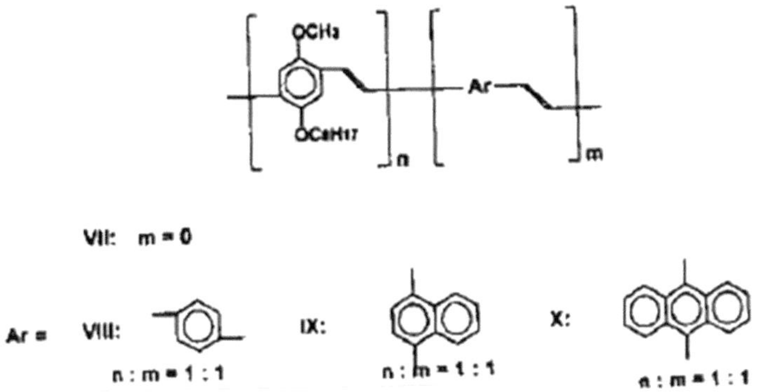

Fig. 35.23 Wittig synthesis of unique poly(arylene vinylenes) (After Ref. [1542], reproduced with permission)

Barashkov et al. [1542] described a Wittig reaction synthesis of poly(arylene vinylenes) incorporating 2-methoxy-5-(2′-ethylhexyloxy)-*p*-phenylene fragments in the polymer chain. The final polymers are depicted in Fig. 35.23. These CPs, which in one sense could be defined as copolymers with P(PV), showed intense green light emission coupled with high solubility and facile film formation, properties of potential use in LEDs.

35.5.3   Poly(phenylene sulfide) (P(PS)), Poly(phenylene oxide) (P(PO)), and Related Poly(phenylene chalcogenide)s

The CPs poly(phenylene sulfide) (P(PS)) and the 2,6-dimethyl derivative of poly (phenylene oxide) (P(PO)) were among the first CPs to be manufactured commercially, as RYTON (originally produced by Phillips Petroleum) and Noryl, respectively. Phillips implemented significant expansion of its P(PS) capacity from 7300 to 10,000 metric tons per year at its Borger, TX, USA, site in the 1990s [1543].

Due to *p*-orbital overlap, the O-atom in P(PO) contributes a continuity in electron delocalization from continued orbital overlap along the polymer chain even if the phenyl rings are non-coplanar; this is even more so for the S-atom in P (PS) due to participation of *d*-orbitals. The absence of a heteroatom in the ring, as in P(T) or P(Py), leads to a higher bandgap, close to that of P(PP). However, these polymers also possess very high thermal stability: P(PS) melt at ca. 270 °C (glass transition temperature, T_g, 92 °C) and can be easily melt processed. P(PS) is also highly soluble in organic solvents, e.g., Cl-naphthalene and NMP, although only at elevated temperatures; for example, it can be cast as films from a 10% solution in diphenyl ether at ca. 200 °C. P(PS) possesses ca. 65% crystallinity, and a single-crystal X-ray structure is available [1544]. Many alkyl-, sulfoxide-, and fluoro-

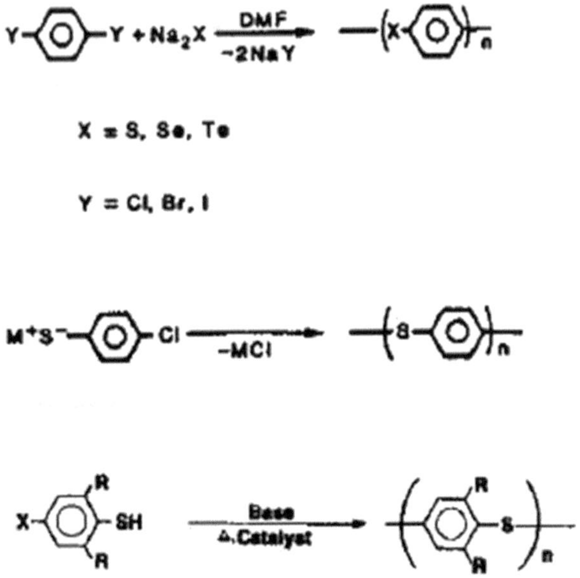

Fig. 35.24 Some common syntheses of P(PS) (After Refs. [1280, 1445], reproduced with permission)

$$Y-\bigcirc-Y + Na_2X \xrightarrow[-2NaY]{DMF} -(X-\bigcirc)_n$$

$$X = S, Se, Te$$

$$Y = Cl, Br, I$$

$$M^+S^- -\bigcirc-Cl \xrightarrow{-MCl} -(S-\bigcirc)_n$$

$$X-\bigcirc_{R}^{R}-SH \xrightarrow[\triangle, Catalyst]{Base} -(\bigcirc_{R}^{R}-S)_n$$

derivatives of P(PS) have been made and characterized [1545]. P(PO) is less easily processible than P(PS) and in its doped form has poorer conductivity; consequently, it has received less attention.

Three common syntheses [1546–1548] for P(PS) are shown in Fig. 35.24.

The MWt of the polymer obtained from the above syntheses is typically ca. 10,000–12,000. Doping generally requires the use of strong dopants due to its high oxidation potential. Chemical doping of cast films with SO_3 [40, 653] yields a conductivity of ca. 80 S/cm, while electrochemical doping by SO_3 can also be carried out, in conc. H_2SO_4. The polymer yields a "doped CP solution" after solubilizing in very concentrated H_2SO_4 (98%) over several days, while with lesser H_2SO_4 concentrations, e.g., 92%, it can be electrochemically cycled and displays a notable spectroelectrochemistry [976].

Similarly, doping is possible with strong oxidants such as AsF_5 (in liquid AsF_3 solution), SbF_5, fluorosulfonic acid, and $NOPF_6$, yielding conductivities in the 0.02–3 S/cm range (similar doping of P(PO) yields conductivities about two orders of magnitude lower). Doping with AsF_5 vapor is however said to yield a modified polymer, with the S-atom part of a fused 5-membered ring [1549]. Interestingly, deep-blue "doped conducting polymer solutions" have also been obtained with doping of P(PS) by AsF_5 in liquid AsF_3 medium [1520]. Removal of the liquid AsF_3 solvent is said to allow deposition of doped P(PS) films with claimed conductivities up to 200 S/cm; under identical conditions, P(PO) is said to yield films of conductivity ca. 100 S/cm. The practical utility of such a toxic and corrosive "solvent" however is in question.

Ding and Hay [1550] recently described a new synthesis of P(PS) from 4-bromobenzenethiol using a free radical initiator, as shown schematically in Eq. 35.11 below: The product was found to almost exclusively possess the 1,4-linkage only. Diaryl sulfide initiators used included bis(4-Br-phenyl) disulfide, 2,2'-dithiobis(benzothiazole) and a cyclic disulfide oligomer (Eq. 35.11).

$$(35.11)$$

Higher hetero-atom analogues of P(PO) and P(PS), viz., P(PSe) and P(PTe), of limited practical utility, have also been synthesized [1534, 1551, 1552]. Conductivities of AsF_5-doped P(PSe), as an example, are in the 10^{-3} S/cm range, while those of P(PTe) are even poorer. More recently, such CPs have seen renewed work. For example, Peulon et al. [1553] described a study of the effect of dopant on the electrochemical and physical properties of poly(selenyl thiophene) (P(SeT)), prepared by electropolymerization of the monomer in acetonitrile. They claimed that dopants such as hexafluorophosphate and perchlorate had a stronger influence on CP properties than for P(T) alone.

35.6 Poly(azulenes)

CPs that are derivatives of poly(azulene) (base structure, Eq. 35.12) had been synthesized and studied in detail by the Bargon group at IBM [1319, 1450], who have studied the parent CP as well as the 1-Me-, 1-Phe-, 4,6,8-TriMe-, 4,6,8-TriMe-1-aldehyde, 1-(Phe-acetylenenyl)-, and the 1,3,-Di(trimethylsilylacetylene) derivatives.

$$(35.12)$$

The most common and practical method of synthesis of these CPs is via electropolymerization from their monomers, whose chemical synthesis is available in the literature. Bargon et al. [1319] elegantly used the electropolymerization of azulene to shed light on the general scheme of electropolymerization in CPs; this has been discussed at some length in an earlier chapter; recent studies [1554] have confirmed and complimented this work. Electrochemical polymerization of azulenes follows the standard procedure for nonaqueous electropolymerizations described in an earlier chapter. All monomers elucidated in the previous paragraph have oxidation potentials in the 0.89–1.04 V (vs. SSCE) range. Free-standing, somewhat flexible films can be peeled off from Pt electrodes. Common nonaqueous dopants, e.g., perchlorate, tetrafluoroborate, and tosylate, are used and yield conductivities of ca. 0.01–0.05 S/cm. Poly(azulene) films can be electrochemically cycled in acetonitrile as well as in other common electrolyte media, such as propylene carbonate. Other than scientific and mechanistic interest, however, this class of CPs appears to have little practical utility.

35.7   Ladder Polymers: BBL, BBB, PBT, and PBO

The structures of the ladder polymers, poly{7-oxo-, 10H–benz[de]imidazo [4′,5′:5,6]benzimidazo[2,1-a]isoquinoline-3,4:10,11-tetrayl)-10-carbonyl} (*BBL*), poly{6,9-dihydro-6,9-dioxobisbenzimidazo[2,1b:1′,2j]benzo[1mn]-[3,8]-phenanth roline-3,12-diyl} (*BBB*), poly(*p*-phenylene-2,6-benzobisthiazolediyl) (*PBT*), and poly(*p*-phenylene-2,6-benzobisoxazolediyl) (*PBO*) are shown schematically in Fig. 35.25.

The reason they are called "ladder" polymers is because they may be considered derivatives of poly(acene), whose structure (Fig. 35.25) is characterized by two linear side frames with interconnecting "rung" bonds, very much like that of a ladder. For our purposes, they *do* qualify as CPs, since they possess a π-conjugated polymeric structure and can indeed also be doped for greatly increased conductivity. The interest in these polymers has been due to their greater thermal and oxidative stability as compared to more conventional CPs, their facile processibility in their undoped form (they may be cast as films and spun into fibers from solutions), and their interesting NLO properties.

The chemical syntheses of BBL and BBT have been established for several decades [1555], much before the advent of the field of CPs. As-synthesized polymer is usually dissolved in acids such as polyphosphoric acid or methanesulfonic acid for further processing.

It has been found that prior to doping, these polymers' conductivity can be enhanced by simple heat treatment – a form of thermomechanical annealing which increases alignment within the polymers [1057]. Thus, Wang et al. [1057] found that 8:1 draw ratio fibers of BBL, when treated at 350 °C, yielded dramatically improved mechanical properties (Young's modulus of 120 GPa) while experiencing an increase in conductivity of four orders of magnitude, to 3×10^{-4} S/cm. Agrawal et al. [1556] found anisotropic conductivity in such heat-treated BBL films. Such treated polymer can then be doped for further enhanced conductivity.

Doping has been carried out chemically [1557] and electrochemically [1558] and via techniques such as ion implantation, a technique adapted from the semiconductor industry. Chemical doping, typically with SO_3 or conc. H_2SO_4 vapor,

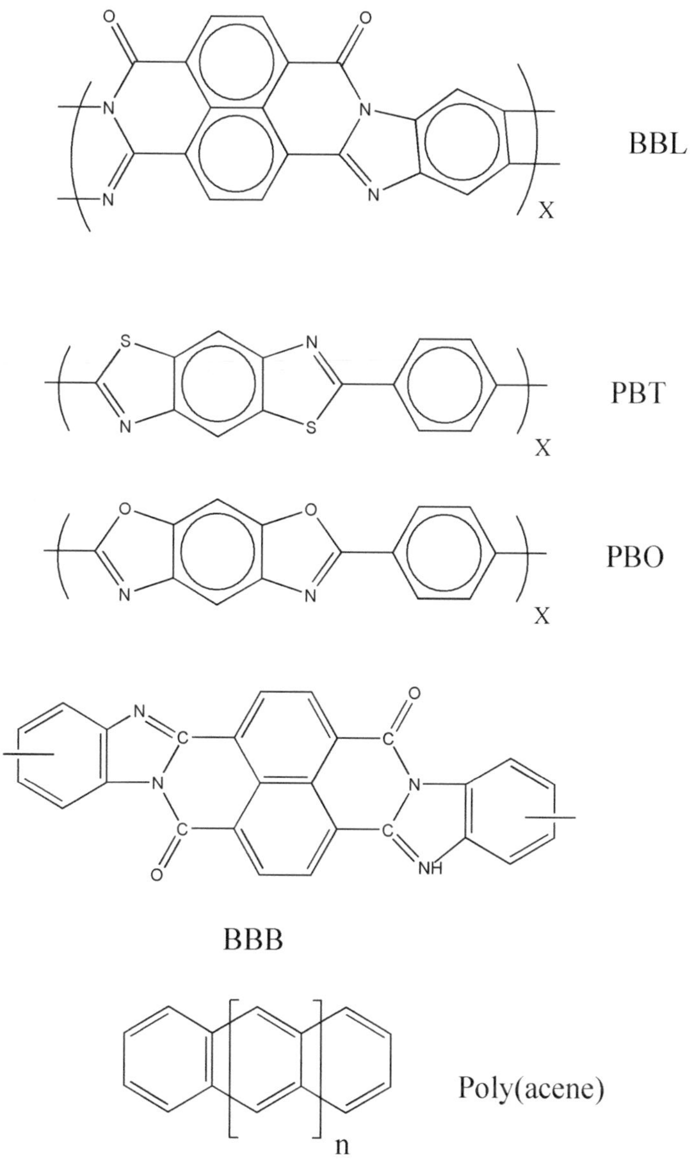

Fig. 35.25 Structures of ladder CPs BBL, PBT, PBO, BBB, and poly(acene)

yields conductivities of ca. 1–2 S/cm for BBL and BBB [1559]. Interestingly, like P (PP), these polymers may also be n-doped. Thus, BBL, on doping with K-naphthalide, yields a conductivity of ca. 1 S/cm [1559]. Burns et al. and Shacklette et al. [891, 1544] adapted ion implantation techniques for Kr$^+$-ion implantation of BBL, PBO, and PBT to obtain claimed conductivities of 220 S/ cm for BBL and ca. 130 S/cm for PBO.

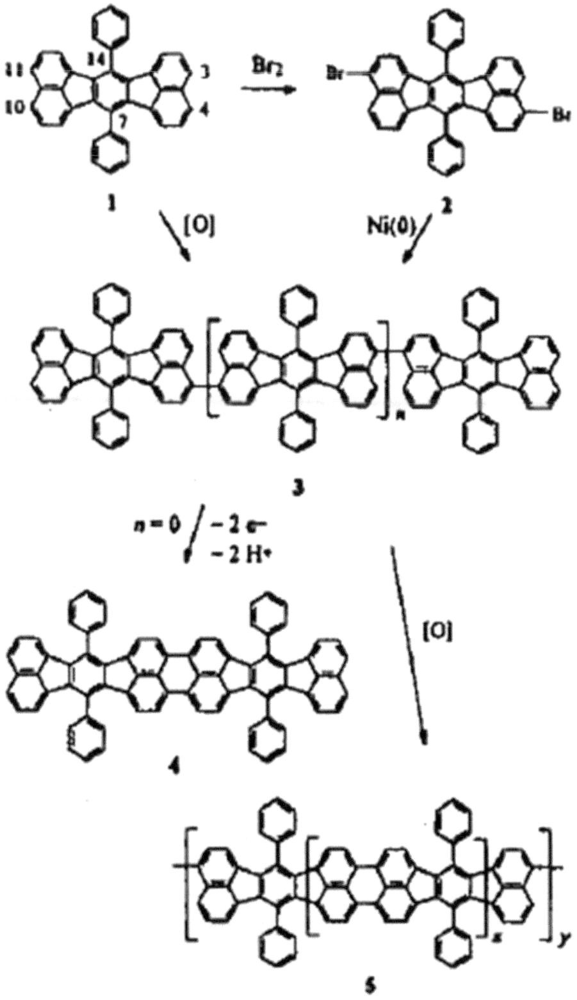

Fig. 35.26 Synthesis of unique fluoranthrene-derivative ladder CP (After Ref. [1560], reproduced with permission)

Debad and Bard [1560] recently described a new ladder polymer which displayed electrochromism. Its synthesis is depicted schematically in Fig. 35.26. The color change observed in this CP in acetonitrile with PF_6^- dopant was pale orange (reduced) through pale green (neutral) to pale gray (oxidized). These polymers have been studied for a variety of applications, where their processibility especially comes in handy. These include third-order NLO effects [1382] and waveguiding [1561].

35.8   Poly(quinolines) and Derivatives

This class of polymers possesses properties similar to those of the ladder polymers, in that their syntheses were known much before the advent of the field of CPs, they are thermally more stable and more easily processible into films and fibers than more conventional CPs, and they can be doped to appreciable conductivity and do possess some semblance of a π-conjugated structure that qualifies them as CPs. Some examples are given in Fig. 35.27. In the early preparations, e.g., that of Stille [1562], an acid-catalyzed Friedlander synthesis was employed.

These polymers may also be chemically or electrochemically doped, with both n- and p-doping found. Conductivities as high as 50 S/cm have been claimed [1563, 1564] for Bu_4N^+-doping (n-type) of the first polymer shown in the schematic above. Methane-sulfonate doping (p-type) of the second polymer above yields a conductivity of ca. 1 S/cm.

The facile processibility and thermal stability of this class of polymers have also led to increased interest in potential applications. Thus, Agrawal and Jenekhe have studied rigid-rod poly(quinolines) for third-order NLO applications [1565], while

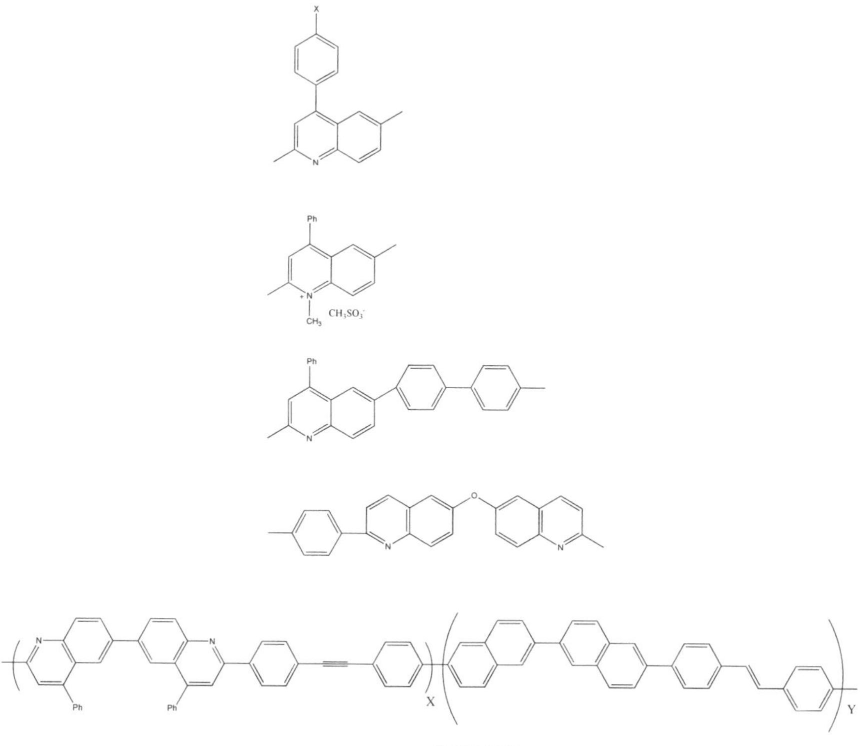

PBAPQ/PSPQ

Fig. 35.27   Structural units of some poly(quinolines) and derivatives

Cai et al. [1393] have looked at poled guest-host poly(quinoline) thin films for second-order NLO applications. Wolf et al. [1506, 1843] documented the use of poly(quinolines) in electrochromic devices. Chandrasekhar et al. [956, 999, 1003, 1004, 1235] have synthesized a series of unique poly(amino quinolines) (below) and demonstrated significant third-order NLO effects in them.

35.9 Other Polymers

Schulz et al. [1368] described a series of poly(oxadiazoles) for potential LED applications. Some of their structures have been depicted earlier in Fig. 35.23. These were prepared by a one-step polycondensation of the aromatic dicarboxylic acids with hydrazine hydrate in polyphosphoric acid, which acted as a solvent as well as dehydration agent.

El-Shekeil et al. [1566] recently described synthesis of novel poly(azomethine) CPs which were claimed to be crystalline as well as conductive. These were synthesized by the polycondensation of p-phenylenediamine and 2-nitro-p-phenylenediamine with terephthalaldehyde and p-diacetylbenzene. The polymers with -NO_2 and other polar functional groups were found to be more soluble in common organic solvents.

González-Tejera et al. [1567] recently reported electropolymerization of poly (furan)/perchlorate from monomer solutions in aprotic media in which dopant concentrations were lower than monomer concentrations. The CP films were claimed to be conductive. Belloncle et al. [1568] claimed production of conductive, electroactive CP films from electropolymerization, in tetrafluoroborate/methylene chloride, of several crown ethers such as bis-binaphtho-22-crown-6 and binaphtho-20-crown-6.

Pandey and Prakash [1569] studied the electropolymerization of indole in perchlorate/−dichloromethane in detail, by potential sweep (−0.2–1.0 V vs. Ag$^+$), potentiostatic (1.0 V vs. Ag$^+$), and galvanostatic (0.2–0.3 mA/cm^2) methods. Conductivities in the region of 1 S/cm were obtained. A Zn/poly(indole) secondary battery was constructed and tested.

Talbi and Billaud [1570] also studied electropolymerization of indole and 5-cyano-indole in standard media (perchlorate/acetonitrile). The electrochemistry of these electroactive films was reported.

35.10 Problems and Exercises

1. Write down the structures of the doped and pristine forms of one member of the following classes (discussed in this chapter), and outline one chemical and one electrochemical (if available) synthesis for it: P(T), P(T) derivatives, P(PP) and related polymers, poly(azulenes), ladder polymers, and poly(quinolines). In

which cases are electrochemical (or chemical) polymerizations unavailable and why?

2. Enumerate the most common dopants for one member of each of the polymer classes listed in problem 1 above, and briefly outline the doping method.

3. Make a list of the classes listed in problem 1 in order of decreasing conductivity. Use conductivity ranges of the highest conductivity observed for a member of each class.

4. Outline one example of methods of processing – whether in pristine or doped form, in melt or in solution – for one member of each of the classes listed in problem 1. Discuss the so-called doped CP solutions and whether and how they may be classified as true solutions.

5. For each class outlined in problem 1, briefly discuss the difference in all possible properties you can enunciate between the parent polymer and common derivatives (e.g., poly(3-alkyl thiophene) and P(T).

6. Of the classes enumerated in problem 1, which are of practical utility? Cite all applications studied thus far for these.

7. Identify at least two ways in which the P(PP)s (Sect. 14.3), ladder polymers (Sect. 14.5), and poly(quinolines) (Sect. 14.6) are different from more "conventional" CPs such as P(Py) or P(ANi). How may they still be classified as CPs? Give structural details, if appropriate.

Part VI
Conducting Polymers, Applications

Chapter 36
Sensors

Contents

36.1 Modes of Sensing with CPs

At the outset of this chapter, and also after going through it to the end, the reader may be justified in asking, "in view of the large volume of publications in this field, as of this writing (2016), are there any sensors or sensing applications based on CPs that have actually been commercialized and are now available on the market?" The answer to this, unfortunately, is "very few, almost none." In spite of the volume of publications in the CP sensor field, the number of actual CP sensor products on the market is less than three as of this writing (2016) [1571]; these are conductometric or fluorescence-based sensors for explosives residues and/or chemical warfare agents (CWAs), and none of them appear to be competitive with other means of sensing for these analytes; indeed, none of them appear to have been commercialized in a successful way. Thus, predictions of "significant impact" of CP sensors [1572] appear not to have been borne out to date.

Sensing has been accomplished using CPs through a number of modes:

1. *Amperometric Sensing*, using change of conductivity (or resistivity, impedance or admittance).
2. *Potentiometric*, monitoring the change of open circuit (rest or equilibrium) potential at a CP electrode.

© Springer International Publishing AG 2018

P. Chandrasekhar, *Conducting Polymers, Fundamentals and Applications*,

https://doi.org/10.1007/978-3-319-69378-1_36

3. *Amperometric*, monitoring the current at a CP electrode.
4. *Voltammetric*, a variant of amperometric, monitoring the change in current while varying the applied potential at a CP electrode.
5. *Gravimetric*, involving the effect of a weight change in a CP and having two primary embodiments – electromechanical actuators and the quartz crystal microbalance.
6. *Optical*, usually involving the effect of doping or analyte binding on optical properties of a CP, including absorption, reflectance, and fluorescence.
7. *pH-based*, a derivative method, using the effect of change in pH in the sensor environment on such parameters as potential, current, or conductivity.
8. *Incorporated receptor (host) based*, another derivative technique, in which an analyte-specific receptor or host, e.g., an enzyme, is incorporated into a CP film during electropolymerization: the effect of sensor contact with analyte on a number of properties of the CP film, e.g., conductivity, open circuit potential, or current, is then monitored; thus, this mode can fall under the conductometric, potentiometric, amperometric, and other classes described above.

Each of the above modes or categories is now discussed separately in the sections below with illustrative examples, which are in fact the best way of understanding them. In addition to the above modes, there are several less widely used modes of sensing using CPs, e.g., pressure/piezoelectric, refractive index, NLO properties, and magnetic properties (e.g., EPR); their practical implementation has been unsuccessful or scarce. These will thus be treated in this chapter only in exemplary mention.

36.2   Conductometric-Mode Sensors

CPs change conductivity over as many as *ten orders of magnitude* with doping, influenced by presence of analyte. We may cite several illustrative examples.

Work from the Contractor group at the Indian Institute of Technology, Mumbai, India, described an elegant glucose sensor, shown in Fig. 36.1a [1267]. A "bread/butter/jam" layer configuration was used, with a Pt electrode (the bread) on which a layer of P(ANi) (the butter) was deposited, on which in turn a layer of P(ANi) incorporating glucose oxidase enzyme (the jam) was deposited. Two Pt disk electrodes 8 µm apart were used, with the first, "butter," layer of P(ANi) electrodeposited to a thickness sufficient to bridge the gap between the electrodes. The sensor was immersed in glucose solutions for at least 20 s, and its conductivity then measured using an AC peak-to-peak method (see elsewhere in this book). The change in resistance between the two electrodes, i.e., the conductivity of the sensor, was found to be linear with glucose concentration up to a concentration of 10 mM (Fig. 36.1b). The sensor was highly selective, as expected due to the presence of the enzyme, for example, not responding at all to mannose. Since the enzymatic reaction is known to produce a lactone which spontaneously hydrolyses to gluconic

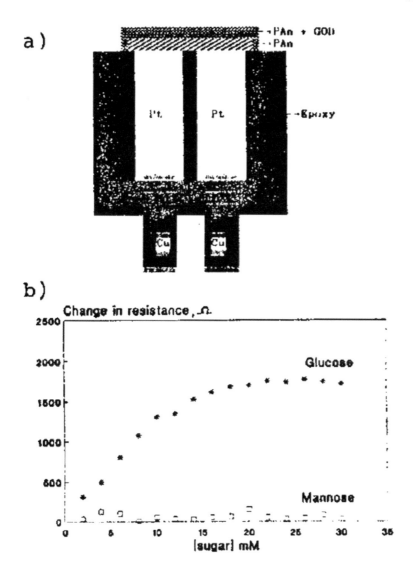

Fig. 36.1 Structure (**a**) and performance data (**b**) for novel glucose biosensor of Contractor group [1717] (After Ref. [1717], reproduced with permission)

acid (pK_a 3.76), and the latter is known to affect pH, it was surmised that this was the likely mode of sensor function, as pH changes are known to dramatically affect P(ANi) conductivity. The authors ruled out other possible modes of function through a number of tests.

Forzani et al. [1573] described a conductometric glucose sensor based on nanojunctions formed by bridging a pair of nanoelectrodes separated by a small gap (20–60 nm) with P(ANi)/glucose oxidase (GOX). Detection was based on the change in the nanojunctions conductance as a result of glucose-oxidation-induced

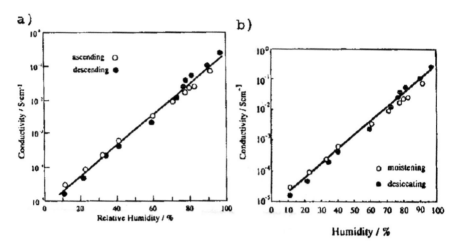

Fig. 36.2 Representative data for a poly(o-phenylenediamine)-/polyvinyl alcohol-based humidity sensor (**a**) first version (**b**) more refined version (After Refs. [1574, 1167], reproduced with permission)

change in the CP redox state. Due to the small size of the nanojunction sensor, the authors claimed that the enzyme was regenerated naturally without the need of redox mediators, thus consuming a minimal amount of oxygen and at the same time giving a faster response (<200 ms).

Ogura et al. [1167] described an interesting conductometric sensor for humidity, based on a composite of poly(o-phenylenediamine) (P(o-PD)) and polyvinyl alcohol (PVA). The composite was prepared from dimethyl sulfoxide (DMSO) solution and deposited on a comb-shaped Pt electrode, used to monitor the conductivity, on a glass substrate. The data for this sensor (Fig. 36.2a) showed a linear relation with relative humidity (RH) in the region 5% through 100%. The mode of function was thought to be the pH affecting the CP composite film's conductivity, much like the glucose sensor described above, through the following equilibrium:

$$-NH^+_2(P(o\text{-}PD)) + -OH(PVA) = -NH(P(o\text{-}PD)) + (PVA)^+ + H_2O \quad (36.1)$$
$$(X^-) \hspace{7cm} (X^-)$$

These authors subsequently refined their sensor [1574] to yield a more linear response with less hysterisis, as seen in Fig. 36.2b.

Ram et al. [1575] described a conductometric sensor for CO based on "ultrathin" layers of P(ANi)/poly(styrene sulfonate) deposited using a self-assembly, layer-by-layer method onto substrates such as SnO_2 and TiO_2. They demonstrated a response in the 10 to 1000 ppm CO range. However, their data did not determine if this was simply a generic response or the effect of interferents or false positives.

An interesting conductometric application described by Baughman et al. [1576] was a radio frequency antenna coated with a CP film and responding to external, remote radio frequency interrogation. As the conductivity of the CP film changed, in response to such external stimuli as change in humidity or oxygen content, the radio frequency response of the antenna would change, through such factors as shorting of the antenna, RF shielding, or capacitance coupling. Such antennae, costing less than US$0.05 each, could be placed inside containers of sensitive or perishable goods, and the effect of such factors as damaged containers or attempted theft could thus be determined remotely, without opening the container.

A number of workers have tested CP-based sensors for detection of gases and vapors. [1577]. CPs tested include P(Py), P(N-Me-Py), and P(ANi), and gases/vapors tested include NO_2, I_2, SO_2, and MeOH. Generally, reduced (de-doped) CP films were used, and an increase in conductivity with exposure to the gas/vapor recorded. However, selectivity was poor or nonexistent, although pattern recognition algorithms applied to the data could be used to enhance it. Boyle et al. [1578] described a P(ANi)-based vapor sensor in which the conductivity was very simply monitored by fabricating the P(ANi) film as part of a standard four-point conductivity probe. The vapors were transported over the sensor in N_2 carrier gas, and the potential difference resulting from the applied current was read directly as conductivity. The typical response of this sensor is shown in Fig. 36.3 for various vapors.

Collins and Buckley [1579] described an interesting conductometric sensing application in which fabrics woven from threads of PET or nylon were coated with P(Py) and P(ANi) (available commercially from Milliken Research Corp., Spartanburg, SC). Analytes tested included vapors of NH_3, NO_2, dimethyl

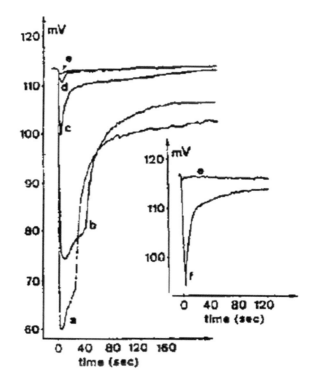

Fig. 36.3 P(ANi)-based vapor sensor of Boyle et al. [1578]. Vapors: (a) MeOH (b) EtOH (c) acetone (d) acetonitrile (e) benzene (f) MeOH 100 ppm, all others 1000 ppm (After Ref. [1578], reproduced with permission)

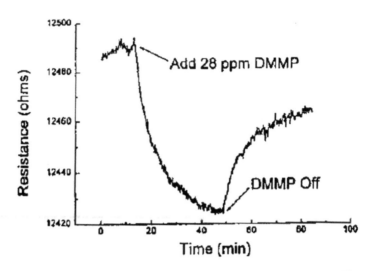

Fig. 36.4 Typical response observed for a NDSA-doped poly(pyrrole) film on PET exposed to 28 ppm DMMP in dry air (After Ref. [1579], reproduced with permission)

methylphosphonate (DMMP, which is a simulant for several chemical warfare agents), and H_2O (i.e., humidity). The ultimately targeted application was sensing of chemical warfare agents using resistivity changes in soldiers' clothing. However, with the exception of NH_3, sensitivity was moderate or poor and response time slow (cf. Fig. 36.4). Since the fabric also turned out to be a good humidity sensor, problems with sensing of chemical agents included the unpredictable effect of humidity on response. In similar work, Kincal et al. [1580] described P(Py)-coated fabrics for sensing of gases such as NH_3 and HCl, which in reality simply functioned as vapor-phase dopants. Although the conductivity change observed was reversible, it was found to decrease in magnitude with time. These authors also demonstrated sensing of CO_2. Figure 36.5 shows typical sensor response obtained by them.

Ellis et al. [1581] described an ultrasensitive (ppt) sensor based on CP films for hydrazine (H) and monomethyl hydrazine (MMH) vapor, rocket fuel components which are themselves highly toxic (USEPA Maximum Permissible Exposure 10 ppb/8 h) and also potential mimics for chemical warfare agents. In this work, very thin films of poly(3-hexyl-T) (P(3HT)) were spin coated on interdigitated Au microelectrodes on deposited on a quartz substrate; the P(3HT) film was partially doped with PF_6^- using $NOPF_6$. The sensing mode was two-point resistivity between the two interdigitated electrodes. H and MMH were detected when flowing over the sensor through their reaction of the undoped P(3HT) film through electron donation and presumed complexation with the dopant. One drawback of this sensor however was its irreversibility of operation, as this reaction of the P(3HT) film was irreversible. For the application envisioned, however – onetime detection of the presence of toxic quantities of the analyte in a work environment – this irreversibility did not present a problem. Concentrations in the range 0.1–100 ppb of H and

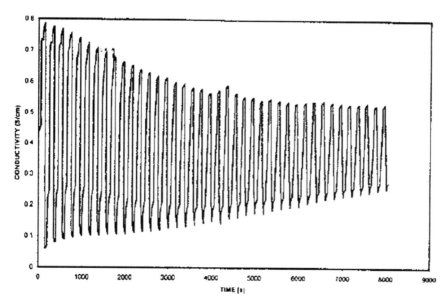

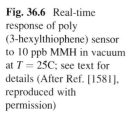

Fig. 36.5 Conductivity change for P(Py)/textile during cycling with HCl/N$_2$/NH$_3$/N$_2$ (After Ref. [1580], reproduced with permission)

Fig. 36.6 Real-time response of poly (3-hexylthiophene) sensor to 10 ppb MMH in vacuum at $T = 25$C; see text for details (After Ref. [1581], reproduced with permission)

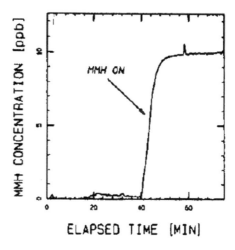

MMH were claimed to be monitorable with an accuracy of $+/-20\%$. Figure 36.6 shows the real-time response for this sensor. It was claimed that interference from such species as ammonia, dimethyl amine, and ethylene diamine at typical concentrations found for these was negligible, ascribed possibly to the much higher affinity of the H and MMH with the P(3HT)/dopant film.

Selampinar et al. [1582] described a sensor based on a P(Py) composite with polyamides for sensing of NH_3 and CO_2 gases. The sensitivity and response time however appeared to be poor, the latter of the order of several minutes.

An interesting conductometric sensor, for detection of *odors* in the form of vapors of a variety of alcohols, was described by De Rossi et al. [1583]. Responses were monitored in the form of steady state percent resistance change, and an artificial neural network that mimicked a natural olfactory system was used.

Ramanathan et al. [1585] reported a functionalized P(Py) "nanowire" for bioaffinity-based sensing. In preliminary work, this group had demonstrated the feasibility of fabricating single and multiple, individually addressable P(Py) and P (ANi) nanowires of controlled dimension (100 nm wide and up to 13 µm long) and location by electrodeposition within a channel between two electrodes on the surface of a Si wafer and their application as pH sensors. They had employed a similar electrode structure with a channel 100 or 200 nm wide by 3 µm long for the entrapment of the model protein, *avidin*, during electrochemical polymerization of P(Py) in a single step. The initial investigations on biomolecular functionalization of Ppy nanowires by entrapment in a single step during electropolymerization were performed using avidin-conjugated ZnSe/CdSe quantum dots (Aqd). In further work to demonstrate the utility of the functionalized P(Py) "nanowires" as sensors, biotin conjugated to a 20-mer DNA oligo (biotin-DNA) was applied as the sensing electrode. The resistance of the 200-nm wide avidin-functionalized nanowires was shown to increase rapidly to a constant value upon addition of 1 nM of the biotin-DNA conjugate, and the resistance change increased with increasing concentrations up to 100 nM. Increasing the concentration further to 1 µM resulted in only a 4% increase over 100 nM, indicating saturation of recognition sites. It was postulated that the changes in resistance were due to the binding of biotin-DNA with avidin in the functionalized nanowires.

Kwon et al. [1586] described a combined conductometric/amperometric sensor for chemical warfare agents (CWAs) based on hydroxylated poly (3,4-ethylenedioxythiophene) (PEDOT) nanotubes with surface substructures such as nanonodules (NNs) and nanorods (NRs). These yielded what were claimed to be "multidimensional CP nanostructures." A detection limit for dimethyl methylphosphonate (DMMP, a CWA simulant) of 10 *ppt* was claimed by these authors. It was not clear, however, if the sensor had simply a generic response to many different analytes and responded to interferents and false positives. Additionally, although the work was reported in 2012, it appears that it did not lead to further development of a successful sensor.

Many CP-based conductometric sensors for common toxic gases such as ammonia have been demonstrated. As an illustrative example, Yoon et al. [1588] demonstrated a P(Py)-based ammonia sensor capable of single-ppm region detection.

Aptamers are nucleic acids (from DNA or RNA) that can selectively bind to small or macromolecules [1589]. Aptamers for specific macromolecules are arrived at using a procedure called SELEX. Aptamer-based sensors not based on CPs have been widely used because they are so specific to a particular protein or other macromolecular analyte [1589, 1590]. It would be expected that researchers would also investigate CP-based aptamer sensors (using CPs as the immobilization matrix) in due course, and so it has been. Just some examples of these: Rahman et al. [1591] demonstrated an Au nanoparticle (NP) + CP aptasensor with a selective response to thrombin, a protein involved in the blood clotting pathway. Olowu et al. [1592] developed an electrochemical DNA aptasensor, based on PEDOT doped with Au-NPs, which exhibited a specific affinity to endocrine-disrupting 17-β-estradiol. And in a claimed demonstration of a direct DNA sensor, Dupont-Filliard et al. [1593] prepared a biotinylated P(Py) film biosensor for DNA detection. A PPy-biotin/avidin/DNA probe was fabricated using intercalated layers of avidin for the immobilization of DNA probes. This sensing layer was claimed to show reproducible, selective responses without loss of activity.

36.3 Potentiometric Sensing

A very common mode of fabricating a highly selective sensor with a CP film is that classified as category (8) above, i.e., incorporation of a selective receptor or host, e.g., an enzyme, into the CP film during electropolymerization, and then monitoring a property of the CP, e.g., conductivity or open circuit potential [1594]. The conductometric glucose sensor developed by the Contractor group described above falls into this category. Another typical such sensor is one described by the Shimidzu group in early work [1576]. In this, such receptors as adenosine triphosphate, poly(adenylic acid), uridine triphosphate, poly(uridylic acid), and sequence-defined oligonucleotide receptors are incorporated into P(Py) or P(T) during electropolymerization. The change in open circuit potential of the P(Py) film on exposure to solutions containing various nucleotides, including DNA/RNA analogues, then provides a means of qualitative and quantitative detection of the latter. Such a sensor is called a *potentiometric* sensor.

Oyama and Hirokawa [2056] described a potentiometric pH sensor that used a film of poly(1-aminopyrene) (P(1APyre)) in which no internal standard solution, as is required in conventional glass pH electrodes, was used. The pH detection was based on the sensor functioning as an ion-sensitive field-effect transistor (ISFET). The sensor had high ion selectivity with respect to Na^+, K^+ and Ca^{2+}, and insensitivity to O_2, much like a conventional pH glass-membrane electrode. Response time was in the 30 to 60 sec region, with excellent linearity of response in the pH 4.0–10.0 region.

An interesting P(ANi)-based sensor for the detection of HCN gas was described by Josowicz and Janata [1596]: into a P(ANi) film electropolymerized on a Pt substrate were incorporated Hg and Ag salts, which formed P(ANi)-Hg and P(ANi)-Ag clusters due to the spontaneous reduction of the metals. These transition metal/CP clusters then adsorbed HCN, causing a change in the open circuit potential of the P(ANi) film. The response time was however very slow, of the order of minutes. Figure 36.7 shows some results from this work. As with the

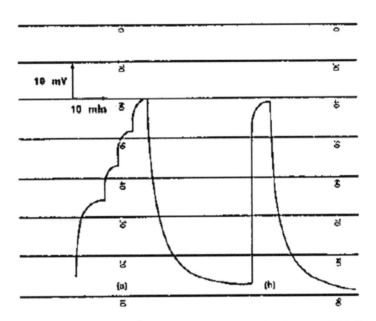

Fig. 36.7 Typical response of PANI*Hg layer to concentration step change of HCN. Concentration steps are (**a**) 0 to 1.50; 3.77; 8.36 and 19.3 ppm HCN. For (**b**) the step is 0 to 19.3 ppm HCN (After Ref. [1596], reproduced with permission)

conductometric sensor work cited earlier, this work appeared to show that CP sensors are not particularly amenable to the detection of gases.

A conductometric P(ANi)/glucose-oxidase sensor was described in Sec. 36.2 above. The same sensor construction could in principle also be used for potentiometric detection of glucose, since the variation of open circuit potential with pH is substantial; such work has been described recently [889, 1597]. The sensing range is claimed to be higher, with an acceptable calibration curve claimed to be obtained over three orders of magnitude [889].

Among other potentiometric or combined potentiometric/amperometric sensors demonstrated for biomolecules are the following: *urea* detection via entrapment of glutamate dehydrogenase in poly(pyrrole)/polyvinyl sulphonate films [1598], *lactate* sensing via glutaraldehyde-mediated linking of lactate dehydrogenase [1599], *cholesterol* sensing using both P(Py) and P(ANi) CPs and based on immobilized cholesterol oxidase/esterase enzymes [1572, 1600–1602], and glucose sensing via binding of glucose oxidase to poly(*o*-amino benzoic acid) [1603].

A number of other potentiometric-mode CP sensors have been described or modeled. Kankare et al. [1604] modeled oxygen sensor response in neutral aqueous solutions based on several CPs, including P(ANi), P(Py), and Poly(3-methyl thiophene). Mazeikiene and Malinauskas [1605] described a nitrate-sensitive electrode comprising P(ANi)/Pt which showed a linear response to nitrate in the 10^{-1} to 10^{-3} M concentration range with a slope of -44 mV/decade. Osaka et al. [1606] which were immobilized with the enzyme creatinine iminohyrolase, for detection of creatinine.

The basis of the response was again found to be pH change in the vicinity of the electrode due to enzymatic reaction, and use of different dopants changed the sensitivity remarkably. Sensitivity, detection range, and detection limit were in the region of 30 mV/decade, 10^{-3} to 10^{-6} M, and 10^{-6} M, respectively.

36.4 Amperometric Sensing

Together with potentiometric and conductometric sensing, amperometric sensing represents one of the most commonly used sensing modes for CP-based sensors [1589].

Glucose, already perhaps the material most commonly sensed with CP-based sensors using other sensing modes, has been the analyte of interest in a number of amperometric sensors. Thus, a sensor described by Iwakura et al. [1576c] used a P(Py) film into which glucose oxidase enzyme and ferrocenecarboxylate (an electron transfer mediator) were incorporated during electropolymerization. After several minutes equilibration time, the current response was measured at fixed applied potential, and acceptable calibration curves were obtained with a glucose concentration in the range of 0 to 20 mM. Work both preceding and following Iwakura et al.'s work however showed [1607–1610] that there were serious problems with a P(Py)-based sensor for glucose, due to rapid degradation of the P(Py) by the H_2O_2 produced during the enzymatic reaction and/or oxidative degradation of the P(Py) at the sensing potentials used. Attempts to circumvent this problem by using ferrocene as a mediator between the Pt substrate and the P(Py), with binding through the pyrrole N-atoms, met with limited success [1611]. Incremental improvements to such amperometric glucose sensors continue to be applied. Khan and Wernet [1612] observed that electrochemical platinization of a flexible film of P(Py) containing the glucose oxidase, covering with gelatin, and cross-linking with glutaraldehyde increased the H_2O_2 oxidation current and the glucose sensor sensitivity. The effect of Pt-black loading that they observed is seen in the response curve in Fig. 36.8.

This problem was circumvented by the use of poly(o-phenylenediamine) (P(o-PD)), which has a large, electroinactive potential window (see Sect. 36.2 above) coinciding with the detection potentials used [1613]. This yielded a sensor with high sensitivity, a 10-day shelf life, a 20 h active-use lifetime, and a insensitivity to interference from ascorbic acid.

Amperometric sensors containing *immobilized enzymes* (similar to the many glucose sensors based on immobilized glucose oxidase described at length elsewhere in this chapter) have also been demonstrated for the following biomolecules [1595]: D-alanine, atrazine, cholesterol, choline, glutamate, fructose, hemoglobin, and phenols.

Genies and Marchesiello [1614] described a P(Py)/glucose-oxidase glucose sensor which used p-benzoquinone as a redox mediator and also a stable P(3-Me-T)/

Fig. 36.8 Effect of
Pt-black loading on the
response current for a P
(Py)-based glucose sensor.
Pt-black loading: μg/cm²,
(Δ) 120, (□) 240, (∘)
300, (·) 600, (■) 1050.
Batch measurement,
applied potential: 0.4 V. no
cover membrane (After Ref.
[1612], reproduced with
permission)

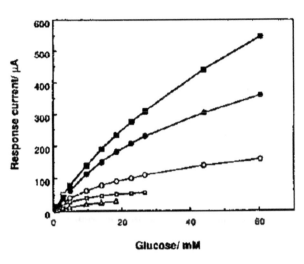

Fig. 36.9 Amperometric P
(Py) glucose sensor:
Response as function of
potential step from open
circuit potential (225 mV,
vs. Ag/AgCl) to 350 mV as
function of glucose
concentration (After Ref.
[1615], reproduced with
permission)

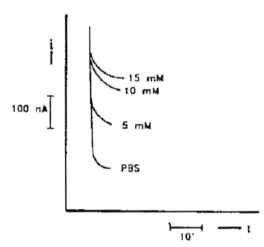

glucose-oxidase sensor. Koopal et al. [1615] demonstrated a sensor constructed
from microtubules of P(Py) which spontaneously adsorbed glucose oxidase. This
sensor, which did not need a mediator, showed accurate sensing in the 1–15 mM
glucose concentration region. The chronoamperometric response of this sensor is
shown in Fig. 36.9.

Gao et al. [1616] prepared a PPy/CNT/glucose oxidase (GOX) sensing electrode
by electrochemical co-deposition of PPy and GOX onto a CNT array; this electrode
was claimed to exhibit high sensitivity toward glucose, nearly one order of magnitude higher than that of the PPy film alone deposited on an Au substrate. The
increase in sensitivity was attributed to increase in the sensing surface area, which
allowed glucose to easily access GOX loaded in the ultrathin PPy film.
Supplementing this work of the Wallace group, many variants of sensors for

glucose incorporating GOX and CPs have been developed [1617–1622]. Other CP sensors using the same principle of an enzyme immobilized on an electrode have been used for sensing of NADH and for pH as well [1623, 1624].

An advanced amperometric CP-based sensor for the detection for the prostate cancer biomarker miR-141 was demonstrated by Tran et al. [288]. The sensing electrodes comprised 3-(5-Hydroxy-1,4-dioxo-1,4-dihydronaphthalen-2(3)-yl) propanoic acid ("JUGA") and 5-hydroxy-1,4-naphthoquinone ("JUG") coupled with MWCNTs, deposited on glassy carbon (GC) surfaces. It was claimed that addition of the miR-141 gave a "signal-on" response, i.e., a current increase due to enhancement of CP electroactivity. Noncomplementary miRNA moieties such as miR-103 and miR-29b-1did not lead to significant change in observed current. A detection limit of ca. 8fM was claimed.

Again, using the same principle of incorporation of an enzyme and mediators into a CP film during electropolymerization, Yabuki et al. [1625] incorporated alcohol dehydrogenase, NAD, and the redox dye Meldola blue into a P(Py) film during electropolymerization. Meldola blue facilitates regeneration of coenzyme NADH and also binds strongly to alcohol dehydrogenase. This amperometric sensor was selective to ethanol at a specific applied potential. Other workers have immobilized receptors such as ATP and lactate in CPs such as P(Py) for potential sensing applications [1578].

Yang et al. [1626] described a P(Py)/galactose oxidase (GalOx) sensor with nearly linear amperometric response in the 0 to 10 mM galactose concentration region, as seen in Fig. 36.10. On the other hand, Jinqing et al. [1627] observed that a GalOx/P(Py) sensor they fabricated showed no response whereas a GalOx/P(ANi) one did.

An amperometric sensor for amino acids based on flow injection analysis (FIA) and using microelectrodes (10 μm diameter) primarily of P(Py) doped with sulfonate dopants such as tosylate and 3-sulfobenzoate was demonstrated by Akhtar et al. [1628, 1629]. Linear response was demonstrated for analytes such as aspartic

Fig. 36.10 P(Py)-based amperometric galactose sensor: data at 0.5 V, in 0.1 M phosphate buffer (pH 6.1) at 25C (After Ref. [1626], reproduced with permission)

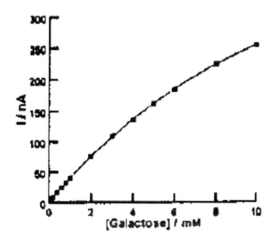

Fig. 36.11 Response of P (Py) sensor in 0.1 M NaNO$_3$ eluent for aspartic acid of the following concentrations: a) 6×10^{-5} M b) 4.5×10^{-5} M c) 3×10^{-5} M d) 1.5×10^{-5} M (After Ref. [1629], reproduced with permission)

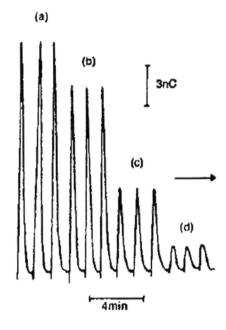

acid and glutamic acid over the concentration range 7.5×10^{-6} to 10^{-4} with sensitivities in the region of 1.5 nC-M and detection limits of ca. 10^{-6} M. These authors also showed the use of a pattern recognition technique using the responses of six detector electrodes. Figure 36.11 shows typical response of one of their sensors.

Cosnier and Innocent [1069] recognized that one of the problems with incorporation of enzymes from solutions into CP films during electropolymerization was that the concentration of the enzyme in the CP film could only be approximately controlled. They took a derivatized pyrrole monomer containing a surfactant group (Eq. 36.2) which has a strong affinity for the enzyme tyrosinase within specific pH ranges. A film of the monomer-tyrosinase complex, with predetermined monomer-tyrosinase concentrations, was cast from solution onto a Pt electrode. The electrode was then removed from this solution and immersed in a second electrolyte for polymerization of the adsorbed monomer-tyrosinase complex, yielding a P(Py)-derivative/tyrosinase film. Tyrosinase catalyzes the oxidation of monophenols and o-diphenols to o-quinones in the presence of O$_2$. The sensor could thus be used for detection of such analytes as catechol, as shown in Fig. 36.12. Response times were of the order of seconds.

$$\text{\raisebox{-0.5em}{\includegraphics}}N-(CH_2)_{12}-\overset{+}{N}Et_3[BF_4]^- \qquad\qquad (36.2)$$

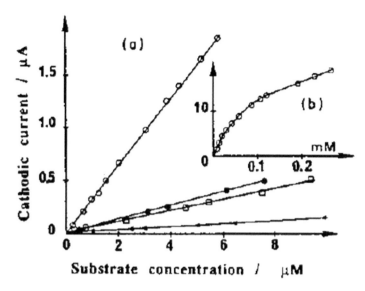

Fig. 36.12   Calibration curves for the poly(pyrrole)/−tyrosinase electrode. (**a**) With low concentrations of various substrates; ∘, catechol; ·, phenol; □, *p*-cresol; ⋆, dopamine. (**b**) With higher concentrations of catechol. (After Ref. [1069], reproduced with permission)

The electrocatalytic properties of CPs, discussed in detail in a subsequent chapter, have also been used for amperometric sensing [1630]. Thus, an amperometric O_2 sensor constructed from P(Ac) has been described which was constructed in the following manner: A BF_4^--doped film of P(Ac) and a piece of Pb are placed in a 48% HBF_4/H_2O solution and connected electrically. Spontaneous dissolution of the Pb occurs, forming de-doped P(Ac) and $Pb(BF_4)_2$. If O_2 is bubbled over the P(Ac)/BF_4 electrode, it is simultaneously oxidized. This oxidation is registered by a current flow directly dependent on the O_2 concentration, yielding an amperometric sensor.

A truly novel amperometric sensor, for detection of *color*, was described by Tada and Yoshino. This was based on poly(3-hexylthiophene) and is schematically depicted in Figs. 36.13a, b. It consisted of two photocells having different CP thicknesses and used the difference in the photocurrent spectrum of the photocells. The color of incident light was detected by measuring the ratio of the short circuit photocurrent of the two photocells. This current/wavelength dispersion is depicted in Fig. 36.13b.

36.5   Conductometric/Amperometric Sensing with Microsensors

A multitude of CP-based sensors for gases were demonstrated early on. For example, Liu et al. [1632] demonstrated a P(ANi) "nanowire"-based sensor for ammonia with a claimed detection limit of 0.5 ppm although coupled with a slow response

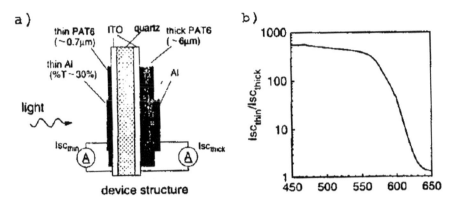

Fig. 36.13 Poly(3-hexylthiophene)-based color sensor: (**a**) Structure. (**b**) Typical performance data; ISC denotes short circuit current (After Ref. [1631], reproduced with permission)

time (10 s to a few minutes); it was however not clear whether this was simply a generic response to a multitude of analyte gases, as no interferents or false-positive data appeared to have been presented. Similarly, a PEDOT "nanowire"-based NO sensor with a claimed detection limit of 10 ppm was reported by Lu et al. [1633], with again no data on interferents, possibility of a generic response, or false positives.

Zhang et al. [1408, 1624] reported a sensor comprised of poly(methyl vinyl ether-alt-maleic acid)-doped P(ANi) nanotubes for oligonucleotide detection. And Peng et al. [212, 217] reported a functionalized poly(thiophene) as an active substrate for a label-free electrochemical geno-sensor.

A variant of conductometric sensing based on using CP films in micro- or nanoscale *field-effect transistors (FETs)* and claimed sensing limits in the femtomolar region was first demonstrated in the mid-1980s by the Wrighton group at MIT [1576d,e] and further developed by many groups [1635, 1636, and Reference therein]. The operation of this type of transistor sensor is shown schematically in Fig. 36.14. An applied gate potential, V_g, produces a gate current, I_g, between the CP film (covering the drain and source) and the counter electrode, also denoted as the gate electrode or gate. An analyte causes a change in conductivity of the CP of as much as six orders of magnitude. This conductivity is monitored as resistivity change between the drain and source electrodes by applying a drain voltage, V_d, across these and monitoring the resulting drain current, I_d (see elsewhere in this book for a description of this measurement method).

Because the dimensions involved are small (e.g., 0.5 nm between the drain and source electrodes in an array) and the total quantity of CP involved also correspondingly small (e.g., 10^{-14} mole), detection limits are very low, with as little as 10^{-10} C of charge producing a response in as little as 100 μs. Such sensors have been implemented with CPs including P(T)s, P(ANi)s, and P(Py). The adaptation of the schematic sensor representation of Fig. 36.14, which is for a liquid electrolyte

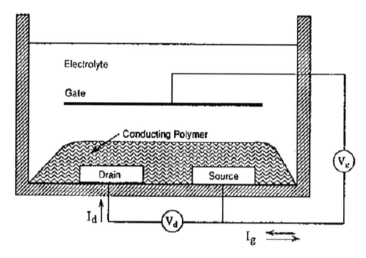

Fig. 36.14 Schematic representation of a CP-based electrochemical sensor analogous to a transistor (After Ref. [1576], reproduced with permission)

medium, to one containing a solid electrolyte and allowing permeation of gaseous analytes to the CP thereby affecting the CP's conductivity, is obvious.

The microsensor configurations first researched by the Wightman, Wrighton, and other groups described above can also be used in an amperometric mode, with some significant advantages. At such dimensions, the "IR drop," an overpotential due to the resistance of the system and one of the bogeys of the electrochemist, is significantly reduced, while the quantity of analyte required is also greatly reduced. These properties, together with developments in microarray electrodes in conventional (i.e., non-CP) electrochemistry and in personal computers, pattern recognition algorithms, and neural networks, have given rise to new developments in sensors based on CP films deposited on microelectrode arrays [1637–1643]. Thus, a typical analytical application has been detection of simple anions with flow injection analysis (FIA) or in electrochemical detection for chromatography. Table 36.1 illustrates the advantages to be gained by detection at a P(Py)-microelectrode vs. a P(Py)-macroelectrode array [1644]. The interrogating waveform is a type of pulsed chronoamperometry (with current monitoring at specific times) causing repeated reduction and oxidation of the electrode, i.e., continued rejuvenation of the CP film for sensing. Some additional selectivity is achievable through modification of the applied potential values (pulsewidths) and times for current sampling.

Field-effect transistors (FETS) based on individual CP nanofibers have been reported by many groups since the early work cited above [1409, 1645–1649]. Pinto et al. [1645] reported an electrospun P(ANi)/polyethylene oxide (PEO) nanofiber FET. Liu et al. [1646] reported a claimed single-nanofiber FET from electrospun poly(3-exylthiophene); this FET exhibited a hole field-effect mobility of 0.03 cm^2/V-s in the saturation regime and a current on/off ratio of 103 in the

Table 36.1 Amperometric anion detection limits using poly(pyrrole) chloride electrodes

Analyte	Macroelectrode[a]	Microelectrode
NO_3^-	1×10^{-5}	5×10^{-6}
Cl^-	1×10^{-5}	5×10^{-8}
CH_3COO^-	1×10^{-4}	5×10^{-8}
CO_3^{2-}	1×10^{-5}	5×10^{-7}
PO_4^{3-}	1×10^{-5}	5×10^{-8}
Ds^-	1×10^{-4}	5×10^{-6}

Anion detection limit (in M): minimum detectable amount when signal is twice noise level. Pulsed potential: E_1, +0.40 V; E_2, −1.0 V. $t_1 = t_2 = 60$ ms. Current sampled for 16 ms at end of pulse. E_1, 1 M glycine eluent. Flow rate, 1 mL/m. Injection volume, 50 µL
[a]Values with glycine eluent. After Ref. [1644], reproduced with permission

accumulation mode. An ultrashort poly(3-hexylthiophene) field-effect transistor with effective channel length of about 6 nm and width of about ~2 nm was also reported by Qi et al. [1647].

Gao et al. [1650] described FETs based on graphene-oxide(GO)/poly (3-hexylthiophene)(P(3HT)) composites with an open-circuit voltage of about 0.62 V, considerably higher than P(3HT)-only devices.

A conventional P(ANi)/glucose-oxidase sensor (see Sect. 36.2) but prepared in such a microelectrode array configuration is claimed to have shown greatly improved sensing capabilities over its macroelectrode counterpart [1638]. Such CP microelectrode arrays have also been used to conductometrically detect such analytes as penicillin by incorporation of the appropriate enzyme, here penicillinase, into the conducting polymer during electropolymerization [1639].

The recent developments in neural networks and pattern recognition algorithms cited above, along with, of course, the near ubiquity of powerful personal computers, have made it possible to get around the relatively poor selectivity frequently found with conductometric sensors not incorporating specific receptors such as enzymes. In such an approach, microelectrode arrays are nearly always used [1637–1643]. Indeed, the technology of such conductometric CP-film-coated microelectrode array sensors has now progressed so far that several commercial "electronic nose" systems are now available with intended application to the food, fragrance, and beverage industry, where human noses are still very much used. These include the "AromaScan" produced by Alphatech International: the "Bloodhound," produced by the University of Leeds Innovations Ltd. in Leeds, England; and the "NOSE" (Neotronics Olfactory Sensing Equipment, by Neotronics, Ltd.) [1641, 1642].

36.6 Voltammetric Sensing

Rather than using a single applied potential or a simple potential program to generate a current as in standard amperometric sensing, the potential can be scanned (swept) using a standard waveform, e.g., cyclic, linear sweep, or square wave voltammetry. This can allow for both qualitative (through location of peaks) and quantitative (through peak current or area) sensing. A sensor based on this principle is a *voltammetric* sensor. We now cite illustrative examples of such voltammetric sensing.

A rather unique and highly specific voltammetric sensor for catechol was reported by Lakshmi et al. [1651]. For this, a monomer that *combines aniline and methacrylamide functionalities* was used. Conducting films were prepared on the surface of electrodes (Au/glass) by electropolymerization of the aniline moiety. A layer of "molecularly imprinted polymers" ("MIP") was then photochemically grafted over the P(ANi), via N,N-diethyldithiocarbamic acid benzyl ester (iniferter) activation of the methacrylamide groups. Detection of catechol by the hybrid-MIP sensor was via cyclic voltammetry between −0.6 V and +0.8 V (vs. Ag/AgCl) at a scan rate of 50 mV/s in PBS buffer solution (pH 7.4). The calibration curve for catechol was found to be linear to 144 μM, with a limit of detection of 228 nM. More importantly, similarly structured compounds and interferents, including phenol, resorcinol, hydroquinone, serotonin, and ascorbic acid, had minimal effect (<3%) on the detection of the analyte. And non-imprinted hybrid electrodes and bare Au/glass electrodes failed to give any response to catechol at concentrations below 0.5 mM.

A voltammetric sensor was reported by Cha et al. [1652] for oligonucleotides based on the unique CP poly(thiophen-3-yl-acetic acid 1,3-dioxo-1,3-dihydro-isoindol-2-yl ester) ("PTAE"). In this work, a direct chemical substitution of the probe oligonucleotide to good leaving group site in the PTAE was carried out on the CP film. Biological recognition could then be monitored by comparison of the cyclic voltammograms of the single- and double-strand oligonucleotides. The sensitivity of this sensor was claimed by the authors to be 0.62 mA/nM, with a detection limit of 1 nM. The oxidation current for the double-strand state of the oligonucleotide was observed to be half that of the single-strand state.

Another, very similar voltammetric sensor, for the detection of the pesticide atrazine, was reported by Pardieu et al. [1653]. In this, a "molecularly imprinted conducting polymer" ("MICP"), specifically poly(3,4-ethylenedioxythiophene-co-thiophene-acetic acid), was electrochemically polymerized onto a Pt electrode using two steps: (i) polymerization of comonomers in the presence of atrazine, already associated with the acetic acid substituent through hydrogen bonding and (ii) removal of atrazine from the resulting polymer, leaving the acetic acid substituents open for association with atrazine. Detection was via cyclic voltammetry. The sensor was claimed to be highly specific toward atrazine. The claimed range of detection was 10^{-9} M to 1.5×10^{-2} M, with a (conflicting!) detection limit of about 10^{-7} M.

A voltammetric sensor for TNT was reported by Aguilar et al. [1655], based on poly(ethylenedioxythiophene) (PEDOT) deposited on a Si chip. A drawback of this sensor however was that it required preheating to 60 °C.

Variations of *combined* voltammetric/potentiometric/amperometric detection have been used for detection of organic and bio-relevant moieties. For example, Vasantha and Chen used PEDOT-modified glassy carbon electrodes for the detection of dopamine and ascorbic acid [1656, 1657]. Manisankar et al. used poly (3,4-ethylenedioxythiophene) (PEDOT)-modified electrodes for the determination of pesticides [1656, 1658]. Pernites et al. developed a novel chemo-sensitive ultrathin film with high selectivity for the detection of naproxen, paracetamol, and theophylline [1656, 1659].

Shiu and Chan [1660] described a voltammetric sensing method for Cu(I/II) in which Alizarin Red S, a specific ligand for Cu, was incorporated into a P(Py) film on an electrode. After equilibration with a Cu(I/II) solution for a specific time period, the CP/ligand film was emersed and characterized by cyclic and square wave voltammetry. A linear calibration curve could be obtained for both oxidation states of Cu, with sensitivity greater but linear range less for Cu(II).

Yano et al. [1661] used a truly novel voltammetric sensing mode for the selective detection of both I^- and Br^- at a Pt electrode, using the CP poly(*o*-phenylenediamine) (P(*o*-PD)). Firstly, they established that a P(*o*-PD) film had a large, electroinactive window (Fig. 36.15, top). Secondly, they established that I^- and Br^- showed reversible voltammetric peaks precisely within this window (Fig. 36.15, bottom). Thirdly, they established that when quinones are progressively incorporated into the CP film via progressive voltammetric cycling in quinone solution, the permeability of the halide ions through the CP film is progressively reduced. Thus, depending on the number of voltammetric cycles in quinone solution, the permeability of I^- and Br^- differed substantially, and they could be selectively detected and distinguished (Fig. 36.15b).

Although glucose sensors have been the most widely studied sensors based on immobilized receptors, similar sensors based on immobilized antibodies have also been suggested [1662].

Emge and Bäuerle [1663] studied several new uracil-substituted poly (bithiophenes), whose synthesis was described in an earlier chapter, as voltammetric-mode sensors (more correctly "molecular recognition modules") for the complementary DNA/RNA bases; the sensitivity of this technique however appeared to be poor.

36.7 Gravimetric-Mode Sensing

A CP film can be deposited on a quartz crystal microbalance (QCM) or a surface acoustic wave (SAW) device, and mass changes due to absorption of analyte interacting either with a dopant or the parent CP itself can then be monitored. A combination of these mass changes with impedance/admittance changes may also

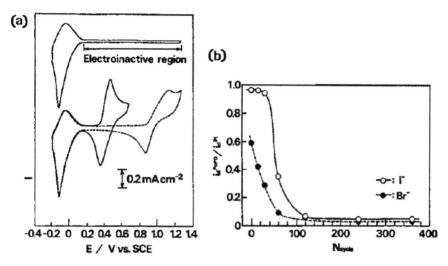

Fig. 36.15 (**a**) (*Top*): Cyclic voltammogram showing the redox activity of a poly(ortho-phenylenediamine) film itself in a Walpole buffer solution (pH 1.1) at 20 mVs^{-1}. (*Bottom*): Cyclic voltammograms for a Walpole buffer solution containing 10 mM KI (solid line) and for that containing 10 nM KBr (broken line) at 20 mVs^{-1} using the poly(ortho-phenylenediamine)-film-coated electrode. Each buffer solution was adjusted to pH 1.1. (**b**) Relationship between the permeability of I^{-} and Br^{-} and the incorporation amount of o-benzoquinone into the film, where $i_d^{P(o-PD)}$ is the limiting current obtained from the RDE voltammogram measured with the poly(o-phenylenediamine)-film-coated electrode, i^{Pt}_d that measured with a bare Pt electrode, and N_{cycle} the number of the potential cycles (After Ref. [1661], reproduced with permission)

be used. Many workers have experimented with such CP sensors [1664–1666], although results have in general not been encouraging. Thus, Fig. 36.16 [1596] shows the results of analysis of methanol vapor using a 500 nm thick P(Py)Tos film deposited on a quartz crystal plate by a combination of mass and admittance change monitoring: the linear mass response occurs only in the region labeled I. Such response cannot generally be used efficiently for sensing applications.

Another mode of sensing which falls into the gravimetric category is use of the electromechanical *actuator* property of CPs to construct "bimetallic" type strips or sensors, whose bending indicates presence of certain analytes. This property of CPs is discussed subsequently in this book, and sensors based on it are also discussed therein.

36.8 Optical-Mode Sensing

The Swager group, originally at the University of Pennsylvania, Philadelphia, USA (since moved to Massachusetts Institute of Technology, Cambridge, MA, USA), described a novel, optical-mode sensor based on a CP which used fluorescence as the sensing mode, much like fluorescent chemosensors [1662]. In the latter, an

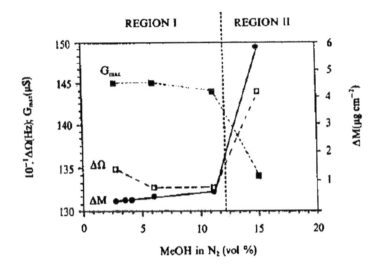

Fig. 36.16 Combined admittance and mass measurements for a 5000 Å thick P(Py)-TOS film deposited on a 5 MHz BT-cut quartz crystal plate. The mass sensitive area equals 0.31 cm^2 and the mass sensitivity is 27.1 ng Hz^{-1} cm^2 (After Ref. [1596], reproduced with permission)

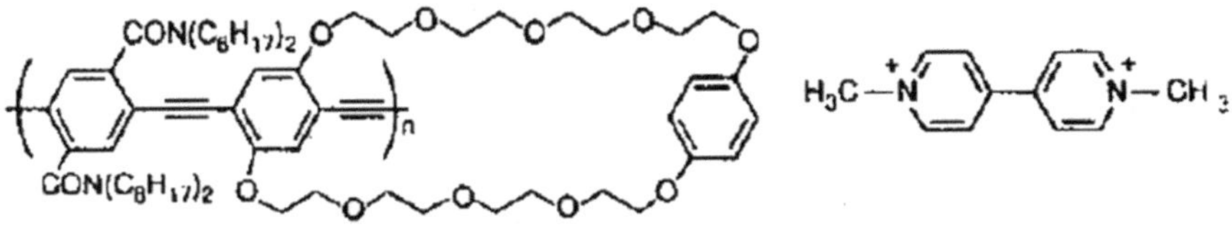

Fig. 36.17 Novel fluorescence-based CP sensor: *Left*: poly(phenylene-ethynylene) with cyclophane receptor (host). *Right*: Paraquat (guest) (After Ref. [832], reproduced with permission)

analyte ("guest") binds selectively and reversibly to an active site ("host") on the sensing species and causes a concomitant change in the fluorescence properties of that species. In their work, a unique derivative of P(PV) was used (Fig. 36.17) in which alternating phenylene rings were part of a large macrocycle called cyclophane, known to be a good receptor for paraquat (the analyte or guest used, see illustration). The paraquat binds to the cyclophane by inserting into its cavity. Without paraquat, the CP fluoresced strongly, the fluorescence thought to be related in some measure to the extended conjugation in the CP. However, when paraquat was bound into the CP, even in very small concentration, the fluorescence was greatly diminished, with the paraquat guest acting much like a "short" in the extended chain CP "wire." In this manner, a sensitivity up to 1000-fold over similar fluorescence sensors was claimed to be achieved.

The Swager group at MIT also reported sensors for explosives and "explosives taggants" based on fluorescence sensing. In early work [1667], they described a chemosensor for TNT based on fluorescence of a porous form of a phenylene/acetylene containing CP, shown in Eq. 36.3.

$$(36.3)$$

More recently, Swager's group described a fluorescence sensor for the "taggant" 2,3-dimethyl-2,3-dinitrobutane (DMNB) using high-bandgap poly(phenylene)s [1668]. However, in that study, response times appeared to be in the tens of seconds, concentrations detected were relatively high, and it was not clear if the sensor would generically respond to other analytes or interferents.

Several workers have attempted to use the absorption, reflectance, or other optical changes resulting from the doping/de-doping process of CPs in sensing applications. Thus, P(ANi), P(T), and P(Py), used in a reduced or semi-reduced state, have been tested in this manner with analytes such as vapors of NH_3, methanol, CH_2Cl_2, and $CHCl_3$ [1596, 1669, 1670] presumed to alter the doping state of the CP or themselves act as chemical dopants. However, as the data in Fig. 36.18 show, sensitivity and selectivity are poor.

A very interesting and novel method of transducing the standard P(ANi)/glucose-oxidase sensor construction (e.g., that of Contractor et al. described in Sec. 36.2 above) to optical response was described recently by Brown et al. [1597]. It is known that besides changes in the conductivity and open circuit potential of the P(ANi) in the presence of the analyte (glucose), presumably due to a change in the pH in the immediate vicinity of the P(ANi), its optical absorption/reflection spectrum of course also changes. The change is substantial, as seen in Fig. 36.19a. Although these data show a nonlinear relationship of pH with the NIR absorbance, if fitted to a 5th power equation, an excellent calibration curve is obtained. Brown et al. fabricated a novel sensor, shown in Fig. 36.19b, in which the cladding on silica and chalcogenide fiber optic cores was replaced by thin films of P(ANi)-glucose-oxidase. This sensor was a variant of an earlier potentiometric pH sensor they developed [1671]. The activity of the sensor persisted for up to 30 days. Response time was of the order of 3 minutes, with detection in the region of glucose concentration of 0 to 20 mM.

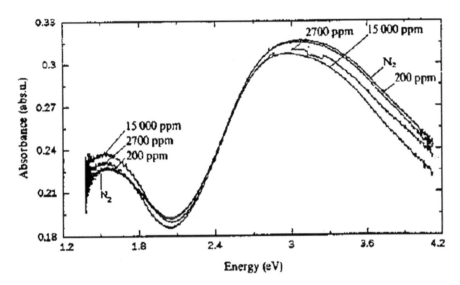

Fig. 36.18 Changes of the absorption spectrum of "as grown" P(Py)-TOS layer obtained during continuous exposure to various concentrations of methanol vapor (After Ref. [1596], reproduced with permission)

36.9 Other Sensing Modes

In addition to the modes described in the above sections, several other sensing modes have been employed with CPs [1672]. Unfortunately, many of these have either been unsuccessful, or have found little or no actual implementation, and are thus cited here only briefly. These modes have included refractive index, magnetism, piezoelectric effect, simple pressure, and NLO properties [1136, 1673, 1674].

Pan et al. [1675] described an "ultrasensitive" resistive pressure sensor based on an elastic, microstructured P(Py)-hydrogel thin film and ultimately based on the piezoelectric effect. The unique synthetic procedure used was claimed to impart the CP film with structure-derived elasticity and a low effective elastic modulus. The detection of pressures of less than 1 Pa, a short response times, reproducibility, good cycling stability, and temperature-stable sensing were claimed.

Boyle et al. [1578] used a P(ANi)sensing film whose EPR spectrum was monitored as various vapors in N_2 carrier were transported above it. This sensor showed selective response depending on the electron affinity and ionization potential of the vapor molecule. These data are shown in Fig. 36.20.

In theory, such effects as the solvatochromic effect observed in many P(3-alkyl-T)'s could be used in sensing applications, although to date no detailed studies of this kind have been produced, to the best of this author's knowledge.

Several sensors based on conductivity (i.e., resistivity) changes in thin CP films have been described. For example, Ellis et al. [1581] described such a sensor for hydrazine and monomethyl hydrazine vapors up to the ppt level which could also

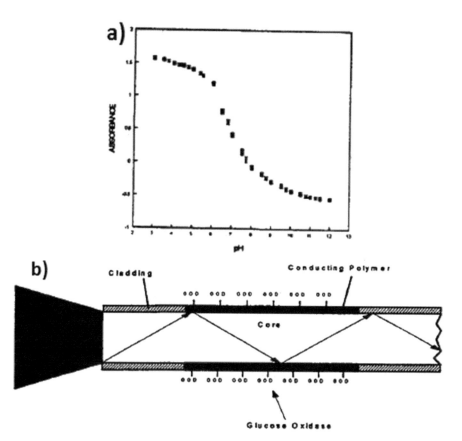

Fig. 36.19 (a) P(ANi)/glucose pH response sensor transduced to optical response (**b**) Schematic construct of an actual sensor using this principle (After Ref. [1597], reproduced with permission)

potentially be applied to nerve agents or other chemical and biological warfare agents. The MacDiarmid group at the University of Pennsylvania, Philadelphia, USA [1676], had earlier also been active in this area.

36.10   Problems and Exercises

1. Outline the principles of working of the six primary and two derivative modes of sensing with CPs identified in Sect. 36.1 and at least two other sensing modes identified in Sect. 36.9.
2. Describe in detail the fabrication and operation of at least one example for each of the modes identified in problem 1.

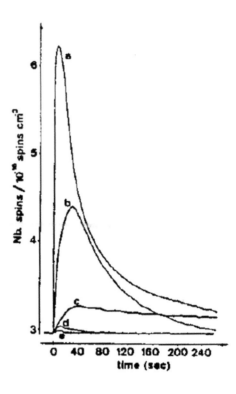

Fig. 36.20 Response of EPR based CP sensor (After Ref. [1578], reproduced with permission)

3. Describe in detail the fabrication and operation of one example for each of the following: (a) A conductometric sensor for the detection of organic vapors. (b) An amperometric sensor for the same. (c) A sensor interfaceable to a PC and usable as a detector in chromatography. (d) A conductometric sensor for glucose. (e) An amperometric sensor for the same. (e) A potentiometric sensor for the same. (f) A sensor whose sensitivity can be controlled electrically. (g) A sensor incorporating an enzyme into the CP film wherein the enzyme concentration is precisely controllable. (h) A sensor using an optical fiber probe and fluorescence-based detection.

4. Besides the commercially marketed sensors described in Sect. 36.5, which sensor(s) do you feel are most likely to be commercially viable and closest to market? What improvements or changes would you make in these to make them commercially viable? You may use specific examples cited in this chapter.

Chapter 37
Batteries and Energy Devices

Contents

37.1 Types of Batteries Incorporating CPs

37.1.1 Summary of Battery Applications

CPs have been investigated for a very wide variety of applications in batteries and other energy devices, although mostly for secondary (rechargeable) batteries. They have been used as the anode as well as the cathode material, although *cathode materials in Li batteries* (especially *secondary* (rechargeable) batteries) have overwhelmingly been the main focus of interest. Applications have included all-CP (anode/cathode) batteries [1677], lead-acid batteries [1678], Zn batteries [1679], and others. Although poly(aniline) (P(ANi)) and poly(pyrrole) (P(Py)) have overwhelmingly been the primary focus of interest, other common CPs studied have included poly(*p*-phenylene) (P(PP)), poly(acetylene) (P(Ac)), and poly(thiophene) (P(T)).

© Springer International Publishing AG 2018   575
P. Chandrasekhar, *Conducting Polymers, Fundamentals and Applications*,
https://doi.org/10.1007/978-3-319-69378-1_37

37.1.2 Advantages of CPs in Batteries

The interest in CPs as battery electrode materials lies in their lightweight, presumed low cost, presumed processibility into odd shapes and sizes (such as ultrathin, flexible or button cells), less corrosive nature, and compatibility with organic liquid and solid electrolytes. They are considered primarily for low-discharge-rate applications, such as in electronic devices, rather than high-rate applications such as transportation systems; besides poor performance under high discharge rates, their specific capacities (see below) are too low for transportation systems. An early attempt at commercialization of Li/CP batteries was that by Bridgestone Corp. in collaboration with Seiko Electronic Components, Ltd., Tokyo, Japan [1309], with a Li/CP battery first introduced in September 1987. To date, however, this has not been further commercialized.

37.1.3 Battery Parameters and Performance

We shall start our discussion then with the focus of the vast majority of the studies, i.e., CPs as cathode materials in Li secondary batteries, whence the principles of CP use in batteries will become familiar. We shall then progress to the other battery applications from there. Before this, however, we must digress briefly to the benchmarks, parameters, and terms used to judge battery performance, so that the reader is familiar with where Li batteries stack in relation to other batteries.

The major parameters used to measure battery performance are *discharge voltage* (or *open-circuit voltage*, e.g., 1.5 V for a common "D" cell); *energy density*, usually expressed in mWh/g (milliWatt-hours/gram); *specific capacity*, usually expressed in mAh/g (A = amperes); number of deep discharge cycles obtainable (*cyclability* or *number of cycles*); *coulombic efficiency* (ratio of charge consumed during charging and discharging), in %; and discharge *current density*. The discharge rate is also sometimes signified by a combination of specific capacity and number of cycles. A battery capable of a high discharge rate is termed "high rate," usable for high-power applications such as transportation, with the converse term being "low rate," usable in low-power applications such as consumer electronics. Most CP-based batteries are low rate.

Table 37.1 lists some comparative parameters for common battery types and for selected batteries using CPs. We shall come back to this Table again later in this chapter. A common characterization parameter for batteries in operation is their charge/discharge curve. Figure 37.1a, b shows such curves for a "well-behaved" and a poor Li/CP battery system [1680]. The useful function of the battery is in the region where its potential is reasonably constant, near ca. 3.0 V.

The electrode–electrolyte configuration of a battery is usually denoted in short-hand, starting with the anode. Thus, a battery consisting of a Li anode, a liquid

Table 37.1 Comparative parameters for various common secondary (rechargeable) batteries and those using CPs

Type (Anode/Cathode)	Discharge voltage	Energy density (m/Wh/g)
Ni/Cd	1.2	27 to 45
Pb/PbO$_2$/H$_2$SO$_4$ (lead-acid)	1.8–2.0	35 to 168
Li/V$_2$O$_5$-V$_6$O$_{13}$	3.2	100
LiC$_6$(graphite)/LiMnO$_2$	3.0	> 100
Li/P(Ac)ClO$_4$	3.3	297
Li/P(Py)BF$_4$	2.8	< 20
Li/P(ANi)Trifl	2.5–3.18	> 200
Li/P(Py)ClO$_4$	2.8	298
Li/P(T)	3.5	> 200
Li/poly(azulene)	3.37	> 100

Type (Anode/Cathode)	Spec. capacity (mAh/g)	Deep cycles
Ni/Cd	160	
Pb/PbO$_2$/H$_2$SO$_4$ (lead-acid)		
Li/V$_2$O$_5$-V$_6$O$_{13}$	3860	300
LiC$_6$(graphite)/LiMnO$_2$	372	150
Li/P(Ac)	90	50–100
Li/P(Py)BF$_4$	95	
Li/P(ANi)Trifl	120	
Li/P(Py)ClO$_4$	90	
Li/P(T)	29	
Li/poly(azulene)	86	

Table 37.1 Continued

Type (Anode/Cathode)	Coulombic efficiency (%)	Current density (mA/cm^2)
Ni/Cd		
Pb/PbO$_2$/H$_2$SO$_4$ (lead-acid)		
Li/V$_2$O$_5$-V$_6$O$_{13}$	ca. 97%	
LiC$_6$(graphite)/LiMnO$_2$		
Li/P(Ac)	91%	1.0
Li/P(Py)BF$_4$	81%	0.1
Li/P(ANi)Trifl	97%	2.0
Li/P(Py)ClO$_4$	91%	0.6
Li/P(T)	87%	0.6
Li/poly(azulene)	90%	0.5

electrolyte of LiClO$_4$ in propylene carbonate (PC), and a P(Py)ClO$_4$ cathode would be denoted as Li/LiClO$_4$-PC-electrolyte/P(Py)ClO$_4$.

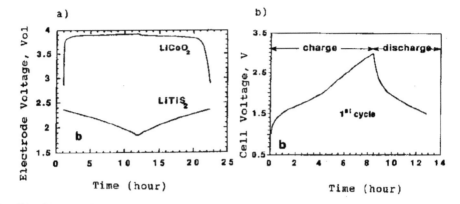

Fig. 37.1 (**a, b**) Illustrative charge/discharge characteristics of (**a**) "well-behaved" and (**b**) "poor" Li secondary battery (After Ref. [1680], reproduced with permission)

37.2   Li Secondary (Rechargeable) Batteries

37.2.1   Principles

The working of a Li secondary battery, as shown schematically in Fig. 37.2a, is simple. During discharge (i.e., use), Li ions migrate from the anode through the Li-ion-conducting electrolyte (liquid or solid) to the cathode, which is a Li-ion acceptor. During charge, the Li ions are retransported to the anode and reduced to Li metal thereon. Because of the high place of Li in the electrochemical series and its low atomic weight, such batteries have high discharge voltages, >3 V, and high energy density. Problems with the simple battery configuration of Fig. 37.1a include severe corrosion due to the highly reactive Li at the anode, circumvented by substitution of Li metal by Li-intercalates and liquid electrolyte by less reactive solid electrolytes with high conductivity at ambient temperature, as shown in the refined schematic of Fig. 37.2b. Acceptability of Li ions at the cathode, another problem, is improved by use of cathode materials such as V and Mo oxides or sulfides. Li_x-graphite is a common intercalate anode material, and a $LiCoO_2$ is a common intercalate cathode material. The latter combination has been commercialized as "Li-ion" batteries, sometimes called "Li rocking-chair batteries"due to the fact that Li ions rock back and forth between anode and cathode during discharge/charge. Many excellent monographs on Li batteries are available [1681].

37.2.2   Li/CP Batteries

To understand how a Li/CP battery operates and why use of the CP is advantageous, we may briefly look at a typical such battery, that combining a Li anode with a *cis*-P

a)

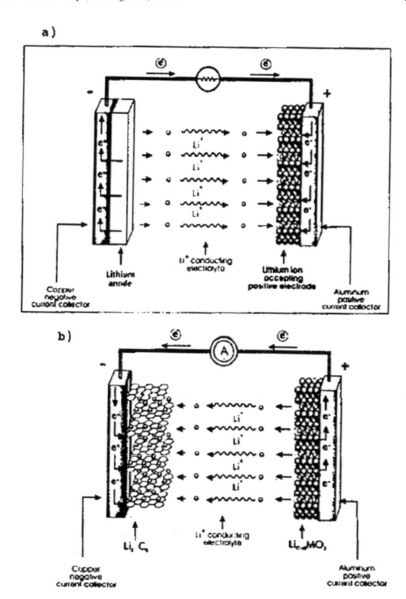

Fig. 37.2 (a, b) Schematics of working of Li battery (After Ref. [1682], reproduced with permission)

(Ac) cathode and using 1 M LiClO$_4$ in propylene carbonate (PC) as liquid electrolyte, first demonstrated by Nigrey et al. [1683]. This battery had an initial discharge voltage of 3.7 V and used 6% doped P(Ac). During discharging, de-doping of the CP occurs, while during charging, doping occurs; the reactions at the anode and cathode are:

$$\text{Positive electrode :} \quad \xrightarrow[\text{discharge}]{\text{charge}}$$

$$\left(CH\right)_x + xyClO_4^{\ -}\text{-}xye^- \xrightarrow[\longleftarrow \text{----------}]{\text{----------} \rightarrow} \left[CH^{+y}\left(ClO_4\right)_y\right]_x$$

$$\text{Negative electrode :} \quad \xrightarrow[\text{discharge}]{\text{charge}}$$

$$xyLi^+ + xye_- \xrightarrow[\longleftarrow \text{----------}]{\text{----------} \rightarrow} xyLi \qquad (37.1)$$

$$\text{Total :} \quad \xrightarrow[\text{discharge}]{\text{charge}}$$

$$\left(CH\right)_X + XYClO_4^{\ -} \xrightarrow[\neg\text{----------}]{\text{----------} \rightarrow} \left[CH^{+y}\left(ClO_4\right)_y\right]_x$$

Thus, it is clear from the above that while at the anode, the Li/Li^+ equilibrium common to most Li batteries is maintained; at the cathode, Li^+ does not need to be accommodated but, rather, the counterion, $ClO_4^{\ -}$, readily and reversibly incorporated by a CP, is the charge carrier. This reduces effects of corrosion and other problems due to the highly reactive Li^+. (It is noted that there is one battery type which uses n-doped CPs (usually P(Ac)) where the de-doping of the CP corresponds to charging, but this is rare and impractical and so is not discussed further here.)

Besides the standard $PEO/LiClO_4$ solid electrolytes, best operated at elevated temperatures (>60 °C), other common solid electrolytes include poly(acrylonitrile) (PAN)–ethylene carbonate (EC)–salt which frequently also includes PC, the EC and PC being plasticizers, and PAN/PMMA/poly(vinylidene difluoride) (PVDF), the latter being composite two-phase electrolytes where a liquid phase is trapped in a matrix obtained by UV cross-linking or gelification [1682, 1684, 1685].

37.2.3 Problems Associated with Li/CP Batteries

Evidently, there are several problems associated with use of CPs as cathode electrode materials in Li batteries, which have limited their commercial exploitation hitherto. Stability of CPs such as P(Ac), especially in contact with leaking air/moisture or liquid electrolytes, is a major issue. Spontaneous de-doping when in contact with electrolyte containing counterions inhibits discharge. Poor conductivity of the CP in its de-doped state, e.g., after full charging, renders it a pseudo-electrode and inhibits battery function. For example, the rapid deterioration of a Li/P(T) battery with $LiClO_4/PC$ liquid electrolyte upon simple storage for 50 hrs is clearly visible in the constant current discharge curve of Fig. 37.3. While the corrosive and reactive effects of liquid electrolytes can be avoided with use of common solid polymer electrolytes, the latter's much lowered conductivity severely inhibits ambient temperature function.

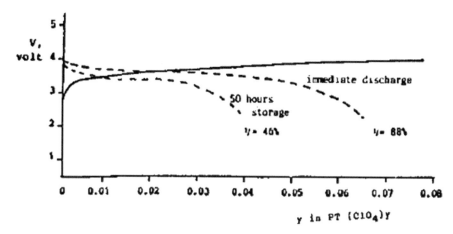

Fig. 37.3 Immediate and delayed (50 h of storage) constant current (0.033 mA/Cm2) discharge of a Li/LiClO4– PC cell at room temperature (After Ref. [985], reproduced with permission)

Yet another drawback of CPs is their maximum dopability, for example, ca. 33% for poly(pyrrole) (P(Py)) or ca. 50% for P(ANi). This inhibits their uptake of counterions during charging, in turn lowering their specific capacity, which is significantly lower than, e.g., that of $LiCoO_2$ or $Li_xVO_{2.17}$, common cathodes in Li-ion batteries. An associated problem is that the optimum counterion (=dopant) concentration desired in the electrolyte in order to maximize CP doping level is considerably higher than achievable. And use of solid electrolytes with CP electrodes greatly diminishes the diffusion coefficient of the counterions involved in CP doping/de-doping, lowering battery efficiency. Indeed, improving physical contact between CP electrode and solid electrolyte layer is contended to be a major problem in solid-state CP batteries.

We will now, in the subsequent sections, discuss illustrative examples of batteries incorporating various CPs as electrode materials. Parameters influencing and improving battery characteristics and performance will become apparent to the reader from these many examples.

37.3 Li Batteries Using Poly(Acetylene) (P(Ac))

As noted above, poly(acetylene) (P(Ac)) batteries, including all-P(Ac) (i.e., non-Li) batteries, were among the first CP-based secondary batteries fabricated [1683, 1686, 1687]. The open circuit potential of a Li/P(Ac) battery is significantly dependent upon the initial P(Ac) doping level. Besides the drawback of the poor stability of P(Ac) in the presence of O_2 or moisture, other drawbacks of Li/P (Ac) batteries include poor stability of the P(Ac) in the presence of most liquid and solid electrolytes and poor shelf life due to spontaneous de-doping; the latter is

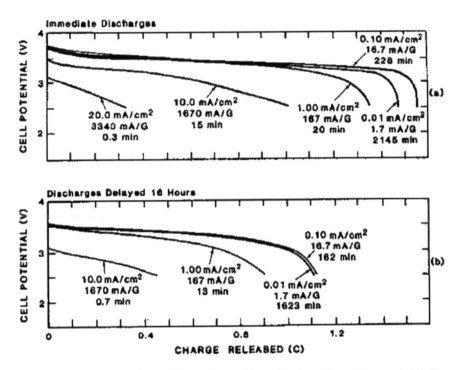

Fig. 37.4 (**a, b**) Poly(acetylene)/lithium cell potential as a function of the oxidation extent during discharge at various constant currents: (**a**) immediate discharge and (**b**) discharge delayed for 16 h (After Ref. [1688], reproduced with permission)

visible in the constant current discharge curves at several current densities shown in Fig. 37.4a,b. The characteristic poor performance of CP-based batteries in high-rate applications is also seen in this figure.

P(Ac)-based batteries follow the characteristic behavior of all CP-based batteries but in a more pronounced fashion due to the greater instability of this CP. Thus, the degradation of capacity with storage, shown in immediate-discharge vs. delayed-discharge curves, is more pronounced for P(Ac) cathodes. As another example, at a discharge current density of 0.1 mA/cm^2, a coulombic efficiency of 86% is obtained, while at 5.0 mA/cm^2, a coulombic efficiency of less than 27% is obtained [1689]. Increasing the doping level of the P(Ac) would be thought to improve battery energy density, capacity, and efficiency. In fact, after an optimal doping level of ca. 2%, coulombic efficiency steadily decreases, going, e.g., from ca. 86% at 2% doping to 60% at 7% doping to ca. 35% at 19% doping [1689, 1690].

37.4 Li Batteries Using Poly(Pyrrole) (P(Py)), Poly (Aniline) (P(ANi)), and Poly(Thiophenes)

P(Py), along with P(ANi), represents the CP most commonly investigated for commercially practical secondary batteries. Li/P(Py) batteries exhibit good discharge characteristics and cyclability, as Table 37.1 reveals. Typical discharge and charge reactions for a standard, Li/LiClO$_4$-liquid or solid electrolyte/P(Py) battery are very similar to those cited for P(Ac) above. We may now briefly discuss illustrative, individual examples of batteries and improvements therein.

Gurunathan et al. [1691] described and demonstrated a cylindrical AA-dimension rechargeable battery using a P(ANi)-TiO$_2$ composite as cathode, Zn outer container as anode, cellulose acetate as the separator, and polyvinyl sulfate and carboxy methyl cellulose as the solid polymer electrolytes (SPE). The battery displayed acceptable performance: open-circuit voltage (OCV) of 1.4 V, current of 250 mA to 1.0 A, 50 recharge cycles, power density of 350 Ah/kg, and power efficiency of 70%.

Cheng et al. [1692] produced P(ANi) nanofibers and nanotubes using a spray technique by wetting a template with a CP solution. These nanofibers and nanotubes showed good electrochemical performance when used as the cathode electrode material in lithium batteries.

The discharge capacity value of the doped P(ANi) nanotubes/nanofibers reached 75.7 mAh/g and retained 72.3 mAh/g 95.5% of the highest discharge capacity) in the 80th cycle, significantly higher than the practical discharge capacity of commercially doped polyaniline powders, about 54.2 mAh/g. The specific discharge energy of the nanostructures was claimed to reach 227 Wh/Kg. The average capacity deterioration of these materials was less than 0.05 mAh/g for one cycle, indicating good cycling capability. In addition, the nanotube electrodes were claimed to exhibit longer charge and discharge plateaus as compared to those for electrodes composed of commercial powders.

In rather unique work, Liu et al. [1694] designed poly(phenylene) derivative CPs, designated PFFO and PFFOMB, whose structures are depicted in Fig. 37.5. These were used with Si in electrode materials in secondary Li batteries. Comparative performance of these as electrode materials in such batteries, vs. conventional P(ANi)-based electrode materials, is also shown in the figure.

Park et al. [1695] reported very unique materials for Li secondary battery cathodes based on a CP-Fe compound, as shown in Fig. 37.6 below. Nevertheless, the electrode capacities observed were somewhat mediocre, about 140 mAh/g after the 10th cycle.

An et al. [1696] reported a highly flexible Li battery electrode based on V$_2$O$_5$ blended with poly(3-hexylthiophene)-*block*-poly(ethylene oxide) ("P3HT-*b*-PEO"), with no cracking or mechanical distortion during normal Li battery function, and doubling of Li-ion diffusion.

The Campos group from Brookhaven National Laboratory, Upton, NY, USA, recently reported on the combination of benzodithiophene-based CP donors and

Fig. 37.5 (a) Structures of the unique poly(phenylene) derivative CPs, designated PFFO and PFFOMB, as designed and synthesized by Liu et al. [1694]. (b) Comparative performance of these as electrode materials, vs. conventional P(ANi)-based electrode materials (After Ref. [1694], reproduced with permission)

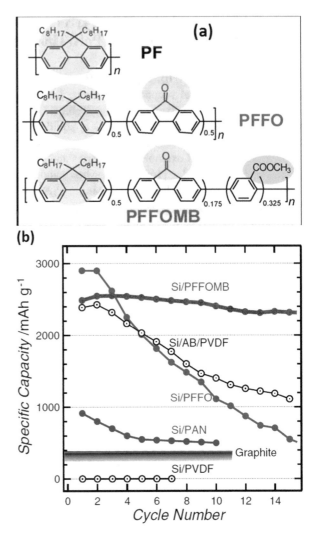

oxidized thiophene-based CP acceptors to improve the efficiency of CP-based solar cells by increasing the formation of multiple excitons [1697]. However, no practical, functional solar cells have as yet resulted from this work.

CPs have been used to coat "nanotubules" and "nanosheets" of conventional Li battery cathode materials, for claimed improved performance. For example, Yoneyama et al. [1698] prepared P(Py)-coated $LiMn_2O_4$ "nanotubules" for superior performance as Li battery cathodes, claimed by the authors to be due to the decrease in real current density and high specific surface area at the tubule electrodes as well as the short diffusion length for Li ions through. In other such work, after that, Torresi et al. [1699] synthesized fibril-shaped P(ANi)/vanadium oxide (VO_x) nanocomposites.

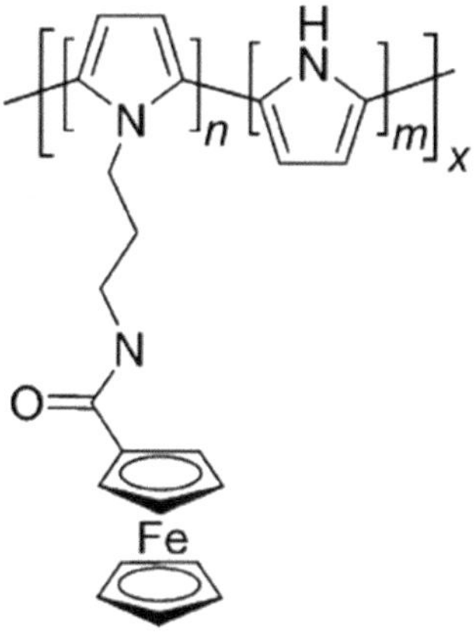

Fig. 37.6 Unique CP-Fe compound for Li secondary battery cathodes as reported by Park et al. [1695] (After Ref. [1695], reproduced with permission)

Zhang et al. [1700, 1701] used PPy "nanowires" and poly(pyrrole-co-aniline) (PPyA) "nanofibers" to fabricate sulfur/CP "nanocomposite" electrode materials for rechargeable Li/S batteries, claiming to have obtained an initial (i.e., at the 0th cycle) discharge capacity of 1222 mAh/g and a capacity at the 20th cycle of 570 mAh/g.

Yang et al. [1702] demonstrated that a thin coating of poly(3,4-ethylenedi-oxythiophene)-poly(styrene sulfonate) (PEDOT/PSS) applied onto a mesoporous carbon/sulfur composite electrode (CMK-3) as used in Li-sulfur batteries led to significantly improved performance. They reasoned that PEDOT/PSS is thermally stable at about 85 °C for over 1000 h with minimal change in electrical conductivity and thus potentially a good candidate for such application. They found that, with a PEDOT/PSS coating, the capacity retention of the sulfur electrode was enhanced from about 70% over 100 cycles to about 80% over 100 cycles, with a 10% increase in delivered discharge capacity. After 80 cycles, the capacity decay was observed to be just 15% over 100 cycles with the PEDOT/PSS coating, while the bare (uncoated) electrode exhibited a capacity decay of 40% over 100 cycles. The coulomb efficiency was also, marginally improved, from 93% to 97%.

Chen et al. [174] demonstrated flexible, "three-dimensionally interconnected" CNT-CP-Hydrogel materials for flexible Li battery electrodes, also based on PEDOT/PSS, along with added TiO_2 and SiNP. They claimed that, unlike previously reported CP-based Li battery electrodes which were mechanically fragile and incompatible with aqueous, their materials exhibited good mechanical properties, high conductivity, and facile ion transport, leading to facile electrode kinetics and high strain tolerance during electrode volume change. They claimed that their flexible TiO_2 electrodes achieved a capacity of 76 mAh/g in 40 s of charge/discharge; an area capacity of 2.2 mAh/cm^2 was claimed for the flexible SiNP-based electrodes at 0.1C discharge rate.

Chao et al. [1703] reported a lightweight, freestanding V_2O_5-based cathodes prepared by growing a V_2O_5 "nanobelt array" (NBA) directly on 3D "ultrathin graphite foam" (UGF), followed by coating the V_2O_5 with a mesoporous thin layer

Fig. 37.7 Coin type Li/P
(Py) cell (After Ref. [1704],
reproduced with
permission)

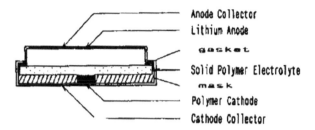

of PEDOT/PSS. The advantages they claimed for their unique electrode included:
(i) The 3D porous UGF acts as both a lightweight scaffold for the growth of V_2O_5
NBAs and an efficient current collector. (ii) The nanoarray architecture is considered more desirable than powder nanostructures in terms of shorter Li + diffusion
paths and more direct electron transport; and the array architecture is favorable for
accommodating the strain caused by the Li-ion insertion/extraction and alleviating
the nanostructure deterioration. (iii) The conductive PEDOT layer coating is
applied without any additional surfactant (e.g., p-toluenesulfonic acid). This thin
homogeneous layer was claimed to facilitate the electron transfer around the V_2O_5
and preserves the whole array structure during long-term cycling.

The Scrosati group described a very standard, Li/P(Py) secondary battery [1685]
using a PAN-EC-PC based solid electrolyte. This battery has predictable performance, with high energy density and coulombic efficiency approaching 90% but,
predictably, very poor shelf life due to spontaneous de-doping and other side
reactions of the P(Py) during storage. A similar, standard solid-state Li/LiClO$_4$-
PEO/P(Py)ClO$_4$ battery was described by Osaka et al. [1606, 1704, 1716, 1911],
with claimed columbic efficiency of 90% at a discharge current density of 0.1 mA/
cm^2 and claimed cyclability to 1400 cycles at high coulombic efficiency. Their
battery, fabricated in the form of a coin, is depicted schematically in Fig. 37.7. A
"rough" P(Py) morphology was claimed to yield increased efficiency due to better
counterion diffusion. Unfortunately, like other solid-electrolyte batteries, it must be
operated at elevated temperature, ca. 80 °C.

Momma et al. [1705] described a Li/P(Py) secondary battery using a standard
LiClO$_4$/poly(ethylene oxide) (PEO) solid electrolyte, in which the cathode was a P
(Py)/Nafion "composite," i.e., during electropolymerization, Nafion was used as the
dopant counterion. This simple modification was claimed to lead to a much
"rougher" CP cathode, yielding better counterion diffusion from the solid electrolyte (through the Nafion) and thus improved battery performance as compared to a
simple P(Py)ClO$_4$ electrode. In a similar study, Shimidzu et al. [1706] noted that
when a large polymeric dopant such as poly(vinyl sulfate) was used with P(Py) as a
cathode material, the resulting "reverse" protonic doping (see discussion elsewhere
in this book) yielded improved battery characteristics.

In the standard Li/lithiated-metal-oxide-composite battery (cf. Fig. 37.2 and Eq.
37.1), lithiated MnO$_2$ is an important and useful composite cathode. It would thus
be logical to consider a composite electrode incorporating this material and a
CP. Such a study, incorporating lithiated MnO$_2$ and P(Py) was carried out by

Kuwabata et al. [1707, 1708], wherein the cathodes were fabricated by electropolymerization of pyrrole in the presence of suspended MnO_2 particles in PC solution. The charge capacity of the cathode, and thus battery energy density, was found to be enhanced because both Li^+ ($Li_xMnO_2 \leftrightarrow MnO_2 + xLi^+$) and the dopant counterion (Cl^-) functioned as charge carriers. In its conductive form, the P (Py) also enhanced conductivity between the Li-MnO_2 particles. In a similar vein, Coffey et al. [1709] described a P(Py)/graphite fiber composite electrode obtained by direct electropolymerization of pyrrole into 8 μm graphite fibers from perchlorate aqueous solution. Specific capacities, reversible over 50 cycles, of 90 mAh/g, with a cell voltage of ca. 2.0 V, were claimed. Again, the charge capacity was enhanced by two, rather than one, charge-transfer processes at the cathode, Li intercalation into the graphite and CP doping/de-doping. In the closest approximation to Li-ion batteries, Panero et al. [1710] recently described a Li/$LiClO_4$-EC-dimethylcarbonate(DMC)-electrolyte/P(Py)ClO_4 battery; this preliminary prototype however had a lower open-circuit voltage and much lower energy density and cyclability than Li-ion batteries.

Li/P(Py) batteries have been the subject of a myriad of studies relating to the effects of all sorts of variables, such as the porosity/morphology of the P(Py), synthesis method, and dopant type/concentration. Thus, for instance, it was found that the porosity of P(Py), and battery efficiency due to improved counterion diffusion, is higher in the order PF_6^- > triflate > perchlorate > tetrafluoroborate [1711]. Similarly, P(Py) chemically polymerized with ferric perchlorate showed improved charge/discharge characteristics as compared to that polymerized using Cu tetrafluoroborate. One of the improved properties of Li/P(Py) batteries, as compared, e.g., to Li/P(Ac) batteries, is the improved storage (shelf) life. This can be seen, e.g., from Fig. 37.8a, b, showing charge/discharge curves for very similarly constructed Li/P(Ac) and Li/P(Py) batteries, respectively [1238][1712].

Nishizawa et al. [1713] studied Li secondary batteries based on cathodes comprised of P(Py)-coated spinel $LiMn_2O_4$ nanotubules of ca. 200 nm outer diameter, prepared by thermal decomposition of an aqueous solution containing Li and Mn nitrates using a nanoporous alumina membrane as a template which was later dissolved off. These electrodes showed superior performance to P(Py)-coated $LiMn_2O_4$ thin films, with capacities claimed to be up to 12 times greater.

Along with P(Py), *P(ANi)* is a CP most studied for battery applications, and the only one, to date, which has been successfully commercialized, as detailed further below. The uniqueness of P(ANi) is, of course, the diffusion of protons, rather than more bulky ions, as the counterions, and the fact that doping levels as high as 50% are attainable. Several examples of and improvements in P(ANi) batteries may now be cited.

Among CPs, P(ANi) appears to yield the best-performing batteries, in terms of such parameters as spontaneous discharge (de-doping, which retards shelf life), coulombic efficiency, and cyclability. Thus, e.g., a typical Li/$LiClO_4$-PC/P(ANi) battery using electrochemically synthesized, fibrous P(ANi) yields a specific capacity of 83 mAh/g and a nearly 100% coulombic efficiency over 500 charge/discharge cycles, with a spontaneous self-discharge rate of less than 2% a day,

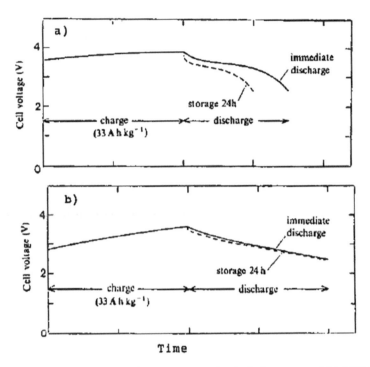

Fig. 37.8 (**a, b**) Comparative charge/discharge curves for similarly constructed (**a**) Li/P(Ac) and (**b**) Li (P(Py)) (After Ref. [1238], reproduced with permission)

[1714, 1715]. In some cases, energy densities as high as 379 mWh/g have been claimed to be observed in simple Li/LiClO$_4$-PC/P(ANi)-triflate cells [1716]. When using chemically synthesized P(ANi), ferric perchlorate and/or cupric tetrafluoroborate are the preferred oxidizing agents due to compatibility of the anion dopants with Li-ion battery electrolytes in liquid media such as PC or γ-butyro-lactone (γ-BL). Combination solvents such as PC + 1,2-dimethoxyethane (DME), EC-DME, and γ-BL-DME appeared to be the best, yielding specific capacities of ca. 90–110 mAh/g at temperatures ranging from −20 °C to room temperature [1717–1721].

Tsutsumi et al. [1722] described a battery using a novel P(ANi) cathode and a composite of P(ANi) with poly(p-styrenesulfonic acid-co-methoxy-oligo(ethylene glycol) acrylate) (PSSA-co-MOEGA), together with a solid electrolyte consisting of cross-linked PEO-grafted-PMMA/LiClO$_4$ (room-temperature conductivity 10^{-4} S/cm). The cathode was prepared by standard electropolymerization of aniline in HClO$_4$, followed by electrochemical doping in aqueous PSSA-co-MOEGA solution, the copolymer being prepared earlier from PSSA and MOEGA. Comparison of this battery with a standard Li/P(ANi)ClO$_4$ using the same solid electrolyte showed much greater cyclability. This was ascribed to two factors, the cation doping of the P(ANi)-PSSA-co-MOEGA electrode vs. anion doping of the P (ANi)ClO$_4$ electrode and the improved interface between the electrode and the solid electrolyte due to the oligo-ethylene oxide side chains in both. A similar improvement in cathode performance due to cation exchange capability and improved compatibility with electrolyte was obtained by Morita et al. [1723], who used a cathode comprising conventional P(ANi)ClO$_4$/Cl with an overlayer of P(ANi)-poly(styrene sulfonate) (PSS), the latter being a cation exchanger; a liquid electrolyte (LiClO$_4$/PC/DME) was used. Although the specific capacity of a battery using this cathode was 84 mAh/g as compared to ca. 140 mAh/g for a conventional P(ANi)ClO$_4$ cathode, the energy density was claimed to be higher, at 116 mWh/g vs. 41 mWh/g.

Oyama et al. [1725] described a novel CP cathode made by mixing chemically polymerized P(ANi) with 2,5-dimercapto-1,3,4-thiadiazole (DMcT) in N-Me-2-pyrrolidinone solvent and spreading and drying the resulting dark, viscous ink on a carbon film to obtain the active cathode. This cathode functioned as a Li$^+$ ion reservoir, much in the tradition of the Li-intercalate composite electrodes and showed the following reactions:

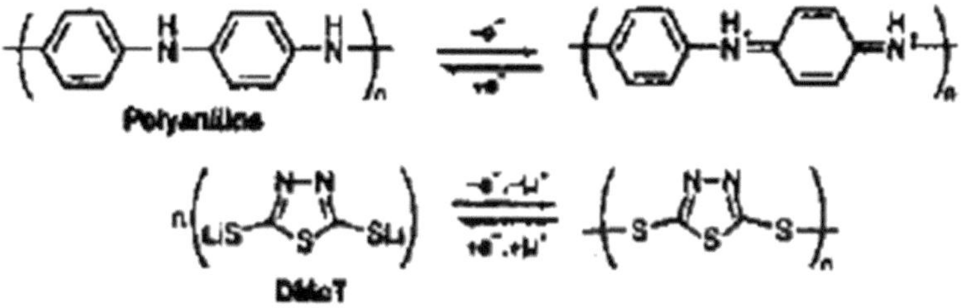

$$(37.2)$$

The battery used a solid polymer electrolyte comprising ethylene carbonate, PC and acrylonitrile-MMA copolymer. Rapid charge transfer between the intimately mixed P(ANi) and DMcT was the key to performance. Although this battery was claimed to have a specific capacity and energy density 50% greater than that of standard rocking-chair Li batteries based on LiCoO$_2$, it still retained a Li anode, with its attendant corrosion problems. For example, charging/discharging currents were required to be 0.05 mA/cm^2 or less to avoid severe cycle life shortening. Addition of a P(Py) derivative to the cathode allowed charging, but not discharging, at higher current densities. In a later refinement of this DMcT/P(ANi) cathode [1725], Oyama et al. "doped" the DMcT with a cupric salt, claiming to observe a higher specific capacity (up to 300 mAh/g over 10 cycles), although the battery ultimately failed beyond ca. 130 cycles even at lower discharge rates (170 mAh/g), and replacement of the Li anode was required. The Oyama group continued studies of these cathode materials with variations including incorporation of acetylene black and current collectors such as Cu, Ni, Al, Ti, and Au-plated Ti [1726].

Leroux et al. [1727] studied P(ANi)/V$_2$O$_5$ "nanocomposite" cathodes derived from sol-gel deposited V$_2$O$_5$ and functioning in Li batteries via efficient Li intercalation, claiming superior battery performance.

With regard to *poly(thiophenes)*, several Li/P(T) batteries have been described. Typically, such batteries have an open circuit potential of 3.8 V, an energy density of 240 mWh/g, and a specific capacity of 17 mAh/g with a coulombic efficiency of 96% [1712].

With regard to 3-alkyl-thiophene polymers, a Li/LiBF$_4$-PC-EC-electrolyte/P (3-MeT)BF$_4$ cell yielded a claimed 100% coulombic efficiency at a charge capacity of 85 mAh/g [1728]. The (spontaneous) self-discharge rate was a low 1.2%/day.

37.5   Li Batteries Using Other CPs

Kawai et al. [971] investigated poly(*trans*-1,2-di(2-thienyl)ethylene), a P(T) deriv-ative, as a cathode material for a secondary Li battery, although they did not actually construct a battery. An open circuit potential and a coulombic efficiency as high as 3.8 V and 99% (over 100 cycles), respectively, were claimed in liquid, PC- or nitrobenzene-based electrolytes.

P(PP) has been an important candidate as both anode and cathode for secondary batteries, due to its *n*-doping (e.g., by Li$^+$) as well as *p*-doping (e.g., by AsF$_6^-$). A Li (anode)/0.5 M LiPF$_6$-PC(electrolyte)/P(PP)PF$_6$(cathode) battery showed the fol-lowing typical characteristics at a ca. 3% CP doping level [1729, 1730]: open circuit potential 4.0 to 4.2 V, coulombic efficiency at 0.5 mA/cm^2 discharge rate 90%, and maximum current density 40 mA/cm^2.

A unique approach, used by the Allied Signal group [1576, 1731, 1732], has been to use a Li-metal alloy, such as Li$_x$Pb, in combination with a CP, as the anode (rather than the cathode). The CP here serves to act as a conductive binder and transfer interface for transfer and reduction of Li$^+$ ions. Thus, using this type of anode, with the CP being P(PP), and an inorganic intercalate cathode such as Li$_x$VO$_{2.17}$, energy densities of 70 mWh/g were achieved with open circuit poten-tials of 1.9 V and claimed improved cyclability and shelf life.

Among other CPs, poly(azulene) (P(Az)) has been among the most promising for battery applications. In a typical configuration, Li/LiClO$_4$-PC-electrolyte/P(Az) ClO$_4$ with the CP at a ca. 20% doping level, an energy density of 150 mWh/g and a coulombic efficiency of 99% at a 1.0 mA/cm^2 current density was claimed [1238, 1733]; a coulombic efficiency of >90% was said to be maintained for 900 cycles.

Other CPs worth mentioning as having been studied for secondary battery applications include poly(3-butyl-thiophene-co-3-methyl-thiophene) and poly (dimethoxy-phenylene vinylene) [1734]. Pandey and Prakash [1735] investigated a Zn/poly(indole) secondary battery and found a maximum capacity of 90 Ah/kg with an open circuit potential of 1.45 V and good coulombic efficiency.

37.6   Non-Li Batteries

Among the first *non-Li* CP-based batteries investigated were all-CP batteries using P(Ac) as both the anode and cathode material. The negative electrode is Al/P (Ac) and the positive electrode Au/P(Ac), with the electrolyte 1 M $LiClO_4$ in PC. The open-circuit voltage and theoretical energy density of such a battery are 2.4 to 2.8 V and 75 mWh/g, respectively [1686, 1687]; current densities can be as high as 100 mA/cm^2. Since no Li is used, the cyclability of this battery is improved, claimed to 2000 cycles; however, shelf life continues to be poor, due to spontaneous de-doping and other effects. The overall battery reaction during discharge is:

$$(CH)_x + nxClO_4^- \rightarrow [(CH^{n+})(ClO_4^-)_n]_x + nxe^-$$

and

$$(CH)_x + nxR_4N^+ + nxe^- \rightarrow [(R_4N^+)_n(CH^{n-})]_x$$

(37.3)

net reaction is:

$$2(CH)_x + nxR_4NClO_4 \rightarrow \left[CH^{n+}(ClO_4^-)_n\right]_x + \left[(R_4N^+)_n(CH^{n-})\right]$$

An Al(anode)/$LiClO_4$-PC-EC-electrolyte/P(Ac) cathode battery was described by Nagatomo et al. [1737], who obtained a coulombic efficiency of 70% through 150 cycles.

Koura et al. [1738] studied an interesting non-Li battery configuration using room-temperature melt chloroaluminate electrolytes, $AlCl_3$-1-butylpyridinium-Cl (BPC) or $AlCl_3$-1-Et-3-Me-imidazolium-Cl (EMIC), with an Al/P(ANi)Cl-powder cathode and anode (Fig. 37.9). The optimum open circuit (initial discharge) poten-tial was 1.6 V, the specific capacity 68 mAh/g-(P(ANi)), and claimed coulombic efficiency 99%.

Several types of non-Li batteries using P(T) have been described. An all-P(T) battery, i.e., with both anode and cathode being P(T), has been described

Fig. 37.9   Unique P(ANi)/P (ANi) battery. Note: PAn = P(ANi) (After Ref. [1738], reproduced with permission)

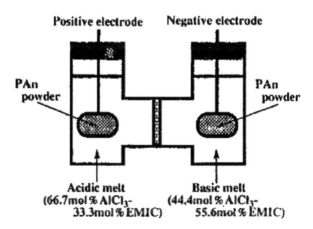

[1739, 1740] with open circuit potential of 3.0 to 3.1 V and energy density of 75 to 93 mWh/g. A Zn/P(T) (anode/cathode) battery has been described using aqueous ZnI$_2$ electrolyte and iodine as dopant for the P(T), with an open circuit potential of 1.4 V and acceptable cyclability [1741]. A similar, Zn(anode)/aqueous-ZnCl$_2$-electrolyte/P(ANi)(cathode) battery and a Pb-acid type battery with a P(PP) cathode have also been described [1678, 1679].

An interesting battery system which appears to be a non-Li system but on closer examination reveals the participation of the Li/Li$^+$ redox couple is that described by Killian et al. [1742]. This uses pyrrole electropolymerized into graphite fibers as both the anode and cathode, with a LiClO$_4$-EC-PAN-PC-acetonitrile gelled electrolyte. The reactions during battery discharge are:

anode:

$$[P(Py)^0/PSS^-] \, LI^+ \rightarrow [P(Py)^+/PSS^-] + Li^+ + e^-$$

(37.4)

cathode:

$$[P(Py)^+] \, ClO_4^- + e^- \rightarrow [P(Py)^0] + ClO_4$$

being opposite during charging. Open circuit potentials are low, in the range 0.9 to 2.0 V, depending on precise cell composition. Specific capacities of 22 mAh/g and cyclabilities to 100 cycles are claimed.

Kumar et al. [1743] reported a Mg/P(ANi) battery and liquid aqueous electrolytes based on Mg perchlorate, chloride, or bromide. The battery exhibited an open-circuit voltage of 1.6–1.8 V, with the nature of the anion having a strong effect. Gofer et al. [1744] reported an all-CP battery based on poly(3-(3,4,5-trifluorophenyl)thiophene) and poly(3(3,5-difluorophenyl)thiophene) films obtained by electropolymerization onto graphite-coated Teflon supports and a polymer gel electrolyte film. Discharge voltages of ca. 2.4 V and storage capacities of 9.5–11.5 mA/h-g were obtained. An advantage claimed was high flexibility of these all-polymer devices.

37.7   Market Implementation of CP-Based Batteries

The first known commercialization of Li/CP batteries was that by Bridgestone Corp. in collaboration with Seiko Electronic Components Ltd., Tokyo, Japan [1309], first introduced in September 1987. This battery had the configuration Li-Al-alloy(anode)/LiBF$_4$-PC-1,2-dimethoxyethane(DME)-electrolyte/P(ANi)BF$_4$. Figure 37.10 shows one configuration of these "button cells" in schematic. In one cell configuration, the P (ANi) used was electropolymerized from aniline/HBF$_4$ solution. The tiny button cells weighed from 0.4 to 2.6 g, had an operating voltage of 2 to 3 V, nominal specific capacity of 0.5–8 mAh, discharge currents in the region of 1 μA to 5 mA, a cyclability of 1000 cycles, and a self-discharge rate of 3.3% per month. The application was

Fig. 37.10 Schematic of
Bridgestone's
commercialized Li/CP
battery (After Ref. [1309],
reproduced with
permission)

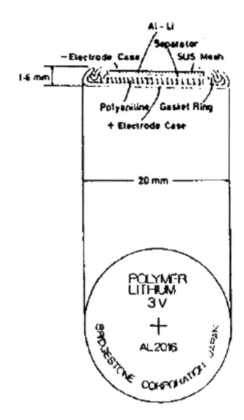

primarily as backup power sources for various electronic and other devices and
watches.

It appears however that these CP-based battery products have been discontinued as
of this writing (2016). And as of this writing, despite many academic studies, no Li
battery commercial products incorporating CPs are known to be on the market as of
this writing (2016). For example, many recent academic studies touted PEDOT/PSS
(poly(ethylene dioxythiophene)-poly(styrene sulfonate)) as a "revolutionary" new
binder material for Li batteries [1745], and others have touted novel CP-based
electrolytes such as poly[(4-styrenesulfonyl)(trifluoromethyl(S-trifluoromethylsul
fonylimino)sulfonyl)imide] (PSsTFSI−), plus high-molecular-weight poly(ethylene
oxide) (PEO) as "revolutionary" electrolytes for Li batteries [1746]; to date (2016),
there is no known Li battery commercial product incorporating these materials.

Although the battery could be damaged irreversibly with high discharge rates, it
was suitable for low-rate use, in the applications cited above, and with a depth of
discharge limited to 30%, 1000 cycles were easily achievable; specific capacity and
energy density were however low.

Initial marketing attempts were made by several German companies, including
BASF and VARTA Batteries of an A size, secondary, low discharge rate, 3.3 V,

Fig. 37.11 BASF's
cylindrical P(Py) battery
(After Ref. [1747],
reproduced with
permission)

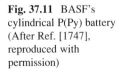

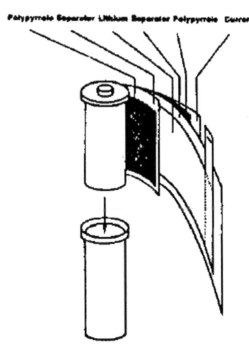

Li/P(Py) battery, claimed to be capable of 500 cycles with an energy density
approaching 300 mWh/g [985, 1747]. This was based on a Li-BF$_4$/ClO$_4$-in-PC
electrolytes. One configuration of this battery is shown in Fig. 37.11. However, the
marketing of this battery appears not to have been successful.

In this respect, the lack of commercialization of a Li/CP battery is in stark
contrast to the wide commercial success of Li-ion (rocking-chair) batteries using
inorganic intercalate cathodes; these have energy densities up to three times those
of Ni-metal-hydride or Ni-Cd batteries, and cyclabilities up to 1000, and are used in
items ranging from camcorders to laptop computers.

37.8 Supercapacitors

Before discussing CP use in supercapacitors, it is useful to get "one's bearings," as
it were, by recounting some numbers with respect to "good" supercapacitors and
"poor" supercapacitors [1748]. Among the best materials for supercapacitors are
(*specific capacitances, in F/g, in parentheses*): *RuO$_2$ (720 to 900)*; *MnO$_2$*
(600–700); *and graphene-P(ANi) nanofiber (490)*. *Among the worst are acti-*
vated-C (200) and *poly(thiophene) nanoparticles (110)*. (Specific capacitances, in
F/g, are generally measured by standard electrochemical techniques such as cyclic
voltammetry.)

CPs have been combined with other conductive materials, specifically CNTs and graphene, for enhance performance in supercapacitors. Thus, e.g., Peng et al. [739, 1749, 2057] describe supercapacitors based on composites of CPs and MWCNTs with specific capacitances of (CP component in parentheses) 506 F/g (P(Py)) and 670 F/g (P(ANi)). And Zhang et al. [1750] reported a very high capacitance of 500 F/g for P(Py)- "nanofiber"/graphene oxide- "nanocomposites."

Liu et al. [1751] demonstrated composites composed of MnO2 nanowires coated with PEDOT (poly(ethylene dioxythiophene)) shells as supercapacitor materials, with acceptable specific capacitance and the ability to maintain capacitance at high current density, preserving 85% of their specific capacitance as the current density increased from 5 to 25 mA/cm^2.

Xia et al. [1752, 1754] reported P(ANi)−/mesoporous carbon composite materials for supercapacitors with claimed specific capacity as high as 900 F/g at a charge–discharge current density of 0.5 A/g (or 1221 F/g based on pure P(ANi) in the composite), to be compared with that of much more expensive, amorphous hydrated RuO2 (840 F/g). They attributed the high claimed capacitance to the growth of ordered whisker-like P(ANi) on the surface of the mesoporous carbon template. The nanoscale P(ANi) was said to form "V-type" nanopores. It was claimed that these nanopores yielded high electrochemical capacitance because the "V-type" channels facilitated faster penetration of the electrolyte and the shorter diffusion length of ions within the electrode during the charge–discharge process. Concomitantly, the high conductivity of polyaniline and mesoporous carbon greatly reduced energy loss and power loss at high charge–discharge current density.

Alvi et al. [1748] described a supercapacitor based on graphene (G)–poly (ethylenedioxythiophene) (PEDOT) nanocomposites. The G-PEDOT nanocomposite was synthesized using a chemical oxidative polymerization technique. An "estimated" specific discharge capacitance of 374 F/g was reported.

Zhang and Zhao [1753] reported on the use as supercapacitor materials of reduced graphene oxide (RGO) sheets coated with the CPs poly(3,4-ethylenedioxythiophene) (PEDOT), polyaniline (PANi), and polypyrrole (PPy) via an in situ polymerization technique. The resulting CP-RGO nanocomposites had varied loadings of the CPs. The authors claimed that the use of ethanol in the coating process led to a uniform coating of the polymers on RGO sheets. The (RGO-PANi) composite exhibited a specific capacitance of 361 F/g at a current density of 0.3 A/g. The (RGO-PPy) and (RGO-PEDOT) composites displayed specific capacitances of 248 and 108 F/g, respectively, at the same current density. More than 80% of initial capacitance was claimed to be retained after 1000 charge/discharge cycles.

Xia et al. [1754] reported supercapacitor materials comprised of 3D porous and thin graphite foams (GF), Co$_3$O$_4$ as the nanowire core, and a composite of PEDOT and MnO$_2$ as the outer shell. Their materials were claimed to exhibit initial specific capacities in the range of about 190 F/g for the worst and about 400 F/g for the best, deteriorating over 20,000 cycles to about 150 F/g and 350 F/g, respectively, over 20,000 cycles.

Kuila et al. [1755] reported dense arrays of ordered, aligned nanorods of P(ANi) showing an incredible specific capacitance value of 3407 F/g; however, it appears that this work could not be duplicated by other groups [1409].

37.9   Solar Cells/Photovoltaics

The use of CPs in solar cells and photovoltaics, especially *dye-sensitized solar cells* (*DSSCs*), is relatively recent, with intense study only after about 2005 [1756]. Nevertheless, CP-based solar cells have shown some promise for commercial implementation recently (since about 2012) [1756]. A typical CP-based solar cell is shown schematically in Fig. 37.12 below. A typical DSSC is shown in Fig. 37.13 below.

One of the earliest CPs used in solar cells was poly[2-methoxy-5-(2'--ethylhexyloxy)-p-phenylene vinylene] (*MEH-PPV*), originally synthesized by Wudl et al. [1758].

Subsequently, Yu et al. [1759] blended MEH-PPV with C_{60} and its derivatives to give the first CP-based solar cell with a high conversion efficiency. Since then, higher performance has been obtained: poly[2,6-(4,4-bis-(2-ethylhexyl)-4H-cyclopenta [2,1-b;3,4-b']dithiophene)-alt-4,7-(2,1,3   benzothiadiazole)]   (PCPDTBT),   a low-bandgap polymer whose absorption

extends up to 900 nm, has shown conversion efficiencies of about 3%. PSCs made from this polymer have showed an initial efficiency of around 3% [1760], which incorporation of alkanedithiol additives was enhanced to about 5.5% [1761]. Leclerc et al. [1762] developed poly[N-9"-hepta-decanyl-2,7-carbazole-alt-5,5-(4',7'-di-2-thienyl-2',1',3'-benzothiadiazole)]   (PCDTBT),   which gave a PCE of 3.6%. Subsequently, others increased this to 6.1% by incorporating a TiOx layer as an optical spacer [1763]. Still later, Liang et al. and Chen et al. [1764–1766] developed a CP comprised of alternating units of thieno[3,4-b]-thiophene (TT) and benzodithiophene (BDT), which yielded conversion efficiencies as

Fig. 37.12 Schematic of construction of a typical CP-based solar cell (After Ref. [1756] reproduced with permission)

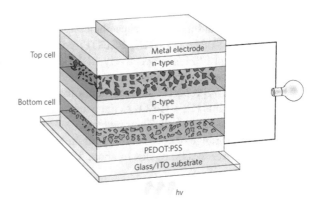

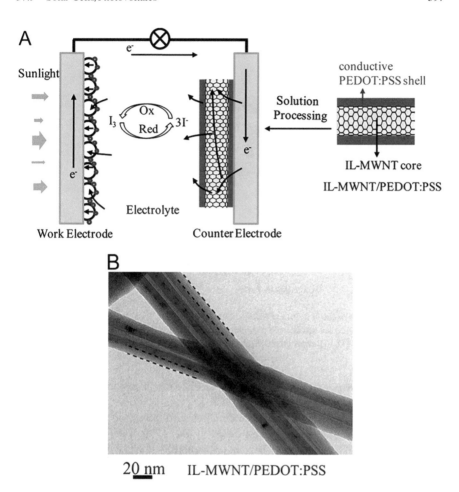

Fig. 37.13 Schematic of typical dye-sensitized solar cell (DSSC), this one based on PEDOT/PSS-coated MWCNTs and utilizing an (I3)-(3I−) electrochemical couple (After Ref. [1757], reproduced with permission)

high as 8%. Subsequent work consistently reported conversion efficiencies of 7% in CP-based solar cells [1767–1773].

Poly(thiophene) derivatives such as P(3-(*hexyl, butyl, or octyl*) thiophene) have been the CPs most commonly used as the *donors* in solar cells, coupled with fullerene (C_{60}) and its soluble derivatives (e.g., [6,6]-phenyl-C61-butyric acid methyl ester (PCBM) and [6,6]-phenyl-C71-butyric acid methyl ester (PC70BM)), as the most common *acceptors* [1774, 1775–1802]. Such solar cells generally have efficiencies only in the 2–6% region. With the conventional or inverted device configurations. [1125, 1186–1189]. Another, commonly used solar cell, which poly[2-methoxy-5-(3',7'-dimethyloctyloxy)-1,4-phenylenevinylene] (MDMO-PPV) as the donor and

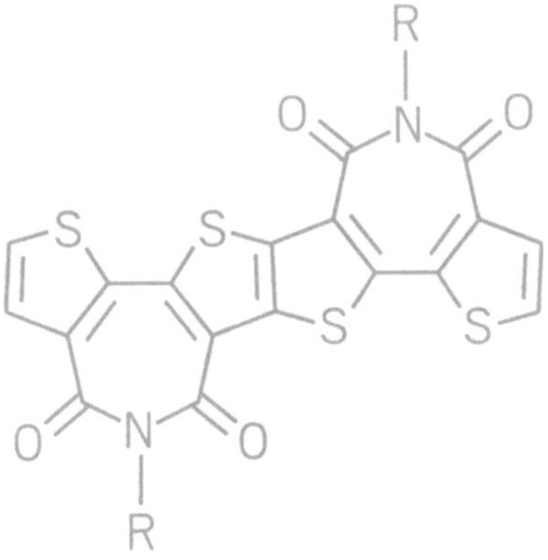

TBI

R = 2-decyltetradecyl
or 2-hexyldecyl

Fig. 37.14 Structure of dithienylthienothiophenebisimide ("TBI"), used as a basis for thiophene-based CPs with potential for use as both p- and n-type semiconductors in photovoltaics, developed by Ritter (After Ref. [1814], reproduced with permission)

and C_{60} derivatives such as PCBM and PC70BM as acceptors generally shows typical efficiencies of <3% [1774, 1780, 1781].

CP-based solar cells have also been coupled with *inorganic* acceptors, such as ZnO, TiO_2, CdS, CdSe, Si, and even CNTs [1794, 1779, 1803–1812]. However, these have generally shown poor efficiencies, generally <3%, with the exception of Si/PEDOT/PSS-based solar cells, which have shown efficiencies as high as 6.35% [1811].

Xia and Ouyang [1813] described PEDOT/PSS (poly(styrene sulfonate)) films with conductivities enhanced from 0.2 S/cm to >100 S/cm via preferential solvation of the hydrophobic PEDOT and hydrophilic PSS with co-solvents, for use in solar cells, although the highest efficiencies obtained were <3%.

In more recent developments (2016), Osaka et al. [1814] developed a new series of thiophene-based CPs with potential for use as both *p*- and *n*-type semiconductors in photovoltaics. Their CPs are based on the dimer dithienylthienothiophenebisimide ("TBI"), depicted in Fig. 37.14 below.

Yoo et al. [1815] described a unique, *thermoelectric* energy device and a Pt-free dye-sensitized solar cell (DSSC), using poly(3,4-ethylenedioxythiophene): poly (4-styrenesulfonate) (PEDOT/PSS)/graphene composites fabricated by in situ polymerization. In the synthesis of the base material, graphene was dispersed in a solution of poly(4-styrenesulfonate) (PSS), and polymerization was directly carried out by addition of 3,4-ethylenedioxythiophene (EDOT) monomer to the dispersion. The content of the graphene was varied and optimized to give high electrical conductivity. The composite solution was ready to use without any reduction process since reduced graphene oxide was used. The fabricated film had a conductivity of 637 S/cm.

The conductive composite films were employed in an organic thermoelectric device, which showed a power factor of 45.7 $\mu W/(mK^2)$, claimed to be 93% higher than a device based on pristine PEDOT/PSS alone. When these PEDOT/PSS−graphene composite films were used in Pt-free DSSCs, they showed an energy conversion efficiency of 5.4%, 21% higher than that of a DSSC based on PEDOT/PSS alone.

37.10 Other Energy Devices

CPs have been cursorily investigated for use in *fuel cells*. For example, P(Py) nanotubes synthesized in alumina templates have been employed in methanol fuel cells as Pt catalyst supports, showing a good electrocatalytic activity and fair stability for the oxidation of methanol [1774, 2059, 2060]. In these, as the Pt loading is increased in the P(Py) anodes, the activity increases from 25.6 mA/cm^2 (Pt = 10 $\mu g/cm^2$) to 302.5 mA/cm^2 (Pt = 140 $\mu g/cm^2$). And tubular Co-P(Py)/MWCNT nanocomposites have been demonstrated as cathode materials for oxygen reduction in polymer electrolyte fuel cells and methanol fuel cells, operating at temperatures ranging from 70 to 90 °C [1774, 2061]. Unique cobalt-porphyrin/P(Py)-nanorod composites have been employed as catalysts for oxygen reduction in fuel cells [1774, 2062]. P(ANi)-CNT composite "nanocables" and P(Py)-CNT "nanocomposites" have been used as the anode material in unique *biological fuel cells* which use *E. coli* [1774, 1816–1818].

37.11 Problems and Exercises

1. Enumerate the principles of operation, using a schematic diagram as an aid and including relevant reactions during charging and discharging, of the following real and hypothetical secondary battery configurations: A Li/perchlorate-liquid-electrolyte/Li$_x$VO$_y$ battery; a Li$_x$C$_6$/solid-electrolyte/Li$_x$MnO$_2$-P(Py); P(Ac)/liquid-electrolyte-(choose any realistic dopant)/P(Ac); P(PP)-Li/liquid electrolyte/P(Py).
2. Identify at least six CPs used in secondary batteries, starting with the most widely used and most successful. Include at least two which could potentially be used as both cathode and anode material.
3. Enumerate the important terms and parameters (cf. Section 37.1) used in defining secondary battery performance, and list comparative values for the following: NiCd, standard Li ion, and at least one battery system using each of the following CPs: P(Py), P(ANi), and P(Ac).

4. What are the continuing problems with Li/CP batteries and which appear to be on the way to solution at the present time? What solutions would you propose for some of these problems? Can you propose a solution for one particular problem, that of high spontaneous (self-) discharge rates?
5. Write down the shorthand configuration–notation for at least one actually studied battery system using each of the following CPs. Also, write out the reactions occurring during charge and discharge at the cathode and anode for them: P(Ac), P(ANi), P(Py), P(T), and P(PP).
6. Carry out the exercise of problem 5 for one example each from two "unconventional" battery systems: the composite–electrode approach of the Allied Signal group and the room-temperature chloroaluminate melt battery described in Sect. 37.6.

Chapter 38
Electrochromics

Contents

38.1   Introduction and Device Types

We must start this chapter by making the important observation that, although there is a surfeit of studies of electrochromic properties of a wide variety of CPs in a laboratory environment, there is a dearth of such studies of *actual, practical* devices. Furthermore, there is an even greater dearth of examples of commercial implementation of CP-based electrochromic devices. The tests in laboratory environments typically involve looking at the spectroelectrochemical characteristics of a CP film deposited chemically or electrochemically on a transparent conductive electrode such as ITO/glass in a liquid electrolyte containing cell. A plethora of such studies have been cited in an earlier chapter, and the reader is referred to these for an appreciation of the multitude of colors, dynamic ranges, and other performance characteristics obtainable with CP-based electrochromics.

Electrochromic devices may be of several types:

1. *Typed by mode of function: reflectance-, transmission-,* and *cumulative*-mode devices.
2. *Typed by spectral region of response: near-UV-visible-only, near-UV-visible-NIR, IR* (infrared), *visible-NIR-IR, microwave–mm-wave,* and *wide spectral.* The last-cited region may potentially encompass all the previous spectral regions cited.
3. *Typed by composition*: *liquid electrolyte* vs. *solid electrolyte (i.e., solid state).*
4. *Typed by electrode mode: two-electrode mode,* the most common, having only an active electrochromic (working) electrode and a counter electrode, and *three-electrode mode* (having an additional reference electrode).

© Springer International Publishing AG 2018
P. Chandrasekhar, *Conducting Polymers, Fundamentals and Applications*,
https://doi.org/10.1007/978-3-319-69378-1_38

The most common electrochromic devices are transmission-mode devices, where incident light passes through the device to an observer on the other side of it. An electrochromic window is the most obvious example. In a reflectance-mode device, incident light is *reflected* from the device to the observer; an electrochromic (e.g., flat-panel) display is the most obvious example. An electrochromic automobile rearview mirror is an example of a fundamentally transmission-mode device modified for reflectance-mode operation. For the want of a better word, "cumulative-mode" device is used to denote one in which transmission, reflectance, and absorption are all important in determining the electrochromic modulation achieved; a new generation of CP-based microwave-energy-modulating devices being developed very recently belongs to this category.

Most electrochromic devices that one envisions, e.g., ones based on P(ANi), modulate the near-UV-visible-NIR spectral region. Very recently, however, a new generation of devices modulating the visible-NIR-IR spectral region in a tailorable manner and in a reflectance mode has been developed. As noted above, very recent work has also seen the development of microwave–mm-wave region modulation by CP-based devices, achieved primarily via conductivity modulation. And most recently, there has been some talk of developing CP-based devices capable of modulating the entire spectrum from the visible through the IR to the microwave regions. Although theoretically possible, in practical terms this may present many difficulties, as the reader will see below.

Electrochromic devices may have liquid electrolytes, which make for rapid switching and sharper color changes due to greater conductivity and ion mobility. However, liquid electrolytes also give rise to faster degradation, due to the physical transport of ions involved, and problems with physical containment and operating temperatures. Solid electrolytes do away with some of the latter problems, but necessarily yield slower devices with less distinct color changes.

All the electrochromic performance parameters cited in an earlier chapter are valid equally for electrochromic devices as for laboratory-cell electrochromic systems: *dynamic range, switching time, cyclability, and open circuit memory.* The reader is referred to this chapter for further reference and definitions.

For practical reasons, most electrochromic devices have a two-electrode mode of operation, with a working electrode containing the active electrochromic material, and a counter electrode, sometimes also containing an active electrochromic. A third reference electrode may be added in devices according to prescribed electrochemical practice [1819]. This yields greater applied potential control at the working electrode, but it also allows the counter electrode to experience more extreme potentials than in a two-electrode mode, which may be detrimental in many situations and is more cumbersome overall.

As will be seen in the sequel, an important factor neglected in the construction of many electrochromic devices is the provision of an adequate counter electrode reaction which would permit reasonable faradaic reversibility of the electrochemical system. Frequently, the only available counter electrode reaction is redox of impurities such as O_2, H_2O, and trace metal ions, leading to poorer cyclability.

In the sequel, we discuss the subjects touched upon in this Introduction in more detail.

38.2   Visible-Region Devices

38.2.1   Structure

The visible-region electrochromic device is the most common electrochromic device based on CPs. A very common, logical, and nearly universal structure for transmission-mode devices is the sandwich structure depicted in Fig. 38.1. Both working and counter electrode substrates are necessarily transparent and conductive: most commonly indium tin oxide (ITO) on glass or NESA glass, less commonly a thin ($<$100 nm), nearly transparent layer of a metal such as Au or Pt on glass. The active electrochromic electrodes, a thin film of CP, comprising the working electrode and the counter electrode, face each other through the medium of an electrolyte. As noted above, in most constructions, the counter electrode is without any electroactive material, and no provision is made for a defined counter electrode reaction to balance the redox of the electrochromic material at the working electrode. With provision for electrical connections, the entire device is sealed, most simply with epoxy or silicone adhesive.

For reflectance-mode operation, the simple expedient of inserting a mirror behind the entire apparatus may be used, or a more complex and unique structure with the working electrode facing the incident light may be used, as described in many issued patents and patent applications [1820–1824].

It is noted that for both transmittance- and reflectance-mode operations, the performance of CP devices is not affected by viewing angle, an important advantage over their closest competitors, liquid crystal devices.

Liquid electrolytes used may be aqueous, for example, aqueous acidic (e.g., 0.2 M HCl or H_2SO_4) electrolytes used typically for P(ANi), or they may be nonaqueous, for example, propylene carbonate (PC) or acetonitrile (ACN) (e.g., 0.2 M $LiClO_4$ in PC, with ClO_4^- the dopant) used typically for P(Py) or P(3MT).

In addition to the possible provision of a non-electrochromic counter electrode reaction, several studies have incorporated an electrochromic counter electrode to supplement the electrochromism of the CP. These have a complementary color change (they darken when the CP lightens, etc.). Examples of this include Prussian blue and WO_3 glasses; the latter use the cation of the dopant electrolyte, e.g., Li^+ of $LiClO_4$, in the electrochromic reaction.

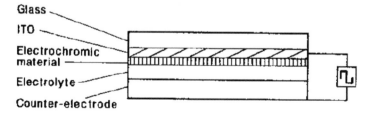

Fig. 38.1 Schematic of typical transmission-mode electrochromic device (After Ref. [948], reproduced with permission)

38.2.2   Function of Devices: Laboratory vs. Actual Devices

Typical electrochromic color changes in the visible region for CPs have been cited in an earlier chapter. We may briefly recapitulate *some* of these:

- *P(ANi)*: nearly colorless (reduced, de-doped state, negative applied potential) ↔ green (intermediate state) ↔ dark green–blue-black (oxidized, highly doped, state, + applied potential)
- *P(Py)*: light yellow (- appl. potl.) ↔ dark blue–black (+ appl. potl.)
- *P(3MT)*: light orange-red (- appl. potl.) ↔ dark blue (+ appl. potl.)
- *Poly(isothianaphthene)*: dark blue–black (- appl. potl.) ↔ very light blue (+ appl. potl.)

Indeed, it may be noted that one of the advantages of CPs in electrochromic devices as compared to such materials as liquid crystals is the wider intrinsic (i.e., filter-less) color range obtainable.

Yang and coworkers [955, 1825–1827] performed detailed studies of the electrochromism of P(ANi) in liquid and solid electrolytes, with polymeric and other dopants, and in multilayer systems. They monitored effects of variables such as pH, film thickness, multilayer configuration, and combination of dopants. Again, however, few of these studies, at least in the public domain, described actual, working devices, as opposed to laboratory studies in electrochemical cells.

The gap between laboratory studies and actual devices is apparent in high-performance data obtained in the laboratory not always reproducible in devices. For example, Lacroix et al. [983] observed 100 µs "switching times" (not clearly defined, Fig. 38.2) for 120 nm thick P(ANi) films in liquid aqueous H_2SO_4, which however have not been duplicated in sealed devices to date. Similarly, very high cyclabilities, claimed up to 10^6 cycles, observed by Kobayashi et al. [1828] or by Foot and Simon [1829] for fairly thick P(ANi) films, could not be duplicated in devices, whether fabricated with liquid or solid electrolyte, e.g., those of Akhtar et al. [1830a].

38.2.3   Examples of Functional Devices

We may now briefly discuss some studies of actual, fully fabricated, CP-based electrochromic devices.

Akhtar et al. [1830a] were one of the first to describe completely assembled, sealed, solid-state electrochromic devices based on CPs. In one set of devices, the fairly common Li-triflate/poly(ethylene oxide) (PEO)/acetonitrile formulation for nonaqueous solid electrolytes was used. However, in another set, the unique combination of poly(ethyleneimines) of different MWt and protonic acids such as hydrochloric, sulfuric, phosphoric, acetic, and poly(styrene sulfonic) was used. Additionally, the films of the CP, P(ANi), were prepared electrochemically as

Fig. 38.2 Data showing
very rapid switching times
claimed for P(ANi) devices
by LaCroix et al. (After Ref.
[983], reproduced with
permission)

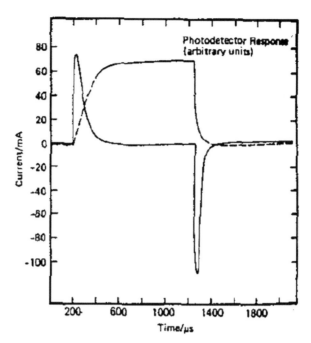

well as by sublimation, and in one set of devices, Fe tungstate was used as a counter
electrode to provide a definitive counter electrode reaction (lithiation). While
cyclabilities to several thousand cycles were claimed, the electrochromic dynamic
range and other parameters were fairly poor, as seen in Fig. 38.3. Very rapid
switching times have been claimed for many P(ANi)- or P(ANi)-derivative-based
devices. For example, Ram et al. [1830b] claimed a 143 ms switching time for
liquid electrolyte devices based on poly(aniline-*co-o*-anisidine).

A number of electrochromic devices have been fabricated and studied which
include a CP working electrode and a counter electrode which is also
electrochromic, coloring in a complementary way to the working electrode. The
most common example of this is that using inorganic electrochromics based on
metal oxides, e.g., WO_3, which go from colorless to colored (blue for WO_3) upon
reduction, complementary to the colored to colorless transition of *p*-type CPs used
as active electrochromics. Toyota Motor Co. was one of the first to study complementary P(ANi)/WO_3 systems for rearview mirrors, with their first systems having
a liquid electrolyte (LiClO$_4$/PC), switching times of the order of seconds, dynamic
ranges of 50% (30–80 %T) in most of the visible region [181–1832], and claimed
cyclability to 10^5 cycles. These devices typically had dimensions of 5 cm × 5 cm,
contained a 0.6 μm P(ANi) film, and had a 0.1 mm-thick liquid electrolyte layer.
Kim et al. [950, 1833] described an entirely solid-state device based on a P(ANi)-*N*-
Bu-sulfonate anode and WO_3 cathode using solid electrolytes based on photo-cured
PEG methacrylate + tripropylene glycol diacrylate + LiClO$_4$. Cyclabilities up to

Fig. 38.3 Data for aqueous
P(ANi) electrochromic
devices of Akhtar et al.
(After Ref. [1830b],
reproduced with
permission)

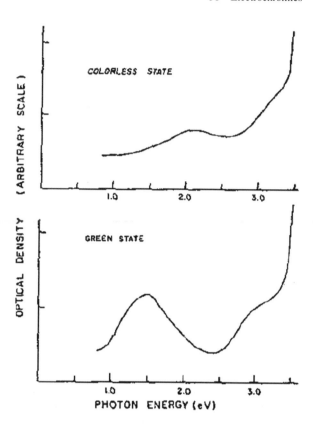

2,000 cycles were claimed. Figure 38.4 shows time drive (%T as function of time and applied voltage) data for one of these devices.

In a similar vein, Leventis and Chung described a novel electrochromic device [1834] which used a P(Py)/Prussian blue (PB) "composite" as the working electrode and, optionally, a complementarily coloring poly(viologen) electrode as the counter electrode. A liquid, commonly aqueous, electrolyte was used. It was found that the P(Py)/PB composite appeared to have enhanced cyclability as compared to either of its components. Figure 38.5 depicts the spectroelectrochemical characterization of the complementary P(Py)/PB-poly(viologen) system. While the dynamic ranges were very substantial throughout the visible-NIR region, the switching times were greater than ca. 5 sec for all systems studied. A similar system, but with the PB incorporated as a complementarily coloring counter electrode and using a solid electrolyte, was described in an earlier patent [1835].

Several devices incorporating P(ANi) as a dispersion in a polymer matrix have been described [1836]. Several reflectance-mode devices incorporating CPs have also been described, some even including such environmentally unstable CPs as poly(acetylene) and reductively coloring ones such as poly(isothianaphthene) [1836–1839]. Nissan Motor Co. and several other manufacturers have described

Fig. 38.4 Electrochromic data for solid-state P(ANi) device (After Ref. [950], reproduced with permission)

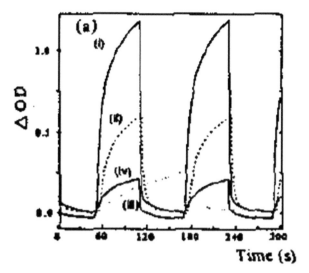

Fig. 38.5 Electrochromic characterization for P(Py)/ Prussian blue/poly (viologen) system (After Ref. [1834], reproduced with permission)

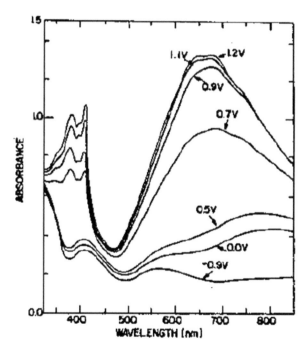

electrochromic devices capable of being incorporated into automobiles in rearview mirrors, displays, or windshields [1840].

In an early work, Corradini et al. [1841] described ca. 2 cm × 1 cm devices based on P(3MT) in which the liquid electrolyte (2 mm thickness) was $LiClO_4$/PC and the

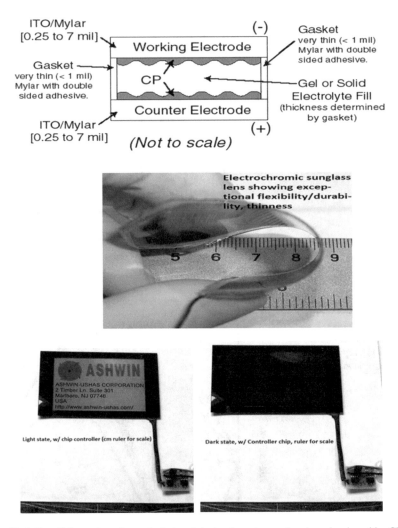

Fig. 38.6 *Top:* Schematic and actual photo of electrochromic sunglass lens developed by Chandrasekhar et al. [1845] *Bottom:* Demo lens showing high light/dark contrast

counter electrode reaction was presumed to involve Li-ion insertion. Charge capacities of ca. 7.5 mC/cm^2 were reported. Arbizzani et al. [1842] described a similar device, with Li-ion insertion into Li/NiO/ITO as the counter electrode reaction. Color changes of purple to very pale blue were claimed. Figure 38.6a, b shows, respectively, the spectroelectrochemical characterization of one of their devices and its spectroelectrochemistry at 530 nm.

In a US patent [1843], Wolf et al. of AlliedSignal, Inc., Morristown, NJ, USA, provided a detailed description of an "adjustable tint" thermal window incorporating an air gap and a CP.

ITO deposited on flexible, plastic substrates such as PMMA or polycarbonate was used as the conductive electrode substrate, with the active electrochromic electrodes producible as a roll which could be attached to window panes with common (e.g., cyanoacrylate) adhesives. This device again optionally used a counter electrode which was also electrochromic, with the difference that it could be not only a metal oxide such as WO_3 but also, interestingly, an n-type CP, which of course displays electrochromism which is complementary to that of the more common p-type CPs. Thus, as cathode materials, the p-type CPs P(ANi)s, P(Py)s, and poly(phenylene vinylene) were listed as usable, with virtually all the common dopants. As anode materials, WO_3, MoO_3, poly(isothianaphthene), and the n-type CPs poly(alkoxythienylene vinylene), poly(p-phenylene), poly(phenyl quinoline), and poly(acetylene) were listed as usable. Liquid nonaqueous electrolytes based on common solvents such as DMSO and THF were used. No electrochromic data were however given in the patent or in subsequent publications.

Chandrasekhar et al. [999, 1844, 1845] described a number of sealed, liquid and gel electrolyte devices based on poly(aromatic amines) similar in structure to P (ANi) and poly(diphenyl amine), operating in two- and three-electrode mode. Color changes were from glass-clear transparent to dark green–blue-black. Figure 38.7a–c shows spectroelectrochemical characterization data for some of these. Figure 38.8a, b shows spectroelectrochemistry data for solid-state devices. (Figs. 38.9 and 38.10, Table 38.1)

The performance of the Chandrasekhar et al. electrochromic lenses cited above may also be compared with the recent work of Ma et al. [1846], who reported an electrochromic device based on the CP poly($(CH_3)_2$-Bz-ProDOT) as the cathodically coloring material (synthesized earlier by Welsh et al. [1847]) and the metal oxide V_2O_5 as the anodically coloring material. They subsequently produced lens-shaped devices and, eventually, retrofitted them into a pair of sunglasses. The comparison with the Chandrasekhar electrochromics can be briefly summarized as follows:

(1) Due to the use of a metal oxide electrochromic as the anodically coloring material, the stability, environmental durability, and shelf-life of their devices are significantly compromised.
(2) Ma et al.'s "transparent" state still has a bluish tinge.
(3) Ma et al.'s light/dark %T spectra of their devices have a large element in the near-IR region (>700 nm) and very little contrast through 500 nm, indicating they are *not* optimized for human vision (400–700 nm).
(4) Ma et al.'s switching times are significantly slower (Figs. 38.11, 38.12, and 38.13).

MacDiarmid et al. [1676] described the deposition of very thin (ca. 7 nm) films of CPs such as P(Py) and P(ANi) in situ on substrates such as poly(ethylene terephthalate) (PET). The deposition was based on a "wetting" of the substrate by the aqueous deposition solution, e.g., $FeCl_3$ + pyrrole + anthraquinone-2-sulfonate (dopant). The films had surface resistivities of ca. 7500 Ω/square. These were then

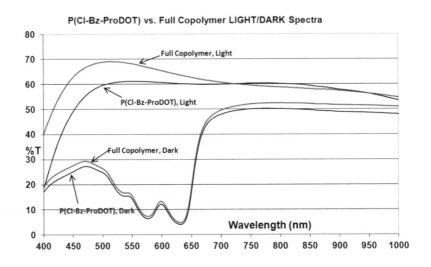

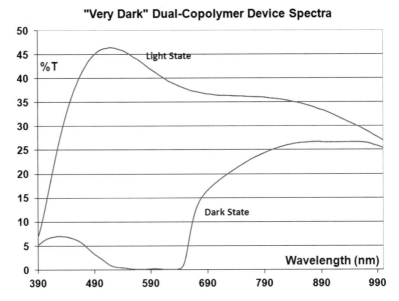

Fig. 38.7 Electrochromic characterization data for the electrochromic sunglass systems developed by Chandrasekhar et al. [1845]. Note that these are actual, "WYSIWYG" (what you see is what you get) data, against air reference rather than substrate reference. Thus, e.g., the substrate (ITO/PET) itself shows a maximum ca. 85% transmission throughout the spectral window shown

used as conductive substrates for polymer dispersed liquid crystals (PDLCs) which functioned as the active electrochromic material; i.e., they were PDLC displays. This technology has since been implemented by a small company based in the USA. MacDiarmid et al. [1676] further refined this technique by generating hydrophilic and hydrophobic micropatterns on PET substrates using a PC-controlled laboratory

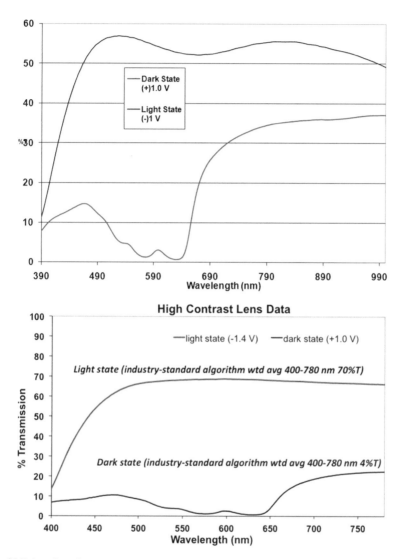

Fig. 38.7 (continued)

plotter. When the substrate treated in this fashion was repeatedly dipped in the aqueous CP-deposition solution, only the hydrophilic areas were wetted and received a CP deposit, yielding a micropattern (line width ca. 70 μm). When this substrate was then coated with a PDLC, the result was a PDLC display showing a computer-generated micropattern.

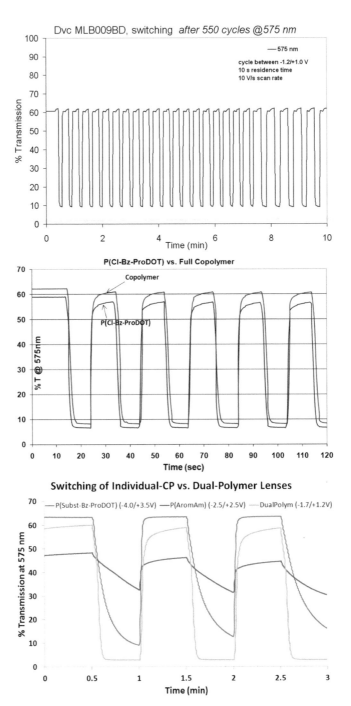

Figs. 38.8 Light/dark switching data for the electrochromic sunglass systems developed by Chandrasekhar et al. [1845]. Note that these are actual, "WYSIWYG" data, against air reference rather than substrate reference. Thus, e.g., the substrate (ITO/PET) itself shows a maximum ca. 85% transmission throughout the spectral window shown

Dual-Polymer Lens,
  Lightest State
  L*= 72 a*= -6, b*=3
                b*=-30

Dual Polymer Lens,
  Darkest State
  L*= 14 a*= 12,

Single-CP Lens, P(Subst-BzProDOT)
                              BzProDOT),
Lightest State
L*= 77 a*= -3, b*=2

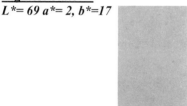

Single-CP Lens, P(Subst

Darkest State
L*= 29 a*= 24, b*=-62

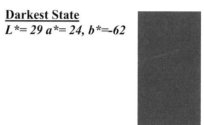

Single-CP Lens, P(AromAmine)
Lightest State
L*= 69 a*= 2, b*=17

Single-CP Lens, P(AromAmine),
Darkest State
L*= 41 a*= -1, b*=10

Fig. 38.9 Representative results for L*,a*,b* values (CIE coordinates, i.e., color profile), and actual colors, for the individual CPs and for dual-polymer and single-CP lenses. After Chandrasekhar et al. [1845]

38.3 IR-Region Devices

For purposes of discussion, we denote IR-region devices as including those operating in broader spectral regions. These include, e.g., visible-IR or IR-microwave regions, encompassing the mid-wave and long-wave (MWIR, LWIR) regions, ca. 2.5–28 μm.

The primary driver behind the development of IR-region devices has been the military need for dynamic (switchable) countermeasures against visible/IR sensors and/or emulation of random noise background in the IR. This need exists for all (land, sea, air vehicles and personnel) components of a modern armed force operating at night. Secondary drivers have included spacecraft thermal control (a spacecraft needs to have very low IR-region solar absorptance when facing the sun and high emissivity when away from the sun, for efficient thermal control); cockpit displays which combine day/night vision for personnel viewing through

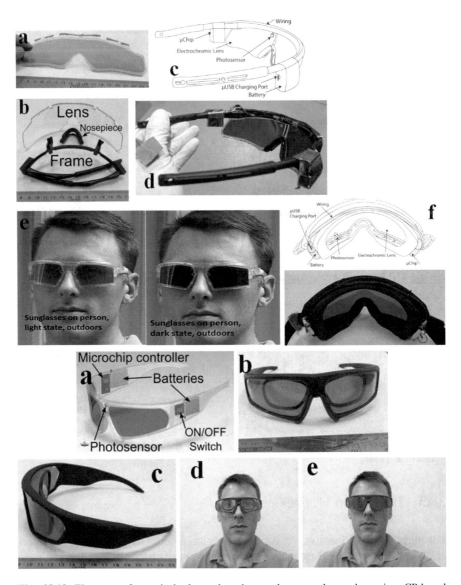

Fig. 38.10 Elements of practical electrochromic sunglasses and goggles using CP-based electrochromics (After Chandrasekhar et al. [1845])

Table 38.1 *Haze* measurements on electrochromic lenses of Chandrasekhar et al. [1845]. Haze is an important property for practical, commercial sunglasses. It should ideally be <2%

Lens#	Light state %T	Light state haze	Dark state %T	Dark state haze	Δ%T
S1D109ED	58.3%	0.98%	12.0%	1.5%	46.3%
S1D109FD	57.0	1.06%	9.0%	1.41%	48.0%

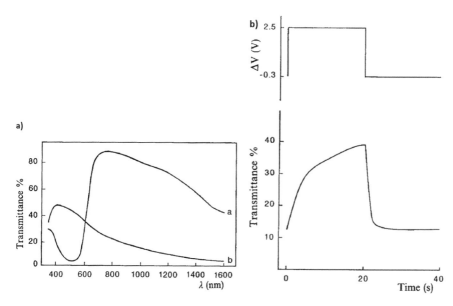

Fig. 38.11 Spectroelectrochemical (electrochromic) characterization (**a**) and time drive (%T as function of time and applied voltage) data (**b**) for poly(3-Me-thiophene) electrochromic device having a complementarily coloring NiO/ITO counter electrode (After Ref. [949], reproduced with permission)

bulky night vision helmets that are difficult to remove; and dynamic microwave absorption (8–40 GHz, X, Ku, and Ka radar bands).

Due to the necessities of construction, most IR-region devices operate in the *reflectance mode*. Before the advent of CPs, painfully few technologies were available that were capable of dynamic (switchable) electrochromism in the IR region. Liquid crystals and materials such as WO_3 are inactive in the IR. Phase change (electro-thermochromic) materials, e.g., poly(vinyl stearate) and poly (octadecene), which show a crystalline–amorphous transition accompanied by electrochromism, and electroluminescents, afforded very poor dynamic range. True dynamic IR-electrochromics, e.g., the semiconductor device described by Daehler [1848], were radiators with high power requirements rather than modulators, although designs for reflectance-mode visible-region devices which could be adapted to the IR region abounded [1821, 1822].

Bennett et al. [1820] described a CP-based electrochromic device using a unique, reflectance-mode configuration. In this, the counter electrode was placed behind the working electrode, which comprised of a CP deposited on a metal and in turn deposited on a porous substrate. Furthermore, both counter and working electrodes were fabricated from the same CP, giving the device high faradaic reversibility and thus high cyclability. The cyclability and dynamic range of this device were nonetheless less than desirable. CPs that were primary candidates in the device included P(ANi) and P(Py). Both working and counter electrode

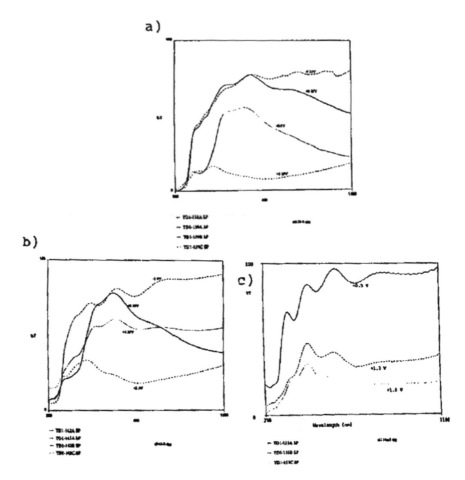

Fig. 38.12 Spectroelectrochemical (electrochromic) characterization data for *early-generation* poly(aromatic amine) electrochromic device. (**a**), (**c**) Three-electrode mode; (**b**) two-electrode mode (After Chandrasekhar [999, 1844])

substrates were flexible, leading to a very thin (< 5 mm), flexible, device that could be draped over odd-shaped and variable area objects. The end application envisioned was military camouflage, as described earlier.

Chandrasekhar et al. [1849] first showed that significant, dynamic IR modulation could be achieved with CPs. The modulation resulted from a combination of changing CP and dopant absorptions in the IR region of interest and morphological changes upon doping/de-doping along the dimensions of the IR wavelengths of interest. They used unique poly(aromatic amines) combined with unique dopants [1824], which were shown to have superior response to conventional CPs such as P (Py) and P(ANi), and several unique device designs, allowing for very efficient reflectance-mode operation.

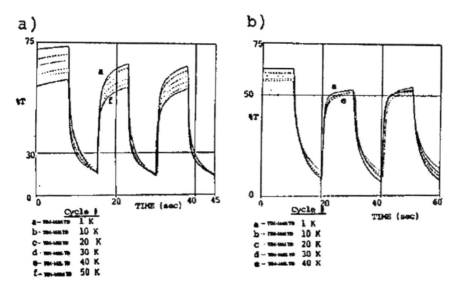

Fig. 38.13 (**a, b**) Time drive (%T as function of time and applied voltage) data for *early-generation* devices of cycle numbers as shown Fig. 38.7 (After Chandrasekhar et al. [999, 1844])

Subsequent to the above-cited work, Chandrasekhar et al. further refined their IR-region electrochromics [968]. Figure 38.14 summarizes some of these newer data. It is important to note that the IR-electrochromics are also designated *variable-emittance* materials, since variation of the IR signature causes variation of the thermal property denoted as *emittance* (which is emissivity integrated over the approximately 2–45 μm IR spectral region) (Fig. 38.15, Table 38.2 and 38.3).

Figures 38.16, 38.17, 38.18, 38.19, and 38.20 show some of the spectroelectrochemical characterization data for Chandrasekhar et al.'s reflectance-mode devices. Figures 38.16 and 38.17 show, respectively, specular and diffuse reflectance data. Figure 38.18 shows data for an optimized system with results of emittance calculation indicated. Figure 38.19 shows comparative data obtained with common CPs other than those used by Chandrasekhar et al., but using one of Chandrasekhar et al.'s unique device configurations; Figure 38.20 shows data obtained with one of their best CPs but with an alternative, undefined configuration; these two sets of data show clearly that both the uniqueness of the poly (aromatic amine) systems used and the uniqueness of the device design are instrumental in producing the significant dynamic ranges and other improved electrochromic performance parameters that Chandrasekhar et al. obtained

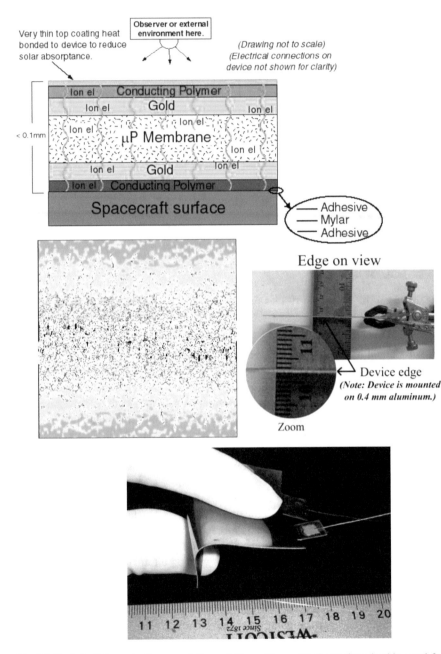

Fig. 38.14 *Left:* Schematic diagram of the variable-emittance IR-electrochromic skin, used for spacecraft thermal control. *Right:* Detail showing that the Au and CP layers are also "porous" and fibrillar in nature, rather than solid surfaces (as in the schematic at left). *Bottom:* Actual photos of the skins (After Chandrasekhar et al. [968])

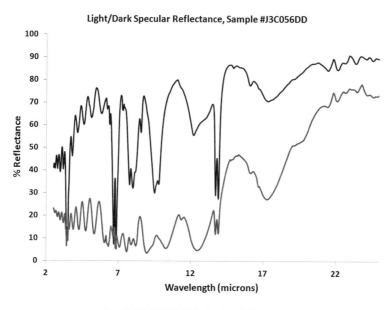

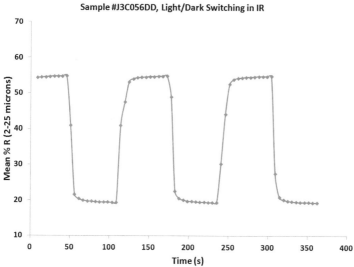

Fig. 38.15 *Top:* Light/dark specular FTIR spectra for the variable-emittance IR-electrochromic skins shown in the previous figure. *Bottom:* Corresponding light/dark switching data (After Chandrasekhar et al. [968])

Table 38.2 Summary data of emissometer-based emittance measurements for various variable-emittance IR-electrochromic ("devices"). Those with and without the a solar absorptance (α(s))-reduction coating are compared, to demonstrate the (minimal) effect of this coating on emittance function (After Chandrasekhar et al. [968])

Device (sample) #	Light ε	Dark ε	$\Delta\varepsilon$
Skins with alpha(s)-reduction coating			
J3C056DD	0.325	0.771	*0.446*
J3A038FD	0.298	0.784	*0.486*
J3A038CD	0.234	0.676	*0.442*
J3A174BD	0.389	0.646	*0.427*
Skins without alpha(s)-reduction coating			
JB_011AD	0.257	0.771	*0.514*
JB_042AD	0.237	0.751	*0.514*
JA_168FD	0.335	0.835	*0.500*

Table 38.3 Space durability/stability data for V-E devices (skins) subject to continuous electrochromic (light/dark) cycling while also thermally cycled between ca. -20 and $+50\ ^{\circ}$C, all in space vacuum (10^{-7} Torr, 10^{-5} Pa), for a period of 212 days (ca. 7 months) (After Chandrasekhar et al. [968])

Sample #	ε's, pretest			ε's, posttest, 64 days			ε's, posttest, 212 days		
	Light ε	Dark ε	$\Delta\varepsilon$	Light ε	Dark ε	$\Delta\varepsilon$	Light ε	Dark ε	$\Delta\varepsilon$
J3A056AD	0.363	0.720	*0.357*	0.397	0.724	*0.327*	0.420	0.747	*0.327*
J3A056ED	0.249	0.613	*0.364*	0.241	0.587	*0.346*	0.291	0.648	*0.357*

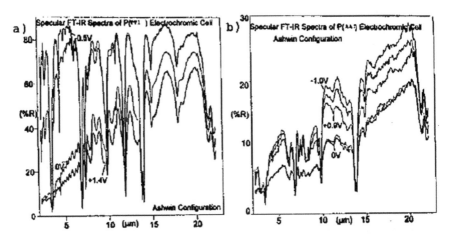

Fig. 38.16 Specular reflectance FTIR electrochromic data, in 2nd Ashwin® configuration for *early-generation* solid-state, thin-film, flexible devices of (**a**) P(TT1), a thiophene-derivative CP, and (**b**) P(AA1), a poly(aromatic amine) (After Chandrasekhar et al.)

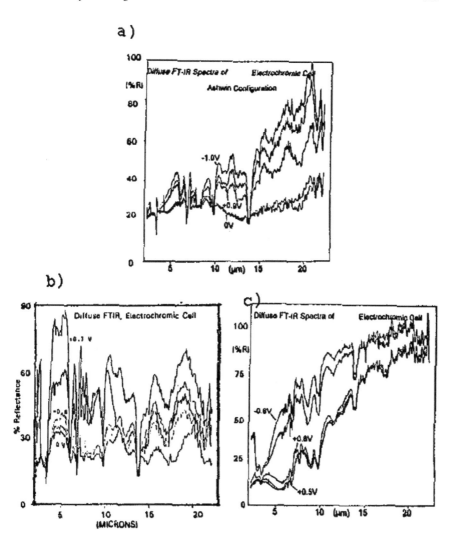

Fig. 38.17 (a–c) Diffuse reflectance FTIR electrochromic data for *early-generation* solid-state, thin-film, flexible devices of P(AA1), a poly(aromatic amine), in three different device configurations, CON 1/2/3 (a through c) (After Chandrasekhar [1850])

38.4 Other Spectral Regions

Of the other spectral regions, that receiving most attention has been the microwave–mm-wave region encompassing the 8–40 GHz region in which the militarily significant X, Ku, and Ka radar bands lie. Unfortunately, much of the work has remained highly classified, unpublished, and undisseminated.

In order to achieve dynamic, microwave region modulation, the primary parameter varied is the conductivity of the CP. In addition to this basic parameter,

a)   b)

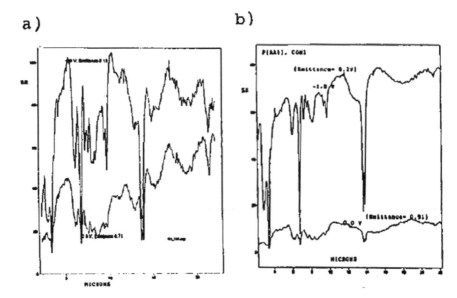

Fig. 38.18 (**a, b**) Diffuse reflectance FTIR electrochromic data, with emittance computations shown, for optimized, *early-generation* P(AA1) system in CON1 device configuration. For the emittance computation, 8 points are taken from these spectra, and 2 points beyond the ca 25 μm limit shown are extrapolated (After Chandrasekhar [1850])

however, other considerations include the geometry of the object doing the modulation, important for radar cross section (RCS) control; the thickness of the CP and dimensions of the underlying conductive electrode grid, important in determining resonance; and whether the modulation is broadband or only for a narrow range of frequencies. Some of these factors were discussed briefly in an earlier chapter, to which the reader is referred.

One of the groups that have been active in dynamic microwave–mm-wave electrochromics using CPs is that of Stenger-Smith et al. at the US Army's China Lake, CA, facility [1349]. From what is known in the public domain, one of the approaches they have used is to lay down a conductive metallic grid (e.g., of Au) on a "low-loss" substrate such as Teflon and then deposit the CP onto this grid. The spaces between the grid then electrochromically modulate between conductive and non-conductive states (the placement of the counter electrode is a proprietary secret). The dimensions of the grid determine resonance and broadband effects. These thin devices can be stacked for greater effect, possibly with interleaving low-loss materials to further break resonance effects and ensure broadband character. It is evident from this that device design is a major consideration in this spectral region. Chandrasekhar et al. [1851] have also been active in this field, but with quite different device designs.

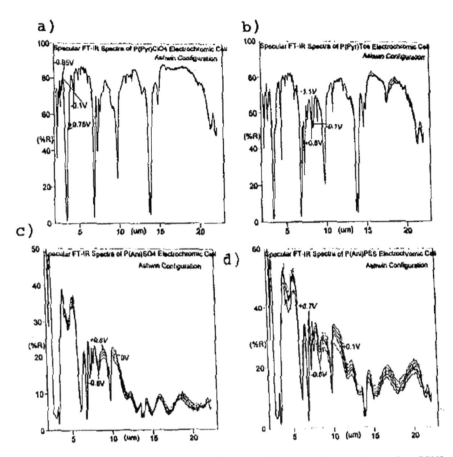

Fig. 38.19 (a–d) Comparative specular reflectance FTIR electrochromic data, using CON2 Ashwin® device configuration, for *early-generation* devices, but with "common" CP systems, as indicated. These data show that choice of the CP system is critical for IR-region electrochromism (After Chandrasekhar)

38.5   Problems and Exercises

1. Describe and differentiate the following terms: reflectance-, transmission-, and cumulative-mode device, wide-spectral and solid-state device, and two-electrode- and three-electrode-mode device. Give advantages and drawbacks of each.
2. Give detailed constructions and one example for each of the device types given in problem 1.
3. Redefine the following terms as they pertain to electrochromic devices: dynamic range, switching time, cyclability, open circuit memory, optical memory, and broadband.

Fig. 38.20 Comparative specular reflectance FTIR electrochromic data for *early-generation* P(AA1) devices, but with an unrefined configuration that permits electrolyte in front of the active electrochromic surface. Compare with Figs. 38.15 and 38.16 (After Chandrasekhar)

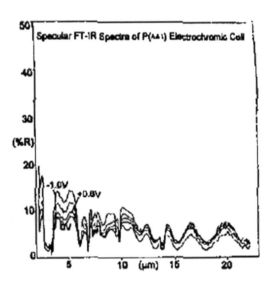

4. Draw out the most common transmission-mode and reflectance-mode electrochromic device constructs. Investigate optimized placements of a third reference electrode in each and alternative placements of the counter electrode.

5. What are the factors limiting cyclability, switching time, and dynamic range in CP devices with liquid electrolyte? With solid electrolyte? How may these be ameliorated?

6. What are the constraints that would limit performance when a laboratory CP electrochromic system is transitioned to a device? How may these be avoided?

7. Describe the construction and function of two examples of electrochromic devices having a complementarily coloring counter electrode; one of these examples should be a CP. What are the redox potential constraints in such devices?

8. Using information available in this chapter and in the public domain, what are the unique device design features required for efficient reflectance-mode operation of IR-region electrochromic devices based on CPs?

9. The placement of the counter electrode in microwave–mm-wave devices based on CPs has been described as one of the crucial design aspects in function of such devices. Where and how would you place the counter electrode(s) in such devices which consist of a stack of 20 thin-film CP working electrodes?

10. Describe the fundamental principles underlying the electrochromic effect of CPs in the visible, IR, and microwave spectral regions. Is there any conflict with respect to these between the different spectral regions?

Chapter 39
Displays, Including Light-Emitting Diodes (LEDs) and Conductive Films

Contents

39.1 Introduction

One of the most attention-grabbing developments in the field of CPs in the early years of CP research was the discovery, first reported by the Cambridge, England, group [1852], that CPs such as poly(phenylene vinylene) (P(PV)) could be used as the emissive layer in *light-emitting diodes (LEDs)*. One of the keys to this discovery was improvements in synthesis and processing of P(PV) in its undoped state that allowed production of CP fairly free of defects which quenched luminescence. Since then, however, this excitement has been somewhat tempered with the realization that, as with all new technologies, practical and commercial implementation still has a number of hurdles before CP-LEDs can begin to take on more established technologies in flat-panel displays for computer and electronic uses, the primary targeted application at present. This is in spite of constant new findings, such as the recent report of CP-LEDs powered by CP FETs based on poly(3-hexylthiophene) from the Cambridge group [1853], in which the performance was claimed to rival that based on inorganic, Si-based semiconductors.

© Springer International Publishing AG 2018
P. Chandrasekhar, *Conducting Polymers, Fundamentals and Applications*,
https://doi.org/10.1007/978-3-319-69378-1_39

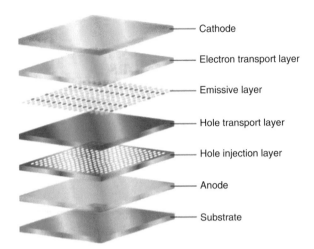

Cathode

Electron transport layer

Emissive layer

Hole transport layer

Hole injection layer

Anode

Substrate

Fig. 39.1 (**a, b**) Schematic of a typical OLED display, e.g., as used in TVs. The substrate can be flexible, e.g., plastic, or rigid, e.g., glass or metal. With a voltage applied across the OLED, current flows from the cathode to the anode. This adds electrons to the emissive layer and creates holes at the anode. At the layers' boundary, electrons and holes combine to give off light. Electron and hole injection and transport layers can be added to further modulate electron movement (After Ref. [1861])

Organic-based LEDs, i.e., *OLEDs*, are not new. Those based on such materials as anthracene, naphthalene, and pyrene, with Ag, liquid, alkali metal, or even ITO electrodes, have been known for several decades [1854–1858]. For example, Fig. 39.1a, b shows the schematic structure and typical luminance data for organic LEDs based on perylene-doped poly(vinyl carbazole) (P(VCz)) observed by Kojima et al. [1859] recently. Indeed, new developments in processing of non-CP organic LEDs have made them more competitive [1860]. Figure 39.2a, b shows a schematic of a typical OLED display.

However, the high voltages frequently needed for single crystals (sometimes as low as 10–20 V as in the above data, but frequently as much as 50–100 V), and low efficiencies in these LEDs, were detrimental to their practical and commercial implementation. The attractiveness of CPs has been that, besides requiring low voltages, of the order of 2– 5 V DC (and recently exhibiting AC operation, see below), they are processible in their undoped state in polymer form into odd shapes/ sizes and films. They are relatively inexpensive and exhibit comparatively higher quantum efficiencies, of the order of 0.01% (initially for the first P(PV) device) to nearly 4% (claimed recently). Their emission colors are possibly "tunable," through structural changes in the CP backbone, combination of multiple CPs, or several other methods, throughout the visible and even NIR region.

An initial, strong hope of commercial potential of CP-LEDs as replacements for LCDs in displays was the initiation of development and commercialization plans by companies such Dow Chemical, DuPont, Hoechst, and Philips [2063]. These companies had collaborative agreements with Cambridge Display Technology, Ltd.,

a)

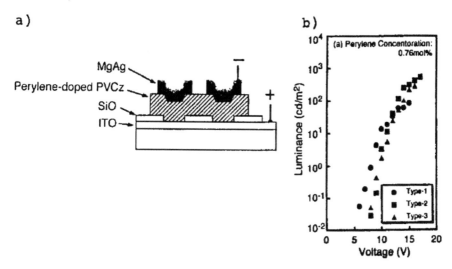

MgAg

Perylene-doped PVCz

SiO

ITO

b)

(a) Perylene Concentoration:
0.76mol%

Luminance (cd/m²)

Type-1
Type-2
Type-3

Voltage (V)

Fig. 39.2 Schematic of construction (**a**) and luminance data (**b**) for P(VCz)-LED, a non-CP-based organic LED (After Ref. [1859], reproduced with permission)

Cambridge, England, the commercial offshoot from the Cambridge group which holds most of the patents on the technology. In the initial years (in the early 2000s), these companies forecast a market of near US$ 10 billion/year if LEDs were able to successfully transplant LCDs in displays. However, it is important to note that, as of this writing (2016) *these hopes of commercial success of CP-LEDs have not been borne out at all: today, CP-LEDs comprise a miniscule proportion of the commercial LEDs (including organic LEDs) market.*

Since the early work cited above, a very wide variety of CP-based LEDs have been studied. Examples include LEDs based on P(ANi), P(Py), and PEDOT nanofibers/nanotubes [1409, 1862–1865] and field emission displays based on the same polymers [1409, 1866–1869]. μm- and nm-size LEDs based on PEDOT nanowires were reported first by Granström et al. [1862].

39.2   Principles of CP-Based LEDs

Figure 39.3a shows a schematic of the simplest CP-based LED, the original device based on P(PV) described by the Cambridge group [1852], while Fig. 39.3b shows a slightly more complicated one, based on two layers of different CPs. As shown in the schematics of Fig. 39.1, normally, one electrode needs to be semitransparent, e.g., ITO/glass, so that the electroluminescence is visible. The function of CP-LEDs is intimately associated with properties such as electroluminescence (EL), upon which the LED's function is based, photoluminescence (PL), and photo-induced absorption (PIA), and the reader is referred to elsewhere in this book for a quick

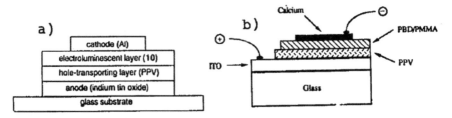

Fig. 39.3 Simple design of one of the first CP-LEDs (left) and variant thereof (After Ref. [1369, 1870], reproduced with permission)

review of these. Indeed, it goes without saying that it is necessary to characterize these properties for any new CP in detail before attempting to construct an LED with it. Besides these properties, other information about the CP highly useful for designing an LED is its equilibrium redox potential or "work function," obtainable from electrochemical data such as voltammetry. Although excitons are still poorly understood, they are thought to be localized within a few repeat units of a single polymer chain, exactly like polarons.

An LED functions on the principle of *double charge injection*. That is, through the medium of an applied electric field, holes are "injected" at one electrode (into the valence band of the active LED material) and electrons at the other (into its conduction band). They are then transported through the body of the active LED material, i.e., the CP, where they combine or "capture" each other, forming excitons (usually mid-gap states). Certain of these excitons, usually the singlets, decay radiatively, emitting light at a wavelength characteristic of the CP's bandgap. It is noted that because singlet:triplet excitons are created in the ratio 1:3 and singlets are primarily responsible for EL, quantum efficiency of EL is found to be only about 25% of PL.

It is evident from this brief description that the determinative processes of LED function are, sequentially, hole and electron injection and transport, formation of excitons, and the latter's radiative decay. Clearly, hole and electron injection must be balanced for highest efficiency, so that any currents produced are utilized for exciton production, rather than simply being carried between electrodes by the carrier which is in excess (whether hole or electron). The thermodynamic work function of the contact electrode thus must match that of the CP (in terms of its overall redox potential or more specifically its ionization potential (IP) for hole injection or electron affinity (EA) for electron injection) in such a way as to facilitate hole or electron injection.

Most CPs studied have low IPs, and thus hole injection is not a problem for them. Even transparent ITO/glass appears to be an acceptable electrode for hole injection. However, since the EA of CPs is generally also quite low, electron injection into them has been found to be more problematic. Figure 39.4 shows schematically the relation between work functions of metals and bandgaps of CPs, here for a two-layer device studied by the Cambridge group. Initially, very low work function metals, such as Ca [1871], were required to be used for acceptable electron

Fig. 39.4 Schematic
energy level diagram for
bilayer CP-LED under
forward bias (After Ref.
[1360], reproduced with
permission)

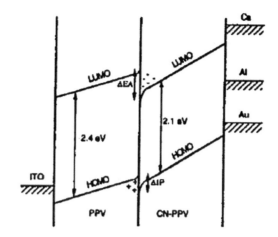

Fig. 39.5 Variation of
current and intensity of
electroluminescence with
forward-bias voltage for a
diode fabricated with layers
of indium tin oxide, P(PV),
and calcium (After Ref.
[1022], reproduced with
permission)

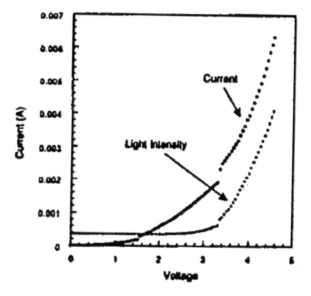

injection, creating problems of air/moisture exclusion, degradation, and practical assembly of devices. Metals such as Al have higher a work function than Ca, resulting in much lowered device efficiency. The primary mode of electron injection in such devices is thought to be tunneling.

Figure 39.5 shows the current and light emission characteristics of one of the simple, early LEDs, consisting of ITO-glass/P(PV)/Ca. The effect of the metal electrode used on the efficiency is shown for the same basic device in Fig. 39.6. Detailed studies of the effect of the electron-injecting metal electrode, for example, with poly(3-octylthiophene) [1362, 1872], have shown that a change of just 0.9 eV in the metal work function (corresponding to a change from Au to Al) results in a

Fig. 39.6 Variation of light output with bias voltage for P(PV) diodes fabricated with indium tin oxide anode and a range of metals as cathode (bias is reversed for Au as indicated). Quantum efficiencies are indicated against each data set (After Ref. [1022], reproduced with permission)

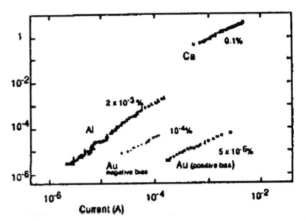

change in the preferred current direction and a change in the LED rectification ratio of six orders of magnitude.

Besides differing efficiencies of hole and electron injection at the electrode/CP interface, hole and electron mobilities within the CP also differ substantially, with electron mobilities being impeded through trapping by impurities such as oxygen. Additionally, if electron-hole recombination occurs near one of the electrodes, quenching is more likely to occur. Thus, it is preferred to somehow be able to keep electron-hole capture, i.e., exciton formation, away from the electrodes and in the interior of the device.

39.3 Varieties of CPs Used

P(PV) was the first CP used in LEDs, due, as noted above, to its easy processibility in defect-free form. In theory, any CP exhibiting singlet exciton formation on charge injection could be used in an LED. In practice, processing difficulties in the fabrication of a device, and the presence of impurities which quench excitonic states, has limited practical implementation to a few CP classes.

These classes include:

- P(PV) and its derivatives, some of which, e.g., the cyano-compounds of the Cambridge group, are substantial modifications of the P(PV) structure.
- Poly(3-alkyl thiophenes), e.g., P(3-octyl-T), again due to their being easily processed in their undoped forms into substantially defect-free thin films, but additionally also because it is relatively facile to alter their structure to achieve significant changes in bandgap and thus color.
- The relatively new CP structures embodied in poly(pyridine) and poly(pyridyl vinylene), used especially for AC-driven LEDs by the Epstein group.
- Poly(*p*-phenylene) (P(P)).

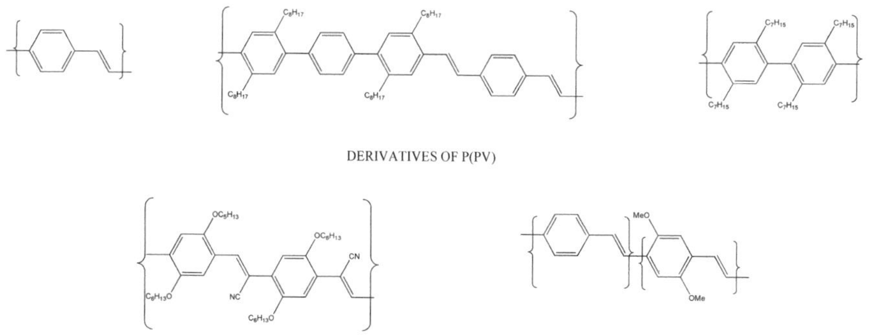

DERIVATIVES OF P(PV)

Fig. 39.7 Structures of some common CPs used in LEDs. As can be seen, most are derivatives of poly(phenylene-vinylene) (P(PV))

- The relatively novel block and other copolymers researched by the Spangler and Hilberer/Hadziioannou groups, which have segments that are EL inactive to tailor emission wavelength; poly(bithiophene) derivatives, such as those probed by the Inganäs group for polarized light emission.
- Poly(thienylene vinylene) derivatives. All these classes are discussed in more detail below.

Figure 39.7 gives structures of some common CPs used in LEDs.

39.4 Exemplary Device Assembly

For illustrative purposes, we may now briefly describe fabrication and assembly of a typical CP-LED, having the configuration ITO-glass/P(PV)/(Al or Ca or Au) and corresponding to the schematic in Fig. 39.7.

P(PV) is deposited on ITO-glass substrates via a precursor polymer route [1874]. The tetrahydro-thiophenium precursor is spin coated onto the ITO substrate and then converted to P(PV) via heating at 220 °C for ca. 10 h in vacuo for a final P (PV) thickness in the region of 100–250 nm. Metal, electron-injecting electrode, contacts (Ca, Al, Au) of several mm^2 area are then formed by vacuum evaporative deposition. In operation, electric fields of ca. 10^6 V/cm are required to obtain current densities of several mA/cm^2. The characterization data for a typical such device, with Ca as the electron-injecting electrode, has been shown in Fig. 39.5.

39.5 Addressing Problems and Tailoring Performance

Among major problems with CP LEDs are adjusting hole and electron currents for balance. This problem can be somewhat dealt with by selecting the metal electrode used, especially that for electron injection. However, this leads to much reduced efficiency of emission.

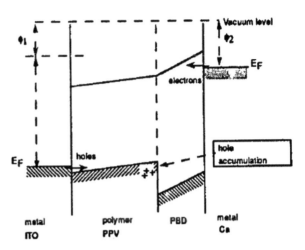

Fig. 39.8 Functioning of typical CP-LED (After Ref. [1022], reproduced with permission)

The Cambridge group first suggested an alternative approach, use of an additional CP layer interfacing between the electroluminescent CP and either or both of the electrodes, functioning much like a photosensitizer. The latter approach has been depicted schematically in the device configuration in Fig. 39.7b. The intervening layer can either be a hole-transporting layer or an electron-transporting layer. The Cambridge group typically used a molecular semiconductor in a matrix such as PMMA as the electron-transporting layer [1875–1879].

The functioning of such a device is illustrated schematically in Fig. 39.8. Here, holes injected from the ITO electrode are blocked at the interface with the electron-transporting polymer layer, which comprises a 1:1 blend of PMMA with 2-(4-biphenylyl)-5-(4-tert-butylphenyl)-1,3,4-oxadiazole) (called butyl-PBD). This blocking causes increased electron injection from the other electrode, forcing a balance in electron and hole currents. Additionally, excitons formed at the PPV/ PBD-PMMA interface are kept away from the other electrode. Such tailoring allows the use of less reactive metal cathodes, such as Mg, in place of Ca, and yields quantum efficiencies as high as 0.4%.

A similar, "blocking" approach was used by Uchiyama et al. [1880], but with alkyl-oligothiophenes (up to sesquithiophene) rather than full-fledged CPs. Although strictly not using CPs, their work confirmed that inhibition of carrier injection at one electrode increases that occurring at the other electrode and increases quantum efficiencies up to a thousandfold. An interesting but poorly understood result obtained by Zyung et al. [1881] when trying a similar "blocking" approach with the use of two CP layers was a substantial difference in PL and EL spectra. These workers used the same CP, a P(PV) derivative, but processed (dialyzed) it differently, preparing two layers in an LED.

A third approach, also used by the Cambridge group [1360, 1877], has been the use of derivatized CPs which have a substantially higher EA than P(PV), such as the cyano-group containing P(PV) derivatives. In such devices, P(PV) itself can be used as a hole-transporting layer. The schematic energy level diagram for the latter

a)

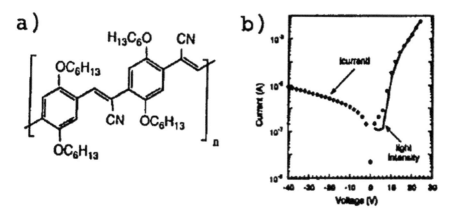

b)

Fig. 39.9 Structure of CN-P(PV) (left) and typical characterization data for LED constructed using this (right) (After Ref. [1360], reproduced with permission)

approach has been presented above in Fig. 39.4. Figure 39.9 shows characterization data for one such LED, using a poly(cyanoterephthalylidene) with a configuration ITO-glass/CP/Al. Among improvements in such a device vis-à-vis one based on P (PV) only are lower drive voltages (of the order of 10^5 vs. 10^6 V/cm^2 for P (PV) only). Good rectifying behavior is exhibited (10^3 forward/reverse ratio at high bias). Light output varies linearly with current over four orders of magnitude. Brightness is 2.6 W/steradian/m^2 at a 50 mA/cm^2 current density.

In a later refinement of this work [1882], the Cambridge group reported that devices having structures of the type shown in Fig. 39.10a showed high luminances, ca. 10^4 cd/m^2 at current densities of 1.5 A/cm^2, when driven with pulsed voltages, turn-on times as low as 3 µs (consistent with transit times of injected charge carriers), and a large range of linear variation of light output with voltage, all said to be favorable for applications in passive-matrix displays. Figure 39.10b shows luminance/voltage data for such devices of various thicknesses, while Fig. 39.10c illustrates the turn-on time for a 62 nm device at various drive voltages.

Östergard et al. [1883] attempted to tailor performance of CP-LEDs by employing Langmuir–Blodgett (LB) films of poly(3-hexylthiophene) derivatives, one of the first of such studies. They claimed that this led to better balancing of injected holes and electrons, leading to improved quantum efficiency.

39.6 Tailoring of Color

Among the challenges for tailoring of performance of CP-LEDs are tailoring the color of emission to encompass the entire visible spectral region.

The most direct route to tailoring the color of emission of CP-LEDs is via alteration of their bandgap, which in turn has been most frequently affected to date via structural alteration of CPs whose use in LEDs has already been

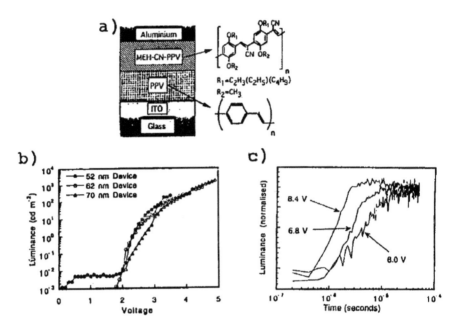

Fig. 39.10 Schematic structure (**a**), luminance (**b**), and turn-on time (**c**) data for LEDs using cyano-P(PV)s (After Ref. [1882], reproduced with permission)

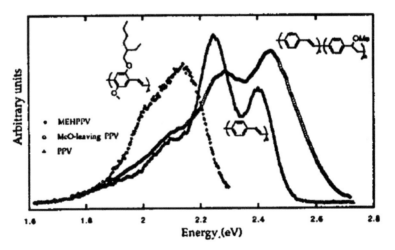

Fig. 39.11 Electroluminescence spectra from P(PV) and two derivatives (After Ref. [1022], reproduced with permission)

established, e.g., P(PV). The cyano-substituted P(PV) derivative in Fig. 39.11 is an illustrative case of this. Such structural tailoring can be more clearly seen in the comparative electroluminescence data of P(PV) and two of its derivatives in

Fig. 39.11. The Cambridge group has used this approach to achieve variations in color of the emitted light from blue-green to orange-red [1022, 1870, 1884].

Perhaps one of the most elegant manifestations of this approach of using substituents or structures within a CP family to tailor bandgap, which in turn tailors color, has been seen in recent work of the Inganäs group [1885, 1886]. The structural variation, in a family of P(T) derivatives, and resultant electroluminescent spectra are represented in Fig. 39.12. These CPs were tested successfully in LEDs with the configuration ITO/CP/Ca/Al (with the added Al layer being used to protect the Ca from atmospheric degradation). In such devices, the polymers abbreviated PCHMT, PCHT, PTOPT, and POPT (see figure) showed emission colors, turn-on voltages, and quantum efficiencies, respectively, as follows: PCHMT (blue, 7 V, and 0.6% (at 24 V)); PCHT (green, 2.4 V, and 0.01% (at 7 V)); PTOPT (orange, 1.6 V, and 0.1% (at 8 V)); and POPT (red/NIR, 1.4 V, and 0.3% (at 6 V)). This CP series thus probably represents the most comprehensive series in terms of spectral/color range achieved in LEDs.

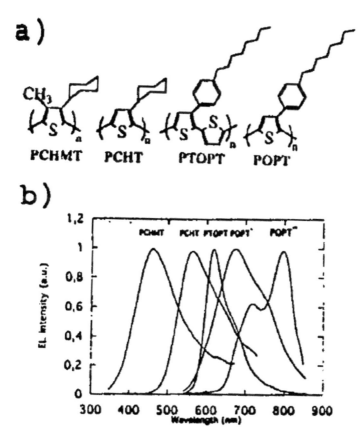

Fig. 39.12 Structures (top) and EL spectra (bottom) for P(T)-derivative LEDs, showing tailorability of color (After Refs. [1885, 1886] reproduced with permission)

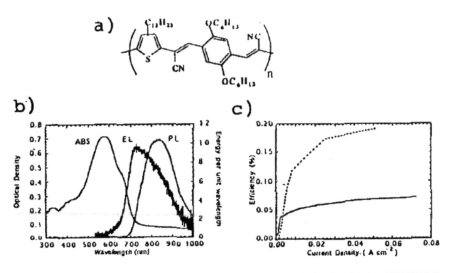

Fig. 39.13 Structure (**a**), EL/PL data (**b**), and EL quantum efficiencies (**c**) for CN-P(PV)-P (TV) derivative (After Ref. [1887], reproduced with permission)

Again, using the same structural modification approach, the Cambridge group [1887] synthesized the CN-thiophene derivative shown in Fig. 39.13a. This very low bandgap material displayed the absorption, EL, and PL spectra shown in Fig. 39.13b and emitted well into the NIR, up to ca. 1000 nm. This CP, whose LUMO lay closer to the Fermi level of Ca than of Al, functioned well only with the former metal, in the configuration ITO/CP/P(PV)(hole-injecting polymer)/Ca, as seen in the comparative data in Fig. 39.13c. Devices showed maximum quantum efficiencies of ca. 0.2%.

Derivatives of P(PV) continue to be the targets of choice for tailoring of LED emission color. For example, Yin et al. [1888] described a unique derivative, poly (2,5-diphenyl-1,3,4-oxadiazolyl)-4,4'vinylene) (structure in Eq. (39.1)), with a peak emission at 483 nm and a turn-on voltage of 6 V in an ITO/CP/Al configuration.

$$\text{(structure)} \tag{39.1}$$

Another approach toward tailoring of color is of course discovery of entirely new, electroluminescent CPs which emit at different wavelengths. Such an approach has been demonstrated in more recent work by the Epstein group at Ohio State University [1889, 1890]. Thus, they have found that poly(p-pyridine) emits blue light and poly(p-pyridyl vinylene) emits orange light. Many of these materials and devices are discussed in the section on AC-driven LEDs below.

Onoda and MacDiarmid [1891] showed that a yellow-orange emission could be obtained with poly(p-pyridyl-vinylene) (P(PyV)) emitting layers in a P(PV)/P

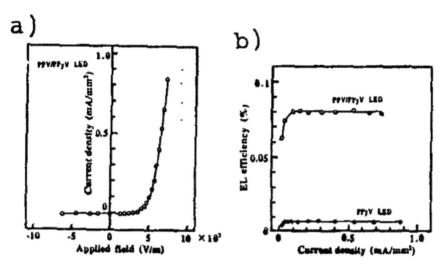

Fig. 39.14 P(PyV)-based CP-LED performance data (After Ref. [1891], reproduced with permission)

(PyV) device using Al as the electron-injecting contact. Representative data for these devices are shown in Fig. 39.14. Wang et al. [1892] showed convincingly that the high electron affinity of the pyridine moiety in CPs with the pyridyl group [1893] allows the use of stabler electron-injecting contacts such as Al or even ITO and better electron-transport properties, allowing use of bilayers with polymers such as poly(9-vinyl carbazole) (PVK). This reduces the turn-on voltage considerably. Figure 39.15 shows representative performance data for a typical such device, with the structure of the CP inset.

The use in LEDs of various poly(3-alkyl thiophenes), most emitting in the red region [1894, 1895], and poly(p-phenylene) prepared using the precursor route, emitting in the blue region [1896], has recently been demonstrated. Indeed, P (3-alkyl-T)s such as poly(3-octylthiophene) are favorite materials for LED studies [1362] due to their high solubility and facile processibility in the undoped state, without the need for going through a precursor step. Figure 39.16 shows characterization data for one such device with two configurations, ITO/P(3-OT)/Al (left) and ITO/P(3-OT)/Au (right); device area is 8 mm^2, thickness (all three layers) 100 nm.

Yet another approach to tailoring the color of CP LEDs is the use of copolymers consisting of EL-active and EL-inactive segments. This approach has been demonstrated by Spangler et al. [1125] and the Hilberer/Hadziioannou group [1130, 1362]. Figure 39.2 (from Spangler et al.'s work), encapsulated the synthetic aspects of this approach, showing four candidate CPs studied, labeled LED1–4. Using Al as the electron-injecting electrode, these authors found that LED1, LED2, and LED4 exhibited emission colors of blue (λ(max) 470 nm, quantum efficiency (QE) 2.07 × 10^{-4}), blue-green (λ(max) 500 nm, QE 6.75 × 10^{-4}), and blue-violet

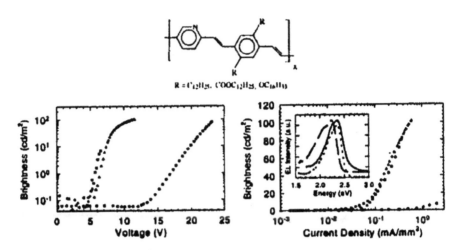

R = C₁₂H₂₅, COOC₁₂H₂₅, OC₁₆H₃₃

Fig. 39.15 Performance data for pyridine-derivative CPs (structure at top) (After Ref. [1893], reproduced with permission)

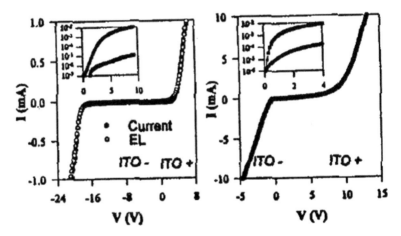

Fig. 39.16 Current voltage and electroluminescence voltage characteristics for a device with ITO/P3OT/Al (**a**) and ITO/P3OT/au (**b**) under forward and reverse bias. The inset shows the same data on a semilogarithmic scale. The device thickness is 100 mm, active area 8 mm³ (After Ref. [1362], reproduced with permission)

(λ(max) 440 nm, QE 2.33 × 10^{-6}), respectively. Although promising for color tunability, their efficiencies were rather poor. Figure 39.17 shows characterization data for LED2 and LED4.

The approach of the Hilberer/Hadziioannou group [1130, 1362] used block copolymers which consisted of π-conjugated sequences, i.e., conventional CPs, "interrupted" by non-π-conjugated sequences. A thiophene derivative, or an oligo (phenylene vinylene), could be used in the CP sequence. These block copolymers were synthesized using a Pd-catalyzed Heck coupling. The schematic structure and

Fig. 39.17 Performance data for LED4 (After Ref. [1125], reproduced with permission)

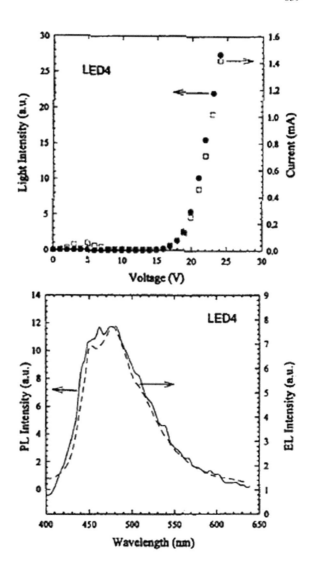

electroluminescence and characterization data for one such LED using an Al electrode are shown in Fig. 39.18a–c. Emission is in the blue or blue-green region for most of these block copolymers and requires an electric field in the region of 2 to 3×10^8 V/m.

Leising et al. [1897] demonstrated highly efficient, full-color electroluminescence from P(PP)-type ladder polymers and fabrication of highly efficient red–green–blue (RGB) and white-light devices. The RGB devices in particular allowed the realization of full-color flat-panel displays. These were fabricated using a new technique which allowed production of virtually any color in the Vis-NIR spectral region. An efficient white-light emission was generated by an internal excitation

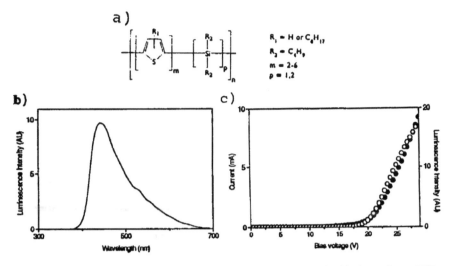

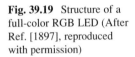

Fig. 39.18 Schematic structure (**a**) and typical performance data (**b, c**) for block copolymer LEDs (After Ref. [1362], reproduced with permission)

Fig. 39.19 Structure of a full-color RGB LED (After Ref. [1897], reproduced with permission)

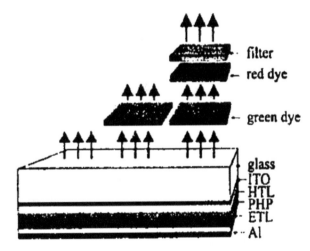

energy transfer from the blue CP component to a red light-emitting CP in a polymer blend, which was used as the active layer in a light-emitting diode. Structures of two of the CPs studied are shown in Eq. (39.2) (Fig. 39.19).

$$(39.2)$$

Jiang et al. [1898] claimed color tunability from blue through close to red for devices based on CN-P(PV)/PVK combinations. Their structure and performance data are shown in Fig. 39.20.

Sun et al. [1899] carried out an elegant study of multicolor EL in LEDs constructed from various P(Ac) derivatives with the configuration Mg-Al/CP/ITO. Red, green, or blue emissions were obtainable, depending on the choice of CP. Structures and EL spectra of LED devices are shown in Fig. 39.21.

Chen et al. [1900] described near-white-light (380–750 nm) emission for an EL diode with P(ANi) as the emitting layer, ITO/glass as the hole injector, and Mg or Al as the electron injector. Turn-on voltages of 13 and 6 V for Al and Mg electrodes were observed using a CP thickness of 80 nm. Figure 39.22 shows typical EL spectra obtained for these LEDs.

39.7 AC-Driven LEDs

More recent developments in CP-based LEDs have seen more specialized applications, in particular devices operated by AC rather than DC drive voltages, and those emitting polarized light or having claimed submicron structures (discussed in the subsequent sections).

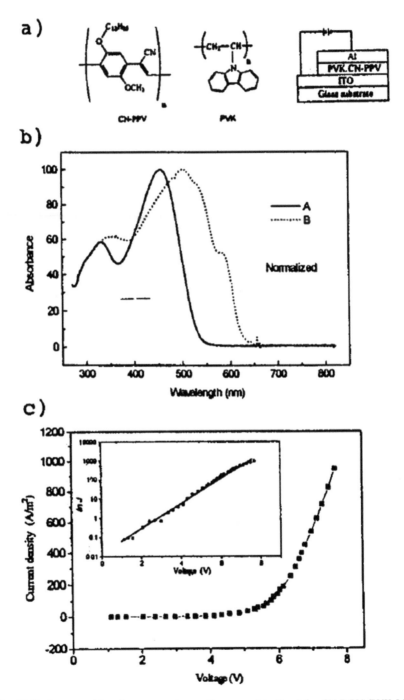

Fig. 39.20 Structure (**a**) and representative performance data (**b**, **c**) for CN-P(PV)/PVK LEDs (After Ref. [1898], reproduced with permission)

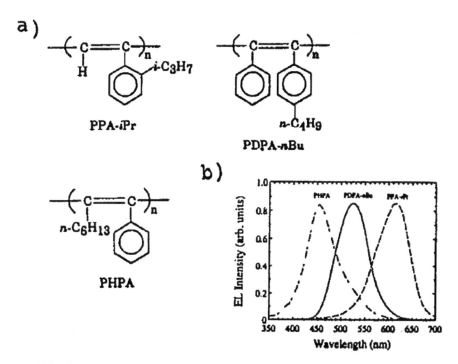

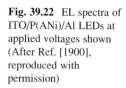

Fig. 39.21 Structure (**a**) and representative performance data (**b, c**) for P(Ac)-derivative electroluminescence (EL) devices (After Ref. [1899], reproduced with permission)

Fig. 39.22 EL spectra of ITO/P(ANi)/Al LEDs at applied voltages shown (After Ref. [1900], reproduced with permission)

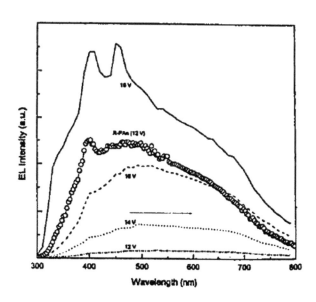

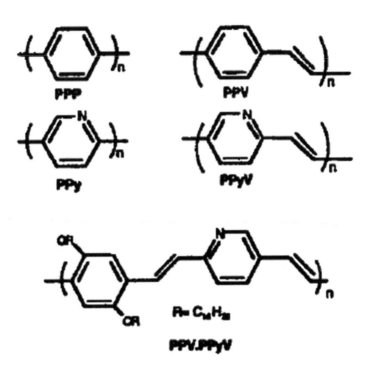

Fig. 39.23 CPs used in AC-driven LEDs, by the MacDiarmid/Epstein groups [1891]

Normally, the rectifying behavior of LEDs entails that they emit light efficiently only under one, "forward-bias" mode, precluding AC operation; AC operation might be considered advantageous in situations where household current is desired to be used. Indeed, AC-LEDs based on Mn/ZnS are available commercially. However, a serendipitous discovery in the laboratories of the Epstein group at Ohio State University [1889, 1890, 1901] showed that a symmetric device given the acronym SCALE (symmetrically configured AC light-emitting) device (see Fig. 39.24) emitted light efficiently under AC operation. The reasons for this were not entirely understood, although it was guessed that charge injection controlled via the insulating polymer/CP interface increased efficiency in the reverse bias mode. These devices used conventional electroluminescent CPs, such as P (PV), as well as more unconventional ones, such as poly(p-pyridine) and poly(p-pyridyl vinylene), and a related copolymer, whose structures are given below. The first two of these emitted in the blue and orange regions, respectively. Figure 39.23 shows structures of some of the CPs used by these groups in AC-driven LEDs. Figure 39.24 shows a schematic of the SCALE structure.

In these SCALE devices, the insulating polymer layer was typically undoped poly(3-hexylthiophene) or P(ANi) (as emeraldine base). Furthermore, these devices functioned well with high work function metals such as Cu or Au and did not need Ca or Al electrodes, for reasons again not fully understood although guessed at being due to "better coordination" between these metal atoms and the insulating

Fig. 39.24 Schematic of SCALE device (After Ref. [1889], reproduced with permission)

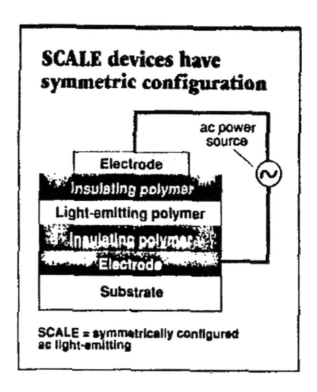

Fig. 39.25 Typical performance data for a SCALE devices (After Ref. [1901], reproduced with permission)

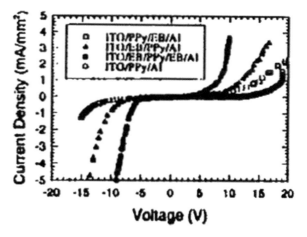

polymer layers used. Yet another unique feature of these devices was that, expectedly, they emitted light pulses twice during one AC cycle but that, unexpectedly, these pulses were not always of the same color. Interestingly, it was found that CP-Nylon-6,6 blends also functioned well as emissive layers in the LEDs [764]. Figure 39.25 shows comparative characterization data for LEDs, based on poly(pyridine) (denoted in the figure as P(Py)), using the standard single-CP-layer configuration and various multilayer configurations with the insulating polymer layer being the emeraldine base form of P(ANi), denoted as EB in the figure.

39.8   LEDs Emitting Polarized Light

The discovery of polarized light-emitting CP-LEDs was less serendipitous and more planned and systematic. Since chains in most CPs normally possess a random orientation, electroluminescence produces randomly oriented (nonpolarized) light. It is well known that stretch orientation of CPs produces oriented CP chains (see description elsewhere in this book) and that other polymers, when stretch-oriented and illuminated with light, give off polarized light. The Inganäs group [1886, 1902] logically surmised that stretch-oriented CPs, if electroluminescent, may give off polarized light. They demonstrated this for poly(3-(4-octylphenyl)-2,2′-bithiophene), a CP which is especially amenable to stretch orientation, and two other P(T) derivatives, although with very low quantum efficiency, ca. 0.1%. A potential application for polarized CP-LEDs is thought to be backlit LCDs, where polarizers are currently used, although cost and quantum efficiency may be major impediments.

39.9   Submicron and Other Specialty Applications

The Inganäs group proposed a unique approach to fabrication of micron- and submicron-sized CP-LEDs, as depicted schematically in Fig. 39.26a, using poly (2,3-ethylene-dioxythiophene) (PEDOT) as a hole-injecting polymer layer coupled to the active electroluminescent CP, poly(3-(4-octylphenyl)-2,2′-bithiophene) (PTOPT). In a first, "Type I" structure (top of figure), the CP was polymerized and doped in the pores of a commercial microporous membrane, e.g., polycarbonate membrane of 100 nm pore diameter, which was in turn attached to Au contacts. CP electropolymerization was controlled to stop when the CP reached the membrane surface. In this way, these workers claimed to achieve light sources of the same size as the pores. In a second, "Type II" structure (bottom of figure), the natural tendency of the CP PTOPT to form micron-scale island-coagulates within a PMMA matrix was used. Figure 39.26b shows characterization data for these devices.

Faraggi et al. [1904] demonstrated fabrication of a CP-LED pixel microarray having the CP sandwiched between ITO and Al, with pixel size as small as 20 μm × 20 μm, and claimed extendable to the nm range. A photo-ablation method employing a 193 nm excimer laser was used. A schematic representation of the procedure is shown in Fig. 39.27. McGehee et al. [1905] also fabricated sub-μm-dimension LEDs using a unique, "shadow evaporation" technique.

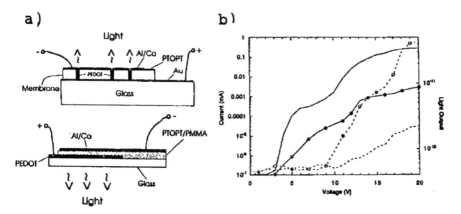

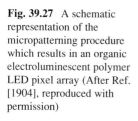

Fig. 39.26 (**a**) Submicron-sized LEDs of the Type I (top) and Type II (bottom) type. (**b**) Performance data for Type I and Type II (with circles) LEDs (After Ref. [1903], reproduced with permission)

Fig. 39.27 A schematic representation of the micropatterning procedure which results in an organic electroluminescent polymer LED pixel array (After Ref. [1904], reproduced with permission)

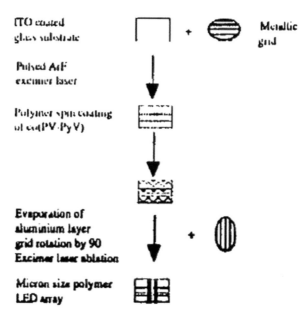

39.10   Device and Commercial Applications

CP-based LEDs or, more accurately, CP-based OLEDs (organic light-emitting diodes) have been talked about as being on the cusp of commercialization since about the year 2000. In the early 2000s, Prof. Alan Heeger of the University of California at Santa Barbara, CA, USA (one of the three persons receiving the Nobel Prize in Chemistry in 2000 for work on CPs), spun off a company for the commercialization of OLEDs based on CPs. This was subsequently purchased by the giant American chemical company DuPont and named DuPont Displays (although the company used other materials as well, besides CPs, as a basis for their displays). In 2012, with funding from a grant from the state of Delaware, USA, and DuPont investing a further US$20 M, a production facility for the manufacture of OLED displays for TVs was launched [1906–1908].

In July 2016, the American Chemical Society's flagship membership publication, *Chemical & Engineering News*, featured a cover story on OLED displays (primarily for TV applications), wherein OLEDs were described as the "rich man's TV" option, i.e., more expensive than standard LEDs for TVs but giving much richer colors and resolution [1861]. More realistically, this article also noted that the market for OLED TVs still remains relatively small compared to the overall LED TVs. It also noted that OLEDs in 2016 were still confined to a niche mobile phone display market (e.g., in some Samsung Galaxy series smartphones), and other types of LEDs, such as those based on (inorganic) quantum dots, were superior in terms of cost and performance to OLEDs generally, even though they consumed much more power due to the need for a backlight. In 2016, the South Korean company LG remained the only marketer of OLED TVs.

Most importantly, however, it is apparent from the references cited above that OLED displays in the year 2016 (i.e., at the time of this writing) are still based on *non*-CP-based organic materials. CP-based OLEDs do not appear to have been able to compete successfully with other organic-based OLEDs. Figure 39.28 shows structures of *non*-CP materials used in OLEDs.

Apart from the above, more recent developments, there have been many *historically important* developments (from about 1999 through about 2005) particularly related to device and commercial applications; these were targeted to making CP-LEDs more commercially viable. We now briefly discuss these as well as the principles behind them.

As an example of one of the earliest, Liedenbaum et al. of the Philips Research Laboratories, Eindhoven, the Netherlands [1909], demonstrated low-voltage operation of large-area CP-LEDs based on P(PV) building blocks, with minimal degradation over several thousand hours of operation. Figure 39.29 shows typical performance data for one such large-area LED.

Platinum octaethylporphyrin   Tris[2-phenylpyridinato-C^2,N]iridium(III)   Diphenylsulfone dimethyldihydroacridine

Fig. 39.28 Structures of exemplary *non*-CP materials used in OLEDs (After Ref. [1861], reproduced with permission)

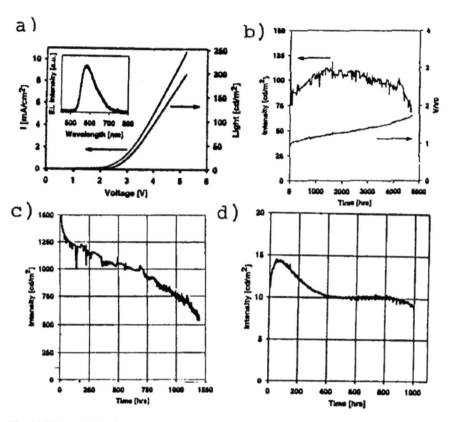

Fig. 39.29 Typical performance data for large-area P(PV)-derivative LEDs (After Ref. [1909], reproduced with permission)

Fukuda et al. [1910] studied poly(quinoxaline-5,8-diyl) (P(Qx), structure shown in Fig. 39.30, as a superior electron injection material. A device configuration such as ITO/P(PV)/P(Qx)/metal was shown to be more efficient than alternative configurations without P(PV) and P(Qx). Figure 39.30b compares these several configurations schematically, while Fig. 39.30c shows representative luminescence characteristics of the first configuration.

In a similar vein, Osaka et al. [1911] showed that use of a blend of nitrile-butadiene rubber with poly(3-alkyl thiophenes) as the emissive layer in CP-LEDs could lead to reduction in leakage current. Figure 39.31 shows characteristics of one of their devices.

Lidzey et al. [1912] showed that photolithography could also be used to pattern and fabricate LEDs and could be applied to fabrication of arrays of micron-sized LEDs, although a significant degradation in LED performance was noted as compared to conventional fabrication methods.

Noting that synthesis of processable CPs such as those based on P(PV) in quantities large enough to be commercially viable remained a major challenge,

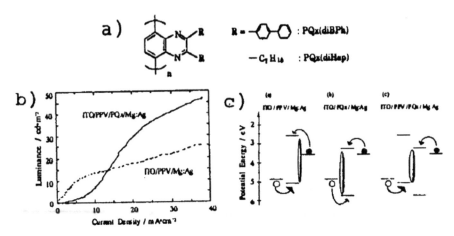

Fig. 39.30 Structure (**a**), representative luminescence data (**b**) for P(Qx) LEDs and schematic comparison of several configurations (**c**) (solid circle = electron, open circle = hole) (After Ref. [1910], reproduced with permission)

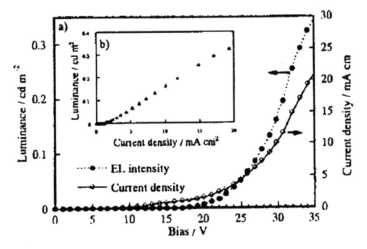

Fig. 39.31 Performance data for rubber/P(3AT) blend LEDs (After Ref. [1911], reproduced with permission)

Hsieh and coworkers [1913] at Xerox, Inc., Rochester, NY, USA, reported successful synthesis of multigram quantities of several such P(PV) derivatives. One such synthesis is depicted in Fig. 39.32.

Brütting et al. [1914] carried out a detailed study showing that control of impurities in P(PV)-based LEDs can dramatically improve their performance.

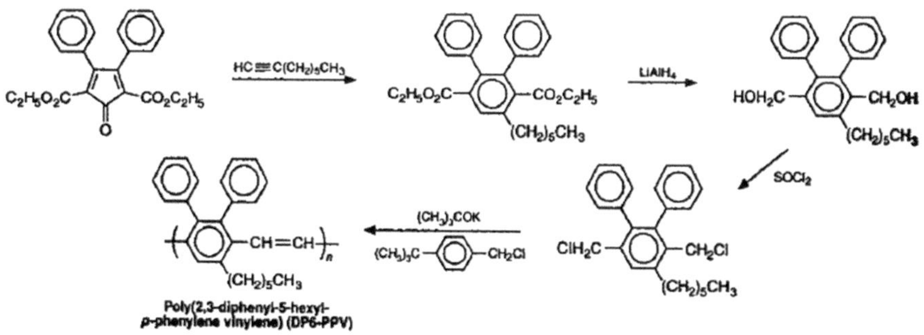

Fig. 39.32 Successful, multigram Diels–Alder-based synthesis of P(PV) derivatives (After Ref. [1913], reproduced with permission)

39.11 "LECs" and Other Device Types

A variant of LEDs which have incorporated CPs have been light-emitting electro-chemical cells, or "LECs," in which an electrolyte of sorts is incorporated into the CP and in which the function may not strictly be that of a diode. Thus, Greenwald et al. [1915] characterized LECs based on poly(3-octylthiophene) (P(3OT)) with P (3OT) + PEO + triflate comprising the CP-electrolyte component, coated onto ITO/glass electrodes. Al was then vacuum deposited on this assembly. Weak orange light emission at a turn-on voltage of 2 V was observed.

A novel device type described recently by Weder et al. [1916] is one which incorporates the photoluminescence properties of P(PV)-type CPs, in the form of thin, polarized photoluminescent layers, with conventional liquid crystal displays (LCDs). This is done to eliminate the LCDs' problems of brightness, contrast, and viewing angle emanating from their use of highly absorbing polarizers and color filters. The three types of devices studied by these workers are shown schematically in Fig. 39.33. These authors used derivatives of poly(2,5-dialkoxy-phenylene-ethynylene). Such devices are the first instance of photoluminescent polarizers being built into LCD devices. In subsequent work, Montali et al. [1917] noted that polarizers in such devices could be operated on a two-step principle. In the first step, randomly oriented "sensitizer" molecules harvested the incident light by isotropic absorption. In the second step, they efficiently transferred the energy to a uniaxially oriented photoluminesent polymer, from which colored light with a high degree of linear polarization was then emitted.

39.12 Conductive Films

CPs have been investigated for use as *transparent conductive films*, to replace such materials as indium tin oxide (ITO), at present the versatile workhorse of conductive films. Xia et al. [1918] studied solution-processed PEDOT/PSS (poly

Fig. 39.33 Schematic structures of (historically important) PL display devices using CP/LC combinations (After Ref. [1916], reproduced with permission)

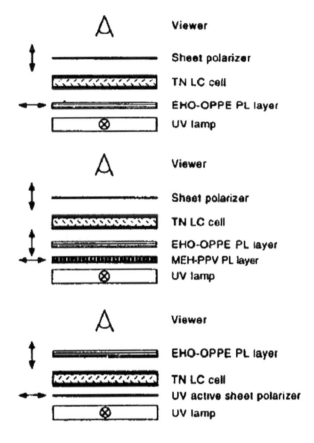

(ethylenedioxythiophene)/poly(styrene sulfonate)) for application as transparent electrodes for optoelectronic devices. After a unique treatment with H_2SO_4 at about 160 °C, their PEDOT/PSS films were claimed to have a conductivity of up to 3065 S/cm or a sheet resistance of 136 Ω/square, comparable to the poorest-conducting ITO. They presented data showing a % transmission of greater than 65% in the 400–700 nm region for five layers of PEDOT/PSS films of thickness about 109 nm. However, it was not clear from their presentation whether this five-layer film also corresponded to a high conductivity. Figure 39.34 shows % transmission data for their films. These authors also then used their PEDOT/PSS films, at a thickness of 70 nm, to replace ITO as the transparent electrode of polymer solar cells, specifically poly(3-hexylthiophene) (P3HT)/[6,6]-phenyl-C 61-butyric acid methyl ester (PCBM) solar cells. They claimed that these solar cells exhibited photovoltaic performance, J_{sc}, of 9.29 mA/cm^2, V_{oc} of 0.59 V, FF of 0.65, and efficiency of 3.56%, comparable to the performance of control devices using ITO anodes.

In other work in the potential use of PEDOT/PSS in transparent conductive films, Vosgueritchian et al. [1919] prepared such films with a fluorosurfactant, to

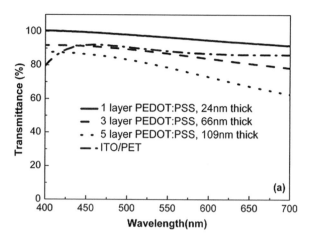

Fig. 39.34 Transparency of PEDOT/PSS films prepared by Xia et al. as a potential substitute for ITO, all on PET (poly(ethylene terephthalate)) substrates (After Ref. [1918], reproduced with permission)

arrive at *flexible* conductive films with claimed sheet resistance of 46 Ω/square at a transmission of 82% at 550 nm. The films were deposited on poly(dimethyl siloxane) (PDMS) substrates to achieve high flexibility. However, to date (2016) since their 2012 publication, these promising transparent conductors based on CPs do not appear to have been further developed or commercialized.

39.13   Problems and Exercises

1. Sketch in detail the structure of a one-CP-layer LED and describe in one paragraph the principles of its operation. Provide an actual example, with details of the chemical, electrochemical, and physical steps used in its fabrication.
2. How are EL, PL, absorption, and fluorescence spectra of a CP related (you may refer to other chapters of this book if you like)? What information does each of these furnish with regard to operation of LEDs?
3. Why does the nature of the (+) electrode have such a strong influence but that of the ($-$) electrode little influence on functioning of CP-LEDs?
4. Summarize the factors responsible for quantum efficiency in CP-LEDs.
5. Carry out the exercise of problem 1 for (a) a two-CP-layer LED, with one being the emitter and one a hole injector; give likely candidate CPs usable for (b) a three-layer LED which may be AC driven.
6. What is the principle of operation of an LED emitting polarized light? Is there any way to specify the plane of polarization of the light?
7. Suggest a means of fabrication of nanoscale LEDs other than the two studied by the Inganäs group described in this chapter.
8. Identify CPs other than P(PV) and P(T) derivatives which have been used successfully in LEDs.

9. Outline the various methods currently available to balance hole and electron injection rates and hole and electron mobilities in CP-LEDs.

10. Identify and describe with examples at least three methods which can be used to tailor the color of emission of CP-LEDs. Can you suggest a way of having a CP-LED emit several, selectable colors?

11. Suggest applications of CP-LEDs and identify limitations still preventing their commercial implementation, including any not suggested in this chapter.

Chapter 40
Microwave- and Conductivity-Based Technologies

Contents

40.1 Introduction, Applications Covered, and Frequencies

Initially, the primary drivers of development for applications of the microwave region and conductivity properties of CPs were the following: electromagnetic impulse (or interference) (EMI) shielding, conductive coatings and composites for electrostatic charge dissipation (ESD) or antistatic applications, and passive absorbers for microwave-region radiation for radar cross-section (RCS) reduction. To these the following have recently been added: dynamic (switchable) microwave absorption and RCS control, microwave-smoke camouflage, and several other applications.

For EMI shielding, the frequencies of interest have ranged from 1 KHz to 1 GHz, determined primarily by regulatory requirements. For microwave absorption and RCS, the frequencies of interest have principally been those of military relevance, e.g., the X, Ku and Ka bands (respectively ca. 8–12 GHz, 12–18 GHz, and 26–40 GHz), and 94 GHz.

© Springer International Publishing AG 2018
P. Chandrasekhar, *Conducting Polymers, Fundamentals and Applications*,
https://doi.org/10.1007/978-3-319-69378-1_40

40.2 Electromagnetic Impulse (Emi) Shielding

Electromagnetic interference shielding effectiveness (EMI-SE) was recognized early as one of the major areas of potential application of CPs [1920]. Required shielding of consumer and commercial electronic equipment from interfering electromagnetic signals can range from a low of 30 to a high of 200 μV/m permissible signal levels. Currently, the need is met by sprayed metal coatings, e.g., Zn flame-sprayed on the thermoplastic composite housings of the electronic components. Clearly, highly conductive CPs would be expected to have greater compatibility with such plastic materials than metals; if they could be incorporated into the thermoplastic before molding of the housing, the expensive step of metal spraying could be obviated.

Thus, most studies of CPs for EMI shielding applications have envisioned this sort of blending within a nonconductive polymer matrix. As an example, many of these studies have evaluated DC and AC conductivity and shielding of CPs as blends in such thermoplastic matrices as poly(vinyl chloride) (PVC), acrylonitrile–butadiene–styrene (ABS), poly(methyl methacrylate) (PMMA), and poly(ethylene terephthalate) (PET).

As mentioned in an earlier chapter, two of the most well-established test methods for EMI shielding effectiveness (EMI-SE) are the ASTM (American Society for Testing and Materials, Conshohocken, PA, USA) method ES 7-83 and the US military standard MIL-STD-462. Frequencies in the 1 KHz to 1 GHz region are most commonly used. However, the cavity perturbation technique and the network analyzer technique, both used for generic microwave measurements and discussed earlier elsewhere in this book, can also be used to generate EMI-SE data.

Some of the earliest work was by the Shacklette group in the 1990s [1921, 1922], wherein P(ANi) was melt-blended with polyvinyl chloride (PVC) to form a composite with conductivity of about 20 S/cm, adequate for some EMI-SE applications.

Subsequent to this work, Chandrasekhar and Naishadham [1923] showed that microwave parameters such as complex permittivity, absorption, reflection, and SE of bulk PANI, which was doped with two proprietary sulfonate dopants, were measured over a broadband of 4–18 GHz for the first time using coaxial line techniques. The permittivity decreased monotonically with increase of frequency. The real part varied from 188 at 4 GHz to ca. 32 at 10 GHz and 10 at 18 GHz; the imaginary part varied from about 35 at 4 GHz to ca. 2 at 18 GHz. The microwave SE of the PANI, calculated on the basis of the measured complex permittivity, was lower than 15 dB over the frequency range and higher than that obtained in ref. [18]. Lower values as 50 dB was estimated for multilayer of PANI. The reflection loss and absorption loss were 5–1 dB and 5 dB over the range, respectively.

Further to this work, Trivedi and Dhawan [1924] grafted PANI on surfaces of fabrics and measured their EMI-SE using the coaxial transmission line method in the frequency range of 1000 kHz to 1 GHz. The results showed that at higher frequencies (0.1 MHz to 1 GHz), the SE is at 16–18 dB, while at lower frequencies, it is more than 40 dB. Mäkelä et al. [1925, 1929] studied EMI-SE of P(ANi) doped with camphorsulfonic acid (CSA), single as well as double layer, double layer, spin coated from m-cresol solution on an insulating substrate, both in the near-field with a dual chamber box and in the far-field with a transmission line method, at a frequency range of 0.1–1000 MHz. In the near-field measurement, the EMI-SE decreased with increasing frequency, indicating reflection from the CP film. In the far-field measurement, the EMI-SE of either the single-layered (10 mm) or double-layered film was independent on frequency, and both increased with the decreasing of surface resistance of the film, which agreed with theoretical values obtained from a good-conductor approximation. The highest EMI-SE of 39 dB was achieved for the three-layered (30 mm) film at 1GHz with surface resistance of 3 Ω/square.

Wojkiewicz et al. [1920] prepared conductive composites by, firstly, dissolving both P(ANi) and PVC in NMP (N-methyl pyrrolidinone) first, then casting the solution on to fiberglass, and finally doping with HCl, and, secondly, by dissolving camphorsulfonic acid (CSA)-doped P(ANi) and polyurethane or styrene-acrylonitrile (SAN) in m-cresol. EMI-SE measurements indicated that the second-type composite had higher conductivity and SE. In the low frequency range (50 MHz to 1 GHz), the SAN composite with PANI loading of 40% exhibited an SE of 50–70 dB and, over the microwave band, an SE of 60–70 dB, higher than that of other composites.

P(Py)-based materials have generally shown poorer EMI-SE performance than P(ANi)-based materials. For example, Kim et al. [1920, 1927] carried out EMI-SE measurements at the frequency range from 50 MHz to 13.5GHz of P(Py)-coated nylon 6 fabric, which was prepared chemically or electrochemically, showed that the multilayer structure was better than single layer for EMI-SE, given the thickness and conductivity of the P(Py) are the same, and the maximum SE achieved was approximately 40 dB.

Naishadham [1347] used the cavity perturbation technique to study the overall microwave characteristics, including EMI-SE, of the CPs poly(acetylene) (P(Ac)) and poly(p-phenylene benzobisthiazole) (P(BT)), at 8.9 and 9.89 GHz and continuously in the 2–20 GHz frequency region. Figure 40.1a–c summarizes some of his results.

Chandrasekhar [1928] carried out EMI-SE studies, in the 1 KHz to 1 GHz region, of composite blends of poly(3-methyl thiophene) (P(3MT)), the thermoplastics acrylonitrile–butadiene–styrene (ABS), polycarbonate and poly(phenylene sulfide) (itself also a CP), and, as an optional additive, graphite microfibers. Figure 40.2a, b show some of his results. The shielding provided is not only somewhat higher than graphite sheet, it is also more broadband than graphite sheet, which shows large shielding gaps between ca. 100 KHz and 1 GHz.

Mäkela et al. [1929] carried out detailed studies of the EMI-SE properties of 1–30 μm thick camphorsulfonic acid-doped P(ANi) films having conductivities in the 10–100 S/cm region. Measurements were carried out in the near-field with a dual chamber, and in the far-field using a transmission line method, in 0.1 MHz to 1 GHz region. A strong correlation with surface film resistivity was found. Multilayered structures were found to enhance shielding considerably, up to 40 dB at

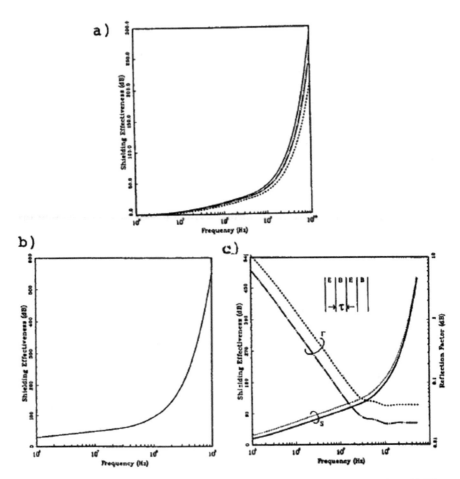

Fig. 40.1 EMI-SE work of Naishadham et al. [1354]. (**a**) Frequency dependence of the shielding effectiveness of 6.25-mm-thick sheets of polymer PBT (dot-dash line, PBT (solid line), polyacetylene cis-(CHI $_{0A045}$)$_x$ (dashed line), and polyacetylene trans-(CHI$_{0A045}$)$_x$ (dotted line), at normal incidence ($\Theta_1 = 0$). (**b**) Frequency dependence of the shielding effectiveness of a 6.25-mm-thick sheet of polymer polyacetylene cis-(CHI0A8)x at normal incidence. (**c**) Frequency dependence of shielding effectiveness and reflection factor at two angles of incidence (perpendicular polarization) for a four-layer absorber consisting of contiguous layers of polymers polyacetylene cis-(CHI0A8)x and PBT. The thickness $\tau = 1.625$ mm. Solid and dotted lines correspond to S, while dashed and dot-dash lines correspond to T, for $\Theta1 = 0$ and 60, respectively (After Ref. [1347], reproduced with permission)

100 MHz to 1 GHz. Figure 40.3 summarizes some of their results in the near- and far-field.

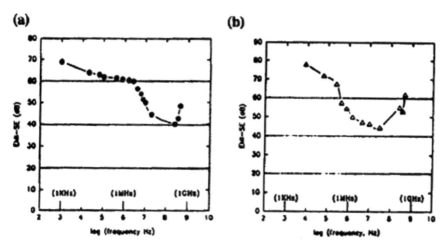

Fig. 40.2 Exemplary EMI-SE data for a poly(3-Me-thiophene) + ABS component at (**a**) 20 w/w% loading of CP in ABS (**b**) 40 w/w% loading (After Chandrasekhar [1928])

Fig. 40.3 Near-field (at 10 MHz) and far-field shielding as a function of surface resistance R $_s$ for thin P(ANi)/CSA films (After Ref. [1929], reproduced with permission)

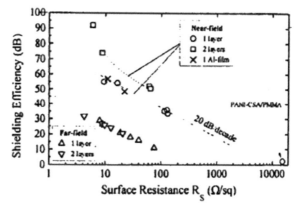

40.3   Electrostatic Discharge and Antistatic

Electrostatic discharge (ESD) or antistatic coatings are currently used in a wide variety of packaging of electronic circuit elements. C-black-filled plastics or plastics coated with a cohesive, non-flaking conductive paint are most commonly used in such applications, since a bulk conductivity of the order of 10^{-4} S/cm suffices. CPs incorporated into plastics as blends or composites could however present many advantages over this current technology, including higher conductivity/weight ratios, lower cost, and potentially more facile processing. As another similar application, most cables in the electrical and electronic industries which require shielding contain a metal braid, usually of Al, as an outer sheath to the cable. Some sheaths have however been fabricated with such plastics as polyethylene or polypropylene containing dispersed C-particles, with less success. These materials could easily be replaced by blends of CPs in sheathing plastics; such blends have been well characterized (see description elsewhere in this book).

Kulkarni and others at the Americhem, Inc., Cleveland, OH, USA, described blends of P(ANi) with thermoplastics such as PVC, nylon, and PMMA, where P (ANi) loadings of 15 to 30 v/v% gave claimed conductivities of 10–20 S/cm [1173]; these blends have been described in an earlier chapter. An alternative method of fabrication of such conductive blends described by these workers was dispersion in a liquid (solvent or sol), usable with such thermoplastics as PET, polycarbonate, acrylics, and PETG. These workers found that the requirements for antistatic or electrostatic discharge (ESD) coatings of ca. 10^9–10^3 Ω/square resistivity and transparency in the >40% region could be met easily with these P(ANi)/thermoplastic blends.

The BASF group has attempted to market antistatic conductive film based on poly(pyrrole) coatings [1930]. Additionally, several Japanese camera and semiconductor companies, e.g., Konica and Hitachi, have patented ESD technology for photofilm and semiconductor applications based on CPs, primarily P(ANi) [1931, 1932]. The semiconductor applications include packaging and handling of semiconductor wafers and magnetic disks.

An additional potential area of application for antistatic materials is photographic films, which currently use polymeric coatings with conductive particles or ionizable groups, to fight dust accumulation and other problems. These coatings however are very environmentally sensitive, especially to humidity. An additional requirement for such photofilm antistatic coatings however is transparency in the visible, initially not thought to be attainable by CP coatings. However, recent work at Americhem, Inc. in the USA, with very thin coatings of P(ANi) [1173, 1215], and at Agfa-Gevaert, Mortsel, Belgium, with poly(isothianaphthene) (P(ITN)) [1219, 1933], which is unique among CPs in being substantially transparent in its oxidized, conductive form, has shown more promise. In the Americhem work, the commercially available, VERSICON (Monsanto Co., St. Louis, MO, USA) form of P(ANi) was coated from a dispersion in an organic solvent containing a film-forming material. Coatings with resistivities of ca. 10^7 and ca. 10^{10} Ω/square had mid-visible (550 nm) % transmittances, respectively, of 50% and 80%. All films had high moisture resistance. The Agfa workers coated P(ITN) from aqueous dispersions containing lambda-carrageenan as a polymeric surfactant. Transparent (visible optical density ca. 0.02) coatings of P(ITN) on PET substrates showed resistivities of ca. 8.48 Ω/square.

40.4 Microwave Absorption and Radar Cross-Section (Rcs) Reduction

A fair estimation of the microwave absorption properties of a material is provided by a quantity known as the loss tangent:

$$\tan \delta = \sigma/(\omega \varepsilon \varepsilon 0) = \varepsilon'' \varepsilon' \tag{40.1}$$

where ε, ε', ε'', ε_o, and σ are, respectively, the complex permittivity of the sample, its real component, its imaginary component, the complex permittivity of free space, and the conductivity of the sample at the frequency of interest, ω. The Epstein group [1934] performed measurements, on P(ANi) samples of various orientations, of tan(δ) at a fixed microwave frequency, 6.5 GHz. Their results are shown in Fig. 40.4. This group has reported tan(δ) values as high as 8.0 at room temperature (cf figure) [1935].

Within the last 5 years or so, several studies have been performed on dynamic (switchable) microwave absorbers. Studies at the US Navy's China Lake, CA facility, although highly secretive, are thought to involve metallic grids deposited on high-microwave loss substrates such as Teflon on which a CP layer is deposited [1936]. This CP layer is switched between its conductive and insulating states, thereby modulating the microwave absorption or attenuation. Stacks or other configurations of such CP devices may be used. More recently, Chandrasekhar et al. [1004] have shown that a configuration used in a switchable, IR-region electrochromic device based on CPs can be modified to achieve microwave-region electrochromic modulation.

An interesting all-solid-state, "microwave shutter," was described by Rose et al. [1937]. The active CP was CSA-doped P(ANi) which was however switched using Li^+ as the mobile ion in conjunction with $Li_xMn_2O_4$ as the counter electrode. Fig. 40.5a, b part (a) shows the structure of this shutter in schematic, while part (b) shows some typical results. Changes in the transmission and reflection of X-band microwave energy were monitored by these authors as a function of time and applied potential. Using potentials of +/− 3 V, modulation of ca. 10 dB was obtained. However, the response time was slow (ca. 10 min), and the cyclability was poor.

40.5 Comprehensive Properties' Studies

Naishadham [1347] carried out comprehensive microwave studies of the CPs P (Ac) and P(BT), as cited earlier in this chapter. Figure 40.5 above has shown the correlation he observed between the EMI-SE and the reflection factor for a multi-layer absorber consisting of several CPs. As predicted, due to less resonance in multilayer configuration, the EMI-SE is more broadband.

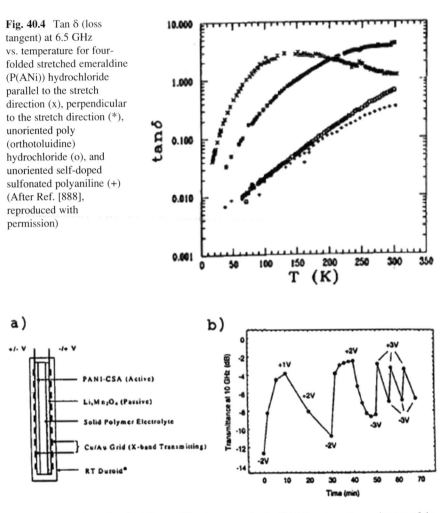

Fig. 40.4 Tan δ (loss tangent) at 6.5 GHz vs. temperature for four-folded stretched emeraldine (P(ANi)) hydrochloride parallel to the stretch direction (x), perpendicular to the stretch direction (*), unoriented poly (orthotoluidine) hydrochloride (o), and unoriented self-doped sulfonated polyaniline (+) (After Ref. [888], reproduced with permission)

Fig. 40.5 (a) Schematic of solid state CP microwave shutter (b) Microwave transmittance of the shutter on applied biases as indicated (After Ref. [1937], reproduced with permission)

In very recent work, Chandrasekhar and Naishadham [1235b, c] studied the comprehensive microwave properties of bulk P(ANi) samples in the 2–18 GHz frequency region using network analyzer/coaxial line techniques. The P(ANi) was chemically polymerized using the standard peroxydisulfate-oxidant synthesis but with unique double-doping using two proprietary sulfonate-group containing dopants. It possessed a conductivity of ca. 11 S/cm. Measurements were carried out on compressed powder samples of several mm thickness.

Figure 40.6a–c shows, respectively, the S-parameters, complex permittivity, and conductivity of the P(ANi) samples as measured by Chandrasekhar and Naishadham. The permittivity data may be compared, e.g., with those obtained by Olmedo et al. [1356] for P(ANi) with Cl^-, Tos^-, and NDSA dopants (Fig. 40.6).

Fig. 40.6 Microwave studies of P(ANi) double doped with two different sulfonate dopants: (**a**) S-parameters, (**b**) complex permittivity, and (**c**) conductivity (After Chandrasekhar and Naishadham [1235d])

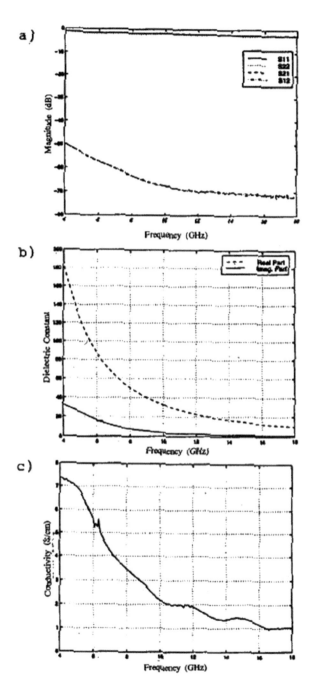

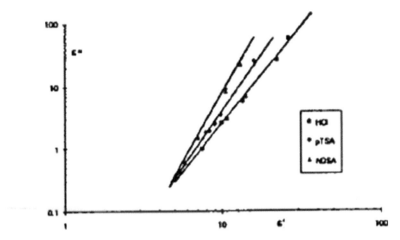

Fig. 40.7 ε"(10GHz) versus ε'(10GHz) for poly(aniline) samples doped with three different acids: (HCl), p-toluene sulfonic acid (PTSA), and naphthalene 1,5-disulfonic acid (NDSA) (After Ref. [1356], reproduced with permission)

Figure 40.7 succinctly summarizes all the relevant measurable parameters for these samples, viz., reflection (return) loss, microwave absorption, and EMI-SE, across the wide frequency range studied.

40.6 Conductive Textiles

This topic merits separate treatment due to the novelty of the technology and wide variety of studies thus far: the applications are primarily passive (unswitchable) camouflage in the microwave and far-IR regions and sensors.

A company called Nagase ChemteX, based in Japan, has collaborated with DuPont of the USA, to attempt to develop and bring to market conductive textiles, especially such items as "smart socks" to the market [1938]. In one of their proposed products, a variant of PEDOT/PSS (poly(ethylene dioxythiophene/poly (styrene sulfonate)) that is claimed to be more flexible is deposited from an ink and screen printed onto textiles. This material is claimed to stand up to 100 wash cycles (which any reader will recognize as still inadequate).

The most well-known and most advanced early work in CP-based conductive textiles has been that of Kuhn and coworkers at the Milliken Research Corp., Spartanburg, SC, USA, a division of the Milliken textiles group [1137, 1138, 1939, 1940]. This work involves the in situ, template-type chemical polymerization of CPs, primarily poly(pyrrole), onto individual textile fibers, yielding a continuous, conductive coating. These fibers are then woven, knitted, or otherwise processed into cloths. Fibers used have included various nylons and Kevlar. Continuous production of up to 1000 linear yards at a time in 80" width has been carried

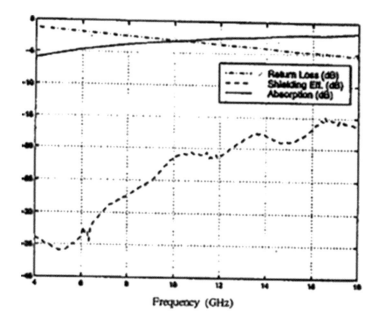

Fig. 40.8 Cumulative microwave parameters (as indicated) for same P(ANi) sample as in Fig. 10. 6 (After Chandrasekhar and Naishadham [1935d])

out. The CP-coated fabric has been given the trademark Contex, and a variant combining the microwave (X-band region) camouflage with visible/IR region camouflage and used in the form of nets has been given the trademark Intrigue.

At typical X-band frequencies between 8 and 10 GHz. Contex fabrics having surface resistivity of ca. 13 Ω/square have shown attenuations of ca. 28 dB. Milliken's Contex and Intrigue products have been tested as camouflage materials and decoys in field conditions. Due to the continuous surface conductivity of CPs as opposed to the island-like, granular conductivity of C-particle filled materials, such textiles have better performance than C-particle coated materials. Fig. 40.8 shows the microwave attenuation properties of Milliken's Contex brand CP P(Py) coated textiles before and after 53 d outdoor exposure. The data reported are however a bit ambiguous, since the loss of microwave attenuation with aging which appears minimal from this figure does not parallel the data on loss of P(Py) conductivity on aging (Fig. 40.9).

In a more recent study [1137], the Milliken group described their in situ polymerization procedure in detail and investigated in detail the effects of such factors as dopant type and level, substrate (fiber) type, degradation, and monomer concentration. They found that the nature of the fiber appeared to have little influence on the polymerization, with hydrophilic fibers such as rayon or cotton working as well as hydrophobic fibers such as polypropylene or polyester. Quartz fibers worked as well as those of Kevlar, and indeed, even high surface area ceramic beads could be coated with CP successfully.

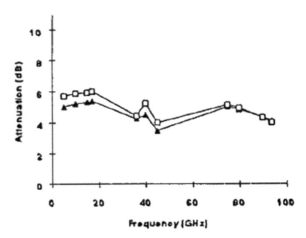

Fig. 40.9 Microwave attenuation of Milliken Research Corp.'s Contex® net as a function of frequency. (□) initial reading, (▲) after 53 days outdoor aging (After Ref. [1940], reproduced with permission)

It is to be noted however that these textiles remain static, passive (unswitchable, non-dynamic) camouflage materials. Applications for such textiles have also been claimed in ESD/antistatic coatings for industrial belts, e.g., in coalmines and personnel uniforms, e.g., for explosion-proofness.

Wiersma et al. at DSM Research, Geleen, the Netherlands [1220], described an interesting way of coating textile fibers with CP/polyurethane and CP/alkyd blends from aqueous colloidal dispersions made in a commercially available aqueous resin dispersion. The dispersions included a polyurethane-in-water dispersion (DSM Resins URAFLEX XP 401 UZ, 40% solids content). A variant of the template-type in situ polymerization was used. After polymerization, the dispersed CP sol could be applied as a coating onto the textile fibers. CPs used included P(Py) and P(ANi).

The Naarmann group at BASF also carried out work on CP-coated textile fibers. Fibers tested included nylon, glass fiber fleeces, and C-fibers [1941].

Genies et al. [1141] described coating of glass textiles via in situ polymerization of aniline thereon, obtaining P(ANi) thicknesses from 0.07 to 1.3 μm and surface resistivities of 1416–29,268 Ω/square. The ready-made glass textile substrates used were manufactured by Porcher Textile, containing glass fiber weaves of 17.5 or 24 threads by warp and 12 or 23 threads by weft. Each thread was made of several 7 or 9 μm glass filaments (Vetrotex St. Gobain). The conductivity of the P(ANi) itself ranged from 0.08 to 1.84 S/cm.

Olmedo et al. [1356] described another typical template-type in situ chemical polymerization of P(ANi) and poly(alkyl thiophenes) on many textile fabrics, with various dopants, but using $K_2Cr_2O_7$ as oxidant. Real and imaginary permittivities for poly(3-octyl/butyl thiophenes) doped with $FeCl_4^-$ lay, respectively, in the 7–30 and 2–80 ranges.

Collins and Buckley [1579] described the use of P(Py) and P(ANi) coatings on PET and nylon threads woven into a fabric, for a very different application – that of conductometric sensing of chemical warfare agents. Conductivity changes would be expected to occur in the CP coatings due to adsorption of the agents. Chemical warfare simulants tested included dimethyl methylphosphonate, ammonia, and NO_2. What was most interesting about this work is that the CP-coated fabrics were procured commercially from Milliken Research. They included: P(Py)/naphthalene-disulfonate on a 150 denier PET, resistivities in the 211–3000 Ω/cm^2 region; P(Py) on an 840 denier nylon with the same dopant, resistivity ca. 100 Ω/cm^2; P(ANi) on a 150 denier PET doped with Cl^-, resistivity ca. 75 Ω/cm^2.

Other microwave-active CP textiles described include a P(Py)/aramid honeycomb [1942] and materials for use in naval aviation applications [1943].

40.7 Microwave Smoke

In the present practice of the art, a passive (unswitchable, non-dynamic) screening of radar signals in the battlefield is achieved by spreading a conductive "smoke," e.g., in the form of a smoke bomb, consisting of conductive microparticles of graphite, metal-coated fiber (e.g., Fe-coated glass), or, the like, essentially a "microwave-smoke." These materials are also denoted as "microwave and mm-wave obscurants." The frequencies of interest here are typically the important military radar frequencies of 94 and 35 GHz, and the protection sought is typically for advancing tanks. Typical graphite fibers currently used are those with large "aspect ratios" (length/diameter), e.g., 0.1 × 200 μm. The current technology, besides making an environmental mess, provides inadequate screening [1944]. Besides more adequate microwave screening, what is also sought is particles that would degrade rapidly even in desert environments.

With this in mind, Jenkins et al. [1945] studied coatings of P(Py) and P(ANi) on environmentally degradable filaments for such applications in detail. Among filaments tested were polyethylene monofilament (Allied Signal "Spectra") plasma treated to enhance coatability, cellulose, and polyester, the latter two very biodegradable. The standard template-type in situ chemical polymerization which by now should be very familiar to the reader was used to prepare the coatings. Freshly coated fiber conductivities obtained varied in the range 0.6–4 S/cm, vs. ca. 0.6 S/cm for 8 μm graphite filaments. Normalized extinction coefficients, in g/m^3 at 35 GHz, obtained for graphite fiber (3.2 mm × 8 μm), uncoated cellulose filament, P(ANi)-coated cellulose filament, and P(Py)-coated cellulose filament were, respectively, 1.7, 0.021, 0.365, and 0.03. For 94 GHz, these figures were, respectively, 0.32, 0.012, 0.11, and 0.011.

In a more recent study, Buckley and Eashoo [1305] investigated commercially procured (ETI Inc., Orlando, FL, USA; Milliken) P(Py)-coated fibers of poly(vinyl alcohol) (PVA) and PET for microwave/mm-wave obscurants. While these fibers possessed the desirable property of environmental instability, they also unfortunately were substantially humidity and heat sensitive, leading to probable rapid degradation in obscuration performance.

A very significant and interesting study of microwave/mm-wave obscurants was produced recently by Carter, working at Lockheed Martin's Tactical Air Systems, Fort Worth, TX, USA [1305]. In this, P(Py)-coated "microballoons" were prepared. The electrically active filler in the radar absorbing material consisted of mineral ash microballoons coated with a P(Py) formulation. This filler was then blended into a commercial resin or spray-on preparation, and the mixture was cured at a temperature determined by the binder. The resulting "composite" (blend) had a density of less than 1. Electrically, it exhibited an unusual amount of dispersion: the real part of the complex permittivity dropped rapidly as a function of increasing frequency, while the imaginary part, which governs absorption, remained relatively strong. These properties translated into advantages in terms of broadband radar absorbing treatments. The variable diameter, void-containing mineral ash microballoon substrates eliminated resonance emphasis on frequencies, yielding the broadband nature.

40.8   Microwave Welding

Epstein et al. apparently first proposed the use of CPs to promote a "microwave-welding" of plastic joints [888, 1935]. In this procedure, the CP is placed between two pieces of the plastic to be welded, e.g., poly(ethylene) (PE). The primary application is remote welding or sealing, where the parts to be joined, once fabricated, are in an inaccessible area. Thus as an example, in one study described by Epstein et al., nylon 6, PETG, polycarbonate, or high-density PE could be welded in a standard, 500 W microwave oven (2.45 GHz). The CP, emeraldine-HCl, was used as a compacted powder or as a gasket (compression molded with HDPE powder, 8 MPa, 190 °C, 1 hr., loss tangent ca. 0.6). In addition to microwave heating, a small pressure was also applied on the joint during welding, via introduction of compressed air into the oven. Microwave exposure time typically was 30 sec. A separate study at 23 MHz showed that the rise in temperature in bulk CP samples was very rapid and reached its maximum, ca. 200 °C, at ca. 45 and 18 secs., respectively, for P(ANi)-HCl and P(ANi)-sulfonate. The joint strength obtained for such welded joints ranged from ca. 15 MPa to close to 25 MPa (the value for HDPE alone) as evaluated on an Instron 4201 machine. The strengths of joints of polycarbonate, nylon, and PETG were, respectively, 59%, 60%, and 86% of the strengths of the neat materials. Similar comparison of strengths of blends of 50% P(ANi)-HCl with HDPE showed that welded samples had joint strengths ca. 82% of those of compression molded samples.

Staicovici et al. [1946] studied the microwave welding and disassembly of HDPE bars using P(ANi). They noted that controlling the amount of CP remaining at the assembly interface after microwave welding was the key to whether a joint could be disassembled or not.

Several patents have also been issued for "compositions" for microwave welding of plastics. Thus, Yamamoto et al. [1947] describe a formulation with 100 parts polyethylene resins and 5–100 parts P(ANi) particles. In a typical application, a sheet of composition 100 parts LDPE +30 parts Versicon (P(ANi)) is wound around a tube of HDPE plastic, which is inserted into another tight-fitting tube as a joint end, and the assembly irradiated at 2.45 GHz and 500 W, to yield a joint with peel strength of 15.5 kg/cm^2.

40.9   Problems and Exercises

1. Give one example of each of the following applications covered in this chapter, including detailed fabrication of materials or devices, if any is involved, and the characterization methods used to determine performance: (a) Broad-spectrum EMI shielding. (b) Transparent antistatic coatings. (c) Microwave attenuation. (d) Radar cross-section (RCS) reduction. (e) Passive camouflage cloths. (f) Passive micro-/mm-wave obscuration. (g) Microwave welding.
2. (a) What in your opinion are the most important properties determining passive RCS reduction provided by CP coatings? (b) Besides the methods described in this chapter, how would you design coatings or devices for dynamic (switchable) RCS reduction using CPs? (c) How would you achieve broadband (wide-spectrum) performance, minimizing the effect of resonance within the devices or coatings, in both (a) and (b)? (d) Are there any other frequencies of military or other interest in the micro-/mm-wave region besides those cited in this chapter? Make a list of all frequencies of interest.
3. Describe an in situ synthesis of a woven, CP-coated, conductive textile fabric. Identify at least two applications of this besides any cited in the answer to question 1 above.
4. Can you suggest a way in which a CP itself can be made biodegradable and fashionable into high-aspect-ratio fibers for use as a microwave smoke directly (i.e., not as a coating)?
5. Suggest a drawback of CP-based microwave welding which would prevent its use in consumer-based end applications.

Chapter 41
Electro-Optic and Optical Devices

Contents

41.1   Device Types and Motivations for Development

Electro-optic (E/O) and optical devices based on CPs which are discussed in this chapter can be usefully classified into three categories:

1. Those based on second-order nonlinear optical (NLO) properties
2. Those based on third-order NLO properties
3. Those based on properties other than NLO
4. CP-based lasers

Waveguides, one of the most studied CP applications, fall into categories (1) and (2).

The driving forces behind the development of E/O devices based on CPs have been several. The increasing speed of common microprocessors such as those used in PCs, with, e.g., 450 MHz processors not uncommon as of the date of this writing, has created severe demands not on the on-chip processing technology but rather on the electrical interconnects between chip elements. Typical problems include cross talk, power dissipation across distances, and EMI shielding. Optical interconnections between electrical components, i.e., "E/O interconnects," are one possible solution. Such E/O interconnects can also be used in telephone and other telecommunication as interfaces between optical fibers.

There has also been much talk of "all-optical computing," presumably viable at some future date, whose speed and "cleanliness" are predicted to be orders of magnitude better than present-day electronic computers. E/O devices and memories

© Springer International Publishing AG 2018   671
P. Chandrasekhar, *Conducting Polymers, Fundamentals and Applications*,
https://doi.org/10.1007/978-3-319-69378-1_41

based on CPs could be expected to be a major component of such optical computing. Specific component targets may include memories, beam splitters, switches,
and routers. CP E/O devices could also be used to encode information into laser
beams via modulation of the beam's intensity or phase.

The major advantages of CPs in such applications are said to be their relatively
easy processibility (in de-doped form), very rapid (sub-ps) response times, and
expected lower cost. Unfortunately, however, two major requirements of E/O
materials for such applications are low optical loss (loss is caused by both absorption and scattering), measured in dB, and high thermal and optical stability (ability
to withstand heating and other effects caused by the high optical intensities required
for most applications): thus far, CPs have failed in being able to meet both these
requirements, as we will see below. Scattering losses can frequently be minimized
by improved processing eliminating such things as air bubbles and inhomogeneities
in the active material, but intrinsic absorption losses are harder to deal with.

Among very specific applications fancied for practical second-order CP materials when they are developed are phase and amplitude modulators and frequency
doublers to produce blue light from semiconductor lasers as applied to data storage
and xerography and parametric amplifiers as applied to communications. Among
very specific applications fancied for practical third-order CP materials, again when
they become available, are all-optical serial information processing at terahertz and
faster rates, parallel processing of optical information, and dynamic holography.

41.2   Waveguides

One of the most studied E/O applications for CPs has been optical waveguides. The
best-known waveguides are of course optical fibers, but waveguides easier to
construct for test purposes are planar or slab and channel waveguides. The kinds
of integrated optic devices that may be constructed from optical waveguides and
used, e.g., in optical computing, have been summarized by Stegeman et al. [1948]
and DeMartino et al. [1949] and are depicted in Fig. 41.1.

Waveguide studies with CPs initially started as an extension of waveguide
studies with other, non-CP, polymers such as PMMA, polyimides, and poly(vinyl
pyridine). It is useful therefore to cite some typical performance parameters for
PMMA, polyimide, and poly(vinyl pyridine) which may be used as a comparison
for the performance of CP-based devices cited in the sequel. For these three
materials in sequence, these are listed in Table 41.1. Table 41.2 gives nonlinear
waveguide parameters for the best-performing CP, the poly(diacetylene) PTS
compared with those of semiconductors and glasses, both on-resonance (i.e., at an
absorption) and off-resonance.

Singh and Prasad [1441] demonstrated optical bistability in a planar device
functioning as a quasi-waveguide at 0.7 and 0.653 μm. The upper switching
threshold power was 3.86 and 2.55 MW/cm^2, respectively, for each of these
frequencies. This corresponded to values of the nonlinear refractive index, n_2 of

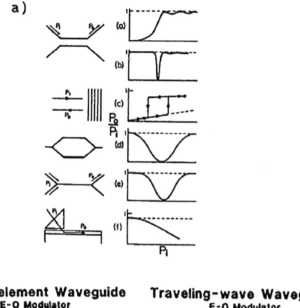

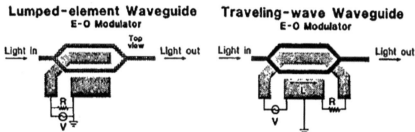

Fig. 41.1 Schematic illustrations of E/O devices based on CPs. (**a**) Standard integrated optics devices and their response to optical power with and without nonlinearities: (*a*) 1/2 beat-length directional coupler, (*b*) 1 beat-length directional coupler, (*c*) distributed feedback grating, (*d*) Mach-Zehnder interferometer, (*e*) mode sorter, and (*f*) prism. (**b**) Two waveguide types (After Ref. [1948], reproduced with permission)

Table 41.1 Reference performance parameters for planar optical waveguides, measured at 632.8 nm.

Polymer	n(TE)	n(TM)	Loss (dB/cm)
PMMA	1.508	1.542	1.8–4.2
Polyurethane	1.56	1.56	<1
Polyimide	1.690	1.645	~10
Poly(vinyl pyridine)	1.572	1.572	1.6

After Ref. [1950–1952]

n(TE) and n(TM) are the refractive indices with the electric vector of the incident beam normal and in-plane to the polymer film

Table 41.2 Typical comparative nonlinear waveguide parameters for various classes of materials

Waveguide material	$n_2(M^2/w)$	$\tau(s)$	$\alpha(cm^{-1})$	Δn_{sat}	W
(a) Semiconductors; e.g., MQW GaAs/GaAlAs					
On-resonance	-10^{-8}	10^{-8}	10^4	0.1	~0.1
Off-resonance	-10^{-12}	10^{-8}	30	$\sim2\times10^{-3}$	~0.9
(b) Organics; e.g., PTS (a poly(diacetylene))					
On-resonance	2×10^{-15}	2×10^{-12}	10^4	~0.1	~0.2
Off-resonance	10^{-16}	$<0.03\times10^{-12}$	$<10^2$	$>10^{-3}$	>0.15

After Ref. [1953]

n_2 nonlinear refractive index, τ switching time, α absorption coefficient, and Δn change in RI

ca. 5×10^{-3} cm^2/MW, indicating a thermal effect. Sasaki et al. [1954] constructed a hybrid waveguide from a Langmuir–Blodgett film of a poly(diacetylene), a glass waveguide, and a silica substrate. The optical bistability observed could not be definitively ascribed to electronic origin and could have been thermal.

Mittler-Neher et al. [1561] reported studies with slab waveguide structures fabricated using the ladder-type CPs poly(p-phenylene benzobisthiazole) (P (BZT)) and poly[2,2'-(1,4-phenylene)-6,6'-bis(4-phenylquinoline)] (P(PPQ)) in pure form and as a 50:50 blend with Nylon. To fabricate the waveguides, the pure polymer or blend was spin coated from a solution of its complexes onto a fused silica substrate to which a grating had been previously ion milled. Chemical treatment to remove the complexed groups yielded insoluble P(BZT) or P(PPQ) blend-based waveguides. Samples were studied at 1.064, 1.319, and 1.535 µm with incident beam polarizations corresponding to TE and TM modes (respectively, the incident light electric vector in-plane with and perpendicular to the CP film). In all samples, refractive indices decreased with increasing wavelength. Table 41.3 summarizes some of these results. Figure 41.2 shows the mode profile (transmission through the waveguide vs. the incidence angle) for a freestanding P(BZT) film. The losses in these waveguides however remained too high for practical device applications.

One of the first classes of CPs studied for optical waveguiding was the poly (diacetylenes), both in single crystal and film form. For single crystals, a special growth technique is needed to be used to obtain films thin enough for optical waveguide fabrication; the complex refractive index and intensity-dependent phase shifts were characterized in detail for several poly(diacetylenes) [1950, 1955]. In many spin-coated thin films of poly(diacetylenes), however, the NLO effects were postulated to be primarily thermal in origin [1956]. Practical devices were fabricated primarily using the poly(diacetylene) known as P (TS) (or PTS), having the formula -CH$_2$OSO$_2$(C$_6$H$_4$)CH$_3$; devices switching at incident intensities of tens of mW were claimed [1956].

However, it is noted that even this power range is rather high as compared to µW ranges which are desirable for practical E/O device application. Thus, the caveat

Table 41.3 Summary results for P(BZT) slab waveguide structures

Material	Thickness [μm]	Wavelength [μm]	n(TE)	n(TM)	Losses [dB/cm] min-max
P(BZT)	2.87	1.064	1.815	1.629	29.8–58.9
		1.319		1.627	15.7–31.0
P(BZT)	2.0	1.064	1.742	1.679	No streak
P(BZT)	1.0	1.064	1.743	1.743	21.4–42.2

(legend) [1561]

Fig. 41.2 Mode profile (transmission through the waveguide vs. incidence angle) for a freestanding P (BZT) film at $\lambda = 1.064$ μm (After Ref. [1561], reproduced with permission)

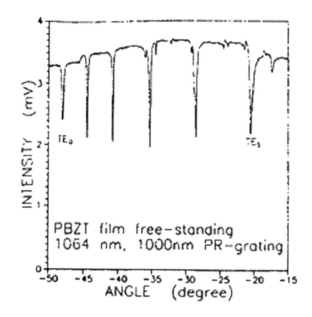

cited earlier that observed nonlinearities are still several orders of magnitude less than those desired for practical device application still stands.

Baker et al. [1957] presented a detailed description of the fabrication of waveguide structures from soluble, high-MWt ($> 10^5$) poly(diacetylenes), poly (4BCMU) and poly(3BCMU). These were spin-cast from cyclopentanone, chlorobenzene, or DMF solutions. Micron-sized patterns were created in the CP films of thickness > 1 μm using deep-UV lithography. In an alternative fabrication method, composite waveguides were constructed from a patterned glass substrate with a CP overlayer. The devices were sealed to protect the CP from the environment. The two procedures are depicted schematically in Fig. 41.3a, b. The latter procedure yielded waveguides with single-moded behavior at 1.3 μm with the light confined to the CP layer immediately above the channel and the patterned glass, when evaluated by end-fire coupling of light into the composite structure. Losses, including coupling losses, for a 5-cm composite waveguide at 1.3 μm were 5–8 dB. In earlier work from the same group [1958], prism coupling techniques were used to evaluate the optical properties of asymmetric slab waveguides fabricated from thin, spin-cast

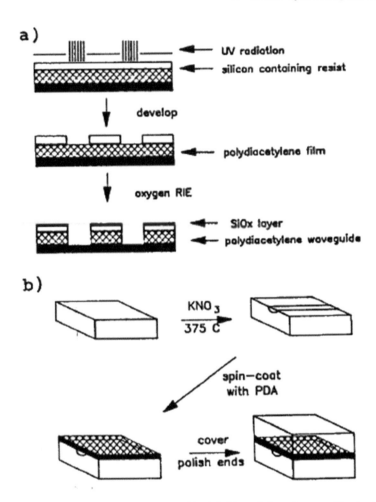

Fig. 41.3 (a) Bilayer scheme for the lithographic definition of channel waveguide structures in poly(diacetylene) films.(b) Scheme for the fabrication of composite channel waveguide structures (After Ref. [1957], reproduced with permission)

films of poly(3BCMU) and poly(4BCMU). This work showed that guiding of 1.06 μm light over distances greater than 3 cm in the CP films could be achieved. Losses were of the order of 1 to 2 dB per cm. The refractive index normal to the plane of the film was found to be lower, evidenced by the fact that the coupling angle observed for the TM mode (incident light electric vector perpendicular to the film) was ca. 6° higher than that for the TE (vector in plane of film) mode. Figure 41.3 shows coupling angle data obtained in this work.

In work of some potential import for practical waveguides, Halvorson and Heeger [1959] measured the two-photon absorption spectrum of oriented *trans*-P (Ac) and found that the two-photon absorption susceptibility ruled out waveguiding devices in this CP. They then cautioned that any CP must be characterized for its two-photon absorption coefficient before attempting to fabricate waveguides from it.

41.3 Other Second-Order NLO Applications

Among the requirements for E/O devices based on second-order NLO effects is the absence of symmetry, frequently achieved by aligning the dipole moments of chromophores (which are responsible for the absorptions creating the nonlinearities), a process known as "poling." A common method of poling is to include a second-order NLO-active CP in a polymer matrix, heat the latter to above its glass transition temperature, apply a strong electric field, and then rapidly cool the matrix to lock in the asymmetry. Such methods with CPs have met with limited success: the locked-in asymmetry degrades over time, and second-order electro-optic coefficients are far short of those achieved with non-CP-based polymers, up to 55 ppm/ V at an incident light energy of 1.3 μm [1371b]. The success obtained with hanging pendant groups (chromophores) having high NLO activity to non-NLO-active polymer chains of non-CP polymers has thus far not been duplicated with CPs.

41.4 Semiconductor/CP (SC/CP) Interfaces

A novel type of device configuration first apparently proposed by Inganäs and Lundström [1960] and developed to a considerable degree of refinement by Chandrasekhar et al. [1961] is one which is based on interfacing a CP to an inorganic semiconductor (SC). This utilizes the well-known electrochromic properties of the CP. The SC may be a material such as CdS or n-Si. In this context, it should be noted that the CP, in its doped form, is also a SC, but its semiconducting properties are not the primary focus of interest in such SC/CP interfaces – rather, its electrochromic properties are.

When the n-type SC is illuminated, it is able to accept charge from a p-type CP, to which class most CPs belong, thus causing its oxidation and electrochromic darkening. If the SC/CP device's electrical configuration is suitably designed, the CP can then either spontaneously revert to its original, de-doped, light-colored state, or it may be switched back by an applied potential. The idea behind using the inorganic SC is to obtain rapidity of response, potentially in the ps regime, and to allow the SC to provide both the trigger and the energy for the CP switching. The preferred medium of SC excitation is a laser beam.

In the Inganäs and Lundström work, an n-Si/poly(N-Me-pyrrole) SC/CP interface was used, with application as a fast-optical memory envisioned. In the write step of this memory, the CP was oxidized by illuminating the SC. In the erase step, a negative (cathodic or reducing) potential was applied to the CP, returning it to its original, de-doped state. This device however required the use of liquid electrolyte,

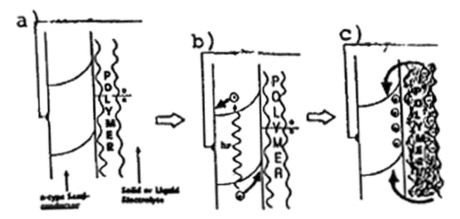

Fig. 41.4 Schematic representation of mode of action of SC/CP interface. The n-type semiconductor bands are bent upward at the polymer-SC interface, as shown. O/R is the electronic level corresponding to the O/R redox potential: rest state (**a**), laser excitation (**b**), and charge transport in polymer (i.e., polymer switching) (**c**). The processes occurring in chronological order are laser excitation of the SC (sub-nanosecond) (**a**), charge transfer to polymer (sub-nanosecond) (**b**), oxidation-reduction of initial polymer layers (sub-nanosecond risetime; falltime from ca. 10 ns to microseconds) (**c**), and charge transport in bulk polymer (slow process) (c)

and switching times were of the order of ms. It thus did not utilize the solid-state and "ultrafast" photonic capabilities of such an interface.

In later work, Chandrasekhar et al. [1961] constructed devices having the interface shown in Fig. 41.4. The sequential steps involved in electrochromic switching of the CP are indicated in the figure legend.

The driver behind this work was the increasing use of pulse and continuous wave (CW) lasers in the battlefield for such uses as range finding, target designation and communication, and the concomitant need to protect sensitive sensors and personnel eyes from accidental (or deliberate) exposure [1953].

For fabrication of a SC/CP interface, commercial thin-film SCs were unavailable at reasonable cost (excepting for ITO), and single-crystal SCs could not be used due to high absorptions. Chandrasekhar et al. [1961] thus deposited thin-film SCs on quartz substrates via thermal evaporation using the appropriate raw materials, e.g., Cd, Se, and CdSe for CdSe thin films. They found that if a near monolayer was deposited and annealed first, subsequent depositions of very thin SC films ($<$70 nm) were epitaxial. The SCs they tested included CdSe, CdS, AlSb, Bi_2S_3 and Se. At moderate or high laser energies (high nJ/cm^2 per pulse), film ablation in such preliminary devices was a serious problem. Subsequently, it was found that deposition via coating from nanoparticulate SC sols prepared via ball milling in inert solvents and coating of the CP in a similar fashion but in an inert polymer matrix such as PMMA or poly(carbonate) yielded SC/CP interface devices which showed minimal damage from the high intensity of lasers.

Samples were characterized by a standard laser pump-and-probe transient absorption (TA) methodology, where the sample (device) is pumped with a laser

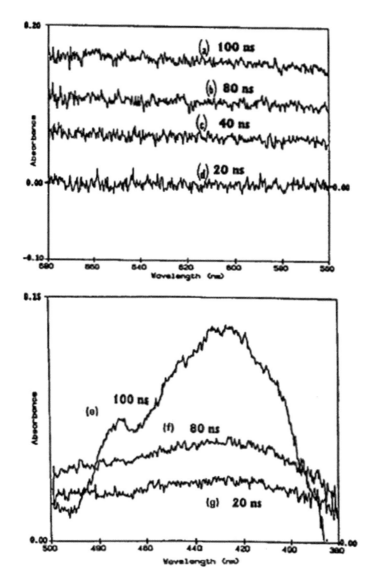

Fig. 41.5 Initial transient absorption data: for (*a–d*), the Se/poly(3-amino-quinoline) system shows transient absorption decay from 100 ns; for (*e–g*), the CdSe/poly(diphenyl benzidine) system shows transient absorption decay from 100 ns (After Chandrasekhar [1961])

and its broad-spectrum absorption then probed as this decays. Figure 41.5 shows initial TA data on two systems, Se/poly(3-amino-quinoline) (Se/P(3AQ)) and CdSe/poly(diphenyl benzidine).

The laser shielding efficiency of the SC/CP interfaces was characterized as *Delta (OD) = OD(laser) - OD(rest)*; here OD = optical density, OD(laser) = transient

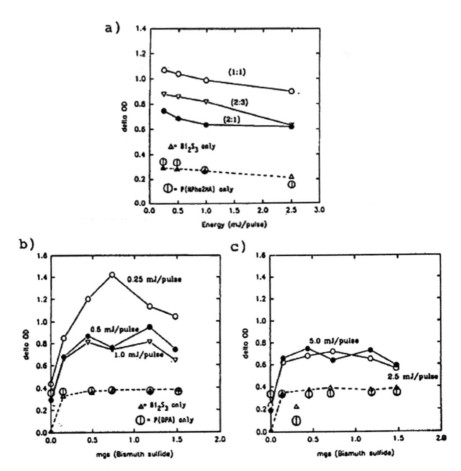

Fig. 41.6 (a) Laser shielding efficiency, ΔOD, as a function of incident laser energy for the Bi_2S_3/ poly(N-phenyl-2-naphthylamine) system, also showing data for SC-only and CP-only references. Molar ratios, CP-SC, are in parentheses. All data in this and subsequent figures are for immobilized fine-particulate configuration.(**b, c**) Laser shielding efficiency ΔOD, as function of proportion of Bi_2S_3, for the Bi_2S_3/poly(diphenyl amine) system. The apparent higher efficiencies at lower incident energies are a combined artifact of sample ablation and saturation effects, (b) and (c), respectively, for lower and higher incident energies. After Chandrasekhar [1961]

OD of sample induced by laser pumping, and OD(rest) = sample OD. Figure 41.6 shows some of Chandrasekhar's results.

Chandrasekhar et al. [1961] concluded from their results that actual charge transfer and switching of the CPs were occurring only in the initial CP layers directly at the interface, to a depth of ca. 10–15 nm, i.e., it was a surface phenomenon. This appeared to confirm hypotheses of conduction mechanisms within CPs, i.e., that intrachain charge propagation is very fast but interchain charge hopping is slow and rate determining. This however meant that the SC/CP interfaces, although highly novel, would not come near to meeting the specifications for efficient military laser shielding devices, which required a broadband (near-UV-NIR)-switched OD of ca. 4.0.

41.5   CP-Based Lasers

Another very new development in CP applications has been CP-based semiconductor lasers. The Friend group at Cambridge, England, first noted [1962] that one way to study whether the photoexcited states in electroluminescent CPs such as P(PV) are fundamentally non-emitting interchain species or emitting intrachain species was to confine solid films of the CP in a microcavity, where spontaneous and stimulated emission of the CP could be studied. What resulted in an early test of such a device [1962] was the observation that upon excitation at 355 nm, stimulated emission, i.e., lasing activity (predominantly at ca. 550 nm), was observed. Figure 41.7a shows a schematic of the microcavity structure used by these authors. Figure 41.7b shows characterization data for such P(PV)-based microcavity (laser) devices [1962, 1963].

Lasing activity in CPs is not surprising, since CPs are semiconductors and semiconductor lasers are well known. Work prior to that cited above had provided indications of stimulated emission in CPs, but it appeared that this was overwhelmed by absorption, resulting in no net emission. Indeed, lasing activity had been reported as early as 1996 in TiO_2 nanocrystals containing very small proportions of a P(PV) derivative [1964]. Net emission was achieved by the Friend group via suitable modification of the CP morphology and microcavity design.

In a more detailed study in sub-µm thick films of a number of CPs including derivatives of P(PV), P(P), and poly(fluorene), Diaz-Garcia et al. [1965] found lasing evidenced by a dramatic collapse of the emission line width (to ca. 7 nm) at pump energy thresholds as low as 10 µJ/cm^2. They found that lasing wavelengths in these CPs spanned the entire visible spectral region and that processing could be used to control lasing through chain packing. Figure 41.8a, b shows lasing data for one candidate CP, a derivative of P(PV).

41.6   Other Optical Devices

Other optical devices based on CPs which have been proposed have included thermo-optic switches and modulators [1953, 1966]. Such devices would typically use a change in refractive index caused by heating a CP to modulate an optical beam. The change in refractive index with such thermal modulation is usually much greater than for electrical or optical modulation, but it is also much slower. Thermal optical bistability was observed with switching power of less than 4 MW by Blau in 1987 [1441c]. More recently, extinction coefficients up to 25 dB and drive powers of ca. 30 mW have been claimed for such thermo-optic devices [1967]. Halvorson

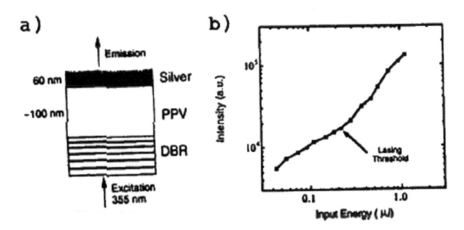

Fig. 41.7 Lasing from P(PV)-based CP microcavities: (**a**) schematic of microcavity structure and (**b**) spectrally integrated intensity of microcavity laser (After Ref. [1962, 1963], reproduced with permission)

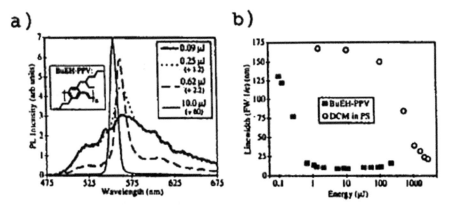

Fig. 41.8 (**a**) Photoluminescence spectra of 210 nm thick film of BuEH-PPV spun on glass at various pump energies both above and below the lasing threshold. (**b**) Line width vs. pump pulse energy for same sample as in (**a**) and for the laser dye DCM suspended in polystyrene (circles) (After Ref. [1965], reproduced with permission)

and Heeger [1959] observed that spatially modulated four-wave mixing could potentially be used in ultrafast (sub-ps) optical computing applications to get around the problem of large two-photon absorption found in many CPs, such as *trans*-P(Ac).

Peyghambarian at the University of Arizona, Tucson, AZ, USA, in collaboration with Marder at CalTech, Pasadena, CA, USA, investigated the use of holographic and phase change properties emanating from third-order NLO effects in CPs for the encryption of credit cards for foolproof identification and to prevent fraud [2064]. To date, however, the technology has not been commercially implemented.

41.7 Practical Implementation

In spite of the large amount of research carried out with the end goal of practical CP-based E/O devices, practical implementation has eluded most researchers. Indeed, in second-order NLO materials, ancient Li niobate still rules the roost in practical uses, despite its fancied drawbacks and the professed superiority of CPs thereagainst.

A more clear indication of practical limitations of CP-based E/O devices is provided by the fact that the research arms of all major industrial groups that may be potential users of the technology – e.g., DuPont, Bellcore, Bell Labs, IBM, Philips, British Telecom, Akzo, and the US Wright-Patterson Air Force Base's Materials Laboratory – quietly stopped work in the area in 1995–1996 [1372]. One of the major drawbacks cited for such polymeric materials, as mentioned earlier, is high optical loss assigned to both scattering and absorption. Other drawbacks have been quirks such as unexplained reduction in transmission when light far from any absorption peaks of the CP is transmitted through waveguides fabricated from the NLO-active material [1372]. Practical incorporation into present-day devices is also a problem, for example, the core diameter of an optical fiber is 8 μm, whereas NLO films thicker than 1 μm require unacceptably high poling electric fields or have unacceptably high absorptions. Thus, butt coupling along the lines of the simplistic representation of Fig. 41.9 may not be practically feasible.

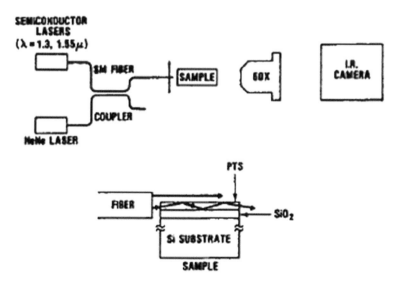

Fig. 41.9 Simplistic representation of butt coupling to a poly(diacetylene) CP single crystal (PTS) which does not take into consideration mismatch between CP film thickness and fiber diameter (After Ref. [1968], reproduced with permission)

41.8   Problems and Exercises

1. Describe the fabrication and function in at least one E/O application of wave-guides fabricated from a poly(diacetylene) and a ladder CP. Compare its expected performance parameters with those of Li niobate and polyimides.
2. Design and describe construction of an ultrafast optical memory using a CP thin film. What properties of the CP would be most important in determining the efficiency and success of such a memory? How would the memory be interfaced to other components of an all-optical computer?
3. Delineate the driving forces for development of CPs (including military applications, if any) in all E/O and optical applications and their limitations therein.
4. Describe the working of an SC/CP interface in the ultrafast mode. Cite its potential applications and potential drawbacks/limitations.

Chapter 42
Electrochemomechanical, Chemomechanical, and Related Devices

Contents

42.1 Introduction, History, and Principles

Electrochemomechanical and *chemomechanical actuators* (also called *electromechanical actuation* and *artificial muscles*) with CPs are one of the more fascinating applications to have emerged very recently. Relatively few groups have been active in this area. Although the Baughman group at Allied-Signal, Inc., Morristown, NJ, USA, claimed to have first conjectured electrochemomechanical actuators and reduced them to initial laboratory practice, the Otero group in San Sebastián, Spain, carried artificial muscles to considerable refinement in the laboratory. The Inganäs group in Linköping, Sweden, studied chemomechanical actuators, and Shahinpoor's group at Albuquerque, NM, USA, has carried the artificial muscle concept to some refinement with the demonstration of "artificial wings" and "artificial propellers" suitable for installation in small remotely piloted vehicles. Nearly all this work has been carried out since 1990.

The principles behind electrochemomechanical and chemomechanical actuation using CPs are fairly simple: CPs are well known to swell on doping (which means electrochemical or chemical oxidation for most CPs, which are *p*-type). This increases volume at least ca. 35% according to measurable parameters and probably quite a bit more. Thus, if a bilayer of a CP and another substance which is flexible but does not swell or contract is created, electrochemical or chemical redox of the CP will lead to bending of this bilayer. The intricacies of this sort of actuation are however more complex, as seen in the sequel.

The reasons for the choice of the defining words should then also be obvious: *electrochemomechanical* implies an applied potential which leads to a chemical redox, which in turn leads to a mechanical deformation. *Electrochemomechanical* in our context is then simply a short form for electrochemomechanical. And *chemomechanical* implies we skip the electrical (applied potential) step and effect redox of the CP by direct chemical means.

42.2 Artificial Muscles

Otero et al. [1446, 1969–1972, 1975, 1983] prepared freestanding P(Py) films on stainless steel electrodes from acetonitrile solution (2% water content) containing LiClO$_4$ electrolyte and monomer, using a double potential step algorithm. They then peeled off these films and applied them to one side of a commercially available, double-sided adhesive tape, inserting a Pt wire for electrical contact. The P(Py)/flexible plastic bilayer so produced, dimensions ca. 3 cm $\times$ 1 cm, constituted the artificial muscle, when clamped on one end (the other end being free). It was used as the working electrode in a laboratory electrochemical cell with Pt counter and SCE reference electrodes. Applying various potentials on the bilayer for periods of about 3 min altered the position of the free end reversibly up to 90° and less reversibly up to 180°. Figure 42.1 shows in rough schematic the action of this artificial muscle. At ca. +0.2 V and ca. -0.7 V vs. SCE, the film was, respectively, vertical and horizontal.

The "muscle" could be held at any one physical position by simply removing the applied potential and thus stopping any current flow. A reproducible correlation of applied potential and angle of bending could be obtained. Time needed to reach a particular, desired potential ranged from ca. 20 s to greater than 2 min. Higher currents were correlated with a faster response time (faster movement). Unfortunately, cyclability appeared to be poor: after ca. 100 cycles, fissures developed in the P(Py) film close to the point of clamping. It was also found that Joule (heating) effects from mechanical movement + current flow caused degradation of the polymer.

The phenomenon was denoted by Otero et al. as "electrochemomechanical" since an electrical force (the applied potential) led to a chemical reaction (CP redox or doping/de-doping) which in turn led to a mechanical reaction (the bending). It was distinguished from the operation of an animal muscle however by the fact that in the latter, the electrical force (nerve impulse) is simply an actuator and not the driving energy behind the chemical reaction.

Otero et al. estimated that the volume change upon extreme oxidation (doping) of the P(Py) film was at least 35%, a value roughly equal to the doping volume changes verified independently for other CPs by such methods as ellipsometry [1974]. However, Otero et al. noted errors in such estimations and postulated that the volume changes were likely to be much larger. The effect was predictably assigned to entry (during doping) and exit (during de-doping) of highly solvated dopant counterions into the CP matrix.

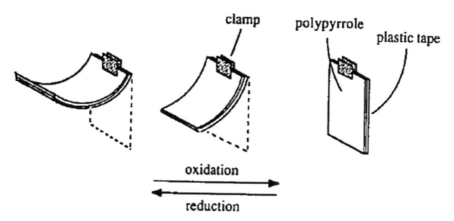

Fig. 42.1 Action of CP-based artificial muscle (After Ref. [1446, 1983], reproduced with permission)

More interestingly, they found that the phenomenon was pronounced only in aqueous systems, and did not occur or occurred negligibly in organic solvent media such as acetonitrile (in which the P(Py) was synthesized) and propylene carbonate. This intriguing effect was ascribed to several factors: the increased solvation of ions in aqueous as opposed to organic media; the greater affinity of the P(Py) for the ions in the more dipolar aqueous media; the retention, possibly, of some organic solvent molecules even in the reduced, de-doped CP, as compared to complete expulsion of water molecules; and a somewhat nebulous "more effective swelling of P(Py) in aqueous solutions."

In subsequent, more detailed studies, Otero et al. [1983] investigated the weight that a P(Py) artificial muscle could carry, the more precise control of movement, and the phenomenon of swelling. In their study of the mechanical carrying capacity, they attached a wire weighing 152 mg to the free end of the bilayer (having a P(Py) mass of ca. 5 mg). They then measured the time it took to elevate the wire-loaded bilayer 1 cm (to a 90° angle) starting at an applied potential of +0.8 V and taking it to various potentials, comparing it with the unloaded bilayer. Their results from this and similar experiments are summarized in Tables 42.1 and 42.2. Otero et al. found that the position of the bilayer could be more precisely controlled by monitoring and controlling the amount of charge – rather than applied potential or current – applied in the form of a pulse. They also more precisely measured the charge consumed per mg of P(Py) required to deflect the device through a constant angle, which they found to be independent of the charge density. They found that the movements of the device are best correlated to the mass of the electroactive material.

Otero et al. conjectured that in addition to electrostatic attraction effects during oxidation (doping) facilitating the flow of solvated ions into the polymer causing swelling, repulsion between the positive charges generated along the disordered CP chain caused it to also "uncoil" or "open up," creating channels for flow of large, solvated counterions.

Table 42.1 Times needed and electrical charges consumed to describe an angular movement of 180° by the free end of a bilayer of P(Py) and nonconductive, flexible polymer when submitted to step potentials from 800 mV to different cathodic potentials in 0.1 M LiClO$_4$ solution

E, mV vs. SCE	−400	−500	−700	−1000	−1200	−1500	−1800	−2000
Time over 180 s	65	59	45	33	29	24	21	19
Charge consumed, mC	520	529	536	545	558	596	633	661

After Ref. [1446]

Table 42.2 Times required to elevate the free end of the bilayer 1 cm by reduction of the P(Py) from 800 mV to different cathodic potentials: empty bilayer and loaded with a wire weighing 152 mg across the end of the bilayer

E$_{red}$, mV	−500	−600	−700	−800	−900	−1000
Time over 90°, s	14.7	15.7	10.4	8.5	8.1	8
Time over 90° (loaded), s	24	21	18	15	13	11
Power/dynes.cm.s^{-1}	16.02	28.11	19.60	22.92	30.40	49.65
Power/mg of P(Py)	3.20	5.62	3.92	4.58	6.08	9.93

After Ref. [1446]

As compared to Otero et al.'s experience with the lack of response in organic electrolytes, Baughman et al. [1976] claimed to have observed strong bending in LiClO$_4$/PC solution in a bilayer constructed from a copolymer of P(3-MeT) and P(3-octyl-T) sputter coated with Au.

42.3 Other Electrochemomechanical Actuation

Baughman et al. [1576, 1976] discussed in detail the merits of CP-based electrochemomechanical actuators as compared to other electromechanical actuators, e.g., those based on piezoelectrics. They noted, for instance, that the voltages required for the CP-based actuators and a piezoelectric polymer such as poly (vinylidene fluoride) were of the order of single volts and tens of volts, respectively, while the fraction length change, $\Delta L/L$, obtained with the former was >10%, whereas that with the latter was ca. 0.1%.

Baughman [1576, 1976] calculated theoretical performance parameters for CP-based actuators, based on the known mechanical properties of undrawn and drawn P(T), P(ANi), and P(Ac). He found that for small changes in doping level, stresses ranging from 40 to 450 MPa could be developed in such CP actuators, being one to two orders of magnitude higher than the tensile stresses developed in the piezoelectric poly(vinylidene fluoride), typically ca. 3 MPa, or in the pH-driven, salt-saturated poly(acrylic acid)/PVA gels (ca. 0.3 MPa). Baughman et al. also postulated [1576, 1976] that even if $\Delta L/L$ for the CP was restricted to 1% to enhance cyclability, it still had an order of magnitude advantage over piezoelectrics in work density per cycle.

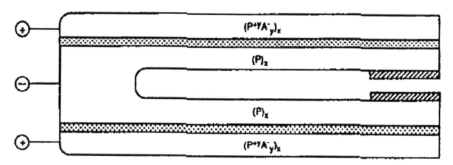

Fig. 42.2 Proposed use of paired bimorph actuators as microelectrochemical tweezers. $(P)_x$ represents any CP, e.g., P(Py). Electrochemical transfer of dopant from the outer layer to the inner layer of each bimorph causes the opening of the tweezers (After Ref. [1576], reproduced with permission)

Baughman et al. described microtweezers that could be fabricated from CP layers (Fig. 42.2); the electrolyte used could be liquid or solid. When the inner layer of CP, denoted $(P)_x$ in the figure, is doped, i.e., dopant is transferred from the outer layer to this inner layer (with polarities reversed from those in the figure), the tweezers open. Potential applications postulated included, besides microtweezers, microvalves, micropositioners, and microsorters. Baughman compared these putative CP tweezers with piezoelectric microtweezers of ca. 200 μm length and noted that over 100 V were required in the latter to achieve the same opening/closing dimensions.

Baughman et al. however also noted the considerable disadvantages of CPs: their cyclability (just 100 cycles or so according to the work of Otero et al. cited above), as compared to possibly millions of cycles obtainable with piezoelectric materials, and response time, of the order of seconds or tens of seconds, as compared to ms or even μs for piezoelectric materials. It would clearly be such factors which would limit commercial application of CP-based actuators.

Smela et al. [1977] constructed CP unimorph hinges, of dimension ca. 90 × 90 μm, and used these to rotate paddles of dimension 900 × 900 μm in solution, as shown in Fig. 42.3. This would be the equivalent of a human arm moving a 30 × 30 m plate in water. They also used a P(Py)/Au bilayer to construct a "microscopic box actuator," as shown in Fig. 42.4.

Actuators with similar action but based on P(ANi) single fibers have also been described [1976]. However, some of the applications envisioned for these, such as steerable optical fiber catheters for medical applications, seem a bit premature for the level of development achieved.

Shahinpoor [1978], working at the "Artificial Muscles Research Institute," University of New Mexico, Albuquerque, NM, USA, fabricated devices for a wide variety of applications based on electrochemomechanical principles, from ion-conducting polymers (not CPs). These polymers included poly(acrylic acid-bisacrylamide) (PAAM), poly(2-acrylamido-2-methyl-1-propane-sulfonic acid) (poly(AMPS)), and polyacrylonitrile (PAN). While these are not CPs, Shahinpoor

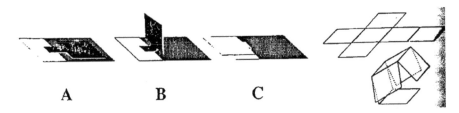

Fig. 42.3 CP unimorph hinges, dimensions ca. 90 μm × 90 μm. After Smela et al. [1977]

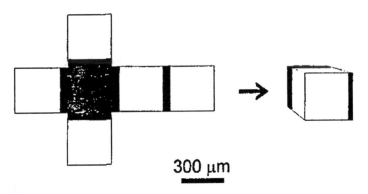

Fig. 42.4 A "microscopic box actuator" based on a P(Py)/An bilayer, with dimensions as shown. The block sections represent the bilayer, the other sections inactive polymer [1976]

also indicated that similar action could be expected, with minor modifications, from CPs such as poly(arylene vinylenes) and poly(thienylene vinylenes). Shahinpoor typically used a metal (e.g., Pt) + ion-conductive polymer composite in place of the customary bilayers. Some of the applications envisioned or demonstrated for ion-conductive polymers included microactuators, motion sensors, accelerometers, oscillating artificial muscles, inchworms, cardiac-circulation assistants, noiseless propulsion swimming robots for military applications, fully constituted contractile artificial muscles, miniature flying machines, and electrically controllable adaptive optical lenses (Fig. 42.5). The potential military applications of these have fueled much interest recently [1979].

In perhaps the first exhibition of its kind in the public domain, Okuzaki et al. [1980] demonstrated the function of a complete rotor based on an electrochemomechanical, P(Py) element. A continuous rotation speed of 21 cm/min was claimed.

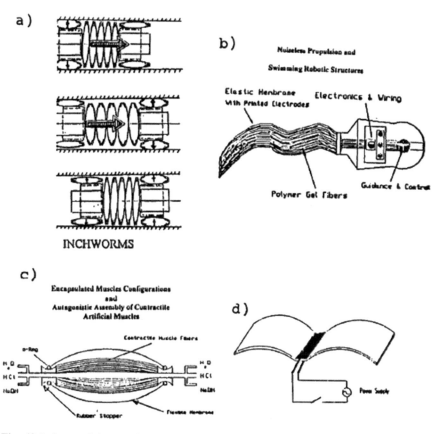

Fig. 42.5 Some of the novel applications envisioned for CP-based electrochemomechanical devices by the Shahinpoor group (After Shahinpoor et al. [1978])

42.4 Chemomechanical Actuators and Sensors

Pei and Inganäs [1981] fabricated bilayer strips similar to those of Otero et al. – CP/Au/poly(ethylene) (PE) or CP/Au/poly(imide) (PI) – but with an entirely different mode of application in mind: no electrochemical potential or current was applied, no electrolyte was used, but rather, the bilayers were used as sensors. Their bending was monitored upon exposure to various analytes. P(Py), P(T), and various P(3-alkyl-T)s were used as the CP. Thicknesses were ca. 20 μm and bilayer dimensions ca. 4 cm × 0.3 cm. Analyte vapors used were ammonia for P(Py) and iodine for the P(T) and P(3-alkyl-T)s. Response time of P(Py) bilayers upon ammonia turn-on was of the order of tens of seconds, but upon turn-off the response decayed slowly over hundreds of seconds (Fig. 42.6). In the case of the P(3-alkyl-T)s, some selectivity could be achieved between the various alkyl substituents. Most significantly, however, the authors were at pain to ascribe the nature of the effect,

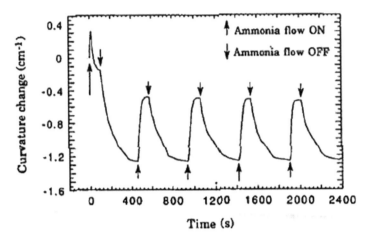

Fig. 42.6 Curvature change of a perchlorate-doped P(Py)/Au/PE strip in response to ammonia gas flow (After Ref. [1981], reproduced with permission)

conjecturing that it could be doping/de-doping, likely in the case of P(Py)/NH$_3$, or simply gas adsorption/desorption and associated mass changes, likely in the case of P(T)/I$_2$.

Okamoto et al. [1982] reported an actuator based on doping/de-doping-induced volume change in anisotropic P(Py) films. And Otero and Cortes [1983] reported the movement of an all-polymeric triple-layer artificial muscle based on P(Py).

42.5   Problems and Exercises

1. Define and differentiate in detail the terms *electrochemomechanical, chemomechanical, and electromechanical.* How do the artificial muscles described in this chapter and natural muscles differ in fundamental terms?
2. Enumerate all possible reasons why a CP-based artificial muscle would work better in aqueous than in nonaqueous environments. Describe in detail its principles of operation on a macroscopic and microscopic (molecular) level.
3. Would a 35% volume change upon doping be expected to give rise to the electrochemomechanical effects observed with CPs? What are the factors that would disguise the actual total volume change during measurements? What do you estimate to be the typical actual volume changes?
4. What are the input parameters responsible for control of the following properties during bending of CP artificial muscles: speed/response time, position, torque (lifting capacity)? Which among the following appears to be the best way to control them: voltage pulses, voltage sweeps, current pulses, or current sweeps?

5. What modifications to the work cited in this chapter could be expected to yield more specificity and selectivity in CP chemomechanical sensors?

6. Compare the advantages and drawbacks of CP-based actuators as compared to those based on a piezoelectric polymer such as poly(vinylidene fluoride). Include quantitative comparisons (e.g., tensile stresses, cyclability, response times) wherever possible.

7. What in your opinion are the primary drawbacks inhibiting commercial exploitation of CP-based actuators and muscles? From what you have garnered from the applications cited in this book thus far, are these similar to the drawbacks of CPs in virtually every field of application?

8. Design a CP bilayer-based swimming robot "octopus," such as that shown in Fig. 42.5, including control circuitry. Describe its principle of operation.

Chapter 43
Miscellaneous Applications

Contents

43.1 Corrosion Protection Applications

43.1.1 Principles of Anti-corrosion Methods

The principles behind corrosion protection based on CP coatings are best understood if we take a quick look at the three other methods most widely used at present for protection of metals from corrosion:

The first method is simply *coating the metal* with a material, such as an epoxy, an enamel, an inorganic silicate polymer, or an alkyd resin, that physically impedes the agents causing corrosion. These agents primarily include a combination of H_2O (as liquid or vapor), O_2, and salt. Such a physical impediment is however powerless against the formation of cracks or pinholes and against vapor phase or other slow diffusion of the corroding agents through the coating matrix.

The second method is the use of chemical "*conversion coatings*" to alter the metal surface for protection. This is illustrated by the most widely used – and the most successful to date – chemical conversion coatings, chromates, for Al protection. Etching agents such as HF are first used to remove native Al oxides. The chromates are then applied, with the overall reaction:

$$2Al + 2H_2O + 2H^+ + Cr_2O_7^{2-} \rightarrow Al_2O_3 + 2Cr(OH)_3 \qquad (43.1)$$

The alumina and the $Cr(OH)_3$ formed together serve as a chemical barrier to corrosion. This barrier is however not entirely immune to pinholes and cracks. Additionally, the chromates are highly toxic, and there is a strong movement at present to eventually replace them due to environmental considerations.

© Springer International Publishing AG 2018
P. Chandrasekhar, *Conducting Polymers, Fundamentals and Applications*,
https://doi.org/10.1007/978-3-319-69378-1_43

Commercially available chromate conversion coatings include Elf-Atochem's "Alumigold" and "Accelagold" and Parker Amchem's "Alodine 1200S."

The third widely used protection method is that of "*cathodic protection*," where a small negative potential is continuously applied to the metal surface to render it passive. Its counterpart, "anodic protection," can also be used to keep a metal in a permanently oxidized state, rendering it passive to corrosion. Quite evidently, this method is more cumbersome and expensive than most methods, although it does find niche uses where it is more practical, e.g., metal pipelines which have periodic control stations on the pipeline.

43.1.2   Driving Forces Behind Development

The development drivers for alternative anti-corrosion coatings should be evident from the above discussion. They address the following: (a) the issue of cracks and pinholes, as well as slow corrodant diffusion to the metal surface; (b) environmental issues, such as the unacceptability of chromates; (c) cost; and (d) ease of application and practicality, e.g., the lack of it in such methods as cathodic protection.

The potential market for metal corrosion protection is, quite evidently, very large. Both military and commercial seagoing vessels, metal structures in seaside environments (e.g., oil rigs), and metal components of seaside buildings are just some of the applications for steel-based metals. For the case of Al, the Al alloys that both military and commercial aircraft are made of are still, as of this writing (1998), protected using chromate conversion coatings for the most part. These coatings must be phased out in the very near future due to environmental considerations. As an example of military aviation use in the USA, at a typical US Air Force base, 60,000 m^2 (ca. 600,000 ft^2) of Al-alloy surface on airplanes must be stripped, recoated for corrosion protection, and repainted. Thus, the market, at a conservative estimate, runs into several billion US$ [1984].

43.1.3   Principles of CP-Based Anti-corrosion Coatings

From the above discussions, it should be apparent that metal corrosion is primarily an oxidative process, involving loss of electrons from the metal to an oxidizing species in the environment. Thus, the fundamental nature of corrosion is electrochemical. It is this fundamental electrochemical, electronic nature of corrosion that CP coatings address.

Two theoretical views have emerged of the action of CP anti-corrosion coatings, both different representations of the same picture (it is pointed out to the reader that most studies have used the doped (oxidized) form of *p*-type CPs as the coating). In the simpler version, for example, that was espoused by Ahmad and MacDiarmid [1985, 1986a], the CP is said to simply serve as an in situ oxidant or "anodic protectant." It oxidizes the surface of the metal it is in contact with, thereby also getting reduced itself. In some cases, re-oxidation of the CP by ambient air is said to "rejuvenate" the CP for further corrosion protection. Quite evidently, such a hypothesis requires that the redox or open circuit (equilibrium) potentials of the CP *in the corrosive environment where protection is sought* should match the Fermi (equilibrium electronic) levels of the metal surface. This hypothesis is simple but is sometimes not rigorous enough to account for anomalies such as the observation that good corrosion protection is provided even by reduced (de-doped) CP [1986] or in cases where there is a CP-metal mismatch in terms of simple redox potentials.

A more theoretically rigorous approach to explain CP anti-corrosion is that first apparently propounded by Jain et al. in a series of papers [1988–1990]. Jain et al. theoretically and experimentally studied metal/semiconductor (M/SC) interfaces, using primarily inorganic SCs such as In-Sn oxides (ITO) and SiO_2-ITO. Their principles are however equally applicable when CPs, which are after all semiconducting polymers, serve as the SCs.

As pictured in Fig. 43.1a, the use of a semiconductor (SC) coating on a metal creates a built-in electric field, resulting from the interfacial, positive dipole space charge layers. This creates a force ($F = -eE$, Fig. 43.1a) that opposes electron transfer from metal to oxidizing species in the external environment (such as O_2), providing an electronic, rather than physical or chemical, barrier to corrosion. A *bending of the SC bands* occurs at the M/SC interface (Fig. 43.1b). The electric field is similar to an actively applied electric field, e.g., as used in cathodic protection of metals from corrosion, but with the distinction that it is passive or intrinsic. Moreover, the dimensions of the field are such that it extends across small pinholes or scratches, up to 250 μm (10 mil) in dimension, providing protection even across these, as prior research has confirmed [1984–1990]. The novelty and clear advantage of this effect is that, unlike physical or chemical barriers to oxidative corrosion provided by conventional anti-corrosion coatings, which are susceptible to problems such as diffusion of oxidative species through the coating to the metal, an electronic basis for corrosion protection is provided. Thus, SC or CP coatings address the issue of corrosion protection from its very fundamental, electrochemical/electronic basis.

In this technology, the choice of the SC (i.e., the CP) is important: higher bandgap SCs generally provide a higher electronic barrier but, in many cases, also a poorer match to the Fermi level (work function) at the interface.

43.1.4 Brief Historical Development

Menholi et al. [1986b] generated metal-S-bridged P(ANi) coatings on iron by electropolymerizing from a basic solution of ANi and $(NH_4)_2S$. Insulating oligomeric films were obtained and cured at 150°C to yield cross-linked, insoluble sulfurized P(ANi) which showed good corrosion resistance. DeBerry [1992] first showed that P(ANi) could impart "anodic protection" to types 410 and

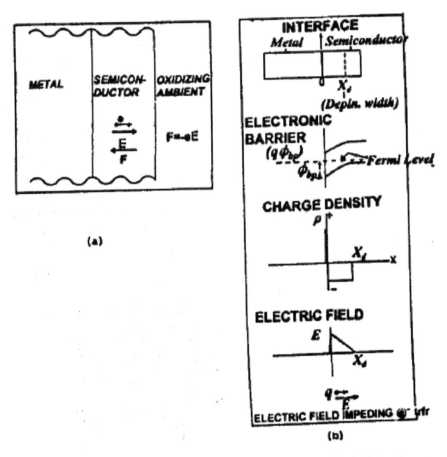

Fig. 43.1 (**a**, *Left*) The metal/semiconductor (M/SC) interface, showing intrinsic electronic barrier to corrosion generated by the SC (After Ref. [931]). In our case, the SC is a coating of a conducting polymer (CP).(**b**, *Right*): The M/SC (M/CP) interface, showing band bending and development of the interfacial space charge and electronic barrier to corrosion (After Ref. [1988])

430 (Cr-containing) stainless steels even in solutions of 0.2 M sulfuric acid + 0.2 M NaCl. He used electron microscopy, open circuit potential, and potentiodynamic scans to measure corrosion protection. Jain et al. studied inorganic SC-metal interfaces, as noted above, using electrochemical methods to demonstrate effective corrosion protection on Al. Their studies however fueled interest in CP-based coatings because of the firm theoretical foundation that they provided. In a NASA-funded collaborative effort, Thompson and Bryan at the NASA Kennedy Space Center, FL, USA, and Benicewicz and Wrobleski at Los Alamos National Lab, Los Alamos, NM, investigated a variety of CPs for corrosion protection [1504, 1991], finding effective corrosion protection for steels. Since their work, a number of studies have emerged and patents issued.

43.1.5 Advantages of CP Coatings

Besides projected lower cost and ease of application, the two major advantages of CP coatings as opposed to other metal corrosion protection methods are seen to be the protection provided over pinholes and scratches, as discussed above, and relative environmental benignness, especially when compared to chromates.

43.1.6 Most-Studied Candidates and Coating Methodology

The most-studied candidate CP to date remains P(ANi), although P(Py) has also been well studied. Wrobleski et al. [1505, 1991], have studied a number of poly (3-alkyl thiophenes) and poly(3-thienylene acetates). Chandrasekhar et al. [1984] have studied a number of poly(aromatic amines), such as derivatives of poly (diphenyl amine).

CP coatings have generally been applied in two ways: (1) as neat CP coatings per se, which are subsequently topcoated with epoxies or other resins for physical durability (i.e., a two-component system is used), and (2) as blends with epoxy or other binders (i.e., in one-component systems).

Neat CP coatings have generally been put down from solutions of de-doped CP (in solvents such as 1-Me-2-pyrrolidinone (NMP)), since doped CPs are mostly insoluble. These undoped coatings are then chemically doped via immersion in concentrated dopant solution or, occasionally, electrochemically. Some doped CP coatings have been put down as sols (colloidal suspensions) [1984] or as "partial" solutions [1986].

One-component coatings generally involve preparing a blend or suspension of the CP, usually in its doped form, with the resin or polymer used as the binder (e.g., an epoxy or an adhesive). The blend is then directly coated on the metal. Coatings have been applied by standard methods, including dip, spray, and brush coating.

43.1.7 Testing Methodology

Electrochemical methods of corrosion testing yield the quickest results. These include the Tafel scan, cyclic polarization (also known as potentiodynamic scan), linear polarization (also known as polarization resistance), and electrochemical impedance spectroscopy (EIS). It cannot be proposed to discuss these in detail here, and reference is made to any one of a number of excellent monographs on corrosion technology [1993]. Many of these have very specific application areas. For example, comparison of linear polarization data for coated vs. bare metal gives quick resistance numbers allowing one to decide whether further studies are worthwhile, while EIS is very useful for studying high-resistance (e.g., undoped CP) coatings.

(a) Another popular test method is salt spray fog testing (in a chamber that generates a heavy, salt-water fog), e.g., using American Society of Testing Materials' (ASTM) method B-117-94 or US Military Standard MIL-C-5541E. Other methods include: (a) Scratch tests and "filiform" tests, e.g., those described in US Military Standards MIL-P-85582B, MIL-P-23377G, or TT-P-2756.

(b) Immersion/weight loss – a popular, quick test whose simplest version is monitoring weight loss and visual damage after extended immersion in 3.5% NaCl solution; its more rigorous version is found in ASTM G-1-90.

(c) Adhesion tests, which are also important, e.g., vide ASTM-D-3359, ASTM-D-4541-95, or US Federal Standard FED-STD-141C-Method-6301.2.

A caveat frequently cited by many experienced corrosion workers may be noted: all above methods are only rough indicators, and none of them tell how a system is eventually going to perform in a corrosive environment.

43.1.8   Results with P(ANi)-Based Coatings

The historically significant work of Mengoli et al. and of DeBerry has already been cited above.

Wrobleski et al. and Thompson et al. [1505, 1991] studied a variety of CPs, including P(ANi) and poly(thiophene) derivatives (see below), for corrosion protection of steels in marine environments. An immediate need they sought to fulfill was protection in the space launch environment, which included HCl fumes, elevated temperatures, and salty air. The Zn chromate coatings in use for the purpose were frequently leached out by the HCl vapors. These workers synthesized P(ANi) chemically using persulfate oxidation; converted it into the soluble (reduced), emeraldine base form; and formed a solution in NMP. The solution was then applied by dip- and spray-coating methods, with the coatings then re-doped chemically by immersion in concentrated dopant solution. Following this, a topcoat of cross-linked epoxy, e.g., Ciba-Geigy Bisphenol A GY 2600 with hardener XU265, was used to impart physical durability to the coating. Their system was thus a two-component system, with P(ANi) applied as the first component followed by the second, epoxy component.

In their patent, Wrobleski et al. [1991] claimed a wide variety of CPs besides P(ANi) and substituted P(ANi)s. They used CP coating thicknesses of 0.5– 5 mils (ca. 12 to 120 µm). Dopants used included BF_3, PCl_5, $AlCl_3$, $SnCl_4$, $Zn(NO_3)_2$, tosylate, WCl_6, and, the dopant that gave the best results, tetracyanoethylene (TCNE). After coating of the P(ANi) film from emeraldine base solution in NMP, chemical doping was performed by prolonged exposure to 0.1–1 M dopant solutions in NMP, THF, methanol, acetonitrile, or water (solvents depending on the

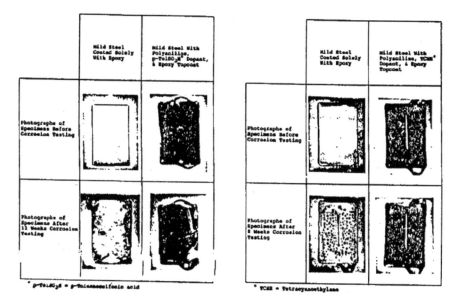

Fig. 43.2   Results of Wrobleski et al.'s [1505] works for P(ANi)-coated steel coupons (After Ref. [1504], reproduced with permission)

dopant used) or in the gas phase for gaseous dopants such as BF_3. Topcoats comprising either epoxy or polyurethane, of thickness 1–125 mils (25 µm to 3 mm), were then applied. Coatings were applied by dip, spray, or roll coating. The conductivities of the coatings were actually quite moderate, ca. 0.1 S/cm. Coated coupons, mainly of mild steel, were tested primarily by immersion in 0.1 M HCl and 3.5% NaCl solutions.

Figure 43.2a shows some of Wrobleski et al.'s results for P(ANi)-Tos before and after 12 weeks of immersion exposure in aerated 3.5% NaCl. More importantly, these workers confirmed the prediction of Jain et al. [1988] that scratches of significant width were protected. This is seen in the results of immersion for 8 weeks in aerated 0.1 M HCl shown in Fig. 43.2. These authors also confirmed their results with other methods such as salt spray fog and electrochemical methods.

In their patent, Kinlen et al. [1994] described the use of one-component systems, comprising a CP dispersed in a non-thermoplastic polymeric binder, and two-component systems, comprising the same first component plus an additional adhesive binder overlayer for added physical durability. The primary purpose of the binders was adhesion to the metal, which the authors claimed was superior to two-component CP + binder systems such as those of Wrobleski et al. described above. In their patent, these authors claimed use of virtually all existing *p*-type CPs, although they used primarily P(ANi), P(Py), and P(3-Me-thiophene) in their examples. As dopants, again, they claimed a wide variety, although using primarily TCNE, Zn nitrate, and tosylate. As binders, they used inorganic compounds, such as silicates and zirconates, or more common resins, such as shellac, phenolic resins, alkyd resins, and epoxies. The preferred binders were epoxies, polyurethanes, and phenolic resins, applied as one-component or two-component (e.g., resin + hardener) system. Preferred concentrations of the CP in the binder were 1–50% by volume (v/v). A typical binder formulation comprised a dispersion of P(ANi) and poly(butyl methacrylate) (a polymer similar to PMMA) in butyrolactone. Coating thicknesses were between 0.1 and 5 mils (2.5 and 125 μm). A typical epoxy topcoat comprised carboline 890. The authors again claimed that nearly all common metals could be corrosion protected by their formulations, including Fe, Ni, Cu, Zn, Co, Pb, steels, Al, and Ag, although citing only steel examples. In the examples cited, steels such as C1018 appeared to be well protected through immersion in aerated 0.1 M HCl.

Ahmad and MacDiarmid [1985, 1986] described simple P(ANi) coatings, coated in the emeraldine base form from NMP, DMSO, or tetramethylurea solutions (0.5 to 5 w/w% P(ANi)) and then chemically doped as in the Wrobleski work above. No further topcoats were however used. Coating thicknesses were small – in the 1–200 μm range (preferably 5–75 μm). These workers also described coating from solutions containing partially doped P(ANi), with dopants such as dodecyl benzene sulfonate and camphor sulfonate, although they noted that the corrosion protection provided was inferior.

These authors noted that adhesion of the neat P(ANi) coating could be improved by pretreating the metal surface with compounds such as phosphoric acid, polyphosphoric acid, and alizarin sulfonate. Targeted metals were primarily steels, e.g., SS-304 and SS-430. According to these authors, P(ANi) was most suited for such steels because its open circuit potential (V_{OC}) matched well with them at moderately acidic pHs found in corrosive environments. Notably, these authors' primary mode of monitoring corrosion protection was the V_{OC} of the CP-coated surface, apparently unsubstantiated by other electrochemical or salt spray methods. The authors then issued the following guidelines for selection of any CP for steel corrosion protection: (a) The potential range for passivation of the metal should be ascertained in the corrosive environment envisioned. (b) A CP film should be selected whose V_{OC} in the same corrosive environment lay a little + of the minimum passivation potential of the metal. (c) If the V_{OC} of the CP was < +0.4 V vs. SCE, then an additional benefit would be that after reduction on contact with the metal surface, the CP would be re-oxidized by air, thus continuing to provide corrosion protection indefinitely. (d) If, however, the V_{OC} of the CP was < +0.4 V vs. SCE, corrosion protection would be provided only until the CP was completely reduced.

In very interesting but somewhat perplexing work not thus far duplicated anywhere else, McAndrew et al. at Air Products and Chemicals, Allentown, PA, USA, described [1987] protection of carbon (mild) steels with *nonconductive* P (ANi), claiming that it performed better than conductive (doped) P(ANi). These authors however noted that oxidation in the corrosive environment or air oxidation of the nonconductive P(ANi) coating appeared to dramatically improve its performance and appeared to be critical in some tests. Thus their P(ANi) coating may after all have been an (O_2-) doped, partially conductive P(ANi). These authors used the CP + binder dispersion (single component) method of coating. Binders used included epoxies, styrene/acrylate copolymers, polyurethanes, and polyimides. P(ANi) was coated from ca. 5 w/w% emeraldine base solutions in NMP to a thickness of ca. 50 μm. Corrosion protection was monitored by EIS, salt spray fog, and immersion methods. Using one method of monitoring, that of "pore resistance," which measures how easily corrodants can access the metal surface, these authors found pore resistance values for coatings of nonconductive P(ANi), conductive P(ANi), and air-re-oxidized P(ANi) of 2×10^8, 2×10^6, and 5×10^8, respectively.

In work that substantially duplicated that of the Wrobleski/Thompson groups cited above, Lu et al. [1995] studied P(ANi) coatings on mild steel (e.g., UNS G10100) in 0.1 N HCl and 3.5 w/w% NaCl solutions. A topcoat of epoxy was used. The samples were scribed (scratched) and the exposed, scratched area monitored. Corrosion rates for the P(ANi)-coated samples, as compared to samples coated with epoxy only, were only two times less in the NaCl solution but as much as 100 times less in the HCl solution. The authors used results from EIS, other electrochemical, ESCA, and Auger studies to postulate that the formation of passivating γ-Fe_2O_3 and Fe_3O_4 layers occurred. They further postulated that galvanic coupling of the P(ANi) to the steel occurs. Results of Tafel scans in HCl and in NaCl are shown in Fig. 43.3a, b. Li et al. [1996] also postulated, based on EIS studies, that P(ANi) coatings on mild steel provide protection primarily via passivation of the metal surface through charge transfer, with concomitant reduction of the P(ANi). The P(ANi) then re-oxidizes itself with atmospheric oxygen. This hypothesis was also put forth by Fahlman et al. [1997] in studies of P(ANi) coatings on cold rolled steel and iron. The latter studies also appeared to indicate that when the metal's top and interfacial oxide layers are removed prior to CP coating deposition (as is done for chromate conversion coatings), corrosion protection provided is superior. These oxide layers were analyzed by X-ray photoelectron spectroscopy to be ca. 1.5 nm of Fe_2O_3 over a thicker layer of Fe_3O_4 in the latter studies.

In some of the first work exclusively on Al alloys, specifically the aerospace grade alloy 7075, the Yang group at the University of Rhode Island, Kingston, RI, USA, [1998] showed substantial corrosion protection. They again confirmed that small, scratched areas continued to be protected and postulated a "passive film formation" protection mechanism. These authors used the template-polymerized P(ANi)/poly(styrene sulfonate) "double-strand" CP-dopant combination described elsewhere in this book, which is substantially water soluble and can be coated directly in the partially doped form from aqueous solution. These authors noted that this direct coating method and the greater thermal stability of their CP-dopant complex gave their technique processing and durability advantages over other techniques for coating P(ANi). They used electrochemical techniques and electron microscopy to verify corrosion protection. Figure 43.4 shows potentiodynamic scan results they obtained for coated and uncoated Al alloy samples in 0.5 M NaCl solution.

Brusic et al. [1999] showed that P(ANi) and P(ANi)-derivative coatings could also be applied to corrosion protection of other metals such as Cu and Ag.

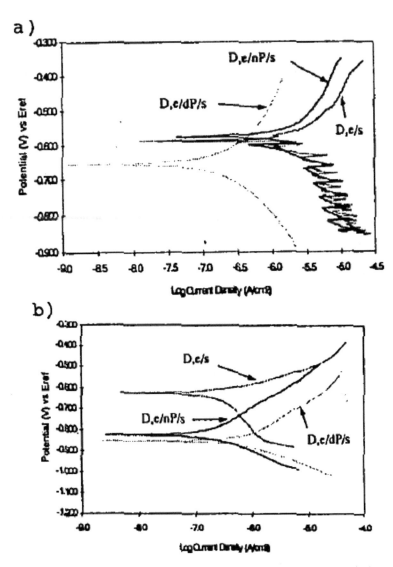

Fig. 43.3 Corrosion-protective P(ANi) coatings on mild steel: Tafel electrochemical measurements in (**a**) HCl and (**b**) NaCl. "D,e/s" indicates drilled epoxy on steel, "D,e/nP/s" drilled epoxy + neutral P(ANi) on steel, and "D,e/αP/s" drilled epoxy + doped P(ANi) on steel (After Ref. [1997], reproduced with permission)

The P(ANi) derivative poly(o-phenetidine) was found to provide corrosion protection superior to that of P(ANi). Figure 43.5 shows representative corrosion data obtained by these authors.

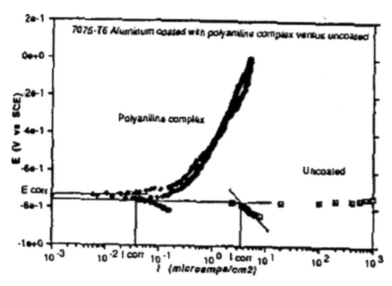

Fig. 43.4 Potentiodynamic scan of 7075-T6 aluminum alloy samples coated with P(ANi)/poly-electrolyte vs. an uncoated aluminum sample, in .5 N NaCl solutions (After Ref. [2004], reproduced with permission)

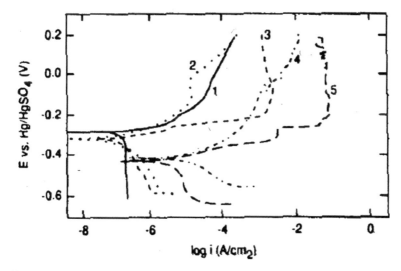

Fig. 43.5 Potentiodynamic polarization curves in a droplet of water for Cu covered with (1) poly (o-phenetidine) base, (2) HCl-doped poly(o-phenetidine), (3) unsubstituted P(ANi) base, (4) DBSA-doped unsubstituted P(ANi), and (5) native oxide only (After Ref. [1998], reproduced with permission)

43.1.9   Other Poly(Aromatic Amines)

Chandrasekhar et al. [1984] carried out a very detailed study of corrosion protection of the aerospace Al alloys 2024 and 7075, the former of which has a substantial Cu content. They used coatings of poly(aromatic amines), primarily derivatives of poly (diphenyl amine) P(DPA)). In this work, these CPs were compared with P(ANi) and with the commercially available chromate conversion coatings "Accelagold" and "Alumigold" (Elf Atochem America). Chandrasekhar et al. chose P(DPA) derivatives because of their higher bandgaps and much better match to Fermi levels at interfaces with Al-based alloys than P(ANi), as also their proven greater oxidative and environmental stability (>400 °C in air as shown by TGA data).

Their coatings were applied from paint-like sols (colloidal suspensions) of nanoparticulate CPs in benign, multicomponent solvents (with components such as ethyl acetate). A poly(ether) was added to the sol to serve as a binder or plasticizer. The poly(ether) served as a binder for better physical durability of the coating, much like the binders described in other work cited above. Notably, coating conductivities were in the region of 1 S/cm, substantially higher than nearly all the prior work with P(ANi), which typically generated coating conductivities less than 0.1 S/cm.

Chandrasekhar et al. [1984] used several independent methods to monitor corrosion, including Tafel scan, cyclic polarization, and polarization resistance among electrochemical methods, immersion methods, salt spray fog techniques vide ASTM B-117-94 and MIL-C-5541E, scratch tests, and adhesion tests. They found that their coatings had excellent adhesion, passing the ASTM tests mentioned above, and excellent durability in salty atmospheres, as evidenced by the samples differing very little in appearance before and after the salt spray fog tests.

Some of these authors' results are shown in Figs. 43.6 and 43.7. Figure 43.6 shows comparative cyclic polarization (also called "pitting," potentiodynamic) scans for uncoated, Accelagold (chromate)-coated, and CP(A) (P(DPA) derivative)-coated Al alloy coupons in aerated 3.5% NaCl. The large loop area and hysteresis observed in the top curves for the uncoated samples are typical of high corrosion; the loop area is frequently used as a direct measure of the extent of corrosion. The complete absence of a loop and the back scan retracing the forward scan (i.e., no hysteresis) in the bottom curves for the CP-coated samples demonstrate corrosion protection superior even to that provided by the Accelagold-coated samples (middle curves). Figure 43.7 shows comparative salt spray fog chamber tests, vide ASTM B-117 and MIL-C-5541E, for the various coatings indicated. It is evident that Accelagold appears to just satisfy the MIL specification threshold, whereas the CP(A) and CP(B) coatings are substantially below this threshold; thus, they provide superior corrosion protection to Accelagold.

Chandrasekhar et al. [1984] were also able to demonstrate that their CP coatings could be sprayed on by techniques commonly used at present in the aircraft industry, such as high-vacuum low-pressure (HVLP) and air-assisted spray coating.

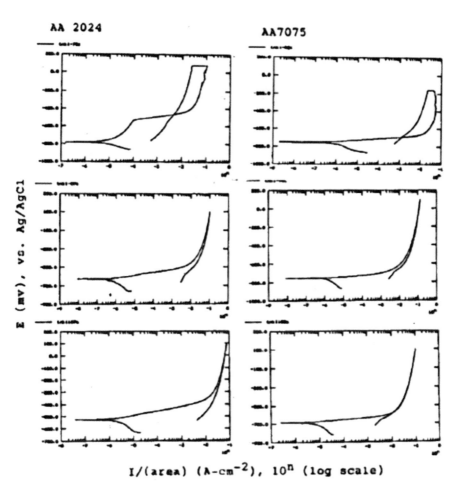

Fig. 43.6 Comparative electrochemical cyclic polarization ("pitting") scans in aerated 3.5% NaCl, for aluminum alloy samples *top*, uncoated substrates; *middle*, substrates coated with Accelagold; *bottom*, substrates coated with the poly(diphenyl amine) derivative designated CP (A). (*Left*, alloy 2024; *right*, alloy 7075). (Ordinate, E, Volts, vs. Ag/AgCl; abscissa, I/area, A-cm^{-2}). Note the large loop areas and hysteresis, as expected, for highly corroding uncoated substrates (*top*), while the areas for CP(A) (*bottom*) are even smaller than those for Accelagold (*middle*) (After Chandrasekhar et al. [1984])

43.1.10   Other CPs

Wrobleski et al. [1505, 1991], in their work described at length above, also tested other CPs, including poly(3-alkyl-thiophenes) (alkyl = hexyl, octyl) and poly (3-thienyl acetates) with an acetate group, i.e., -CH$_2$(C = O)R, (R = Me, Eth), at the 3-position of thiophene. They however found P(ANi) superior in anti-corrosion performance. Similarly, Kinlen et al. [1994], in their work also described above,

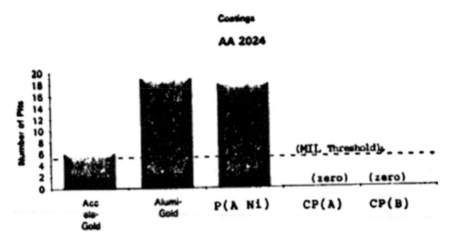

Fig. 43.7 Comparative salt spray fog chamber tests for various samples as indicated. CP(A) and CP(B) are two poly(diphenyl amine) derivatives (After Chandrasekhar, et al. [1984])

claimed the use of virtually all *p*-type CPs and actively tested P(Py) and P(3-Me-T). Again, however, they found P(ANi) the most versatile. Among the many reasons for the superior performance of P(ANi) were the better match with the metal substrates, expressed through its V_{OC}, and more facile processibility.

Zarras et al. [2000] described the use of poly(2,5-bis(*N*-Me-*N*-alkylamino)) phenylene vinylenes) coatings for corrosion protection of Al alloys. They studied these coatings via simple potentiostatic and galvanostatic electrochemical methods and found a significant corrosion–inhibition effect.

43.1.11   Anti-corrosion Coatings Based on CPs: Looking Ahead

While the work carried out thus far, as cited above, has appeared promising, and several patents to many parties have been issued on the technology, to date (2016), there is no commercialization of such coatings. Indeed, research and development in this area has tapered off drastically since about 2008, if the number of publications in the field is any indicator.

43.2   Biomedical Applications, Including Drug Delivery

CPs have been studied for a variety of biomedical applications [2001–2007].

Green et al. [2008] discussed the *potential* application of CPs as coatings for metal electrodes used in implantable devices. The rationale for such use was that

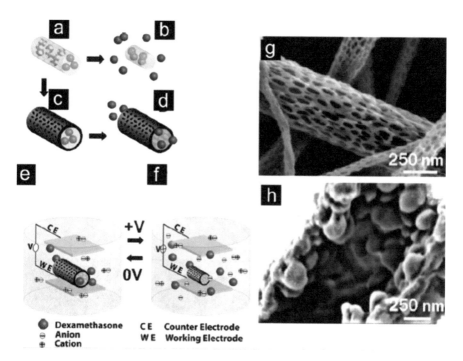

Fig. 43.8 Schematic illustration of the delivery (controlled release) of the drug dexamethasone using PEDOT as an agent: (**a**) dexamethasone-loaded electrospun poly(lactide-co-glycolide) (PLGA); (**b**) hydrolytic degradation of PLGA fibers leading to release of the drug; (**c**) electrochemical deposition of PEDOT around the dexamethasone-loaded electrospun PLGA fiber, which slows down the release of dexamethasone as shown in (**d**). (**e**) PEDOT nanotubes in a neutral electrical condition. (**f**) External electrical stimulation controls the release of dexamethasone from the PEDOT nanotubes due to contraction or expansion of the PEDOT. Shrinkage causes the drug to come out of the ends of tubes. SEM images of (**g**) PLGA nanoscale fibers and (**h**) a single PEDOT nanotube which was polymerized around a PLGA nanoscale fiber (After Ref. [2011], reproduced with permission)

they would result in improvement of the neural tissue–electrode interface and thus increase of the effective lifetime of such implants. The metal electrodes used include gold, platinum, iridium oxide, and glassy carbon. CP coatings would presumably improve electrode–tissue communication through providing a high surface-area material more conducive to cell and tissue integration. Charge transfer would presumably be improved through reduced impedance and greater selectivity for both recording and stimulating applications. Since polymers are typically softer materials, it was also hypothesized that inflammation would reduce due to reduction in strain mismatch between tissue and electrode surface. However, to date, these have remained only predictions, with no well-known experimental studies, either in vitro or in vivo.

Also with regard to *neural interfaces*, Kang et al. [2009] fabricated a porous nerve growth factor (NGF)-doped PPy "nanoarchitecture" using self-assembly of

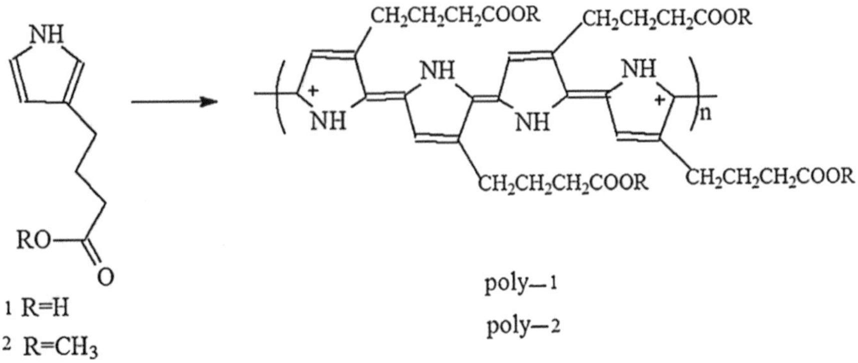

CH$_2$CH$_2$CH$_2$COOR   CH$_2$CH$_2$CH$_2$COOR

NH   NH

NH   NH

NH   NH

CH$_2$CH$_2$CH$_2$COOR   CH$_2$CH$_2$CH$_2$COOR

RO—

O

1 R=H

2 R=CH$_3$

poly—1

poly—2

Fig. 43.9 Zelikin et al.'s "unconventionally biodegradable" CPs (After Ref. [2019], reproduced with permission)

Fig. 43.10 SEM image of C2C12 myoblast cells cultured on the surface of hydrogels based on poly (3-thiophene acetic acid) (After Mawad et al. [2004, 2045], reproduced with permission)

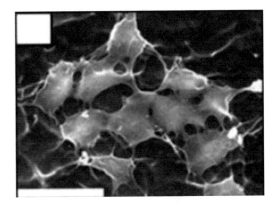

polystyrene beads as a sacrificial template. The open-pore structure of the P(Py) resulted in amplified actions in NGF release profiles, when combined with electrical fields. The structure provided free diffusion of neurotrophic factors, favorable cell adherence, and an electrically conductive surface, leading to enhanced cellular behavior.

Xie et al. [2029] developed conductive PCL-PPy and polylactate (PLA)-P(Py) core-sheath NFs that facilitated neurite growth and extension (Fig. 7.131). The conductive core-sheath NFs were prepared by combining electrospinning with aqueous polymerization.

Abidian et al. [2011] reported delivery of the drug dexamethasone using PEDOT as an agent. This is illustrated schematically in Fig. 43.9 below.

CPs have been studied for use as scaffolds for *nerve tissue engineering* [2002, 1588] and for other applications requiring in vivo use. Proposed advantages of CPs over other materials in these applications include flexibility and compatibility with body tissues combined with acceptable conductivity.

Mawad et al. [2004] described the use of CP hydrogels based on poly (3-thiopheneacetic acid) covalently cross-linked with the polymer with 1,1-′-carbonyldiimidazole (CDI). The mechanical properties of the resulting networks were characterized as a function of the cross-linking density and were claimed to be comparable to those of muscle tissue. Hydrogels were found to be electroactive and conductive at physiological pH. Fibroblast and myoblast cells cultured on the hydrogel substrates were shown to adhere and proliferate.

Chen et al. [2006] prepared well-ordered nanofibers composed of polyaniline (PANi) and poly(e-caprolactone) (PCL) via electrospinning. They then cultured mouse C2C12 myoblasts on random PCL (R-PCL), aligned PCL (APCL), random PCL/PANi-3 (R-PCL/PANi), and aligned PCL/PANi-3 (A-PCL/PANi) nanofibers. Their results appeared to show that the aligned nanofibers (A-PCL and A-PCL/ PANi) could guide myoblast orientation and promote myotube formation (i.e., approximately 40% and 80% increases in myotube numbers) compared with

R-PCL scaffolds. In addition, electrically conductive A-PCL/PANi nanofibers further enhanced myotube maturation (i.e., approximately 30% and 23% or 15% and 18% increases in the fusion and maturation indices) compared with nonconductive A-PCL scaffolds or R-PCL/PANi.

Biocompatibility and cytotoxicity are major concerns when considering such materials as CPs for use in vivo. In this respect, many researchers have attempted to introduce *biodegradability* to CPs. Thus, e.g., Shi et al. [2012] introduced biodegradability into CPs by blending them with suitable biodegradable polymers. Huang et al. [2013] fabricated an electroactive, biodegradable composite containing polylactide as the biodegradable component and the low molecular weight aniline pentamer as the electroactive component. A novel block copolymer of polyglycolide and aniline pentamer that is electroactive and degradable was also synthesized by Ding and coworkers [2014]. Zhang et al. [2015] synthesized an electrically conductive and biodegradable polyphosphazene polymer containing aniline pentamer with glycine ethyl ester as a side chain; they evaluated its biocompatibility using Schwann cells, with results showing that the polymer exhibited no cytotoxicity. Huang et al. [2016] synthesized a multiblock copolymer (PLAAP) through condensation polymerization of hydroxyl-capped poly(L-lactide) (PLA) and carboxyl-capped aniline pentamer (AP) and showed it had good electroactivity and biodegradability, suitable for tissue engineering applications. And Jiang et al. [2017] coated polyester fabrics with P(Py) and investigated the in vivo biocompatibility and biostability of P(Py)-coated polyester fabrics. Their results showed in vivo biocompatibility of the P(Py)-coated and non-coated polyester fabrics. And Olayo et al. [2018] synthesized semiconductor biomaterials of iodine-doped P(Py) and P(Py)-polyethylene glycol. Their results showed good compatibility of these materials after implantation into the transectioned spinal cord tissue.

A unique strategy for synthesis of biodegradable CPs was the attachment of skeletal groups that yield high *solubility* to the resulting CPs, the rationale being that the polymers would simply dissolve away in body fluids and eventually be expelled by the body, rather than "biodegrade" in the conventional sense. One of the first to report this strategy was Zelikin et al. [2019]. The structure of these polymers is illustrated in Fig. 43.9 below.

Degradable and conductive hydrogels represent another strategy for eventual use of CP derivatives in vivo. In this respect, Liu et al. [2020] synthesized degradable and conductive hydrogels by joining together the photopolymerizable macromer acrylated poly(d,l-lactide)-poly(ethyleneglycol)-poly(d,l-lactide) (AC-PLA-PEG-PLA-AC), glycidylmethacrylate (GMA), and ethylene glycol dimethacrylate (EGDMA), along with aniline tetramer, yielding a degradable, conductive network.

Valle et al. [2021] investigated the interaction of PEDOT films with epithelial cells and showed the biocompatibility of PEDOT with epithelial cells. Luo et al. [2022] synthesized PEDOT films and investigated their biocompatibility by seeding NIH3T3 fibroblasts on PEDOT films and carried out subcutaneous implantation of PEDOT films by in vivo study. Their results showed that PEDOT films exhibit very low intrinsic cytotoxicity and that their inflammatory response upon implantation was good. Richardson–Burns et al. [2023] polymerized PEDOT around living cells and described a neural cell-templated conducting polymer coating for microelectrodes and a hybrid CP–live neural cell electrode.

Ramanaviciene et al. [2024] claimed to demonstrate the biocompatibility of P(Py) nanomaterials, specifically P(Py)-cellulose "nanocomposites" which they claimed showed good biocompatibility and stability in mammalian cells over 48 h. Oh et al. [2025] reported that the cytotoxicity of and immune response to PEDOT nanomaterials in IMR90 and J774A.1 cells were within acceptable limits.

With regard to drug delivery, Sanchvi et al. [2003] used phage display to select peptides that specifically bind to P(Py) and which can subsequently be used to modify the surface of P(Py). The use of selected peptides for P(Py) by phage display can be extended to encompass a variety of therapies and devices, such as PPy-based drug delivery vehicles, nerve guidance channel conduits, and coatings for neural probes. Selection of peptides using phage display is claimed to represent a simple alternative to methods based on electrostatic and hydrophobic interactions between two moieties to achieve adsorptive modification of surfaces.

Oh et al. [1589, 2025] demonstrated the use of *carbonized* P(Py) nanoparticles (NPs) for drug delivery, prepared by pyrolysis of P(Py) NPs. The NPs so obtained displayed uniform sizes, large micropore volumes, and high surface areas. Magnetic phases formed during pyrolysis were useful for selective NP separation. Hydrophobic medicines, such as ibuprofen, were incorporated into the carbonized PPy NPs by surface adsorption and pore filling, and further plasma modification enhanced drug sustainability via surface covalent coupling. Drug-loaded NPs demonstrated sustained release properties. Additionally, the carbonized PPy NPs showed low toxicity and were readily incorporated into cells.

With regard to *biocompatibility studies*, Lee et al. [2027] synthesized carboxy-endcapped polypyrrole (P(Py)-α-COOH), composed of a P(Py) surface modified with pyrrole-α-carboxylic acid, and further grafted it with the cell-adhesive Arg-Gly-Asp (RGD) motif. Human umbilical vein endothelial cells (HUVECs) cultured on RGD-modified P(Py)-α-COOH demonstrated significantly higher adhesion and spreading on the RGD-modified PPy-α-COOH than negative P(Py)-α-COOH without RGD motif and unmodified P(Py). Cullen et al. [2028] fabricated small-diameter (<400 μm) fibers consisting of a blend of electrically conductive P(ANi) and P(Py) and carried out bioadhesive surface modification using either poly-L-lysine (PLL) alone or PLL combined with collagen type I.

With regard to nerve tissue engineering, Ghasemi et al. [2002] carried out the electrospinning of a blend of poly(ε-caprolactone) (PCL), gelatin, and P(ANi) to obtain conductive nanofiber scaffolds for nerve tissue engineering. Xie et al. [2029] fabricated a new type of scaffold comprised of conductive core-sheath nanofibers prepared by in situ polymerization of pyrrole on electrospun PCL or poly(L-lactide) (PLA) nanofibers. Thompson et al. [2030] showed that neurite outgrowth from cochlear neural explants grown on the PPy-containing NT-3 and BDNF was significantly improved when the polymer containing both neurotrophins was electrically stimulated.

Li et al. prepared P(ANi)-DNA hybrid "nanowires" with a conductivity of about (10^{-2} S/cm) [2031]; in these, DNA was claimed to be the dopant as well as template. However, other than academic curiosity, the precise utility of these constructs was not clarified by these authors. Cheng et al. [2032] polymerized aniline monomer in the presence of chitosan as a steric stabilizer, to obtain surface-functionalized P(ANi)-chitosan "nanocomposites," with a view to increase biocompatibility; these composites had an electrical of about 0.25 S/cm.

The biocompatibility of various films and composites containing CPs has been investigated at length by many researchers. For example, P(ANi) composites and films were investigated by culturing with PC12 cells, in comparison with bare Si substrates by Liu et al. [2033]. It was found that the P(ANi) constructs are appropriate for proliferation of PC12 cells; after 2 days of culture, cells continue to proliferate well on PAni film, but not on a reference Si wafer, suggesting that the rougher P(ANi) film provides a better surface for cell growth. Several similar studies demonstrated cell and tissue compatibility of P(Py)-based constructs in vitro and in vivo, as tested with endothelial cells, osteoblasts, nerve tissue, PC-12 cell, neuro2a, and neuroblastoma cells [2002, 2034–2037]. The implants of these P(Py) constructs in rat muscle, hypodermal tissue, guinea pig brain, rat peripheral nerve, and mouse peritoneum were also studied in vivo over a significant period of implantation, with, however, mixed results [2002, 2037–2042]. The in vitro and in vivo biocompatibility of P(Py), both chemically and electrochemically synthesized, was investigated in depth by Wang et al. [2041]. The extracted solution of chemically prepared P(Py) powders was assessed using systematic toxicity tests according to ISO 10993 and ASTM 1748–82 standards. They found that extract solution possessed neither acute or subacute toxicity, it did not change body temperature, and it did not cause hemolysis of blood red cells; allergic responses such as nose scratching, piloerection, dyspnea, and spasm; or mutagenesis of cells. In addition, the growth of Schwann cells cultured on electrochemically polymerized P(Py) was enhanced as compared to glass substrates.

Xiao et al. [2043] and Green et al. [2044] showed that electropolymerization of poly(3,4-ethylenedioxythiophene) (PEDOT) films in the presence of collagen or laminin improved the adhesion of PC12 cells thereto. Mawad et al. [2045] showed that C2C12 myoblast cells culture well on the surface hydrogels based on poly (3-thiophene acetic acid), as shown in Fig. 43.10.

Del Valle et al. [2046] investigated the surface of the PEDOT films generated by anodic polymerization regarding the adhesion and proliferation of epithelial cells. The results showed that epithelial cells Hep-2 presented significant activity on the surface of PEDOT electrodeposited on stainless steel electrodes and no cytotoxicity was detected due to this polymer. The electrochemical characteristics of PEDOT covered with cells were determined in different biological media using cyclic voltammetry. It was found that the electroactivity of PEDOT films is significantly *enhanced* by the attached cellular monolayer.

Che et al. [2047] electropolymerized PEDOT onto Au electrodes with glutamate (Glu) as dopant. The resultant PEDOT/Glu coating had lower impedance over a wide range of frequencies than bare gold. The coating exhibited good electroactivity in buffer solution, implying that it can establish a bridge between the modified metal electrode and brain tissue. Incorporation of Glu promoted the attraction and anchoring of neural cells on the coating, showing better cell attachment and growth than a reference PEDOT-Na-tosylate coating.

43.3   Miscellaneous Applications

Bergren et al. [2048] soaked up PEDOT:PSS into a rose plant using various means, including applying vacuum to the leaves. They then applied a small voltage and claimed that the plant glowed. These authors however did not elaborate on the exact scientific or practical utility of this experiment.

Samori et al. [2050] presented preliminary results on the potential use of CPs in flexible memory (as in computer memory) devices. They developed a blend of light-sensitive diarylethene and P(3HT) (poly(3-hexyl thiophene). When excited by UV light, the diarylethene moieties change from an open to a closed configuration, allowing "writing" of information with ns-duration laser pulses. These moieties can then be reopened with visible light to "erase" the written information.

43.4   Problems and Exercises

1. Describe in detail the principles of corrosion protection with CPs. Use diagrams if necessary. How does it differ from the most common physical and chemical anti-corrosion methods?
2. Name at least three key advantages of CP corrosion protection coatings. Name any drawbacks preventing commercialization at present. Besides the naval, aviation, and structural markets mentioned in this chapter, are there any other potential markets (hint – e.g., for industrial processes)?

3. What are the advantages and drawbacks of two-component vis-à-vis one-component coating methods for direct coating of doped CPs, e.g., from sols, vs. in situ doping of de-doped polymer coatings?
4. What properties would you look for in a binder for CP coatings on Al alloys? On mild steel? Which among the following general classes would be better for each: polyimides, epoxies, and polyurethanes?
5. Briefly outline the major corrosion test methods used today.
6. Describe in detail the characterization of the anti-corrosion properties of a two-component P(ANi) coating using TCNE dopant, starting from the synthesis of the P(ANi) and including comparative characterization methods and expected results.
7. Outline methods you would use to select a CP for proper matching – in terms of energy levels, pretreatment, adhesion, and all other conceivable factors – to a Cu-rich Al alloy and to Cr- and Ta-containing steels.
8. What are the factors most likely to be hindering the use of CPs other than P(ANi) and poly(3-alkyl thiophenes) for corrosion protection?

Literature Cited

1. Iijima, S.: Helical microtubules of graphitic carbon. Nature. **354**(6348), 56–58 (1991). Bibcode:1991Natur.354...56I. doi: https://doi.org/10.1038/354056a0
2. Hata, K., Futaba, D.N., Mizuno, K., Namai, T., Yumura, M., Iijima, S.: Water-assisted highly efficient synthesis of impurity-free single-walled carbon nanotubes. Science. **306** (5700), 1362–1365 (2004). Bibcode:2004Sci...306.1362H. doi: https://doi.org/10.1126/science.1104962. PMID 15550668
3. Iijima, S.: High resolution electron microscopy of some carbonaceous materials. J. Microsc. **119**(1), 99–111 (1980)
4. Радушкевич, Л.В.: О Структуре Углерода, Образующегося При Термическом Разложении Окиси Углерода На Железном Контакте. Журнал Физической Химии. **26**, 88–95 (1952) (in Russian). (see http://nanotube.msu.edu/HSS/2006/4/2006-4.pdf)
5. Schützenberger, P., Schützenberger, L.: C. R. Acad. Sci. **111**, 774–780 (1890)
6. Oberlin, A., Endo, M., Koyama, T.: Filamentous growth of carbon through benzene decomposition. J. Cryst. Growth. **32**(3), 335–349 (1976). Bibcode:1976JCrGr..32..335O. doi: https://doi.org/10.1016/0022-0248(76)90115-9
7. Baker, R.T.K., Harris, P.S.: The formation of filamentous carbon. Chem. Phys. Carbon. **14**, 83–165 (1978)
8. Abrahamson, J., Wiles, P.G., Rhoades, B.L.: Structure of carbon fibers found on carbon arc anodes. Carbon. **37**(11), 1873–1874 (1999). https://doi.org/10.1016/S0008-6223(99)00199-2
9. Izvestiya Akademii Nauk SSSR. Metals. **3**, 12–17 (1982)
10. Tennent, H.G.: Carbon fibrils, method for producing same and compositions containing same (1987)
11. Ebbesen, T.W., Ajayan, P.M.: Large-scale synthesis of carbon nanotubes. Nature. **358**, 220–222 (1992)
12. Mintmire, J.W., Dunlap, B.I., White, C.T.: Are fullerene tubules metallic? Phys. Reve. Lett. **68**, 631–634 (1992)
13. Dresselhaus, M.S., Dresselhaus, G., Saito, R.: Carbon fibers based on C_{60} and their symmetry. Phys. Rev B Condens Matter. **45**, 6234–6242 (1992)
14. Jones, D.E.H.: New Sci. **110**, 1505 (1986)
15. Ajayan, P.M.: Nanotubes from Carbon. Chem. Rev. **99**(7), 1787–1800 (1999)
16. Tans, S.J., Devoret, M.H., Dai, H., Thess, A., Smalley, R.E., Geerligs, L.J., Dekker, C.: Individual single-wall carbon nanotubes as quantum wires. Nature. **386**, 474–477 (1997)

© Springer International Publishing AG 2018
P. Chandrasekhar, *Conducting Polymers, Fundamentals and Applications*,
https://doi.org/10.1007/978-3-319-69378-1

17. Terrones, M.: Science and technology of the twenty-first century: synthesis, properties, and applications of carbon nanotubes. Annu. Rev. Mater.Res. **33**, 419–501 (2003)

18. De Volder, M.F.L., Tawfick, S.H., Baughman, R.H., Hart, A.J.: Carbon nanotubes: present and future commercial applications. Science. **339**(535), 535–539 (2013)

19. Balasubramanian, K., Burghard, M.: Chemically functionalized carbon nanotubes. Small. **2**, 180–192 (2005)

20. Hu, H., Bhowmik, P., Zhao, B., Hamon, M.A., Itkis, M.E., Haddon, R.C.: Chem. Phys. Lett. **108**(4), 227 (2006)

21. Zhao, X., Liu, Y., Inoue, S., Suzuki, T., Jones, R., Ando, Y.: Smallest carbon nanotube is 3 Å in diameter. Phys. Rev. Lett. **92**(12), 125502 (2004). Bibcode:2004PhRvL..92l5502Z. doi: https://doi.org/10.1103/PhysRevLett.92.125502. PMID 15089683

22. Zhang, R., Zhang, Y., Zhang, Q., Xie, H., Qian, W., Wei, F.: Growth of half-meter long carbon nanotubes based on Schulz–Flory distribution. ACS Nano. **7**(7), 6156–6161 (2003). https://doi.org/10.1021/nn401995z. PMID 23806050

23. Davenport, M.: Twist and shouts: a nanotube story. Chem. Eng. News. **93**(23), 10–15 (2016.) Ouyang, M., Huang, J.-L.: Fundamental electronic properties and applications of single-walled carbon nanotubes. Acc. Chem. Res. **35**, 1018–1025 (2002)

24. Wilder, J.W.G., Venma, L.C., Rinzler, A.G., Smalley, R.E., Dekker, C.: Electronic structure of atomically resolved carbon nanotubes. Nature. **391**, 59–62 (1998). https://doi.org/10.1038/34139

25. Odom, T.W., Huang, J.-L., Kim, P., Lieber, C.M.: Structures and electronic properties of carbon nanotubes. J. Phys. Chem. B. **104**, 2794–2809 (2000)

26. Kim, P., Odom, T.W., Huang, J.L., Lieber, C.M.: Electronic density of atomically resolved single-walled carbon nanotubes: van Hove singularities and end states. Phys. Rev. Lett. **82**, 1225–1228 (1999)

27. Sgobba, V., Guldi, D.M.: Carbon nanotubes—electronic/electrochemical properties and application for nanoelectronics and photonics. Chem. Soc. Rev. **38**, 165–184 (2009)

28. Hong, S., Myung, S.: Nanotube electronics: a flexible approach to mobility. Nat. Nanotechnol. **2**(4), 207–208 (2007). Bibcode:2007NatNa...2..207H. https://doi.org/10.1038/nnano.2007.89. PMID 18654263

29. Wei, B.Q., Vajtai, R., Ajayan, P.M.: Reliability and current carrying capacity of carbon nanotubes. Appl. Phys. Lett. **79**, 1172 (2001)

30. Schnorr, J.M., Swager, T.M.: Emerging applications of carbon nanotubes. Chem. Mater. **23**(3), 646–657 (2011) https://doi.org/10.1021/cm102406h

31. Tans, S.J., Devoret, M.H., Dai, H.J., Thess, A., Smalley, R.E., et al.: Individual single-wall carbon nanotubes as quantum wires. Nature. **386**, 474–477 (1997)

32. Bachtold, A., Strunk, C., Salvetat, J.-P., Bonard, J.-M., Forró, L., Nussbaumer, T., Schönenberger, C.: Aharonov-Bohm oscillations in carbon nanotubes. Nature. **397**, 673–675 (1999)

33. Iijima, S.: Helical microtubules of graphitic carbon. Nature. **354**, 56–58 (1991)

34. Iijima, S., Ichihashi, T.: Single-shell carbon nanotubes of 1-nm diameter. Nature. **363**, 603–605 (1993.) Bethune, D.S., Kiang, C.H., de Vries, M.S., Gorman, G., Savoy, R., et al.: Cobalt-catalysed growth of carbon nanotubes with single-atomic-layer walls. Nature **363**, 605–6 (1993)

35. Ishigami, N., Ago, H., Imamoto, K., Tsuji, M., Iakoubovskii, K., Minami, N.: Crystal plane dependent growth of aligned single-walled carbon nanotubes on sapphire. J. Am. Chem. Soc. **130**(30), 9918–9924 (2008)

36. Syangdev, N., Ishwar, K.P.: A model for catalytic growth of carbon nanotubes. J. Phys. D Appl. Phys. **41**(6), 065304 (2008)

37. Meyer-Plath, A., Ortis-Gil, G., Petrov, S., et al.: Plasma-thermal purification and annealing of carbon nanotubes. Carbon. **50**(10), 3934–3942 (2012)

38. Ren, Z.F., Huang, Z.P., Xu, J.W., Wang, J.H., Bush, P., Siegal, M.P., Provencio, P.N.:
Synthesis of large arrays of well-aligned carbon nanotubes on glass. Science. **282**(5391),
1105–1107. Bibcode:1998Sci...282.1105R. (1998). https://doi.org/10.1126/science.282.
5391.1105

39. Neupane, S., Lastres, M., Chiarella, M., Li, W.Z., Su, Q., Du, G.H.: Synthesis and field
emission properties of vertically aligned carbon nanotube arrays on copper. Carbon. **50**(7),
2641–2650 (2012)

40. Richard, S.E., Li, Y., Moore, V.C., Price, B.K., Colorado, R., Schmidt, H.K., Hauge, R.H.,
Barron, A.R., Tour, J.M.: Single wall carbon nanotube amplification: en route to a type-
specific growth mechanism. J. Am. Chem. Soc. **128**(49), 15824–15829 (2006)

41. Hata, K., Futaba, D.N., Mizuno, K., Namai, T., Yumura, M., Iijima, S.: Water-assisted
highly efficient synthesis of impurity-free single-walled carbon nanotubes. Science. **306**
(5700), 1362–1365 (2004). Bibcode:2004Sci...306.1362H. https://doi.org/10.1126/science.
1104962. PMID 15550668

42. Cao, Q., Rogers, J.A.: Ultrathin films of single-walled carbon nanotubes for electronics and
sensors: a review of fundamental and applied aspects. Adv. Mater. **21**, 29 (2009)

43. Hata, K., et al.: Water-assisted highly efficient synthesis of impurity-free single-walled
carbon nanotubes. Science. **306**, 1362 (2004.) Kozio, L.K., et al. High-performance carbon
nanotube fiber. Science. 318, 1892 (2007)

44. Chang, H., Lin, C., Kuo, C.: Iron and cobalt silicide catalysts-assisted carbon nanostructures
on the patterned Si substrates. Thin Solid Films. **420–421**, 219–224 (2002)

45. Ting, G., Nikolaev, P., Rinzler, A.G., Tomanek, D., Colbert, D.T., Smalley, R.E.: Self-
assembly of tubular fullerness. J. Phys. Chem. **99**(27), 10694–10697 (1995)

46. Ting, G., Nikolaev, P., Thess, A., Colbert, D., Smalley, R.: Catalytic growth of single-
walled nanotubes by laser vaporization. Chem. Phys. Lett. **243**(1-2), 49–54 (1995)

47. Chiang, M., Liu, K., Lai, T., Tsai, C., Cheng, H., Lin, I.: Electron field emission properties
of pulsed laser deposited carbon films containing carbon nanotubes. J Vac Sci Technol B. **19**
(3), 1034–1039 (2001)

48. Srivastava, D., Brenner, D.W., Schall, J.D., Ausman, K.D., Yu, M., Ruoff, R.S.: Predictions
of enhanced chemical reactivity at regions of local conformational strain on carbon
nanotubes: kinky chemistry. J. Phys. Chem. B. **103**(21), 4330–4337 (1999.) [in press]

49. Soundarrajan, P., Patil, A., Liming, D.: Surface modification of aligned carbon nanotube
arrays for electrochemical sensing applications. Am. Vac. Soc. **21**, 1198–1201 (2002)

50. Thess, A., Lee, R., Nikolaev, P., Dai, H., Petit, P., et al.: Crystalline ropes of metallic carbon
nanotubes. Science. **273**, 483–487 (1996)

51. Smiljanic, O., Stansfield, B.L., Dodelet, J.-P., Serventi, A., Désilets, S.: Gas-phase synthesis
of SWNT by an atmospheric pressure plasma jet. Chem. Phys. Lett. **356**(3–4), 189–193
(2002). Bibcode:2002CPL...356..189S. https://doi.org/10.1016/S0009-2614(02)00132-X

52. Kim, K.S., Cota-Sanchez, G., Kingston, C., Imris, M., Simard, B., Soucy, G.: Large-scale
production of single-wall carbon nanotubes by induction thermal plasma. J. Phys. D Appl.
Phys. **40**(8), 2375–2387. Bibcode:2007JPhD...40.2375K (2007). https://doi.org/10.1088/
0022-3727/40/8/S17

53. Hsu, W.K., Hare, J.P., Terrones, M., Kroto, H.W., Walton, D.R.M., Harris, P.J.F.:
Condensed-phase nanotubes. Nature. **377**, 687 (1995)

54. Hsu, W.K., Hare, J.P., Terrones, M., Kroto, H.W., Walton, D.R.M.: Electrlytic formation of
carbon nanostructures. Chem. Phys. Lett. **262**, 161–166 (1996)

55. Laplaze, D., Bernier, P., Master, W.K., Flamant, G., Guillard, T., Loiseau, A.: Carbon. **36**,
685–688 (1998)

56. Guillard, T., Flamant, G., Robert, J.F., Rivoire, B., Olalde, G., et al.: J. Phys. **IV9**, 59–64
(1999)

57. Alvarez, L., Guillard, T., Olalde, G., Rivoire, B., Robert, J.F., et al.: Synth. Met. **103**,
2476–77 (1999)

58. Chen, J., Hamon, M.A., Hu, H., Chen, Y., Rao, A.M., Eklund, P.C., Haddon, R.C.: Solution properties of single-walled carbon nanotubes. Science. **282**, 95 (1998)

59. Graupner, R., Abraham, J., Wunderlich, D., Vencelová, A., Lauffer, P., Röhrl, J., Hundhausen, M., Ley, L., Hirsch, A.: Nucleophilic−alkylation−reoxidation: a functionalization sequence for single-wall carbon nanotubes. J. Am. Chem. Soc. **128**(20), 6683–6689 (2006). https://doi.org/10.1021/ja0607281

60. Syrgiannis, Z., Hauke, F., Röhrl, J., Hundhausen, M., Graupner, R., Elemes, Y., Hirsch, A.: Covalent sidewall functionalization of SWNTs by nucleophilic addition of lithium amides. Eur. J. Org. Chem. **15**, 2544–2550 (2008). https://doi.org/10.1002/ejoc.200800005

61. Balaban, T.S., Balaban, M.C., Malik, S., Hennrich, F., Fischer, R., Rösner, H., Kappes, M.M.: Polyacylation of single-walled nanotubes under Friedel–Crafts conditions: an efficient method for functionalizing, purifying, decorating, and linking carbon allotropes. Adv. Mater. **18**(20), 2763–2767 (2006). https://doi.org/10.1002/adma.200600138

62. Karousis, N., Tagmatarchis, N., Tasis, D.: Current progress on the chemical modification of carbon nanotubes. Chem. Rev. **110**(9), 5366–5397 (2010). https://doi.org/10.1021/cr100018g

63. Yang, H., Wang, S.C., Mercier, P., Akins, D.L.: Diameter-selective dispersion of single-walled carbon nanotubes using a water-soluble, biocompatible polymer. Chem. Commun. 1425–1427 (2006). https://doi.org/10.1039/B515896F

64. Chen, R., Radic, S., Choudhary, P., Ledwell, K.G., Huang, G., Brown, J.M., Chun Ke, P.: Formation and cell translocation of carbon nanotube-fibrinogen protein corona. Appl. Phys. Lett. **101**(13), 133702 (2012). https://doi.org/10.1063/1.4756794. PMC 3470598

65. Wang, Z., Li, M., Zhang, Y., Yuan, J., Shen, Y., Niu, L., Ivaska, A.: Thionine-interlinked multi-walled carbon nanotube/gold nanoparticle composites. Carbon. **45**(10), 2111–2115 (2007). https://doi.org/10.1016/j.carbon.2007.05.018

66. Campidelli, S., Sooambar, C., Lozano Diz, E., Ehli, C., Guldi, D.M., Prato, M.: Dendrimer-functionalized single-wall carbon nanotubes: synthesis, characterization, and photoinduced electron transfer. J. Am. Chem. Soc. **128**(38), 12544–12552 (2006). https://doi.org/10.1021/ja063697i

67. Ballesteros, B., de la Torre, G., Ehli, C., Aminur Rahman, G.M., Agulló-Rueda, F., Guldi, D.M., Torres, T.: Single-wall carbon nanotubes bearing covalently linked phthalocyanines − photoinduced electron transfer. J. Am. Chem. Soc. **129**(16), 5061–5068 (2007). https://doi.org/10.1021/ja068240n

68. Georgakilas, V., Bourlinos, A.B., Zboril, R., Trapalis, C.: Synthesis, characterization and aspects of superhydrophobic functionalized carbon nanotubes. Chem. Mater. **20**(9), 2884–2886 (2008). https://doi.org/10.1021/cm7034079

69. Fabre, B., Hauquier, F., Herrier, C., Pastorin, G., Wu, W., Bianco, A., Prato, M., Hapiot, P., Zigah, D.: Covalent assembly and micropatterning of functionalized multiwalled carbon nanotubes to monolayer-modified Si(111) surfaces. Langmuir. **24**(13), 6595–6602 (2008). https://doi.org/10.1021/la800358w

70. Bianco, A., Kostarelos, K., Partidos, C.D., Prato, M.: Biomedical applications of functionalized carbon nanotubes. Chem. Commun. (5), 571–577 (2005)

71. Shin, W.H., Jeong, H.M., Kim, B.G., Kang, J.K., Choi, J.W.: Nitrogen-doped multiwall carbon nanotubes for lithium storage with extremely high capacity. Nano Lett. **12**(5), 2283–2288 (2012). Bibcode:2012NanoL..12.2283S. https://doi.org/10.1021/nl3000908. PMID 22452675

72. Yin, L.-W., Bando, Y., Li, M.-S., Liu, Y.-X., Qi, Y.-X.: Unique single-crystalline beta carbon nitride nanorods. Adv. Mater. **15**(21), 1840–1844 (2003). https://doi.org/10.1002/adma.200305307

73. Glerup, M., Steinmetz, J., Samaille, D., Stephan, O., Enouz, S., Loiseau, A., Roth, S., Bernier, P.: Synthesis of N-doped SWNT using the arc-discharge procedure. Chem. Phys. Lett. **387**, 193 (2004)

74. Sen R., Satishkumar, B.C., Govindaraj S., Harikumar K.R., Renganathan M.K., Rao C.N R.: Mater. Chem. **7**, 2335 (1997)

75. Bahr, J.L., Yang, J., Kosynkin, D.V., Bronikowski, M.J., Smalley, R.E., Tour, J.M.: Functionalization of carbon nanotubes by electrochemical reduction of aryl diazonium salts: a bucky paper electrode. J. Am. Chem. Soc. **123**, 6536 (2001)

76. Knez, M., Sumser, M., Bittner, A.M., Wege, C., Jeske, H., Kooi, S., Burghard, M., Kern, K.: Electrochemical modification of individual nano-objects. J. Electroanal. Chem. **522**, 70 (2002)

77. Vlandas, A., Kurkina, T., Ahmad, A., Kern, K., Balasubramanian, K.: Enzyme-free sugar sensing in microfluidic channels with an affinity-based single-wall carbon nanotube sensor. Anal. Chem. **82**, 6090 (2010)

78. (a) Kalyanasundaram, K.: Photochemistry of polypyridine and porphyrin complexes. Academic Press, London (1997); (b) Dolphin, D.: The porphyrins. Academic Press, New York (1978); (c) Kadish, K.M., Smith, K.M., Guilard, R.: The porphyrin handbook. Academic Press, New York (2003)

79. (a) Baskaran, D., Mays, J.W., Zhang, X.P., Bratcher, M.S.: Carbon nanotubes with covalently linked porphyrin antennae: photoinduced electron transfer. J. Am. Chem. Soc. **127**, 6916 (2005); (b) Li, H., Martin R.B., Harruff, B.A., Carino, R.A., Allard, L.F., Sun, Y.P: Single-walled carbon nanotubes tethered with porphyrins: synthesis and photophysical properties. Adv. Mater. **16**, 896 (2004)

80. O'Reagan, B.; Grätzel, M.: Nature **353**, 737 (1991)

81. Wang, P., Moorefiled, C.N., Li, S., Hwang, S. H., Shreiner, C.D., Newkome, G.R.: Chem. Commun. **10**, 1091 (2006)

82. Lim, J.K., Yoo, B.K., Yi, W., Hong, S., Paik, H.Y., Chun K., Kim, S.K., Joo, S.W.: J. Mater. Chem. **16**, 2374 (2006)

83. Guo, Z., Du, F., Ren, D., Chen, Y., Zheng, J., Liu, Z., Tian, J.: Covalently porphyrin-functionalized single-walled carbon nanotubes: A novel photoactive and optical limiting donor-acceptor nanohybrid. J. Mater. Chem. **16**, 3021 (2006)

84. Cui, J.B., Burghard, M., Kern, K.: Reversible sidewall osmylation of individual carbon nanotubes. Nano Lett. **3**, 613 (2003)

85. Banerjee, S., Wong, S.S.: Selective metallic tube reactivity in the solution-phase osmylation of single-walled carbon nanotubes. J. Am. Chem. Soc. **126**, 2073 (2004.) Liu J., Rinzler A. G., Dai H.J., Hafner J H., Bradley R.K., et al.: Fullerene pipes. Science **280**, 1253–56 (1998)

86. Chen, J., Hamon, M.A., Hu, H., Chen, Y.S., Rao, A.M., et al.: Solution properties of single-walled carbon nanotubes. Science. **282**, 95–98 (1998)

87. Arnold, M.S., Green, A.A., Hulvat, J.F., Stupp, S.I., Hersam, M.C.: Nat. Nanotechnol. **21**, 29 (2009)

88. Kim, W.-J., Usrey, M.L., Strano, M.S.: Selective functionalization and free solution electrophoresis of single-walled carbon nanotubes: separate enrichment of metallic and semiconducting SWNT. Chem. Mater. **19**(7), 1571 (2007). https://doi.org/10.1021/cm061862n

89. Maultzsch, J., Reich, S., Thomsen, C., Webster, S., Czerw, D.L., Carroll, D.L., Vieira, S.M. C., Birkett, P.R., Rego, C.A.: Raman characterization of boron-doped multiwalled carbon nanotubes. Appl. Phys. Lett. **81**, 2647 (2002)

90. Weisman, R.B., Bachilo, S.M.: Dependence of optical transition energies on structure for single-walled carbon nanotubes in aqueous suuspension: an empirical Kataura plot. Nano Lett. **3**(9), 1235–1238 (2003). Bibcode:2003NanoL...3.1235W. https://doi.org/10.1021/nl034428i

91. Reisch, M.S.: Molecular rebar design unravels carbon nanotubes. Chem. Eng. News. **93**(9), 25 (2015)

92. Zhao, Q., Nardelli, M.B., Bernholc, J.: Ultimate strength of carbon nanotubes: a theoretical study. Am. Phys. Soc. **65**(144105), 1–6 (2002)

93. Thess, A., Lee, R., Nikolaev, P., Dai, H.J., Petit, P., Robert, J., Xu, C.H., Lee, Y.H., Kim, S. G., Rinzler, A.G., Colbert, D.T., Scuseria, G.E., Tomanke, D., Fischer, J.E., Smalley, R.E.: Crystalline ropes of metallic carbon nanotubes. Science. **273**(5274), 483–487 (1996)

94. Ruoff, R.S., Tersoff, J., Lorents, D.C., Subramoney, S., Chan, B.: Radial deformation of carbon nanotubes by van der Waals forces. Nature. **364**(6437), 514–516 (1993). Bibcode:1993Natur.364..514R. https://doi.org/10.1038/364514a0

95. Palaci, I., Fedrigo, S., Brune, H., Klinke, C., Chen, M., Riedo, E.: Radial elasticity of multiwalled carbon nanotubes. Phys. Rev. Lett.. **94**(17) (2005). arXiv:1201.5501. Bibcode:2005PhRvL..94q5502P. https://doi.org/10.1103/PhysRevLett.94.175502

96. Ruoff, R.S., Qian, D., Liu, W.K.: Mechanical properties of carbon nanotubes: theoretical predictions and experimental measurements. C. R. Phys. **4**, 993–1008 (2003)

97. Paradise, M., Goswami, T.: Carbon nanotubes – production and industrial applications. Mater. Des. **28**, 1477–1489 (2007)

98. Yu, M.-F., Lourie, O., Dyer, M.J., Moloni, K., Kelly, T.F., Ruoff, R.S.: Strength and breaking mechanism of multiwalled carbon nanotubes under tensile load. Science. **287** (5453), 637–640 (2000). Bibcode:2000Sci...287..637Y. https://doi.org/10.1126/science. 287.5453.637. PMID 10649994

99. Peng, B., Locascio, M., Zapol, P., Li, S., Mielke, S.L., Schatz, G.C., Espinosa, H.D.: Measurements of near-ultimate strength for multiwalled carbon nanotubes and irradiation-induced crosslinking improvements. Nat. Nanotechnol. **3**(10), 626–631 (2008). https://doi. org/10.1038/nnano.2008.211. PMID 18839003; Demczyk, B.G., Wang, Y.M., Cumings, J., Hetman, M., Han, W., Zettl, A., Ritchie, R.O.: Direct mechanical measurement of the tensile strength and elastic modulus of multiwalled carbon nanotubes. Mater. Sci. Eng. A **334**(1–2), 173–178 (2002). doi:https://doi.org/10.1016/S0921-5093(01)01807-X

100. Stainless Steel – 17-7PH (Fe/Cr17/Ni 7) Material Information, Archived from the original on July 19, 2011

101. Wagner, H.D.: Reinforcement. Encycl. Polym. Sci. Technol. John Wiley & Sons. doi: https://doi.org/10.1002/0471440264.pst317

102. Yamabe, T.: Recent development of carbon nanotubes. Synthetic Met. 1511–1518 (1995)

103. Yu, M.F., Files, B.S., Arepalli, S., Ruoff, R.S.: Tensile loading of ropes of single wall carbon nanotubes and their mechanical properties. Phys. Rev. Lett. **84**(24), 5552–5555 (2000)

104. Filleter, T., Bernal, R., Li, S., Espinosa, H.D.: Ultrahigh strength and stiffness in cross-linked hierarchical carbon nanotube bundles. Adv. Mater. **23**(25), 2855–2860 (2011). https://doi.org/10.1002/adma.201100547

105. Treacy, M., Ebbesen, T.W., Gibson, J.M.: Exceptionally high Young's modulus observed for individual carbon nanotubes. Nature. **381**, 678–680 (1996)

106. Wong, E.W., Sheehan, P.E., Lieber, C.M.: Nanobeam mechanics: elasticity, strength, and toughness of nanorods and nanotubes. Science. **277**, 1971–1975 (1997)

107. Popov, M., Kyotani, M., Nemanich, R., Koga, Y.: Superhard phase composed of single-wall carbon nanotubes. Phys. Rev. B. **65**(3), 033408 (2002). Bibcode:2002PhRvB..65c3408P. https://doi.org/10.1103/PhysRevB.65.033408

108. Kim, P., Shi, L., Majumdar, A., McEuen, P.L.: Thermal transport measurements of individual multiwalled nanotubes. Phys. Rev. Lett. **87**, 215502 (2001)

109. Chem. Eng. News 25 (2016)

110. Kelly, B.T.: Physics of graphite. Applied Science, London (1981)

111. Hone, J.: Carbon Nanotubes **80**, 273–86 (2001)

112. Yi, W., Lu, L., Zhang, D.L., Pan, Z.W., Xie, S.S.: Linear specific heat of carbon nanotubes. Phys. Rev. **59**(14), R9015–R9018 (1999)

113. Hone, J., Llaguno, M.C., Biercuk, M.J., Johnson, A.T., Batlogg, B., Benes, Z., Fischer, J.E.: Thermal properties of carbon nanotubes and nanotube-based materials. Appl. Phys. A. **74**, 339–343 (2002)

114. Pradhan, N.R., Duan, H., Liang, J., Iannacchione, G.S.: The specific heat and effective thermal conductivity of composites containing single-wall and multi-wall carbon nanotubes. Nanotechnol. **20**, 1–7 (2009)
115. Yang, D.J., Zhang, Q., Chen, G., Yoon, S.F., Ahn, J., Wang, S.G., Zhou, Q., Wang, Q., Li, J. Q.: Thermal conductivity of multiwalled carbon nanotubes. Phys. Rev. B. **66**, 165440-1–165540-6 (2002)
116. Kolosnjaj, J., Szwarc, H., Moussa, F.: Toxicity studies of carbon nanotubes. Adv. Exp. Med Biol. **620**, 181–204 (2007). https://doi.org/10.1007/978-0-387-76713-0_14. ISBN 978-0-387-76712-3. PMID 18217344
117. Corredor, C., Hou, W.C., Klein, S.A., Moghadam, B.Y., Goryll, M., Doudrick, K., Westerhoff, P., Posner, J.D.: Disruption of model cell membranes by carbon nanotubes. Carbon. **60**, 67–75 (2013). https://doi.org/10.1016/j.carbon.2013.03.057
118. Lam, C.W., James, J.T., McCluskey, R., Arepalli, S., Hunter, R.L.: A review of carbon nanotube toxicity and assessment of potential occupational and environmental health risks. Crit Rev Toxicol. **36**(3), 189–217 (2006). https://doi.org/10.1080/10408440600570233. PMID 16686422
119. Firme III, C.P., Bandaru, P.R.: Toxicity in the application of carbon nanotubes to biological systems. Nanomed. Nanotechnol. Biol. Med. **6**, 245–256 (2010)
120. Yang, S.T., Wang, X., Jia, G., Gu, Y., Wang, T., Nie, H., et al.: Long-term accumulation and low toxicity of single-walled carbon nanotubes in intravenously exposed mice. Toxicol Lett. **181**, 182–189 (2008)
121. Byrne, J.D., Baugh, J.A.: The significance of nano particles in particle-induced pulmonary fibrosis. McGill J. Med. **11**(1), 43–50 (2008)
122. Porter, A., Gass, M., Muller, K., Skepper, J.N., Midgley, P.A., Welland, M.: Direct imaging of single-walled carbon nanotubes in cells. Nat. Nanotechnol. **2**(11), 713–717 (2007). Bibcode:2007NatNa...2..713P. https://doi.org/10.1038/nnano.2007.347. PMID 18654411
123. Fatkhutdinova, L.M., Khaliullin, T.O., Vasil'yeva, O.L., Zalyalov, R.R., Mustafin, I.G., Kisin, E.R., Birch, M.E., Yanamala, N., Shvedova, A.A.: Fibrosis biomarkers in workers exposed to MWCNTs. Toxicol. Appl. Pharmacol. **299**, 125–131 (2016). https://doi.org/10.1016/j.taap.2016.02.016
124. Shvedova, A.A., Castranova, V., Kisin, E.R., Schwegler-Berry, D., Murray, A.R., Gandelsman, V.Z., et al.: Exposure to carbon nanotube material: assessment of nanotube cytotoxicity using human keratinocyte cells. J Toxicol Environ Health. **66**, 1909–1926 (2003)
125. Lee, J.S., Choi, Y.C., Shin, J.H., Lee, J.H., Lee, Y., Park, S.Y., Baek, J.E., Park, J.D., Ahn, K.: Health surveillance study of workers who manufacture multi-walled carbon nanotubes. Nanotoxicol. **9**(6), 802–811 (2015). https://doi.org/10.3109/17435390.2014.978404. ISSN 1743-5390. PMID 25395166
126. Liou, S.-H., Tsai, C.S.J., Pelclova, D., Schubauer-Berigan, M.K., Schulte, P.A.: Assessing the first wave of epidemiological studies of nanomaterial workers. J. Nanopart. Res.. **17**(10), 1–19 (2015). https://doi.org/10.1007/s11051-015-3219-7. ISSN 1388-0764. PMC 4666542. PMID 26635494
127. Grosse, Y.: Carcinogenicity of fluoro-edenite, silicon carbide fibres and whiskers, and carbon nanotubes. Lancet Oncol. **15**(13), 1427–1428 (2014)
128. Schulte, P.A., Kuempel, E.D., Zumwalde, R.D., Geraci, C.L., Schubauer-Berigan, M.K., Castranova, V., Hodson, L., Murashov, V., Dahm, M.M.: Focused actions to protect carbon nanotube workers. Am. J. Ind. Med. **55**(5), 395–411 (2012). https://doi.org/10.1002/ajim. 22028. ISSN 1097-0274
129. Current intelligence bulletin 65: occupational exposure to carbon nanotubes and nanofibers. Natl. Inst. Occup.Saf. Health. **65**, 1–156 (2013)
130. Lacerda, L., Bianco, A., Prato, M., Kostarelos, K.: Carbon nanotubes as nanomedicines: from toxicology to pharmacology. Adv. Drug Deliv. Rev. **58**, 1460–1470 (2006)

131. Pacurari, M., Yin, X.J., Zhao, J., Ding, M., Leonard, S., Schwegler-Berry, D., et al.: Raw single-wall carbon nanotubes induce oxidative stress and activate MAPKs, AP-1, NF-kB, and Akt in normal and malignant human mesothelial cells. Environ Health Perspect. **116**, 1211–1217 (2008)

132. Jacobsen, N.R., Pojana, G., White, P., Moller, P., Cohn, C.A., Korsholm, K.S., et al.: Genotoxicity, cytotoxicity, and reactive oxygen species induced by single-walled carbon nanotubes and C60 fullerenes in the FE1-Muta mouse lung epithelial cells. Environm. Mol. Mutagen. **49**, 476–487 (2008)

133. Guo, L., Bussche, A.V.D., Buechner, M., Yan, A., Kane, A.B., Hurt, R.H.: Adsorption of essential micronutrients by carbon nanotubes and the implications for nanotoxicity testing. Small. **4**(6), 721–727 (2008)

134. Glenn, H.: U.S. Launches Inquiry Into Plastic resin Imports. Chem. Eng. News. **93**(12), 28 (2015)

135. Zyvex Technologies. http://zyvextech.com

136. Edwards, B.C.: The space elevator: a revolutionary earth-to-space transportation system. BC Edwards (2003)

137. Miaudet, P., Badaire, S., Maugey, M., Derré, A., Pichot, V., Launois, P., Poulin, P., Zakri, C.: Hot-drawing of single and multiwall carbon nanotube fibers for high toughness and alignment. Nano Lett. **5**(11), 2212–2215 (2005). Bibcode:2005NanoL...5.2212M. https://doi.org/10.1021/nl051419w. PMID 16277455

138. Li, Y.-L., Kinloch, I.A., Windle, A.H.: Direct spinning of carbon nanotube fibers from chemical vapor deposition synthesis. Science. **304**(5668), 276–278 (2004). Bibcode:2004Sci...304..276L. https://doi.org/10.1126/science.1094982. PMID 15016960

139. Pötschke, P., Andres, T., Villmow, T., Pegel, S., Brünig, H., Kobashi, K., Fischer, D., Häussler, L.: Liquid sensing properties of fibres prepared by melt spinning from poly(lactic acid) containing multi-walled carbon nanotubes. Compos. Sci. Technol. **70**(2), 343–349 (2010). https://doi.org/10.1016/j.compscitech.2009.11.005

140. Chen, P., Kim, H.S., Kwon, S.M., Yun, Y.S., Jin, H.J.: Regenerated bacterial cellulose/multi-walled carbon nanotubes composite fibers prepared by wet-spinning. Curr. Appl. Phys. **9**(2), e96. Bibcode:2009CAP.....9...96C (2009). https://doi.org/10.1016/j.cap.2008.12.038

141. Coleman, J.N., Khan, U., Blau, W.J., Gun'Ko, Y.K.: Small but strong: a review of the mechanical properties of carbon nanotube–polymer composites. Carbon. **44**(9), 1624–1652 (2006). https://doi.org/10.1016/j.carbon.2006.02.038

142. Alimohammadi, F., Parvinzadeh, M., Shamei, A.: Carbon nanotube embedded textiles. US20110171413 A1, 14 July 2011

143. Alimohammadi, F., Parvinzadeh Gashti, M., Shamei, A.: Functional cellulose fibers via polycarboxylic acid/carbon nanotube composite coating. J. Coat. Technol. Res. **10**, 123–132 (2012). https://doi.org/10.1007/s11998-012-9429-3

144. Alimohammadi, F., Gashti, M.P., Shamei, A.: A novel method for coating of carbon nanotube on cellulose fiber using 1,2,3,4-butanetetracarboxylic acid as a cross-linking agent. Prog. Org. Coat. **74**(3), 470–478 (2012). https://doi.org/10.1016/j.porgcoat.2012.01.012

145. Zhu, H.W., Xu, C.L., Wu, D.H., Wei, B.Q., Vajtai, R., Ajayan, P.M.: Direct synthesis of long single-walled carbon nanotube strands. Science. **296**, 884 (2002)

146. Zhang, M., Atkinson, K.R., Baughman, R.H.: Multifunctional carbon nanotube yarns by downsizing an ancient technology. Science. **306**, 1358 (2004)

147. Zhang, M., Fang, S., Zakhidov, A.A., Lee, S.B., Aliev, A.E., Williams, C.D., Atkinson, K. R., Baughman, R.H.: Strong, transparent, multifunctional, carbon nanotube sheets. Science. **309**, 1215 (2005)

148. Yildirim, T., Gülseren, O., Kılıç, Ç., Ciraci, S.: Pressure-induced interlinking of carbon nanotubes. Phys. Rev. B. **62**(19), 19 (2001). arXiv:cond-mat/0008476. Bibcode: 2000PhRvB..6212648Y. https://doi.org/10.1103/PhysRevB.62.12648

149. Zhang, M., Atkinson, K.R., Baughman, R.H.: Multifunctional carbon nanotube yarns by downsizing an ancient technology. Science. **306**, 1358 (2004)

150. Behabtu, N., et al.: Strong, light, multifunctional fibers of carbon nanotubes with ultrahigh conductivity. Science. **339**, 182 (2013)

151. Thoteson, E., Ren, Z., Chou, T.: Advances in the science and technology of carbon nanotubes and their composites: a review. Compos. Sci. Technol. **61**(13), 1899–1912 (2001)

152. Delmotte, J.P., Rubio, A.: Mechanical properties of carbon nanotubes: a fiber digest for beginners. Carbon. **40**(10), 1729–1734 (2002)

153. Chou, T.-W., Gao, L., Thostenson, E.T., Zhang, Z., Byun, J.-H.: Compos. Sci. Technol. **64**, 2363 (2004)

154. Gojny, F.H., Wichmann, M.H.G., Kopke, U., Fiedler, B., Schulte, K.: Carbon Nanotube-reinforced epoxy composites: enhanced stiffness and fracture toughness at low nanotube content. Compos. Sci. Technol. **64**, 2363 (2004)

155. Yao, Z., Braidy, N., Botton, G.A., Adronov, A.: Polymerization from the surface of single-walled carbon nanotubes - preparation and characterization of nanocomposites. J. Am. Chem. Soc. **125**, 16015 (2003)

156. Veedu, V.P., et al.: Multifunctional composites using reinforced laminae with carbon-nantube forests. Nat. Mater. **5**, 457 (2006)

157. Garcia, E.J., Wardle, B.L., Hard, A.J., Yamamoto, N.: Fabrication and multifunctional properties of a hybrid laminate with aligned carbon nanotubes grown in situ. Compos. Sci. Technol. **68**, 2034 (2008)

158. Seeger, T., Köhler, T., Frauenheim, T., Grobert, N., Rühle, M., et al.: Chem. Commun. **35**, 34–35 (2002)

159. Kashigawa, T., et al.: Nanoparticle networks reduce the flammability of polymer nanocomposites. Nat. Mater. **4**, 928 (2005)

160. Smith, J.: Slicing it extra thin, Tireview, 2005. Available from: http://www.tireview.com

161. Krishnamoorti, R., Dyke, C.A., Tour, J.M.: To be submitted for publication

162. Bauhofer, W., Kovacs, J.Z.: A review and analysis of electrical percolation in carbon nanotube polymer composites. Compos. Sci. Technol. **69**, 1486 (2009)

163. Yu, K., Ganhua, L., Zheng, B., Shun, M., Junhong, C.: Carbon nanotube with chemically bonded graphene leaves for electronic and optoelectronic applications. J. Phys. Chem. Lett. **2**(13), 1556–1562 (2011). https://doi.org/10.1021/jz200641c

164. Bourzac, K.: Nano paint could make airplanes invisible to radar, Technology Review. Mit, 5 December 2011

165. Dai, L., Chang, D.W., Baek, J.-B., Lu, W.: Carbon nanomaterials for advanced energy conversion and storage. Small. **8**, 1130 (2012)

166. Evanoff, J., et al.: Adv. Mater. **24**, 433 (2012)

167. Sotowa, C., et al.: The reinforcing effect of combined carbon nanotubes and acetylene blacks on the positive electrode of lithium-ion batteries. ChemSusChem. **1**, 911 (2008)

168. Wu, G.T., Wang, C.S., Zhang, X.B., Yang, H.S., Qi, Z.F., Li, W.Z.: Lithium insertion into CuO/carbon nanotubes. J. Power Sources. **75**, 175–179 (1998)

169. Sakamoto, J.S., Dunn, B.: Vanadium oxide-carbon nanotube composite electrodes for use in secondary lithium batteries. Electrochem Soc. **149**, A26–A30 (2002)

170. Gao, B., Bower, C., Lorentzen, J.D., Fleming, L., Kleinhammes, A., Tang, X.P., McNeil, L. E., Wu, Y., Zhou, O.: Enhanced saturation lithium composition in ball-milled single-walled carbon nanotubes. Chem. Phys. Lett. **327**, 1–2, (69–75) (2000). Bibcode:2000CPL...327... 69G. https://doi.org/10.1016/S0009-2614(00)00851-4

171. See, Beyond Batteries: Storing Power in a Sheet of Paper, at https://www.eurekalert.org/ pub_releases/2007-08/rpi-bbs080907.php (2007). Accessed 2016

172. Hu, L., Choi, J.W., Yang, Y., Jeong, S., Mantia, F.L., Cui, L.-F., Cui, Y.: Highly conductive paper for energy-storage devices. Proc. Natl. Acad Sci. **106**(51), 21490–21494 (2009). https:// doi.org/10.1073/pnas.0908858106. ISSN 0027-8424. PMC 2799859. PMID 19995965

173. Hu, L., Wu, H., La Mantia, F., Yang, Y., Cui, Y.: Thin, flexible secondary Li-ion paper batteries. ACS Nano. **4**(10), 5843–5848 (2010). https://doi.org/10.1021/nn1018158. ISSN 1936-0851

174. Chen, Z., To, J.W.F., Wang, C., Lu, Z., Liu, N., Chortos, A., Pan, L., Wei, F., Cui, Y., Boa, Z.: A three-dimensionally interconnected carbon nanotube-conducting polymer hydrogel network for high-performance flexible battery electrodes. Adv. Energy Mater. **4**, 1400207 (2014). https://doi.org/10.1002/aenm.201400207

175. Lee, S.W., Yabuuchi, N., Gallant, B.M., Chen, S., Kim, B., Hammond, P.T., Shao-Horn, Y.: High-power lithium batteries from functionalized carbon-nanotube electrodes. Nat. Nanotechnol. **5**, 531–537 (2010). https://doi.org/10.1038/NNANO.2010.116

176. Guo, J., Xu, Y., Wang, C.: Sulfur-impregnated disordered carbon nanotubes cathode for lithium-sulfur batteries. Nano Lett. **11**, 4288–4294 (2011)

177. Endo, M., Hayashi, T., Kim, Y.A., Terrones, M., Dresselhaus, M.S.: Applications of carbon nanotubes in the twenty-first century. R. Soc. **362**, 2223–2238 (2004)

178. Baughman, R.H., Zakhidov, A.A., De Heer, W.A.: Carbon nanotubes–the route toward applications. Science. **297**, 787–792 (2002)

179. Frackowiak, E., Beguin, F.: Electrochemical storage of energy in carbon nanotubes and nanostructured carbons. Carbon. **40**, 1775–1787 (2002)

180. Ma, R.Z., Liang, J., Wei, B.Q., Zhang, B., Xu, C.L., Wu, D.H.: Processing and performance of electric double-layer capacitors with block-type carbon nanotube electrodes. Bull. Chem. Soc. Jpn. **72**, 2563–2566 (1999)

181. Jurewicz, K., Delpeux, S., Bertagna, V., Be-guin, F., Frackowiak, E.: Supercapacitors from nanotubes/polypyrrole composites. Chem. Phys. Lett. **347**, 36–40 (2001)

182. Schnorr, J.M., Swager, T.M.: Emerging applications of carbon nanotubes. Chemistry of Materials. **23**(2), 646–657 (2011)

183. Wee, G., Mak, W.F., Phonthammachai, N., Kiebele, A., Reddy, M.V., Chowdari, B.V.R., Gruner, G., Srinivasan, M., Mhaisalkar, S.G.: J. Electrochem. Soc. **157**, A179 (2010)

184. Xie, X., Gao, L.: Characterization of a manganese dioxide/carbon nanotube composite fabricated using an in situ coating method. Carbon. **45**, 2365 (2007)

185. Zhou, Y.-k., He, B.-l., Zhou, W.-j., Huang, J., Li, X.-h., Wu, B., Li, H.-L.: Electrochemical capacitance of well-coated single-walled carbon nanotube with polyaniline composites. Electrochim. Acta. **49**, 257 (2004)

186. Khomenko, V., Frackowiak, E., Béguin, F.: Electrochim. Acta. **50**, 2499 (2005)

187. M. H. van der Veen et al.: Paper presented at the 2012, IEEE international interconnect technology conference, San Jose, CA, 4 to 6 June 2012

188. Liu, C., Bard, A.J., Wudl, F., Weitz, I., Heath, J.R.: Electrochemical characterization of films of single-walled carbon nonotubes and their possible application in supercapacitors. Electrochem. Solid-State Lett. **2**(11), 577–578 (1999)

189. Gong, K., Du, F., Xia, M., Durstock, M., Dai, L.: Nitrogen-doped carbon nanotube arrays with high electrocatalytic activity for oxygen reduction. Science. **323**, 760 (2009)

190. Matsumoto, T., et al.: Science **2004**, 840 (2011)

191. Le Goff, A., et al.: From hydrogenases to noble metal-free catalytic nanomaterials for H2 production and uptake. Science. **326**, 1384 (2009)

192. Lee, J.M., et al.: Selective electron- or hole-transport enhancement in bulk-heterojunction organic solar cells with N- or B-doped carbon nanotubes. Adv. Mater. **23**, 629 (2011)

193. Gabor, N.M., Zhong, Z., Bosnick, K., Park, J., McEuen, P.L.: Extremely efficient multiple electron-hole pair generation in carbon nanotube photodiodes. Science. **325**, 1367 (2009)

194. Ajayan, P., Zhou, O.: Applications of carbon nanotubes, carbon nanotubes. Top. Appl. Phys. **80**, 391–425 (2001)

195. Pederson, B. M.: J. Phys. Rev. Lett. **69**(2689), 405 (1992)

196. Dillon, A.C., Jones, K.M., Bekkedahl, T.A., Kiang, C.H., Bethune, D.S., Heben, M.J.: Storage of hydrogen in single-walled carbon nanotubes. Nature. **386**(6623), 377–379. Bibcode:1997Natur.386..377D (1997). https://doi.org/10.1038/386377a0

197. Jhi, S.H., Kwon, Y.K., Bradley, K., Gabriel, J.C.P.: Hydrogen storage by physisorption: beyond carbon. Solid State Commun. **129**(12), 769–773. Bibcode:2004SSCom.129..769J (2004). https://doi.org/10.1016/j.ssc.2003.12.032

198. Vohrer, U., et al.: Carbon nanotube sheets for the use as artificial muscles. Carbon. **42**, 1159 (2004). https://doi.org/10.1016/j.carbon.2003.12.044

199. Chang, T., Guo, Z.: Temperature-induced reversible dominoes in carbon nanotubes. Nano Lett. **10**(1021), 101623 (2010)

200. Baughman, R.H., et al.: Carbon nano actuators. Science. **284**(5418), 1340 (1999)

201. Spinks, G.M., et al.: Pneumatic carbon nanotube actuators. Adv. Mater. **14**, 1728 (2002)

202. Aliev, A.E., et al.: Giant-stroke, superelastic carbon nanotube aerogel muscles. Science. **323** (5921), 1575–1578 (2009)

203. Madden, D.W.: Materials science: stiffer than steel. Science. **323**(5921), 1571–1572 (2009)

204. Yuzvinsky, T.: Nanotube nanomotor, tailoring carbon nanotubes. https://users.soe.ucsc.edu/ ~yuzviknsy/research/nanomotor.php. Accessed 30 Sept 2016

205. Berger, M.: Speeding up catalytic nanomotors with carbon nanotubes. Nano Werk [Online]. http://www.nanowerk.com/spotlight/spotid=5553.php (2008). Accessed 30 Sept 2016

206. Laocharoensuk, R., Burdick, J., Wang, J.: Carbon-nanotube-induced acceleration of catalytic nanomotors. ACS Nano. **2**(5), 1069–1075 (2008)

207. Bailey, S.W.D., Amanatidis, I., Lambert, C.J.: Carbon nanotube electron windmills: a novel design for nanomotors. Phys. Rev. Lett. **100**, 256802 (2008)

208. Baugham, R., Zakhidov, A., Heer, W.: Carbon nanotubes – the route toward applications. Science. **297**, 787–792 (2002)

209. Tang, Z.K., et al.: Science. **292**, 2462 (2001)

210. Tennent, H.G.: Carbon fibrils, method for producing same and compositions containing same. US 4663230 A, 5 May 1987

211. Kong, N.R., Franklin, C., Zhou M.C., Chapline, S., Peng, K., Cho, H.D.: Science **287**(622), 406–15 (2000)

212. Kong, J., Franklin, N.R., Zhou, C.W., Chapline, M.G., Peng, S., et al.: Nanotube molecular wires as chemical sensors. Science. **287**, 622–625 (2000)

213. Varghese, O.K., Kichambre, P.D., Gong, D., Ong, K.G., Dickey, E.C., Grimes, C.A.: Gas sensing characteristics of multi-wall carbon nanotubes. Sens. Actuators B. **81**, 32–41 (2001)

214. Valentini, L., Cantalini, C., Lozzi, L., Armanetano, I., Kenny, J. M., Santucci, S.: Mater. Sci. Eng. C **23**, 523 (2003)

215. Valentini, L., Cantalini, C., Lozzi, L., Armanetano, I., Kenny, J. M., Santucci, S.: Sens. Actuators B **93**, 333 (2003)

216. Valentini, L., Cantalini, C., Lozzi, L., Armanetano, I., Kenny, J.M., Santucci, S.: J. Eur. Ceram. Soc. **24**,1405 (2004)

217. Qi, P., Vermesh, O., Grecu, M., Javey, A., Wang, Q., Dai, H., Peng, S., Cho, K.J.: Toward large arrays of multiplex functionalized carbon nanotube sensors for highly sensitive and selective molecular detection. Nano Lett. **3**, 347 (2003)

218. Ahn, K.S., Kim, J.H., Lee, K.N., Kim, C.O., Hong, J.P.: Multi-wall carbon nanotubes as a high-efficiency gas sensor. J. Korean Phys. Soc. **45**, 158 (2004)

219. Valentini, L., Bavastrello, V., Stura, E., Armanetano, I., Nicolini, C., Kenny, J.M.: Chem. Phys. Lett. **383**, 617 (2004)

220. He, J.-B., Chen, C.-L., Li, J.-H.: Sens. Actuators B **99**, 1 (2004)

221. Wong, Y.M., Kang, W.P., Davidson, J.L., Wisitsora-at, A., Soh, K.L.: Sens. Actuators B **93**, 327 (2003)

222. Suehiro, J., Zhou, G.B., Hara, M.: J. Phys. D **36**, L109 (2003)

223. Varghese, O.K., Kichambre, P.D., Gong, D., Ong, K.G., Dickey, E.C., Grimes, C.A.: Sens. Actuators B **81**, 32 (2001)

224. Snow, E.S., Perkins, F.K., Houser, E.J., Badescu, S.C., Reinecke, T.L.: Chemical detection with a single-walled carbon nanotube capacitor. Science. **307**, 1942 (2005)

225. Esser, B., Schnoor, J.M., Swager T.M.: Angew Chem. Int. Ed. **51**, 5752 (2012)

226. Novak, J.P., Snow, E.S., Houser, E.J., Park, D., Stepnowski, J.L., McGill, R.A.: Nerve agent detection using networks of singlewalled carbon nanotubes. Appl. Phys. Lett. **83**, 4026 (2003)

227. Li, J., Lu, Y., Ye, Q., Cinke, M., Han, J., Meyyappan, M.: Carbon nanotube sensors for gas and organic vapor detection. Nano Lett. **3**, 929 (2003)

228. Valentini, L., Armentano, I., Kenny, J.M., Cantalini, C., Lozzi, L., Santucci, S.: Sensors for sub-ppm NO 2 gas detection based on carbon nanotube thin films. Appl. Phys. Lett. **82**, 961 (2003)

229. Mubeen, S., Zhang, T., Yoo, B., Deshusses, M.A., Myung, N.V.: Palladium nanoparticles decorated single-walled carbon nanotube hydrogen sensor. J. Phys. Chem. C. **111**, 6321 (2007)

230. Sun, Y., Wang, H.: High-performance, flexible hydrogen sensors that use carbon nanotubes decorated with palladium nanoparticles. Adv. Mater. **19**, 2818 (2007)

231. Sun, Y., Wang, H.: Electrodeposition of Pd nanoparticles on single-walled carbon nanotubes for flexible hydrogen sensors. Appl. Phys. Lett. **90**, 213107 (2007)

232. Lu, Y.J., Li, J., Han, J., Ng, H.T., Binder, C., Partridge, C., Meyyappan, M.: Room temperature methane detection using palladium loaded single-walled carbon nanotube sensors. Chem. Phys. Lett. **391**, 344 (2004)

233. Star, A., Han, T.R., Joshi, V., Gabriel, J.C.P., Gruner, G.: Nanoelectronic carbon dioxide sensors. Adv. Mater. **16**, 2049 (2004)

234. Star, A., Joshi, V., Skarupo, S., Thomas, D., Gabriel, J.C.P.: Gas sensor array based on metal-decorated carbon nanotubes. J. Phys. Chem. B. **110**, 21014 (2006)

235. Lee, H., Naishadham, K., Tentzeris, M.M., Shaker, G.: A novel highly-sensitive antenna-based 'smart skin' bas sensor utilizing carbon nano tubes and inkjet printing, pp. 1593–1596 (2011)

236. Chopra, S., Pham, A., Gaillard, J., Parker, A., Rao, A.M.: Appl. Hys. Lett. **80**, 4632–34 (2002)

237. Rubianes, M.D., Rivas, G.A.: Electrochem. Commun. **5**, 689 (2003)

238. Farajian, A., Yakobson, B., Mizeseki, H., Kawazoe, Y.: Electronic transport through bent carbon nanotubes: nanoelectromechanical sensors and switches. Phys Rev. **67**, 1–6 (2003)

239. Gao, M., Dai, L.M., Wallace, G.G.: Biosensors based on aligned carbon nanotubes coated with inherently conducting polymers. Electroanalysis. **15**, 1089 (2003)

240. Star, A., et al.: Label-free detection of DNA hybridization using carbon nanotube network field-effect transistors. Proc. Natl. Acad. Sci. U.S.A. **103**, 921 (2006)

241. Trojanowicz, M., Mulchandani, A., Mascini, M.: Carbon nanotubesmodified screen-printed electrodes for chemical sensors and biosensors. Anal. Lett. **37**, 3185 (2004)

242. Britto, P.J., Santhanam, K.S.V., Ajayan, P.M.: Carbon nanotube electrode for oxidation of dopamine. Bielectrochem. Bioenerg. **41**, 121 (1996)

243. Zhang, M., Smith, A., Gorski, W.: Anal. Chem. **76**, 1083 (2004)

244. Xu, J.Z., Zhu, J.J., Wu, Q., Hu, Z., Chen, H.Y.: An amperometric biosensor based on the coimmobilization of horseradish peroxidase and methylene blue on a carbon nanotubes modified electrode. Electroanalysis (NY). **15**, 219 (2003)

245. Jiang, L.-C., Zhang, W.-D.: A highly sensitive nonenzymatic glucose sensor based on CuO nanoparticles-modified carbon nanotube electrode. Biosens. Bioelectron. **25**, 1402 (2010)

246. Pérez López, B., Merkoçi, A.: Improvement of the electrochemical detection of catechol by the use of a carbon nanotube based biosensor. Analyst. **134**, 60 (2009)

247. Zhao, Y., Gao, Y., Zhan, D., Liu, H., Zhao, Q., Kou, Y., Shao, Y., Li, M., Zhuang, Q., Zhu, Z.: Selective detection of dopamine in the presence of ascorbic acid and uric acid by a carbon nanotubes-ionic liquid gel modified electrode. Talanta. **66**, 51 (2005)

248. Guo, M., Chen, J., Li, J., Tao, B., Yao, S.: Anal. Chem. Acta **532**, 71 (2005)

249. Mancuso, S., Marras, A.M., Magnus, V., Baluska, F.: Noninvasive and continuous record-ings of auxin fluxes in intact root apex with a carbon nanotube-modified and self-referencing microelectrode. Anal. Biochem. **341**, 344 (2003)

250. Hun, C.G., Wang, W.L., Wang, S.X., Zhu, W., Li, Y.: Investigation on electrochemical properties of carbon nanotubes. Diamond Relat. Mat. **12**, 1295 (2003)

251. Wang, S.-F., Xu, Q.: Square wave voltammetry determination of brucine at multiwall carbon nanotube-modified glassy carbon electrodes. Anal. Lett. **38**, 657 (2005)

252. Gong, K.P., Dong, Y., Xiong, S.X., Chen, Y., Mao, L.: Novel electrochemical method for sensitive determination of homocysteine with carbon nanotube-based electrodes. Biosens. Bioelectron. **20**, 253 (2004)

253. Deo, R.P., Wang, J.: Electrochemical detection of carbohydrates at carbon-nanotube modified glassy-carbon electrodes. Electrochem. Commun. **6**, 284 (2004)

254. Ye, J.S., We, Y., De Zhang, W., Gan, L.M., Xu, G.Q., Sheu, F.S.: Electroanalysis (NY) **15**, 1693 (2003)

255. Zhao, G., Zang, S.Q., Liu, K.Z., Lin, S., Liang, J., Guo, X.Y., Zhang, Z.J.: Determination of trace xanthine by anodic stripping voltammetry with carbon nanotube modified glassy carbon electrode. Anal. Lett. **35**, 2233 (2002)

256. Zhao, G., Liu, K.Z., Lin, S., Liang, J., Guo, X.Y., Zhang, Z.J.: Application of a carbon nanotube modified electrode in anodic stripping voltammetry for determination of trace amounts of 6-benzylaminopurine. Microchim. Acta. **143**, 255 (2003)

257. Zeng, B.Z., Huang, F.: Electrochemical behavior and determination of fluphenazine at multi-walled carbon nanotubes/(3-mercaptopropyl)trimethoxysilane bilayer modified gold electrodes. Talanta. **64**, 380 (2004)

258. Yang, C.H.: Microchem. Acta **148**, 87 (2004)

259. Britto, P.J., Santhanam, K.S.V., Ajayan, P.M.: Carbon nanotube electrode for oxidation of dopamine. Bioelectrochem. Bioenerg. **41**, 121 (1996)

260. Wang, Z.H., Liu, J., Liang, Q.L., Wang, Y.M., Luo, G.: Carbon nanotube-modified electrodes for the simultaneous determination of dopamine and ascorbic acid. Analyst. **127**, 653 (2002)

261. Rubianes, M.D., Rivas, G.A.: Carbon nanotubes paste electrode. Electrochem. Commun. **5**, 689 (2003)

262. Wang, J.X., Li, M.X., Shi, Z.J., Li, N.Q., Gu, Z.N.: Electrocatalytic oxidation of norepinephrine at a glassy carbon electrode modified with single wall carbon nanotubes. Electroanalysis. **14**, 225 (2002)

263. Moore, R.R., Banks, C.E., Compton, R.G.: Basal plane pyrolytic graphite modified electrodes: comparison of carbon nanotubes and graphite powder as electrocatalysts. Anal. Chem. **76**, 2677 (2004)

264. Valentini, F., Amine, A., Orlanducci, S., Terranova, M.L., Palleschi, G.: Carbon nanotube purification: preparation and characterization of carbon nanotube paste electrodes. Anal. Chem. **75**, 5413 (2003)

265. Wang, J., Musameh, M.: Carbon nanotube/teflon composite electrochemical sensors and biosensors. Anal. Chem. **75**, 2075 (2003)

266. Wang, J., Musameh, M.: Solubilization of carbon nanotubes by nafion toward the preparation of amperometric biosensors. J. Am. Chem. Soc. **125**, 2408 (2003)

267. Rubianes, M.D., Rivas, G.A.: Carbon nanotubes paste electrode. Electrochem. Commun. **5**, 689 (2003)

268. Wang, J.: Carbon-nanotube based electrochemical biosensers: a review. Electroanalysis. **17** (1), 7–14 (2005)

269. Wang, J., Musameh, M.: Carbon nanotube screen-printed electrochemical sensors. Analyst. **129**, 1 (2004)

270. Lin, Y., Lu, F., Wang, J.: Disposable carbon nanotube modified screen-printed biosensor for amperometric detection of organophosphorus pesticides and nerve agents. Electroanalysis. **16**, 145 (2004)

271. Patolsky, F., Weizmann, Y., Willner, I.: Long-range electrical contacting of redox enzymes by SWCNT connectors. Angew. Chem., Int. Ed. Engl. **43**, 2113 (2004)

272. Wang, J., Liu, G., Jan, M.R.: Ultrasensitive electrical biosensing of proteins and DNA: carbon-nanotube derived amplification of the recognition and transduction events. J. Am. Chem. Soc. **126**, 3010 (2004)

273. Cheng, G., Zhao, J., Tu, Y., He, P., Fang, Y.: A sensitive DNA electrochemical biosensor based on magnetite with a glassy carbon electrode modified by muti-walled carbon nanotubes in polypyrrole. Anal. Chim. Acta. **533**, 11 (2005)

274. Koehne, J.E., Chen, H., Cassell, A.M., Ye, Q., Han, J., Meyyappan, M., Li, J.: Miniaturized multiplex label-free electronic chip for rapid nucleic acid analysis based on carbon nanotube nanoelectrode arrays. Clin. Chem. **50**, 1886 (2004)

275. Zhang, X., Jiao, K., Liu, S., Hu, Y.: Readily reusable electrochemical DNA hybridization biosensor based on the interaction of DNA with single-walled carbon nanotubes. Anal. Chem. **81**, 6006 (2009)

276. Staii, C., Johnson, A.T., Chen, M., Gelperin, A.: DNA-decorated carbon nanotubes for chemical sensing. Nano Lett. **5**, 1774 (2005)

277. Wang, J., Kawde, A., Mustafa, M.: Carbon-nanotube-modified glassy carbon electrodes for amplified label-free electrochemical detection of DNA hybridization. Analyst. **128**, 912 (2003)

278. Pedano, M., Rivas, G.A.: Adsorption and electrooxidation of nucleic acids at carbon nanotubes paste electrodes. Electrochem. Commun. **6**, 10 (2004)

279. Wang, J., Li, M., Shi, Z., Li, N., Gu, Z.: Electroanalysis **16**, 140 (2004)

280. Gooding, J.J.: Nanostructuring electrodes with carbon nanotubes: A review on electrochemistry and applications for sensing. Electrochimica Acta. **50**, 3049–3060 (2005)

281. Li, J., Ng, H.T., Cassell, A., Fan, W., Chen, H., Ye, Q., Koehne, J., Han, J., Meyyappan, M.: Carbon nanotube nanoelectrode array for ultrasensitive dna detection. Nano Lett. **3**, 597 (2003)

282. Nguyen, C.V., Delzeit, L., Cassell, A.M., Li, J., Han, J., Meyyappan, M.: Preparation of nucleic acid functionalized carbon nanotube arrays. Nano Lett. **2**, 1079 (2002)

283. Koehne, J., Chen, H., Li, J., Cassell, A.M., Ye, Q., Ng, H.T., Han, J., Meyyappan, M.: Ultrasensitive label-free DNA analysis using an electronic chip based on carbon nanotube nanoelectrode arrays. Nanotechnol. **14**, 1239 (2003)

284. Koehne, J., Li, J., Cassell, A.M., Chen, H., Ye, Q., Ng, H.T., Han, J., Meyyappan, M.: The fabrication and electrochemical characterization of carbon nanotube nanoelectrode arrays. J. Mater. Chem. **14**, 676 (2004)

285. Johnston, D.H., Glasgow, K.C., Thorp, H.H.: Electrochemical measurement of the solvent accessibility of nucleobases using electron transfer between DNA and metal complexes. J. An. Chem. Soc. **117**, 8933 (1995)

286. Thorp, H.H.: TIBTECH **16**, 117 (1998)

287. Gooding, J.J.: Electrochemical DNA hybridization biosensors. Electroanalysis. **14**, 1149 (2002)

288. Tran, H.V., Piro, B., Reisberg, S., Tran, L.D., Duc, H.T., Pham, M.C.: Label-free and reagentless electrochemical detection of microRNAs using a conducting polymer nanostructured by carbon nanotubes: Application to prostate cancer biomarker miR-141. Biosens. Bioelectro. **49**, 164–169 (2013)

289. Star, A., Gabriel, J.C.P., Bradley, K., Gruner, G.: Electronic detection of specific protein binding using nanotube FET devices. Nano Lett. **3**, 459–463 (2003)

290. Boussaad, S., Tao, N., Zhang, N.J., Zhang, R., Hopson, T., Nagahara, L.A.: In Situ detection of cyto chrome adsorption with single walled carbon nanotube device. Chem. Commun. **9**, 1502–1503 (2003)

291. Forzani, E.S., Li, X.L., Zhang, P.M., Tai, N.J., Zhang, R., Amlani, I., Tsui, R., Nagahara, L. A.: Turning the chemical selectivity of SWNT-FETs for detection of heavy-metal ions. Small. **2**, 1283–1291 (2006)

292. Gooding, J.J., Wibowo, R., Liu, J.Q., Yang, W.R., Losic, D., Orbons, S., Mearns, F.J., Shapter, J.G., Hibbert, D.B.: Protein electrochemistry using aligned carbon nanotube arrays. J. Am. Chem. Soc. **125**, 9006–9007 (2003)

293. Yu, X., et al.: Carbon nanotube amplification strategies for highly sensitive immunodetection of cancer biomarkers. J. Am. Chem. Soc. **128**, 11199–11205 (2006)

294. Besterman, K., Lee, J.O., Wiertz, F.G.M., Heering, H.A., Dekker, C.: Nano Lett. **3**, 727 (2003)

295. So, H.M., Won, J., Kim, Y.H., Kim, B.K., Ryu, B.H., Na, P.S., Kim, H., Lee, J.O.: Single-walled carbon nanotube biosensors using aptamers as molecular recognition elements. J. Am. Chem. Soc. **127**, 11906 (2007)

296. So, H.M., Park, D.W., Jeon, E.K., Kim, Y.H., Kim, B.S., Lee, C.K., Choi, S.Y., Kim, S.C., Chang, J., Lee, J.O.: Detection and titer estimation of Escherichia coli using aptamer-functionalized single-walled carbon-nanotube field-effect transistors. Small. **4**, 197 (2008)

297. Yoon, H., Kim, J.H., Lee, N., Kim, B.G., Jang, J.: A novel sensor platform based on aptamer-conjugated polypyrrole nanotubes for label-free electrochemical protein detection. ChemBioChem. **9**, 634 (2008)

298. Wohlstadter, J.N., Wilbur, J.L., Sigal, G.B., Biebuyck, H.A., Billadeau, L.W., Dong, L.W., Fischer, A.B., Gudbande, S.R., Jamieson, S.H., Kenten, J.H., Leginus, J., Leland, J.K., Massey, R.J., Wohlstadter, S.J.: Carbon nanotube-based biosensor. Adv. Mater. **15**, 1184 (2003)

299. Sánchez, S., Pumera, M., Fabregas, E.: Biosens. Biolectron. **22**, 332 (2007)

300. Sánchez, S., Roldán, M., Pérez, S., Fabregas, E.: Toward a fast, easy, and versatile immobilization of biomolecules into carbon nanotube/polysulfone-based biosensors for the detection of hCG hormone. Anal. Chem. **80**, 6508 (2008)

301. Pumera, M.: The electrochemistry of carbon nanotubes: fundamentals and applications. Chem. Eur. J. **15**, 4970–4978 (2009)

302. Musameh, M., Wang, J., Merkoci, A., Lin, Y.: Low-potential stable nadh detection at carbon-nanotube-modified glassy carbon electrodes. Electrochem. Commun. **4**, 743 (2002)

303. Campbell, J.K., Sun, L., Crooks, R.M.: Electrochemistry using single carbon nanotubes. J. Am. Chem. Soc. **121**, 3779 (1999)

304. Heller, I., Kong, J., Heering, H.A., Williams, K.A., Lemao, S.G., Dekker, C.: Individual single-walled carbon nanotubes as nanoelectrodes for electrochemistry. Nano Lett. **5**, 137 (2005)

305. Zhang, C., Wang, G., Liu, M., Feng, Y., Zhang, Z., Fang, B.: Individual single-walled carbon nanotubes as nanoelectrodes for electrochemistry. Electrochim. Acta. **55**, 2835 (2010)

306. Kong, L., Wang, J., Luo, T., Meng, F., Chen, X., Li, M., Liu, J.: Novel pyrenehexafluoroisopropanol derivative-decorated single-walled carbon nanotubes for detection of nerve agents by strong hydrogen-bonding interaction. Analyst. **135**, 368 (2010)

307. Novak, J.P., Snow, E.S., Houser, E.J., Park, D., Stepnowski, J.L., McGill, R.A.: Nerve agent detection using networks of singlewalled carbon nanotubes. Appl. Phys. Lett. **83**, 4026 (2003)

308. Cherukuri, P., Bachilo, S.M., Litovsky, S.H., Weisman, R.B.: Near-infrared fluorescence microscopy of single-walled carbon nanotubes in phagocytic cells. J. Am. Chem. Soci. **126**, 15638–15639 (2004)

309. Welsher, K., Sherlock, S.P., Dai, H.: Deep-tissue anatomical imaging of mice using carbon nanotube fluorophores in the second near-infrared window. Proceedings of the National Academy of Sciences. **108**(22), 8943–8948 (2011)

310. Heller, D.A., Baik, S., Eurell, T.E., Strano, M.S.: Single-walled carbon nanotube spectroscopy in live cells: towards long-term labels and optical sensors. Adv. Mater. **17**, 2793 (2005)

311. Yang, W., Thordarson, P., Gooding, J.J., Ringer, S.P., Braet, F.: Carbon nanotubes for biological and biomedical applications. Nanotechnol. **18**(412001), 1–12 (2007)

312. Kam, N.W.S., O'Connell, M., Wisdom, J.A., Dai, H.J.: Carbon nanotubes as multifunctional biological transporters and near-infrared agents for selective cancer cell destruction. Proc. Natl Acad. Sci. USA. **102**, 11600–11605 (2005)

313. Pantarotto, D., Briand, J.P., Prato, M., Bianco, A.: Translocation of bioactive peptides across cell membranes by carbon nanotubes. Chem. Commun. (Camb). (1), 16–17 (2004)

314. Bhandavat, R., Feldman, A., Cromer, C., Lehman, J., Singh, G.: Very high laser-damage threshold of polymer-derived Si(B)CN- carbon nanotube composite coatings. ACS Appl. Mater Interfaces. **5**(7), 2354–2359 (2013). https://doi.org/10.1021/am302755x

315. Dai, H.J., Hafner, J.H., Rinzler, A.G., Colbert, D.T., Smalley, R.E.: Nanotubes as nanoprobes in scanning probe microscopy. Nature. **384**, 147–150 (1996)

316. Yumura, M.: Carbon nanotube industrial applications. Res. Cent. Adv. Carbon Mater. **10**, (2004)
317. Kim, P., Lieber, C.M.: Nanotube nanotweezers. Science. **286**, 2148 (1999)
318. Chen, H.-W., Wu, R.-J., Chan, K.-H., Sun, Y.-L., Su, P.-G.: Sens. Actuators B **104**, 80 (2005)
319. Penza, M., Antolini, F., Antisari, M.V.: Sens. Actuators B **100**, 47 (2004)
320. Sumanasekera, G.U., Pradham, B.K., Adu, C.K.W., Romero, H.E., Foley, H.C., Eklund, P. C.: Mol. Cryst. Liq. Cryst. **387**, 255 (2002)
321. Modi, A., Koratkar, N., Lass, E., Wei, B.Q., Ajayan, P.M.: Miniaturized gas ionization sensors using carbon nanotubes. Nature (London). **424**, 171 (2003)
322. Dharap, P., Zhiling, L., Nagarajaiah, S., Barrera, E.V.: Nanotube film based on single-wall carbon nanotubes for strain sensing. Nanotechnol. **15**, 379–382 (2004)
323. Lee, C., Liu, X., Zhou, C.: Carbon nanotube field-effect inverters. Appl. Phys. Lett.**79**(20), (2001)
324. De La Zerda, A., et al.: Carbon nanotubes as photoacoustic molecular imaging agents in living mice. Nat. Nanotechnol. **3**, 557 (2008)
325. Shi Kam, N.W., Jessop, T.C., Wender, P.A., Dai, H.: Nanotube molecular transporters: internalization of carbon nanotube-proteinconjugates into mammalian cells. J. Am. Chem. Soc. **126**, 6850–6851 (2004)
326. Cai, D., Mataraza, J.M., Qin, Z.H., Huang, Z., Huang, J., Chiles, T.C., Carnahan, D., Kempa, K., Ren, Z.: Highly efficient molecular delivery into mammalian cells using carbon nanotube spearing. Nat. Methods. **2**, 449–454 (2005)
327. Bianco, A., Kostarelos, K., Prato, M.: Applications of carbon nanotubes in drug delivery. Curr. Opin. Chem. Biol. **9**, 674–679 (2005)
328. Kam, N.W.S., Dai, H.J.: Carbon nanotubes as intracellular protein transporters: generality and biological functionality. J. Am. Chem. Soc. **127**, 6021–6026 (2005)
329. Kam, N.W.S., Liu, Z.A., Dai, H.J.: Carbon nanotubes as intracellular transporters for proteins and DNA: an investigation of the uptake mechanism and pathway. Angew. Chem. Int. Ed. **45**, 577–581 (2006)
330. Shao, N., Lu, S.X., Wickstrom, E., Panchapakesan, B.: Integrated molecular targeting of IGF1R and HER2 surface receptors and destruction of breast cancer cells using single wall carbon nanotubes. Nanotechnol. **18**, 315101 (2007)
331. Guiseppi-Elie, A., Lei, C.H., Baughman, R.H.: Direct electron transfer of glucose oxidase on carbon nanotubes. Nanotechnol. **13**, 559–564 (2002)
332. Pantarotto, D., Partidos, C.D., Graff, R., Hoebeke, J., Briand, J.P., Prato, M., Bianco, A.: Synthesis, Structural characterization and immunological properties of carbon nanotubes functionalized with peptides. J. Am. Chem. Soc. **125**, 6160–6164 (2003)
333. Liu, Z., Winters, M., Holodniy, M., Dai, H.J.: siRNA delivery into human T cells and primary cells with carbon-nanotube transporters. Angew. Chem. Int. Ed. **46**, 2023–2027 (2007)
334. Dean, D.A., Strong, D.D., Zimmer, W.E.: Nuclear entry of nonviral vectors. Gene. Ther. **12**, 881–890 (2005)
335. Luo, D.: A new solution for improving gene delivery. Trends Biotechnol. **22**, 101–103 (2004)
336. Schmidt-Wolf, G.D., Schmidt-Wolf, I.G.: Non-viral and hybrid vectors in human gene therapy: an update. Trends Mol. Med. **9**, 67–72 (2003)
337. Chaudhuri, A. (ed.): Special issue: cationic transfection lipids. Curr. Med. Chem. **10**, 1185–1315 (2003)
338. Pantarotto, D., Singh, R., McCarthy, D., Erhardt, M., Briand, J.P., Prato, M., Kostarelos, K., Bianco, A.: Functionalised carbon nanotubes for plasmid DNA gene delivery. Angew Chem. Int. Ed. Engl. **43**, 5242–5246 (2004)
339. Singh, R., Pantarotto, D., McCarthy, D., Chaloin, O., Hoebeke, J., Partido, C.D., Briand, J. P., Prato, M., Bianco, A., Kostarelos, K.: Binding and condensation of plasmid DNA onto functionalized carbon nanotubes: towards the construction of nanotube-based gene delivery vectors. J. Am. Chem. Soc. **127**, 4388–4396 (2005)

340. Bianco, A., Hoebek, E.J., Godefroy, S., Chaloin, O., Pantarotto, D., Briand, J.P., Muller, S., Prato, M., Partidos, C.D.: Cationic carbon nanotubes bind to CpG oligodeoxynucleotides and enhance their immunostimulatory properties. J. Am. Chem. Soc. **127**, 58–59 (2005)

341. Lu, Q., Moore, J.M., Huang, G., Mount, A.S., Rao, A.M., Larcom, L.L., Ke, P.C.: RNA polymer translocation with single-walled carbon nanotubes. Nano. Lett. **4**, 2473–2477 (2004)

342. Wu, W., Wieckowski, S., Pastorin, G., Klumpp, C., Benincasa, M., Briand, J.P., Gennaro, R., Prato, M., Bianco, A.: Targeted delivery of amphotericin B to cells using functionalised carbon nanotubes. Angew Chem. Int. Ed. Engl. **44**(39), 6358–6352 (2005., in press). https://doi.org/10.1002/anie.200501613

343. Yinghuai, Z., Peng, A.T., Carpenter, K., Maguire, J.A., Hosmane, N.S., Takagaki, M.: Substituted carborane-appended water-soluble single-wall carbon nanotubes: new approach to boron neutron capture therapy drug delivery. J. Am. Chem. Soc. **127**, 9875–9880 (2005)

344. Murakami, T., Ajima, K., Miyawaki, J., Yudasaka, M., Iijima, S., Shibe, K.: Drug-loaded carbon nanohorns: adsorption and release of dexamethasone in vitro. Mol. Pharm. 399–405 (2004)

345. Liu, Z., Sun, X., Jakayama-Ratchford, N., Dai, H.: Supramolecular chemistry on water-soluble carbon nanotubes for drug loading and delivery. ACS Nano. **1**, 50 (2007)

346. Matson, M.L., Wilson, L.J.: Nanotechnology and MRI contrast enhancement. Future Med. Chem. **2**(3), 491–502 (2010). https://doi.org/10.4155/fmc.10.3. PMID 21426177

347. Wang, J.X., Li, M.X., Shi, Z.J., Li, N.Q., Gu, Z.N.: Direct electrochemistry of cytochrome c at a glassy carbon electrode modified with single-wall carbon nanotubes. Anal. Chem. **74**, 1993 (2002)

348. Davis, J.J., Coles, R.J., Hill, H.A.O.: Protein electrochemistry at carbon nanotube electrodes. J. Electroanal. Chem. **440**, 279 (1997)

349. Wang, G., Xu, J.J., Chen, H.Y.: Interfacing cytochrome c to electrodes with a DNA—carbon nanotube composite film. Electrochem. Commun. **4**, 506 (2002)

350. Zhao, G.C., Zhang, L., Wei, X. W., Yang, Z.S.: Electrochem. Commun. **5**, 825 (2003)

351. Yu, X., Chattopadhyay, D., Galeska, I., Papadimitrakopoulos, F., Rusling, J.F.: Peroxidase activity of enzymes bound to the ends of single-wall carbon nanotube forest electrodes. Electrochemical. Commun. **5**, 408 (2003)

352. Wang, L., Wang, J.X., Zhou, F.M.: Direct electrochemistry of catalase at a gold electrode modified with single-wall carbon nanotubes. Electroanalysis. **16**, 627 (2004)

353. Cao, Q., Rogers, J.A.: Ultrathin films of single-walled carbon nanotubes for electronics and sensors: a review of fundamental and applied aspects. Adv. Mater. **21**, 29–53 (2009)

354. Kocabas, C., Pimparkar, N., Yesilyurt, O., Kang, S.J., Alam, M.A., Rogers, J.A.: Experimental and theoretical studies of transport through large scale, partially aligned arrays of single-walled carbon nanotubes in thin film type transistors. Nano Lett. **7**(5), 1195–1202 (2007)

355. Javey, A., Guo, J., Wang, Q., Lundstrom, M., Dai, H.: Ballistic carbon nanotube transistors. Nature. **424**, 654–657 (2003)

356. Javey, A., Guo, J., Farmer, D.B., Wang, Q., Yenilmez, E., Gordon, R.G., Lundstrom, M., Dai, H.: Self-aligned ballistic molecular trasnisters and electrically parallel nanotube arrays. Nano Lett. **4**(7), 1319–1322 (2004)

357. Gabriel, J.-C.P.: 2d Random networks of carbon nanotubes. C.R. Phys. **11**(5–6), 362–374 (2010)

358. Gabriel, J.-C.P.: Large scale production of carbon nanotube transistors: a generic platforms for chemical sensors. Mat. Res. Soc. Symp. Proc. **762**, Q.12.7.1 (2003)

359. Nanomix Elab System: Fast and fully automated point of care diagnostic system. http://www.nano.com. Accessed 30 Sept 2016

360. Gabriel, J.-C.P.: Dispersed growth of nanotubes on a substrate, Patent WO 2004040671A2, 10 Aug 2005

361. Bradley, K., Gabriel, J.-C.P., Grüner, G.: Flexible nanotube transistors. Nano Lett. **3**(10), 1353–1355 (2003)

362. Armitage, P. N., Bradley, K., Gabriel, J. -C. P., Grüner, G.: Flexible nanostructure electronic devices. US 8456074, 25 Aug 2005.
363. Franklin, A.D., et al.: Sub-10 nm carbon nanotube transistor. Nano Lett. **12**, 758 (2012)
364. Cao, Q., et al.: Medium-scale carbon nanotube thin-film integrated circuits on flexible plastic substrates. Nature. **454**, 495 (2008)
365. Park, H., et al.: High-density integration of carbon nanotubes via chemical self-assembly. Nat. Nanotechnol. **7**, 787 (2012)
366. McCarthy, M.A., et al.: Low-voltage, low-power, organic light-emitting transistors for active matrix displays. Science. **332**, 570 (2011)
367. Cao, Q., Kim, H.S., Pimparkar, N., Kulkarni, J.P., Wang, C.J., Shim, M., Roy, K., Alam, M. A., Rogers, J.A.: Medium-scale carbon nanotube thin-film integrated circuits on flexible plastic substrates. Nature. **454**, 495 (2008)
368. Kocabas, C., Kim, H.S., Banks, T., Rogers, J.A., Pesetski, A.A., Baumgardner, J.E., Krishnaswamy, S.V., Zhang, H.: Radio frequency analog electronics based on carbon nanotube transistors. Proc. Natl. Acad. Sci. U.S.A. **105**, 1405 (2008)
369. Chimot, N., Derycke, V., Goffman, M.F., Bourgoin, J.P., Happy, H., Dambrine, G.: Gigahertz frequency flexible carbon nanotube transistors. Appl. Phys. Lett. **91**, 153111 (2007)
370. Kang, S.J., Kocabas, C., Kim, H.S., Cao, Q., Meitl, M.A., Khang, D.Y., Rogers, J.A.: Printed multilayer superstructures of aligned single-walled carbon nanotubes for electronic applications. Nano. Lett. **7**, 3343 (2007)
371. Katz, E., Willner, I.: Biomolecule-functionalized carbon nanotubes: applications in nanobioelectronics. ChemPhysChem. **5**(8), 1084–1104 (2004)
372. Rinzler, A., Hafner, J.H., Nikolaev, P., Lou, L., Kim, D.G., et al.: Unraveling nanotubes: field emission from an atomic wire. Science. **269**, 1550–1553 (1995)
373. Saito, Y., Hamaguchi, K., Hata, K., Uchida, K., Tasaka, Y., et al.: Conical beams from open nanotubes. Nature. **389**, 554–555 (1997)
374. De Vita, A., Charlier J.C., Blase, X., Car, R.: Appl. Phys. A **68**, 283–86 (1999)
375. Saito, Y., Uemura, S., Hamaguchi, K.: Jpn. J. Appl. Phys. **37**, L346–48 (1998)
376. Sugie, H., Tanemura, M., Filip, V., Iwata, K., Takahashi, K., Okuyama, F.: Carbon nanotubes as electron source in an x-ray tube. Appl. Phys. Lett. **78**, 2578–2580 (2001)
377. Lee, N.S., Chung, D.S., Han, I.T., Kang, J.H., Choi, Y.S., Kim, H.Y., Park, S.H., Jin, Y.W., Yi, W.K., Yun, M.J., Jung, J.E., Lee, C.J., You, J.H., Jo, S.H., Lee, C.G., Kim, J.M.: Application of carbon nanotubes to field emission displays. Diamond Relat. Mater. **10**, 265–270 (2001)
378. Choi, W.B., Chung, D.S., Kang, J.H., Kim, H.Y., Jin, Y.W., Han, I.T., Lee, Y.H., Jung, J.E., Lee, N.S., Park, G.S., Kim, J.M.: Fully sealed, high-brightness carbon-nanotube field-emission display. App. Phys. Lett. **75**(20), 3129–3131 (1999)
379. Lee, Y.H., Lim, S.C., An, K.H., Kim, W.S., Jeong, H.J., et al.: New Diamond Front. Carbon Technol. **12**, 181–207 (2002)
380. Endo, M., Kim, C., Nishimura, K., Fujino, T., Miyashita, K.: Carbon **38**, 183–197 (2000)
381. Rotman, D.: Tech. Rev. **38**–45 (2002)
382. Wang, Q.H., Setlur, A.A., Lauerhaas, J.M., Dai, J.Y., Seelig, E.W., Chang, R.H.: Appl. Phys. Lett. **72**(2912), 499–400.1998,
383. Yue, G.Z., Qiu, Q., Gao, B., Cheng, Y., Zhang, J., et al.: Generation of continuous and pulsed diagnostic imaging x-ray radiation using a carbon-nanotube-based field-emission cathode. Appl. Phys. Lett. **81**, 355–357 (2002)
384. Lee, N.S., Chung, d.S., Han, I.T., Kang, J.H., Choi, Y.S., et al.: Application of carbon nanotubes to field emission displays. Diamond Rel. Mater. **10**, 265–270 (2001)
385. Wu, Z., et al.: Transparent, conductive carbon nanotube films. Science. **305**, 1273 (2004)
386. De, S., Coleman, J.N.: The effects of percolation in nanostructured transparent conductors. MRS Bull. **36**, 774 (2011)
387. Polytechnic University of Catalonia: Transparent conductive coatings based on carbon nanotubes. https://upcommons.upc.edu/bitstream/handle/2099.1/6114/Transparent%20Conductive%20C...%20based%20on%20Carbon%20Nanotubes.pdf;sequence=1.   Accessed 4 Oct 2016

388. Glatkowski, P.J.: Carbon nanotube based transparent conductive coatings", Pennsylvania State University. http://citeseerx.ist.psu.edu/viewdoc/download?doi=10.1.1.123.5774&rep=rep1& type=pdf. (Accessed February 2018)

389. Berkei, M.: Conductive coatings using carbon nanotubes: a fascinating material for the coating producer's toolbox. Chemicals. **7–8**, 10 (2011)

390. Akhmadishina, K.F., Bobrinetskii, I.I., Ibragimov, R.A., Komarov, I.A., Malovichko, A.M., Nevolin, V.K., Petukhov, V.A.: Fabrication of flexible transparent conductive coatings based on single-walled carbon nanotubes. Inorganic Materials. **50**(1), 23–28 (2014)

391. Janas, D., Herman, A.P., Boncel, S., Koziol, K.K.: Iodine monochloride as a powerful enhancer of electrical conductivity of carbon nanotube wires. Carbon. **73**, 225–233 (2014)

392. Yao, Z., Wei, J., Vajtai, R., Ajayan, P.M., Barrera, E.V.: Iodine doped carbon nanotube cables exceeding specific electrical conductivity of metals. Sci. Rep. (Nature). **1**, 83 (2011)

393. Subramaniam, C., Yamada, T., Kobashi, K., Sekiguchi, A., Futaba, D.N., Yumura, M., Hata, K.: One hundred fold increase in current carrying capacity in a carbon nanotube-copper composite. Nat. Commun. **4**, (2013)

394. Hamada, N., Sawada, A., Oshiyama, A.: A new one-dimensional conductor: graphitic microtubules. Phys. Rev. Lett. **68**, 1579–1581 (1992)

395. Geerligs, L.J., Harmans, C.J., Kouwenhoven, L.P.: The physics of few-electron nanostructures. North-Holland, Amsterdam (1993)

396. Ando, T.: Theory of Electronic States and Transport in Carbon Nanotubes. J. Phys. Soc. Jpn. **74**(3), 777–817 (2005)

397. Akera, H., Ando, T.: Phys. Rev. B. **43**, 11676 (1991)

398. Dekker, C.: Carbon nanotubes as molecular quantum wires. Phys. Today. **52**(5), 22–28 (1999)

399. White, C.T., Todorov, T.N.: Carbon nanotubes as long ballistic conductors. Nature. **393**, 240 (1998)

400. Kreupl, F., Graham, A.P., Duesberg, G.S., Steinhögl, W., Liebau, M., Unger, E., Hönlein, W.: Carbon nanotubes in interconnect applications. Microelectron. Eng. **64**, 399 (2002)

401. Tang, Z.K., Zhang, L., Wang, N., Zhang, X.X., Wen, G.H., Li, G.D., Wang, J.N., Chan, C. T., Sheng, P.: Superconductivity in 4 angstrom single-walled carbon nanotubes. Science. **292**, 2462–2465 (2001)

402. Camilli, L., Pisani, C., Gautron, E., Scarselli, M., Castrucci, P., D'Orazio, F., Passacantando, M., Moscone, D., De Crescenzi, M.: A three-dimensional carbon nanotube network for water treatment. Nanotechnol. **25**(6), 1–3 (2014)

403. Hashim, D.P., Narayanan, N.T., Romo-Herrera, J.M., Cullen, D.A., Hahm, M.H., Lezzi, P., Suttle, J.R., Kelkhoff, D., Muñoz-Sandoval, E., Ganguli, S., Roy, A.K., Smith, D.J., Vajtai, R., Sumpter, B.G., Meunier, V., Terrones, H., Terrones, M., Ajayan, P.M.: Covalently bonded three-dimensional carbon nanotube solids via boron induced nanojunctions. Sci. Rep.. **2**, (2012)

404. Zhang, S., Shao, T., Selcen Kose, H., Karanfil, T.: Adsorption of aromatic compounds by carbonaceous adsorbents: a comparative study on granular activated carbon, activated carbon fiber, and carbon nanotubes. Environ. Sci. Technol. **44**(12), 6377–6383 (2010)

405. Fasano, M., Chiavazzo, E., Asinari, P.: Water transport control in carbon nanotube arrays. Nanoscale Res. Lett. **9**(1), 559 (2014)

406. Gethard, K., Sae-Khow, O., Mitra, S.: Water deslination using carbon-nanotube-enhanced membrane distillation. ACS Appl. Mater. Interfaces. **3**(2), 110 (2011)

407. Corry, B.: Designing carbon nanotube membranes for efficient water desalination. J. Phys. Chem. B. **112**, 1427 (2008)

408. Dumée, L.F., Sears, K., Schütz, J., Finn, N., Huynh, C., Hawkins, S., Duke, M., Gray, S.: Characterization and evaluation of carbon nanotube Bucky-Paper membranes for direct contact membrane distillation. J. Membraine Sci. **351**, 36 (2010)

409. Gao, G., Vecitis, C.D.: Electrochemical carbon nanotube filter oxidative performance as a function of surface chemistry. Environ. Sci. Technol. **45**, 9726 (2011)

410. Rahaman, M.S., Vecitis, C.D., Elimelech, M.: Electrochemical carbon-nanotube filter performance toward virus removal and inactivation in the presence of natural organic matter. Environ. Sci. Technol. **46**, 1556 (2012)

411. Srivastava, A., Srivastava, O.N., Talapatra, S., Vajtai, R., Ajayan, P.M.: Carbon nanotube filters. Nat. Mater. **3**, 610 (2004)

412. Brady-Estévez, A.S., Nguyen, T.H., Gutierrez, L., Elimelech, M.: Impact of solution chemistry on viral removal by a single-walled carbon nanotube filter. Water Res. **44**, 3773 (2010)

413. Brady-Estéves, A.S., Kang, S., Elimelech, M.: A single-walled-carbon-nanotube filter for removal of viral and bacterial pathogens. Small. **4**, 481 (2008)

414. Seldon Water: Carbon nanotube technology: the science behind our products. Accessed June 2016

415. Herrera-Herrara, A.V., González-Curbelo, M.Á., Hérnandez-Borges, J.: Carbon nanotubes applications in separation science: a review. Analytica Chimica Acta. **734**, 1–30 (2012)

416. Wang, S.: Optimum degree of functionalization for carbon nanotubes. Curr. Appl. Phys. **9**, 1146–1150 (2009)

417. Safavi, A., Maleki, N., Doroodmand, M.M.: Single-walled carbon nanotubes as stationary phase in gas chromatographic separation and determination of argon, carbon dioxide and hydrogen. Anal. Chim. Acta. **675**, 207–212 (2010)

418. Speltini, A., Merli, D., Quartarone, E., Profumo, A.: Separation of alkanes and aromatic compounds by packed column gas chromatography using functionalized multi-walled carbon nanotubes as stationary phases. J. Chromatogr. A. **1217**, 2918–2924 (2010)

419. Hussain, C.M., Saridara, C., Mitra, S.: Self-assembly of carbon nanotubes via ethanol chemical vapor deposition for the synthesis of gas chromatography columns. Anal. Chem. **82**, 5184–5188 (2010)

420. Merli, D., Speltini, A., Ravelli, D., Quartarone, E., Costa, L., Profumo A.: J. Chromatogr. A **1218**, 7275–7281 (2010)

421. Zhao, L., Ai, P., Duan, A.H., Yuan, L.M.: Single-walled carbon nanotubes for improved enantioseparations on a chiral ionic liquid stationary phase in GC. Anal. Bioanal. Chem. **399**, 143–147 (2011)

422. Na, N., Cui, X., De Beer, T., Li, T., Tang, T., Sajid, M., Ouyang, J.: The use of silica nanoparticles for gas chromatographic separation. J. Chromatogr. A. **1218**, 4552–4558 (2011)

423. Hussain, C.M., Saridara, C., Mitra, S.: Altering the polarity of self-assembled carbon nanotubes stationary phase via covalent functionalization. RSC Adv. **1**, 685–689 (2011)

424. Speltini, A., Merli, D., Dondi, D., Paganini, G., Profumo, A.: Improving selectivity in gas chromatography by using chemically modified multi-walled carbon nanotubes as stationary phase. Anal. Bioanal. Chem. **403**, 1157–1165 (2012) https://doi.org/10.1007/s00216-011-5606-y

425. André, C., Gharbi, T., Guillaume, Y.C.: A novel stationary phase based on amino derivatized nanotubes for hplc separations: Theoretical and practical aspects. J. Sep. Sci. **32**, 1757–1764 (2009)

426. Liang, X., Liu, S.H., Liu, X., Jiang, S.: J. Sep. Sci. **33**, 3304–3312 (2010)

427. Zhong, Y., Zhou, W., Zhang, P., Zhu, Y.: Preparation, characterization, and analytical applications of a novel polymer stationary phase with embedded or grafted carbon fibers. Talanta. **82**, 1439–1447 (2010)

428. Chambers, S.D., Svec, F., Frechet, J.M.J.: Incorporation of carbon nanotubes in porous polymer monolithic capillary columns to enhance the chromatographic separation of small molecules. J. Chromatogr. A. **1218**, 2546–2552 (2011)

429. André, C., Aljhani, R., Gharbi, T., Guillaume, Y.C.: Incorporation of carbon nanotubes in a silica HPLC column to enhance the chromatographic separation of peptides: theoretical and practical aspects. J. Sep. Sci. **34**, 1221–1227 (2011)

430. André, C., Agiovlasileti, D., Guillaume, Y.C.: Peculiarities of a novel bioenzymatic reactor using carbon nanotubes as enzyme activity enhancers: application to arginase. Talanta. **85**, 2703 (2011)

431. Yoo, J.T., Ozawa, H., Fujigaya, T., Nakashima, N.: Evaluation of affinity of molecules for carbon nanotubes. Nanoscale. **3**, 2517–2522 (2011)

432. Chen, J.L.: Multi-wall carbon nanotubes bonding on silica-hydride surfaces for open-tubular capillary electrochromatography. J. Chromatogr. A. **1217**, 715–721 (2010)

433. Chen, J.L., Lu, T.L., Lin, Y.C.: Multi-walled carbon nanotube composites with polyacrylate prepared for open-tubular capillary electrochromatography. Electrophoresis. **31**, 3217–3226 (2010)

434. Chen, J.L., Hsieh, K.H.: Polyacrylamide grafted on multi-walled carbon nanotubes for open-tubular capillary electrochromatography: comparison with silica hydride and polyacrylate phase matrices. Electrophoresis. **31**, 3937–3948 (2010)

435. Chen, J.L., Lin, Y.C.: The role of methacrylate polymerized as porous-layered and nanoparticle-bound phased for open-tubular capillary electrochromatography: substitution of a charged monomer for a bulk monomer. Electrophoresis. **31**, 3949–3958 (2010)

436. Stege, P.W., Sombra, L.L., Messina, G., Martinez, L.D., Silva, M.F.: Determination of melatonin in wine and plant extracts by capillary electrochromatography with immobilized carboxylic multi-walled carbon nanotubes as stationary phase. Electrophoresis. **31**, 2242–2248 (2010)

437. Jiménez-Soto, J.M., Moliner-Martínez, Y., Cárdenas, S., Valcárcel, M.: Evaluation of the performance of single walled carbon nanohorns in capillary electrophoresis. Electrophoresis. **31**, 1681–1688 (2010)

438. Yu, J., Dushu, H., Kelong, H., Yong, H.: Preparation of hydroxypropyl-β-cyclodextrin cross-linked multi-walled carbon nanotubes and their application in enantioseparation of clenbuterol. J. Chem. **29**, 893–897 (2011)

439. Reid, V.R., Stadermann, M., Bakajin, O., Synovec, R.E.: High-speed, temperature programmable gas chromatography utilizing a microfabricated chip with an improved carbon nanotube stationary phase. Talanta. **77**, 1420–1425 (2009)

440. Goswami, S., Bajwa, N., Asuri, P., Ci, L., Ajayan, P.M., Cramer, S.M.: Aligned carbon nanotube stationary phases for electrochromatographic chip separations. Chromatographia. **69**, 473–480 (2009)

441. Wu, R.G., Yang, C.S., Wang, P.C., Tseng, F.G.: Electrophoresis. **30**, 2024–2031 (2009)

442. Moigensen, K.B., Chen, M., Molhave, K., Boggild, P., Kutter, J.P.: Lab Chip 2116–2118 (2011)

443. Shrivas, K., Wu, H.-F.: Multifunctional nanoparticles composite for MALDI-MS: Cd2+ doped carbon nanotubes with CdS nanoparticles as the matrix, preconcentrating and accelerating probes of microwave enzymatic digestion of peptides and proteins for direct MALDI-MS analysis. J. Mass Spectrom. **45**, 1452–1460 (2010)

444. Yang, H.-J., Lee, A., Lee, M.-Y., Kim, W., Kim, J.: Bull. Korean Chem. Soc. **31**, 35–40 (2010)

445. Gholipour, Y., Giudicessi, S.L., Nonami, H., Erra-Balsells, R.: Diamond, titanium dioxide, titanium silicon oxide, and barium strontium titanium oxide nanoparticles as matrixes for direct matrix-assisted laser desorption/ionization mass spectrometry analysis of carbohydrates in plant tissues. Anal. Chem. **82**, 5518–5526 (2010)

446. Wei, Y.-L., Zhou, W., Liu, F., Ren, Z.-Y., Zhang, L.-Y., Du, Y.-P., Guo, Y.-L., Chem. J. Chinese U. **31**, 1729–1733 (2010)

447. Li, X.-S., Wu, J.-H., Xu, L.-D., Zhao, Q., Luo, Y.-B., Yuan, B.-F., Feng, Y.-Q.: Chem. Commun. **47**, 9816–9818 (2011)

448. Tang, H.-W., Ng, K.-M., Lu, W., et al.: Ion desorption efficiency and internal energy transfer in carbon-based surface-assisted laser desorption/ionization mass spectrometry: desorption mechanism(s) and the design of SALDI substrates. Anal. Chem. **81**, 4720–4729 (2009)

449. Hsu, W.-Y., Lin, W.-D., Hwu, W.-L., Lai, C.-C., Tsai, F.-J.: Anal. Chem. **82**, 6814–6820 (2010)

450. Lee, J., Kim, Y.-K., Min, D.-H.: J. Am. Chem. Soc. **132**, 14714–14717 (2010)

451. Kim, Y.-K., Na, H.-K., Kwack, S.-J., Ryoo, S.-R., Lee Y., Hong, S., Hong, S., Jeong, Y., Min, D.-H.: ACS Nano **5**, 4550–4561 (2011)

452. Pumera, M.: Carbon nanotubes contain residual metal catalyst nanoparticles even after washing with nitric acid at elevated temperature because these metal nanoparticles are sheathed by several graphene sheets. Carbon. **23**, 6453 (2007)

453. Lordi, V., Yao, N., Wei, J.: Method for supporting platinum on single-walled carbon nanotubes for a selective hydrogenation catalyst. Chem. Mater. **13**, 733 (2001)

454. Jiang, L., Gu, H., Xu, X., Yan, X.: J. Mol. Catal. A-Chem. **310**, 144 (2009)

455. Pawelec, B., La Parola, V., Navarro, R.M., Murcia-Mascaros, S., Fierro, J.L.G.: On the origin of the high performance of MWNT-supported PtPd catalysts for the hydrogenation of aromatics. Carbon. **44**, 84 (2006)

456. Yoon, B., Pan, H.-B., Wai, C.M.: Relative catalytic activities of carbon nanotube-supported metallic nanoparticles for room-temperature hydrogenation of benzene. J. Phys. Chem. C. **113**, 1520 (2009)

457. Tan, X., Deng, W., Liu, M., Zhang, Q., Wang, Y.: Carbon nanotube-supported gold nanoparticles as efficient catalysts for selective oxidation of cellobiose into gluconic acid in aqueous medium. Chem. Commun. (46), 7179 (2009)

458. Sullivan, J.A., Flanagan, K.A., Hain, H.: Suzuki coupling activity of an aqueous phase Pd nanoparticle dispersion and a carbon nanotube/Pd nanoparticle composite. Catal. Today. **145**, 108 (2009)

459. Baleizao, C., Gigante, B., Garcia, H., Corma, A.: Vanadyl salen complexes covalently anchored to single-wall carbon nanotubes as heterogeneous catalysts for the cyanosilylation of aldehydes. J. Catal. **221**, 77 (2004)

460. Carol, L.: Carbon SUPER-SPRINGS, Mechanical Engineering (2010)

461. Novoselov, K.S., Geim, A.K., Morosov, S.V., Jiang, D., Zhang, Y., Dubonos, S.V., Grigorieva, I.V., Firsov, A.A.: Electric field effect in atomically thin carbon films. Science. **306**(5696), 666–669 (2004)

462. Geim, A.K., Novoselov, K.S.: The rise of graphene. Nat. Mater. **6**, 183–191 (2007)

463. Fitzer, E., Kochling, K.-H., Boehm, H.P., Marsh, H.: Recommended terminology for the description of carbon as a sold (IUPAC Recommendations 1995). Pure Appl. Chem. **67**(3), 473–506 (2009)

464. Mouras et al.: Synthesis of first stage graphite intercalation compounds with fluorides. Revue de Chimie Minérale. **24**, 572 (1987)

465. The Official Web Site of the Nobel Prize: Graphene – the perfect atomic lattice. http://www. nobelprize.org/nobel_prizes/physics/laureates/2010/press.html. Accessed June 2016

466. Boehm, H.P., Clauss, A., Fischer, G.O., Hofmann, U.: Das Adsoprtionsverhealten sehr dünner Kohlenstoff-Folien. Zeitschrift für anorganische und allgemeine Chemie. **316** (3-4), 119–127 (1962)

467. Boehem, H.P., Clauss, A., Fischer, G., Hofmann, U., Surface properties of extremely thin graphite lamellae. In: Proceedings of the Fifth Conference on Carbon (1961)

468. Graphene Times: Boehm's 1961 isolation of graphene. Acessed June 2016

469. Brodie, B.C.: On the atomic weight of graphite. Philos. Trans. R. Soc. London. **149**, 249–259 (1859)

470. Wallace, P.R.: The band structure of graphite. Phys. Rev. **71**, 622–634 (1947)

471. Ruess, G., Vogt, F.: Höchstlamellarer Kohlenstoff aus Graphitohydroxyd. Monatshefte für Chemie. **78**, 222–242 (1948)

472. Oshima, C., Nagashima, A.: Ultra-thin epitaxial films of graphite and hexagonal boron nitride on solid surfaces. J. Phys. Condens. Matter. **9**, 1–20 (1997)

473. Jang, B.Z., Huang, W.C.: Nano-scaled graphene plates. US 7071258 B1, 4 July 2006

474. Affoune, A.M., Prasad, B.L.V., Sato, H., Enoki, T., Kaburagi, Y., Hishiyama, Y.: Experimental evidence of a single nano-graphene. Chem. Phys. Lett. **348**, 17 (2001)

475. Lee, C., Wei, X., Kysar, J.W., Hone, J.: Measurement of elastic properties and intrinsic strength of monolayer graphene. Science. **321**(5887), 385–388 (2008)

476. Terrones, M., Botello-Méndez, A.R., Campos-Delgado, J., López-Urías, F., Vega-Cantú, Y. I., Rodríguez-Macías, F.J., Elías, L.A., Muñoz-Sandoval, E., Cano-Márquez, A.G., Charlier, J.-C., Terrones, H.: Graphene and graphite nanoribbons: morphology, properties, synthesis, defects and applications. Nano Today. **5**, 351–372 (2010)

477. Zhang, P., Ma, L., Fan, F., Zeng, Z., Peng, C., Loya, P.E., Liu, Z., Gong, Y., Zhang, J., Zhang, X., Ajayan, P.M., Zhu, T., Lou, J.: Fracture toughness of graphene. Nat. Commun. **5**, 3782 (2014)

478. Rozhkov, A.V., Giavaras, G., Bliokh, Y.P., Freilikher, V., Nori, F.: Electronic properties of mesoscopic graphene structures: change confinement and control of spin and charge transport. Phys. Rep. **503**, 77–114 (2011)

479. Rao, C.N.R., Sood, A.K., Subrahmanyam, K.S., Govindaraj, A.: Graphene: the new two-dimensional nanomaterial. Angew. Chem. Int. Ed. **48**, 7752–7777 (2009)

480. Edwards, R.S., Coleman, K.S.: Graphene synthesis: relationship to applications. Nanoscale. **5**, 38 (2013)

481. Singh, V., Joung, D., Zhai, L., Das, S., Khondaker, S.I., Seal, S.: Graphene based materials: past, present and future. Prog. Mater. Sci. **56**, 1178–1271 (2011)

482. (a) Meyer, J.C., Geim, A.K., Katsnelson, M.I., Novoselov, K.S., Booth, T.J., Roth, S.: The structure of suspended graphene sheets. Nature. **446**, 60 (2007); (b) Pop, E., Roy, A.K., Varshney, V.: Thermal properties of graphene: Fundamentals and applications. MRS Bull. **37**, 1273 (2012); (c) Thermal Conductivity. http://ndl.ee.ucr.edu/MRS-Talk-09.pdf. Accessed June 2016

483. (a) Novoselov, K.S., Jiang, D., Schedin, F., Booth, T.J., Khotkevich, V.V., Morosov, S.V., Geim, A.K.: Two-dimensional atomic crystals. Proc. Natl. Acad. Sci. USA **102**, 10451 (2005); (b) Blake, P., Hill, E.W., Neto, A.H.C., Novoselov, K.S., Jiang, D., Yang, R., Booth, T. J., Geim, A.K.: Appl. Phys. Lett. 91, 063124 (2007)

484. Stolyarova, E., Taegh, R.K., Ryu, S., Maultzsch, J., Kim, P., Brus, L.E., Heinz, T.F., Hybertsen, M.S., Flynn, G.W.: High-resolution scanning tunneling microscopy imaging of mesoscopic graphene sheets on an insulating surface. Proc. Natl. Acad. Sci. USA. **104**, 9209 (2007)

485. Meyer, J.C., Kisielowski, C., Erni, R., Rossell, M.D., Crommie, M.F., Zettl, A.: Direct imaging of lattice atoms and topological defects in graphene membranes. Nano Lett. **8**, 3582 (2008)

486. Talukdar, Y., Rashkow, J.T., Lalwani, G., Kanakia, S., Sitharaman, B.: The effects of graphene nanostructures on mesenchymal stem cells. Biomaterials. **35**(18), 4863–4877 (2014)

487. News from Brown: Jagged Graphene can slice into cell membranes. https://news.brown. edu/articles/2013/07/graphene. Accessed 4 Oct 2016

488. Li, Y., Yuan, H., von Dem Bussche, A., Creighton, M., Hurt, R.H., Kane, A.B., Gao, H.: Graphene microsheets enter cells through spontaneous membrane penetration at edge asperities and corner sites. Proc. Natl. Acad. Sci. **110**(30), 12295–12300 (2013)

489. Mullick Chodhury, S., Lalwani, G., Zhang, K., Yang, J.O., Neville, K., Sitharaman, B.: Cell specific cytotoxicity and uptake of graphene nanoribbons. Biomaterials. **34**(1), 283–293 (2013)

490. Wang, K., Ruan, J., Song, H., Zhang, J., Wo, Y., Guo, S., et al.: Biocompatibility of graphene oxide. Nanoscale Res. Lett. **6**, 1 (2010)

491. Akhavan, O., Ghaderi, E.: Toxicity of graphene and graphene oxide nanowalls against bacteria. ACS Nano. **4**, 5731 (2010)

492. Lee, C., Wei, X., Kysar, J.W., Hone, J.: Measurement of the Elastic Properities and Intrinsic Strength of Monolayer Graphene. Science. **5887**, 385–388 (2008.) Frank, I.W., Tanenbaum, D. M., Van Der Zande, A.M., McEuen, P.L.: Mechanical properties of suspended graphene sheets. J. Vac. Sci. Technol. B , **25**(6), 2558–2561 (2007)

493. Hill, E.W., Vijayaragahvan, A., Novoselov, K.: Graphene sensors. IEEE Sens. J. **11**(12), 3161–3170 (2011)

494. Lee, C., et al.: Measurement of the elastic properties and intrinsic strength of monolayer graphene. Science. **321**(5887), 385–388 (2008)

495. Poot, M., van der Zant, H.S.J.: Nanomechanical properties of few-layer graphene membranes. Appl. Phys. Lett. **92**(6), 063111–063113 (2008)
496. Scott, A.: Graphene's global race to market. Chem. Eng. News. **94**(15), 28–33 (2016)
497. Graphene Flagship: http://graphene-flagshup.eu/. Accessed June 2016
498. IDTechEx: Graphene & 2D Materials Europe. http://www.idtechex.com/graphene-europe/show/en/. Accessed June 2016
499. IDtechEx: Event Highlights. http://www.idtechex.com/events/summary/E16/. Accessed June 2016
500. EV and More: Production of revolutionary graphene batteries begins in Spain. http://blog.evandmore.com/production-of-revolutionary-graphene-batteries-begins-in-spain/. Accessed June 2016
501. Vorbeck: RFID. http://vorbeck.com/pages/products-rfid. Accessed June 2016
502. Vorbeck: Antennas. http://vorbeck.com/pages/products-antennas. Accessed June 2016
503. Saito, R., Dresselhaus, G., Dresselhaus, M.S.: Phyical properies of carbon nanotubes. Imperial College Press (1998)
504. Nika, D.L., Pokatilov, E.P., Askerov, A.S., Balandin, A.A.: Phonon thermal conduction in graphene: role of Umklapp and edge roughness scattering. Phys. Rev. B. **79**, 155413 (2009)
505. Savchenko, A.: Transforming graphene. Science. **323**, 589–590
506. Novoselov, K.S., Geim, A.K., Morosov, S.V., Jiang, D., Katsnelson, M.I., Grigorieva, I.V., Dubonos, S.V., Firsov, A.A.: Two-dimensional gas of massless Dirac fermions in graphene. Nature. **438**, 197 (2005)
507. Zhang, Y., Tan, Y., Stormer, H.L., Kim, P.: Experimental observation of the quantum Hall effect and Berry's phase in graphene. Nature. **438**, 201 (2005)
508. Novoselov, K.S., McCann, E., Morosov, S.V., Fal'ko, V.I., Katsnelson, M.I., Zeitler, U., et al.: Unconventional quantum Hall effect and Berry's phase of 2 pi in bilayer graphene. Nat. Phys. **2**, 177 (2006)
509. Akturk, A., Goldsman, N.: Electron transport and full-band electron-phonon interactions in graphene. J. Appl. Phys. **103**(5), 053702 (2008)
510. Morosov, S.V., Novoselov, K., Katsnelson, M., Schedin, F., Elias, D., Jaszczak, J., Geim, A.: Giant intrinsic carrier mobilities in graphene and its bilayer. Phys. Rev. Lett. **100**(1), 016602 (2008)
511. Novoselov, K.S., Geim, A.K., Morozov, S.V., Jiang, D., Katsnelson, M.I., Grigorieva, I.V., Dubonos, S.V., Firsov, A.A.: Two-dimensional gas of massless Dirac fermions in graphene. Nature. **438**(7065), 197–200 (2005)
512. Chen, J.H., Jang, C., Xiao, S., Ishigami, M., Fuhrer, M.S.: Intrinsic and extrinsic performance limits of graphene devices on SiO 2. Nat. Nanotechnol. **3**(4), 206–209 (2008)
513. Neto, A.C., Peres, N.M.R., Novoselov, K.S., Geim, A.K., Geim, A.K.: The electronic properties of graphene. Rev. Mod. Phys. **81**, 109–162 (2009)
514. Sagade, A.A., et al.: Highly air stable passivation of graphene based field effect devices. Nanoscale. **7**, 3558–3564 (2015)
515. Evaldsson, M., Zozoulenko, I.V., Xu, H., Heinzel, T.: Edge-disorder-induced Anderson localization and conduction gap in graphene nanoribbons. Phys. Rev. B. **78**, 161407 (2008)
516. Cervantes-Sodi, F., Csanyi, G., Piscanec, S., Ferrari, A.C.: Edge-functionalized and substitutionally doped graphene nanoribbons: electronic and spin properties. Phys. Rev. B. **77**, 165427 (2008)
517. Zhang, Y., Tang, T.-T., Girit, C., Hao, Z., Martin, M.C., Zettl, A., Crommie, M.F., Shen, Y.R., Wang, F.: Direct observation of a widely tunable bandgap in bilayer graphene. Nat. Lett. **459**, 820–823 (2009)
518. Lui, C.H., Zhiqiang, L., Mak, K.F., Cappelluti, E., Heinz, T.F.: Observation of an electrically tunable band gap in trilayer graphene. Nat. Phys. **7**, 944–947 (2011)
519. Oostinga, J.B., Heersche, H.B., Liu, X., Morpurgo, A.F., Vandersypen, L.M.K.: Gate-induced insulating state in bilayer graphene devices. Nat. Mater. **7**, 151 (2008)

520. Allen, M.J., Tung, V.C., Kaner, R.B.: Honeycomb Carbon: A Review of Graphene. Chem. Rev. **110**, 132–145 (2010)

521. Geim, A.K., Novoselov, K.S.: The rise of graphene. Nat. Mater. **6**, 183 (2007)

522. Ohta, T., et al.: Controlling the electronic structure of bilayer graphene. Science. **313**(5789), 951–954 (2006)

523. Kaiser, A.B., Gómez-Navarro, C., Sundaram, R.S., Burghard, M., Kern, K.: Electrical conduction mechanism in chemically derived graphene monolayers. Nano Lett. **9**(5), 1787–1792 (2009)

524. Wang, Y., Huang, Y., Song, Y., Zhang, X., Ma, Y., Lang, J., Chen, Y.: Room-temperature ferromagnetism of graphene. Nano Lett. **9**(1), 220–224 (2009)

525. Ghosh, S., Nika, D.L., Pokatilov, E.P., Balandin, A.A.: Heat conduction in graphene: experimental study and theoretical interpretation. New J. Phys. **11**, 1–18 (2009)

526. Cai, W., Moore, A.L., Zhu, Y., Li, X., Chen, S., Shi, L., Ruoff, R.S.: Thermal transport in suspended and supported monolayer graphene grown by chemical vapor deposition. Nano Lett. **25**(10), 1645–1651 (2010)

527. Xu, X., Periera, L.F.C., Wang, Y., Wu, J., Zhang, K., Zhao, X., Bae, S., Tinh Bui, C., Xie, R., Thong, J.T.L., Hong, B.H., Loh, K.P., Donadio, D., Li, B., Özyilmaz, B.: Length-dependent thermal conductivity in suspended single-layer graphene. Nat. Commun. **5**, 3689 (2004)

528. Seol, J.H., Jo, I., Moore, A.L., Lindsay, L., Aitken, Z.H., Pettes, M.T., Li, X., Yao, Z., Huang, R., Broido, D., Mingo, N., Ruoff, R.S., Shi, L.: Two-dimensional phonon transport in supported graphene. Science. **328**, 213–216 (2010)

529. Balandin, A., Wang, K.L.: Significant decrease of the lattice thermal conductivity due to phonon confinement in a free-standing semiconductor quantum well. Phys. Rev. B. **58**, 1544 (1998)

530. Zou, J., Balandin, A.: Phonon heat conduction in a semiconductor nanowires. J. Appl. Phys. **89**, 2932 (2001)

531. Ghosh, S., Nika, D.L., Pokatilov, E.P., Balandin, A.A.: Heat conduction in graphene: experimental study and theoretical interpretation. New J. of Phys. **11**, 1–18 (2009)

532. Geim, A.K., MacDonald, A.H.: Graphene: exploring carbon flatland. Phys. Today. **60**(8), 45–41 (2007)

533. Alzari, V., Nuvoli, D., Scognamillo, S., Piccinini, M., Gioffredi, E., Malucelli, G., Marceddu, S., Sechi, M., Sanna, V., Mariana, A.: Graphene-containing thermoresponsive nanocomposite hydrogels of ply(N-isopropylacrylamide) prepared by frontal polymerization. J. Mater. Chem. **21**(24), 8727 (2011)

534. Nuvoli, D., Valentini, L., Alzari, V., Scognamillo, S., Bon, S.B., Piccinini, M., Illescas, J., Mariani, A.: High concentration few-layer graphene sheets obtained by liquid phase exfoliation of graphite in ionic liquid. J. Mater. Chem. **21**(10), 3428–3431 (2011)

535. Vallés, C., Drummond, C., Saadaoi, H., Furtado, C.A., He, M., Roubeau, O., Ortolani, L., Monthioux, M., Penicaud, A.: Solutions of negatively charged graphene sheets and ribbons. J. Am. Chem. Soc. **130**, 15802 (2008)

536. Kamali, A.R., Fray, D.J.: Molten salt corrosion of graphite as a possible way to make carbon nanostructures. Carbon. **56**, 121–131 (2013)

537. Kamali, A.R., Fray, D.J.: Large-scale preparation of graphene by high temperature insertion of hydrogen into graphite. Nanoscale. **7**, 11310–11320 (2015)

538. Hamilton, C.E., Lomeda, J.R., Sun, Z., Tour, J.M., Barron, A.R.: High-yield organic dispersions of unfunctionalized graphene. Nano Lett. **9**, 3460 (2009)

539. Bourlinos, A.B., Georgakilas, V., Zboril, R., Steriotis, T.A., Stubos, A.K.: Liquid-phase exfoliation of graphite towards solubilized graphenes. Small. **5**, 1841 (2009)

540. Woltornist, S.J., Oyer, A.J., Carrillo, J.-M.Y., Dobrynin, A.V., Adamson, D.H.: Conductive thin films of pristine graphene by solvent interface trapping. ACS Nano. **7**(8), 7062–7066 (2013)

541. Dhakate, S.R., Chauhan, N., Sharma, S., Tawale, J., Singh, S., Sahare, P.D., Mathur, R.B.: An approach to produce single and double layer graphene from re-exfoliation of expanded graphite. Carbon. **49**, 1946–1954 (2011)

542. Safavi, A., Tohidi, M., Mahyari, F.A., Shahbaazi, H.: J. Mater. Chem. **22**, 3825–3831 (2012)

543. Pu, N.-W., Wang, C.-A., Sung, Y., Liu, Y.-M., Ger, M.-D.: Mater. Lett. **63**, 1987–1989 (2009)

544. Li, X.L., Wang, X.R., Zhang, L., Lee, S.W., Dai, H.J.: Chemically derived, ultrasmooth graphene nanoribbon semiconductors. Science. **319**, 1229 (2008)

545. JAyasena, B., Sathyan, S.: A novel mechanical cleavage method for synthesizing few-layer graphenes. Nanoscale Res. Lett. **6**, 95 (2011)

546. Niyogi, S., Bekyarova, E., Itkis, M.E., McWilliams, J.L., Hamon, M.A., Haddon, R.C.: Solution properties of graphite and graphene. J. Am. Chem. Soc. **128**(24), 7720–7721 (2006)

547. Schniepp, H.C., Li, J.L., McAllister, M.J., Sai, H., Herrera-Alonso, M., Adamson, D.H., Prud'homme, R.K., Car, R., Saville, D.A., Aksay, I.A.: Functionalized single graphene sheets derived from splitting graphite oxide. J. Phys. Chem. B. **110**, 8535 (2006)

548. Hofmann, M., Chiang, W.-Y., Nguyễn, T.D., Hiseh, Y.-P.: Controlling the properties of graphene produced by electrochemical exfoliation – IOPscience. Nanotechnol. **26**, 334607 (2015)

549. Wang, G., Wang, B., Park, J., Wang, Y., Sun, B., Yao, J.: Highly efficient and large-scale synthesis of graphene by electrolytic exfoliation. Carbon. **47**, 3242–3246 (2009)

550. Su, C.-Y., Lu, A.-Y., Xu, Y., Chen, F.-R., Khlobystov, A.N., Li, L.-J.: High-quality thin graphene films from fast electrochemical exfoliation. ACS Nano. **5**, 2332–2339 (2011)

551. Wang, J., Manga, K.K., Bao, Q., Loh, K.P.: High-yield synthesis of few-layer graphene flakes through electrochemical expansion of graphite in propylene carbonate electrolyte. J. Am. Chem. Soc. **133**, 8888–8891 (2011)

552. Suslick, K.S., Price, G.J.: Applications of ultrasound to materials chemistry. Annu. Rev. Mater. Sci. **29**, 295–326 (1999)

553. Paton, K.R., et al.: Scalable production of large quantities of defect-free few-layer graphene by shear exfoliation in liquids. Nat. Mater. **13**, 624–630 (2014)

554. Sutter, P.W., Flege, J., Sutter, E.A.: Epitaxial graphene on ruthenium. Nat. Mater. **7**, 406 (2008)

555. Wei, D., Liu, Y., Zhang, H., Huang, L., Wu, B., Chen, J., Yu, G.: Scalable synthesis of few-layer graphene ribbons with controlled morphologies by a template method and their applications in nanoelectromechanical switches. J. Am. Chem. Soc. **131**, 11147–11154 (2009)

556. Reina, A., Jia, X., Ho, J., Nezich, D., Son, H., Bulovic, V., Dressel Haus, M.S., Kong, J.: Large area, few-layer graphene films on arbitrary substrates by chemical vapor deposition. Nano Lett. **9**, 30 (2009)

557. Caltech: Caltech scientists develop cool process to make better graphene. https://www.caltech.edu/news/caltech-scientists-develop-cool-process-make-better-graphene-45961. Accessed June 2016

558. Electronics Weekly: Good graphene over square centimetres without high temperatures. http://www.electronicsweekly.com/news/research-news/good-graphene-square-centimetres-without-high-temperatures-2015-03/. Accessed June 2016

559. Boyd, D.A., Lin, W.-H., Hsu, C.-C., Teague, M.-L., Chen, C.-C., Lo, Y.-Y., Chan, W.-Y., Su, W.-B., Cheng, T.-C., Chang, C.-S., Wu, C.-I., Yeh, N.-C.: Single-step deposition of high-mobility graphene at reduced temperatures. Nat. Commun. **6**, 6620 (2015)

560. Bointon, T.H., Barnes, M.D., Russo, S., Carciun, M.F.: High quality monolayer graphene synthesized by resistive heating cold wall chemical vapor deposition. Adv. Mater. **27**(28), 4200–4206 (2015)

561. Malesevic, A., Vitchev, R., Schouteden, K., Volodin, A., Zhang, L., Tendeloo, G.V., Vanhulsel, A., Haesendonck, C.V.: Synthesis of few-layer graphene via microwave plasma-enhanced chemical vapour deposition. Nanotechnol. **19**, 305604 (2008)

562. Wang, J.J., Zhu, M.Y., Outlaw, R.A., Zhao, X., Manos, D.M., Holoway, B.C.: Free-standing subnanometer graphite sheets. Appl. Phys. Lett. **85**, 1265 (2004)

563. Wang, J.J., Zhu, M.Y., Outlaw, R.A., Zhao, X., Manos, D.M., Holoway, B.C.: Synthesis of carbon nanosheets by inductively coupled radio-frequency plasma enhanced chemical vapor deposition. Carbon. **42**, 2867 (2004)

564. Wei, D., Liu, Y., Wang, Y., Zhang, H., Huang, L., Yu, G.: Synthesis- of N-doped graphene by chemical vapor deposition and its electrical properties. Nano Lett. **9**, 1752 (2009)

565. Georgakilas, V., Otyepka, M., Bourlinos, A.B., Changdra, V., Kim, N., Kemp, K.C., Hobza, P., Zboril, R., Kim, K.S.: Functionalization of graphene: covalent and non-covalent approaches, derivatives and applications. Chem. Rev. **112**, 6156–6214 (2012)

566. Chpucair, M., Thordarson, P., Stride, J.A.: Gram-scale production of graphene based on solvothermal synthesis and sonication. Nat. Nanotechnol. **4**, 30 (2009)

567. Staudemaier, L.: Verfahren zur Darstellung der Graphitsäure. Ber. Dtsch. Chem. Ges. **31**, 1481 (1898)

568. Fan, X., Peng, W., Li, Y., Li, X., Wang, S., Zhang, G., Zhang, F.: Deoxygenation of exfoliated graphite oxide under alkaline conditions: a green route to graphene preparation. Adv. Mater. **20**, 4490 (2008)

569. Li, D., Muller, M.B., Gilje, S., Kaner, R.B., Wallace, G.G.: Processable aqueous dispersions of graphene nanosheets. Nat. Nanotechnol. **3**, 101 (2008)

570. Eigler, S., Enzelberger-Heim, M., Grimm, S., Hofmann, P., Kroener, W., Geworski, A., Dotzer, C., Röckert, M., Xiao, J., Papp, C., Lytken, O., Steinrück, H.-P., Hirsch, A.: Wet chemical synthesis of graphene. Adv. Mater. **25**(26), 3583–3587 (2013)

571. Athanasios, B., Bourlinos, D.G., Petridis, D., Szabo, T., Szeri, A., Dekany, I.: Graphite oxide: chemical reduction to graphite and surface modification with aliphatic amines and amino acids. Langmuir. **19**, 6050 (2003)

572. Schniepp, H.c., Li, J.L., McAllister, M.J., Sai, H., Herrera-Alonso, M., Adamson, D.H., et al.: Functionalized single graphene sheets derived from splitting graphite oxide. J. Phys. Chem. B. **110**, 8535 (2006)

573. McAllister, M.J., Li, J.-L., Adamson, D.H., Schniepp, H.C., Abdala, A.A., Liu, J., et al.: Single sheet functionalized graphene by oxidation and thermal expansion of graphite. Chem. Mater. **19**, 4396 (2007)

574. Gilje, S., Han, S., Wang, M., Wang, K.L., Kaner, R.B.: A chemical route to graphene for device applications. Nano Lett. **7**, 3394 (2007)

575. Cote, L.J., Kim, F., Huang, J.: Langmuir–Blodgett assembly of graphite oxide single layers. J. Am. Chem. Soc. **131**, 1043 (2009)

576. Xu, Y., Bai, H., Lu, G., Li, C., Shi, G.: Flexible graphene films via the filtration of water-soluble noncovalent functionalized graphene sheets. J. Am. Chem. Soc. **130**, 5856 (2008)

577. Somani, P.R., Somani, S.P., Umeno, M.: Planer nano-graphenes from camphor by CVD. Chem. Phys. Lett. **430**, 56 (2006)

578. Yang, X., Dou, X., Rouhanipour, A., Zhi, L., Rader, H.J., Mullen, K.: Two-dimensional graphene nanoribbons. J. Am. Chem. Soc. **130**, 42167 (2008)

579. Wang, X., Zhi, L., Tsao, N., Tomović, Ž., Li, J., Müllen, K.: Transparent carbon films as electrodes in organic solar cells. Angew. Chem. Int. Ed. **47**, 2990–2992 (2008)

580. Simpson, C.D., Brand, J.D., Berresheim, A.J., Przybilla, L., Rader, H.J., Mullen, K.: Synthesis of a giant 222 carbon graphite sheet. Chem.-Eur. J. **8**, 1424–1429 (2002)

581. Chen, L., Hernandez, Y., Feng, X., Müllen, K.: From nanographene and graphene nanoribbons to graphene sheets: chemical synthesis. Angew. Chem. Int. Ed. **51**, 7640–7654 (2012)

582. Tang, L., Li, X., Ji, R., Teng, K.S., Tai, G., Ye, J., Wei, C., Lau, S.P.: Bottom-up synthesis of large-scale graphene oxide nanosheets. J. Mater. Chem. **22**(12), 5676 (2012)

583. Lara-Avila, S., Kalaboukhov, A., Paolillo, S., Syväjä, M., Yakimova, R., Fal'ko, V., Tzalenchuk, A., Kubatkin, S.: SiC graphene suitable for quantum hall resistance metrology, Science Brevia (2009)

584. Sutter, P.: Epitaxial graphene: how silicon leaves the scene. Nat. Mater. **8**(3), 171–172 (2009)

585. Choucair, M., Thordarson, P., Stride, J.A.: Gram-scale production of graphene based on solvothermal synthesis and sonication. Nat. Nanotechnol. **4**(1), 30–33 (2008)
586. Sutter, P.: Epitaxial graphene: how silicon leaves the scene. Nat. Mater. **8**, 171–172 (2009)
587. Rolings, E., Gweon, G.-H., Zhou, S.Y., Mun, B.S., McChesney, J.L., Hussain, B.S., Fedorov, A.V., First, P.N., de Heer, W.A., Lanzara, A.: Synthesis and characterization of atomically thin graphite films on a silicon carbide substrate. J. Phys. Chem. Solids. **67**, 2172 (2006)
588. Virojanadara, C., Syväjarvi, M., Yakimova, R., Johansson, L.I.: Homogeneous large-area graphene layer growth on 6H -SiC(0001). Phys. Rev. B. **78**, 245403 (2008)
589. Emtsev, K.V., Bostwick, A., Horn, K., Jobst, J., Kellog, G.L., Ley, L., McChesney, J.L., Ohta, T., Reshanov, S.A., Rohrl, J., Rotenberg, E., Schmid, D., Rotenberg, E., Schmid, A. K., Waldmann, D., Weber, H.B., Seyller, T.: Towards wafer-size graphene layers by atmospheric pressure graphitization of silicon carbide. Nat. Mater. **8**, 203 (2009)
590. Kim, D.Y., Sinha-Ray, S., Park, J.-L., Lee, J.-G., Cha, Y.-H., Bae, S.-H., Ahn, J.-H., Jung, Y.C., Kim, S.M., Yarin, A.L., Yoon, S.S.: Supersonic spray creates high-quality graphene layer. Adv. Funct. Mater. **24**(31), 4986–4995 (2014)
591. MIT Technology Review: How to make graphne using supersonic buckyballs. https://www. technologyreview.com/s/539911/how-to-make-graphene-using-supersonic-buckyballs/. Accessed June 2016
592. Chiu, P.L., Mastrogiovanni, D.D.T., Wei, D., Lous, C., Jeong, M., Yu, G., Saad, P., Flach, C. R., Mendelsohn, R.: Microwave- and notrionium ion-enabled rapid and direct production of highly conductive low-oxygen graphene. J. Am. Chem. Soc. **134**(13), 5850–5856 (2012)
593. Patel, M., Feng, W., Svaram, K., Khoshi, M.R., Huang, R., Sun, J., Rabie, E., Flach, C., Mendelsohn, R., Garfunkel, E., He, H.: Microwave enabled one-pot, one-step fabrication and notrogen doping of holey graphene oxide for catalytic applications. Small. **11**(27), 3358–3368 (2015)
594. Kim, J., Lee, G., Kim, J.: Wafer-scale synthesis of multi-layer graphene by high-temperature carbon ion implantation. Appl. Phys. Lett. **107**(3), 033104 (2015)
595. Campos-Delgado, J., Romo-Herrera, J.M., Jia, X., Cullen, D.A., Muramatsu, H., Kim, Y.A., Hayashi, T., Ren, Z., Smith, D.J., Okuno, Y., Ohba, T., Kanoh, H., Kaneko, K., Endo, M., Terrones, H., Dresselhaus, M.S., Terrones, M.: Bulk production of a new form of sp (2) carbon: crystalline graphene nanoribbons. Nano Lett. **8**, 2773 (2008)
596. Kula, P., Pietrasik, R., Dybowski, K., Atraszkiewicz, R., Szymanski, W., Kolodziejczyk, L., Niedzielski, P., Nowak, D.: Single and multilayer growth of graphene from the liquid phase. Appl. Mech. Mater. **510**, 8–12 (2014)
597. Subrahmanyam, K.S., Panchakarla, L.S., Govindaraj, A., Rao, C.N.R.: Simple method of preparing graphene flakes by an arc-discharge method. J. Phys. Chem. C. **113**, 4257–4259 (2009)
598. Chen, Y., Zhao, H., Sheng, L., Yu, K., An, J., Xu, Y., Ando, Y., Zhao, X.: Mass-production of highly-crystalline few-layer graphene sheets by arc discharge in various H2-inert gas mixtures. Chem. Phys. Lett. **538**, 72–76 (2012)
599. Shen, B., Ding, J., Yan, X., Feng, W., Li, J., Xue, Q.: Influence of different buffer gases on synthesis of few-layered graphene by arc discharge method. Appl. Surf. Sci. **258**, 4523–4531 (2012)
600. Wang, X.K., Lin, X.W., Mesleh, M., JArrold, M.F., Dravid, V.P., Ketterson, J.B., Chang, R. P.H.: J. Mater. Res.I **10**, 1977 (1995)
601. Wang, X.K., Lin, X.W., Dravid, V.P., Ketterson, J.B., Chang, R.P.H.: Carbon nanotubes synthesized in a hydrogen arc discharge. Appl. Phys. Lett. **66**, 2430 (1995)
602. Panchokarla, L.S., Subrahmanyam, K.S., Saha, S.K., Govindaraj, A., Krishnamurthy, H.R., Waghmare, U.V., Rao, C.N.R.: Synthesis, structure, and properties of boronand nitrogen-doped graphene. Adv. Mater. **21**, 4726 (2009)
603. Ci, L., Song, L., Jin, C., Jariwala, D., Wu, D., Li, Y., Srivastava, A., Wang, Z.F., Storr, K., Balicas, L., Liu, F., Ajayan, P.M.: Atomic layers of hybridized boron nitride and graphene domains. Nat. Mater. **9**, 430 (2010)

604. Deng, D., Pan, X., Yu, L., Cui, Y., Jiang, Y., Qi, J., Li, W.X., Fu, Q., Ma, X., Xue, Q., Sun, G., Bao, X.: Toward N-doped graphene via solvothermal synthesis. Chem. Mater. **23**, 1188 (2011)

605. Brumfiel, G.: Nanotubes cut to ribbons new techniques open up carbon tubes to create ribbons. Nature (2009)

606. Kosynkin, D.V., Higginbotham, A.L., Sinitskii, A., Lomeda, J.R., Dimiev, A., Price, B.K., Tour, J.M.: Longitudinal unzipping of carbon nanotubes to form graphene nanoribbons. Nature. **458**(7240), 872–876 (2009)

607. Jiao, L., Zhang, L., Wang, X., Diankov, G., Dai, H.: Narrow graphene nanoribbons from carbon nanotubes. Nature. **458**, 877–880 (2009)

608. Chakrabarti, A., Lu, J., Skrabutenas, J.C., Xu, T., Xiao, Z., Maguire, J.A., Hosmane, N.S.: Conversion of carbon dioxide to few-layer graphene. J. Mater. Chem. **21**(26), 9491 (2011)

609. Kurzweil: Carbon nanotubes as reinforcing bars to strengthen graphene and increase conductivity. http://www.kurzweilai.net/carbon-nanotubes-as-reinforcing-bars-to-streng then-graphene-and-increase-conductivity. Accessed June 2016

610. Blake, P., Hill, E.W., Neto, A.H.C., Novoselov, K.S., Jiang, D., Yang, R., Booth, T.J., Geim, A.K.: Making graphene visible. Appl. Phys. Lett. **91**, 063124 (2007)

611. Treossi, E., Melucci, M., Liscio, A., Gazzano, M., Samori, P., Palermo, V.: High-contrast visualization of graphene oxide on dye-senstized glass, quartz, and silicon by fluorescence quenching. J. Am. Chem. Soc. **131**, 15576 (2009)

612. Park, J.S., Reina, A., Saito, R., Kong, J., Dresselhaus, G., Dresselhaus, M.S.: G' band Raman spectra of single, double and triple layer graphene. Carbon. **47**, 1303 (2009)

613. Kim, J., Cote, L.J., Kim, F., Huang, J.: Visualizing graphene based sheets by fluorescence quenching microscopy. J. Am. Chem. Soc. **132**, 260 (2009)

614. Ferrari, A.C., Meyer, J.C., Scardaci, V., Casiraghi, C., Lazzeri, M., Mauri, F., et al.: Raman spectrum of graphene and graphnen layers. Phys. Rev. Lett. **97**, 187401 (2006)

615. Kosynkin, D.V., Higginbotham, A.L., Sinitskii, A., Lomeda, J.R., Dimiev, A., Price, B.K., Tour, J.M.: Nature. **458**, 872 (2009)

616. Sinitskii, A., Dimiev, A., Corley, D.A., Fursina, A.A., Kosynkin, D.V., Tour, J.M.: ACS Nano. **4**, 1949 (2010)

617. Niyogi, S., Bekyarova, E., Itkis, M.E., Zhang, H., Shepperd, K., Hicks, J., Sprinkle, M., Berger, C., Ning Lau, C., de Heer, W.A., Conrad, E.H., Haddon, R.C.: Nano Lett. **10**, 4061 (2010)

618. Fang, M., Wang, K., Lu, H., Yang, Y., Nutt, S.: J. Mater. Chem. **19**, 7098 (2009)

619. Vadukumpully, S., Gupta, J., Zhang, Y., Xu, C.Q., Valiyaveettil, S.: Nanoscale. **3**, 303 (2011)

620. He, H., Gao, C.: Chem. Mater. **22**, 5054 (2010)

621. Zhong, X., Jin, J., Li, S., Niu, Z., Hu, W., Li, R., Ma, J.: Aryne cycloaddition: highly efficient chemical modification of graphene. Chem. Commun. **46**, 7340 (2010)

622. Liu, Y., Zhou, J., Zhang, X., Liu, Z., Wan, X., Tian, J., Wang, T., Chen, Y.: Synthesis, characterization and optical limiting property of covalently oligothiophene-functionalized graphene material. Carbon. **47**, 3113 (2009)

623. Yu, D., Yang, Y., Durstock, M., Baek, J.B., Dai, L.: Soluble P3HT-grafted graphene for efficient bilayer-heterojunction photovoltaic devices. ACS Nano. **4**, 5633 (2010)

624. Liu, Q., Liu, Z., Zhang, X., Yang, Y., Zhang, N., Pan, G., Yin, S., Chen, Y., Wei, J.: Adv. Funct. Mater. **19**, 894 (2009)

625. Xu, Y., Liu, Z., Zhang, X., Wang, Y., Tian, J., Huang, Y., Ma, Y., Zhang, X., Chen, Y.: A graphene hybrid material covalently functionalized with porphyrin: synthesis and optical limiting property. Adv. Mater. **21**, 1275 (2009)

626. Karousis, N., Sandanayaka, A.S.D., Hasobe, T., Economopoulos, S.P., Sarantopoulou, E., Tagmatarchis, N.: J. Mater. Chem. **21**, 109 (2011)

627. Liu, Z.B., Xu, Y.F., Zhang, X.Y., Zhang, X.L., Chen, X.L., Tian, J.G.: J. Phys. Chem. B. **113**, 9681 (2009)

628. Zhang, X., Feng, Y., Huang, D., Li, Y., Feng, W.: Carbon. **48**, 3236 (2010)
629. Ramanathan, T., Abdala, A.A., Stankovich, S., Dikin, D.A., Herrera-Alonso, M., Piner, R. D., Adamson, D.H., Schniepp, H.C., Chen, X., Ruoff, R.S., Nguyen, S.T., Aksay, I.A., Prud'Homme, R.K., Brinson, L.C.: Nat. Nanotechnol. **3**, 327 (2008)
630. Das, B., Eswar Prasad, K., Ramamurty, U., Rao, C.N.R.: Nanotechnol. **20** (125705) (2009)
631. Liu, Z., Robinson, J.T., Sun, X., Dai, H.: J. Am. Chem. Soc. **130**, 10876 (2008)
632. Salavagione, H.J., Gomez, M.A., Martınez, G.: Macromol. **42**, 6331 (2009)
633. Cheng, H.C., Shiue, R.J., Tsai, C.C., Wang, W.H., Chen, Y.T.: ACS Nano. **5**, 2051 (2011)
634. Liang, Y.Y., Wu, D.Q., Feng, X. L., Müllen, K.: Adv. Mater. **21**, 1679 (2009)
635. Zhang, X.Y., Li, H.P., Cui, X.L., Lin, Y.: J. Mater. Chem. **20**, 2801 (2010)
636. Noble 3D Printers: Graphenite™ WX™ For Lost Wax Casting. https://www.noble3dprinters.com/product/graphenite-wx-lost-wax-casting/. Accessed June 2016
637. Head: Graphenext. http://www.head.com/us/sports/tennis/technology/graphene-xt/. Accessed June 2016
638. Huntsman: Huntsman advanced materials and Haydale devlop graphnen enhanced Araldite® resin. http://www.huntsman.com/advanced_materials/Applications/itemrenderer?p_render title=no&p_renderdate=no&p_renderteaser =no&p_item_id=997718783&p_item_caid=1223. Accessed June 2016
639. Composites World: Haydale, Huntsman to work together on graphene enhanced polymer resins. http://www.compositesworld.com/news/haydale-huntsman-to-work-together-on-graphene-enhanced-polymer-resins. Accessed June 2016
640. Thomas Swan: Elicarb® Graphene. http://www.thomas-swan.co.uk/advanced-materials/ elicarb%C2%AE-graphene. Accessed June 2016
641. Yoon, H.J., Jun, D.H., Yang, J.H., Zhou, Z., Yang, S.S., Cheng, M.C.-C.: Carbon dioxide gas sensor using a graphene sheet. Sen. Actuators B. **157**, 310–313 (2011)
642. Schedin, F., Geim, A.K., Morozov, S.V., Hill, E.W., Blake, P., Katsnelson, M.I., Novoselov, K.S.: Detection of individual gas molecules adsorbed on graphene. Nat. Mater. **6**, 652–655 (2007)
643. Schedin, F., Geim, A.K., Morozov, S.V., Hill, E.W., Blake, P., Katsnelson, M.I., Novoselov, K.S.: Nat. Mater. **6**, 652 (2007)
644. Novoselov, K., Geim, A.: Mater. Technol. **22**, 178–179 (2007)
645. Sun, J., Muruganathan, M., Mizuta, H.: Room temperature detection of individual molecular physisoprtion using suspended bilayer graphene. Sci. Adv. **2**(4), e1501518 (2016)
646. Fowler, J.D., Allen, M.J., Tung, V.C., Yang, Y., Kaner, R.B., Weiller, B.H.: ACS Nano. **3**, 301 (2009)
647. Ganhua, L., Ocola, L.E., Chen, J.: Reduced graphene oxide for room-temperature gas sensors. Nanotechnol. **20**, 445502 (2009)
648. Robinson, J.T., Perkins, F.K., Snow, E.S., Wei, Z., Sheehan, P.E.: Reduced graphene oxide molecular sensors. Nano Lett. **8**, 3137 (2008)
649. Jung, I., Dikin, D., Park, S., Cai, W., Mielke, S.L., Ruoff, R.S.: Effect of water vapor on electrical properties of individual reduced graphene oxide sheets. J. Phys. Chem. C. **112**, 20264 (2008)
650. Shafiei, M., Spizzirri, P.G., Arsat, R., Yu, J., du Plessis, J., Dubin, S., et al.: Platinum/ graphene nanosheet/SiC contacts and their application for hydrogen gas sensing. J. Phys. Chem. C. **114**, 13796 (2010)
651. Arsat, R., Breedon, M., Shafiei, M., Spizziri, P.G., Gilje, S., Kaner, R.B., et al.: Graphene-like nano-sheets for surface acoustic wave gas sensor applications. Chem. Phys. Lett. **467**, 344 (2009)
652. Joshi, R.K., Gomez, H., Alvi, F., Kumar, A.: Graphene films and ribbons for sensing of O_2 and 100 ppm of CO and NO_2 in practical conditions. J. Phys. Chem. C. **114**, 6610 (2010)
653. Fowler, J.D., Allen, M.J., Tung, V.C., Yang, Y., Kaner, R.B., Weiller, B.H.: Practical chemical sensors from chemically derived graphene. ACS Nano. **3**(2), 301–306 (2009)

654. Nallon, E.C., Schnee, V.P., Bright, C., Polcha, M.P., Qilang, L.: Chemical discrimination with an unmodified graphene chemical sensor. ACS Sens. **1**(1), 26–31 (2016)

655. Rangel, N.L., Seminario, J.M.: Vibronics and plasmonics based graphene sensors. J. Chem. Phys. **132**, 125102 (2010)

656. Shao, Y., Wang, J., Wu, H., Liu, J., Aksay, I.A., Lin, Y.: Graphene based electrochemical sensors and biosensors: a review. Electroanalysis. **22**(10), 1027–1036 (2010)

657. Liu, Y., Dong, X., Chen, P.: Biological and chemical sensors based on graphene materials. Chem. Soc. Rev. **41**, 2283–2307 (2012)

658. McCreery, R.L.: Chem. Rev. **108**, 2646 (2008)

659. Niwa, O.L, Jia, J., Sato, Y., Kato, D., Kurita, R., Maruyama, K., Suzuki, K., Hirono, S.: J. Am. Chem. Soc. **128**, 7144 (2006)

660. Zhou, M., Zhai, Y.M., Dong, S.J.: Anal. Chem. **81**, 5603 (2009)

661. Shan, C.S., Yang, H.F., Song, J.F., Han, D.X., IVaska, A., Niu, L.: Anal. Chem. **81**, 2378 (2009)

662. Kang, X.H., Wang, J., Wu, H., Aksay, A.I., Liu, J., Lin, Y.H.: Biosens. Bioelectron. **25**, 901 (2009)

663. Tehrani, Z.: Generic epitaxial graphene biosensors for ultrasnensitive detection of cancer risk biomarker. 2D Mater. **1**, 025004 (2014)

664. Park, D.-W., Schendel, A.A., Mikael, S., Brodnick, S.K., Roichner, T.J., Ness, J.P., Hayat, M.R., Atry, F., Frye, S.T., Pashaie, R., Thongpang, S., Ma, Z., Williams, J.C.: Graphene-based carbon-layered electrode array technology for neural imaging and optogenetic applications. Nat. Commun. **5**, 5258 (2014)

665. He, Q.Y., Sudibya, H.G., Yin, Z.Y., Wu, S.X., Li, H., Boey, F., et al.: Centimeter-long and large-scale micropatterns of reduced graphene oxide films: fabrication and sensing applications. ACS Nano. **4**, 3201 (2010)

666. Mao, S., Lu, G., Yu, K., Bo, Z., Chen, J.: Specific protein detection using thermally reduced graphene oxide sheet decorated with gold nanoparticle-antibody conjugates. Adv. Mater. **22**, 3521 (2010)

667. Lee, H., Choi, T.K., Lee, Y.B., Cho, H.R., Ghaffari, R., Wang, L., Choi, H.J., Chung, T.D., Lu, N., Hyeon, T., Choi, S.H., Kim, D.-H.: A graphene-based electrochemical device with thermoresponsive microneedles for diabetes monitoring and therapy. Nature Nanotechnol. **11**, 566–572 (2016)

668. Wang, Y., Shao, Y., Matson, D.W., Li, J., Lin, Y.: ACS Nano. **4**, 1790 (2010)

669. Zhang, B., Cui, T.: An ultrasenstivie and low-cost graphene sensor based on layer-by-layer nano self-assembly. Appl. Phys. Lett. **98**, 073116 (2011)

670. Mohanty, N., Berry, V.: Graphene-based single-bacterium resolution biodevice and DNA transistor: interfacing graphene derivatives with nanoscale and microscale biocomponents. Nano Lett. **8**(12), 4468–4476 (2008)

671. Keeley, G.P., et al.: Electrochemical ascorbic acid sensor based on DMF-exfoliated graphene. J. Mater. Chem. **20**(36), 7864–7869

672. Kang, X., et al.: A graphene-based electrochemical sensor for sensitive detection of paracetamol. Talanta. **81**(3), 754–759

673. Zhou, M., Zhai, Y.M., Dong, S.J.: Anal. Chem. **81**, 5603 (2009)

674. Tang, L.H., Wang, Y., Li, Y.M., Feng, H.B., Lu, J., Li, J.H.: Adv. Funct. Mater. **19**, 2782 (2009)

675. Wang, Y., Li, Y.M., Tang, L.H., Lu, J., Li, J.H.: Electrochem. Commun. **11**, 889 (2009)

676. Alwarappan, S., Erdem, A., Liu, C., Li, C.Z.: J. Phys. Chem. C. **113**, 8853 (2009)

677. Shang, N.G., Papalonstantinou, P., McMullan, M., Chu, M., Stamboulis, A., Potenza, A., Dhesi, S.S., Marchetto, H.: Adv. Funct. Mater. **18**, 3506 (2008)

678. Zhou, M., Zhai, Y.M., Dong, S.J.: Anal. Chem. **81**, 5603 (2009)

679. Li, J., Guo, S.J., Zhai, Y.M., Wang, E.K.: Anal. Chim. Acta. **649**, 196 (2009)

680. Li, J., Guo, S.J., Zhai, Y.M., Wang, E.K.: Electrochem. Commun. **11**, 1085 (2009)

681. Kawano, Y.: Wide-band frequency-tunable terhertz and infrared detection with graphene. Nanotechnol. **24**(21), 214004 (2013)
682. Petruk, O., Szewczyk, R., Ciuk, T., Strupiński, W., Salach, J., Nowvicki, M., Pasternak, I., Winiarski, W., Tszcinka, K.: Sensitivity and offset voltage testing in the hall-effect sensors made of graphene. Adv. Intell. Syst. Comput. **267**, 631–640 (2014)
683. Dauber, J., Sagade, A.A., Oellers, M., Watanabe, K., Taniguchi, T., Neumaier, D., Stampfer, C.: Ultra-senstive Hall sensors based on graphnen encapsulated in hexagonal boron nitride. Appl. Phys. Lett. **106**, 193501 (2015)
684. Li, X., Zhu, M., Du, M., Lv, Z., Zhang, L., Li, Y., Yang, Y., Yang, T., Li, X., Wang, K., Zhu, H., Fang, Y.: High detectivity graphene-silicon heterojunction photodetector. Small. **2**(5), 595–601 (2015)
685. Ghosh, S., Sarker, B.K., Chunder, A., Zhai, L., Khondaker, S.I.: Position dependent photodetector from large area reduced graphenen oxide thin films. Appl. Phys. Lett. **96**, 163109 (2010)
686. R & D Magazine: Bosch announces breakthrough in graphene sensor technology. http://www.rdmag.com/news/2015/06/bosch-announces-breakthrough-graphene-sensor-technol ogy. Accessed June 2016
687. Kazakova, O., et al.: Optimization of 2 DEG InAs/GaSb hall sensors for single particle detection. IEEE Trans. Magn. **44**(11), 4480–4483 (2008)
688. Pisana, S., et al.: Graphene magnetic field sensors. IEEE Trans. Mag. **46**(6), 1910–1913 (2010)
689. Boone, T.D., et al.: Temperature dependence of magnetotransport in extraordinary magne-toresistance devices. IEEE Trans. Magn. **42**(10), 3270–3272 (2006)
690. Kuzmenko, A.B., et al.: Universal optical conductance of graphite. Phys. Rev. Lett. **100**(11), 4 (2008)
691. Booth, T.J., et al.: Macroscopic graphene membranes and their extraordinary stiffness. Nano Lett. **8**(8), 2442–2446 (2008)
692. Boland, C.S., Khan, U., Backes, C., O'Neill, A., McCauley, J., Duane, S., Shanker, R., Liu, Y., Jurewicz, I., Dalton, A.B., Coleman, J.N.: Sensitive, high-Strain, high-rate bodily motion sensors based on graphene-rubber composites. ACS Nano. **8**(9), 8819–8830 (2014)
693. Yao, F., Güneş, F., Ta, H.Q., Lee, S.M., Chae, S.J., Sheem, K.Y., Cojocaru, C.S., Xie, S.S., Lee, Y.H.: Diffusion mechanism of lithium ion through basal plane of layered graphene. J. Am. Chem. Soc. **134**(20), 8646–8654 (2012)
694. IEEE Spectrum: Faster and cheaper for graphene in Li-ion batteries. http://spectrum.ieee.org/nanoclast/semiconductors/nanotechnology/faster-and-cheaper-process-for-graphene-in-liion-batteries. Accessed June 2016
695. Brownson, D.A.C., Kampouris, D.K., Banks, C.E.: An overview of graphnen in energy production and storage applications. J. Power Sources. **196**, 4873–4885 (2011)
696. Paek, S.-M., Yoo, E., Honma, I.: Nano Lett. **9**, 72 (2009)
697. Lian, P., Zhu, X., Liang, S., Li, Z., Yang, W., Wang, H.: Electrochim. Acta. **55**, 3909 (2010)
698. Guo, P., Song, H., Chen, X.: Electrochem. Commun. **11**, 1320 (2009)
699. Yang, S., Feng, X., Ivanovici, S., Müllen, K.: Fabrication of graphene-encapsulated oxide nanoparticles: towards high-performance anode materiels for lithium storage. Angew Chem. Int. Ed. **49**, 8408 (2010)
700. Wang, H., Cui, L.-F., Yang, Y., Sanchez Casalongue, H., Robinson, J.T., Liang, Y., et al.: Mn_3O_4 graphene hybrid as a high-capacity anode material for lithium ion batteries. J. Am. Chem. Soc. **132**, 13978 (2010)
701. Wu, Z.-S., Ren, W., Wen, L., Gao, L., Zhao, J., Chen, Z., et al.: Graphene anchored with Co_3O_4 nanoparticles as anode of lithium ion batteries with enhanced reversible capacity and cyclic performance. ACS Nano. **4**, 3187 (2010)
702. Larcher, D., Beattie, S., Morcrette, M., Edstrom, K., Jumas, J.-C., Tarascon, J.-M.: Recent findings and prospects in the field of pure metals as negative electrodes for Li-ion batteries. J. Mater. Chem. **17**, 3759 (2007)

703. Zhou, G., Wang, D.-W., Li, F., Zhang, L., Li, N., Wu, Z.-S., et al.: Graphen-wrapped Fe_3O_4 anode material with improved reversible capacity and cyclic stability for lithium ion batteries. Chem. Mater. **22**, 5306 (2010)

704. Lee, J.K., Smith, K.B., Hayner, C.M., Kung, H.H.: Silicon nanoparticles-graphene paper composites for Li ion batter anodes. Chem. Commun. **46**, 2025 (2010)

705. Wang, H., Maiyalagan, T., Wang, X.: Review on recent progress in nitrogen-doped graphene: synthesis, characterization, and its potential applications. ACS Catal. **2**, 781–794 (2012)

706. Pan, D., Wang, S., Zhao, B., Wu, M., Zhang, H., Wang, Y., Jiao, Z.: Chem. Mater. **21**, 3136 (2009)

707. Wu, Z.S., Ren, W., Xu, L., Li, F., Cheng, H.M.: ACS Nano. **5**, 5463 (2011)

708. Reddy, A.L.M., Srivastava, A., Gowda, S.R., Gullapalli, H., Dubey, M., Ajayan, P.M.: ACS Nano. **4**, 6337 (2010)

709. Shao, Y., Zhang, S., Engelhard, M.H., Li, G., Shao, G., Wang, Y., Liu, J., Aksay, I.A., Lin, Y.: Nitrogen-doped graphene and its electrochemical applications. J. Mater. Chem. **20**, 7491–7496 (2010)

710. David, L., Bhandavat, R., Barrera, U., Singh, G.: Silicon oxycarbide glass-graphene composite paper electrode for long-cycle lithium-ion batteries. Nature Comm. **7**, (2016)

711. Liu, C., Alwarappan, S., Chen, Z., Kong, X., Li, C.-Z.: Biosens. Bioelectron. **25**, 1829 (2010)

712. Jafri, R.I., Rajalakshmi, N., Ramaprabhu, S.: J. Mater. Chem. **20**, 7114 (2010)

713. Seger, B., Kamat, P.V.: Electrocatyltically active graphene-platinunm nanocomposites. Role of 2-D carbon support for PEM fuel cells. J. Phys. Chem. C. **113**, 7990 (2009)

714. Kou, R., Shao, Y., Wang, D., Engelhard, M.H., Kwak, J.H., Wang, J., et al.: Enhanced activity and stability of Pt catalysts on functionalized graphnen sheets for electrocatalytic oxygen reduction. Electrochem Commun. **11**, 954 (2009)

715. Yumura, T., Kimura, K., Kobayashi, H., Tanaka, R., Okumura, N., Yamabe, T.: The use of nanometer-sized hydrographene species for support material for fuel cell electrode catalyists: a theoretical proposal. Phys. Chem. Chem. Phys. **11**, 8275 (2009)

716. Hu, S., Lozado-Hidalgo, M., Wang, F.C., Mishchenko, A., Schedin, F., Nair, R.R., Hill, E. W., Boukhvaolv, D.W., Katsnelson, M.I., Dryfe, R.A.W., Grigorieva, I.V., Wu, H.A., Geim, A.K.: Proton transport through one-atom-thick crystals. Nature. **516**, 227–230 (2014)

717. Shao, Y., Zhang, S., Engelhard, M.H., Li, G., Shao, G., Wang, Y., Liu, J., Aksay, I.A., Lin, Y.: J. Mater. Chem. **20**, 7491 (2010)

718. Luo, Z., Lim, S., Tian, Z., Shang, J., Lai, L., MacDonald, B., Fu, C., Shen, Z., Yu, T., Lin, J.: J. Mater. Chem. **21**, 8038 (2011)

719. Sheng, Z.H., Shao, L., Chen, J.J., Bao, W.J., Wang, F.B., Xia, X.H.: ACS Nano **5**, 4350 (2011); Huang, Y., Liang, J., Chen, Y.: An overview of the applications of graphene-based materials in supercapacitors. Small **8**(12), 1805–1834 (2012)

720. Vivekchand, S.R.C., Rout, C.S., Subrahmanyam, K.S., Govindaraj, A., Rao, C.N.R.: J. Chem. Sci. **120**, 9 (2008)

721. Kim, T.Y., Lee, H.W., Stoller, M., Dreyer, D.R., Bielawski, C.W., Ruoff, R.S., Suh, K.S.: ACS Nano. **5**, 436 (2011)

722. Liu, C., Yu, Z., Neff, D., Zhamu, A., Jang, B.Z.: Nano Lett. **10**, 4863 (2010)

723. Stoller, M.D., Park, S., Zhu, Y., An, J., Ruoff, R.S.: Graphene-Based Ultracapacitors. Nano Lett. **8**(10), 3498–3502 (2008)

724. Wu, Z.-S., Wang, D.-W., Ren, W., Zhao, J., Zhou, G., Li, F., et al.: Anchoring hydrous RuO_2 on graphene sheets for high-performance electrochemical capacitors. Adv. Funct. Mater. **20**, 3595 (2010)

725. Zhu, Y., Stoller, M.D., Cai, W., Velamakanni, A., Piner, R.D., Chen, D., et al.: Exfoliation of graphite oxide in propylene carbonate and thermal reduction of the resulting graphene oxide platelets. ACS Nano. **4**, 1227 (2010)

726. Wang, Y., Shi, Z., Huang, Y., Ma, Y., Wang, C., Chen, M., et al.: Supercapacitor devices based on graphene materials. J. Phys. Chem. C. **113**, 13103 (2009)

727. Murugan, A.V., Muraliganth, T., Manthiram, A.: Rapid, facile microwave-solvothermal synthesis of graphene nanosheets and th4eir polyaniline nanocomposites for energy storage. Chem. Mater. **21**, 5004 (2009)

728. Chen, S., Zhu, J.W., Wu, X.D., Han, Q.F., Wang, X.: Graphene Oxide-MnO$_2$ Nanocomposites for Supercapacitors. ACS Nano. **4**, 2822 (2010)

729. Yu, D., Dai, L.: Self-assembled graphene/carbon nanotube hybrid films for supercapacitors. J. Phys. Chem. Lett. **1**, 467 (2009)

730. Wu, Z.S., Ren, W.C., Wang, D.W., Li, F., Liu, B.L., Cheng, H.M.: High-energy MnO$_2$ nanowire/graphene and graphene asymmetric electrochemical capacitors. ACS Nano. **4**, 5835 (2010)

731. Zhu, Y.W., Murali, S., Stoller, M.D., Velamakanni, A., Piner, R.D., Ruoff, R.S.: Microwave assisted exfoliation and reduction of graphite oxide for ultracapcitors. Carbon. **48**, 2118 (2010)

732. Du, X., Guo, O., Song, H., Chen, X.: Electrochim. Acta. **55**, 4812 (2010)

733. Yan, J., Wei, T., Shao, B., Fan, Z., Qian, W., Zhang, M., Wei, F.: Carbon. **48**, 487 (2010)

734. Chen, Y., Zhang, X., Yu, P., Ma, Y.: J. Power Sources. **195**, 3031 (2010)

735. Lu, T., Zhang, Y., Li, H., Pan, L., Li, Y., Sun, Z.: Electrochim. Acta. **55**, 4170 (2010)

736. Chen, S., Zhu, J., Wang, X.: J. Phys. Chem. C. **114**, 11829 (2010)

737. R & D Magazine: Laser-induced graphene "super" for electronics. http://www.rdmag.com/news/2015/01/laser-induced-graphene-super-electronics. Accessed June 2016

738. R & D Magazine: Quick-charging hybrid supercapacitors. http://www.rdmag.com/news/2015/04/quick-charging-hybrid-supercapacitors. Accessed June 2016

739. Peng, Z., Ye, R., Mann, J.A., Zakhidov, D., Li, Y., Smalley, P.R., Lin, J., Tour, J.M.: Flexible boron-doped laser-induced graphene microsupercapacitors. ACS Nano. **9**(6), 5868–5875 (2015)

740. Yan, J., Wei, T., Shao, B., Fan, Z., Qian, W., Zhang, M., et al.: Preparation of a graphene nanosheet/polyaniline composite with high specific capacitance. Carbon. **48**, 487 (2010)

741. Wang, H., Hao, Q., Yang, X., Lu, L., Wang, X.: ACS Appl. Mater. Inter. **2**, 821 (2010)

742. Wang, H., Hao, Q., Yang, X., Lu, L., Wang, X.: Electrochem. Commun. **11**, 1158 (2009)

743. Zhang, K., Zhang, L.L., Zhao, X.S., Wu, J.: Chem. Mater. **22**, 1392 (2010)

744. Zhang, L.L., Zhao, S., Tian, X.N., Zhao, X.S.: Langmuir. **26**, 17624 (2010)

745. Mini, P.A., Balakrishnan, A., Nair, S.V., Subramanian, K.R.V.: Chem. Commun. **47**, 5753 (2011)

746. Chen, S., Zhu, J., Wu, X., Han, Q., Wang, X.: ACS Nano. **4**, 2822 (2010)

747. Wang, X., Zhi, L., Müllen, K.: Transparent, conductive graphene electrodes for dye-sensitized solar cells. Nano Lett. **8**(1), 323–327 (2008)

748. Wang, Y., Chen, X., Zhong, Y., Zhu, F., Loh, K.P.: Large area, continuous, few-layered graphene as anodes in organic photovoltaic devices. Appl. Phys. Lett. **95**, 063302 (2009)

749. Li, X., Zhu, H., Wang, K., Anyuan, C., Wei, J., Li, C., Jia, Y., Li, Z., Li, X., Wu, D.: Graphene-on-silicon Schottky junction solar cells. Adv. Mater. **22**(25), 2743–2748 (2010):

750. Wang, J.T.-W., Ball, J.M., Barea, E.M., Abate, A., Alexander-Webber, J.A., Huang, J., Saliba, M., Mora-Sero, I., Bisqurt, J., Snaith, H.J., Nicholas, R.J.: Low-temperature processed electron collection layers of graphene/TiO$_2$ nanocomposites in thin film perovskite solar cells. Nano Lett. **14**(2), 724–730 (2014)

751. De Arco, L.W., Zhang, Y., Schlenker, C.W., Ryu, K., Thompson, M.E., Zhou, C.: Continuous highly flexible, and transparent graphene films by chemical vapor deposition for organic photovoltaics. ACS Nano. **4**(5), 2865 (2010)

752. Wu, J., Agrawal, M., Becerril, H.A., Bao, Z., Liu, Z., Chen, Y., Peumans, P.: Organic light-emitting diodes on solution-processed graphene transparent electrodes. ACS Nano. **4**(1), 43–48 (2010)

753. Lin, Y., Norman, C., Srivastava, D., Azough, F., Wang, L., Robbins, M., Simpson, K., Freer, R., Kinloch, I.A.: Thermoelectric power generation from lanthanum strontium titanium oxide at room temperature through the addition of graphene. ACS Appl. Mater. Interfaces. **7**(29), 15898–15908 (2015)

754. Geng, X., Niu, L., Xing, Z., Song, R., Liu, G., Sun, M., et al.: Aqueous-processable noncovalent chemically converted graphene – quantum dot composites for flexible and transparent optoelectronic films. Adv. Mater. **22**, 638 (2010)

755. Lin, Y., Zhang, K., Chen, W., Liu, Y., Geng, Z., Zeng, J., et al.: Dramatically enhanced photoresponse of reduced graphene oxide with linker-free anchored CdSe nanoparticles. ACS Nano. **4**, 3033 (2010)

756. Xiang, Q., Yu, J., Jaroniec, M.: J. Am. Chem. Soc. **134**, 6575 (2012)

757. Roy-Mayhew, J.D., Bozym, D.J., Punckt, C., Aksay, I.A.: Functionalized graphene as a catalytic counter electrode in dye-sensitized solar cells. ACS Nano. **4**(10), 6203–6211 (2010)

758. Inhabitat: Graphene-based solar cells could yield 60% efficiency. http://inhabitat.com/graphene-based-solar-cells-could-yield-60-efficiency/. Accessed June 2016

759. Zhu, S., Li, T.: Hydrogenation-assisted graphene origami and its application in program-mable molecular mass uptake, storage, and release. ACS Nano. **8**(3), 2864–2872 (2014)

760. Burress, J., Gadipelli, S., Ford, J., Simmons, J., Zhou, W., Yildirim, T.: Angew. Chem. Int. Ed. **49**, 8902 (2010)

761. Ao, Z.M., Jiang, Q., Zhang, R.Q., Tan, T.T., Li, S.: Al doped graphene: a promising material for hydrogen storage at room temperature. J. Appl. Phys. **105**, 074307 (2009)

762. Liang, J., Wang, Y., Huang, Y., Ma, Y., Liu, Z., Cai, J., et al.: Electromagnetic interference shielding of graphene/epoxy composites. Carbon. **47**, 922 (2009)

763. Yousefi, N., Sun, X., Lin, X., Shen, X., Jia, J., Zhang, B., Tang, B., Chan, M., Kim, J.-K.: Highly aligned graphene/polymer nanocomposites with excellent dielectric properties for high-performance electromagnetic interference shielding. Adv. Mater. **26**, 5480–5487 (2014)

764. Liang, J., Wang, Y., Huang, Y., Ma, Y., Liu, Z., Cai, J., Zhang, C., Gao, H., Chen, Y.: Electromagnetic interference shielding of graphene/epoxy composites. Carbon. **47**, 922–925 (2009)

765. Wu, B., Tuncer, H.M., Naeem, M., Yang, B., Cole, M.T., Milne, W.I., Hao, Y.: Experi-mental demonstration of a transparent graphene millimeter wave absorber with 28% fraction bandwidth at 140 GHz. Sci. Rep. **4**, 4130 (2014)

766. Britnell, L., Gorbachev, R.V., Geim, A.K., Ponomarenko, L.A., Mishchenko, A., Greena-way, M.T., Fromhold, T.M., Novoselov, K.S., Eaves, L.: Resonant tunneling and negative differential conductance in graphene transistors. Nat. Commun. **4**, 1794 (2013)

767. Yang, J.W., Lee, G., Kim, J.S., Kim, K.S.: J. Phys. Chem. Lett. **2**, 2577 (2011)

768. Jin, Z., Yao, J., Kittrell, C., Tour, J.M.: ACS Nano. **5**, 4112 (2011)

769. Park, J., Lee, W.H., Huh, S., Sim, S.H., Kim, S.B., Cho, K., Hong, B.H., Kim, K.S.: J. Phys. Chem. Lett. **2**, 841 (2011)

770. Park, J., Jo, S.B., Yu, Y.J., Kim, Y., Yang, J.W., Lee, W.H., Kim, H.H., Hong, B.H., Kim, P., Cho, K., Kim, K.S.: Adv. Mater. **24**, 407 (2012)

771. Novoselov, K.S., Jiang, D., Schedin, F., Booth, T.J., Khotkevich, V.V., Morozov, S.V., Geim, A.K.: Proc. Natl. Acad. Sci. U.S.A. **102**, 10451 (2005)

772. Morozov, S.V., Novoselov, K.S., Katsnelson, M.I., Schedin, F., Elias, D.C., Jaszczak, J.A., Geim, A.K.: Phys. Rev. Lett. **100**, 016602 (2008)

773. Bolotin, K.I., Sikes, K.J., Jiang, Z., Klima, M., Fudenberg, G., Hone, J., Kim, P., Stormer, H. L.: Solid State Commun. **146**, 351 (2008)

774. Geim, A.K., MacDonald, A.H.: Graphene: exploring carbon flatland. Phys. Today. **60**, 35 (2007)

775. Chen, J.-H., Ishigami, M., Jang, C., Hines, D.R., Fuhrer, M.S., Williams, E.D.: Printed graphene circuits. Adv. Mater. **19**(21), 3623–3627 (2007)

776. Ponomarenko, L.A., Schedin, F., Katsnelson, M.I., Yang, R., Hill, E.W., Novoselov, K.S., Geim, A.K.: Chaotic Dirac billiard in graphene quantom dots. Science. **320**(5874), 356–358 (2008)

777. Jia, C., Migliore, A., Xin, N., Huang, S., Wang, J., Yang, Q., Wang, S., Chen, H., Wang, D., Feng, B., Liu, Z., Zhang, G., Qu, D.-H., Tian, H., Ratner, M.A., Xu, H.Q., Nitzan, A., Guo, X.: Covalently bonded single-molecule junctions with stable and reversible photoswitched conductivity. Science. **352**(6292), 1443–1445 (2016)

778. Halford, B.: Diarylethene molecular switch has staying power. Chem. Eng. News. **94**(25), 5 (2016)

779. Georgia Tech Research News: Carbon-based electronics: researchers develop foundation for circuitry and devices based on graphite. http://gtresearchnews.gatech.edu/newsrelease/graphene.htm. Accessed June 2016

780. Lemme, M.C., Echtermeyer, T.J., Baus, M., Kurz, H.: A graphene field-effect device. IEEE Electron Device Lett. **28**(4), 1–12 (2007)

781. Kedzierski, J., Hsu, P.-L., Healey, P., Wyatt, P., Keast, C., Sprinkle, M., Berger, C., de Heer, W.: Epitaxial graphene transistors on SiC substrates. IEEE Trans Electron Devices. **55**, 2078 (2008)

782. Moon, J.S., Curis, D., Hu, M., Wong, D., McGuire, C., Campbell, P.M., Jernigan, G., Tedesco, J.L., Vanmil, B., Myers-Ward, R., Eddy, C., Gaskill, D.K.: Epitaxial-graphene RF field-effect transistors on Si-Face 6H-SiC substrates. IEEE Electron Device Lett. **30**(6), 650–652 (2009)

783. Sordan, R., Traversi, F., Russo, V.: Logic gates with a single graphene transistor. Appl. Phys. Lett. **94**, 073305 (2009)

784. Chang, H., Sun, Z., Yuan, Q., Ding, F., Tao, X., Yan, F., et al.: Thin film field-effect phototransistors from bandgap-tunable, solution-processed, few-layer reduced graphene oxide films. Adv. Mater. **22**, 4872 (2010)

785. Xia, F., Mueller, T., Lin, Y.-M., Valdes-Garcia, A., Avouris, P.: Ultrafast graphene photo-detector. Nat. Nanotechnol. **4**, 839 (2009)

786. (a) Lee, W.H., Park, J., Sim, S.H., Lim, S., Kim, K.S., Hong, B.H., Cho, K.: J. Am. Chem. Soc. **133**, 4447 (2011); (b) Lee, W.H., Park, J., Sim, S.H., Jo, S.B., Kim, K.S., Hong, B.H., Cho, K.: Adv. Mater. **3**, 1752 (2011)

787. Torrisi, F., Hasan, T., Wu, W., Sun, Z., Lombardo, A., Kulmala, T., Hshieh, G.W., Jung, S. J., Bonaccorso, F., Paul, P.J., Chu, D.P., Ferrari, A.C.: Ink-jet printed graphene electronics. ACS Nano. **6**, 2992 (2012)

788. MIT Technology Review: Printed graphene transistors promise a flexible electronic future. https://www.technologyreview.com/s/518606/printed-graphene-transistors-promise-a-flexible-electronic-future/. Accessed June 2016

789. Eda, G., Fanchini, G., Chhowalla, M.: Large-area ultrathin films of reduced graphene oxide as a transparent and flexible electronic material. Nat. Nanotechnol. **3**(5), 270–274 (2008)

790. The physics arXiv blog: Graphene transistors clocked at 26 GHz. http://arxivblog.com/?p=755. Accessed June 2016

791. Traversi, F., Russo, V., Sordan, R.: Integrated complementary graphene inverter. Appl. Phys. Lett. **94**, 223312 (2009)

792. Lin, Y.-M., Dimitrakopoulos, C., Jenkins, K.A., Farmer, D.B., Chiu, H.-Y., Grill, A., Avouris, P.: 100-GHz transistors from wafer-scale epitaxial graphene. Science. **327**(u), 662 (2010)

793. Lin, Y.-M., Valdes-Garcia, A., Han, S.-J., Farmer, D.B., Meric, I., Sun, Y., Wu, Y., Dimitrakopoulos, C., Grill, A., Avouris, P., Jenkins, K.A.: Wafer-Scale Graphene Integrated Circuit. Science. **332**(6035), 1294–1297 (2011)

794. Physics World: Graphene circuit breaks the gigahertz barrier. http://physicsworld.com/cws/article/news/2013/jun/17/graphene-circuit-breaks-the-gigahertz-barrier. Accessed June 2016

795. Skrypnychuk, V., Boulanger, N., Yu, V., Hilke, M., Mannsfeld, S.C.B., Toney, M.F., Barbero, D.R.: Enhanced verticle charge transport in a semiconducting P3HT thin film on single layer graphene. Adv. Funct. Mat. **25**(5), 664–670 (2014)

796. Ming, L., Xin, X., Ulin-Avila, E., Geng, B., Zentgraf, T., Ju, L., Wang, F., Zhang, X.: A graphene-based broadband optical modulator. Nature. **474**, 64–67 (2011)

797. Phys.org: Graphene could lead to faster chips. phys.org/news/2009-03-graphene-faster-chips.html. Accessed June 2016

798. Zhou, Q., Zheng, J., Onishi, S., Crommie, M.F., Zettl, A.K.: Graphene electrostatic microphone and ultrasonic radio. PNAS. **112**(29), 1–5 (2015)

799. Mikhailov, S.A., Ziegler, K.: Nonlinear electromagnetic response of graphene: frequency multiplication and the self-consistent-field effects. J. Phys.: Condens. Matter. **20**, 1–13 (2008)

800. Wang, X., Zhi, L.J., Mullen, K.: Nano Lett. **8**, 323 (2008)

801. Wang, X., Zhi, L., Müllen, K.: Transparent, conductive graphene electrodes for dye-sensitzed solar cells. Nano Lett. **8**(1), 323–327 (2008)

802. Wu, J., Agrawal, M., Becerril, H.A., Bao, Z., Liu, Z., Chen, Y., Peumans, P.: Organic light-emitting diodes on solution-processed graphene transparent electrodes. ACS Nano. **4**(1), 43–48 (2010)

803. New Atlas: First flexible graphene-based display created. http://www.gizmag.com/graphene-flexible-electrophoretic-display/33765/. Accesed June 2016

804. Optics.org: Flexible, inorganic LEDs and solar cells grown on graphene. http://optics.org/news/5/9/45. Accessed June 2016

805. Anguita, J.V., Ahmad, M., Haq, S., Allam, J., Silva, S.R.P.: Ultra-broadband light trapping nusing nanotextured decoupled graphene multilayers. Sci. Adv. **2**(2), e1501238 (2016)

806. Fabro, A., et al.: Graphene-based interfaces do not alter target nerve cells. ACS Nano. **10**(1), 615–623 (2016)

807. Lalwani, G., Henslee, A.M., Farshid, B., Lin, L., Kasper, F.K., Qin, Y.-X., Mikos, A.G., Sitharaman, B.: Two-dimensional nanostructure- reinforced biodegradable polymeric nanocomposites for bone tissue engineering. Biomacromolecules. **14**(3), 900–909 (2013)

808. Kanakia, S., Toussaint, J.D., Choudhury, S.M., Lalwani, G., Tembulkar, T., Button, T., Shroyer, K.R., Moore, W., Sitharaman, B.: Physicochemical characterization of a novel graphene-based magnetic resonance imaging contrast agent. Int. J. Nanomed. **8**, 2821–2833 (2013)

809. Lalwani, G., Sundaraj, J.L., Schaefer, K., Button, T., Sitharaman, B.: Synthesis, characterization, in vitro phantom imaging, and cytotoxicity of a novel graphene-based multimodal magnetic resonance imaging – X-ray computed tomography contrast agent. J. Mater. Chem. B Mater Biol. Med. **2**(22), 3519–3530 (2014)

810. Abdul, K.R., Kafafy, R., Salleh, H.M., Faris, W.F.: Enhancing the efficiency of polymerase chain reaction using graphene nanoflakes. Nanotechnol. **23**(45), 455106 (2012)

811. Min, S.K., Kim, W.Y., Cho, Y., Kim, K.S.: Nat. Nanotechnol. **6**, 162 (2011)

812. Zhu, S., Zhang, J., Qiao, C., Tang, S., Li, Y., Yuan, W., Li, B., Tian, L., Liu, F., Hu, R., Gao, H., Wei, H., Zhang, H., Sun, H., Yang, B.: Chem. Commun. **47**, 6858 (2011)

813. Hu, S.H., Chen, Y.W., Hung, W.T., Chen, I.W., Chen, S.Y.: Adv. Mater. **24**, 1748 (2012)

814. Mao, X., Tian, D., Li, H.: Chem. Commun. **48**, 4851 (2012)

815. Das, T.K., Prusty, S.: Graphene-based polymer composites and their applications. Polym.-Plast. Technol. Eng. **52**, 319–331 (2013)

816. Yan, Z., et al.: Rebar Graphene. ACS Nano. **8**(5), 5061–5068 (2014)

817. Kim, T., Park, J., Sohn, J., Cho, D., Jeon, S.: Bioinspired, highly stretchable, and conductive dry adhesives based on 1D-2D hybrid carbon nanocomposites for all-in-one ECG electrodes. ACS Nano. **10**(4), 4770–4778 (2016)

818. Cohen-Tanugi, D., Grossman, J.C.: Water desalination across nanoporous graphene. Nano Lett. **12**(7), 3602–3608 (2012)

819. Romanchuk, A.Y., Slesarev, A.S., Kalmykov, S.N., Kosynkint, D.V., Tour, J.M.: Graphene oxide for effective radionuclide removal. Phy. Chem. Chem. Phy. **15**, 2321–2327 (2013)

820. Rice University News & Media: Another tiny miracle: graphene oxide soaks up radioactive waste. http://news.rice.edu/2013/01/08/another-tiny-miracle-graphene-oxide-soaks-up-radioactive-waste/. Accessed June 2016

821. Xu, J., Wang, L., Zhu, Y.: Langmuir. **28**, 8418 (2012)

822. Ohta, T., Bostwick, A., Seyller, T., Horn, K., Rotenberg, E.: Science. **313**, 951 (2006)
823. Zhao, G., Li, J., Ren, X., Chen, C., Wang, X.: Environ. Sci. Technol. **45**, 10454 (2011)
824. Zhao, G., Jiang, L., He, Y., Li, J., Dong, H., Wang, X., Hu, W.: Adv. Mater. **23**, 3959 (2011)
825. Chandra, V., Kim, K.S.: Chem. Commun. **47**, 3942 (2011)
826. Geng, Z., Lin, Y., Yu, X., Shen, Q., Ma, L., Li, Z., Pan, N., Wang, X.J.: Mater. Chem. **22**, 3527 (2012)
827. Chandra, V., Park, J., Chun, Y., Woo Lee, J., Hwang, I.C., Kim, K.S.: ACS Nano. **4**, 3979 (2010)
828. Zhang, J., Xiong, Z., Zhao, X.S.J.: Mater. Chem. **21**, 3634 (2011)
829. Chandra, V., Yu, S.U., Kim, S.H., Yoon, Y.S., Kim, D.Y., Kwon, A.H., Meyyappan, M., Kim, K.S.: Chem. Commun. **48**, 735 (2012)
830. Mishra, A.K., Ramaprabhu, S.J.: Mater. Chem. **22**, 3708 (2012)
831. Phys.org: Graphene proves a long-lasting lubricant. http://phys.org/news/2014-10-graphene-long-lasting-lubricant.html. Accessed June 2016
832. Wei, Y., Xie, H.: Significant thermal conductivity enhancement for nanofluids containing graphene nanosheets. Phys. Lett. A. **375**(10), 1323–1328 (2011)
833. R & D Magazine: Graphene-copper sandwich may improve, shrink electronics. http://www.rdmag.com/news/2014/03/graphene-copper-sandwich-may-improve-shrink-electronics. Accessed June 2016
834. Bomgardner, M.M.: Biobased firms win funding. Chem. Eng. News. **93**(9), 15 (2015)
835. Jia, C., Migliore, A., Xin, N., Huang, S., Wang, J., Yang, Q., Shuopei, W., et al.: Covalently bonded single-molecule junctions with stable and reversible photoswitched conductivity. Science. **352**(6292), 1443–1445 (2016)
836. Cambridge Dictionary: Graphene. http://dictionary.cambridge.org/dictionary/english/graphene?a=british. Accessed June 2016
837. Wessling, B.: Synth. Met. **93**, 143 (1998)
838. Wallace, G.G., Spinks, G.M., Teasdale, P.R.: Conductive electroactive polymers: intelligent materials systems. Technomic (1996)
839. Malinauskas, A., Malinauskiene, J., Ramanavičius, A.: Conducting polymer-based nanostructurized materials: electrochemical aspects. Nanotechnol. **16**, R51–R62 (2005)
840. Li, C., Bai, H., Shi, G.: Conducting polymer nanomaterials: electrosynthesis and applications. Chem. Soc. Rev. **38**(8), 2397–2409 (2009)
841. Ito, T., Shirakawa, H., Ikeda, S.: J. Plym. Sci. Plym. Chem. Edu. **12**, 11 (1974)
842. Letheby, H.: J. Chem. Soc. **15**, 161 (1862)
843. Mohlner, D.M., Adams, R.N., Arlgersinger Jr., W.J.: J. Am. Chem. Soc. **84**, 3618 (1962)
844. Gardini, G.P.: Adv. Heterocycl. Chem. **15**, 67 (1973)
845. Angeli, A.: Gazz. Chim. Ital. 46, 279 (1916) (1916)
846. Lund, H.: Acta Chem. Scand. **11**, 1323 (1957)
847. Peover, M.E., White, B.S.: J. Electroanal. Chem. **13**, 93 (1967)
848. Osa, T., Yildiz, A., Kuwana, T.: J. Am. Chem. Soc. **91**, 3994 (1969)
849. Armour, M., Gavies, A.G., Upadhyay, J., Wasserman, A.: J. Polym. : Sci. **A1**, 1527 (1967)
850. Jozefowicz, M., Yu, L.T., Belorgey, G., Buvet, R.: J. Polym. Sci. Part C. **16**, 2943 (1969)
851. Dall'Ollio, A., Dascola, Y., Varacca, V., Bocchi, V.: Comptes Rendus. **C267**, 433 (1968)
852. (a) Chiang, C.K., Fincher, C.R., Park, Y.W., Heeger, A.J., Shirakawa, H., Louis, E.J., Gau, S.C., MacDiarmid, A.G.: Phys. Rev. Lett. **39**, 1089 (1977); (b) Genies, W.M., Bidan, G., Diaz, A.: J. Electroanal. Chem. **149**, 101 (1983)
853. (a) Diaz, A.F., Kanazawa, K.K., Gardini, G.P.: J. Chem. Soc. Chem. Commun. 635 1979; (b) Scott, J.C., Pfluger, P., Krounbi, M.T., Street, G.B.: Phys. Rev. B **28**, 2140 (1983); (c) Chen, J., Heeger, A.J.: Synth. Met. **24**, 311 (1988)
854. Salzner, U.: Electronic structure of conducting organic polymers: insights from time-dependent density functional theory. WIREs Comput. Mol. Sci. **4**(6), 601–622 (2014)
855. Brazovskii, S., Kirova, N.: Physical theory of scitons in conducting polymers. Chem. Soc. Rev. **39**, 2453–2465 (2010)

856. Brédas, J.L., Street, G.B.: Acc. Chem. Res. **18**, 309 (1985)

857. Brédas, J.L., Scott, J.C., Yakushi, K., Street, G.B.: Phys. Rev. **30**, 1023 (1984)

858. Guay, J., Leclerc, M., Dao, L.H.: J. Electroanal. Chem. **251**, 31 (1988)

859. Kitani, A., Yano, J., Kunai, A., Sasaki, K.: J. Electroanal. Chem. **221**, 69 (1987)

860. Basescu, N., Liu, Z.-X., Moses, D., Heeger, A.J., Naarmann, H., Theophilou, N.: Long mean free path coherent transport in doped polyacetylene. In: Kuzmany, H., Mehring, M., Roth, S. (eds.) Electronic properties of conjugated polymers, vol. 76, p. 18. Springer, Berlin/Hedelberg (1987)

861. Laks, B., Galvao, D.S.: Phys. Rev. B: Condens. Matter. **56**, 967 (1997)

862. Chance, R.R., Brédas, J.L., Silbey, R.: Phys. Rev. B. **29**, 4491 (1984)

863. Schimmel, T., Gläser, D., Schwoerer, M., Naarmann, H.: Properties of highly conducting polyacetylene. In: Brédas, J.L., Silbey, R. (eds.) Conjugated polymers: the novel science and technology of highly conducting and nonlinear optically active materials, p. 49. Kluwer Academic Publishers, Norwell (1991)

864. Kuwabara, M., Abe, S., Ono, Y.: Synth. Met. **85**, 1109 (1997)

865. Roth, S., Kaiser, M., Reichenbach, J.: Physica Scripta. **T45**, 230 (1992)

866. Bott, D.C.: Structural basis for semiconducting and metallic polymers. In: Skotheim, T.A. (ed.) Handbook of conducting polymers, vol. 2, p. 1191. Marcel Dekker, Inc., New York (1986)

867. Roth, S.: Conductive polymers in molecular electronics: conductivity and photoconductivity. In: Salaneck, W.R., Clark, D.T., Samuelsen, E.J. (eds.) Science and applications of conducting polymers, p. 129. Adam Hilger, New York (1991)

868. Kivelson, S., Heeger, A.J.: Synth. Met. **22**, 371 (1988)

869. Pietronero, L.: Synth. Met. **8**, 225 (1983)

870. Mao, H., Pickup, P.G.: J. Am. Chem. Soc. **112**, 1776 (1990)

871. Aldissi, M.: Transport properties-semiconductor to metal transition. In: Inherently conducting polymers: processing, fabrication, applications, limitations, p. 43. Noyes Data Corporation, New Jersey (1989)

872. Rehwald, W.; Kiessm H.G. Charge transport in polymers. In Conjugated conducting polymers, Springer Ser. Solid-State Sci.; Kiess, H. Ed.; Springer, New York, 1992; Vol. 102, p. 158.

873. Ochmanska, J., Pickup, P.G.: J. Electroanal. Chem. **297**, 211 (1991)

874. Mott, N.F., Davis, E.A.: Electronic processes in non-crystalline materials, 3rd edn. Clarendon, Oxford (1979)

875. Schäfer-Siebert, D., Roth, S.: Limitation of the conductivity of polyacetylene by conjugational defects. Synthetic Metals. **28**(3), (1989)

876. Hauser, J.J.: J. Non-Cryst. Solids **23**, 21 (1977); (b) Singh, R., Narula, A.K., Tandon, R.P., Mansingh, A., Chandra, S.: J. Appl. Phys. **81**, 3726 (1997); (c) Sheng, P.: Phys. Rev. B **21**, 2180 (1980)

877. Paasch, G., Lehmann, G., Wuckel, H.: Properties of highly conducting polyacetylene. In: Brédas, J.L., Silbey, R. (eds.) Conjugated polymers: the novel science and technology of highly conducting and nonlinear optically active materials, p. 49. Kluwer Academic Publishers, Dordrecht (1991)

878. Kivelson, S.: Phys. Rev. B. **21**, 3798 (1982)

879. Kivelson, S., Heeger, A.J.: Phy. Rev. Lett. **55**, 308 (1985)

880. Heeger, A.J.: Polyacetylene, new concepts and new phenomena. In: Skotheim, T.A. (ed.) Handbook of conducting polymers, vol. 2, p. 729. Marcel Dekker, Inc., New York (1986)

881. Yoon, C.O., Reghu, M., Moses, D., Heeger, A.J., Cao, Y., Chen, T.-A., Wu, X., Rieke, R.D.: Synth. Met. **75**, 229 (1995)

882. Plocharski, J.: Mechanisms of conductivity in conjugated polymers and relations to moprhology. In: Plocharski, J., Roth, S. (eds.) Material science forum, vol. 42, p. 17. Trans Tech Publications, Switzerland (1989)

883. Roth, S.: Conducting polymers- present state of physical understanding. In: Plocharski, J., Roth, S. (eds.) Materials science forum, vol. 42, p. 1. Trans tech Publications, Switzerland (1989)

884. Pfluger, P.: Electronic structure and transport in the organic 'amorphous semi conductor' polypyrrole. In: Skotheim, T.A. (ed.) Handbook of conducting polymers, vol. 2, p. 1369. Marcel Dekker, Inc., New York (1986)

885. Kanazawa, K.K., Diaz, A.F., Gill, W.D., Grant, P.M., Street, G.B., Gardini, G.P., Kwak, J. F.: Synth. Met. **1**, 329 (1979)

886. Epstein, A.J.: AC conductivity of polyacetylene: distinguishing mechanisms of charge transport. In: Skotheim, T.A. (ed.) Handbook of conducting polymers, vol. 2, p. 1047. Marcel Dekker, Inc., New York (1986)

887. Sariciftci, N.S., Kobryanskii, V.M., Reghu, M., Smilowitz, L., Halvorson, C., Hagler, T.W., Mihailovic, D., Heeger, A.J.: Synth. Met. **53**, 161 (1993)

888. Epstein, A.J., MacDiarmid, A.G.: The controlled electromagnetic response of polyanilines and its application to technologies. In: Salaneck, W.R., Clark, D.T., Samuelsen, E.J. (eds.) Science and applications of conducting polymers, p. 141. Adam Hilger, New York (1991)

889. (a) Wang, Z.H., Ray, A., MacDiarmid, A.G., Epstein, A.J.: Phys. Rev. B **43**, 4373 (1991); (b) Courves, L.D., Porter, S.J.: Synth. Met. **28**, C761 (1989)

890. Wang, Z.H., Scherr, E.M., MacDiarmid, A.G., Epstein, A.J.: Phys. Rev. B. **45**, 4190 (1992)

891. Burns, A., Wang, Z.H., Du, G., Joo, J., Epstein, A.J., Osaheni, J.A., Jenekhe, S.A., Wang, C. S.: In: Chiang, L.Y., Garito, A.F., Sandman, D.J. (eds.) Electrical, optical, and magnetic properties of organic solid state materiels. IN Mat. Res. Soc. Symp, vol. 247, p. 735. Materials Reasearch Society, Pittsburgh (1992)

892. Heeger, A.J.: Conducting polymers: The route from fundamental science to technology. In: Salaneck, W.R., Clark, D.T., Samuelsen, E.J. (eds.) Science and applications of conducting polymers, p. 1. Adam Hilger, New York (1991)

893. Epstein, A.J., Rommelmann, H., Abkowitz, M., Gibson, H.W.: Phys. Rev. Lett. **47**, 1549 (1981)

894. Javadi, H.H.S., Cromack, K.R., MacDiarmid, A.G., Epstein, A.J.: Phys. Rev. B. **39**, 3579 (1989)

895. Lee, K., Reghu, M., Yuh, E.L., Sariciftci, N.S., Heeger, A.J.: Synth. Met. **68**, 287 (1995)

896. Sieger, K., Gill, W.D., Clark, T.C., Street, G.B.: Am. Phys. Soc. (1978)

897. Reghu, M., Yoon, C.O., Moses, D., Cao, Y., Heeger, A.J.: Synth. Met. **69**, 329 (1995)

898. (a) Basescu, N., Liu, Z.-X., Moses, D., Heeger, A.J., Naarmann, H., Theophilou, N.: Nature **327**, 403 (1987); (b) Maddison, D.S., Unsworth, J.: Synth. Met. **22**, 257 (1988)

899. Ferraris, J.P., Webb, A.W., Weber, D.C., Fox, W.B., Carpenter, E.R., Brant, P.: Solid State Commun. **35**, 15 (1980)

900. Fukuhara, T., Masubuchi, S., Kazama, S.: Synth. Met. **92**, 229 (1998)

901. Bao, Z.-X., Liu, C.X., Pinto, N.J.: Synth. Met. **87**, 147 (1997)

902. Heeger, A.J., Smith, P.: Solution processing of conducting polymers: opportunities for science and technology. In: Brédas, J.L., Silbey, R. (eds.) Conjugated polymers: the novel science and technology of highly conducting and nonlinear optically active materials, p. 141. Kluwer Academic Publishers, Norwell (1991)

903. MacDiarmid, A.G., Epstein, A.J.: The polyanilines: potential technology based on new chemistry and new properties. In: Salaneck, W.R., Clark, D.T., Samuelsen, E.J. (eds.) Science and applications of conducting polymers, p. 117. Adam Hilger, New York (1991)

904. Burroughes, J.H., Friend, R.H.: The semiconductor device physics of polyacetylene. In: Brédas, J.L., Silbey, R. (eds.) The novel science and technology of highly conducting and nonlinear optically active materials, p. 555. Kluwer Academic Publishers, Norwell (1991)

905. Aldissi, M.: Doping of conjugated polymers: conducting polymers. In: Inherently conducting polymers: processing, fabrication, applications, limitations, vol. 989, p. 40. Noyes Data Corporation, New Jersey (1989)

906. (a) Paloheimo, J., Punkka, E., Stubb, H., Kuivavainen, P.: Polymer field-effect transistors for transport property studies. In: Metzger, R.M. (ed.) Proceedings of NATO ASI conference on Lower Dimensional Systems and Molecular Electronics, Soetses, Greece, June 12–23, p. 989. Plenum Press, New York (1990);(b) Punkka, E.; Isotalo, M.; Ahlskog, M.; Subb, H. Effects of humidy and heat on the conductivity of Poly(3-Alkylltheriophenes), Espoo. Finland.

907. Sato, M.-A., Tanaka, S., Kaeriyama, K.: J. Chem. Soc., Chem. Commun. **713**, (1985)
908. Sato, M.-A., Tanaka, S., Kaeriyama, K.: Synth. Met. **18**, 229 (1987)
909. Peierls, R.E.: Quantum theory of solids, p. 108. Clarendon, Oxford (1955)
910. Dai, Y., Chodhury, S., Blaisten-Barojas, E.: Density functional theory study of the structure and energetics of negatively charged pyrrole oligomers. Quantum Chem. **111**(10), 2295–2305 (2011)
911. Tretiak, S., Igumenshchev, K., Chernyak, V.: Exciton sizes of conducting polymers predicted by time-dependent density functional theory. Phys. Rev. B. **71**, 033201 (2005)
912. Whangbo, M.-H., Hoffmann, R., Woodward, R.B.: Proc. Royal Soc. Lond. A. **366**, 23 (1979)
913. Hoffmann, R.: J. Chem Phys. **39**, 1397 (1963)
914. Delhalle, J., Delhalle, S., André, J.M., Pireaux, J.J., Saudano, R., Verbist, J.J.: J. Electron. Spectrosc. Relat. Phenom. **12**, 293 (1977)
915. Elsenbaumer, R.L., Marynick, D.S., Seong, S., Meline, R.L.: Sulfur containing conjugated polymers with interesting electronic properties. In: Garito, A.F., Jen, A.K.-Y., Lee, C.Y.-C., Dalton, L.R. (eds.) Mat. Res. Soc. Symp. Proc., Electrical, electrical, optical, and magnetic properties of organic solid state materials, vol. 328, p. 221. Materials Research Society, Pittsburgh (1994)
916. Pomerantz, M., Wang, J., Seong, S., Starkey, K.P., Nugyen, L., Marynick, D.S.: A new dithiophene fused p-phenylene vinylene vonducting polymer. Synthesis and study. In: Mat. Res. Soc. Symp. Proc., Electrical optical, and magnetic properties of organic solid state materials, vol. 328, p. 227. Materials Research Society, Pittsburgh (1994)
917. Beck, F.: Metalloberflaeche. **46**, 177 (1992)
918. Dewar, M.J.S., Zoebisch, E.G., Healy, E.F., Stewart, J.F.: J. Am. Chem. Soc. **107**, 3092 (1985)
919. Ford, W.K., Duke, C.B., Paton, A.: J. Chem. Phys. **77**, 4564 (1982)
920. Ford, W.K., Duke, C.B., Salaneck, W.R.: J. Chem. Phys. **77**, 5030 (1982)
921. Su, W.P., Schrieffer, J.R., Heeger, A.J.: Phys. Rev. B. **22**, 2099 (1980)
922. Su, W.P., Schrieffer, J.R., Heeger, A.J.: Phys. Rev. Lett. **42**, 1698 (1979)
923. (a)Kiess, H.G.: Conjugated conducting polymers. In: Kiess, H. (ed.) Springer Ser. Sold-State Sci, vol. 102, p. 1. Springer (1992.); (b) Baeriswyl, D., Campbell, D.K., MAzumdar, S: An overview of the theory of pi-conjugated polymers. In: Conjugated conducting polymers, Springer Ser. Solid-State Sci., Vol. 102, p. 7. Springer, New York (1992)
924. Mazumdar, S., Chandross, M.: Theory of photoexcitations in phenylene-based polymers. In: Yang, S.C., Chandrasekhar, P. (eds.) Proc. SPIE, 2428: optical and photonic applications of electroactive and conducting polymers, vol. 1995, p. 62. SPIE-The International Society for Optical Engineering, Bellingham (2528)
925. Grant, P.M., Batra, I.P.: Solid State Commun. **29**, 225 (1979)
926. Brédas, J.L., Chance, R.R., Silbey, R.: Phys. Rev. B. **26**, 5843 (1982)
927. Kertesz, M., Hughbanks, T.R.: Synth. Met. **69**, 699 (1995)
928. Brédas, J.L.: Electronic structure of highly conducting polymers. In: Skotheim, T.A. (ed.) Handbook of conducting polymers, vol. 2, p. 859. Marcel Dekker, Inc., New York (1986)
929. Pollak, M., Knotek, M.L.: J. Non-Cryst. Solids. **32**, 141 (1979)
930. (a) Yamamoto, T., Sanechika, K., Yamamoto, A.: J. Polym. Sci. Polym. Lett. Ed. **18**, 9 (1980); (b) Lin, J.W.P., Dudek, L.P.: J. Polym. Sci. Polym. Lett. Ed. **18**, 2869 (1980)
931. Brédas, J.L., Heeger, A.J., Wudl, F.: J. Chem. Phys. **85**, 4673 (1986)
932. Tanaka, C., Tanaka, J.: Energy band structure for metallic polyacetylene. In: Chiang, L.Y., Garito, A.F., Sandman, D.J. (eds.) Mat. Res. Soc. Symp. Proc., Electrical, optical, and magnetic properties of organic solid state materials, vol. 247, p. 577. Materials Research Society, Pittsburgh (1992)
933. Brédas, J.L., Thémans, B., Fripiat, J.G., André, J.M., Chance, R.R.: Phys. Rev. B. **29**, 6761 (1984)

934. Lazzaroni, R., Rachidi, S., Brédas, J.L.: Theoretical investigation of chain flexibility in polythiophene and polypyrrole. In: Salaneck, W.R., Clark, D.T., Samuelsen, E.J. (eds.) Science and applications of conducting polymers, p. 13. Adam Hilger, New York (1991)

935. Mele, E.J.: Phonmons and the Peierls instability in polyacetylene. In: Skotheim, T.A. (ed.) Handbook of conducting polymers, vol. 2, p. 795. Marcel Dekker, Inc., New York (1986)

936. Stafström, S.: Electronic properties of heavily doped trans-polyacetylene. In: Brédas, J.L., Silbey, R. (eds.) Conjugated polymers: the novel science and technology of highly conducting and nonlinear optically active materials, p. 113. Kluwer Academic Publishers, Norwell (1991)

937. Chandross, M., Shimoi, Y., Mazumdar, S.: Synth. Met. **85**, 1001 (1997)

938. Gallagher, F.B., Spano, F.C.: Synth. Met. **85**, 1007 (1997)

939. Senevirathne, M.S., Nanayakkara, A., Senadeera, G.K.R.: A theoretical investigation of band gaps of conducting polymers with heterocycles. J. Natn. Sci. Foundation Sri Lanka. **39** (2), 183–185 (2011)

940. Thémans, B., Salaneck, W.R., Brédas, J.L.: Synth. Met. **28**, C359 (1989)

941. (a) Nicholas, G., Durand, P.: J. Chem. Phys. **70**, 2020 (1979); **72**, 453 (1980); (b) André, J. M., Burke, L.A., Delhalle, J., Nicolas, G., Durand, P.: Int. J. Quantum Chem. Symp. **13**, 283 (1979)

942. Brédas, J.L., Chance, R.R., Silbey, R., Nicolas, G., Durand, P.: J. Chem. Phys. **75**, 255 (1981)

943. Delamer, M., Lacaze, P.C., Dumousseau, J.Y., Dubois, J.E.: Electrochim. Acta. **27**, 61 (1982)

944. Tanaka, J., Tanaka, M.: Optical spectra of conducting polymers. In: Skotheim, T.A. (ed.) Handbook of conducting polymers, vol. 2, p. 1269. Marcel Dekker, Inc., New York (1986)

945. Mazumdar, S., Chandross, M.: Theory of photoexcitations in phenylene-based polymers. In: Yang, S.C., Chandrasekhar, P. (eds.) Proc. SPIE, 2528: optical and photonic applications of electroactive and conducting polymers, p. 62. SPIE-The International Society for Optical Engineering, Bellingham (1995)

946. Furukawa, Y., Tazawa, S., Fuji, Y., Harada, I.: Synth. Met. **24**, 329 (1988)

947. Yano, J.L J. Electrochem. Soc. **144**, 477 (1997)

948. Inganäs, O.: Electroactive polymers in large area chromogenics. In: Lampert, C.M., Granqvist, C.G. (eds.) Proc. SPIE, Large-area chromogenics: materials and devices for transmittance control, vol. IS4, p. 328. SPIE Optical Engineering Press, Bellingham (1990)

949. Mastragostino, M.: Electrochromic devices. In: Scrosati, B. (ed.) Applications of electroactive polymers, p. 223. Chapman & Hall, New York (1993)

950. Kim, E., Lee, K.-Y., Lee, M.-H., Shin, J.-S., Rhee, S.B.: Synth. Met. **85**, 1367 (1997)

951. Yamasaki, S., Terayama, K., Yano, J.: J. Electrochem. Soc. **143**, L212 (1996)

952. Tourillon, G.: Polythiophene and its dervatives. In: Skotheim, T.A. (ed.) Handbook of conducting polymers, vol. 1, p. 293. Marcel Dekker, Inc., New York (1986)

953. Neugebauer, R., Neckel, A., Brinda-Konopil, N.: In situ infrared spectro-electrochemical investigations of polythiophenes. In: Kuzmany, H., Mehring, M., Roth, S. (eds.) Electronic properties of polymers and related compounds, vol. 69, p. 226. Springer, New York

954. Meador, M.A., Gaier, J.R., Good, B.S., Sharp, G.R., Meador, M.A.: A review of properties and potential aerospace applications of electrically conducting polymers. In: Internal report national aeronautics and space administration, pp. 1–21. Lewis Research Center, Cleveland (1989)

955. Yang, S.C.: Conducting polymer as electrochromic material: polyaniline. In: Lampert, C.M., Granqvist, C.G. (eds.) Proc. SPIE, Large-area chromogenics, materials and devices for transmittance control, vol. IS4, p. 335. SPIE Optical Engineering Press, Washington, DC (1990)

956. Chandrasekhar, P., Gumbs, R.W.: J. Electrochem. Soc. **138**, 1337 (1991)

957. Onoda, M., Iwasa, T., Kawai, T., Nakayama, J., Nakahara, H., Yoshino, K.: J. Electrochem. Soc. **140**, 397 (1993)

958. Guay, J., Paynter, R., Dao, L.H.: Macromolecules. **23**, 3598 (1990)

959. Duffie, J.A., Beckman, W.A.: Radiation characteristics of opaque materials. In: Solar engineering of thermal processes, p. 184. Wiley, New York (1991)

960. Wake, L.V.: Principles and formulation of solar reflecting and low IR emitting coatings for defense use, p. AD-A218429. Avail. Fr. Defense Technical Information Center, Washington, DC (1989)

961. Inganäs, O., Gustafsson, G., Salaneck, W.R.: Synth. Met. **28**, C377 (1989)

962. Gustafsson, G., Inganäs, O., Salaneck, W.R., Laakso, J., Loponen, M., Taka, T., Österholm, J.-E., Stubb, H., Hjertberg, T.: Processable conducting poly (3-alkylthiopenes). In: Brédas, J.L., Silbey, R. (eds.) Conjugated polymers: the novel science and technology of highly conducting and nonlinear optically active materials, p. 315. Kluwer Academic Publishers, Norwell (1991)

963. Inganäs, O., Salaneck, W.R., Österholm, J.-E., Laakso, J.: Synth. Met. **22**, 395 (1988)

964. Hirota, N., Hisamatsu, N., Maeda, S., Tsukahara, H., Hyodo, K.: Synth. Met. **80**, 67 (1996)

965. Patil, A.O.: Synth. Met. **28**, C495 (1989)

966. Liu, M., Gregory, R.V.: Synth. Met. **72**, 45 (1995)

967. Lanzi, M., Bizzarri, P.C., Casa, C.D.: Synth. Met. **89**, 181 (1997)

968. Chandrasekhar, P., Zay, B.J., Lawrence, D., Caldwell, E., Sheth, R., Stephan, R., Cornwell, J.: Variable-emittance IR-electrochromic skins combining unique conducting polymers, ionic liquid electrolytes, microporous polymer membranes and semiconductor/polymer coatings, for spacecraft thermal control. Appl. Polymer. **131**(19), 40850 (2014)

969. Quill Work V.4

970. Diaz, A.F., Castillo, J.I., Logan, J.A., Lee, W.-Y.: J. Electroanal. Chem. **129**, 115 (1981)

971. Kawai, T., Iwasa, T., Onada, M., Sakamoto, T., Yoshino, K.: J. Electrochem. Soc. **139**, 3404 (1992)

972. Diaz, A.F., Bargon, J.: Electrochemical synthesis of conducting polymers. In: Skotheim, T.A. (ed.) Handbook of conducting polymers, vol. 1, p. 81. Marcel Dekker, Inc., New York (1986)

973. Gopal, J., Vanhouten, K., Cowan, D.O., Poehler, T.O., Madsen, P.V., Searson, P.C.: Diether substitued polyanilines: Novel electrode materials. In: Chiang, L.Y., Garito, A.F., Sandman, D.J. (eds.) Mat. Res. Soc. Symp. Proc.: electrical, optical, and magnetic properties of organic solid state materials, vol. 247, p. 607. Materials Research Society, Pittsburgh (1992)

974. Goldenberg, L.M., Petty, M.C., Monkman, A.P.: J. Electrochem. Soc. **141**, 1573 (1994)

975. Guay, J., Dao, L.H.: J. Electroanal. Chem. **274**, 135 (1989)

976. Arbizzani, C., Mastragostino, M., Dellepiane, G., Piaggio, P.: Chemical and electrochemical doping of PPS in sulfuric acid. In: Chiang, L.Y., Garito, A.F., Sandman, D.J. (eds.) Mat. Res. Soc. Symp. Roc.: electrical, optical, and magnetic properties of organic solid State materials, vol. 247, p. 717. Materials Research Society, Pittsburgh (1992)

977. (a) Oyama, N., Sato, M., Ohsaka, T.: Synth. Met. **29**, E501 (1989). (b) Helbig, M., Hörhold. Elec. Prop. of Polymers **107**, 321 (1992)

978. Chandrasekhar, P., Zay, B.J., Cai, C., Chai, Y., Lawrenced, D.: Matched-dual-polymer elecctrochromic lesnes, using new cathodically coloring conducting polymers, with exceptional performance and incorporated into automated sunglasses. J. Appl. Polym. Sci. **131**, 547–557 (2014.) 41043-1 – 41043-21

979. (a) Camurlu, P., Toppare, L.: J. Macromol. Sci. Pure Appl. Chem. **43**, 449 (2006); (b) Gazotti, W.A., Casalbore-Miceli, G., Geri, A., De Paoli, M.-A: Adv. Mater. **10**, 60 (1998)

980. Pickup, P.G., Osteryoung, R.A.: J. Am. Chem. Soc. **195**, 271 (1985)

981. LaCroix, J.C., Diaz, A.F.: Makromol. Chem., Macromol. Symp. **8**, 17 (1987)

982. LaCroix, J.C., Diaz, A.F.: J. Electrochem. Soc. **135**, 1457 (1988)

983. LaCroix, J.C., Kanazawa, K.K., Diaz, A.: J. Electrochem. Soc. **136**, 1308 (1989)

984. Oyama, N., Hirabayashi, K., Ohsaka, T.: Bull. Chem. Soc. Jpn. **59**, 2071 (1986)

985. Bedioui, F., Bernard, P., Moisy, P., Bied-Charreton, C., Devynck, J.: Poly(Pyrrole-Cobaltpoprhyrin) film modified rlectrodes: preparation and catalytic application. In: Plocharski, J., Roth, S. (eds.) Materials science forum, vol. 42, p. 221. Trans Tech Publications, Switzerland (1989)

986. Levi, M.D.: Mechanism and kinetics of dark redox reactions at polythiophene film electrodes. In: Plocharski, J., Roth, S. (eds.) Materials science forum, vol. 42, p. 101. Trans Tech Publications, Switzerland (1989)

987. Gardini, G.P.: The oxidation of monocyclic pyrroles. Adv. Heterocycl. Chem. **15**, 67–98 (1973)

988. Zahradnik, R.: In: Snyder, J.P. (ed.) Nonbenzenoid aromatic compounds, pp. 1–80. Academic Press, Inc., New York (1971)

989. Bargon, J., Mohmand, S., Waltman, R.J.: IBM J. Res. Develop. **27**, 330 (1983)

990. Tourillon, G., Garnier, F.: J. Electroanal. Chem. **161**, 51 (1984)

991. Tourillon, G., Garnier, F.: J. Electroanal. Chem. **135**, 173 (1982)

992. Street, G.B., Clarke, T.C.: IBM J. Res. Dev. **25**, 51 (1981)

993. Delamer, M., Lacaze, P.C., Dumousseau, J.Y., Dubois, J.E.: Electrochim. Acta. **27**, 61 (1982)

994. Walker, D.G., Wilson, N.E., Jr.: U.S. Patent 3,437,569, (1969); Wisdon, N.E., Jr.: U.S. Patent 3,437,57, 1969

995. Street, G.B., Clarke, T.C., Krounbi, M., Kanazawa, K.K., Lee, V., Pfluger, P., Scott, J.C., Weiser, G.: Mol. Cryst. Liq. Cryst. **83**, 253 (1982)

996. Street, G.B., Geiss, R.H., Lindsey, S.E., Nazzal, A., Pfluger, P.: In: Reineker, P., Haken, H., Wolf, H.C. (eds.) Proc. Conf. Electronic excitation interaction processes Org. Molec. aggregates, p. 242. Springer, New York (1983)

997. Diaz, A.F., Crowley, J.I., Bargon, J., Gardini, G.P., Torrance, J.B.: J. Electroanal. Chem. **121**, 355 (1981)

998. Ohsaka, T., Hirabayashi, K., Oyama, N.: Bull. Chem. Soc. Jpn. **59**, 3423 (1986)

999. (a) Chandrasekhar, P.: Flexible electrochromic window materials based on Poly(Diphenyl Amine) and related conducting polymers, Final Technical Report, Grant No. DE-FG05-93ER81631/A00(1,2,3) for the U.S. Department of Energy, Oak Ridge/Washinton, DC (1998); (b) Chandrasekhar, P.: Flexible electrochromic window materials based on Poly (Diphenyl Amine) and related conducting polymers, Final Report, Grant No. DE-FG05-93ER81631 for the U.S. Department of Energy, Washington, DC (1994)

1000. Hoier, S.N., Park, S.-M.: J. Electrochem. Soc. **140**, 2454 (1993)

1001. Johnson, B.J., Park, S.-M.: J. Electrochem. Soc. **143**, 1277 (1996)

1002. Pickup, P.G., Osteryoung, R.A.: J. Am. Chem. Soc. **106**, 2294 (1984)

1003. (a) Chandrasekhar, P., Masulaitis, A.M., Gumbs, R.W.: Synth. Met. **36**, 303 (1990); (b) Wudl, F., Ikenoue, Y., Patil, A. O. In: Prasad, P.N., Ulrich, D.R. (eds.) Nonlinear optical and electroactive polymers. p. 393 Plenum, New York (1988)

1004. Chandrasekhar, P.: Flexible, visible/IR flat panel electrochromics based on poly (Diphenyl Amine) and related conducting polymers, Final Report, Contract No. N00014-95-C-0069 Office of Naval Research, Arlington, Virginia (1995)

1005. Heinze, J., Dietrich, M.: Cyclic voltammetry as a tool for characterizing conducting polymers. In: Plocharski, J., Roth, S. (eds.) Materials science forum, vol. 42, p. 79. Trans Tech Publications, Switzerland (1989)

1006. Segawa, H., Shimidzu, T., Honda, K.: J. Chem. Soc. Chem. Commun. 132 (1989)

1007. Smyrl, W.H., Lien, M.: Electrical and electrochemical properties of electronically conducting polymers. In: Srosati, B. (ed.) Applications of electroactive polymers, p. 29. Chapman & Hall, New York (1993)

1008. Han, J.H., Motobe, T., Whang, Y.E., Miyata, S.: Synth. Met. **45**, 261–1991

1009. Street, G.B.: Polyrrole from powders to plastics. In: Skotheim, T.A. (ed.) Handbook of conducting polymers, vol. 1, p. 265. Marcel Dekker, Inc., New York (1986)

1010. Krische, B., Zagorska, M.: In: Plocharski, J., Roth, S. (eds.) Materials science forum, vol. 42, p. 79. Transtech Publications, Switzerland (1989)

1011. (a) Kossmehl, G., Chatzitheodorou, G.: Mol. Cryst. Liq. Cryst. **83**, 291 (1982); (b) Kossmehl, G., Chatzitheodorou, G.: Makromol. Chem. Rap. Commun. **2**, 551 (1981)

1012. Angelopoulos, M., Shaw, J.M., Kaplan, R.D., Perrault, S.: J. Vac. Sci. Tech. **B7**, 1519 (1989)

1013. Elsenbaumer, R.L., Jen, K.Y., Oboodi, R.: Synth. Met. **15**, 169 (1986)

1014. Machida, S., Miyata, S., Techagumpuch, A.: Synth. Met. **31**, 311 (1989)

1015. Whang, Y.E., Han, J.H., Nalwa, H.S., Watanabe, T., Miyata, S.: Synth. Met. **41–43**, 3043 (1991)

1016. (a) Katz, T.J., Lee, S.J.: J. Am. Chem. Soc. **102**, 422 (1980); (b) Katz, T.J., Lee, S.J., Shippey, M.A.: J. Mol. Cat. **8**, 219 (1980)

1017. Rubner, M., Deits, W.: J. Polym. Sci. Polym. Chem. Ed. **20**, 2043 (1982)

1018. Masuda, J., Takahashi, T., Higashimura, T.: J. Chem. Soc. Chem. Commun. 1297 (1982)

1019. Thakur, M., Lando, J.B.: Macromolecules **16**, 143 (1983)

1020. Hotta, S., Soga, M., Sonoda, N.: Synth. Met. **26**, 267 (1988)

1021. Feast, W.J.: Synthesis of conducting polymers. In: p (ed.) Handbook of conducting olymers, vol. 1, p. 1. Marcel Dekker, Inc., New York (1986)

1022. Brown, A.R., Greenham, N.C., Gymer, R.W., Pichler, K., Bradley, D.D.C., Friend, R.H., Burn, P.L., Kraft, A., Holmes, A.B.: Conjugated polymer light-emitting diodes. In: Aldissi, M. (ed.) Intrinsically conducting polymers: an emerging technology, p. 87. Kluwer Academic Publisdhers, Boston (1993)

1023. Bao, Z., Yu, L.: In: Yang, S.C., Chandrasekhar, P. (eds.) Optical and photonic applications of electroactive and conducting polymers, vol. 2528. SPIE Optical Engineering Press, Bellingham; Proc. SPIE (1995)

1024. Freund, M.S., Karp, C., Lewis, N.S.: Curr. Sep. **13**, 6 (1994)

1025. Aldissi, M.: J. Polym. Sci. Polym. Lett. Ed. **23**, 167 (1985)

1026. Alva, K.S., Kumar, J., Marx, K.A., Tripathy, S.K.: Macromolecules **30**, 4024 (1997)

1027. Cruz, G.J., Morales, J., Castillo-Ortega, M.M., Olayo, R.: Synth. Met. **88**, 213 (1997)

1028. Schäfer, O., Greiner, A., Pommerehne, J., Guss, W., Vestweber, H., Tak, H.Y., Bässler, H., Schmidt, C., Lüssem, G., Schartel, B., Stümpflen, V., Wendorff, J.H., Spiegel, S., Möller, C., Spiess, H.W.: Synth. Met. **82**,1 (1996)

1029. Francois, B., Mermilliod, N., Zuppiroli, L.: Synth. Met. **4**, 131 (1981)

1030. Liu, J.-M., Sun, L., Hwang, J.-H., Yang, S.C.: Novel template guided synthesis of polyaniline. In: Chiang, L.Y., Garito, A.F., Sandman, D.J. (eds.) Electrical, optical, and magnetic properties of organic solid state materials, vol. 247, p. 601. Materials, Reasearch Society, Pittsburgh (1992)

1031. Shimidzu, T., Iyoda, T., Segawa, H., Fujitsuka, M.: Functionalizations of conducting polymers by mesoscopically structural control and by molecular combination of reactive moiety. In: Aldissi, M. (ed.) Intrinsically conducting polymers: an emerging technology, p. 13. Kluwer Academic Publishers, Boston (1993)

1032. Chiang, C.K., Druy, M.A., Gau, S.C., Heeger, A.J., Louis, E.G., MacDiarmid, A.G., Park, Y.W., Shirakawa, H.: J. Am. Chem. Soc. **100**, 1013 (1978)

1033. Selig, H., Holloway, J.H., Pron, A., Billaud, D.: J. Phys. Paris **C3**, 179 (1983)

1034. Shirakawa, H., Kobayashi, T.: J. Phys. Paris **C3**, 3 (1983)

1035. Billaud, D., Kulszewicz, I., Pron, A., Bernier, P., Lefrant, S.: J. Phys. Paris **C3**, 33 (1983)

1036. Pekker, S., Jánossy, A.: Chemistry of doping and distribution of dopants in polyacetylene. In: Skotheim, T.A. (ed.) Handbook of conducting polymers, vol. 1, p. 45. Marcel Dekker, Inc., New York (1986)

1037. Boils, D., Schue, F., Sledz, J., Giral, L.: J. Phys. Paris **C3**, 189 (1983)

1038. André, J.J., Bernard, M., Francois, B., Mathis, C.: J. Phys. Paris **C3**, 199 (1983)

1039. Francois, B., Mathis, C.: J. Phys. Paris **C3**, 21 (1983)

1040. Pron, A.: Synth. Met. **46**, 277 (1992)

1041. Bidan, G., Blohorn, B., Ehui, B., Lapkowski, M., Kern, J.M., Sauvage, J.P.: Electrochemical behaviour of some hybridized conducting polymers: recent developments. In: Plocharski, J., Roth, S. (eds.) Materials science forum: electrochemistry of conductive polymers, pp. 51–62. Trans Tech Publications, Brookfield (1989)

1042. Girard, F., Ye, S., Bélanger, D.: J. Electrochem. Soc. **142**, 2296 (1995)

1043. Reynolds, J.R., Pyo, M., Qiu, Y.-J.: J. Electrochem. Soc. **141**, 35 (1994)

1044. Hotta S., Hosaka, T., Shimotsuma, W.: Synth. Met. **6**, 69 (1983)

1045. Sung, H., Lee, T., Paik, W.-K.: Synth. Met. **69**, 485 (1995)
1046. Lim, J.Y., Paik, W.-K., Yeo, I.-H.: Synth. Met. **69**, 451 (1995)
1047. Kudoh, Y., Tsuchiya, S., Kojima, T., Fukuyama, M., Yoshimura, S.: Synth. Met. **41–43**, 1133 (1991)
1048. Cao, Y., Smith, P.: Polymer papers. UNIAX Coproration, Santa Barbara (1992)
1049. Armes, S.P., Vincent, B.: J. Chem Soc. Chem. Commun. 288 (1987)
1050. Pryzyluski, J., Zagórska, M., Conder, K.: Pron, A., Polymer **23**, 1872 (1982)
1051. Zagórska, M., Pron, A., Pryzyluski, J., Krische, B., Ahlgren, G.: J.C.S. Chem. Comm. 1125 (1983)
1052. Liu, Y., Xu, Y., Zhu, D.: Synth. Met. **90**, 143 (1997)
1053. Tong, Z.S., Wu, M.Z., Pu, T.S., Zhou, F., Liu, H.Z.: Synth. Met. **68**, 125 (1995)
1054. Clarke, T.C., Krounbi, M.T., Lee, V.Y., Street, G.B.: J. Chem. Soc. Chem. Commun. 384 (1981)
1055. Pitchuman, S., Willig, F.: J. Chem. Soc. Chem. Commun. 809 (1983)
1056. Laakso, J., Österholm, J.-E., Nyholm, P.: Synth. Met. **28**, C467 (1989)
1057. Wang, C.S., Lee, C.Y.-C., Arnold, F.E.: Mechanical and electrical properties of heat-treated ladder polymer fiber. In: Chiang, L.Y., Garito, A.F., Sandman, D.J. (eds.) Electrical, optical, and magnetic properties of organic solid state materials, vol. 247, p. 747. Materials Research Society, Pittsburgh (1992)
1058. Angelopoulos, M., Shaw, J.M., Lee, K.L.: Mat. Res. Soc. Symp. Proc. **214**, 137 (1991)
1059. Hotta, S., Rughooputh, S.D.D.V., Heeger, A.J.: Synth. Met. **22**, 79 (1987)
1060. MacDiarmid, A.G., Epstein, A.J.: Faraday Discuss. Chem. Soc. **88**, 317 (1989)
1061. (a) Galvin, M.E., Wnek, G.E.: Polym. Commun. **23**, 795 (1982); (b) Galvin, M.E., Dandreux, G.F., Wnek, G.E. In: Davidson, T. (ed.) Polymers in electronics. American Chemical Society, Washington, DC, p. 507 (1984)
1062. Wessling, B. In: Electronic properties of conjugated polymers (Kirchberg II). p. 407
1063. Wnek, G.E.: Electrically conductive polymer composites. In: Skotheim, T.A. (ed.) Handbook of conducting polymers, vol. 1, p. 205. Marcel Dekker, Inc., New York (1986)
1064. Yin, X.H., Yoshino, K., Hashizume, K., Isa, I.: Jpn. J. Appl. Phys. Part 1, **36**, 3537 (1997)
1065. MacDiarmid, A.G., Epstein, A.J.: Polyaniline: interrelationships between molecular weight, morpholoy, Donnan potential and conductivity. In: Chiang, L.Y., Garito, A.F., Sandman, D.J. (eds.) Electrical, optical, and magnetic properties of organic solid state materials, vol. 247, p. 545. Materials Research Society, Pittsburgh (1992)
1066. Malhotra, B.D., Ghosh, S., Chandra, R.: J. Appl. Polym. Sci. **40**, 1049 (1990)
1067. Wang, Y., Rubner, M.F.: Fabrication of an electrically conducting full-interpenetrating polymer network. In: Chiang, L.Y., Garito, A.F., Sandman, D.J. (eds.) Electrical, optical, and magnetic properties of organic solid state materials, vol. 27, p. 759. Materials Research Society, Pittsburgh (1992)
1068. Concentrates. Chem. Eng. News, **72**(39), 24 (1994)
1069. Cosnier, S., Innocent, C.: J. Electroanal. Chem. **328**, 361 (1992)
1070. Stanke, D., Hallensleben, M.L., Toppare, L.: Synth. Met. **72**, 95 (1995)
1071. Stanke, D., Hallensleben, M.L., Toppare, L.: Synth. Met. **72**, 89 (1995)
1072. Nazzal, A.I., Street, G.B.: J. Chem. Soc. Chem. Commun. 375 (1985)
1073. Bidan, G., Divisia-Blohorn, B., Kern, J.M., Sauvage, J.P.: J. Chem. Soc. Chem. Commun. 723 (1988)
1074. Demoustier-Champagne, S., Ferain, E., Jerome, C., Jerome, R., Legras, R.: Eur. Polym. J. **34**, 1767 (1998)
1075. Demoustier-Champagne, S., Stavaux, P. Y.: Chem. Mater. **11**, 829 (1999)
1076. Delvaux, M., Duchet, J., Stavaux, P. Y., Legras, R., Demoustier-Champagne, S:, Synth. Met. **113**, 275 (2000)
1077. Mazur, M., Tagowska, M., Palys, B., Jackowska, K.: Electrochem. Commun. **5**, 403 (2003)
1078. Duvail, J. L., Retho, P., Garreau, S., Louarn, G., Godon, C., Demoustier-Champagne, S.: Synth. Met. **131**, 123 (2002)

1079. Jerome, C., Labaye, D., Bodart, I., Jerome, R.: Synth. Met. **101**, 3 (1999)
1080. Qu, L.T., Shi, G.Q., Chen, F., Zhang, J.X.: Macromolecules **36**, 1063–1067 (2003)
1081. Qu, L.T., Shi, G.Q., Yuan, J.Y., Han, G.Y., Chen, F.: J. Electroanal. Chem. **561**, 149–156 (2004)
1082. Qu, L.T., Shi, G.Q.: J. Polym. Sci. Polym. Chem. **42**, 3170–3177 (2004)
1083. Li, X.H., Lu, M., Li, H.L.: J. Appl. Polym. Sci. **81**, 3002 (2001)
1084. Lu, M., Li, X.H., Guo, X.Y., Li, H.L.: Chem. J. Chin. Univ. **12**, 2331 (2002)
1085. Fu, M.X., Chen, F., Zhang, J.X., Shi, G.Q.: J. Mater. Chem. **12**, 2331 (2002)
1086. Lee, J. Y., Park, S.M.: Electrochem. Soc. **147**, 4189 (2000)
1087. Choi, S.J., Park, S.N.: Adv. Mater. **12**, 1547 (2000)
1088. Park, S.M., Lee, J.Y., Choi, S.J.: Synth. Met. **121**, 1297 (2001)
1089. Martin, C.R.: Acc. Chem. Res. **28**, 61 (1995)
1090. Bein, T., Enzel, P.: Inclusion of conducting polymers in inorganic hosts: towards conducting nanostructures. In: Aldissi, M. (ed.) Intrinsically conducting polymers: an emerging technology, p. 51. Kluwer Academic Publishers, Boston (1993)
1091. Wu, C.-G., Bein, T.: Stud. Surf. Sci. Catal. **84**, 2269 (1994)
1092. Mehrotra, V., Giannelis, E.P. In: Schaefer, D.W., Mark, J.E. (eds.) Polymer based molecular composites; Mat. Res. Soc. Symp. Roc., p. 171 (1990)
1093. Brandt, P., Fischer, R.D., Martinez, E.S., Calleja, R.D.: Angew. Chem. Int. Ed. Engl. **28**, 1265 (1989)
1094. Wu, C.-G., Bein, T.: Science **264**, 1757 (1994)
1095. Wu, C.-G., Chen, J.-Y.: Chem. Mater. **9**, 399 (1997)
1096. Qi, Z., Bruce, L.R.: Proc. – Electrochem. Soc. **97–5**, 173 (1997)
1097. Tang, Z.Y., Liu, S.Q., Tang, Z.X., Dong, S.J., Wang, E. K.: Electrochem. Commun. **2**, 32 (2000)
1098. Liu, J., Lin, Y.H., Liang, L., Voigt, J.A., Huber, D.L., Tian, Z.R., Coker, E., McKenzie, B., McDermott, M.J.: Chem. Eur. J. **9**, 605 (2003)
1099. Ge, D.T., Wang, J.X., Wang, Z., Wang, S.C.: Synth. Met. **132**, 93 (2002)
1100. Noll, J.D., Nicholson, M.A., Van Patten, P.G., Chung, C.W., Myrick, M.L.: J. Electrochem. Soc. **145**, 3320 (1998)
1101. Yang, Y.S., Liu, J., Wan, M.X.: Nanotechnol. **13**, 771 (2002)
1102. Cai, X.W., Gao, J.S., Xie, Z.X., Xie, Y., Tian, Z.Q., Mao, B.W.: Langmuir **14**, 2508 (1998)
1103. Jahromi, S., Dijkstra, J., Van der Vegte, E., Mostert, B.: Chem. Phys. Chem. **3**, 693 (2002)
1104. Nyffenegger, R.M., Penner, R.M.: J. Phys. Chem. **100**, 17041 (1996)
1105. Zhang, H.P., Luo, J., Huang, H.G., Wu, L.L., Lin, Z.H.: Chem. Phys. Lett. **326**, 169 (2000)
1106. Smith, J.A., Josowicz, M., Janta, J.: J. Electrochem. Soc. **150**, 384 (2003)
1107. Henry, M.C., Hsueh, C.C., Timko, B.P., Freund, M.S.: J. Electrochem. Soc. **148**, D155 (2002)
1108. Cioffi, N., Torsi, L., Sabbitini, L., Zambonin, P.G., Bleve-Zacheo, T.: J. Electroanal. Chem. **488**, 42 (2000)
1109. Cioffi, N., Torsi, L., Losito, I., Sabbitini, L., Zambonin, P.G.,Bleve-Zacheo, T.: Electrochim. Acta **46**, 4205 (2001)
1110. Tsakova, V., Winkels, S., Schultze, J.W.: J. Electroanal. Chem. **500**, 574 (2001)
1111. Hepel, M.: J. Electrochem. Soc. **145**, 124 (1998)
1112. Grzeszczuk, M., Poks, P.: Electrochim. Acta **45**, 4171 (2000)
1113. Choi, J.H., Park, K.W., Lee, H.K., Kim, Y.M., Lee, J.S., Sung, Y.E.: Electrochim. Acta **48**, 2781 (2003)
1114. Chen, G.Z., Shaffer, M.S.P., Coleby, D., Dixon, G., Zhou, W.Z., Fray, D.J., Windle, A.H.: Adv. Mater. **12**, 522 (2000)
1115. Chen, J.H., Huang, Z.P., Wang, D.Z., Yang, S.X., Wen, J.G., Ren, Z.R.: Appl. Phys. A **73**, 129 (2001)
1116. Chen, J.H., Huang, Z.P., Wang, D.Z., Yang, S.X., Li, W.Z., Wen, J.G, Ren, Z.F.: Synth, Met. **125**, 289 (2001)
1117. Huang, J.E., Li, X.H, Xu, J.C., Li, H.L.: Carbon, **41**, 2731 (2003)

1118. Tahhan, M., Truong, V.T., Spinks, G.M., Wallace, G.G.: Smart Mater. Struct. **12**, 626 (2003)
1119. Raspopov, L.N., Matkovskii, P.Ye., Belov, G.P., Noskova, V.N., Russiyan, L.N., Davydova, G.I., Shtarkin, V.A., Rudakov, V.M., Yusupbekov, A.Kh.: Vysokomol. Soedinen. **33a**, 425 (1991)
1120. Kminek, I., Trekoval, J.: Makromol. Chem., Rapid Commun. **5**, 53 (1984)
1121. Aldissi, M.: J. Chem. Soc. Chem. Commun. 1347 (1984)
1122. Malhotra, B.D., Kumar, N., Ghosh, S., Singh, K.K., Chandra, S.: Synth. Met. **31**, 155 (1989)
1123. Iyoda, T., Ohtani, A., Shimidzu, T., Honda, K.: Synth. Met. **18**, 725 (1987)
1124. Shimidzu, T.: Functionalized conducting polymer membranes/films. In: Scrosati, B. (ed.) Applications of Electroactive Polymers, vol. 283. Chapman & Hall, New York, USA (1993)
1125. Spangler, X.W., Thurmond, J.W., Li, H., He, M., Ghosal, S., Zhang, Y., Casstevens, M.K., Burzynski, R.: New copolymers for applications as organic LEDs. In: Yang, S.C., Chandrasekhar, P. (eds.) Optical and photonic applications of electroactive and conducting polymers, vol. 2528, p. 46. SPIE-The International Society for Optical Engineering, Bellingham (1995)
1126. Yang, Z., Geise, H.J.: Synth. Met. **47**, 105 (1992)
1127. Ochmanska, J., Pickup, P.G.: J. Electroanal. Chem. **297**, 197 (1991)
1128. Cho, H.N., Kim, D.Y., Kim, J.K., Kim, C.Y.: Synth. Met. **91**, 293 (1997)
1129. Beggiato, G., Casalbore-Miceli, G., Geri, A., Berlin, A., Pagani, G.: Synth. Met. **82**, 11 (1996)
1130. Malliaras, G.G., Herrema, J.K., Wildeman, J., Wieringa, R.H., Gill, R.E., Lampoura, S.S., Hadziioannou, G.: Adv. Mater. **5**, 721 (1993)
1131. Millan Rodriguez, J., Martinez Albillos, G., Gomez-Elvira Gonzales, J. M.: Conducting polymers based on copolymers of indole, thiophene, and pyrrole for use in electric applications. Appl. 930015528, Jan 1993, p.8
1132. Mazeikiene, R., Malinauskas, A.: Synth. Met. **92**, 259 (1998)
1133. Yang, C.-H., Wen, T.-C.: J. Electrochem. Soc. **144**, 2078 (1997)
1134. 380. Benjamin, I., Faraggi, E.Z., Cohen, G., Chayet, H., Davidov, D., Neumann, R., Avny, Y.: Synth. Met. **84**, 401 (1997)
1135. (a) Wessling, R.A., Zimmerman, R.G., Ray, G: Polyelectrolytes from bis sulfonium salts. US 3401152 A, 10 Sept 1968; (b) Lenz, R.W., Han, C., Smith, J.S., Karasz, S.E.: J. Polym. Sci., Polym. Chem. **26**, 3241 (1988)
1136. Genies, E.: Intrinsically conducting polymers from fundamental to applied research. In: Aldissi, M. (ed.) Intrinsically conducting polymers: an emerging technology, p. 75. Kluwer Academic Publishers, Boston (1993)
1137. Gregory, R.V., Kimbrell, W.C., Kuhn, H.H.: Synth. Met. **28**, C823 (1989)
1138. Kuhn, H.H., Kimbrell, W.C., Fowler, J.E., Barry, C.N.: Synth. Met. **55–57**, 3707 (1993)
1139. Kuhn, H.H.: Characterization and application of polypyrrole-coated textiles. In: Aldissi, M. (ed.) Intrinsically conducting polymers: an emerging technology, p. 25. Kluwer Academic Publishers, Boston (1993)
1140. Naarmann, H: BASF Plastics, Research and Development, Report No. KVX 8611e, 10.1986 p. 37/38, USP: 4738757 v. 12.02.81 BASF AG/FRG, (1986)
1141. Genies, E.M., Petrescu, C., Olmedo, L.: Synth. Met. **41–43**, 665 (1991)
1142. Li, C., Song, Z.: Synth. Met. **40**, 23 (1991)
1143. Zhang, H., Li, C.: Synth. Met. **44**, 143 (1991)
1144. Ojio, T., Miyata, S.: Polymer J. **18**, 95 (1986)
1145. Rabek, J.F., Lucki, J., Kereszti, H., Krische, B., Qu, B.J., Shi, W.F.: Synth. Met. **45**, 335 (1991)
1146. Jousse, F., Hourquebie, P., Deleuze, C., Olmedo, L.: Synthesis and microwave characterization of polypyrrole-PVC blends. In: Chiang, L.Y., Garito, A.F., Sandman, D.J. (eds.) Electrical, optical, and magnetic properties of organic solid state materials, vol. 247, p. 705. Materials Research Society, Pittsburgh (1992)

1147. Zinger, B., Kijel, D.: Synth. Met. **41–43**, 1013 (1991)
1148. Wiersma, A.E., vd Steeg, L.M.A., Jongeling, T.J.M.: Synth. Met. **71**, 2269 (1995)
1149. Lafosse, X.: Synth. Met. **68**, 227 (1995)
1150. Sun, Y., Ruckenstein, E.: Synth. Met. **72**, 261 (1995)
1151. Morita, M., Hashida, I., Masato, N.: J. Appl. Polym. Sci. **36**, 1639 (1989)
1152. Galvin, M.E., Wnek, G.E.: Polym. Comm. **23**, 795 (1982)
1153. (a) Rubner, M.F., Tripathy, S.K., Geroger, J., Jr., Cholewa, P.: Macromolecules, **16**, 870 (1980); (b) Rubner, M.F., Tripathy, S.K., Georger, J., Jr., Cholewa, P.: Macromolecules **16**, 870 (1983)
1154. Przyluski, J., Zukowska, G.: Polimery **42**, 229 (1997)
1155. Xie, H.-Q., Liu, H., Liu, Z.-H., Guo, J.-S.: Die Angewandte Makromolekulare Chemie **243**, 117 (1996)
1156. Oh, S.Y., Koh, H.C., Choi, J.W., Rhee, H.-W., Kim, H.S.: Polym. J. (Tokyo) **29**, 404 (1997)
1157. Ramachandran, K., Lerner, M.M.: J. Electrochem. Soc. **144**, 3739 (1997)
1158. Bao, J.-S., Xu, C.C., Cai, W., Bi, X.-T.: Electrically conductive composite of polypyrrole and liquid crystalline aromatic copolyamide. In: Chiang, L.Y., Garito, A.F., Sandman, D.J. (eds.) Electrical, optical, and magnetic properties of organic solid state materials, vol. 247, p. 699. Materials Research Society, Pittsburgh (1992)
1159. Niwa, O., Hikita, M., Tamamura, T.: Synth. Met. **18**, 677 (1987)
1160. De Paoli, M.A., Waltman, R.J., Diaz, A.F., Bargon, J.: J. Polym. Sci.: Polym. Chem. Ed. **23**, 1687 (1985)
1161. Lindsey, S.E., Street, G.B.: Synth. Met. **10**, 67 (1984/85)
1162. Selampinar, F., Akbulut, U., Yildiz, E., Güngör, A., Toppare, L.: Synth. Met. **89**, 111 (1997)
1163. Bi, X., Pei, Q.: Synth. Met. **22**, 145 (1987)
1164. Bozkurt, A., Akbulut, U., Toppare, L.: Synth. Met. **82**, 41 (1996)
1165. Stockton, W.B., Rubner, M.F.: Electrically conducting compatible blends of polyaniline/ poly (Vinyl Pyrrolidone). In: Garito, A.F., Jen, A.K.-Y., Lee, C.Y.-C., Dalton, L.R. (eds.) Electrical, optical, and magnetic properties of organic solid state materials, vol. 328, p. 257. Materials Research Society, Pittsburgh (1994)
1166. Yang, Z., Geise, H.J.: Synth. Met. **47**, 105 (1992)
1167. Ogura, K., Kokura, M., Nakayama, M.: J. Electrochem. Soc. **142**, L152 (1995)
1168. Pron, A., Österholm, J-E., Smith, P., Heeger, A.J., Laska, J., Zagorska, M.: Synth. Met. **55–57**, 3520 (1993)
1169. Laska, J., Pron, A., Zagórska, M., Lapkowski, S., Lefrant, S.: Synth. Met. **69**, 113 (1995)
1170. Gonçalves, D., Waddon, A., Karasz, F.E., Akcelrud, L.: Synth. Met. **74**, 197 (1995)
1171. Banerjee, P., Mandal, B.M.: Synth. Met. **74**, 257 (1995)
1172. Tan, L.-S., Simko, S.R., Bai, S.J., Vaia, R.A., Spry, R.J.: Polym. Prepr. **38**, 239 (1997)
1173. Kulkarni, V.G.: Processing of polyanilines. In: Aldissi, M. (ed.) Intrinsically conducting polymers: an emerging technology, vol. 45. Kluwer Academic Publishers, Boston (1993)
1174. Shacklette, L.W., Han, C.C., Luly, M.H.: Synth. Met. **55–57**, 3532 (1993)
1175. Heeger, A.J., Smith, P.: Solution processiong of conducting polymers: opportunities for science and technology. In: Brédas, J.L., Silbey, R. (eds.) Conjugated polymers: the novel science and technology of highly conducting and nonlinear optically active materials, p. 141. Kluwer Academic Publishers, Norwell (1991)
1176. Yoshino, K., Onoda, M., Sugimoto, R.: Jpn. J. Appl. Phys. **27**, L2034 (1988)
1177. Yang, J.P., Rannou, P., Planès, J., Pron, A., Nechtschein, M.: Synth. Met. **93**, 169 (1998)
1178. Mandal, T.K., Mandal, B.M.: Synth. Met. **80**, 83 (1996)
1179. Chiang, L.Y., Wang, L.Y., Kuo, C.S., Lin, J.G., Huang, C.Y.: Synth. Met. **84**, 721 (1997)
1180. Liao, D.C., Hsieh, K.H., Chern, Y.C., Ho, K.S.: Synth. Met. **87**, 61 (1997)
1181. Li, S., White, H.S.: J. Electrochem. Soc. **140**, 2473 (1993)
1182. Andreatta, A., Heeger, A.J., Smith, P.: Polym. Commun. **31**, 275 (1990)
1183. Hsu, C.-H., Vaca-Segonds, P., Epstein, A.J.: Synth. Met. **41–43**, 1005 (1991)
1184. Nemoto, H., Marks, T.J., De Groot, D.C., Kannewurf, C.R.: Chem. Mater. **2**, 349 (1991)

1185. Chiang, C.K., Druy, M.A., Gau, S.C., Heeger, A.J., Louis, E.J., MacDarmid, A.G., Park, Y. W., Shirakawa, H.: J. Am. Chem. Soc. **100**, 1014 (1980)

1186. Heffner, G.W., Pearson, D.S.: Synth. Met. **44**, 341 (1991)

1187. Kathirgamanathan, P., Qayyum, M.M.B.: J. Electrochem. Soc. **141**, 147 (1994)

1188. Kathirgamanathan, P., Boland, B.: J. Electrochem. Soc. **140**, 2815 (1993)

1189. Naarmann, H., Kohler, G., Schlag, J.: Continuous production of polypyrrole films. US 4468291 A, 28 Aug 1984

1190. Andreatta, A., Cao, Y., Chiang, J.C., Heeger, A.J., Smith, P.: Synth. Met. **26**, 383 (1988)

1191. Frommer, J.E.: Acc. Chem. Res. **19**, 2 (1986)

1192. Cao, Y., Smith, P., Heeger, J.: Synth. Met. **32**, 263 (1989)

1193. Kim, I.W., Lee, J.Y., Lee, H.: Synth. Met. **78**, 177 (1996)

1194. Sun, L., Yang, S.C., Liu, J.: Conducting polymer with improved long-time stability: plyanline-polyelectrolyte complex. In: Garito, A.F., Jen, A.K.-Y., Lee, C.Y.-C., Dalton, L.R. (eds.) Electrical, optical, and magnetic properties of organic solid state materials, vol. 328, p. 167. Materials Research Society, Pittsburgh (1994)

1195. Sun, L., Yang, S.C.: Solution processable conducting polymer: polyaniline-polyelectrolyte complexes. In: Garito, A.F., Jen, A.K.-Y., Lee, C.Y.-C., Dalton, L.R. (eds.) Electrical, optical, and magnetic properties of organic solid state materials, vol. 328, p. 209. Materials Research Society, Pittsburgh (1994)

1196. Angelopoulos, M., Patel, N., Shaw, J.M.: Water soluble polyanilines: properties and applications. In: Garito, A.F., Jen, A.K.-Y., Lee, C.Y.-C., Dalton, L.R. (eds.) electrical optical, and magnetic properties of organic solid state materials, vol. 328, p. 173. Materials Research Society, Pittsburgh (1994)

1197. Bates, N., Cross, M., Lines, R., Walton, D.: J. Chem. Soc. Chem. Commun. 871 (1985)

1198. Fan, F.-R., Bard, A.J.: J. Electrochem. Soc. **133**, 301 (1986)

1199. Shi, S., Wudl, F.: Macromolecules **23**, 2119 (1990)

1200. Reynolds, J.R., Child, A.D., Ruiz, J.P., Musfeldt, J.L., Sankaran, B., Larmat, F., Balanda, P., Tanner, D.B.: Optical absorption, luminescence, and redox switching properties of polyphenylene derivatives. In: Garito, A.F., Jen, A.K.-Y., Lee, C.Y.-C., Dalton, L.R. (eds.) Electrical, optical, and magnetic properties of organic solid state materials, pp. 328–191. Materials Resaerch Society, Pittsburgh (1994)

1201. Yue, J., Epstein, A.J.: J. Am. Chem. Soc. **112**, 2800 (1990)

1202. Kim, E., Lee, M.-H., Moon, B.S., Lee, C., Rhee, S.B.: J. Electrochem. Soc. **141**, L26 (1994)

1203. Yang, C.-H., Wen, T.-C.: J. Electrochem. Soc. **141**, 2624 (1994)

1204. (a) Patil, A.O., Ikenoue, Y., Basescu, N., Colaneri, N., Chen, J., Wudl, F., Heeger, A.J.: Synth. Met. **20**, 151 (1987); (b) Patil, A.O., Ikenoue, Y., Wudl, F., Heeger, A.J.: J. Am. Chem. Soc. **109**, 1858 (1987); (c) Ikenoue, Y., Chiang, J., Patil, A.O., Wudl, F., Heeger, A. J.: J. Am. Chem. Soc. **110**, 2983 (1988)

1205. Sato, M-A., Tanaka, S., Kaeriyama, K.: J. Chem. Soc. Chem. Commun. 873 (1986)

1206. Jen, K-Y., Miller, G.G., Elsenbaumer, R.L.: J. Chem. Soc. Chem. Commun. 1346 (1986)

1207. Österholm, J.-E., Laakso, J., Nyholm, P., Isotalo, H., Stubb, H., Inganäs, O., Salaneck, W.R.: Synth. Met. **28**, C435 (1992)

1208. Patel, G.N., Chance, R.R., Witt, J.D.: J. Polym. Sci. Polym. Lett. Ed. **16**, 607 (1978)

1209. Patel, G.N., Chance, R.R., Witt, J.D.: J. Chem. Phys. **70**, 4387 (1979)

1210. Plachetta, C., Schulz, R.C.: Makromol. Chem. Rapid. Commun. **3**, 815 (1982)

1211. Liu, M., Gregory, R.V.: Synth. Met. **72**, 45 (1995)

1212. Yoshino, K., Nakajima, S., Fuji, M., Sugimoto, R.-I.: Polym. Commun. **28**, 309 (1987)

1213. Isotalo, H., Laakso, J., Kuivalainen, P., Stubb, H., Österholm, J.-E., Yli-Lahti, P.: Physica Status Solidi (b), **154**, 305 (1989)

1214. Virtanen, E., Laakso, J., Ruohonen, H., Väkiparta, K., Järvinen, H., Jussila, M., Passiniemi, P., Österholm, J.-E.: Synth. Met. **84**, 113 (1997)

1215. Kulkarni, V.G.: Synth. Met. **71**, 2129 (1995)

1216. Bjorklund, R.B., Liedberg, B.: J. Chem. Soc. Chem. Connun. 1293 (1986)

1217. Armes, S.P.: Potential applications of conducting polymer colloids. In: Aldissi, M. (ed.) Intrinsically conducting polymers: an emerging technology, p. 35. Kluwer Academic Publishers, Boston Massachusetts, USA (1993)

1218. Mattes, B.R., Knobbe, E.T., Fuqua, P.D., Nishida, F., Chang, E.W., Pierce, B. M., Dunn, B., Kaner, R.B.: Synth. Met. **41–43**, 3183 (1991)

1219. (a) Rose, T.L., Liberto, M.C.: Synth. Met. **31**, 395 (1989); (b) Defieuw, G., Samijn, R., Hoogmartens, I., Vanderzande, D., Gelan, J.: Synth. Met. **55-57**, 3702 (1993)

1220. Wiersma, A.E., vd Steeg, L.M.A., Jongeling, T.J.M.: Synth. Met. **71**, 1995 (2269)

1221. Andreatta, A., Tokito, S., Moulton, J., Smith, P., Heeger, A.J.: Processing of high-performance conducting polymers. In: Salaneck, W.R., Clark, D.T., Samuelsen, E.J. (eds.) Science and applications of conducting polymers, p. 105. Adam Hilger, New York (1991)

1222. Yoshino, K., Nakajima, S., Fuji, M., Sugimoto, R.-I.: Polym. Commun. **28**, 310(1987)

1223. Tokito, S., Smith, P., Heeger, A.J.: Synth. Met. **36**, 183 (1990)

1224. Mattes, B.R., Wang, H.L., Yang, D., Zhu, Y.T., Blumenthal, W.R., Hundley, M.F.: Synth. Met. **84**, 45 (1997)

1225. Proceedings of the DARPA Active Polymers Workshop hosted by Institute for Defense Analyses, Baltimore, Maryland, USA, p.C7, 15 Nov 1996

1226. Hagiwara, T., Hirasaka, M., Sato, K., Yamaura, M.: Synth. Met. **36**, 241 (1990)

1227. Rubner, M.F., Skotheim, T.A.: Controlled molecular assemblies of electrically conductive polymers. In: Brédas, J.L., Silbey, R. (eds.) Conjugated polymers: the novel science and technology of highly conducting and nonlinear optically active materials, p. 363. Kluwer Academic Publishers, Norwell (1991)

1228. (a) Hong, K., Rubner, M.F.: Thin Solid Films **160**, 187 (1988); (b) Hong, K., Rubner, M.F.: Thin Solid Films **179**, 215 (1989); (c) Hong, K., Rosner, R.B., Rubner, M.F.: Chem. Mater. **2**, 82 (1990)

1229. Logsdon, P.B., Pfleger, J., Prasad, P.N.: Synth. Met. **26**, 369 (1988)

1230. Era, M., Kamiyama, K., Yoshiura, K., Momii, T., Murata, H., Tokito, S., Tsutsui, T., Saito, S.: Thin Solid Films. **179**, 1 (1989)

1231. Cheung, J.H., Rosner, R.B., Rubner, M.F.: New strategies for preparing electrically conductive Langmuir-Blodgett films. In: Chiang, L.Y., Garito, A.F., Sandman, D.J. (eds.) Electrical, optical, and magnetic properties of organic solid state materials, vol. 247, p. 859. Materials Research Society, Pittsburgh (1992)

1232. (a) Watanabe, I., Hong, K., Rubner, M.F.: J. Chem. Soc. Chem. Commun. **123** (1989) (b) Watanabe, I., Hong, K., Rubner, M.F., Loh, I.H.: Synth. Met. **28**, C473 (1989); (c) Watanabe, I., Hong, K., Rubner, M.F.: Thin Solid Films **179**, 199 (1989); (d) Watanabe, I., Hong, K., Rubner, M.F.: Langmuir: **6**, 1164 (1990)

1233. Plank, R.V., DiNardo, N.J., Vohs, J.M.: Synth. Met. **89**, 1 (1997)

1234. Zhou, J., Wipf, D.O.: J. Electrochem. Soc. **144**, 1202 (1997)

1235. Jang, J., Oh, J.H., Stucky, G.D.: Fabrication of ultrafine conducting polymer and graphite nanoparticles. Angew. Chem. Int. Ed. **41**(21), 4016–4019 (2002); (a) Grossman, C., Heflin, J. R., Wong, K.Y., Zamani-Khamiri, O., Garito, A.F. In: Messier, J., Kajzar, F., Prasa, D.P., Ulrich, D.R. (eds.) Nonlinear optical effects in organic polymers. Kluwer, Dordrecht, p. 61–781989; (b) Chandrasekhar, P., Thorne, J.R.G., Hochstrasser, R.M.: Appl. Phys. Lett. **59**, 1661 (1991), and references therein; (c) Chandrasekhar, P., Thorne, J.R.G., Hochstrasser, R.M.: Synth Met. **53**, 175 (1993), and references therein. (d) Chandrasekhar, P., Naishadham, K. (1999)

1236. Leclerc, M., Diaz, F.M., Wegner, G.: Makromol. Chem. **190**, 3105 (1989)

1237. Liao, H., Y., S.C.: The 1.5eV polaron transition of polyaniline: the spectra electrochemical resolution into sub-bands. In: Chiang, L.Y., Garito, A.F., Sandman, D.J. (eds.) Electrical, optical, and magnetic properties of organic solid state materials, vol. 247, p. 741. Materials Research Society, Pittsburgh (1992)

1238. Furukawa, N., Nishio, K.: Lithium batteries with polymer electrodes. In: Scrosati, B. (ed.) Applications of electroactive polymers, p. 150. Chapman & Hall, New York (1993)

1239. Kanamura, K., Kawai, Y., Yonezawa, S., Takehara, Z.: J. Electrochem. Soc. **142**, 2894 (1995)

1240. Naarman, H.: Snythesis, properties and applications of perconjugated systems. In: Aldissi, M. (ed.) Intrinsically conducting polymers: an emerging technology. Kluwer Academic Publishers, Boston (1993)

1241. Oh, E.J., Min, Y., Wiesinger, J.M., Manohar, S.K., Scherr, E.M., Prest, P.J., MacDiarmid, A. G., Epstein, A.J.: Synth. Met. **55–57**, 977 (1993)

1242. Lazzaroni, R., Rachidi, S., Brédas, J.L.: Theoretical investigation of chain flexibility in polythiophene and polypyrrole. In: Salaneck, W.R., Clark, D.T., Samuelsen, E.J. (eds.) Science and applications of conducting polymers, p. 13. Adam Hilher, New York (1991)

1243. Bueche, F.: Physical properties of polymers, pp. 6–8. Interscience Publishers, New York (1962)

1244. Nazzal, A., Street, G.B.: J. Chem. Soc., Chem. Commun. 83 (1983)

1245. Avlyanov, J.K., Josefowicz, J.Y., MacDiarmid, A.G.: Synth. Met. **73**, 205 (1995)

1246. Rolland, M. Abadie, M. Cadene, M. J. Phys. Orsay Fr. **44**, C3 (1984)

1247. Aldissi, M., Schue, F., Giral, L., Rolland, M.: Polymer. **23**, 246 (1982)

1248. Angelopoulous, M., Liao, Y.-H., Furman, B., Graham, T.: Solvent and salt effects on the morphological structure of polyaniline. In: Yang, S.C., Chandrasekhar, P. (eds.) Optical and photonic applications of electroactive and conducting polymers. SPIE-The International Society for Optical Engineering, Bellingham, vol. 2528, p. 230 (1995)

1249. MacDiarmid, A.G., Epstein, A.J.: Synth. Met. **65**, 103 (1994)

1250. Buckley, L.J., Joseflowicz, J.Y., Xie, L.: Polyaniline surface morpholoy during the eDoping process using atomic force microscopy. In: Garito, A.F., Jen, A.K.-Y., Lee, C.Y.-C., Dalton, L.R. (eds.) Electrical, optical, and magnetic properties of organic solid state materials, vol. 328, p. 197. Materials Research Society, Pittsburgh (1994)

1251. Aldissi, M.: Stabiliy and stabilization of pristine and doped polymers. In: Inherently conducting polymers: processing fabrication, applications, limitations, p. 52. Noyes Data Corporations, New Jersey (1989)

1252. Silk, T., Hong, Q., Tamm, J., Compton, R.G.: Synth. Met. **93**, 59 (1998)

1253. a) Jenekhe, S.: Priv. Commun. (1985); (b) Jenekhe, S.A.: Macromolecules **19**, 2663 (1986); (c) Jenekhe, S.A.: Nature **322**, 345 (1986)

1254. Patil, A.O., Wudl, F.: Macromolecules. **21**, 540 (1988)

1255. Bräunling, H., Jira, R.: Synth. Met. **20**, 375 (1987)

1256. Street, G.B., Lindsey, S.E., Nazzai, A.I., Wynne, K.J.: Mol. Cryst. Liq. Cryst. **118**, 137 (1985)

1257. Fukuda, T., Takezoe, H., Ishikawa, K., Fukuda, A., Woo, H.S., Jeong, S.K., Oh, E.J., Suh, J. S.: Synth. Met. **69**, 175 (1995)

1258. Pouget, J.P., Jozefowicz, M.E., Tang, X., MacDiarmid, A.G., Epstein, A.J.: Macromolecules. **24**, 779 (1991)

1259. Pouget, J.B., Laridjani, M., Jozefowicz, M.E., Epstein, A.J., Scherr, E.M., MacDiarmid, A. G.: Structural aspects of polyaniline family of electronic polymers. In: Chiang, L.Y., Garito, A.F., Sandman, D.J. (eds.) Electrical, optical, and magnetic properties of organic solid state materials, vol. 247, p. 589. Materials Research Society, Pittsburgh (1992)

1260. Tsukamooto, J., Takahashi, A.: Structure and morphology of metallic-conductive polyacetylene. In: Electrical, optical and magnetic properties of organic solid state materials, vol. 247, p. 711. Materials, Research Society, Pittsburgh (1992)

1261. Wegner, G.: Molecular metals, vol. 1, p. 209. Plenum, New York (1979)

1262. van der Pauw, L.J.: Phillips Tech. Rev. **20**, 220 (1958-1959)

1263. Montgomery, H.C.: J. Appl. Phys. **42**, 2971 (1971)

1264. Logan, B.F., Rice, S.O., Wick, R.F.: J. Appl. Phys. **42**, 2975 (1971)

1265. Hermann, G.J., Resetar-Racine, T.M., Hale, J., Stevens, W.C., Del Vecchio, J.A., Sturm, E. A.: Metallized fiber smoke material with tailored degradability. In: Proceedings of the smoke/obscurants smyposium XV, vol. 1, p. 63. Chemical Research, Development & Engineering Center, Aberdeen (1991)

1266. Shibuya, M., Nishina, T., Matsue, T., Uchida, I.: J. Electrochem. Soc. **143**, 3157 (1996)

1267. Hoa, D.T., Kumar, T.N.S., Punekar, N.S., Srinivasa, R.S., Lal, R., Contractor, A.Q.: Anal. Chem. **64**, 2645 (1992)

1268. Schiavon, G., Sitran, S., Zotti, G.: Synth. Met. **32**, 209 (1989)

1269. Paul, E.W. Ricco, A.J., Wrighton, M.S.: J. Phys. Chem. **89**, 1441 (1985), and references therein

1270. Thackeray, J.W., Wrighton, M.S.: J. Phys. Chem. **90**, 6674 (1986), and references therein

1271. Guillaud, G., Rosenberg, N.: J. Phys. E. **13**, 1287 (1980)

1272. Buravov, L.I., Shchegolev, I.F.: Prib. Tekh. Eksp. **4**, 171 (1971)

1273. (a) Epstein, A.J., Rommelmann, H., Bigelow, R., Gibson, H.W., Hoffman, D., Tanner, D.B.: Phys. Rev. Lett. **50**, 1866 (1983); (b) Epstein, A.J.; Rommelmann, H.; Bigelow, R., Gibson, H.W., Hoffman, D., Tanner, D.B.: Phys. Rev. Lett. **51**, 2020 (1983)

1274. Electrochemistry and Corrosion Overview and Techniques, Application Note Corr 4, EG&G Princeton Applied Research Corp., USA (1997)

1275. Nogueira, J.S., Mattoso, L.H.C., Lepienski, C.M., Faria, R.M.: Synth. Met. **69**, 259 (1995)

1276. Amemiya, T., Hashimoto, K., Fujishima, A.: Dynamics of electrochromic phenomena in organic conducting polypyrrole films. In: Chiang, L.Y., Garito, A.F., Sandman, D.J. (eds.) Electrical, optical, and magnetic properties of organic solid state materials, vol. 247, p. 613. Materials Research Society, Pittsburgh (1992)

1277. Passiniemi, P., Väkiparta, K.: Synth. Met. **69**, 237 (1995)

1278. Singh, R., Narula, A.K., Tandon, R.P.: Synth. Met. **82**, 63 (1996)

1279. (a) Bernier, P.: The magnetic properties of conjugated polymers: ESR studies of undoped and doped systems.; In Handbook of conducting polymers; Skotheim, T. A. Ed. Marcel Dekker, Inc., New York, 1986; Vol. 2, p.1099.; (b) Catellani, M., Porzio, W., Musco, A., Pontellini, R: Synthesis and characterization of soluble alkyl substitued poly (2,5-Thienylene Vinylenes). In: Chiang, L.Y., Garito, A.F., Sandman, D.J. (eds.) Electrical optical and magnetic properties of organic solid state materialsMaterials Research Society, Pittsburgh, Vol. 247, p. 681 (1992);

1280. Aldissi, M.: Synthesis of conjugated polymers. In: Inherently conducting polymers: processing, fabrication, applications, limitations, p. 2. Noyes Data Corporation, New Jersey (1989)

1281. Kiess, H.G., Harbeke, G.: Optical properties of conducting polymers. In: Kiess, H.G. (ed.) Conjugated conducting polymers, vol. 102, p. 175. Springer, New York

1282. Hotta, S., Rughooputh, S.D.D.V., Heeger, A.J.: Synth. Met. **22**, 79 (1987)

1283. Hayes, W., Pratt, F.L., Wong, K.S., Kaneto, K., Yoshino, K.: J. Phys. **C18**, L555 (1985)

1284. Shacklette, L.W., Chance, R.R., Ivory, D.M., Miller, G.G., Baughman, R.H.: Synth. Met. **1**, 307 (1980)

1285. Nguyên, M.T., Dao, L.H.: J. Chem. Soc. Chem. Commun. 1221 (1990), and references therein

1286. Jones, M.B., Kovacic, P., Lanska, D.: J. Polym. Sci. Polym. Letts. **19**, 89 (1981)

1287. Chien, J.C.W., Capistran, J.D., Karasz, F.E., Dickinson, L.C., Schen, M.A.: J. Polym. Sci. Lett. **21**, 93 (1983)

1288. Wegner, G.: Angew, Chem. Int. Ed. **20**, 361 (1981)

1289. Soga, K., Kawakami, S., Shirakawa, H., Ideda, S.: Makromol. Chem. Rap. Commun. **1**, 523 (1980)

1290. Davied, S., Nicolau, Y.F., Melis, F., Revillon, A.: Synth. Met. **69**, 125 (1995)

1291. Cao, Y., Andreatta, A., Heeger, A.J., Smith, P.: Polymer. **30**, 2305 (1989)

1292. Plachetta, C., Schulz, R.C.: Makromol. Chem. Rapid Commun. **3**, 815 (1982)

1293. Hoffman, H. Krömer, H. Kuhn, R. Polymeranalytik I.: Georg Thieme Verlag Stuttgart S. 303 (1977)

1294. Seery, T.A.P., Angelopoulos, M., Levon, K., Seghal, A.: Synth. Met. **84**, 79 (1997)

1295. Mulazzi, E., Bivio, G.P., Lefrant, S., Faulques, E., Perrin, E.: Synth. Met. **17**, 325 (1987)

1296. Yong, C., Renyan, Q.: Solid State Commun. **54**, 211 (1985)

1297. Steigmeier, E.F., Auderset, H., Kobel, W., Baeriswyl, D.: Synth. Met. **17**, 219 (1987)

1298. Botta, C., Luzzati, S., Bolognesi, A., Tubino, R., Borghesi, A.: Photoinduced absorption spectroscopy of poly-3-alkylthiophenes. In: Chiang, L.Y., Garito, A.F., Sandman, D.J. (eds.) Electrical, optical, and magnetic properties of organic solid state materials, vol. 247, p. 669. Materials Research Society, Pittsburgh (1992)

1299. Hankin, S.H.W., Sandman, D.J.: Recent studies of Raman spectroscopy of polydiacetylene crystals: poly-Ipudo. In: Chiang, L.Y., Garito, A.F., Sandman, D.J. (eds.) Electrical, optical, and magnetic properties of organic solid state materials, vol. 247, p. 661. Materials Research Society, Pittsburgh (1992)

1300. Fong, Y., Chen, C., Schlenoff, J.B.: In-situ montoring of the kinetics and mechanism of conducting polymer synthesis. In: Chiuang, L.Y., Garito, A.F., Sandman, D.J. (eds.) Electrical, optical, and magnetic properties of organic solid state materials, vol. 246, p. 693. Materials Research Society, Pittsburgh (1992)

1301. (a) Lefrant, S., Mévellec, J.Y., Buisson, J.P., Perrin, E., Eckhardt, H., Han, C.C., Jen, K.Y.: In: Kuzmany, H., Mehring, M., Roth, S. (eds.) Electronic properties of conjugated conducting polymers III: basica models and application, vol. 91, p. 123. Springer, New York (1989); (b) Kuzmany, H., Genies, E.M., Syed, A.: Resonance Raman scattering from polyaniline. In: Kuzmany, H., Mehring, M., Roth, S. (eds.) Electronic properties of polymers and related compounds, vol. 69, p. 223 . Springer, New York (1989)

1302. Sakamoto, A., Furukawa, Y., Tasumi, M., Noguchi, T., Ohnishi, T.: Synth. Met. **69**, 439 (1995)

1303. Kastner, J., Kuzmany, H., Vegh, D., Landl, M., Cuff, L., Kertesz, M.: Synth. Met. **69**, 593 (1995)

1304. Jen, A.K.-Y., Drzewinski, M., Chin, H.H., Boara, G.: Highly conducting and thermally stable conjugated polymers. In: Chiang, L.Y., Garito, A.F., Sandman, D.J. (eds.) Electrical, optical, and magnetic properties of organic solid state materials, vol. 247, p. 687. Materials Research Society, Pittsburgh (1992)

1305. Buckley, L.J., Eashoo, M.: Synth. Met. **78**, 1 (1996)

1306. Kulkarni, V.G., Mathew, W.R., Wessling, B., Merkle, H., Blaettner, S.: Synth. Met. **41–43**, 1009 (1991)

1307. Mathys, G.I., Truong, V.-T.: Snyth. Met. **89**, 103 (1997)

1308. Yue, J., Epstein, A.J.: Macromolecules. **24**, 4441 (1991)

1309. Nakajima, T. Kawagoe, T. Synth. Met. **28**, C629 (1989), and references therein

1310. Salaneck, W.R., Thomas, H.R., Duke, C.B., Paton, A., Plummer, E.W., Heeger, A.J., MacDiarmid, A.G.: J. Chem. Phys. **71**, 2044 (1979)

1311. Hsu, S.L., Signorelli, A.J., Pez, G.P., Baughman, R.H.: J. Chem. Phys. **69**, 106 (1978)

1312. Salaneck, W.R., Thomas, H.R., Bigelow, R.W., Duke, C.B., Plummer, E.W., Heeger, A.J., MacDiarmid, A.G.: J. Chem. Phys. **72**, 3674 (1980)

1313. Nguyen, T.P., Amgaad, K., Cailler, M., Tran, V.H., Lefrant, S.: Synth. Met. **69**, 495 (1995)

1314. Jira, R. Braunling. H.: Synth. Met. **17**, 691 (1987), and references therein

1315. Kahol, P.K., Clark, W.G., Mehring, M.: Magnetic properties of conjugated polymers. In: Kiess, H. (ed.) Conjugated conducting polymers, vol. 102, p. 217. Springer, New York (1992.) and references therein

1316. Yannoni, C.S.: Accts. Chem. Res. **15**, 201 (1982)

1317. Gustafsson, G., Inganas, O., Salaneck, W.R., Laasko, J., Loponen, M., Taka, T., Osterholm, J.-E., Stubb, H., Hjertberg, T.: Processable conducting poly(3-Alkylthiopenes). In: Bredas, J.L., Silbey, R. (eds.) Conjugated polymers, p. 315. Kluwer Academic Publishers, Dordrecht (1991)

1318. Elsenbaumer, R.L., Jen, K.Y., Miller, G.G., Eckhardt, H., Shacklette, L.W., Jow, R.: Springer Ser. Solid- State Sci. **76**, 400 (1987)

1319. Bargon, J., Mohmand, S., Waltman, R.J.: IBM J. Res. Develop. **27**, 330 (1983)

1320. (a) Clarke, T.C., Scott, J.C.: NMR studies of polyacetylene. In: Skotheim, T.A. (ed.) Handbook of conducting polymers, vol. 2, p. 1127. Marcel Dekker, Inc., New York (1986); (b) Thomann, H. ENDOR studies of polyacetylene. In: Skotheim, T. A. (ed.) Handbook of conducting polymers, vol. 2, p. 1157. Marcel Dekker, Inc., New York (1986); and references therein

1321. Yannoni, C.S., Clarke, T.C.: Phys. Rev. Lett. **51**, 1191 (1983)

1322. Fincher Jr., C.R., Chen, C.-E., Heeger, A.J., MacDiarmid, A.G., Hastings, J.B.: Phys. Rev. Lett. **48**, 100 (1982)
1323. Heinmaa, I., Alla, M., Vainrub, V., Lippmaa, E., Khidelkel, M.L., Kotov, A.I., Kozub, G.I.: J. Phys. Colloq. **44**, C3–357 (1983)
1324. Keiser, H., Beccu, U.D., Gutjahr, M.A.: Electrochim. Acta. **21**, 539 (1976)
1325. Wnek, G.E., Chien, J.C.W., Karasz, F.E., Lillya, C.P.: Polymer. **20**, 144 (1979)
1326. Scott, J.C. Pfluger, P. Krounbi, M.T. Street, G.B. Phys. Rev. B **28**, 2140 (1983), and references therein
1327. Tanaka, K., Shichiri, T., Yoshizawa, K., Yamabe, T., Hotta, S., Shimotsuma, W., Yamanchi, J., Deguchi, Y.: Solid State Commun. **51**, 565 (1984)
1328. Kobayashi, M., Chen, J., Chung, T.C., Moraes, F., Heeger, A.J., Wudl, F.: Synth. Met. **9**, 77 (1984)
1329. Chien, J.C.W., Karasz, F.E., Wnek, G.E., MacDiarmid, A.G., Heeger, A.J. J. Polym. Sci. Polym. Lett. Ed. **18**, 45 (1980) and references therein
1330. Su, W.P.: Solid State Commun. **35**, 899 (1980)
1331. Thomann, H. Electronic resonance of the solid state. In: Weil, J. (ed.) Can. Che. Soc. Symp. Ser. 1986
1332. Cailleau, H., Girad, A., Moussa, F., Zeyen, C.M.E.: Solid State Commun. **29**, 259 (1979)
1333. Heeger, A.J. MacDiarmid, A.G. Mol. Cryst. Liq. Cryst. **77**, A1 (1981) , and references therein
1334. Genoud, F., Ménardo, C., NEchtschein, M.: Electrochemical cycling of polyaniline: proton transfer and ESR susceptibility. In: Plocharski, J., Roth, S. (eds.) Materials science forum, vol. 42, p. 85. Trans Tech Publications, Switzerland (1989)
1335. Javadi, H. H.S. Laversanne, R. Epstein, A.J. Kohli, R.K. Scheer, E.M. MacDiarmid, A.G. Synth. Met. **29**, E-439 (1989)
1336. Wang, Z.H., Javadi, H.H.S., Ray, A., MacDiarmid, A.G., Epstein, A.J.: Phys. Rev. B. **42**, 5411 (1990)
1337. Flood, J.D., Heeger, A.J.: Phys. Rev. B. **28**, 2356 (1983)
1338. Moraes, F., Schaffer, H., Kobayashi, M., Heeger, A.J., Wudl, F.: Phys. Rev. B. **30**, 2948 (1984)
1339. Kispert, L.D., Files, L.A., Frommer, J.E., Shacklette, L.W., Chance, R.R.: J. Chem. Phys. **78**, 4858 (1983)
1340. Raynor, J.B.: Mater. Sci. Forum. **21**, 11 (1987)
1341. Thomann, H., Dalton, L.R., Tomkiewicz, Y., Shiren, N.S., Clarke, T.C.: Phys. Rev. Lett. **50**, 533 (1983)
1342. Kuroda, S.-I., Noguchi, T., Ohnishi, T.: Synth. Met. **69**, 423 (1995)
1343. Crecelius, G.: Electron energy loss spectroscopy in the study of conducting polymers. In: Skotheim, T.A. (ed.) Handbook of conducting polymers, vol. 2, p. 1233. Marcel Dekker, Inc., New York (1986)
1344. (a) Ritsko, J.J.: Mater. Sci. **7**, 337 (1981); (b) Ritsko, J.J.: Phys. Rev. **B26**, 2192 (1982)
1345. Ritsko, J.J., Crecelius, G., Fink, J.: Phys. Rev. **B27**, 4902 (1983)
1346. Ritsko, J.J., Fink, J., Crecelius, G.: Solid State Commun. **46**, 477 (1983)
1347. Naishadham, K.: Microwave characterizatioln of polymeric materials with potential applications in electromagnetic interference (EMI) shielding. Int. SAMPE Electron. Conf. **7**, 252 (1994), and references therein
1348. Naishadham, K.: Private Communication (1988)
1349. Stenger-smith, J. J.: NAWCP, China Lake, Califronia, USA, Private Communication (1998)
1350. Radar Cross Section Handbook. Plenum Press, New York (1970); Several Volumes
1351. Naishadham, K., Kadaba, P.K.: IEEE Trans. Microwave Theor. Techn. **39**, 1158 (1991), and refs. therein
1352. Epstein, A.J., MacDiarmid, A.G.: The controlled electromagnetic response of playanilines and its application to technologies. In: Salaneck, W.R., Clark, D.T., Samuelsen, E.J. (eds.) Science and applications of conducting polymers, p. 141. Adam Hilger, New York (1991)

1353. Epstein, A.J., Joo, J., Kohlman, R.S., MacDiarmid, A.G., Weisinger, J.M., Min, Y., Pouget, J.P., Tsukamoto, J.: The metallic state of conducting polymers: microwave dielectric response and optical conductivity. In: Garito, A.F., Jen, A.K.-Y., Lee, C.Y.-C., Dalton, L.R. (eds.) Electrical, optical, and magnetic properties of organic solid state materials, vol. 328, p. 145. Materials Research Society, Pittsburgh (1994)

1354. Hourquebie, P., Olmedo, L., Deleuze, C.: Microwave properties of conductive polymer composites. In: Kuzmany, H., Mehring, M., Roth, S. (eds.) Electronic properties of polymers: orientation and dimensionality of conjugated systems, vol. 107, p. 125. Springer, New York (1992)

1355. Henry, F., Pichot, C., Kamel, A., Aaser, M.S.E.: Colloid Polym. J. **267**, 48 (1989), and references therein

1356. Olmedo, L., Hourquebie, P., Jousse, F.: Synth. Met. **69**, 205 (1995)

1357. Hourquebie, P.: Influence of structural parameters of conducting polymers on their microwave properties. In: Garito, A.F., Jen, A.K.-Y., Lee, C.Y.-C., Dalton, L.R. (eds.) Electrical, optical, and magnetic properties of organic solid state materials, vol. 328, p. 239. Materials research Soceity, Pittsburgh (1994)

1358. Shim, H.-K., Kim, H.-J., Ahn, T., Kang, I.-N., Zyung, T.: Synth. Met. **91**, 289 (1997)

1359. Tubino, R., Botta, C., Destri, S., Porzio, W., Rossi, L.: Optical properties and photoexcitations in regularly alternating conjugated copolymers. In: Garito, A.F., Jen, A.K.-Y., Lee, C.Y.-C., Dalton, L.R. (eds.) Electrical, optical, and magnetic properties of organic solid state materials, vol. 328, p. 673. Materials Research Society, Pittsburgh (1994)

1360. Greenham, N.C., Cacialli, F., Bradley, D.D.C., Friend, R.H., Moratti, S.C., Holmes, A.B.: Cyano-derivative of poly (P-Phenylene Vinylene) for use in thin-film light-emitting diodes. In: Electrical, optical, and magnetic properties of organic solid-state materials, vol. 328, p. 351. Materials Research Society, Pittsburgh (1994)

1361. Orenstein, J.: Photoexcitations of conjugated polymers. In: Skotheim, T.A. (ed.) Handbook of conducting polymers, vol. 2, p. 1297. Marcel Dekker, Inc, New York (1986)

1362. Garten, F., Vrijmoeth, J., Schlatmann, A.R., Gill, R.E., Klapwijk, T.M., Hadziioannou, G.: Effect of the top electrode work function on the rectification ratio of Light Emitting Diodes (LEDs) based on poly(3-octylthiophene). In: Yang, S.C., Chandrasekhar, P. (eds.) Optical and photonic applications of electroactive and conducting polymers, vol. 2528, p. 81. SPIE-The International Society for Optical Engineering, Bellingham (1995.) and reference therein

1363. Vardeny, Z., Ehrenfreund, E., Brafman, O. Heeger, A.J. Wudl, F.: Synth. Met. **18**, 183 (1987), and references therein

1364. Moraes, F., Schaffer, H., Kobayashi, M., Heeger, A.J.: Wudl Phys. Rev. B **30**, 2948 (1984)

1365. Masters, J.G., MacDiarmid, A.G., Kim, K., Ginder, J.M., Epstein, A.J.: Bull. Am. Phys. Soc. **36**, 377 (1991)

1366. Hilberer, A., Wildeman, J., Brouwer, H.-J., Garten, F., Hadziioannou, G.: Conjugated block copolymers for light-emitting diodes. In: Yang, S.C., Chandrasekhar, P. (eds.) Optical and photonic applications of electroactive and conducting polymers, vol. 2528, p. 74. SPIE-The International Society for Optical Engineering, Bellinghma (1995)

1367. Epstein, A.J.: The polyaniliens: model systems for diverse electronic phenomena. In: Brédas, J.L., Silbey, R. (eds.) Conjugated polymers: the novel science and technology of highly conducting and nonlinear optically active materials, p. 211. Kluwer Academic Publishers, Norwell (1991)

1368. Schulz, B., Kaminorz, Y., Brehmer, L.: Synth. Met. **84**, 449 (1997)

1369. Dridi, C., Chaieb, A., Hassen, F., Majdoub, M., Gamoudi, M.: Synth. Met. **90**, 233 (1997)

1370. Kim, H.-K., Ryu, M.-K., Kim, K.-D., Lee, J.-H., Park, J.-W.: Synth. Met. **91**, 297 (1997)

1371. (a) Dagani, R.: Devices based on electro-optic polymers begin to enter marketplace. Chem. Eng. News. **74**(10), 22–27 (1996); (b) Marder, Perry, Staehelin, M., Zysset, B., et al.: Science **271**, 335 (1996)

1372. Spence, R.: Aluminum with fluorinated ligands. Chem. Eng. News. **74**(21), 4–1996

1373. Sauteret, C., Hermann, J.-P., Frey, R., Pradère, F., Ducuing, J., Baughman, R.H., Chance, R. R.: Phys. Rev. Lett. **36**, 956 (1976)

1374. Drury, M.R.: Solid State Commun. **68**, 417 (1988)

1375. Dennis, W.M., Blau, W., Bradley, D.J.: Appl. Phys. Lett. **47**, 200 (1985)

1376. Sinclair, M., McBranch, D., Moses, D., Heeger, A.J.: Synth. Met. **28**, D645 (1989)

1377. Chollet, P.-A., Kajzar, F., Messier, J.: Synth. Met. **18**, 459 (1987)

1378. Kanetake, T., Ishikawa, K., Hasegawa, T., Koda, T., Takeda, K., Hasegawa, M., Kubodera, K., Kobayashi, H.: Appl. Phys. Lett. **54**, 2287 (1989)

1379. Neher, D., Wolf, A., Bubeck, C., Wegner, G.: Chem. Phys. Lett. **163**, 116 (1989)

1380. Byrne, H.J., Blau, W., Jen, K.-Y.: Synth. Met. **32**, 229 (1989)

1381. Ghoshal, S.K.: Chem. Phys. Lett. **158**, 65 (1989)

1382. Lindle, J.R., Bartoli, F.J., Hoffman, C.A., Kim, O.-K., Lee, Y.S., Shirk, J.S., Kafafi, Z.H.: Appl. Phys. Lett. **56**, 712 (1990)

1383. Cao, X.F., Jiang, J.P., Bloch, D.P., Hellwarth, R.W., Yu, L.P., Dalton, L.: J. Appl. Phys. **65**, 5012 (1989)

1384. Agrawal, A.K., Jenekhe, S.A., Vanherzeele, H., Meth, J.S.: Third-order nonlinear optical properties of a series of systematically designed conjugated rigid-rod polyquinolines. In: Chiang, L.Y., Garito, A.F., Sandman, D.J. (eds.) Electrical, optical, and magnetic properties of organic solid state materials, vol. 247, p. 253. Materials Research Society, Pittsburgh

1385. Guo, D., Mazumadar, S., Stegeman, G.I., Cha, M., Neher, D., Aramaki, S., Torruellas, W., Zanoni, R.: Nonlinear optics of linear conjugated polymers. In: Electrical, optical, and magnetic properties of organic solid state materials, vol. 147, p. 151. Materials Research Society, Pittsburgh (1992)

1386. Samoc, M., Samoc, A., Luther-Davies, B., Scherf, U.: Synth. Met. **87**, 197 (1997)

1387. Moon, K.-J., Lee, K.-S., Shim, H.-K.: Synth. Met. **71**, 1719 (1995)

1388. Spangler, C.W., He, M.Q., Laquindanum, J., Dalton, L., Tang, N., Partanen, J., Hellwarth, R.: Bipolaron formation and nonlinear optical properties in bis-thienyl polyenes. In: Electrical, optical, and magnetic properties of organic solid state materials, vol. 328, p. 655. Materials Research Society, Pittsburgh (1994)

1389. Okawa, H., Wada, T., Sasabe, H.: Synth. Met. **84**, 265 (1997)

1390. Meyer, R.K., Benner, R.E., Vardeny, Z.V., Liess, M., Ozaki, M., Yoshino, K., Ding, Y., Barton, T.: Synth. Met. **84**, 549 (1997)

1391. Matsuda, H., Shimada, S., Takeda, H., Masaki, A., Van Keuren, E., Yamada, S., Hayamizu, K., Nakanishi, F., Okada, S., Nakanishi, H.: Synth. Met. **84**, 909 (1997)

1392. Tripathy, S.K., Kim, W.H., Masse, C.E., Bihari, B., Kumar, J.: Novel polydiacetylenes as materials for second and third order nonlinear optics. In: Garito, A.F., Jen, A.K.-Y., Lee, C. Y.-C., Dalton, L.R. (eds.) Electrical, optical, and magnetic properties of organic solid state materials, vol. 328, p. 667. Materials Research Society, Pittsburgh (1994.) and references therein

1393. Cai, Y.M., Jen, A.K.-Y., Liu, Y.J., Chen, T.A.: Highly active and thermally stable nonlinear optical polymers for electro-optical applications. In: Optical and photonic applications of electroactive and conducting polymers, vol. 2528, p. 128. SPIE-The International Society for Optical Engineering, Bellingham (1995)

1394. Shank, C.V., Yen, R., Fork, R.L., Orenstein, J., Baker, G.L.: Phys. Rev. B. **49**, 1660 (1982)

1395. Zheng, L.X., Feng, Z.G., Knopf, F.C.: Polymer Prepr. **37**, 107 (1996), and references therein

1396. Vardeny, Z., Strait, J., Moses, D., Chung, T.-C., Heeger, A.J.: Phys. Rev. Lett. **49**, 1657 (1982)

1397. Kobayashi, T.: Synth. Met. **71**, 1663 (1995)

1398. Oamg, Y., Prasad, P.N.: J. Chem. Phys. **93**, 2201 (1990)

1399. Weinberger, B.R., Kaufer, J., Heeger, A.J., Pron, A., MacDiarmid, A.G.: Phys. Rev. B. **20**, 223 (1979)

1400. Epstein, A.J., Rommelmann, H., Druy, M.A., Heeger, A.J., MacDiarmid, A.G.: Solid State Commun. **38**, 683 (1981)

1401. Shirakawa, H., Ito, T., Ikeda, S.: Makromol. Chem. **179**, 1565 (1978)

1402. Mizoguchi, K., Kachi, N., Sakamoto, H., Kume, K., Yoshioka, K., Masubuchi, S., Kazama, S.: Synth. Met. **84**, 695 (1997)
1403. Kitao, S., Matsuyama, T., Seto, M., Maeda, Y., Masubuchi, S., Kazama, S.: Synth. Met. **69**, 371 (1995)
1404. Suwalski, J., Pron, A., Zucharski, Z.: Mater. Sci. Forum. **21**, 125 (1987)
1405. Hepel, M., Chen, Y.-M., Stephenson, R.: J. Electrochem. Soc. **143**, 498 (1996)
1406. Thyssen, A., Hochfeld, A., Schultze, J.W.: Investigations of the anodic polymerisation of ortho- and meta-toludine with the EQMB. In: Plocharski, J., Roth, S. (eds.) Materials science forum, vol. 42, p. 151. Trans Tech Publications, Switzerland (1989)
1407. Shilov, V.E., Shilova, A.S.: Application of neutron spectoscopy in investigation of magnetic state of pi-electron systems. In: Gordeev, M.E. (ed.) Fiz.-Khim. Metody Issled. Strukt. Din. Mol. Sist. Mater. Vseros. Sovesch, vol. 2, p. 134 (1994)
1408. Wei, Z.X., Zhang, L.J., Yu, M., Yang, Y.S., Wan, M.X.: Self-assembling sub-micrometer-sized tube junctions and dendrites of conducting polymers. Adv. Mater. **15**, 1382–1385 (2003)
1409. Long, Y.-Z., Meng-Meng, L., Changzhi, G., Meixiang, W., Jean-Luc, D., Zongwen, L., Zhiyong, F.: Recent advances in synthesis, physical properties and applications of conducting polymer nanotubes and nanofibers. Progress in Polymer Science. **36**, 1415–1442 (2011)
1410. Purcell, K.F., Kotz, J.C.: Inorganic chemistry, p. 973. WB Saunders Co, Philadelphia (1977), and refs therein
1411. Nigrey, R.J., Heeger, A.J., MacDiarmid, A.G.: Mol. Crist. Liq. Crist. **83**, 309 (1982)
1412. Kuzmany, H.: Pure Appl. Chem. **57**, 235 (1985)
1413. Bernier, P., Rolland, M., Linaya, C., Aldissi, M.: Polymer. **21**, 7 (1980)
1414. Chien, J.C.W., Karasz, F.E., Wnek, G.E., Heeger, A.J., MacDiarmid, A.G.: J. Poly. Sci. Polym. Lett. Ed. **18**, 45 (1980)
1415. Daniels, W.E.: J. Org. Chem. **29**, 2936 (1964)
1416. Luttinger, L.B.: Chem. And Ind. (1960)
1417. Aldissi, M., Linaya, C., Sledz, J., Schue, F., Giral, L., Fabre, J.M., Rolland, M.: Polymer. **23**, 243 (1982)
1418. MacInnes Jr., D., Druy, M.A., Nigrey, P.J., Nairns, D.P., MacDiarmid, A.G., Heeger, A.J.: J. Chem. Soc. Chem. Commun. 317 (1981)
1419. Kaner, R.B., MacDiarmid, A.G.: Synth. Met. **14**, 3 (1986)
1420. Schacklette, L.W., Toth, J.E., Murthy, N.S., Baughman, R.H.: J. Electrochem. Soc. **132**, 1529 (1985)
1421. Farrington, G.C., Scrosati, B., Frydrych, D., de Nuzzio, J.: J. Electrochem. Soc. **131**, 7 (1984)
1422. Chiang, C.K., Blubaugh, E.A., Yap, W.T.: Polymer. **25**, 1112 (1984)
1423. Kietz, K.H., Beck, F.: J. Appl. Electrochem. **15**, 159 (1985)
1424. Nigrey, P.J., MacDiarmid, A.G., Heeger, A.J.: J.C.S. Chem. Comm. 594 (1979)
1425. Will, F.G.: J. Electrochem. Soc. **132**, 743 (1985)
1426. Begin, D., Billaud, D., Goulon, G.: Synth. Met. **11**, 29 (1985)
1427. Jánossy, A., Pogány, L, Pekker, S. Swietlik, R.: Mol. Cryst. Liq. Cryst. **77**, 185(1981)
1428. Benoit, C., Rolland, M., Aldissi, M., Rossi, A., Cadene, M., Bernier, P.: Phys. Stat. Sol. A. **68**, 209 (1981)
1429. Yamamoto, T., Yamamoto, A.: Chem. Soc. Jap. Chem. Lett. 353 (1977)
1430. Shirakawa, H., Ikeda, S.: Synth. Met. **1**, 175 (1979/80)
1431. Lugli, G., Pedretti, U., Perego, G.: J. Polym. Sci.: Polym. Lett. **23**, 129 (1985)
1432. Leising, G., Uitz, R., Ankele, B., Ottinger, W., Stelzer, F.: Mol. Cryst. Liq. Cryst. **117**, 327 (1985)
1433. Naarmann, H., Theophilou, N.: Synth. Met. **22**, 1 (1987)
1434. Swager, T.M., Dougherty, D.A., Grubbs, R.H.: J. Am. Chem. Soc. **110**, 2973 (1988)
1435. Masuda, T., Higashimura, T.: Adv. Polym. Sci. **81**, 121 (1986)
1436. Wagner, G.: Pure Appl. Chem. **49**, 443 (1997)

1437. Aime, J.-P.: Structural characterization of conjugated polymer solutions in the undoped and doped state. In: Brédas, J.L., Silbey, R. (eds.) Conjugated polymers: the novel science and technology of highly conducting and nonlinear optically active materials, p. 229. Kluwer Academic Publishers, Norwell (1991)

1438. (a) Patel, G.N.: J. Polym. Sci. Letters Ed. **16**, 607 (1978); (b) Patel, G.N., Walsh, E.K.: J. Polym. Sci. Lett. Ed. **17**, 203 (1979)

1439. Patel, G.N., Witt, J.D., Khanna, Y.P.: J. Polym. Sci. Polym. Phys. Ed. **18**, 1383 (1980)

1440. Yang, Y., Lee, J.Y., Li, L., Kumar, J., Jain, A.K., Tripathy, S.K.: Polarization dependent photocurrent in thin film polydiacetylene single crystals. In: Chiang, L.Y., Garito, A.F., Sandma, D.J. (eds.) Electrical, optical, and magnetic properties of organic solid state, vol. 247, p. 729. Materials Research Society, Pittsburgh (1992)

1441. (a)Kajzar, F., Messier, J.: Third order nonlinear optical effects in conjugated polymers. In: Brédas, J.L., Silbey, R. (eds.) Conjugated polymers: the novel science and technology of highly conducting and nopnlinear optically active materials, p. 509. Kluwer Academic Publishers, Norwell (1991); (b) Singh, B. P., Prasad, P. N.: J. Opt. Soc. Am. **B5**, 453 (1988); (c) Blau, W.: Opt. Commun. **64**, 85 (1987)

1442. Ohnuma, H., Hasegawa, K., Se, K., Kotaka, T.: Macromolecules. **18**, 2341 (1985)

1443. Winter, M., Grupp, A., Mehring, M., Sixl, H.: Chem. Phys. Lett. **133**, 482 (1987)

1444. Elsenbaumer, R.L., Shacklette, L.W.: Pheneylene-based conducting polymers. In: Skotheim, T.A. (ed.) Handbook of conducting polymers, vol. 1, p. 213. Marcel Dekker, Inc., New York (1986)

1445. Diaz, A.F., Hall, B.: IBM J. Res. Dev. **27**, 342 (1983)

1446. Otero, T.F., Rodríguez, J.: Electrochemomechanical and electrochemopositioning devices: artificial muscles. In: Aldissi, M. (ed.) Intrinsically conducting polymers: an emerging technology, p. 179. Kluwer Academic Publishers, Boston (1993.) and references therein

1447. Salmon, M., Diaz, A.F., Logan, A.J., Krounbi, M., Bargon, J.: Mol. Cryst. Liq. Cryst. **83**, 1297 (1983)

1448. Ouyang, J., Li, Y.: Polymer. **38**, 3997 (1997)

1449. Diaz, A.F., Kanazawa, K.K.: In: Miller, J.S. (ed.) Extended linear chain compounds, p. 417. Plenum, New York (1982)

1450. Bargon, J., Mohmand, S., Waltman, R.J.: IBM, J. Res. Dev. **27**, 330 (1983)

1451. Salmon, M., Diaz, A., Goitia, J.: Chemically modified surfaces. In: Miller J. S. (ed.) Catalysis and electrocatalysis; Am. Chem. Soc. Symp. Ser. vol. 192, p. 65(1982)

1452. (a) Epstein, A.J., Ginder, J.M., Zuo, F., Bigelow, R.W., Woo, H.-S., Tanner, D.B., Richter, A.F., Huang, W.- S., MacDiarmid, A.G.: Synth. Met. **18**, 303 (1987); (b) Ginder, J.M., Richter, A.F., MacDiarmid, A.G., Epstein, A.J.: Solid State Commun. **63**, 97 (1987)

1453. Shacklette, L.W., Han, C.C.: Solubility and dispersion characteristics of polyaniline. In: Dalton, L.R. (ed.) Electrical, optical, and magnetic propperties of organic solid state materials, vol. 328, p. 157. Materials Research Society, Pittsburgh (1994)

1454. MacDiarmid, A.G., Epstein, A.J.: Conducting polymers: past, present, and future.... In: Garito, A.F., Jen, A.K.-Y., Lee, C.Y.-C., Dalton, L.R. (eds.) Electrical, optical, and magnetic properties of organic solid state materials, vol. 328, p. 133. Materials Research Society, Pittsburgh (1994)

1455. (a) Andreatta, A., Cao, Y., Chiang, J.C., Heeger, A.J., Smith, P.: Synth. Met. **26**, 383 (1988); (b) Cao, Y., Smith, P., Heeger, J.: Synth. Met. **32**, 263 (1989)

1456. Tang, H., Kitani, A., Shiotani, M.: J. Electrochem. Soc. **143**, 3079 (1996)

1457. Pham, M.C., Piro, B., Bassaoui, E.A., Hedayatullah, M., Lacroix, J.-C., Novak, P., Haas, O.: Synth. Met. **92**, 197 (1998)

1458. Hwang, G.-W., Wu, K.-Y., Hua, M.-Y., Lee, S.-T., Chen, S.-A.: Synth. Met., **92**, 39 1998

1459. Naoi, K., Kawase, K.-I., Mori, M., Komiyama, M.: J. Electrochem. Soc. **144**, L173 (1997)

1460. Dao, L.H., Nguyen, M.T., Paynter, R.: Synth. Met. **41–43**, 649 (1998)

1461. Wang, X., Yang, Q., Luo, X.: Shandong Jiancai Xueyuan Xuebao. **10**, 19 (1996)

1462. Pham, M.-C., Oulahyane, M., Mostefai, M., Chehimi, M.M.: Synth. Met. **93**, 89 (1998)

1463. Mostefai, M., Pham, M.-C., Marsault, J.-P., Aubard, J., Lacaze, P.-C.: J. Electrochem. Soc. **143**, 2116 (1996)

1464. Yamamoto, T., Sanechika, K., Yamamoto, A.: J. Polm. Sci. Polym. Lett.Ed. **18**, 9 (1980)

1465. Gorman, C.B., Grubbs, R.H.: Conjugated polymers: the interplay between synthesis, structure, and properties. In: Brédas, J.L., Silbey, R. (eds.) Conjugated polymers: the novel science and technology of highly conducting and nonlinear optically active materials, vol. 1. Kluwer Academic Publishers, Norwell (1991)

1466. Sugimoto, R., Takeda, S., Gu, H.B., Yoshino, K.: Chem. Express. **1**, 635 (1986)

1467. Yamamoto, T., Morita, A., Maruyama, T., Zhou, Z., Kanbara, T., Sanechika, K.: Polym. J. **22**, 187 (1990)

1468. Amer, A., Zimmer, H., Mulligan, K.J, Mark, H.B., Pons, S., McAleer, J.F.: J. Polym. Sci. Polym. Lett. Ed. **22**, 77(1984)

1469. Zhang, Q.T., Tour, J.M.: J. Am. Chem. Soc. **119**, 5065 (1997)

1470. (a) Hotta, S., Hosaka, T., Shimotsuma, W.: Synth. Met. **6**, 317 (1983); (b) Druy, M.S., Seymour, R.J.: J. Phys. Paris Colloq. C3 **44**, 595 (1983)

1471. Yamamoto, T., Sanachika, K., Yamamoto, A.: Bull. Chem. Soc. Jpn. **56**, 1497 (1983)

1472. Tourillon, G., Garnier, F.: J. Phys. Chem. **87**, 2289 (1983)

1473. Waltman, R.J., Bargon, J., Diaz, A.F.: J. Phys. Chem. **87**, 2289 (1983)

1474. Torsi, L., Giglio, E.D., Sabbatini, L., Zambonin, P.G.: J. Electrochem. Soc. **141**, 2608 (1994)

1475. Welzel, H.-P., Kossmehl, G., Boettcher, H., Engelmann, G., Hunnius, W.-D.: Macromolecules. **30**, 7419 (1997)

1476. Zhang, X., Shen, X., Yang, S., Lu, W., Zhang, J.: Chin. J. Polym. Sci. **14**, 330 (1996)

1477. Satoh, M., Imanishi, K., Yasuda, Y., Tsushima, R., Yamasaki, H., Aoki, S., Yoshino, K.: Synth. Met. **30**, 33 (1989)

1478. Ng, S.C., Fu, P., Yu, W.-L., Chan, H.S.O., Tan, K.L.: Synth. Met. **87**, 119 (1997)

1479. Guillerez, S., Bidan, G.: Synth. Met. **93**, 123 (1998)

1480. Dogbéavou, R., El-Mehdi, N., Naudin, E., Breau, L., Bélanger, D.: Synth. Met. **84**, 207 (1997)

1481. Hide, F., Greenwald, Y., Wudl, F., Heeger, A.J.: Synth. Met. **85**, 1255 (1997)

1482. Poverenov, E., Li, M., Bitler, A., Bendikov, M.: Major effect of electropolymerization solvent on morphology and electrochromic properties of PEDOT films. Chem. Mater. **22**, 4019–4025 (2010)

1483. Luo, S., Ali, E., Tansil, N., Yu, H., Gao, S., Kantchev, E., Ying, J.: Poly (3,4-ethylenedioxythiophene) (PEDOT) nonbiointerfaces: thin, ultrasmooth, and functionalized PEDOT films with in vitro and in vivo biocompatability. Langmuir. **24**, 8071–8077 (2008)

1484. Xiao, Y., Lin, J., Tai, S., Chou, S., Yue, G., Wu, J.: Pulse electropolymerization of high performance PEDOT/MWCNT counter electrodes for Pt-free dye-sensitized solar cells. J. Mater. Chem. **22**, 19919–19925 (2012)

1485. McFarlane, S., Deore, B., Svenda, N., Freund, M.: A one-step, organic-solvent processable synthesis of PEDOT thin films via in situ metastable chemical polymerization. Macromolecules. **43**, 10241–10245 (2010)

1486. Cho, W., Im, S., Kim, S., Kim, S., Kim, J.: Synthesis and characterization of PEDOT: P (SS-co-VTMS) with hydrophobic properties and excellent thermal stability. Polymers. **8**, 189–199 (2016)

1487. Zhang, X., Lee, J., Lee, G., Cha, D., Kim, M., Yang, D., Manohar, S.: Chemical synthesis of PEDOT nanotubes. Macromolecules. **39**, 470–472 (2006)

1488. Zhang, X.: Chemical synthesis of PEDOT nanofibers. Chem. Comm. **42**, 5328–5330 (2005)

1489. Winther-Jensen: High rates of oxygen reduction over a vapor phase-polymerized PEDOT. Electrode. Science. 321 (2008)

1490. Paradee, N., Sirivat, A.: Synthesis of poly(3,4-ethylenedioxythiophene) nanoparticles by chemical oxidation polymerization. Polym. Int. **63**, 106–113 (2013)

1491. Krishnamoorthy, K., Ambade, A.V., Kanungo, M., Contractor, A.Q., Kumar, A.J.: Mat. Chem. **11**, 2909–2911 (2001)

1492. Chandrasekhar, P., Zay, B.J., Cai, C., Chai, Y., Lawrence, D.: Matched-dual-polymer electrochromic lenses, using new cathodically-coloring conducting polymers, with exceptional performance and incorporated into automated sunglasses, Ashwin-Ushas Corporation, 1–47

1493. Kobayashi, M., Colaneri, N., Boysel, M., Wudl, F., Heeger, A.J.: J. Chem. Phys. **82**, 5717 (1985)

1494. Yashima, H., Kobayashi, M., Lee, K.-B., Chung, D., Heeger, A.J., Wudl, F.: J. Electrochem. Soc. **134**, 46 (1987)

1495. (a) Cava, M.P., Deana, A.A.: J. Am. Chem. Soc. **81**, 4266 (1959); (b) Cava, M.P., Pollack, N.M.: J. Am. Chem. Soc. **88**, 4112 (1966); (c) Cava, M.P., Pollack, N.M., Mamer, O.A., Mitchell, M.J.: J. Org. Chem. **36**, 3932 (1971)

1496. (a) Okiver, J.A., Ongley, P.A.: Chem. Ind. **12**, 1024 (1965); (b) Iddon, B. Adv. Heterocycl. Chem. **14**, 331 (1972)

1497. Jen, K.Y., Elsenbaumer, R.: Synth. Met. **16**, 379 (1986)

1498. (a) Wudl, F., Kobayashi, M., Heeger, A.J.: J. Org. Chem. **49**, 3382 (1984); (b) Chen, S.-A., Lee, C.-C., Method for preparing processable polyisothianaphthene. US 5510457 A, 23 Apr 1996, and references therein

1499. Schlick, U., Teichert, F., Hanack, M.: Synth. Met. **92**, 75 (1998)

1500. Kossmehl, G.: Ber. Bunsenges. Phys. Chem. **83**, 417 (1979)

1501. Jen, K.-Y., Maxfield, M., Shacklette, L.W., Elsenbaumer, R.L.: J. Chem. Soc., Chem. Commun. **309** (1987)

1502. Catellani, M., Porzio, W., Musco, A., Pontellini, R.: Synthesis and characterization of solluble alkyl substituted poly(2,5-Thienylene Vinylenes). In: Chiang, L.Y., Garito, A.F., Sandman, D.J. (eds.) Electrical, optical, and magnetic properties of organic solid state materials, vol. 247, p. 681. Materials Research Society, Pittsburgh (1992)

1503. Chandrasekhar, P.:Nonlinear optical properties of conducting polymer/semiconductor interfaces. In: Proc. 38th Sagamore Army Mat. Res. Conf.; Plymouth, MA, USA, 19991

1504. Thompson, K.G., Bryan, C.J., Benicewicz, B.C., Wrobleski, D.A.: Corrosion-protective coatings from electrically conducting polymers, Los Alamos National Laboratory, Internal Report #LA-UR-92-360, 338–347 (1992)

1505. Wrobleski et al.: Corrosion resistant coating. US5658649 A1, 19 Aug 1997, and references therein

1506. Wolf, J. F. et al.: Adjustable tint window with electro chromic conductive polymer. US 5042923 A, 27 Aug 1991, and references therein

1507. Ng, S.C., Chan, H.S.O., Miao, P., Tan, K.L.: Synth. Met. **90**, 25 (1997)

1508. Inaoka, S., Collard, D.M.: Synth. Met. **84**, 193 (1997)

1509. Pomerantz, M., Gu, X.: Synth. Met. **84**, 243 (1997)

1510. Lambert, T.L., Ferraris, J.P.: J. Chem. Soc. Chem. Commun. 752 (1991)

1511. Ferraris, J.P., Lambert, T.L.: J. Chem. Soc. Chem. Commun. 1268 (1991)

1512. Gunatunga, S.R., Jones, G.W., Kalaji, M., Murphy, P.J., Taylor, D.M., Williams, G.O.: Synth. Met. **84**, 973 (1997)

1513. Beyer, R., Kalaji, M., Kingscote-Burton, G., Murphy, P.J., Pereira, V.M.S.C., Taylor, D.M., Williams, G.O.: Synth. Met. **92**, 25 (1998)

1514. Lima, A., Schottland, P., Sadki, S., Chevrot, C.: Synth. Met. **93**, 33 (1998)

1515. Akoudad, S., Roncali, J.: Synth. Met. **93**, 111 (1998)

1516. Ng, S.C., Xu, J.M., Chan, H.S.O.: Synth. Met. **92**, 33 (1998)

1517. Emge, A., Bäuerle, P.: Synth. Met. **84**, 213 (1997)

1518. Kovacic, P., Kyriakis, A.: J.A.C.S. **85**, 454 (1963)

1519. Fauvarque, J.F., Petit, M.A., Pfluger, F., Jutand, A., Chevrot, C., Troupel, M.: Macromol. Chem. Rapid. Comm. **4**, 455 (1983)

1520. Ballard, D.G.H., Courtis, A., Shirley, I.M., Taylor, S.C.: Macromolecules. **21**, 294 (1988)

1521. McKean, D.R., Stille, J.K.: Macromolecules. **20**, 1787 (1987)
1522. Chem. Abstr. **63**: Fobe (1965); Netherlands Patent Appl. 6,402,650 (1964)
1523. Aeiyach, S., Soubiran, P., Lacaze, P.C., Froyer, G., Pelous, Y.: Synth. Met. **32**, 103 (1989)
1524. Goldenberg, L.M., Pelekh, A.E., Krinichnyi, V.L., Roshchupkina, O.S., Zueva, A.F., Lyubovskaya, R.N., Efimov, O.N.: Synth. Met. **36**, 217 (1990)
1525. Tsuchida, E., Yamamoto, K., Asada, T., Nishide, Y.: Chem. Lett. 1541 (1987)
1526. Brilmyer, G., Jasinski, R.: J. Electrochem. Soc. **129**, 1950 (1982)
1527. Rubinstein, I.: J. Electrochem. Soc. **130**, 1506 (1983)
1528. Li, C., Shi, G., Liang, Y., Ye, W., Sha, Z.: Polymer. **38**, 5023 (1997)
1529. Santos, D.A.D., Galvao, D.S., Laks, B., Dezotti, M.W.C., DePaoli, M.A.: Chem. Phys. **144**, 103 (1990), and references therein
1530. Maafi, M., Lion, C., Aaron, J.J.: Synth. Met. **83**, 167 (1996)
1531. Martínez, Y., Hernández, R., Kalaji, M., Márquez, J., Márquez, O.P.: Synth. Met. **93**, 9 (1998)
1532. Larmat, F., Soloducho, J., Katritzky, A.R., Reynolds, J.R.: J. Electrochem. Soc. **143**, L161 (1996)
1533. Wessling, R.A.: J. Polym. Sci. Polym. Symp. **72**, 55 (1985)
1534. Bradley, D.D.C., Evans, G.P., Friend, R.H.: Synth. Met. **17**, 651 (1987)
1535. Wudl, F.: Proc. of the Int. Conf. on Synth. Met., Santa Fe 1988
1536. Shi, S., Wudl, F.: Conjugated polymeric materials: opportunities in electronics, optoelectronics and molecular electronics. In: Brédas, J.L., Chance, R.R. (eds.) NATO ASI Series, Series E: applied sciences, vol. 82, p. 83. Kluwer Academic Publishers, Dordrecht (1990)
1537. Hilberer, A., Brouwer, H.-J., van der Scheer, B.-J., Wildeman, J., Hadziioannou, G. Macromolecules **18** (23) (1995)
1538. Heck, R.F.: Org. Reac. **27**, 345 (1982)
1539. Remmers, M., Schulze, M., Wegner, G.: Macromolecular Rapid Commun. **17**, 239 (1996)
1540. Stephens, J.A., Friend, R.H., Remmers, M., Neher, D.: Synth. Met. **84**, 645 (1997)
1541. Ueda, M., Hayakawa, T., Haba, O., Kawaguchi, H., Inoue, J.: Macromolecules. **30**, 7069 (1997)
1542. Barashkov, N.N., Novikova, T.S., Ferraris, J.P.: Synth. Met. **83**, 39 (1996)
1543. Concentrates. Chem. Eng. News **76**(11), 14–15 (1998)
1544. Shacklette, L.W., Elsenbaumer, R.L., Chance, R.R., Eckhardt, H., Frommer, J.E., Baughman, R.H.: J. Chem. Phys. **75**, 1919 (1980)
1545. Elsenbaumer, R.L., Shacklette, L.W., Sowa, J.M., Baughman, R.H.: Mol. Cryst. Liq. Cryst. **83**, 229 (1982)
1546. Lenz, R.W., Handlovits, C.E., Smith, H.A.: J. Polym. Sci. **58**, 351 (1962)
1547. Schoch, K.F.: Polym. Prepr. **25**, 278 (1984), and references therein
1548. Edmonds, J. T., Jr., Hill, H.W.: JR. Selective cleavage of monoami-dotriphosphates to produce orthophosphoramidates. US 3345129 A, 3 Oct 1967
1549. (a) Aldissi, M., Liepins, R.: J. Chem. Soc. Chem. Commun. 255 (1984); (b) Hopf, K.: Agitator for a drill and related methods. US 8025111 B2, 27 Sept 2011
1550. Ding, Y., Hay, A.S.: Macromolecules. **30**, 1849 (1997)
1551. Hanack, M., Datz, A., Fay, R., Fischer, K., Keppeler, U., Koch, J., Metz, J., Mezger, M., Schneider, O., Schulze, H.-J.: Synthesis and properties of conducting bridged acrocyclic metal complexes. In: Kotheim, T.A. (ed.) Handbook of conducting polymers, vol. 1, p. 133. Marcel Dekker, Inc., New York (1986)
1552. Acampora, L.A., Dugger, D.L., Emma, T., Mohammed, J., Rubner, M., Samuelson, L., Sandman, D.J., Tripathy, S.: Symp. Ser. **242**, 461 (1984)
1553. Peulon, V., Barbey, G., Malandain, J.-J.: Synth. Met. **82**, 111 (1996)
1554. Shim, Y.-B. Park, S.-M. J. electrochem. Soc. **144**, 3027 (1997)
1555. (a) Van Deusen, R.L.: J. Polym. Sci. B **14**, 211 (1966); (b) Arnold, F.E., Van Deusen, R.L.: Macromol. **2**, 497 (1969)

1556. Agrawal, A. K. Wang, C., Song, H. H: High temperature electrical conductivity of solution-cast bbl films. In Electrical, optical, and magnetic properties of organic solid state materials; Garito, A. F.; Jen, A. K.-Y.; Lee, C.Y.-C.; Dalton, L. R. Eds.; Materials Research Society, Pittsburgh, 1994; Vol. 328, p. 279.

1557. Kim, O.-K.: Mol. Cryst. Liq. Cryst. **105**, 161 (1984)

1558. Wilbourn, K., Murray, R.W.: Macromol. **21**, 89 (1988)

1559. Kim, O.K.: J. Polym. Sci. Polym. Lett. Ed. **20**, 663 (1982)

1560. Debad, J.D., Bard, A.J.: J. Am. Chem. Soc. **120**, 2476 (1998)

1561. Mittler-Neher, S., Otomo, A., Stegeman, I., Lee, C.Y.-C., Mehta, R., Agrawal, A.K., Jenekhe, S.A.: Appl. Phys. Lett. **62**, 115 (1993)

1562. Stille, J.K.: Macromol. **14**, 870 (1981)

1563. Papir, Y.S., Kurkov, V.P., Current, S.P.: Extended Abstracts, 83-1, Abs. 544, Electrochemical Society Meeting, San Francisco (1983)

1564. Schroeder, A.H., Papir, Y.S., Kurkov, V.P.: Extended Abstracts, 83-1, Abs. 544, Electrochemical Society Meeting, San Francisco (1983)

1565. Agrawal, A.K., Jenekhe, S.A., Vanherzeele, H., Meth, J.S.: Chem. Mater. **3**, 765 (1991)

1566. El-Shekeil, A.G., Al-Saady, H.A., Al-Yusufy, F.A.: Polym. Int. **44**, 78 (1997)

1567. González-Tejera, M.J., Carrillo, I., Hernández-Fuentes, I.: Synth. Met. **92**, 187 (1998)

1568. Belloncle, C., Fabre, B, Cauliez, P., Simonet, J.: Synth. Met. **93**, 115(1998)

1569. Pandey, P.C., Prakash, R.: J. Electrochem. Soc. **145**, 999 (1998)

1570. Talbi, H., Billaud, D.: Synth. Met. **93**, 105 (1998)

1571. Fennell, J.F., Liu, S.F., Azzarelli, J.M., Weis, J.G., Rochat, S., Mirica, K.A., Ravnsbæk, J. B., Swager, T.M.: Nanowire chemical/biologiccal sensors: status and a roadmap for the future. Angew. Chem. Int. Ed. **55**, 1266–1281 (2016).) (b) 908 Devices: Safety & Security. http://908devices.com/markets/safety-security/.Accessed Sept 2016

1572. Guimard, N.K., Gomez, N., Schmidt, C.E.: Conducting polymers in biomedical engineering. Prog. Polym. Sci. **32**, 876–882 (2007)

1573. Forzani, E.S., Zhang, H., Nagahara, L.A., Amlani, I., Tsui, R., Tao, N.: A conducting polymer nanojunction sensor for glucose detection. Nano Lett. **4**(9), 1785–1788 (2004)

1574. Ogura, K., Shiigi, H., Nakayama, M.: J. Electrochem. Soc. **143**, 2925 (1996)

1575. Ram, M.K., Yavuz, Ö., Lahsangah, V., Aldissi, M.: CO gas sensing from ultrathin nano-composite conducting polymer film. Sens. Actuators B. **106**, 750–757 (2005)

1576. (a)Baughman, R.H., Shacklette, L.W.: Application of dopant-induced structure-property changes of conducting polymers. In: Salaneck, W.R., Clark, D.T., Samuelsen, E.J. (eds.) Science and applications of conducting polymers, p. 47. Adam Hilger, New York (1991) (b) Shimidzu, T.: React. Polym. **6**, 221 (1987); (c)Iwakura, C., Kajiya, Y., Yoneyama, H.: J. Chem. Soc. Chem. Chem. Commun. 1019 (1988); (d) Thackeray, J.W., White, H.S., Wrighton, M.S.: J. Chem. **89**, 5133 (1985); (e) Jones, E.T.T., Chyan, O.M., Wrighton, M.S.: J. Am. Chem. Soc. **109**, 55526 (1987)

1577. Ogura, K., Shiigi, H., Nakayama, M.: J. Electrochem. Soc. **143**, 2925 (1996)

1578. Boyle, A., Geniés, E.M., Lapkowski, M.: Synth. Met. **28**, C769 (1989)

1579. Collins, G.E., Buckley, L.J.: Synth. Met. **78**, 93 (1996)

1580. Kincal, D., Kumar, A., Chid, A.D., Reynolds, J.R.: Synth. Met. **92**, 43 (1998)

1581. Ellis, D.L. Zakin, M.R., Bernstein, L.S., Rubner, M.F.: Anal. Chem. **68**, 817 (1996)

1582. Selampinar, F., Toppare, L., Akbulut, U., Yalçin, T., Süzer, S.: Synth. Met. **68**, 109 (1995)

1583. De Rossi, D. Gestri, G. Stella. R. Stussi, E. Proc. Ital. Conf. 1st, Sens. Microsyst. 64 (1996)

1584. Ramanthank, K., Bangar, A.M., Yun, M., Chen, W., Myung, N.V., Mulchandani, A.: Bioaffinity sensing using biologically functionalized conducting-polymer nanowire. J.A.C.S. **127**, 496–497 (2005)

1585. (a) Ramanathan, K., Bangar, M., Yun, M., Chen, W., Mulchandani, A.; (b) Myung, N.V.: Nano Lett. **4**, 1237–1239 (2004); (c) Yun, M., Myung, N.V., Vasquez, R.P.: Lee, C., Menke, E., Penner, R.M.: Nano Lett. **4**, 419–422 (2004)

1586. Kwon, O.S., Park, S.J., Lee, J.S., Park, E., Kim, T., Park, H.-W., You, S.A., Yoon, H., Jang, J.: Multidimensional conducting polymer nanotubes for ultrasensitive chemical nerve agent sensing. Nano Lett. **12**, 2797–2802 (2012)
1587. Yoon, H., Jang, J.: Conducting-polymer nanomaterials for high-performance sensor applications: issues and challenges. Adv. Funct. Mater. **19**, 1567–1576 (2009)
1588. Yoon, H., Chang, M., Jang, J.: J. Phys. Chem. B. **110**, 14074 (2006)
1589. Oh, W.-K., Kwon, O.S., Jang, J.: Conducting polymer nanomaterials for biomedical applications: cellular interfacing and biosensing. Polym. Rev. **53**, 407–442 (2013)
1590. Willner, I., Zayats, M.: Electronic aptamer-based sensors. Angew. Chem. Int. Ed. **46**, 6408–6418 (2007)
1591. Rahman, M.A., Son, J.I., Won, M.S., Shim, Y.B.: Gold nanoparticles doped conducting polymer nanorod electrodes: ferrocene catalyzed aptamer-based thrombin immunosensor. Anal. Chem. **81**, 6604–6611 (2009)
1592. Olowu, R.A., Arotiba, O., Mailu, S.N., Waryo, T.T., Baker, P., Iwuoha, E.: Electrochemical aptasensor for endocrine disrupting 17β-estradiol based on a poly (3, 4-ethylenedioxylthiopene)- gold nanocomposite platform. Sensors. **10**, 9872–9890 (2010)
1593. Dupont-Filliard, A., Billon, M., Livache, T., Guillerez, S.: Biotin/avidin system for generation of fully renewable DNA sensor based on biotinylated polypyrrole film. Anal. Chim. Acta. **414**, 271–277 (2004)
1594. Gerard, M., Chaubey, A., Malhotra, B.D.: Application of conducting polymers to biosensors. Biosens. Bioelectron. **17**, 345–349 (2002)
1595. Oyama, N., Hirokawa, T.: Anal. Chem. **59**, 258 (1987)
1596. Josowicz, M., Janata, J.: Electroactive polymers in chemical sensors. In: Scrosati, B. (ed.) Applications of electroactive Polymers, p. 310. Chapman & Hall, New York (1993)
1597. Brown, C.W., Chen, C.-S., Li, Y.: Near- and mid-infrared chemical and biological sensors. In: Yang, S.C., Chandrasekhar, P. (eds.) Optical and photonic applications of electroactive and conducting polymers, vol. 2528, p. 243. SPIE-The International Society for Optical Engineering, Bellingham (1995)
1598. Gambhir, A., Gerard, M., Mulchandani, A., Malhotra, B.D.: Co-immobilization of urease and glutamate dehydrogenase in electrochemically prepared polypyrrole-polyvinyl sulphonate films. Appl. Biochem. Biotechnol. (2001), in press
1599. Chaubey, A., Gerard, M., Singhal, R., Singh, V.S., Malhora, B.D.: Immobilization of lactate dehydrogenase on electrochemically prepared polypyrrole-polyvinyl sulphonate composite films for application to lactate biosensors. Electrochim. Acta. **46**, 732–729 (2000)
1600. Vidal, J.C., Garcia-Ruiz, E., Espuelas, J., Aramendia, T., Castillo, J.R.: Comparison of biosensors based in entrapment of cholesterol oxidase and cholesterol esterase in electro-polymerized films of polypyrrole and diaminonaphtalene derivative for amperometric determination of cholesterol. Anal. Bioanal. Chem. **377**, 273–280 (2003)
1601. Kumar, A., Rajesh, C., Chaubey, A., Grover, S.K., Malhotra, B.D.: Immobilization of cholesterol oxidase and potassium ferricyanide on dodecylbenzen sulfonate ion doped poly=pyrrole film. J. Appl. Polym. Sci. **82**, 3486–3491 (2001)
1602. Singh, S., Solanki, P.R., Pandey, M.K., Malhotra, B.D.: Cholesterol biosensor based on cholesterol esterase, cholesterol oxidase and peroxidase immobilized onto conducting polyaniline films. Sens. Actuators B. **B115**, 534–541 (2006)
1603. Ramanthan, K., Pandey, S.S., Kumar, R., Guati, A., Murthy, A.S.N., Malhotra, B.D.: Covalent immobilization of glucose oxidase to poly(o-amino benzoic acid) for application to glucose biosensor. J. Appl. Polym. Sci. **78**, 662–667 (2000)
1604. Kankare, J., Vinokurov, I.A.: Anal. Chem. **69**, 2337 (1997)
1605. Mazeikiene, R., Malinauskas, A.: Synth. Met. **89**, 77 (1997)
1606. Osaka, T., Komaba, S., Amano, A.: J. Electrochem. Soc. **145**, 406 (1998)
1607. Degani, Y., Heller, A.: Anal. Chem. **91**, 1285 (1987)
1608. Umana, M., Waller, J.: Anal. Chem. **58**, 2979 (1986)
1609. Bartlett, P.N., Witaker, R.G.: J. Electroanal. Chem. **224**, 37 (1987)

1610. Belanger, D., Nadreau, J., Fortier, G.: J. Electroanal. Chem. **274**, 143 (1989)
1611. Folds, N. C., Lowe, C. R. Anal. Chem. **60**, 2473 (1988)
1612. Khan, G.F., Wernet, W.: J. Electrochem. Soc. **143**, 3336 (1996)
1613. Malitesta, C., Palmisano, F., Torsi, L., Zambonin, P. G.: Anal. Chem., **62**, 2735 (1990)
1614. Genies, E.M., Marchesiello, M.: Synth. Met. **44–47**, 3677 (1993)
1615. Koopal, C.G.J., Eijsma, B., Nolte, R.J.M.: Synth. Met. **55–57**, 3689 (1993)
1616. Gao, M., Dai, L.\M., Wallace, G.G.: Electroanalysis. **15**, 1089–1094 (2003)
1617. Arrigan, D.W.M., Bartlett, P.N.: Biosens. Bioelectron. **13**, 293 (1998)
1618. Ju, H.X., Zhou, D.M., Xiao, Y., Chen, H.Y.: Electroanalysis **10**, 541 (1998)
1619. Miao, Y.Q., Qi, M., Zhan, S. Z., He, N.Y., Wang, J., Yuan, C.W.: Anal. Lett. **32**, 1287 (1999)
1620. Ram, M.K., Adami, M., Paddeu, S., Nicolini, C.: Nanotechnol. **11**, 112 (2000)
1621. Kros, A., Nolte, R.J.M., Sommerdijk, N.A.: J. M. Adv. Mater. **14**, 1779 (2002)
1622. Gao, M., Dai, L., Wallace, G.G.: Synth. Met. **137**, 1393 (2003)
1623. Tian, S.J., Liu, J.Y., Zhu, T., Knoll, W.: Chem. Commun. 2738 (2003)
1624. Zhang, X.J., Ogorevc, B., Wang, J.: Anal. Chim. Acta. **452**, 1 (2002)
1625. Yabuki, S., Shinohara, H., Ikariyama, Y., Aziawa, M.: J. Electroanal. Chem. **277**, 179 (1990)
1626. Yang, Y. Mu, S. Chen, H.: Synth. Met. **92**, 173 (1998)
1627. Jinqing, K. Huaiguo, X., Shaolin, M., Hong, C.: Synth. Met. **87**, 2051997)
1628. Akhtar, P., Too, C.O., Wallace, G.G.: Anal. Chim. Acta. **339**, 211 (1997)
1629. Akhtar, P., Too, C.O., Wallace, G.G.: Anal. Chim. Acta. **339**, 201 (1997)
1630. Mammone, R.J., MacDiarmid, A.G.: J. Chem. Soc. Faraday Trans. I. **81**, 105 (1985)
1631. Tada, K., Yoshino, K.: Jpn. J. Appl. Phys. **36**, L1351 (1997)
1632. Liu, H.Q., Kameoka, J., Czaplewski, D.A., Craighead, H.G.: Polymeric nanowire chemical sensor. Nano Lett. **4**, 671–675 (2004)
1633. Lu, H.h., Lin, C.Y., Hisao, T.C., Fang, Y.Y., Ho, K.C., Yang, D.F., Lee, C.K., Hsu, S.M., Lin, C.W.: Electrical properties of single and multiple poly(3,4-ethylenedioxythiophene) nanowires for sensing nitric oxide gas. Anal. Chim. Acta. **640**, 68–74 (2009)
1634. Gao, M., Dai, L., Wallace, G.G.: Biosensors based on aligned carbon nanotubes coated with inherently conducting polymers. Electroanalysis. **15**, 1089–1094 (2003)
1635. Wightman, R.M.: Science. **240**, 415 (1988)
1636. Chidsey, E.D., Murray, R.W.: Science. **231**, 25 (1986)
1637. Sadik, O.A., Wallace, G.G.: Elecroanalysis. **6**, 860 (1994)
1638. Shinohara, H., Chiba, T., Aziawa, M.: Sens. Actuators. **13**, 79 (1988)
1639. Nishizawa, M., Matsue, T., Uchida, I.: Anal. Chem. **64**, 2642 (1992)
1640. Gardner, J.W., Bartlett, P.N.: Sens. Actuator B. **18–19**, 211 (1994)
1641. Stuart, N.: Chem. Ind. **1**, 587 (1994)
1642. Parker, P.S., Chen, J.R., Agber N.E., Monkman, A.P., Mars, P., Petty, M.C: Sens. Actuators B **17**, 143 (1994)
1643. Imisides, M., John, R., Wallace, G.G.: Chemtech. May, 19 (1996)
1644. Moreau, W.M. (ed.): Semiconductor lithography: principle, practices and materials. Plenum, New York (1988)
1645. Pinto, N.J., Johnson Jr., A.T., Muelller, C.H., Theofylaktos, N., Robinson, D.C., Miranda, F. A.: Electrospun polyaniline/polyethylene oxide nanofiber field-effect transistor. Appl. Phys Lett. **83**, 4244–4246 (2003)
1646. Liu, H.Q., Reccius, C.H., Craighead, H.G.: Single electrospun regioregular poly (3-hexylthiophene) nanofiber field-effect transistor. Appl. Phys. Lett. **87**, 253106/1–253106/3 (2005)
1647. Qi, P.F., Javey, A., Rolandi, M., Wang, Q., Yenilmez, E., Dai, H.J.: Miniature organic transistors with carbon nanotubes as quasi-one-dimensional electrodes. J. Am. Chem. Soc. **126**, 11774–11775 (2005)

1648. Alam, M.M., Wang, J., Guo, Y.Y., Lee, S.P., Tseng, H.R.: Electrolyte-gated transistors based on conducting polymer nanowire junction arrays. J. Phys. Chem. B. **109**, 12777–12784 (2005)

1649. Lee, S.Y., Choi, G.R., Lim, H., Lee, K.M., Lee, S.K.: Electronic transport characteristics of electrolyte-gated conducting polyaniline nanowire field-effect transistors. Appl. Phys. Lett. **95**, 013113/1-3 (2009)

1650. Gao, Y., Yip, H.-L., Chen, K.-S., O'Malley, K.M., Acton, O., Sun, Y., Ting, G., Chen, H., Jen, A.K.-Y.: Surface doping of conjugated polymers by graphene oxide and its application for organix electronic devices. Adv. Mat. **XX**, 1–6 (2011)

1651. Lakshmi, D., Bossi, A., Whitcombe, M.J., Chianella, I., Fowler, S.A., Subrahmanyam, S., Piletska, E.V., Piletsky, S.A.: Electrochemical sensor for catechol and dopamine based on a catalytic molecularly imprinted polymer-conducting polymer hybrid recognition Element. Anal. Chem. **81**, 3576–3584 (2009)

1652. Cha, J., Han, J.I., Choi, Y., Yoon, D.S., Oh, K.W., Lim, G.: DNA hybridization electrochemical sensor using conducting polymer. Biosensors and Bioelectronics. **18**, 1241–1247 (2003)

1653. Pardieu, E., Cheap, H., Vedrine, C., Lazerges, M., Lattach, Y., Garnier, F., Remita, S., Pernelle, C.: Molecularly imprinted conducting polymer based celectrochemical sensor for detection of atrazine. Anal. Chim. Acta. **649**, 236–245 (2009)

1654. Yoon, H.: Current trends in sensors based on conducting polymer nanomaterials. Nanomaterials. **3**, 524–549 (2013)

1655. Aguilar, A.D., Forzani, E.S., Leright, M., Tsow, F., Cagan, A., Iglesias, R.A., Nagahara, L. A., Amlani, I., Tsui, R., Tao, N.J.: A hybrid nanosensor for TNT vapor detection. Nano Lett. **10**, 380–384 (2010)

1656. Ates, M.: A review study of (bio)sensor systems based on conducting polymers. Materials Science and Engineering C. **33**, 1853–1859 (2013)

1657. Vasantha, V.S., Chen, S.M.: J. Elecroanal. Chem. **592**, 77–87 (2006)

1658. Mansiankar, P., Viswanatha, S., Pusphalatha, A.M., Rani, C.: Anal. Chim. Acta. **528**, 157–163 (2005)

1659. Pernites, R., Ponnapati, R., Felipe, M.J., Advincula, R.: Biosens. Bioelectron. **26**, 2766–2771 (2011)

1660. Shiu, K.-K., Chan, O.-Y.: J. Electroanal. Chem. **388**, 45 (1995), and references therein.

1661. Yano, J., Shimoyama, A., Ogura, K.: J. Electrochem. Soc. **139**, L52 (1992)

1662. Dagani, R.: Chemosensor shows enhanced sensitivity. Chem. Eng. News. **73**(29), 63–64 (1995.) and references therein

1663. Emge, A., Bäuerle, P.: Synth. Met. **84**, 213 (1997)

1664. Buttry, D.A.: In: Bard, J. (ed.) Electroanalytical chemistry, vol. 17, pp. 1–136. Marcel Dekker, New York (1991)

1665. Frye, G.C., Martin, S.J., Ricco, A.J., Brinker, C.J.: In: Murray, R.W., Dessy, R.E., Heineman, W.R., et al. (eds.) Chemical sensors and microinstrumentation. ACS Symposium Series 403, Washington, DC (1989)

1666. Grate, J.W., Snow, A., Ballantiene, D.S., et al.: Anal. Chem. **60**, 869 (1986)

1667. Concentrates. Chem. Eng. News **76** (21), 42 (1998)

1668. Thomas III, S.W., Amara, J.P., Bjork, R.E., Swager, T.M.: Amplifying fluorescent polymer sensor for the explosives taggant 2,3-dimethyl-2,3-dinitrobutane (DMNB). Chem. Commun. 4572–4574 (2005)

1669. Blackwood, D., Josowicz, M.: J. Phys. Chem. **95**, 493 (1991)

1670. Gustafsson, G., Lundström, I., Liedberg, B., et al.: Synth. Met. **31**, 163 (1989)

1671. Ge, Z., Brown, C.W., Yang, S.C., Sun, L.: Anal. Chem. **65**, 2335 (1993), and references therein

1672. Nambiar, S., Yeow, J.T.W.: Conductive polymer-based sensors for biomedical applications. Biosens. Bioelectron. **26**, 1825–1832 (2011)

1673. Wolfbeis, O.: Fiber optic chemical sensors and biosensors, vol. I and II. CRC Pres, Inc., London (1992)

1674. Epstein, A.J., MacDiarmid, A.G.: In: Salaneck, W.R., Clark, D.T., Samuelsen, E.J. (eds.) Science and applicatios of conducting polymers, p. 141. Adam Hiler, Bristol (1991)

1675. Pan, L., Chrotos, A., Yu, G., Wang, Y., Isaacson, S., Allen, R., Shi, Y., Dausardt, R., Bao, Z.: An ultra-sensitive resistive pressure sensor based on hollow-sphere microstructure induced elasticity in conducting polymer film. Nat. Comm. 5(3002), 1–8 (2014)

1676. MacDiarmid, A.G.: Proceedings of the DARPA active polymers workshop, p. D3. Institute of Defense Analyses, Baltimore (1996)

1677. Nigrey, P.J., MacDiarmid, A.G., Heeger, A.J.: J. Chem. Soc., Chem. Commun. 594 (1979)

1678. Pruss, A., Beck, F.: J. Electroanal. Chem. 172, 281 (1984)

1679. MacDiarmid, A.J. Mu, S.-L., Somasiri, N.L.D., Wu, W.: Mol. Cryst. Liq. Cryst. 121, 187 (1985)

1680. Croce, F., Passerini, S., Scrosati, B.: J. Electrochem. Soc. 141, 1405 (1994)

1681. Venkatasetty, H.V: Lithium batteries. Electrochem. Soc. Monogr. (1984), and references therein

1682. Megahed, S., Scrosati, B.: J. Electrochem. Soc. Interface Winter, 34(1995)

1683. Nigrey, P.J., MacInnes Jr., D., Nairns, D.P., MacDiarmid, A.J.: J. Electrochem. Soc. 128, 1651 (1981)

1684. Jiang, Z., Alamgir, M., Abraham, K.M.: J. Electrochem. Soc. 142, 333 (1995)

1685. Kakuda, S., Momma, T., Osaka, T., Appetecchi, G.B., Scrosati, B.: J. Electrochem. Soc. 142, L1 (1995), and references therein

1686. Broich, B., Hocker, J.: Bunsenges. Phys. Chem. 88, 497 (1984)

1687. Nagatomo, T., Honma, T., Yamamoto, C., Negishi, K., Omoto, O.: Jpn. J. Appl. Phys. 22, L255 (1983)

1688. Aldissi, M.: Future technologica applications. In: Inherently conducting polymers: processing, fabrication, applications, limitations, p. 66. Noyes Data Corporation, New Jersey (1989)

1689. Maxfield, M., Mu, S.L., MacDiarmid, A.G.: J. Electrochem. Soc. 132, 838 (1985)

1690. Shinozaki, K., Timizuka, Y., Nojiri, A.: Jpn. J. Appl. Phys. 23, L892 (1984)

1691. Gurunathan, K., Amalnerkar, D.P., Trivedi, D.C.: Synthesis and characterization of conducting polymer composite (Pan/TiO₂) for cathode material in rechargeable battery. Mater. Lett. 57, 1642–1648 (2003)

1692. Cheng, F.Y., Tang, W., Li, C.S., Chen, J., Liu, H.K., Shen, P.W., Dou, S.X.: Conducting poly(aniline) nanotubes and nanofibers: controlled synthesis and application in lithium/poly (aniline) rechargeable batteries. Chem. -Eur. J. 12, 3082–3088 (2006)

1693. Pan, L., Qiu, H., Dou, C., Li, Y., Pu, L., Xu, J., Shi, Y.: Conducting polymer nanostructures: template synthesis and applications in energy storage. Int. J. Mol. Sci. 11, 2636–2657 (2010)

1694. Liu, G., Xun, S., Vukmirovic, N., Song, X., Olalde-Velasco, P., Zheng, H., Battaglia, V.S., Wang, L., Yang, W.: Polymers with tailored electronic structure for high capacity lithium battery electrodes. Adv. Mater. 23, 4679–4683 (2011)

1695. Park, K.S., Schousgaard, S.B., Goodenough, J.B.: Conducting-polymer/iron-redox-couple composite cathodes for lithium secondary batteries. Adv. Mater. 19, 848–851 (2007)

1696. An, H., et al.: Flexible lithium-ion battery electrodes. Chem. Eng. News. 93(39), 26 (2015)

1697. Jacoby, M.: Two-for-one deal in solar cells. Chem. Eng. News. 93(9), 30 (2015)

1698. Nishizawa, M., Mukai, K., Kuwabata, S., Martin, C.R., Yoneyama, H.: J. Electrochem. Soc. 144, 1923 (1997)

1699. Malta, M., Louarn, G., Errien, N., Torresi, R.M.: Electrochem. Commun. 5, 1011 (2003)

1700. Sun, M., Zhang, S., Jiang, T., Zhang, L., Yu, J.: Electrochem. Commun. 10, 1819 (2008)

1701. Qiu, L., Zhang, S., Zhang, L., Sun, M., Wang, W.: Electrochim. Acta. 55, 4632 (2010)

1702. Yang, Y., Yu, G., Cha, J.J., Wu, H., Vosgueritchian, M., Yao, Y., Bao, Z., Cui, Y.: Improving the performance of lithium-sulfur batteries by conductive polymer coating. ACS Nano. 5(11), 9187–9193 (2011)

1703. Chao, D., Xia, X., Liu, J., Fan, Z., Ng, C.F., Lin, J., Zhang, H., Shen, Z.X., Fan, H.J.: A V_2O_5/conductive-polymer core/shell nanobelt array on three-dimensional graphite foam: a high-rate, ultrastable, and freestandig cathode for lithium-ion batteries. Adv. Mater. **26**(33), 5794–5800 (2014)

1704. Osaka, T., Momma, T., Nishimura, K., Kakuda, S., Ishii, T.: J. Electrochem. Soc. **141**, 1994 (1994)

1705. Momma, T., Kakuda, S., Yarimizu, H., Osaka, T.: J. Electrochem. Soc. **142**, 1766 (1995)

1706. Shimidzu, T., Ohtani, A., Iyoda, T., Honda, K.: J. Chem. Soc., Chem. Commun. **327** (1987)

1707. Gemeay, A.H., Nishiyama, H., Kuwabata, S., Yoneyama, H.: J. Electrochem. Soc. **142**, 4190 (1995)

1708. Kuwabata, S., Kishimoto, A., Tanaka, T., Yoneyama, H.: J. Electrochem. Soc. **141**, 10 (1994)

1709. Coffey, B., Madsen, P.V., Peohler, T.O., Searson, P.C.: J. Electrochem. Soc. **142**, 321 (1995)

1710. Panero, S., Spila, E., Scrosati, B.: J. Electrochem. Soc. **143**, L29 (1996)

1711. Abraham, K.M., Pasquariello, D.M., Willstaedt, E.B.: J. Electrochem. Soc. **137**, 1956 (1990)

1712. Nishio, K., Fujimoto, M., Yoshinaba, N., et al.: J. Power Sources. **34**, 153 (1991)

1713. Nishizawa, M., Mukai, K., Kuwabata, S., Martin, C.R., Yoneyama, H.: J. Electrochem. Soc. **144**, 1923 (1997)

1714. Taguchi, S., Tanaka, T.: J. Power Sources. **20**, 249 (1987)

1715. Goto, F., Abe, K., Oabayashi, K., et al.: J. Power Sources **20**, 243(1987)

1716. Osaka, T., Nakajima, T., Naoi, K., Owens, B.B.: J. Electrochem. Soc. **137**, 2139 (1990)

1717. Genies, E.M., Hany, P., Santier Ch.: Synth. Met. **18**, 751 (1988)

1718. Mizumoto, M., Namba, M., Nishimura, S., et al.: Synth. Met. **18**, C639(1989)

1719. Nishio, K., Fujimoto, M., Yoshinaga, N., et al. 40th ISE Meeting; Extended Abstracts, p. 553 (1989)

1720. Nishio, K., Fujimoto, M., Yoshinaga, N., et al.: 29th Battery Symposium in Japan; Abstracts, p. 227 (1988)

1721. Nishio, K., Fujimoto, M., Yoshinaga, N., Furukawa, N.: 30th Battery Symposium in Japan; Abstracts, p. 127 (1989)

1722. Tsutsumi, H., Fukuawa, S., Ishikawa, M., Morita, M., Matsuda, Y.: J. Electrochem. Soc. **142**, L168 (1995)

1723. Morita, M., Miyazaki, S., Ishikawa, M., Matsuda, Y., Tajima, H., Adachi, K., Anan, F.: J. Electrochem. Soc. **142**, L3 (1995)

1724. Freemantle, M.: Organic cathode spurs battery energy storage. Chem. Eng. News. **73**(8), 5 (1995)

1725. Oyama, N, Pope, J.M., Sotomura, T.: J. Electrochem. Soc. **144**, L47 (1997), and references therein

1726. Sotomura, T., Tatsuma, T., Oyama, N.: J. Electrochem. Soc. **143**, 315 (1996)

1727. Leroux, F., Koene, B.E., Nazar, L.F.: J. Electrochem. Soc. **143**, L181 (1996)

1728. Nagamoto, T., Omoto, O.: J. Electroche. Soc. **135**, 21214 (1988)

1729. Shacklette, L.W., Elsenbaumer, R.L., Chance, R.R. Sowa, J.M. Ivory, D.M. Miller, G.G. Baughman, R.H. J. chem. Soc. Chem. Commun. **1982** 361 (1982)

1730. Kitamura, T., Hasumi, K. In: Electrochemical society in Japan, Spring Meeting Abstracts, p. 76 (1985)

1731. Shacklette, L.W., Maxfield, M., Gould, S., Wolf, J.F., Jow, T.R., Baughman, R. H.: Synth. Met. **18**, 611 (1987)

1732. Shacklette, L.W., Jow, T.R., Maxfield, M., Hatami, R.: Synth. Met. **28**, C655 (1989)

1733. Ivory, D.M., Miller, G.G., Sowa, J.M., et al.: J. Chem. Phys. **71**, 1506 (1979)

1734. Shacklette, L.W., Maxfield, M., Gould, S., Wolf, J.F., Jow, T.R., Baughman, R.H.: Synth. Met. **18**, 611 (1987)

1735. Pandey, P.C., Prakash, R.: J. Electrochem. Soc. **145**, 999 (1998)

1736. Nagatomo, T., Kakehata, H., Ichikawa, C., Omoto, O.: J. Electrochem. Soc. **132**, 1380 (1985)

1737. Nagatomo, T., Ichikawa, C., Omoto, O.: Synth. Met. **18**, 649 (1987)
1738. Koura, N., Ejiri, H., Takeishi, K.: J. Electrochem. Soc. **140**, 602 (1993)
1739. Kaneto, K., Yoshino, K., Inuishi, Y.: Jpn. J. Appl. Phys. **22**, L567 (1983)
1740. (a) Su, W.P., Schrieffer, J.R., Heeger, A.J.: Phys. Rev. B **22**, 2209 (1980); (b) Su, W.P., Schrieffer, J.R., Heeger, A.J.: Phys. Rev. B **28**, 1138€ (1982)
1741. Yamamoto, T. In: Spring electrochemical society meeting extended abstracts; Electrochemical Society, Vol. 82-1, p. 987 (1982)
1742. Killian, J.G., Coffey, B.M., Gao, F., Poehler, T.O., Searson, P.C.: J. Electrochem. Soc. **143**, 936 (1996)
1743. Kumar, G., Sivashanmugam, A., Muniyandi, N., Dhawan, S.K., Trivedi, D.C.: Synth. Met. **80**, 279 (1996)
1744. Gofer, Y., Sarker, H., Killian, J.G., Poehler, T.O., Searson, P.C.: Appl. Phys. Lett. **71**, 1582 (1997)
1745. Higgins, T.M., Park, S.-H., King, P.J., Zhang, C., McEvoy, N., Berner, N.C., Daly, D., Shmeliov, A., Khan, U., Duesberg, G., Nicolosi, V., Coleman, J.N.: A commercial conducting polymer as both binder and conductive additive for silicon nanoparticle-based lithium-ion battery negative Electrodes. ACS Nano. **10**(3), 3702–3713 (2016)
1746. Ma, Q., Zhang, H., Zhou, C., Zheng, L., Cheng, P., Nie, J., Feng, W., Hu, Y.-S., Li, H., Huang, X., Chen, L., Armand, M., Zhou, Z.: Single lithium-ion conducting polymer electrolytes based on a super-delocalized polyanion. Ange. Chem. **55**(7), 2521–2525 (2016)
1747. Münstedt, H., Köhler, G., Möhwald, H., Naegele, D., Bitthin, R., Ely, G., Meissner, E.: Synth. Met. **18**, 259 (1987)
1748. Alvi, F., Ram, M.K., Basnayaka, P.A., Stefanakos, E., Goswami, Y., Kumar, A.: Graphene-polyethylenedioxythiophene conducting polymer nanocomposite based supercapacitor. Electrochimica Acta. **56**, 9406–9412 (2011)
1749. Peng, C., Zhang, S., Jewell, D., Chen, G.Z.: Carbon nanotube and conducting polymer composites for supercapacitors. Prog. Nat. Sci. **18**, 777–778 (2008)
1750. Zhang, L.L., Zhao, S., Tian, X.N., Zhao, X.S.: Langmuir. **26**, 17624 (2010)
1751. Liu, R., Lee, S.B.: MnO2/Poly(3,40-ethylenedioxythiophene) coaxial nanowires by one-step coelectrodeposition for electro chemical energy storage. J. Am. Chem. Soc. **130**, 2942–2943 (2008)
1752. Wang, Y.G., Li, H.Q., Xia, Y.Y.: Ordered whiskerlike polyaniline grown on the surface of mesoporous carbon and its electrochemical capacitance performance. Adv. Mater. **18**, 2619–2623 (2006)
1753. Zhang, Z., Zhao, X.S.: Conducting polymers directly coated on reduced graphene oxide sheets as high-performance supercapacitor electrodes. J. Phys. Chem. C. **116**, 5420–5426 (2012)
1754. Xia, X., Chao, D., Fan, Z., Guan, C., Cao, X., Zhang, H., Fan, H.J.: A new type of porous graphite foams and their integrated composites with oxide/polymer core/shell nanowires for supercapacitors: structural design, fabrication, and full supercapacitor Demonstrations. Nano Lett. **14**(3), 1651–1658 (2014)
1755. Kuila, B.K., Nandan, B., Böhme, M., Janke, A., Stamm, M.: Vertically oriented arrays of polyaniline nanorods and their super electrochemical properties. Chem. Commun. 5749–5751 (2009)
1756. Li, G., Zhu, R., Yang, Y.: Polymer solar cells. Nat. Photonics. **6**, 153–161 (2012)
1757. Zhao, L., Li, Y., Liu, Z., Shimizu, H.: Chem. Mater. **22**, 5949 (2010)
1758. Wudl, F., Srdanov, G.: Conducting polymer formed of poly(2-methoxy-5-(2'--ethylhexyloxy)-p-phenylene vinylene). US5189136, 1993
1759. Yu, G., Gao, J., Hummelen, J.C., Wudl, F., Heeger, A.J.: Polymer photovoltaic cells – enhanced efficiencies via a network of internal donor-acceptor heterojunctions. Science. **270**, 1789–1791 (1995)
1760. Muhlbacher, D., et al.: High photovoltaic performance of a low-bandgap polymer. Adv. Mater. **18**, 2884–2889 (2006)

1761. Peet, J., et al.: Efficiencey enhancement in low-bandgap polymer solar cells by processing with alkane dithiols. Nature Mater. **6**, 497–500 (2007)

1762. Blouin, N., Michaud, A., Leclerc, M.: A low-bandgap poly(2,7-carbazole) derivative for use in high-performance solar cells. Adv. Mater. **19**, 2295–2700 (2007)

1763. Park, S.H., et al.: Bulk heterojunction solar cells with internal quantum efficiency approaching 100%. Nature Photon. **3**, 297–302 (2009)

1764. Liang, Y.Y., et al.: Development of new semiconducting polymers for high performance solar cells. J. Am. Chem. Soc. **131**, 56–57 (2009)

1765. Liang, Y.Y., et al.: Highly efficient solar cell polymers developed via fine-tuning of structural and electronic properties. J. Am. Chem. Soc. **131**, 7792–7799 (2009)

1766. Chen, H.Y., et al.: Polymer solar cells with enhanced open-ciruit voltage and efficiency. Nature Photon. **3**, 649–653 (2009)

1767. Price, S.C., Stuart, A.C., Yang, L.Q., Zhou, H.X., You, W.: Fluorine substituted conjugated polymer of medium band gap yileds 7% efficiency in polymer-fulleren solar cells. J. Am. Chem. Soc. **133**, 4625–4631 (2011)

1768. Zhou, H.X., et al.: Development of fluorinated benzothiadiazole as a structural unit for polymer solar cell of 7% efficiency. Angew. Chem. Int. Ed. **50**, 2995–2998 (2011)

1769. Su, M.S., et al.: Improving device efficiency of polymer/fullerene bulk hetero junction solar cells through enhanced crystallinity and reduced grain boundaries induced by solvent additivies. Adv. Mater. **23**, 3315–3319 (2011)

1770. Yang, J., et al.: A robust inter-connecting layer for achieving high performance tandem polymer solar cells. Adv. Mater. **23**, 3465–3470 (2011)

1771. Sun, Y.M., et al.: Efficient, air-stable bulk heterojunction polymer solar cells using MoO_x as the anode interfacial layer. Adv. Mater. **23**, 2226–2230 (2011)

1772. Chu, T.Y., et al.: Bulk heterjunction solar cells using thieno[3,4-c] pyrrole-4,6-dione and dithieno[3,2-b:2',3'-d] silole copymer with a power conversion efficiency of 7.3%. J. Am. Chem. Soc. **133**, 4250–4253 (2011)

1773. Amb, C.M., et al.: Dithienogermole as a fused electron donor in bulk heterojunction solar cells. J. Am. Che. Soc. **133**, 10062–10065 (2011)

1774. Yin, Z., Zheng, Q.: Controlled synthesis and energy applications of one-dimensional conducting polymer nanostructures: an overview. Adv. Energy Mater. **2**, 179–218 (2012)

1775. Irwin, M.D., Buchholz, B., Hains, A.W., Chang, R.P.H., Marks, T.J.: Proc. Natl. Acad. Sci. U. S. A. **105**, 2783 (2008)

1776. Zhao, G., He, Y., Li, Y.: Adv. Mater. **22**, 4355 (2010)

1777. Brabec, C.J., Gowrisanker, S., Halls, J.J.M., Laird, D., Jia, S.;, Williams, S. P.: Adv. Mater. **22**, 3839 (2010)

1778. Chen, L.-M., Hong, Z., Li, G., Yang, Y.: Adv. Mater. **21**, 1434 (2009)

1779. Huang, J., Yin, Z., Zheg, Q.: Energy Environ. Sci. **4**, 3861 (2011)

1780. Hoppe, H., Sariciftci, N.S.: J. Mater. Chem. **16**, 45 (2006)

1781. Dennler, G., Scharber, M.C., Brabec, C.J.: Adv. Mater. **21**, 1323 (2009)

1782. Kim, J.-H., Park, J.H., Lee, J.H., Kim, J.S., Sim, M., Shim, C., Cho, K.: J. Mater. Chem. **20**, 7398 (2010)

1783. Kim, J.S., Lee, J.H., Park, J.H., Shim, C., Sim, M., Cho, K.: Adv. Funct. Mater. **21**, 480 (2011)

1784. Kim, J.-H. Park, J.H. Lee, J.H. Kim, J.S. Sim, M., Shim, C., Cho, K.: J. Mater. Chem. **20**, 7398 (2010)

1785. Bavel, S.S.V., Soutry, E., With, G.D., Loos, J.: Nano Lett. **9**, 507 (2009)

1786. Jo, J., Kim, S.-S., Na, S.-I., Yu, B.-K., Kim, D.-Y.: Adv. Funct. Mater. **19**, 866 (2009)

1787. Jo, J., Na, S.-I., Kim, S.-S., Lee, T.-W., Chung, Y., Kang, S.-J., Vak, D., Kim, D.-Y.: Adv. Funct. Mater. **19**, 2398 (2009)

1788. Berson, S., De Bettignies, R., Bailly, S., Guillerez, S.: Adv. Funct. Mater. **17**, 1377 (2007)

1789. Sun, S., Salim, T., Wong, L.H., Foo, Y.L., Boey, F., Lam, Y.M.: J. Mater. Chem. **21**, 377 (2011)

1790. Zhao, Y., Shao, S., Xie, Z., Geng, Y., Wang, L.: J. Phys. Chem. C. **113**, 17235 (2009)
1791. Park, J.H., Kim, J.S., Lee, J.H., Lee, W.H., Cho, K.: J. Phys. Chem. C. **113**, 17579 (2009)
1792. Salim, T., Sun, S., Wong, L.H., Xi, L., Foo, Y.L., Lam, Y.M.: J. Phys. Chem. C. **114**, 9459 (2010)
1793. Lee, J.H., Kim, D.W., Jang, H., Choi, J.K., Geng, J., Jung, J.W., Yoon, S.C., Jung, H.-T.: Small. **5**, 2139 (2009)
1794. Kim, J.S., Park, Y., Lee, D.Y., Lee, J.H., Park, J.H., Kim, J.K., Cho, K.: Adv. Funct. Mater. **20**, 540 (2010)
1795. (a) Wang, H.-S., Lin, L.-H., Chen, S.-Y., Wang, Y.-L.; Wei, K.-H.: Nanotechnol. **20**, 075201 (2009); (b) Wang, H.-S., Chen, S.-Y., Su, M.-H., Wangand, Y.-L., Wei, K.-H.: Nanotechnol. **21**, 145203 (2010)
1796. He, X., Gao, F., Tu, G., Hasko, D., Hüttner, S., Steiner, U., Greenham, N.C., Friend, R.H., Huck, W.T.S.: Nano Lett. **10**, 1302 (2010)
1797. Xin, H., Ren, G., Kim, F.S., Jenekhe, S.A.: Chem. Mater. **20**, 6199 (2008)
1798. Xin, H., Kim, F.S., Jenekhe, S.A.: J. Am. Chem. Soc. **130**, 5424 (2008)
1799. Xin, H., Reid, O.G., Ren, G., Kim, F.S., Ginger, D.S., Jenekhe, S.A.: ACS Nano. **4**, 1861 (2010)
1800. Wu, P.-T., Xin, H., Kim, F.S., Ren, G., Jenekhe, S.A.: Macromolecules. **42**, 8817 (2009)
1801. Chen, H.-C., Wu, I.C., Hung, J.-H., Chen, F.-J., Chen, I.W.P., Peng, Y.-K., Lin, C.-S., Chen, C.-H., Sheng, Y.-J., Tsao, H.-K., Chou, P.-T.: Small **7**, 1098 (2011)
1802. Ren, G., Wu, P.-T., Jenekhe, S.A.: ACS Nano. **5**, 376 (2011)
1803. Briseno, A.L., Holcombe, T.W., Boukai, A.I., Garnett, E.C., Shelton, S.W., Fréchet, J.J.M., Yang, P.: Nano Lett. **10**, 334 (2010)
1804. Wang, H.-S., Chen, S.-Y., Wang, Y.-L., Wei, K.-H.: J. Nanosci. Nanotechnol. **11**, 3229 (2011)
1805. Ravirajan, P., Haqu, S.A., Durrant, J.R., Poplavskyy, D., Bradley, D.D.C., Nelson, J.: J. Appl. Phys. **95**, 1473 (2004)
1806. Tepavcevic, S., Darling, S.B., Dimitrijevic, N.M., Rajh, T., Sibner, S.J.: Small **5**, 1776 (2009)
1807. Mor, G.K., Kim, S., Paulose, M., Varghese, O.K., Shankar, K., Basham, J., Grimes, C.A.: Nano Lett. **9**, 4250 (2009)
1808. Guo, Y., Zhang, Y., Liu, H., Lai, S.-W., Li, Y., Li, Y., Hu, W., Wang, S., Che, C.-M., Zhu, D.: J. Phys. Chem. Lett. **1**, 327 (2010)
1809. Sun, B., Greenham, N.C.: Phys. Chem. Chem. Phys. **3**, 3631 (2006)
1810. Shiu, S.-C., Chao, J.-J., Hung, S.-C., Yeh, C.-L., Lin, C.-F.: Chem. Mater. **22**, 3108 (2010)
1811. Lu, W. Wang, C., Yue, W., Chen, L.: Nanoscale **3**, 3631 (2011)
1812. Tezuka, N., Umeyama, T., Matano, Y., Shishido, T., Yoshida, K., Ogawa, T., Isoda, S., Stranius, K., Chukharev, V., Tkachenko, N.V., Lemmetyinen, H., Imahori, H.: Energy Environ. Sci. **4**, 741(2011)
1813. Xia, Y., Ouyang, J.: PEDOT: PSS films with significantly enhanced conductivities induced by preferential solvation with coslovents and their application in polymer photovoltaic cells. J. Mater. Chem. **21**, 4927 (2011)
1814. Ritter, S.K.: Decked-out thiophene adds versatility to semiconducting polymers. Chem. Eng. News. **94**(23), 10 (2016)
1815. Yoo, D., Kim, J., Kim, J.H.: Direct synthesis of highly conductive poly(3,4-ethylenedi-oxythiophene):poly(4-styrenesulfonate)(PEDOT:PSS)/graphene composites and their applications in energy harvesting systems. Nano Res. **7**(5), 717–730 (2014)
1816. Zhou, M., Chi, M., Luo, J., He, H., Jin, T.: J. Power Sources. **196**, 4427 (2011)
1817. Qiao, Y., Li, C.M., Boa, S.-J., Boa, Q.-L.: J. Power Sources. **170**, 79 (2007)
1818. Zou, Y., Xiang, C., Yang, L., Sun, L.-X., Xu, F., Cao, Z.: Int. J. Hydrogen Energy **33**, 4856 (2008)
1819. Ito, T., Mori, T., Kato, M., Uemura, T. Electrochromic device with a reference electrochromic element. US 5073011A, 17 Dec 1991

1820. Bennett, R.N., Kokonaski, W.K., Hannan, M.J., Boxall, L.G. Electrode for display devices, US 5446577 A, 29 Aug 1995
1821. Baranowski, H.-P., Knoblauch, W.: Steam sterilizing apparatus, US 3884636 A, 20 May 1975
1822. Castellion, G. Electrochromic (ec) mirror which rapidly changes reflectivity, US 380732 A, 30 Apr 1974
1823. Baucke, F.G.K., Drause, D., Metz, B., Paquet, V., Zauner, J. Electrochromic mirrors, US 4465339 A, 14 Aug 1984
1824. Beall, G.H., Fehlner, F.P. Electrochromic devices including a mica layer electrolyte, US 4416517 A, 22 Nov 1983
1825. Yang, S.C., Hwang, J.-H. Solid electrolytes for conducting polymer-based color switchable windows for electronic display services. USA 5253100 A, 12 Oct 1993
1826. Yang, S.C., Durand R.R., Jr. Electronic display element. US 4586792 A, 6 May 1986
1827. Yang, S.C., Durand, R.R.Jr.: Variable color transparent panels. US 4749260 A, 6 June 1988
1828. Kobayashi, T.; Yonegama, H., Tamura, H.: J. Electroanal. Chem. **161**, 419 (1984)
1829. Foot, P.J.S., Simon, R.: J. Phys. D. Appl. Phys. **22**, 1598 (1989)
1830. (a) Akhtar, M., Weakliem, H.A., Paiste, R.M., Gaughan, K.: Synth. Met. **26**, 203 (1988); (b) Ram, M.K., Maccioni, E., Nicolini, C.: Thin Solid Films **303**, 27 (1997)
1831. Yoshida, T., Okabayaski, K., Asaoka, T., Katsushi, A.: Electrochem. Soc. Extended Abstracts. **86**, 552 (1988)
1832. Asaoka, T., Okabayashi, K., Abe, T., Yoshida, T.: 40th ISE Meeting (Kyoto), Extended Abstracts, I, 245–246 (1989)
1833. Kim, E., Lee, K., Rhee, S.B.: J. Elecrochem. Soc. **144**, 227 (1997)
1834. Leventis, N., Chung, Y.C. Complementary surface confined polymer electrochromic materials, systems, and methods of fawbrication therefor. US 5457564 A, 10 Oct 1995
1835. Ho, K.-C. Solid-state electrochromic device with proton-conducting polymer electrolyte and Prussian blue counterelectrode. US 5215821 A, 1 June 1993
1836. Coleman, J.P. Electrochromic materials and displays. US 5413739 A, 9 May 1995
1837. Uemachi, H., Sotomura, T., Takeyama, K., Koshida, N. Reversible electrochemical electrode. US 5413883 A, 9 May 1995
1838. Jin, S.-J. Kang, S.-W. Soluble conductive polymer manufacturing method thereof and display device employing the same. US 5616669 A, 1Apr 1997
1839. Wudl, F., Heeger, A., Kobayashi, M. Polymer having isothianaphthene structure and electrochromic display. US 4772940 A, 20 Sept 1988
1840. Ishikawa, M. et al. Electrochromic display device capable of display in plural colors. US 4983957 A, 8 Jan 1991
1841. Corradini, A., Marinangeli, A.M., Mastragostino, M.: Electrochimica Acta. **35**, 1757 (1990)
1842. Arbizzani, C., Mastragostino, M., Passerini, S., et al.: Electrochimica Acta **36**, 837 (1991)
1843. Wolf, J. F. Meller: G. G., Shacklet, L., Elsenbaumer, R.L., Baughman, R.H.: Adjustable tint window with electrochromic conductivev polymer. US 5042923 A, 27 Aug 1991
1844. Chandrasekhar, P.: Flexible electrochromic window materials based on poly(Diphenyl Amine) and related conducting polymers, Final Technical Report, Gront No. DE-FG05-93ER81631 for the U.S. Department of Energy, Oak Ridge/Washington, DC (1998)
1845. (a) Chandrasekhar, P.: Complimentary polymer electrochromic device. US 20130120821 A1. 16 May 2013; (b) Chandrasekhar, P.: Method and apparatus for control of electrhromic devices. US 8902486 B1. 2 Dec 2014; (c) Chandreskhar, P.: Elecrochromic display device. US 5995273 A. 30 Nov 1999; (d) Chandrasekhar, P. Electrolytes. US 6033592 A. 7 Mar 2000; (e) Chandrasekhar, P.: Elechrochromic display device. CA 2321894 A1. 2 Sept 1999
1846. (a) MA, C., Taya, M., Xu, C.: Electrochimica Acta **54**, 598–605 (2008); (b) Ma, C., Taya, M., Xu, C.: Plym. Eng. Sci. **48**, 2224–2228 (2008)
1847. (a) Welsh, D.M., Kumar, A., Meijer, E.W., Reynolds, J.R.: Adv. Mater. **11**, 1379–1382 (1999); (b) Amb, C.M., Kerszulis, J.A., Thompson, E.J., Dyer, A.L., Reynolds, J.R.: Polymer Chem. **2**(4), 812–814 (2011); (c) Shen, D.E., Abboud, K.A., Reynolds, J.R.: J. Macromolec. Sci. Part A Pure Appl. Chem. **47**(1), 6–11. (2010)
1848. Daehler, M. Infrared display device. US 4724356 A, 9 Feb 1988

1849. Chandrasekhar, P., Dooley: FAR-IR transparency and dynamic infrared signature control with novel conducting polymer systems. In: Yang, S.C., Chandrasekhar, P. (eds.) Optical and photonic applications of electroactive and conducting polymers, vol. 2528, p. 169. SPIE-The International Society for Optical Engineering, Bellingham (1995.) and references therein

1850. Chandrasekhar, P. Conducting polymer based actively IR modulating electrochromics, Monthly Report Numbers 1-7 towards Contract No. F29601-98-C-0142 for Kirtland Airforce Base, New Mexico, 1998

1851. Chandrasekhar, P.: Proceedings of the DARPA active Polymers Workshop hosted by the Institute for Defense Analyses, Baltimore, p. C6–C1 1996

1852. Burroughes, J.H., Bradley, D.D.C., Brown, A.R., Marks, R.N., Friend, R.H., Burn, P.L., Holmes, A.B.: Nature. **347**, 539 (1990)

1853. Nanotechnology lessons from a defect-tolerant computer. Chem. Eng. News **76**(24), 24 (1998)

1854. Pope, M., Kallmann, H., Magnante, P.: J. Chem. Phys. **38**, 2042 (1963)

1855. Helfrich, W., Schneider, W.G.: Phys. Rev. Lett. **14**, 229 (1965)

1856. Helfrich, W., Schneider, W.G.: J. Chem. Phys. **44**, 2902 (1966)

1857. Lohmann, F., Mehl, W.: J. Chem. Phys. **50**, 500 (1969)

1858. Gonzalez Basurto, J., Burshtein, Z.: Mol. Cryst. Liq Cryst. **31**, 211 (1975)

1859. Kojima, H., Ozawa, A., Takashi, T., Nagaoka, M., Homma, T., Nagatomo, T., Omoto, O.: J. Electrochem. Soc. **144**, 3628 (1997)

1860. Dagani, R.: Simpler method for organic pixel patterning. Chem. Eng. News. **75**(18), 56–57 (1997)

1861. Tremblay, J.-F.: The rise of OLED displays. Chem. Eng. News. **94**(28), 30–34 (2016)

1862. Granström, M., Berggren, M., Inganäs, O.: Micrometer-and nanometer-sized polymeric light-emitting diodes. Science. **267**, 1479–1481 (1995)

1863. Granström, M., Berggren, M., Inganäs, O.: Polymeric light-emitting diodes of submicron size – Structures and developments. Synth. Met. **76**, 141–143 (1996)

1864. Boroumand, F.A., Fry, P.W., Lidzey, D.G.: Nanoscale conjugated-polymer light-emitting diodes. Nano Lett. **5**, 67–71 (2005)

1865. Grimsdale, A.C., Chan, K.L., Martin, R.E., Jokisz, P.G., Holmes, A.B.: Synthesis of light-emitting conjugated polymers for applications in electro-luminescent devices. Chem. Rev. **109**, 897–1091 (2009)

1866. Kim, B.H., Park, D.H., Joo, J., Yu, S.G., Lee, S.H.: Synthesis, characteristics, and field emission of doped and de-doped polypyrrole, polyaniline, poly(3,4-ethylenedioxy-thiophene) nanotubes and nanowires. Synth. Met. **150**, 279–284 (2005)

1867. Wang, C.W., Wang, Z., Li, M.K., Li, H.L.: Well-aligned polyaniline nano-fibril array membrane and its field emission property. Chem. Phys. Lett. **341**, 431–434 (2001)

1868. Kim, B.H., Kim, M.S., Park, K.T., Lee, J.K., Park, D.H., Joo, J., Yu, S.G., Lee, S.H.: Characteristics and field emission of conducting poly(3,4-ethylenedioxythiophene) nanowires. Appl. Phys. Lett. **83**, 539–541 (2003)

1869. Yan, H.L., Zhang, L., Shen, J.Y., Chen, Z.J., Shi, G.Q., Zhang, B.L.: Synthesis, property and field-emission behavior of amorphous polypyrrole nanowires. Nanotechnol. **17**, 3446–3450 (2006)

1870. Burn, P.L., Holmes, A.B., Kraft, A., Brown, A.R., Bradley, D.D.C., Friend, R.H.: Light-emitting diodes based on conjugated polymers: control of colour and efficiency. In: Chiang, L.Y., Garito, A.F., Sandman, D.J. (eds.) Electrical, optical, and magnetic properties of organic solid state materials, vol. 247, p. 647. Materials Research Society, Pittsburgh (1992)

1871. (a) Braun, D., Heeger, A.J.: Appl Phys. Lett. **58**, 1982 (1991); (b) Braun, D., Heeger, A.J.: Appl. Phys. Lett. **59**, 878 (1991)

1872. Garten, F., Vrijmoeth, J., Schlatmann, A.R., Gill, R.E., Klapwijk, T.M., Hadziioannou, G. Synth. Met. **76**, 85 (1996), and references therein

1873. AlSalhi, M.S., Alam, J., Dass, L.A., Raja, M.: Recent advances in conjugated polymers for light emitting devices. Int. J. Mol. Sci. **12**, 2036–2054 (2011)

1874. Burn, P.L., Bradley, D.D.C., Friend, R.H., Halliday, D.A., Holmes, A.B., Jackson, R.W., Kraft, A.M.: J. Chem. Soc. Perkin Trans. **1**, 3225 (1992)

1875. Brown, A.R., Burroughes, J.H., Greenham, N., Friend, R.H., Bradley, D.D.C., Burn, P.L., Kraft, A., Holmes, A.B.: Appl. Phys. Lett. **61**, 2793 (1992)

1876. Burn, P., Holmes, A.B., Kraft, A., Brown, A.R., Bradley, D.D.C., Friend, R.H.: Mat. Res. Soc. Proc. **247**, 647 (1992)

1877. Moratti, S.C., Bradley, D.C., Friend, R.H., Greenhan, N.C., Holmes, A.B.: Molecularly engineered polymer LEDs. In: Garito, A.F., Jen, A.K.-Y., Lee, C.Y.-C., Dalton, L.R. (eds.) Electrical, optical, and magnetic properties of organic solid state materials, vol. 328, p. 371. Materials Research Society, Pittsubrgh (1994)

1878. Antoniadis, H., Abkowitz, M., Hsieh, B.R., Jenehe, S.A., Stolka, M.: Space-charge-limited charge injextion from ito/ppv into a trap-free molecularly doped polymer. In: Garito, A.F., Jen, A.K.-Y., Lee, C.Y.-C., Dalton, L.R. (eds.) Electrial, optical, and magnetic properties of organic solid state materials, vol. 328, p. 377. Materials Research Society, Pittsburgh (1994)

1879. Brown, A.R., Burroughes, J.H., Greenham, N., Friend, R.H., Bradley, D.D.C., Burn, P.L., Kraft, A., Holmes, A.B.: Appl. Phys. Lett. **61**, 2793 (1992)

1880. Uchiyama, K., Akimichi, H., Hotta, S., Noge, H., Sakaki, H.: Light-emitting diodes using semiconducting oligothiophenes. In: Garito, A.F., Jen, A.K.-Y., Lee, C.Y.-C., Dalton, L.R. (eds.) Electrical, optical, and magnetic properties of organic solid state materials, vol. 328, p. 389. Materials Research Society, Pittsburgh (1994)

1881. Zyung, T., Kang, I.-N., Hwang, D.-H., Shim, H.-K.: Obersvation of new wavelength electroluminescence from multilayer structure device using poly(P-Phenylenevinylene) derivative. In: Yang, S.C., Chandrasekhar, P. (eds.) Optical and photonic applications of electroactive and conducting polymers, vol. 2528, p. 89. The International Society for Optical Engineering, Bellingham (1995)

1882. Baigent, D.R., May, P.G., Friend, R.H.: Synth. Met. **76**, 149 (1996)

1883. Östergård, T., Paloheimo, J., Pal, A.J., Stubb, H.: Synth. Met. **88**, 171 (1997)

1884. Burn, P.L., Holmes, A.B., Kraft, A., Bradley, D.D.C., Brown, A.R., Friend, R.H., Gymer, R. W.: Nature **356**, 47 (1992)

1885. Berggren, M., Inganäs, O., Gustafsson, G., Andersson, M.R., Hjertberg, T., Wennerström, O.: Synth. Met. **71**, 2185 (1995)

1886. Inganäs, O., Berggren, M., Andersson, M.R., Gustafsson, G., Hjertberg, T., Wennerström, O., Dyreklev, P., Granström, M.: Synth. Met. **71**, 2121 (1995)

1887. Baigent, D.R., Hamer, P.J., Friend, R.H., Moratti, S.C., Holmes, A.B.: Synth. Met. **71**, 2175 (1995)

1888. Yin, S. Peng, J. Li, C. Huang, W. Liu, X., Li, W., He, B.: Synth. Met. **93**, 193 (1998)

1889. Chem. Eng. News **73**(17), 35 (1995)

1890. MacDiarmid, A.G., Wang, H.-L., Park, J.-W., Fu, D.-K., Marsella, M.J., Swager, T.M., Wang, Y., Gebler, D.D., Epstein, A.J.: Novel light emitting diodes involving heterocyclic aromatic conjugated polyersm. In: Yang, S.C., Chandrasekhar, P. (eds.) Optical and photonic applications of electroactive and conducting polymers, vol. 2528, p. 2. SPIE-The International Society for Optical Engineering, Bellingham (1995.) and references therein

1891. Onoda, M., MacDiarmid, AG.: Synth. Met. **91**, 307 (1997)

1892. Wang, Y.Z., Gebler, D.D., Lin, L.B., Blatchford, J.W., Jessen, S.W., Wang, H.L., Epstein, A.J.: Appl. Phys. Lett. **68**, 894 (1996)

1893. Wang, Y.Z., Gebler, D.D., Fu, D.K., Swager, T.M., MacDiarmid, A.G., Epstein, A.J.: Synth. Met. **85**, 1179 (1997)

1894. (a) Ohmori, Y., Uchida, M., Muro, K., Yoshino, K.: Solid State Commun. **80**, 605 (1991), (b) Braun, D., Gustaffson, G., McBranch, D., Heeger, A.J.: J. Apply. Phys. **72**, 564 (1992)

1895. Greenham, N.C., Brown, A.R., Bradley, D.D.C., Friend, R.H.: Synth. Met. **55–57**, 4134 (1993)

1896. Grem, G., Leditzky, G., Ulrich, B., Leising, G.: Adv. Mater. **4**, 36 (1992)

1897. Leising, G., Tasch, S., Brandstätter, C., Graupner, W., Hampel, S., List, E.J.W., Meghdadi, F., Zenz, C., Schlichting, P., Rohr, U., Geerts, Y., Scherf, U., Müllen, K.: Synth. Met. **91**, 41 (1997)

1898. Jiang, X.Z., Liu, Y.Q., Song, X.Q., Zhu, D.B.: Synth. Met. **91**, 311 (1997)

1899. Sun, R., Masuda, T., Kobayashi, T.: Synth. Met. **91**, 301 (1997)

1900. Chen, S.-A., Chuang, K.-R., Chao, C.-I., Lee, S.-T.: Synth. Met. **82**, 207 (1996)

1901. Epstein, A.J., Wang, Y.Z., Jessen, S.W., Blatchford, J.W., Gebler, D.D., Lin, L.B., Gustafson, T.L., Swager, T.M., MacDiarmid, A.G.: Poltm. Prepr. **37**, 133 (1996)

1902. Dagani, R.: Polarized light-emitting diode based on polymer. Chem. Eng. News. **73**(4), 28–29 (1995)

1903. Granström, M., Berggren, M., Inganäs, O.: Synth. Met. **76**, 141 (1996)

1904. Faraggi, E.Z., Davidov, D., Cohen, G., Noach, S., Golosovsky, M., Avny, Y., Neumann, R., Lewis, A.: Synth. Met. **85**, 1187 (1997)

1905. McGehee, M.D., Vacar, D., Lemmer, U., Moses, D., Heeger, A.J.: Synth. Met. **85**, 1233 (1997)

1906. DuPont: Dupont displays opens OLED materials scale-up facility for next generation TVs. http://www.dupont.com/products-and-services/display-lighting-materials/oled-organic-ligh t-emitting-diodes/press-releases/20150929-dupont-opens-oled-scale-up-facility.html. Accessed Sept 2016

1907. DuPont: OLED materials and solution process technology for advanced displays and lighting. http://www.dupont.com/products-and-services/display-lighting-materials/oled-organic-light-emitting-diodes.html. Accessed Sept 2016

1908. DuPont: DuPont displays and Kateeva collaborate to optimize inkjet printing for mass production of OLED TVs. http://www.dupont.com/products-and-services/display-lighting-materials/oled-organic-light-emitting-diodes/press-releases/20150601-dupont-displays-kateeva-collaborate.html. Accessed Sept 2016

1909. Liedenbaum, C., Croonen, Y., van de Weijer, P., Vleggaar, J., Schoo, H.: Synth. Met. **91**, 109 (1997)

1910. Fukuda, T., Kanbara, T., Yamamoto, T., Ishikawa, K., Takezoe, H., Fukuda, A.: Synth. Met. **85**, 1195 (1997)

1911. Osaka, T., Komaba, S., Fujihana, K., Okamoto, N., Momma, T., Kaneko, N.: J. Electrochem. Soc. **144**, 742 (1997)

1912. Lidzey, D.G., Pate, M.A., Weaver, M.S., Fisher, T.A., Bradley, D.D.C.: Synth. Met. **82**, 141 (1996)

1913. Dagani, R.: Light-emitting polymer synthesis: versatile new route makes potentially important materials available for R&D. Chem. Eng. News. **76**(3), 9–10 (1998)

1914. Brütting, W., Meier, M., Herold, M., Karg, S., Schwoerer, M.: Synth. Met. **91**, 163 (1997)

1915. Greenwald, Y., Hide, F., Heeger, A.J.: J. Electrochem. Soc. **144**, L241 (1997)

1916. Weder, C., Sarwa, C., Montali, A., Bastiaansen, C., Smith, P.: Science. **279**, 835 (1998)

1917. Montali, A., Bastiaansen, C., Smith, P., Weder, C.: Nature. **392**, 261 (1998)

1918. Xia, Y., Sun, K., Ouyang, J.: Solution-processed metallic conducting polymer films as transparent electrode of optoelectronic devices. Adv. Mater. **24**, 2436–2440 (2012)

1919. Vosgueritchian, M., Lipomi, D.J., Bao, Z.: Highly conductive and transparent PEDOT:PSS films with a fluorosurfactant for stretchable and flexible transparent electrodes. Adv. Funct. Mater. **22**, 421–428 (2012)

1920. Wang, Y., Jing, X.: Intrinsically conducting polymers for electromagnetic interference shielding. Polym. Adv. Technol. **16**, 344–351 (2005)

1921. Colaneri, N.F., Schlette, L.W.: EMI shielding measurements of conductive polymer blends. IEEE Trans. Instrum. Meas. **41**, 291 (1992)

1922. Shacklette, L.W., Colaneri, N.F., Kulkarni, V.G., Wessling, B.: EMI shielding of intrinsically conductive polymers. J. Vinyl Technol. **14**, 118 (1992)

1923. Chandrasekhar, P., Naishadham, K.: Broadband microwave absorption and shielding properties of a poly(aniline). Synth. Met. **105**, 115 (1999)

1924. Trivedi, D.C., Dhawan, S.K.: Shielding of electromagnetic interference using polyaniline. Synth. Met. **59**, 267 (1993)

1925. Mäkelä, T., Pienimaa, S., Taka, T., Jussila, S., Isotalo, H.: Thin polyaniline films in EMI shielding. Synth. Met. **85**, 1335 (1997)

1926. Wojkiewicz, J.L., Fauveaux, S., Miane, J.L.: Electromagnetic shielding properties of polyaniline composites. Synth. Met. **135–136**, 127 (2003)

1927. Kim, S. H.; Jang, S. H.; Byun, S. W.; Lee, J. Y.; Joo, J. S.; Jeong, S.H.;Park, M. J., Electrical properties and EMI shielding characteristics of polypyrrole-nylon 6 composite fabrics, J. Appl. Polym. Sci., 2003, 87,1969.

1928. Chandrasekhar, P. (Ashwin-Ushas Corp. Inc.), Final Reports towards U.S. Contract Nos.: N00024-91-C-4045 (Naval Sea Systems Divn.), DAAH01-92-C-R120, DAAH-01-91-C-R151, DAAH01-90-C-0556 (DARPA), NAS3-25889 (NASA), MML-TSC-93-05 (Lockheed Martin), N00014-95-C-0069, N00014-96-C-0229 (Office of Naval Research), DEFG05-93ER81631 (Dept. of Energy)

1929. Mäkelä, T., Pienimaa, S., Taka, T., Jussila, S., Isotalo, H.: Synth. Met. **85**, 1335 (1997)

1930. Naarman, H.: Correlations between active agents and electrically conducting polymers. In: Salaneck, W.R., Clark, D.T., Samuelsen, E.J. (eds.) science and applications of conducting polymers, p. 81. Adam Hilger, New York (1991)

1931. Kurachi, I.; Ezure, H. JP 09111128. 28 Apr 1997

1932. Amada, H., Ishikawa, T., Misuishi, M., Inoe, T., Suezaki, M. JP 08288190. 1 Nov 1996

1933. De Winter, W., Agfa, N.V.: Aqueous application technology of conductive layers for photographic materials. In: Salaneck, W.R., Clark, D.T., Samuelsen, E.J. (eds.) Science and applications of conducting polymers, p. 179. Adam Hilger, New York (1991)

1934. Wang, Z.H., Javadi, H.H.S., Ray, A., MacDiarmid, A.G., Epstein, A.J.: Phys. Rev. B. **42**, 5411 (1990)

1935. Epstein, A.J., Joo, J., Wu, C.-Y., Benatar, A., Faisst Jr., C.F., Zegarski, J., MacDiarmid, A. G.: Polyanilines: recent advances in processing and applications to welding of plastics. In: Aldissi, M. (ed.) Intrinsically conducting polymers: an emerging technology, p. 165. Kluwer Academic Publishers, Boston (1993.) and references therein

1936. Brupbacher, J.: Lockheed Martin Corp., Private Commun. 1987

1937. Rose, T.L., D'Antonio, S., Jillson, M.H., Kron, A.B., Suresh, R., Wang, F.: Synth. Met. **85**, 1439 (1997)

1938. Reisch, M.S.: Fitness clothing gets smart: Textile technology combines with flexiblmagasee electronic sensors at IDTechEx Conference. Chem. Eng. News. **93**(48), 28–29 (2015)

1939. Kuhn, H.H.: Characterization and application of polypyrrole-coated textiles. In: Aldissi, M. (ed.) Intrinsically conducting polymers: an emerging technology, p. 25. Kluwer Academic Publishers, Boston (1993)

1940. Kuhn, H.H., Child, A.D., Kimbrell, W.C.: Synth. Met. **71**, 2139 (1995), and references therein

1941. Segawa, H., Kunimoto, K., Nakamoto, A. Shimidzu, T.: J. Chem. Soc. Perkin Trans. **1**, 939 (1992)

1942. Jousse, F. Microwave properties of conductive honeycombs. In: Evolving technologies for the competitive edge; Int. SAMPE Symp. Exhib. Book2, p. 1552–1558 (1997)

1943. Stenger-Smith, J D., Zarras, P., Miles, M.H., Hollins, R.A., Chafin, A.P., Lindsay, G.A.: Optical and electronic materials for naval aviation. In: evolving technologies for the competitive edge; Int. SAMPE Sym. Exhib. Book 1, Issue 42, p. 652–66 (1997)

1944. Gerard, S., Klimek, W.: Proceedings of the smoke/obscurants symposium XV, vol. I. Chemical Research, Development & Engineering Center, Aberdeen (1991)

1945. Jenkins, L., Mattson, G., Clausen, C.A., Morgan, P.W.: The Evaluation of conductive polymer filaments for use as an obscurant. In: Proceedings of the smoke/obscurants symposium XV, vol. 1, p. 109. Chemical Research Development & Engineering Center, Aberdeen (1991)

1946. Staicovici, S., Wu, C.-Y., Benatar, A.: Proc. 55 th Annu. Tech. Conf.-Soc. Plast. Eng. **1**, 1140 (1997)

1947. Yamamoto, K., Murashima, M.: JP 09241588. 16 Sept 1997

1948. Stegeman, G.I., Zanoni, R., Seaton, C.T.: Nonlinear organic materials in integrated opticcs structures. In: Heeger, A.J., Orenstein, J., Ulrich, D.R. (eds.) Materials research society symposium proceedings, vol. 109, p. 53. Nonlinear Optical Properties of Polymers, Materials Research Society, Pittsburgh (1988)

1949. DeMartino, R., Haas, D., Khanarian, G., Leslie, T., Man, H.T., Riggs, J., Sansone, M., Stamatoff, J., Teng, C., Yoon, H.: Nonlinear optical polymers for electrooptical devices. In: Heeger, A.J., Orenstein, J., Ulrich, D.R. (eds.) Materials research society symposium proceedings, vol. 109, p. 65. Nonlinear Optical Properties of Polymers, Materials Research Society, Pittsburgh (1988)

1950. (a) Thakur, M., Krol, D.M.: Appl. Phys. Lett. **56**, 1406 (1990); (b) Rochford, K.B., Zanoni, R., Gong, Q., Stegeman, G.I.: Appl. Phys. Lett. **55**, 1161 (1989); (c) Patel, J.S., Lee, S.-D., Baker, G.L., Shelburne, J.A. III, Appl. Phys. Lett. **56**, 131 (1990); (d) Mann, S., Oldroyd, A. R., Bloor, D., Ando, D.J., Wells, P.J.: Proc. SPIE **971**, 245 (1988); (e) Drake, A.F., Udvarhelyi, P., Ando, D.J., Bloor, D., Obhi, J.S., Mann, S.: Polym. **30**, 1063 (1989); (f) Oldroyd, A.R., Mann, S., McCallion, K.J.: Electr. Lett. **25**, 1476 (1989); (g) Driemeier, W., Brockmeyer, A.: Appl. Opt. **25**, 2960 (1986)

1951. Svatinova, I., Tonchev, S., Todorov, R., Venkova, E., Liarokapis, E., Anastassakis, E.: J. Appl. Phys. **67**, 2051 (1990)

1952. Bloor, D.: Conjugated and non-conjugated polymers in integrated optics. In: Salaneck, W. R., Clark, D.T., Samuelsen, E.J. (eds.) Science and applications of conducting polymers, p. 23. Adam Hilger, New York (1991)

1953. Kushner, B.G., Neff, J.A.: Nonlinear optical materials & DoD device requirements. In: Heeger, A.J., Orenstein, J., Ulrich, D.R. (eds.) Nonlinear optical properties of polymers, vol. 109, p. 3. Materials Research Society, Pittsburgh (1988)

1954. Sasaki, K., Fuii, K., Tomioka, T., Kinoshika, T.: J. Opt. Soc. Am. **B5**, 457 (1988)

1955. Thankur, M., Verbeek, B., Chi, G.C., O'Brien, K.J.: Some fundamental aspects of the thin film organization and device-structure fabrication of polydiacetylines. In: Heeger, A.J., Orenstein, J., Ulrich, D.R. (eds.) Nonlinear optical properties of polymers, vol. 109, p. 41. Materials Research Society, Pittsburgh (1998)

1956. Etemad, S., Fann, W.-F., Townsend, P.D., Baker, G.L., Jackel, J.: In: Bredas, J.L., Chance, R.R. (eds.) Conjugated polymeric materials, p. 341. Kluwer Academic Publ., Dordrecht (1990)

1957. Baker, G.L., Klausner, C.F., Shelburne III, J.A., Schlotter, N.E., Jackel, J.L., Townsend, P. D., Etemad, S.: Synth. Met. **28**, D639 (1989)

1958. Townsend, P.D., Baker, G.L., Schlotter, N.E., Etemad, S.: Synth. Met. **28**, D633 (1989)

1959. Halvorson, C., Heeger, A.J.: Synth. Met. **71**, 1649 (1995)

1960. Inganas, O., Lundstrom, I.: J. Electrochem. Soc. **131**, 1129 (1984)

1961. Chandrasekhar, P.: Polymers for activated laser switching. In: Arshady, R. (ed.) Desk reference of functional polymers: syntheses and applications, p. 529, American Chemical Society, Washington, DC (1997)

1962. Tessler, N., Denton, G.J., Friend, R.H.: Nature. **382**, 695 (1996)

1963. Tessler, N., Denton, G.J., Friend, R.H.: Synth. Met. **84**, 475 (1997)

1964. Dagani, R.: Semiconducting polymers: thin films show laser potential. Chem. Eng. News. **74**(36), 4 (1996)

1965. Díaz-García, M.A., Hide, F., Schwartz, B.J., Anderson, M.R., Pei, Q., Heeger, A.J.: Synth. Met. **84**, 455 (1997)

1966. Messier, J. (ed.) In: NATO ASI Series, vol. 162 (1989)

1967. Forrest, S.R., Burrows, P.E., Haskal, E.I., Zhang, Y.: Optoelectronic and strutural properties of vacuum-deposited crystalline organic thin films. In: Electrical, optical, and magnetic properties of organic solid state materials, vol. 328, p. 37. Materials Research Society, Pittsburgh (1994)

1968. Thakur, M., Verbeek, B., Chi, G.C., O'Brien, K.J.: In: Heeger, A.J., Orenstein, J., Ulrich, D.R. (eds.) Nonlinear Optical Properties of Polymers, vol. 109, p. 48. Materials Research Society, Pittsburgh, Pennsylvania, USA (1998)

1969. Otero, T.F., de Larreta-Azelain, E., Tejada, R.: Makromol. Chem. Makromol. Symp. **20**(21), 615 (1988)

1970. Otero, T.F., de Larreta-Azelain, E.: Synth. Met. **26**, 79 (1988)

1971. Otero, T.F., Angulo, E., Rodríguez, J., Santamaría, C.: J. Electroanal. Chem. **341**, 369 (1992)

1972. Otero, T.F., Rodríguez, J., Angulo, E., Santamaría, C.: Synth. Met. **55–57**, 3713 (1993)

1973. Product Literature for "Versicon". Allied-signal Inc., Morristown

1974. Barbero, C., Kötz, R.: J. Electrochem. Soc. **141**, 859 (1994)

1975. Otero, T.F., Sansinena, J.M.: Bioelectroche. Bioenerg. **42**, 117 (1997), and references therein

1976. Baughman, R.H. In: Proceedings of the DARPA active polymers workshop hosted by the Institute for Defense Analyses, Baltimore, Maryland, USA, p. D2 (1996)

1977. Smela, Inganas, Lundstrom: Science (1995)

1978. Shahinpoor, M. In: Proceedings of the DARPA active polymers workshop hosted by the Institute for Defense Analyses, Baltimore, Maryland, USA, p. D7 (1996

1979. Electroactive Polymers and Devices, DARPA (USA) Broad Agency Announcement No. 98-06 (April 15, 1998)

1980. Okuzaki, H., Kuwaraba, T., Kunugi, T.: Polym. **38**, 5491 (1997)

1981. Pei, Q., Inganäs, O.: Synth. Met. **55–57**, 330 (1993)

1982. Okamoto, T., Kato, Y., Tada, K., Onoda, M.: Actuator based on doping/undoping-induced volume change in anisotropic polypyrrole film. Thin Solid Films. **393**, 383–387 (2001)

1983. Otero, T.F., Cortes, M.T.: Artificial muscle: movement and position control. Chem. Commun. 284–285 (2004)

1984. Chandrasekhar, P. Corrosion prevention coatings based on soluble conducting polymers, Final Report, Contract No. F34601-95-C-0542 for Tinker Air Force Base, Oklahoma

1985. Ahmad, J., MacDiarmid, A.G.: Synth. Met. **78**, 103 (1996)

1986. (a) MacDiarmid, A.G., Ahmad, N.: Prevention of corrosion with polyaniline. US 5645890 A. 8 July 1997; (b) Menholi, et al.: J. Appl. Poly. Sci. **28**, 1125 (1983)

1987. (a) McAndrew, T.P., Gilicinski, A.G., Robeson, L.M.: Protecting carbon steel form corrosion with nonconducting poly(aniline). US 5441772 A, 15 Aug 1995, and references therein; (b) Huang, F.C., et al.: Polym. Mater. Sci. Eng. **61**, 895 (1989)

1988. Jain, F.C., Rosato, J.J., Kalonia, K.S., Agarwala, V.S.: Corrosion. **42**, 700 (1986)

1989. Jain et al.: Mat. Res. Soc. Symp. Proc. **125**, 329 (1988)

1990. Jain, et al.: Corrosion prevention in metals using layered semi-conductor/insulator structures forming an interfacial electronic barrier. In: Lee, L.H. (ed.) Adhesive, sealants, and coatings for space and harsh environments, pp. 381–405. Plenum Publishing corp., New York (1988)

1991. Wrobleski, D.A. Benicewicz, B.C., Thompson, K.G. Bryan, C.J. Corrosion resistant coating. US 5658649 A, 19 Aug 1997, and references therein

1992. DeBerry, D.W.: J. Elecrochem. Soc. 1022 (1985)

1993. Basics of Corrosion Measurements: Application Note Corri, EG&G Princeton Applied Research Corp. (1997)

1994. Kinlen, P.J., Silverman, D.C., Tokas, E.F., Hardiman, C.J.: Corrosion inhibiting compositions. US 5532025 A, 2 July 1996

1995. Lu, W.-K., Elsenbaumer, R.L., Wessling, B.: Synth. Met. **71**, 2163 (1995)

1996. Li, P., Tan, T.C., Lee, J.Y.: Synth. Met. **88**, 237(1997)

1997. Fahlman, M., Jasty, S., Epstein, A.J.: Synth. Met. **85**, 1323 (1997)

1998. Racicot, R., Clark, R.L., Liu, H.-B., Yang, S.C., Alias, M.N., Brown, R.: Thin film conductive polymers on aluminum surfaces: interfacial charge-transfer and anti-corrosion aspects. In: Yang, S.C., Chandreskhar, P. (eds.) Optical and photonic applications of electroactive and conducting polymers, vol. 2528, p. 251. SPIE-The International Society for Optical Engineering, Bellingham (1995., and references therein)

1999. Brusic, V., Angelopoulos, M., Graham, T.: J. Electrochem. Soc. **144**, 436 (1997)

2000. Zarras, P., Stenger-Smith, J.D., Miles, M.H.: Polym. Mater. Sci. Eng. **76**, 589 (1997)

2001. Guo, B., Glavas, L., Albertsson, A.-C.: Biodegradable and electrically conducting polymers for biomedical applications. Progress in Polymer Science. **38**, 1263–1286 (2013)

2002. Ghasemi-Mobarakeh, L., Prabhakaran, M.P., Morshed, M., Nasr-Esfahani, M.H., Baharvand, H., Kiani, S., Al-Deyab, S., Ramakrishna, S.: Application of conductive polymers, scaffolds and electrical stimulation for nerve tissue engineering. J. Tissue Eng. Regen. Med. **5**, e17–e35 (2011)

2003. Bendrea, A.-D., Cianga, L., Cianga, I.: Progress in the filed of conducting Polymers for tissure engineering applications. J. Biomater. Appl. **26**, 3 (2011)

2004. Mawad, D., Stewart, E., Officer, D.L., Romeo, T., Wagner, P., Wagner, K., Wallace, G.G.: A single component conducting polymer hydrogel as a scaffold for tissue engineering. Adv. Funct. Mater. 1–8 (2012)

2005. Hardy, J.G., Lee, J.Y., Schmidt, C.E.: Biomimetic conducting polymer-based tissue scaffolds. Current opinion in biotechnology. **24**(5), 847–854 (2013)

2006. Chen, M.-C., Sun, Y.-C., Chen, Y.-H.: Electrically conductive nanofibers with highly oriented structures and their potential application in skeletal muscle tissue engineering. Acta Miomaterialia. **9**, 5562–5572 (2013)

2007. Guimard, N.K., Gomez, N., Schmidt, C.E.: Conducting polymers in biomedical engineering. Prog. Polym. Sci. **32**, 876–921 (2007)

2008. Green, R. A.; Lovell, N. H.; Wallace, G. G.; Poole-Warren, L.A., Conducting polymers for neural interfaces: challenges in developing an effective long-term implant, Biomater., 2008, 29, 3393-3399.

2009. Kang, G., Borgens, R.B., Cho, y.: Well-ordered porous conductive polypyrrole as a new platform for neural interfaces. Langmuiri. **27**, 6179–6184 (2011)

2010. Huang, J., Kaner, R.B.: Flash welding of conducting polymer nanofibers. Nat. Mater. **3**, 783–786 (2004)

2011. Abidian, M.R., Kim, D.H., Martin, D.C.: Conducting-polymer nanotubes for controlled drug release. Adv. Mater. **18**, 405–409 (2006)

2012. Shi, G., Rouabhia, M., Wang, Z., et al.: A novel electrically conductive and biodegradable composite made of polypyrrole nanoparticles and polylactide. Biomater. **25**, 2477–2488 (2004)

2013. Huang, L., Hu, J., Lang, L., et al.: Synthesis and characterization of electroactive and biodegradable ABA block copolymer of polylactide and aniline pentamer. Biomaterials. **28**, 1741–1751 (2007)

2014. Ding, C., Wang, Y., Zhang, S.: Synthesis and characterization of degradable electrically conducting copolymer of aniline pentamer and polygycolide. Eur. Polym. J. **43**, 4244–4252 (2007)

2015. Zhang, Q., Yan, Y., Li, S., et al.: The Synthesis and characterization of a novel biodegradable and electroactive polyphosphazene for nerve regeneration. Mater. Sci. Eng. C. **30**, 160–166 (2010)

2016. Huang, L., Zhuang, X., Hu, J., et al.: Synthesis of biodegradable and electroactive multiblock polylactide and aniline pentamer copolymer for tissue engineering applications. Biomacromolecules. **9**, 850–858 (2008)

2017. Jiang, X., Marois, Y., Traore, A., et al.: Tissue reaction to polypyrrole-coated polyester fabrics: an in vivio study in rats. Tissue Eng. **8**, 634–647 (2002)

2018. Olayo, R., Rios, C., Salgado-Ceballos, H., et al.: Tissue spinal cord response in rats after implants of polypyrrole and polyethylene glycol obtained by plasma. J. Mater. Sci Mater. Med. **19**, 817–826 (2008)

2019. Zelikin, A.N., Lyn, D.M., Farhadi, J., Martin, I., Shastri, V., Langer, R.: Erodible conducting polymers for potential biomedical applications. Ange. Chem. Int. Ed. **41**, 141–144 (2002)

2020. Liu, Y.D., Hu, J., Zhuang, X.L., Zhang, P.B.A., Wei, Y., Wang, x.H., Chen, X.S.: Synthesis and characterization of novel biodegradable and electroactive hydrogel based on aniline oligomer and gelatin. Macromol. Biosci. **12**, 241–250 (2012)

2021. Valle, L.J., Aradilla, D., Olver, R., et al.: Cellular adhesion and proliferation on poly (3,4-ethylenedioxythiophene): benefits in the electroactivity of the conducting polymer. Eur. Polym. J. **43**, 2342–2349 (2007)

2022. Luo, S., Ali, E.M., Tansil, N.C., et al.: Poly(3,4-ethylenedioxythiophene) (PEDOT) nanobiointerfaces: thin, ultrasmooth, and functionalized PEDOT Films with in vitro, and in vivo biocompatibility. Langmuir. **24**, 8071–8077 (2008)

2023. Richardson-Burns, S.M., Hendricks, J.L., Foster, B.: Polymerization of the conducting polymer poly(3,4-ethylenedioxythiphene) (PEDOT) around living neural cells. Biomater. **28**, 1539–1552 (2007)

2024. Ramanaviciene, A., Kausaite, A., Tautkus, S., Ramanavicius, A.: Biocompatibility of polypyrrole particles: An in vivo study in mice. J. Pharm. Pharmacol. **59**, 311–315 (2007)

2025. Oh, W.-K., Kim, S., yoon, H., Jang, J.: Shape-dependent cytotoxicity and proinflammatory response of poly(3,4-ethylenedioxythiophene) nanomaterials. Small. **6**, 8720879 (2010)

2026. Sanchvi, A.B., Miller, K.P.H., Belcher, A.M., et al.: Biomaterials functionalization using a novel peptide that selectively binds to a conducting polymer. Nat. Mater. **4**, 496–502 (2005.)
Oh, W.K., Yoon, H., Jang, J.: Size control of magnetic carbon nanoparticles for drug delivery. Biomater. 31, 1342–1348 (2010)

2027. Lee, J.W., Serna, F., Schmidt, C.E.: Carboxy-endcapped conductive polypyrrole: biomimetic conducting polymer for cell scaffolds and electrodes. Langmuir. **22**, 9816–9819 (2006)

2028. Cullen, D.K., Patel, A.R., Doorish, J.F., et al.: Developing a tissue-engineered neural-electrical relay using encapsulated neuronal constructs on conducting polymer fibers. J. Neural. Eng. **5**, 374–384 (2008)

2029. Xie, J., MacEwan, M.R., Willerth, S.M., et al.: Conductive core-sheath nanofibers and their potential application in neural tissue engineering. Adv. Funct. Mater. **19**, 2312–2318 (2009)

2030. Thompson, B.C., Richardson, R.T., Moulton, S.E., et al.: Conducting polymers, Dual neurotrophins and pulsed electrical stimulation – dramatic effects on neurite outgrowth. J. Control Release. **141**, 161–167 (2010)

2031. Li, X., Wan, M., Li, X., Zhao, G.: The role of DNA in Pani-DNA Hybrid Template and Dopant. Polymer. **50**, 4529–4534 (2009)

2032. Cheng, D., Xia, H., Chan, H.S.: Synthesis and characterization of surface-functionalized conducting polyaniline-chitosan nanocomposites. J. Nanosci. Nanotechnol. **5**, 466–473 (2005)

2033. Liu, S., Wang, J., Zhang, D., Zhang, P., Ou, J., Liu, B., et al.: Investigation of cell biocompatible behaviors of polyaniline film fabricated via electroless surface polymerization. Appl. Surf. Sci. **256**, 3427–3431 (2010)

2034. Wong, J.Y., Langer, R., Ingber, D.E.: Electrically conducting polymers can non-invasivley control the shape and growth of mammalian cells. Proc. Natl. Acad. Sci. USA. **91**, 3201–3204 (1994)

2035. Garner, B., Georgevich, A., Hodgson, A.J., Liu, L., Wallce, G.G.: Polypyrrole heparin composites as stimulus-responsive substrates for endothelial cell growth. J. Biomed. Mater. Res. **44**, 121–129 (1999)

2036. De Giglio, E., Sabbatini, L., Colucci, S., Zambonin, G.: Synthesis, analyitcal characterization and osteoblasts adhesion properties on RGD-grafter polypyrrole coatings on titanium substrates. J. Biomater. Sci. Polym. Ed. **11**, 1073–1083 (2001)

2037. Schmidt, C.E., Shastri, V.R., Vacanti, J.P., Langer, R.: Stimulation of neurite outhrowth using an electrically conducting polymer. Proc. Natl. Acad. Sci. USA. **94**, 8949–8953 (1997)

2038. Wulliams, R.L., Doherty, P.J.: A preliminary assessment of poly(Pyrrole) in nerve guide studies. J. Mater. Sci. Mater. Med. **5**, 429–433 (1994)

2039. Cui, X., Wiler, J., Dzaman, J., Altschuler, M.R.A., Martin, D.C.: Vivo studies of polypyrrole/peptide coated neural probe. Biomaterials. **24**, 777–787 (2003)

2040. Jiang, X., Marois, Y., Traore, A., Tessier, D., Dao, L.H., Guidoin, R., et al.: Tissue reaction to polypyrrole-coated polyester fabrics: an in vivo study in rats. Tissue. Eng. **8**, 635–647 (2002)

2041. Wang, X., Gu, X., Yuan, C., Chen, S., Zhang, P., Zhang, T., et al.: Evaluation of biocompatibility of polypyrrole in vitro and in vivo. J. Biomed. Mater. Res. A. **68**, 411–422 (2004)

2042. Ramanviciene, A., Kausaite, A., Tautkus, S., Ramanivcius, A.: Biocompatibility of polypyrrole particles: an in vivo study in mice. J. Pharm. Pharmacol. **59**, 311–315 (2007)

2043. Xiao, Y., Li, C.M., Wang, S., Shi, J., Ooi, C.P.: Incorporation of collagen in poly (3,4-ethylenedioxythiophene) for a biofunctional film with high bio- and electrochemical activity. J. Biomed. Mater. Res. A. **92A**, 766–772 (2010)

2044. Green, R.A., Lovell, N.H., Poole-Warren, L.A.: Cell attachment functionality of bioactive conducting polymers for neural interfaces. Biomater. **30**, 3637–3644 (2009)

2045. Mawad, D., Gilmore, K., Molino, P., Wagner, K., Wagner, P., Officer, D.L., Wallace, G.G.: An erodible polythiophene-based composite for biomedical applications. J. Mater. Chem. **21**, 5555–5560 (2011)

2046. delValle, L.J., Aradilla, D., Olver, R., Sepulcre, F., Gamez, A., Armmelin, E., et al.: Cellular adhesion and proliferation on poly(3,4-Ethylenedioxythiphene): Benefits in the Electroactivity of Conducting Polymer. Eur. Polym. J. **43**, 2342–2349 (2007)

2047. Che, J., Xiao, Y., Zhu, X., Sun, X.: Electro-Synthesized PEDOT/Glutamate Chemically Modified Elecrode: A Coombination of Electrical and Biocompatible Features. Polym. Int. **57**, 750–755 (**2008**)

2048. Bergren, M.: Cyborg rose carries a current. Chem. Eng. News. **93**(46), 26 (2015)

2049. Davenport, M.: Light and organic chemistry could make smarter, flexible devices. Chem. Eng. News. **94**(26), 7 (2016)

2050. Eureka: Innovation across borders. Amroy Oy. http://www.eurekanetwork.org/content/amroy-oy-0. Accessed June 2016

2051. Frackowiak, E., Lota, G., Cacciaguerra, T., Béguin, F.: Electrochem. Commun. **8**, 129 (2006)

2052. Sun, X., Li, R., Villers, D., Dodelet, J.P., Désilets, S.: Chem. Phys. Lett. **379**, 99 (2003)

2053. Song, Y., Li, X., Mackin, C., Zhang, X., Fang, W., Palacios, T., Zhu, H., Kong, J.: Role of interfacial oxide in high-efficiency graphene-silicon schottky barrier solar cells. Nano. Lett. **15**(3), 2104–2110 (2015)

2054. (a) Hubbard, A.T., Anson, F.C.: Anal. Chem. **58** (1966); (b) Brown, A.P., Anson, F.C.: Anal. Chem 49, **1589** (1977)

2055. (a) Laviron, E.: Bull. Soc. Chem. Fr. **3717** (1967); (b) Laviron, E.: J. Electroanal. Chem. **39**, 1 (1972)

2056. Oyama, N., Hirokawa, T.: Anal. Chem. **59**, 258 (1987)

2057. Peng, C., Zhang, S., Jewell, D., Chen, G.Z.: Carbon nanotube and conducting polymer composites for supercapacitors. Prog. Nat. Sci. **18**, 777–778 (2008)

2058. Rajesh, B., Ravindranatha Thampi, K., Bonard, J.-M., Jorg Mathieu, H., Xanthopoulos, N., Viswanahan, B.: Chem. Commun. 2022 (2003)

2059. Ma, Y., Jiang, S., Jian, G., Tao, H., Yu, L., Wang, X., Wang, X., Zhu, J., Hu, Z., Chen, Y.: Energy Environ. Sci. **2**, 224 (2009)

2060. Mohana Reddy, A.L., Rajalkshmi, N., Ramaprabhu, S.: Carbon. **46**, 2 (2008)

2061. Zhou, Q., Li, C.M. Li, J. Cui, X. Gervasio, D. J. Phys. Chem. C **111**, 11216 (2007)

2062. Will light-emitting polymers outshine LCDs? Chem. Week **42** (1998)

2063. Active Polymers Workshop osted by the Institute for Defense Analyses. Proceedings of the Defense Advance Research Projects Agency Active (DARPA); Baltimore, p. D4(1996)

2064. Brodie, B.C.: On the atomic weight of graphite. Phil. Trans. R. Soc. London. **149**, 249–259 (1859)

2065. Lourie, O., Wagner, H.D.: Evaluation of Young's modulus of carbon nanotubes by micro-Raman spectroscopy. J. Mater. Res. **13**(9), 2418–2422 (1998)

2066. Krishnan, A., Dujardin, E., Ebbesen, T.W., Yianilos, P.N., Treacy, M.M.J.: Young's modulus of single-walled nanotubes. Phys. Rev. B. **58**(20), 14013–14019 (1998)

2067. Mohanty, N., Berry, V.: Graphene-based single-bacterium resolution biodevice and DNA transistor: interfacing graphene derviatives with nanoscale and microscale biocomponents. Nano. Lett. **8**, 4469 (2008)

2068. Cai, H., Cao, X., Jiang, Y., He, P., Fang, Y.: Anal. Bioanal. Chem. **375**, 287 (2003)

2069. Cai, H., Cao, X.N., Jiang, Y., He, P.G., Fang, Y.Z.: Carbon nanotube-enhanced electrochemical DNA biosensor for DNA hybridization detection. Anal. Bioanal. Chem. **375**, 287–293 (2003)

2070. Pantarotto, D., Singh, R., McCarthy, D., Erhardt, M., Briand, J.P., Prato, M., Kostarelos, K., Bianco, A.: Functionalised carbon nanotubes for plasmid DNA gene delivery. Angew Chem. Int. Ed. Engl. **43**, 5242–5246 (2004)

2071. Singh, R., Pantarotto, D., McCarthy, D., Chaloin, O., Hoebeke, J., Partidos, C.D., Briand, J. P., Prato, M., Bianco, A., Kostarelos, K.: Binding and condensation of plasmid DNA onto functionalized carbon nanotubes: towards the construction of nanotube-based gene delivery vectors. J. Am. Chem. Soc. **127**, 4388–4396 (2005)

2072. Liang, Y.T., Vijayan, B.K., Gray, K.A., Hersam, M.C.: Nano Lett. **11**, 2865 (2011)

2073. Ng, Y.J., Iwase, A., Kuda, A., Amal, R.: J. Phys. Chem. Lett. **1**, 2607 (2010); Min, S.K., Kim, W.Y., Cho, Y., Kim, K.S.: Nat. Nanotechnol. **6**, 162 (2011)

2074. Baum, R.: Nanotube characterization. Chem. Eng. News. **6**, (1998)

2075. Xu, B., Yin, J., Liu, Z.: Phonon scattering and electron transport in single wall carbon nanotube. In: Sazuki, S. (ed.) Physical and chemical properties of carbon nanotbues. InTech (2013)

2076. A carbon nanotube page: Carbon nanotube science and technology. http://www.personal. reading.ac.uk/~scsharip/tubes.htm. Accessed June 2016

2077. Karisruhe Institute of Technology: Graphitic materials. https://www.int.kit.edu/1745.php. Accessed June 2016; Wikipedia: Stone-wales defect. https://en.wikipedia.org/wiki/Stone% E2%80%93Wales_defect. ccessed June 2016

2078. Ishii, S., Takano, Y.: High-conductivity boron-doped carbon nanotubes, SPIE (2007)

2079. Cai, L., Wang, C.: Carbon nanotube flexible and stretchable. Nanoscale Res. Lett. **10**(320), 1–21 (2015)

2080. Chen, H., Cao, Y., Zhang, J., Zhou, C.: Large-scale complementary microelectronics using hybrid integration of carbon nanotubes and IGZO thn-film transistors. Nature Comm. **5**, 4097 (2014)

2081. Cao, Q., Kim, H.-S., Pimparkar, N., Kulkarni, J.P., Wang, C., Shim, M., Roy, K., Alam, M. A., Rogers, J.A.: Medium-scale carbon nanotube thin-film integrated circuits on flexible plastic substrates. Nature. **454**, 4955–4500 (2008)

2082. Bookofjoe: World's smallest radio consists of 1 carbon nanotube – listen to it play 'Layla'. http://www.bookofjoe.com/2009/02/nanotube-radio.html. Accessed June 2016

2083. what-when-how: Carbon nanotubes: thermal properties Part 1 (Nanotechnology). http:// what-when-how.com/nanoscience-and-nanotechnology/carbon-nanotubes-thermal-proper ties-part-1-nanotechnology/. Accessed June 2016

2084. Pykal, M., Jurečka, P., Karlický, F., Otyepka, M.: Modelling of graphene functionalization. Phys. Chem. Chem. Phys. **18**, 6351–6372 (2016)

2085. AzoOptics: Using NTEGRA instruments to conduct comprehensive characterization and analysis of graphene flakes by NT-MDT. http://www.azooptics.com/Article.aspx? ArticleID=223http://www.ntmdt-si.com/. Accessed June 2016

2086. Liu, Y., Liu, Z., Lew, W.S., Wang, Q.J.: Temperature dependence of the electrical transport properties in few-layer graphene interconnects. Nanoscale Res. Lett. **8**(1), 335 (2013)

2087. Wang, J., Rathi, S., Singh, B., Lee, I., Maeng, S., Joh, H.-I., Kim, G.-H.: Dielectrophoretic assembly of Pt nanoparticle-reduced graphene oxide nanohybrid for highly-sensitive multiple gas sensor. Sens. and Act. B: Chem. **220**, 755–761 (2015)
2088. Tung, T.T., Castro, M., Kim, T.Y., Suh, K.S., Feller, J.-F.: Graphene quantum resistive sensing skin for the detection of altercation biomarkers. J. Mater. Chem. **22**, 21754–21766 (2012)
2089. Tomalia, D.A.: In quest of a systematic framework for unifying and defining nanoscience. J. Nanopart Res. **11**(6), 1251–1310 (2009)
2090. Chang-Jian, S.-K., Ho, J.-R. Laser patterning of carbon-nanotubes thin films and their applications. In: Marulanda, J.M. (ed.) Carbon nanotubes applications on electron devices. IntTech, **20**

Index

Printed in the United States
By Bookmasters